Quantitative Stratigraphy

Quantitative Stratigraphy

by

F. M. GRADSTEIN
Geological Survey of Canada and Dalhousie University, Halifax

F. P. AGTERBERG
Geological Survey of Canada and Ottawa University

J. C. BROWER
Heroy Geological Laboratory, Syracuse University, N.Y.

W. S. SCHWARZACHER
Department of Geology, Queen's University of Belfast

D. Reidel Publishing Company

A MEMBER OF THE KLUWER ACADEMIC PUBLISHERS GROUP

Dordrecht / Boston / Lancaster / Tokyo

UNESCO / PARIS

Library of Congress Cataloging in Publication Data
Main entry under title:

Quantitative stratigraphy.

Includes index.
1. Stratigraphic correlation–Statistical methods. I. Gradstein, F.M.
QE651.Q27 1985 551.7′01 85-18331
Reidel ISBN 90-277-2116-5 (hardbound)
Unesco ISBN 92-3-102336-5 (limpbound)

Published by The United Nations Educational, Scientific and Cultural Organization, 7 Place de Fontenoy, 75700 Paris, France and D. Reidel Publishing Company, P.O. Box 17, 3300 AA Dordrecht, Holland

Hardbound edition sold and distributed in the U.S.A. and Canada by Kluwer Academic Publishers, 190 Old Derby Street, Hingham, MA 02043, U.S.A.

Hardbound edition in all other countries, sold and distributed by Kluwer Academic Publishers Group, P.O. Box 322, 3300 AH Dordrecht, Holland

Limpbound edition sold throughout the world by Unesco, 7 Place de Fontenoy, 75700 Paris, France

Printed in The Netherlands

CONTENTS

FOREWORD

It gives me great pleasure to introduce this textbook on 'Quantitative Stratigraphy', which is one of the first of its kind.

Modern geological exploration and analysis of oceanic and continental basins make extensive use of stratigraphic techniques. Such techniques form the cornerstone of geological synthesis from the first field and seismic observations to the final understanding of the development and depositional history of a sedimentary basin. Geological correlations as well as subsidence and sedimentation models must necessarily be based on biozonation, requiring accurate knowledge of fossil distribution, meticulous lithological description in time and space, and the erection of linear timescales in years.

It is widely recognized that the present day practice of erecting biozones and lithologically defined rock units must necessarily be separate and distinct from the abstract linear timescale. The use and meaning of these building blocks of geology have been greatly clarified and extended over the last 150 years, although the concepts have not been fundamently altered.

What has changed is the number of data collected, the detail required, and the introduction of a dynamic probabilistic approach to geological and stratigraphic modelling. Greatly improved linear timescales for the last 200 million years, supported by advances in computer technology, as well as the successful cooperation between the statistician and geologist, have lead to powerful quantitative techniques in solving stratigraphic problems. The qualitative and descriptive nature of conventional stratigraphic models have been both their attraction and their weakness. Although highly flexible, they do not provide for rigorous stratigraphic correlation, nor for an array of answers as a function of well defined thresholds.

This carefully conceived book includes original and stimulating contributions to quantitative biostratigraphy and lithostratigraphy. Its four authors are foremost in this branch of science and have pioneered its application. Over the years, they have closely cooperated in international courses and symposia on the subject of quantitative stratigraphy, within the framework of the International Geological Correlation Program (IGCP). Project 148 has played an important role, and continues to be a focal point for scientists in all countries of the world, to share and develop new knowledge and ideas on natural resources, environment and geological events in space and time.

This book, by presenting new methods for geological basin analysis, represents a major advance in the science of stratigraphy.

Ottawa, Canada
March, 1985

Digby J. McLaren

PREFACE

This book introduces concepts and methods of quantitative stratigraphy and the impact of the results and interpretations on the geology of sedimentary basins. We have aimed at a practical, rather than theoretical approach to this interdisciplinary science, which has its roots in stratigraphy on the one hand, mathematics, statistics and computer sciences on the other hand. The fifteen Chapters offer a pathway through the 'jungle of methods' for graduate students in geomathematics and geology, exploration paleontologists and geologists, and for university teachers. Virtually all these methods can now be executed with relatively inexpensive microcomputers.

'Quantitative Stratigraphy' starts with an introduction to selected techniques and a philosophical essay on paleontological - stratigraphical background. Part II contains nine Chapters with detailed reviews and applications of the more common and practical techniques for biostratigraphy, including index fossil coefficients, multivariate zonations, seriations, sequencing methods like ranking and scaling (RASC) and the method of unitary associations. The last Chapter compares the quality of 5 methods, composite standard, ranking, seriation, lateral tracing and clustering for zonation and correlation. Part III reviews the CASC method of quantitative biostratigraphic correlation. Quantitative aspects of lithostratigraphic analysis, correlation and sedimentation models are considered in Part IV. An introduction to modern methods for construction of linear timescales and organisation of stratigraphic data in linear time for the purpose of subsidence and sedimentation models is in Part V. Appendix I lists nine large data bases for further study and shows results of ranking and scaling. In Appendix II we provide a brief description of characteristics of computer routines. The scientific vocabulary is supported by a Glossary and for quick reference there is an Index.

This book, which originated from an invitation by the IGCP secretariat in Paris to the leader of Project-148, F.P. Agterberg to prepare a scientific textbook on accomplishments of the Project, could not have been written without the enthusiasm, criticism and unflagging support of my co-authors. Valuable criticism on individual Chapters was provided by G. Bonham Carter, P. Baumgartner, A. Grant, J. Guex, P. Doeven, C. Drooger, P. Moore, D. Kent, W. Berggren and my wife W. Gradstein. She, together with G.M. Murney, also prepared the Glossary and Index and carefully edited versions of the Chapters. Scientific and computing assistance of S.N. Lew, J. Oliver, F. Thomas, M. Williamson, M. Heller, P. Issler, B. Stam and A. Jackson is gratefully acknowledged. The excellent technical assistance and pleasant cooperation of Chris Archibald, Sharon Hiltz, Olive Ross and Gary Cook made it possible to prepare the camera-ready typescript. The Geological Survey of Canada and Unesco, Paris, contributed financially and provided technical support for this undertaking.

February 1985 Felix Gradstein

UNESCO PREFACE

Unesco's active involvement in studies on quantitative stratigraphy dates back to 1972 when quantitative methods and data processing in geological correlation were identified as one of the four major scientific fields of research to be covered by the International Geological Correlation Programme (IGCP).

A specific project on quantitative stratigraphic correlation was launched in 1976 under the leadership of Dr. J.C. Brower and later Dr. J.M. Cubitt. The scope of this project (IGCP Project No. 148) was enlarged in 1981 to encompass not only biostratigraphic and lithostratigraphic correlation but also chronostratigraphy and large scale application.

In response to the growing interest of the international scientific community in the development of concepts and methods of quantitative stratigraphy, Unesco invited Dr. F.P. Agterberg, since 1979 leader of IGCP Project No. 148, to prepare a guidebook on the subject.

The 15 chapters of this book together represent an up-to-date study on the complex and interdisciplinary field of quantitative stratigraphy and constitute a major contribution towards dissemination of geological information.

Unesco wishes to express its thanks to the four authors who undertook the task of writing the various sections of the book. The contents of their respective chapters reflect their personal views and are not necessarily those of the Organization.

Part I

INTRODUCTION

Contents

I.1

QUANTITATIVE STRATIGRAPHIC CORRELATION TECHNIQUES-IGCP PROJECT 148

FREDERIK P. AGTERBERG

INTRODUCTION

The rapid growth of information in the applied and academic geological sciences has led to an increasing need for numerical models to organize and explain specific geological problems. Studies in the fields of biostratigraphy, lithostratigraphy (especially well logs) and sedimentology make successful use of the quantitative modelling approach. Statistical and other numerical techniques can be used in correlation of biostratigraphic events, biozonations, classification and matching of lithofacies in well logs or sections, lithofacies pattern recognition, and determination of the rate or magnitude of geological processes relative to the numerical time scale.

IGCP Project No. 148 "Evaluation and Development of Quantitative Stratigraphic Correlation Techniques" was initiated in 1976 for the purpose of developing computer-based mathematical theory and analysis of geological information which could be applied to automated correlation techniques in stratigraphy (see also Agterberg, 1983). About 170 participants in 25 countries have been conducting research mainly in the fields of biostratigraphy and lithostratigraphy, and National Working Groups in Canada, India, USA and USSR have met regularly. Specific problems were solved by establishing regional standards of ordered stratigraphic events and performing correlations on the basis of these standards. Comprehensive descriptions and computer programs have been prepared for different techniques which were then applied to the same data sets in order to evaluate their respective advantages and drawbacks. The purpose of these evaluations was to select those techniques which are relatively simple and easily understood, achieve maximum resolution, also in comparison with traditional methods of stratigraphic correlation, and can be implemented on computers of different types. This chapter reviews the results of Project 148 in general and provides examples of some applications of the new methods.

OBJECTIVES AND DEVELOPMENT OF THE PROJECT

Special attention was given to the performance of computer-based quantitative techniques in comparison with the results obtained by conventional qualitative stratigraphic correlation methods. During the

first few years of existence (1976 to 1980), the emphasis within the Project was on method development. The statistical problems encountered when attempting to describe quantitative methods of stratigraphic correlation in a cohesive manner are far more complex and difficult to solve than one might expect. Some of the studies made under the auspices of Project No. 148 would not have been possible without recent advances in the theory of mathematical statistics. During the last few years (from 1980), the primary activity in the Project shifted from method development to application, for solving specific stratigraphic problems using large data bases for regions in North America, Europe and India. Deep Sea Drilling Project data sets in the Atlantic and Pacific Oceans were also analyzed. Except for subprojects on the Silurian in the Baltic region and the Cambrian in Texas, the participants have been working mostly on Cenozoic and Cretaceous stratigraphy.

Research on the following major problems is mostly completed:

- Creation and definition of the mathematical theory of stratigraphic relationships.
- Establishment of standards and codes for the biostratigraphic, lithological and environmental information attainable from well logs, cores, and surface sections.
- Development of a theory for mathematical correlation.
- Development of practical methods of biostratigraphic correlation concentrating on quantification of assemblage zones, sequencing methods, set theoretical approaches, morphometric chronoclines and multivariate methodology.
- Development of practical methods of correlation concentrating on methods of spectral analysis (frequency domain), methods of stretching and zonation (time domain), methods of stratigraphic interpolation and multivariate statistical analysis.

Publications emanating from the Project, including computer programs, have been listed annually in Geological Correlation. The IGCP Catalogue 1973-1979 published by UNESCO in 1980 contains 18 indexed entries for Project No. 148. The bulk of the output (over 90 additional publications) were listed in the second, 1980-1982 IGCP Catalogue. This includes 15 papers presented at the International Working Group meeting in Paris, 1980, and published in the book entitled "Quantitative Stratigraphic Correlation" published by J. Wiley and Sons in 1982. The proceedings of the International Symposium in Geneva, 1982, on "Theory, Application and Comparison of Stratigraphic Correlation Methods" (14 papers) were published in 1984 as a Special Issue in the journal Computers and Geosciences.

One of the objectives of the IGCP is transfer of knowledge to developing countries. India has been an especially active participant since 1980, when a number of quantitative stratigraphic correlation projects were initiated there. An international workshop was organized in Kharagpur, West Bengal, during November 1980, and the final international meeting for Project No. 148, consisting of a symposium and short course, was convened at the Indian Institute of Technology, Kharagpur, in December 1983. The proceedings of this meeting are to be published by the Association of Exploration Geophysicists, Hyderabad (India).

In the International Stratigraphic Guide published in 1976 by International Subcommission on Stratigraphic Classification of IUGS Commission on Stratigraphy, a clear distinction is made between

(1) Lithostratigraphy in which strata are organized into units based on their lithologic character;

(2) Biostratigraphy with units based on fossil content; and

(3) Chronostratigraphy with units based on the age relations of the strata.

For convenience, these three topics will be discussed more or less separately in the next three sections of this introduction. A similar separate approach is used in this book, with biostratigraphy followed by chronostratigraphy, which in turn is followed by lithostratigraphy. It should be noted that Project No. 148 was initiated in 1976 as a project on quantitative biostratigraphic correlation. Later in that same year, the initial proposal was combined with equivalent aspects of lithostratigraphic correlations. Aspects of chronostratigraphic correlation were added in 1981 so that the Project finally has covered all aspects of quantitative stratigraphic correlation.

QUANTITATIVE BIOSTRATIGRAPHY

Numerical methods in biostratigraphy make use of the quantified fossil record in sedimentary rock sections for precise recording of biological events in space and time. They can be grouped into six basic categories:

(1) Sampling of environments with fossils that occur in patches (instead of displaying random spatial distribution);

(2) Automated microfossil recognition;

(3) Analysis of evolutionary sequences;

(4) Measurement of the attributes of index fossils (Chapter II.1);

(5) Determination of the optimum sequence of biostratigraphic events as recorded in different stratigraphic sections (Chapters II.4, 5, 6, 8, 9); and

(6) Analysis of assemblage zones (Chapters II.2, 4).

Emphasis in Project No. 148 has been on subjects (5) and (6). This includes the construction of range charts depicting periods of existence for different fossils in comparison with one another.

Several methods, as discussed in Chapters II.3, 4, 5, 6, 8 and 9, are now available for determining the most likely sequence of biostratigraphic events recorded in different stratigraphic sections and for the construction of quantitative range charts. The resulting zonations can be of either the average or conservative types. In general, average zonations will underestimate the position of the highest occurrence of a range zone at a given place while they overestimate its base. Conservative zonations are produced by sequencing methods designed to give the stratigraphically highest possible estimate of the top of a range zone and the stratigraphically lowest estimate of the base of a range zone. Their drawback is that they are sensitive to anomalous situations arising when, locally, fossils were moved upwards or downwards in a stratigraphic section due to mixing of sediments later in geologic time. When a fossil was poorly preserved, misidentification may also be a reason that its range of occurrence in a section is overestimated. Assemblage zones, overlap zones and other types of zones are easily derived from dissecting the sequence of all events. Assemblage zones can also be determined by means of multivariate statistical methods such as cluster analysis. In the latter methods, the order of successive events in time is not used but zonations are obtained from co-occurrences of different species in the samples (Chapter II.2).

As in other fields, the rapid growth of information in biostratigraphy has led to an increased demand for quantification of information. Quantitative stratigraphy is useful in this because it helps to organize the information in novel ways. The following practical example provides an illustration.

Special properties of the paleontological record form the basis of biostratigraphy. The properties include first appearance datum (entry), range, peak occurrence, and last appearance datum (exit) of fossil taxa. These events differ from physical or chemical events in that they are unique in time, nonrecurrent and that their order is irreversible. Paleontological correlation for geological studies depends on comparing similar fossil occurrences in or between regions and is commonly referred to as a paleontological zonation.

The observed order of biostratigraphic events is generally different from place to place. In correlating wells drilled for oil, occurrences of the same event in different wells normally are connected by straight lines in stratigraphic profiles or fence diagrams. If there is a reversal in order for two events in two wells, these lines will cross. The cross-over frequency for pairs of events therefore provides a measure of inconsistency.

In order to cope with numerous inconsistencies in a data base, for instance the one discussed in Chapter II.4 consisting of foraminiferal events in wells along the Canadian Atlantic margin, a computer program for the ranking and scaling of events (RASC program) has been developed which produces two types of biostratigraphical answers:

(a) The optimum (or average) sequence of stratigraphic 'events' along a relative time scale.

(b) The clustering in time, of these events, based on the cross-over frequencies of the events, weighted for the number of occurrences, using the optimum sequence of (a).

Sequencing and Clustering: A Practical Example

The RASC method is illustrated here by means of a simple example. Figure 1 shows a highest occurrence and nine lowest occurrences of Eocene calcareous nannofossils in nine sections from the California Coast Range. The number of levels per section varies from 2 (section B) to 8 (section E). The names of the 10 biostratigraphic events are shown in Table 1. On the lowest level of section A, there are 6 coeval events. Proceeding from left to right in Figure 1, these are labelled 1 to 6 in Table 1. Moving stratigraphically upwards, 3 other events were observed in A, each on a separate level; these are labelled 7 to 9 in Table 1. The triangle Δ which does not occur in A is represented by the number 10. In the bottom part of Table 1 the occurrence of the events 1 to 10 in the nine sections have been coded as input for the RASC computer program. Coeval events are preceded by a hyphen or minus sign.

The following method of sequencing the 10 events in this example was proposed 12 years ago, before a computer program was available. First a preliminary ranking was carried out with the result shown in column 1 on the righthand side of Figure 1. Next, each event was compared with all other events by counting how often it occurred (a) above and (b) below the other events. Suppose that these counts or frequencies are called f_a and f_b. The positions of events were then changed until the 'optimum sequence' of column 2 was found. The positions of the events in column 2 are such that $f_a \geq f_b$ for each event in comparison with all other events. Thus an event that falls below any other event in the optimum sequence was never observed to occur more frequently above the other event in the sections that could be used for comparison. This method can easily be programmed for a digital computer using the information of Table 1 as input. The optimum sequence obtained by means of the RASC program is shown in Figure 2. If $f_a = f_b$ for a pair of successive events in Figure 2, the events belonging to this pair do not occupy distinct positions in the optimum sequence. Their order as it results from the computer program is arbitrary. Such uncertainty is shown in Figure 2 by means of a 'statistical range.' For example, the events in positions 1 and 2 both range from 0 to 3. This means that they could occur anywhere between positions 0 (above 1) and

Table 1

Stratigraphic information shown in Figure 1 was coded as input for RASC computer program. Upper part of table shows names of events; LO - Lowest Occurrence; HO - Highest Occurrence. Lower part of table contains sequence data ordered stratigraphically upwards starting at the left side of each row. Numbers 1 to 6 correspond to symbols in bottom row of section A in Figure 1; 7 to 9 correspond to other three symbols in A moving upward; 10 represents triangle in Figure 1 (from Gradstein and Agterberg, 1982).

1	LO DISCOASTER DISTINCTUS
2	LO COCCOLITHUS CRIBELLUM
3	LO DISCOASTER GERMANICUS
4	LO COCCOLITHUS SOLITUS
5	LO COCCOLITHUS GAMMATION
6	LO RHABOOSPHAERA SCABROSA
7	LO DISCOASTER MINIMUS
8	LO DISCOASTER CRUCIFORMIS
9	HO DISCOASTER TRIBRACHIATUS
10	LO DISCOLITHUS DISTINCTUS

Section	Observed events								
A	1	-2	-3	-4	-5	-6	7	8	9
B	2	-3	-7	-4	-5	-6	-10	9	
C	2	5	1	9					
D	2	1	7	5	8	9	10		
E	2	-5	1	3	7	8	4	6	9
F	1	-3	4	-5	2	7	-8	9	10
G	7	3	-4	1	-2	-5	10	-8	9
H	7	10	-1	-5	9	4			
I	2	3	-1	5	4	6	9	10	

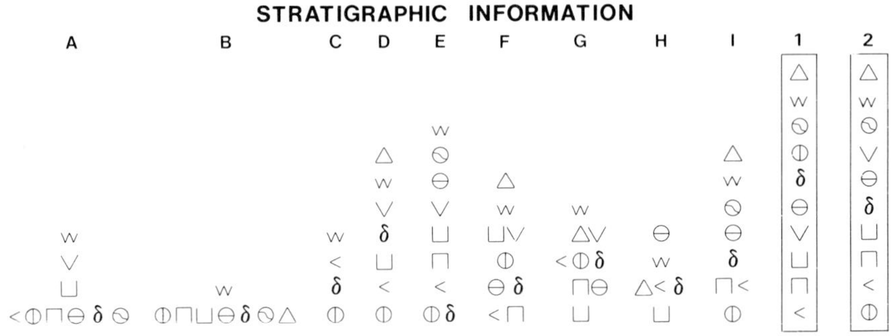

STRATIGRAPHIC INFORMATION

A B C D E F G H I 1 2

Figure 1 *Example of occurrence of 10 biostratigraphic events in 9 sections (A-I). Explanation of symbols is given in text and Table 1. Subjective ranking resulted in order of column 1. Original optimum sequence is shown in column 2 on right side. Other rankings of some 10 events are shown in Figures 2 and 3 (after Hay, 1972).*

SEQUENCE POSITION	FOSSIL NUMBER	RANGE	FOSSIL NAME
1	9	0 — 3	HO DISCOASTER TRIBRACHIATUS
2	10	0 — 3	LO DISCOLITHUS DISTINCTUS
3	8	2 — 5	LO DISCOASTER CRUCIFORHIS
4	6	1 — 5	LO RHABOOSPHAERA SCABROSA
5	4	4 — 6	LO COCCOLITHUS SOLITUS
6	7	5 — 8	LO DISCOASTER MINIMUS
7	5	5 — 8	LO COCCOLITHUS GAMMATION
8	1	7 — 10	LO DISCOASTER DISTINCTUS
9	3	7 — 11	LO DISCOASTER GERMANICUS
10	2	8 — 11	LO COCCOLITHUS CRIBELLUM

Figure 2 *Optimum stratigraphic sequence of 10 events for the example of Figure 1 and Table 1. Range (statistical) indicates uncertainty of sequence position. For further explanation, see text.*

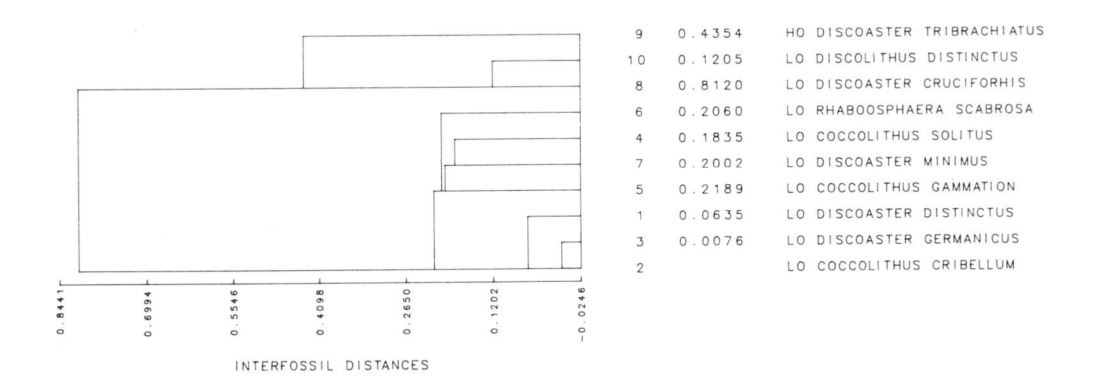

Figure 3 *Optimum clustering of the 10 events. Distances from each event to the next one were also plotted horizontally and connected by vertical lines. Note that events 1 to 6 form a cluster because they tend to be coeval. For explanation of distances, see text.*

2. These events (10 and 9) are coeval on average. Going back to Figure 1 it can be seen that Δ and W occur together in 6 sections, each three times above the other.

Other applications of the RASC program have shown that a unique solution to the problem of finding the optimum sequence with $f_a \geq f_b$ for each event compared to all others normally does not exist because of cyclical inconsistencies in the data base involving 3 or more events. For example, if in pairwise comparison an event 1 occurs more frequently above another event 2, 2 above event 3, and 3 above 1, an optimum sequence cannot be obtained without modifications of the preceding method. The method used for the RASC program is based on graph theory and provides a solution in situations of this type too. The relative position of an event in an optimum sequence is an average of all relative positions encountered. In practice, this may mean that stratigraphical ranges in range charts constructed from an optimum sequence are shorter than those constructed by other methods.

An example of optimum clustering which may lead to the definition of biostratigraphic assemblage-zones is shown in Figure 3. A biostratigraphic assemblage-zone is a body of strata whose content of fossils constitutes a natural association that distinguishes it from adjacent strata. In Figures 1 and 2, the events in the lower parts of the sections are more frequently coeval and show more inconsistencies than those in the upper parts. The cross-over frequency $P = f_a/(f_a+f_b)$ can be computed for each pair of events. Coeval events may be scored as 0.5 toward both f_a and f_b. Note that the cross-over frequency of Δ and W in Figure 1 satisfies $P = 3/(3+3) = 0.5$. By using the Gaussian distribution function, cross-over frequencies can be transformed into 'distances' D which can be plotted along a linear scale. For example, if $P = 0.5$, $D = 0$, or if $P = 0.95$, $D = 1.645$. All possible pairwise comparisons can be made and the frequencies weighted according to sample size. Estimates of average distances between successive events eventually obtained by means of the RASC program are shown in Figure 3. For example, the distance D between events 9 and 10 amounts to 0.4354. Because all possible pairwise comparison were considered, this distance differs from the $D = 0$ that resulted from the direct comparison ($f_a = f_b$, see before). The distances in Figure 3 were plotted in the horizontal direction and connected to each other as is commonly done in biometric cluster analysis. Events 1 to 7 cluster relatively strongly while 8 to 10 tend to occur above the others. Clearly the dendrogram of Figure 3 provides more information than the optimum sequence of Figure 2. In large scale applications, the RASC computer program has produced range charts and assemblage zonations which superseded micropaleontological resolution previously available (see Chapter II.4 and Appendix I).

QUANTITATIVE CHRONOSTRATIGRAPHY

An approach in which biostratigraphy, paleoecology, lithostratigraphy, and geochronology are combined with one another is called burial

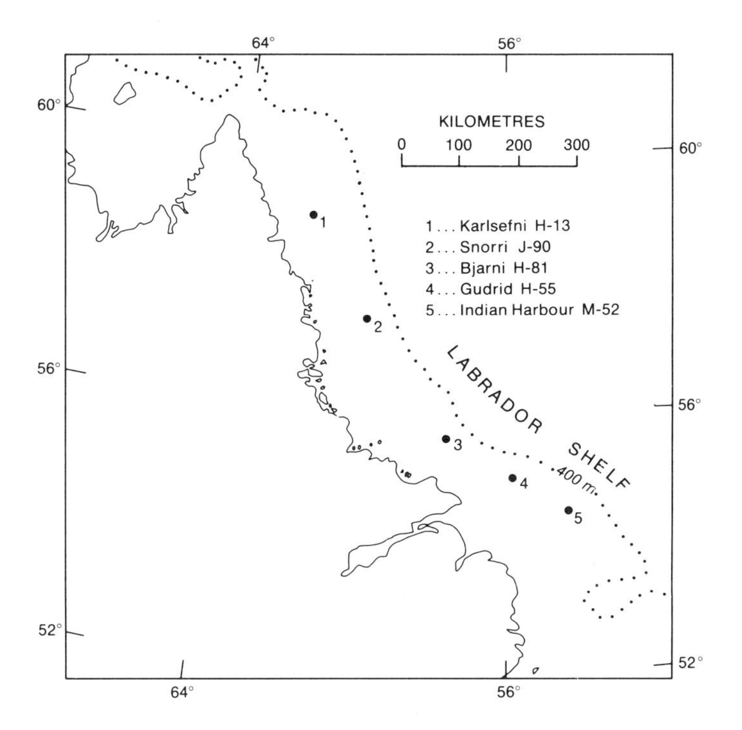

Figure 4 Locations of five wells in Labrador Shelf to be used for example in Figures 5 and 6.

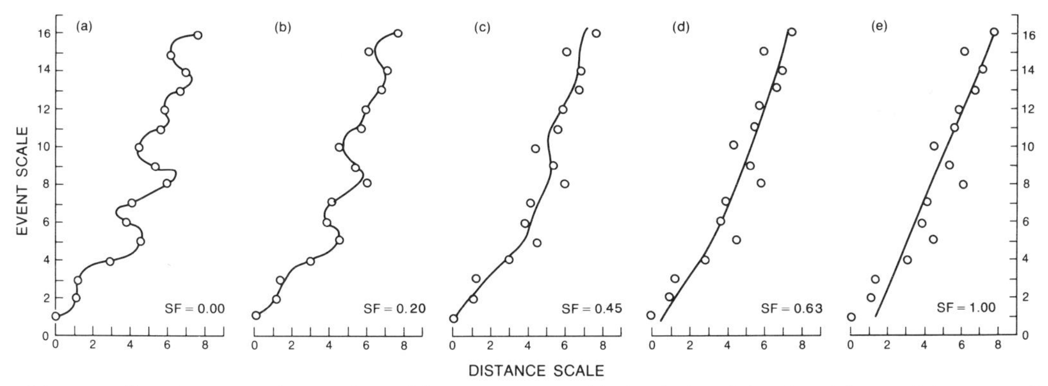

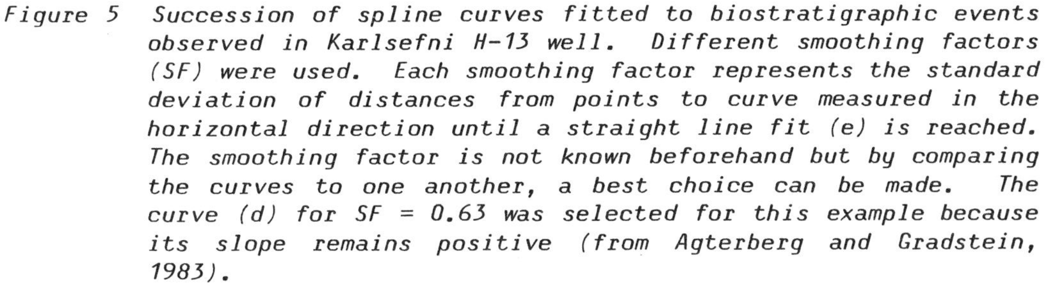

Figure 5 Succession of spline curves fitted to biostratigraphic events observed in Karlsefni H-13 well. Different smoothing factors (SF) were used. Each smoothing factor represents the standard deviation of distances from points to curve measured in the horizontal direction until a straight line fit (e) is reached. The smoothing factor is not known beforehand but by comparing the curves to one another, a best choice can be made. The curve (d) for SF = 0.63 was selected for this example because its slope remains positive (from Agterberg and Gradstein, 1983).

history. It deals with subsidence and sedimentation in time. Data from wells or sections are organized linearly with the rates of subsidence, sedimentation, thermal maturation of organic matter, etc., expressed in years, thousands of years, or larger time units. For an in-depth discussion of burial analysis, the reader is referred to Chapters III.1 and V.1.

The prerequisite of this approach would be a good calibration of zonations with respect to the geochronologic scale, but since the determination of trends is the primary objective, individual errors in calibration are less important. This is because the trends can be generalized and used for extrapolation, whereas errors in calibration produce localized 'noise' which should be eliminated if possible.

Information on rates of sedimentation, change in paleo-water depth, unconformities, and other factors can be integrated in time with sediment thickness data and paleo-waterdepth plots. Refinements are corrections for compaction and loading which will provide information on seafloor or basement subsidence, evaporite movements, undercompaction phenomena and exact timing of important changes in geological history. The linear time perspective significantly clarifies geologic history and therefore exploration geology, primarily because it allows more precise chrono-correlation, using intervals of time in millions of years or the boundaries of chronostratigraphic units. For example, the occurrence of hydrocarbons in a region may be related to these boundaries.

'Explorationists' can also establish a numeric chronostratigraphy for well sections and calculate estimates for the extent in time of the missing section at unconformities. Consequently, a new kind of cross-section can be constructed that shows isochrons imaging chronostratigraphic depositional patterns just like the seismic record does. As their geochronologic resolution normally will be higher than that of seismic sections, isochron cross-sections are most useful in the calibration and the interpretation of the seismic record.

Recently, as a follow-up to the RASC program, a computer-based method of quantitative correlation has been proposed, which uses the numerical geologic time scale resulting from RASC. The computer program is called CASC (Correlation And SCaling in time). It is an interactive program with the user typing in responses to prompts that accompany diagrams appearing on a monitor screen. At present CASC provides two types of displays of which redrafted versions are shown in Figures 5 and 6, respectively, for an application that will now be discussed in more detail.The locations of five offshore wells on the Labrador Shelf are shown in Figure 4. A separate set of biostratigraphic events (exits of microfossils only) was observed in each well. For the well Karlsefni (No. 1 in Figure 4) these events are plotted against their estimated RASC distances in Figure 5. (The RASC distance of an event is usually measured from the first event near the beginning of a section; it is the sum of all distances, in time, between successive events preceding it in

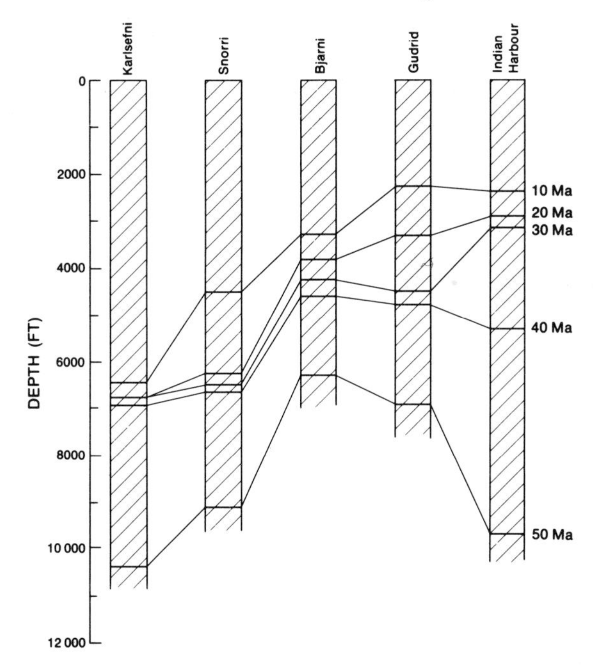

Figure 6 *Distance scale of Figure 5 was changed into age (in Ma) and event scale into depth (in ft). Similar analysis was performed for all 5 wells of Figure 4. Probable ages which are multiples of 10 Ma were plotted at depths read from spline curves and connected by lines of correlation (from Agterberg and Gradstein, 1983).*

Figure 7 *Simulated cycles consisting of limestone (black) and shale (white) layers. Each column represents one cycle with alternating shale and limestone layers reduced to unit thickness. (The true thickness of the natural cycles simulated here varies between 1 and 3 metres.) This diagram was originally displayed on the screen of an Apple microcomputer (from Schwarzacher, 1982).*

an optimum clustering as in Figure 3). If the order of events in Karlsefni were the same as that of the standard consisting of the RASC distances estimated from data for 18 wells, the curve connecting observed values in Figure 4 would never decrease with respect to distance. Any observed reversals are due to biostratigraphic inconsistencies similar to those discussed in the previous section. A spline curve was fitted in Figures 5a-e with variable smoothing factors (SF). SF represents the standard deviation of the distances from the points to the curve measured in the horizontal direction. It can be chosen before a curve is actually fitted. A spline curve consists of a sequence of cubic polynomial curves which pass into one another without abrupt changes of slope at their boundary points. Such a curve can be fitted by means of the statistical method of least squares allowing a variable amount of differences between the curve and the observed values. If SF = 0.0 (Figure 5a) the fitted curve goes through all observed values; the degree of smoothing increases until a best-fitting straight line can actually be estimated (see Chapter III). The slope of the fitted curve is related to rate of sedimentation. If measured in the anticlockwise direction from the distance scale, this slope should lie between 0° and 90°. A horizontal segment of the curve (0° slope) may reflect an unconformity; a vertical segment (90° as in Figure 5c) would represent an infinitely large rate of sedimentation which, of course, is physically not possible. Likewise, slopes exceeding 90° as observed in Figures 5a and b are not realistic. The user may therefore select the pattern for SF = 0.63 (Figure 5d) for further work.

It is possible to replace the RASC distance by geologic age (in Ma) and the event scale by depth. Then a probable age can be determined for all samples taken along a well. Identical probable ages can be connected by lines of correlation as shown in Figure 6, for the five wells of Figure 4, using multiples of 10 Ma. A long hiatus at about 6700 ft. in the northern part of the study region is clearly illustrated in Figure 6. It corresponds to a broad shelf regression in the Oligocene (37.5-23 Ma) probably accentuated by eustatic sea-level lowering. Current research on this methodology is aimed at estimating error bars for probable ages as shown in Figure 6.

QUANTITATIVE LITHOSTRATIGRAPHY

Lithostratigraphic correlation, as treated in Chapters IV.1 and 2, can be defined as the correct identification of lithological boundaries in different locations. When the correlated points are connected, they reproduce the shape of the rock body (lithosome). There is nothing probabilistic about this type of correlation and, in the stratigraphic sense, it is not even measurable. It can only be right or wrong. By establishing quantitative methods, a probability measure of whether a proposed correlation is right or wrong may be found. The similarity between two sections is a measurable quantity. If two portions in the sections are identical, this can be called a match and the number of matches is used as a measure of the similarity. An example of a simple matching technique for estimating the similarity between two successions of lithologies is to divide the number of matches by the total number of comparisons made.

A fundamental prerequisite for such quantitative approach is the meaningful numerical coding of lithologies. In addition, most quantitative modelling studies require interpolation between equal intervals. This can be accomplished by linear interpolation between irregularly spaced points along sections or by using more sophisticated tools such as the cubic spline function. Smoothing factors in spline interpolation can be determined by interactively using a computer terminal, a technique that was discussed in the previous section of this introduction. Regular series can also be created by calculating 'matching percentage' values of lithologies for line segments of equal length. A digitized well-log is an excellent source of information.

In quantitative lithostratigraphic correlation, two possible methods of approach have been recognized. The first is the well-established matching procedure, which determines the similarity between two sections. The second approach consists of prediction of an unknown section by extrapolating from known profiles. To make the second method quantitative, data on the spatial and time-domain correlation structure of stratigraphic profiles are urgently needed. The extrapolation then can be performed by computer simulation. The simulated profiles contain both known and unknown sections. The known data are exactly reproduced, but different computer simulation experiments, which have different conditional limits, produce different lithological sequences for the unknown sections. From these differences it is possible to determine the uncertainty of the predictions.

Sedimentation models in the study of stratigraphic correlation have been developed of which the objective is to predict an unknown section from known profiles by means of computer simulation. The correlation structure is used as a link between reality and hypothesis. So far, only one- or two-dimensional simulation models have been considered; e.g. with distance between profiles plotted in one direction and time in the other. There should be no difficulty in expanding this approach to three-dimensional space-time models. The question of how much can be gained by treating real situations with vastly simplified models based on the correlation structure has not yet been investigated. Nevertheless, the new simulation approach yields practical results which are complementary to the matching procedures commonly used in lithostratigraphic correlation. An example of a number of lithostratigraphic sections simulated on a microcomputer, using known profiles, is shown in Figure 7. Each section shows a single cycle of sedimentary rocks obtained from the same one-dimensional random process-model.

Because of differences in the rate of sedimentation, stretching or shrinking of sections is normally required before lithostratigraphic correlation is possible. An example of a new technique is the slotting method for pairwise comparison of sections (cf. Gordon and Reyment, 1979). Suppose that two sections with observed lithological parameters, A_1, A_2, ..., A_n and B_1, B_2, ..., B_n are to be slotted. One series, e.g. A_1, A_2, B_1, A_3, B_2, A_4, A_5, ..., can be created in which the successive data points show a minimum of dissimilarity. This method works best with continuous lithological variables as obtained in well logging.

SUMMARY

From 1976 to 1985, an international group of stratigraphers and mathematical geologists collaborated on the development and evaluation of quantitative stratigraphic correlation techniques under the auspices of the International Geological Correlation Program (Project No. 148). The present volume reflects the results of this 10-year effort. New concepts and techniques in the field of quantitative biostratigraphy as well as new methods of quantitative chronostratigraphy and lithostratigraphy have been developed. Especially during the closing stages of the project new attempts have been made to link the relative time-scales, which can now be constructed automatically on a regional basis, to the numerical geological time-scale, in order to allow isochron contouring and burial history analysis. This introductory Chapter provides a general outline of the developments leading up to these attempts, with examples to illustrate them, while the next introductory Chapter will concentrate on the stratigraphic concepts that formed the basis for these developments.

REFERENCES

Agterberg, F.P., 1983, Quantitative stratigraphic correlation techniques. Nature and Resources 19, (4), 20-26.

Agterberg, F.P., and Gradstein, F.M., 1983, Interactive system of computer programs for stratigraphic correlation. Current Research, Geol. Survey of Canada, Paper 83-1A, 83-87.

Gordon, A.D., and Reyment, R.A., 1979, Slotting of borehole sequences. Math. Geology 11, 309-327.

Gradstein, F.M., and Agterberg, F.P., 1982, Models of Cenozoic foraminiferal stratigraphy - Northwestern Atlantic Margin. In Cubitt, J.M., and Reyment, R.A., eds., Quantitative Stratigraphic Correlation, John Wiley, Chichester, 119-173.

Hay, W.W., 1972, Probabilistic stratigraphy. Eclogae Geol. Helv. 65, (2), 255-266.

Schwarzacher, W., 1982, Quantitative correlation of a cyclic limestone-shale formation. In Cubitt, J.M., and Reyment, R.A., eds., Quantitative Stratigraphic Correlation, John Wiley, Chichester, 275-286.

I.2

STRATIGRAPHY AND THE FOSSIL RECORD

F.M. GRADSTEIN

INTRODUCTION

Stratigraphy, like most geological sciences, is essentially a natural philosophy. This implies that stratigraphy is rooted in a body of organized cumulative observations, governed by a series of widely accepted principles and rules. The three main attributes of this historical philosophy are:

(1) the irreversible flow of time, often called the arrow of time;

(2) superposition of successively younger strata

(3) the results of events as fossilized in the Earth sedimentary record, and their spatial and temporal relations.

The reconstruction of the likely order and geographic extent of events and their placement within the Geological Timescale provide a framework called Earth geological history. There are many categories of events, each with special properties and values to geological history. Lithostratigraphy, magnetostratigraphy and biostratigraphy contain well used classification systems of such categories. Reasonable rules for the systems are provided in the International Stratigraphic Guide (Hedberg ed 1976).

As a geological discipline, stratigraphy over the years has developed four major components: Lithostratigraphy, biostratigraphy, magnetostratigraphy, and chronostratigraphy. Lithostratigraphy, the framework of older and younger rock units recognized mainly by their physical character, is the older of the four and its principles are generally well understood.

Seismostratigraphy is a special type of lithostratigraphy. During an acoustic shock wave experiment, gradual or abrupt density contrasts between superimposed rock units and acoustic velocity-loss at bedding contacts, lead to acoustic cross-sections of subsurface strata. Unfortunately, classification and mapping of such cross-sections lacks a standardized philosophy. As a result, seismostratigraphy and lithostratigraphy are in some applications used together.

Biostratigraphy, the global or regional record of paleontological events or zones and their limits, which are used to correlate rock sections, is the common link between lithostratigraphy and chronostratigraphy. The combination of corresponding rock units (formations) and corresponding fossil units (zones) provides independent correlations of strata. The presence of a well-defined biozone or a fossil event in several rock-sections is commonly used as a basis to postulate synchroneity in geological time.

Magnetostratigraphy uses the reversals of the earth magnetic field and their spacing through time, as fossilized in the sedimentary and igneous rocks. In combination with biostratigraphy the reversal levels provide a powerful correlation tool, with a theoretical accuracy better than that of most biozonal limits.

Chronostratigraphy, which has led to development of the commonly used scale of geological stages, is entirely relative. As a measure of relative age in geological history, reference is made to the standard chronostratigraphic scheme made up of successive stages, like Cenomanian, Turonian, Coniacian, etc., in the Cretaceous System. The stage unit is a well-delimited body of rocks of an assigned and agreed upon relative age, younger than typical rocks of the next older stage, and older than the typical rocks of the next younger stage.

The principles of this chronostratigraphy were slowly laid down in over a century or more of research in many discontinuous and incomplete sections. Facies changes and a lack of agreement on criteria, particularly fossils, to arrive at a relative age for the rocks in which they occur, have always resulted in a considerable amount of confusion on stage nomenclature and stage use.

The accurate portrayal of the geological history demands that relative and subjective scales be modified into a numerical, linear one. The conversion of a relative to a so-called absolute scale, measured in units of linear time, like one second, one year, or one million years, is embodied in the science coined Geochronology. Geochronometry, sometimes referred to as radiometry, deals with the actual measurements of - radiometrically derived - absolute ages in rocks. Stratigraphically meaningfull radiometric numbers (ages) are needed to produce a good geochronology or geological time scale. A linear time scale, expressed in millions of years (Ma) allows to calculate the rate of a geological process, like rate of sedimentation - of evolution, or of continental drift, or calculate the duration of a zone, a stage, or a glaciation. Numerical time scales and aspects of their underlying methodology are discussed in Chapter V.

THE PALEONTOLOGICAL RECORD

The properties of the paleontological record form the basis of biostratigraphy. In this record we recognize taxa and in the continuous flow of taxa through time we reconstruct events. A taxon is defined as a stable unit consisting of all individuals (fossils) considered to be morphologically sufficiently alike to be given the same (Linnean) name. On the basis of evolutionary theories we know that a taxon may imperceptibly grade into another. For our stratigraphical purpose a taxon (species, or unit of different rank) is recognized by a qualified paleontologist, whether based on single specimens or 'populations.' Commonly, we do not use categories intermediate between such taxa.

A paleontological event is the presence of a taxon in its time context, derived from its position in a rock sequence. For stratigraphic purposes we apply certain events only, such as the first occurrence (appearance, entry), the last occurrence (disappearance, exit), and the range between an entry and an exit. These palontological events are the result of the continuing evolutionary trends of Life on Earth. They differ from physical events in that they are unique, nonrecurrent, and that their order is irreversible. Deviations from the axiom of irreversibility have been found to be on the order of no more than one 'species unit,' along the trend of single evolutionary lineages (Drooger et al., 1979).

Often, first and last occurrences of fossil taxa are relatively poorly recognizable events, based on a few specimens, which are stratigraphically scattered. Particularly with time-wise scattered last occurrences, one may be suspicious that through geological reworking the observed stratigraphic record exceeds the (unknown) natural record. For this reason it is useful to distinguish between stratigraphically first and last occurrences and stratigraphically first and last regular (or consistent) occurrences (Figure 1). An observed first or last occurrence is called consistent when such range end points are part of an observed continuous stratigraphic range. Doeven et al. (1982) successfully applied the concept of the two different types of first and of last occurrences to erect a quantitative range chart using Ranking (see Chapter II.4) on 119 Late Cretaceous nannofossil taxa in 10 Atlantic margin wells.

If the fossil record encountered in the stratigraphic sections which we want to correlate would be ubiquitous and perfect, i.e. if only time would control the appearance, range and disappearance of taxa, then biostratigraphy would be a straightforward excercise. The science of biochronology, as developed for the evolutionary first and last occurrence datums of ocean plankton in conjunction with geomagnetic reversals in Deep Sea Drilling Sites would be a matter of systematic bookkeeping on a worldwide scale, only constrained by taxonomic deliberations. Unfortunately, the geological record is highly imperfect. Such a noisy record, which can make biostratigraphic decision making complex or even murky, is due to a number of factors, including those mentioned in Table 1. These are the principal uncertainty factors which bear on the accuracy of a stratigraphic correlation or other type of biostratigraphic analysis. These factors are both the result of personal bias in observation and classification and of inherent properties of the

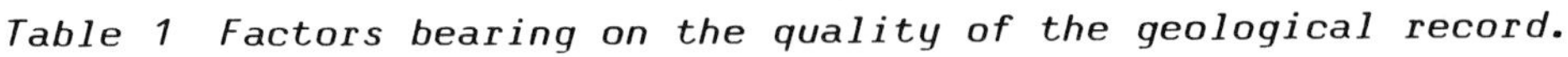

Table 1 Factors bearing on the quality of the geological record.

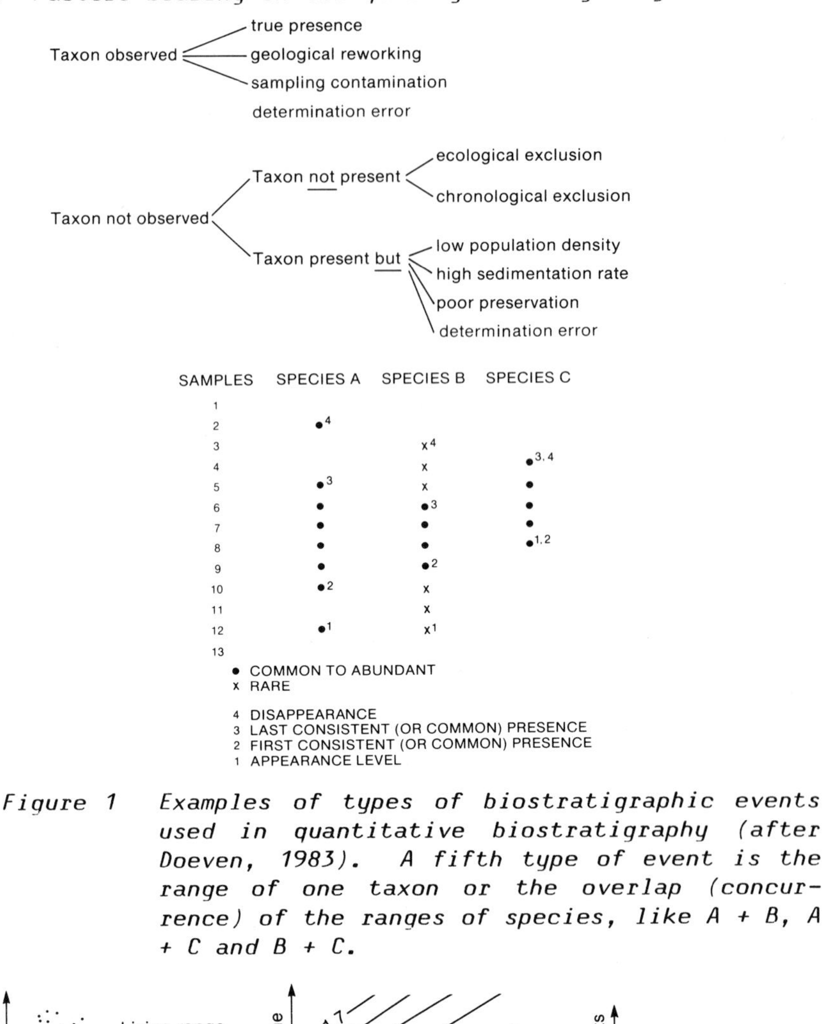

Figure 1 Examples of types of biostratigraphic events used in quantitative biostratigraphy (after Doeven, 1983). A fifth type of event is the range of one taxon or the overlap (concurrence) of the ranges of species, like A + B, A + C and B + C.

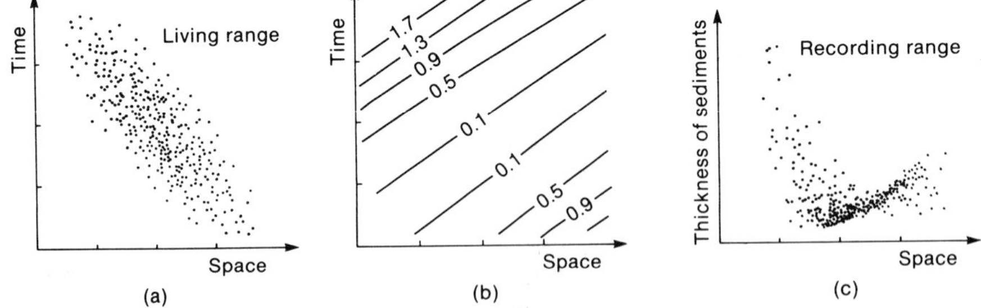

Figures 2a-c Simple, theoretical example which illustrates how the unknown geological living range of a species changes to the actual observed, recording range in sediments (after Davaud 1982):
a. Living range of a species in time.
b. Sedimentation rate gradient during burial.
c. Recording range after burial.

paleontological record. Together these uncertainty factors can be summarized as follows:

(a) Sampling details and the frequency of fossil taxa

(b) Confidence of taxonomic identification

(c) Influence of environmental change on the stratigraphic range of fossils

(d) Differential rate of taxon evolution in different parts of the world

I will briefly review the first two in the context of some recent studies. Influence of environment on stratigraphic range and differential rates of taxon evolution are geological factors and outside the realm of simple modelling.

SAMPLING AND THE FREQUENCY OF FOSSIL TAXA

As shown in Table 1, the presence or absence of a taxon depends on multiple causes. Population and sampling densities directly influence the chance that the total stratigraphic range of a given species has been observed. If it is reasonable to assume that the frequency of specimens in any sample is a function of random variables, statistical inferences may be drawn on the effects of sample size to detect an event. Hay (1972) and Zachariasse et al. (1978) are among those who give useful information on these effects and on the probability of event detection. Unfortunately, it is probably not reasonable to assume that random mixing of fossils has taken place in the sediment sampled. Linked co-occurrences or exclusions due to ecological factors and sampling problems are non-random factors. These are likely to result in poor predictive power of specimen frequency, using a theoretical frequency distribution only.

In the case of geological factors it will not be possible to evaluate the effects of errors which help to calculate the chance that a species at a locality was detected. Davaud (1982) gives a simple theoretical example which illustrates how changing rates of sedimentation, both in time and in space affect the probability of discovery of a particular taxon.

In Figure 2a is a so called time-space domain of the geological living range of a particular species. This entity is abstract and in the words of Davaud (1982) can be thought of as a four dimensional volume, outside of which the probability that the species is present is nil. Within this volume the probability that the particular species occurs is greater than zero, but fluctuates considerably as a function of population density. In the example this density is suggested by the number and spacing of points.

During the same period of time and in the same geographic region, sedimentation rate changed regularly, as shown in Figure 2b, where higher numbers denote a higher rate. As a result, after burial and fossilisation the specimens are in a three-dimensional sediment domain. The lithostratigraphic axis or thickness relates to geological time in proportion to the sedimentation rate. S (x, y, t) is a continuous function showing the local and temporal variation of sedimentation rate. Individuals which lived at time t^0 and at point x^0, y^0 will be located in the sedimentary record at point x^0, y^0, z, where z is equal to the integral of S for each increment of time.

In Figure 2c Davaud (1982) shows the simulated recording range of the species. Clearly, the sedimentary record distorts the simple time - space information. At the base of the recording range the chance to find the species is much greater than higher up the stratigraphic section, where the higher sedimentation rate stretched and thinned the record.

The rate and effects of geological factors on the chance of event detection will generally remain obscure and cannot be modelled prior to extensive sampling and stratigraphic analysis. On the other hand, it is widely known from repeated observations, that for many groups of organisms the majority of taxa is found at few sampling sites and with few specimens. The number of taxa which is common to abundant throughout decreases in proportion to the number of sections studied. Also, most taxa occur in few sites only and few taxa occur in all sites. A practical and reasonable question to ask is, if in fact there is some kind of simple relationship between frequency of taxon recognition and number of sample sites. Figure 3 shows (a) the cumulative number of stratigraphic last occurrences for 169 taxa of Cenozoic foraminifers, as distributed in 24 wells (Gradstein and Agterberg 1982), (b) the same for 116 taxa of Mesozoic foraminifers in 16 wells (Williamson, in press), North Atlantic margin, and (c) the same for first and last occurrences of 220 Late Jurassic to Early Cretaceous radiolarian taxa in 76 outcrop sections and wells in the Mediterranean and Atlantic realms (Baumgartner, 1984). The graphs show that the number of entries or exits of taxa which occur in at least 1, 2, 3 .... n sites steadily decreases. Several authors (in Buzas et al. 1982) have shown previously that the distribution of species occurrence can be modelled, preferably with logarithmic conversions of frequencies. Such graphs, as fitted to a particular dataset predict the number of taxa occurring 1, 2 or ...... n times and help the biostratigrapher to decide on threshold values for the number of taxa and sections needed for a reliable quantitative zonation (see Chapter II.4 on "Ranking and Scaling in Exploration Micropaleontology"). In Atlantic margin well sites, 95% of the taxa were detected after study of 10-15 wells, and few taxa (< 15) occur in 10-20 or more sites. Over 50% of all taxa in the three examples quoted, occur in less than 10 sites.

These numbers should not be construed to indicate that zonations will not increase in quality after studying more than 20 or so sites.

Particularly in quantitative methodology an increase in the number of observations on the order of events increases the relative precision of the results. Such precision means a greater predictive power on relative order and distance in time, and is not necessarily tied to resolution in time.

This account is not finished without some words on the absences of fossil events in the sediments. It is true, as Hay (1972) has clearly expressed that "although biostratigraphers are universally loathe to admit it, biostratigraphy depends as much on the absence as it does on the presence of a certain fossil group," This remark is particularly tailored to microfossils, which are generally abundant, have many widely occurring taxa, and show many stratigraphically significant events. Only if non-existence is recognized in many stratigraphically well-sampled sections, may absences be construed as affirmative for stratigraphic interpretations. If only small samples are available over a long stratigraphic interval, the chance to find long-ranging taxa considerably exceeds the chance to find short ranging forms, unless the latter are abundant. In actual practice, so-called index fossils are generally less common, and have a short range.

CONFIDENCE OF TAXONOMIC IDENTIFICATION

Stories abound of specialists on a group of fossils who, when on different occasions asked to identify a rare taxon, will give different answers. In general, confidence of taxonomic identification is directly proportional to the number of specimens found; it also is inversely proportional to the number of observers used to identify all taxa in an assemblage. In a published experiment, four observers, one American and three Europeans, all reasonably familiar with Pliocene benthic foraminifers were presented with the same set of 200 specimens in a Mediterranean Pliocene sample (Zachariasse et al. 1978, p. 57-62). It turned out that none of the scientists had more than 10 out of 30 or so generic names in common with his colleagues. With respect to species names the differences exceeded 75%. The two participants from the same 'school,' had the higher similarity score, but still showed considerable difference in appreciation of taxonomy.

Although this was only one experiment and a harsh one because specimens were generally undersized, it underscores the danger of treating literature data to further analysis. Particularly in quantitative stratigraphic and in quantitative paleoecologic studies, standardization of taxonomy by one or a single group of observers, working with the same species concept is essential. Standardization also involves a ruling on taxonomic aberrations, for example inclusion of specimens identified as affinis (aff) with the nominate taxon, and treatment of confer (cf) identifiers as separate entities.

Methods like numerical taxonomy and biometrical studies of morphology, using population concepts, are cumbersome to execute. This

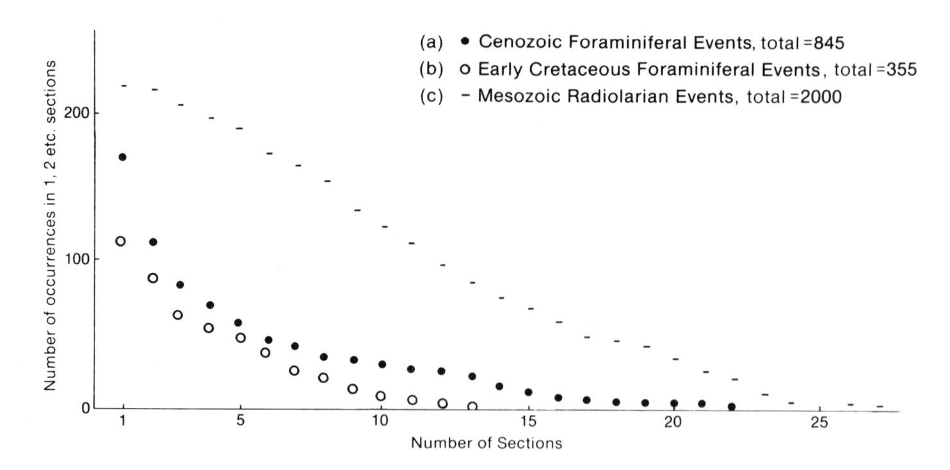

Figure 3 *Curve a - Cumulative frequency of stratigraphic last occurrences for 169 taxa of Cenozoic foraminifers, as distributed in 24 northwestern Atlantic margin wells.*
Curve b - Cumulative frequency of stratigraphic last occurrences for 116 taxa of Late Jurassic and Early Cretaceous foraminifers, as distributed in 16 northwestern Atlantic margin wells (after Williamson, in press).
Curve c - Cumulative frequency of stratigraphic first and last occurrences for 220 taxa of Late Jurassic and Early Cretaceous radiolarians in 76 sample sites in the Mediterranean and Atlantic realms (after Baumgartner 1984).

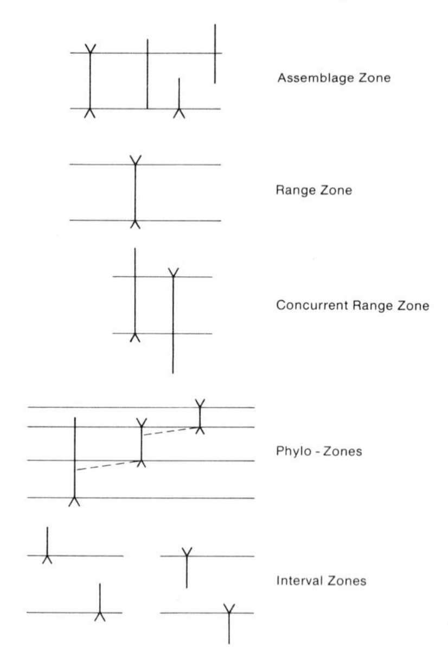

Figure 4 *Common type of zones used for paleontological correlations of strata.*

makes these otherwise more objective methods less attractive for routine examination and classification of large numbers of microfossil taxa in many sites. The latter is a prerequisite for quantitative stratigraphy where routine examination and identification of hundreds of taxa in dozens of sites may have to be performed.

More objective type taxonomies have not gained widespread acceptance. For one reason its applications run counter to the conventional typological, Linnean nomenclature of a central morphotype. Nevertheless, quantitative taxonomy has become essential in selected stratigraphic applications, for instance, to describe the evolutionary trends in the internal apparatus of larger benthic foraminifers. The evolution of Senonian **Orbitoides** (Van Hinte 1965; Van Gorsel 1975), mid-Tertiary **Lepidocyclina** (Adams 1970), and **Miogypsina** (Drooger 1984) are good examples. For those who need further introduction in the many applications of biometrics to stratigraphic micropaleontology there is a comprehensive review in Scott (1974).

Numerical taxonomy has been successfully applied to fish debris (mostly teeth) in Tertiary deep sea samples (Doyle and Riedel, 1979). The system for describing the debris is based on a string of letters and numbers. The morphologic descriptor corresponds to the two-dimensional character of the image as seen in a transmitted-light microscope. The stratigraphic distribution of the fish-debris morpho-species has been used to arrive at a preliminary Tertiary chronostratigraphy of the red clays in the central North Pacific.

The inroads of high speed, high resolution image processors which convert an analog to a digital image and the development of two-dimensional shape descriptors (based on size independent shape formulas) is finding practical application in stratigraphy (Healy-Williams 1983). The new techniques, which make shape analysis more objective, may transform micropaleontological taxonomy from an art into a science.

WEIGHTING OF STRATIGRAPHIC EVENTS

W.R. Riedel and his colleagues have given thought to the problem of weight as a measure of the quality or reliability of events (Riedel 1981; Westberg and Riedel 1982; Riedel and Westberg 1982). In conventional biostratigraphy, some weighting of observations is taken for granted, although rarely accounted for in a satisfactory manner. In quantitative stratigraphy it is equally desirable to index events according to quality, as for example in the methods developed to objectively detect index fossils using Relative Biostratigraphic Value (RBV), (Millendorf et al. 1978, and Chapter II.1).

The original weighting method as proposed is particularly concerned with the vagaries of the stratigraphic record of radiolarians. I reiterate the method in more general terms, as it can be easily adopted to specific fossil groups and situations.

Cornerstone of the method is the index of reliability for each fossil event occurrence in a sample of standardized size. The index takes the format of a score, obtained by multiplying individual scores for each factor and expressing the resulting fraction as a percentage.

Eight different factors are considered (Table 2):

(1) Abundance of taxa is rated in three ranges. One specimen observed in a sample of standardized size scores 0.25; 2-5 specimens 0.50 and more than 5 specimens 1.00. Particularly when dealing with first or last occurrence events a rating of less than 1.00 indicates with a lower probability that the stratigraphic endpoints of a taxon have been detected;

(2) Average relative abundances of ancestor and descendant above and below evolutionary transitions is also rated in three steps. If the average of descendant/ancestor count above the event and ancestor/descendant count below the event is less than 2 x, the score is 0.50. For a factor of 2-5, the score is 0.75, and for more than 5 x it is 1.00.

(3) There is a three step rating for the situation where the ancestor or the descendant of a taxon is known or not, and also the type of morphological transition. A cryptogenetic origin or plain (? local) extinction scores 0.50; evolutionary extinctions and morphotypic limits which bracket an evolutionary transition score 0.75 and evolutionary transitions 1.00.

(4) The facility, using a proper differential diagnosis, of distinguishing a taxon from co-occurring species is relatively subjective. Difficult scores 0.10, moderately easy 0.50, easy 1.00.

(5) The preservation of the assemblage relative to the taxon scores 0.10 for poor, 0.50 for some specimen poor and 1.00 for good. If it is suggested that poor preservation in the assemblages above the upper limit, (or below the lower limit) of a taxon prevented detection of the upper limit, the observed event is downgraded.

(6) Constancy of occurrence in the sequences above or below the event (not applied to evolutionary transitions) rates as follows: Interrupted 0.75, constant 1.00. This weighting compares to the first or last regular or consistent occurrence as opposite to just plain first or last occurrence.

(7) Relation of the sample locality to the periphery of the area of distribution of the taxon scores: Near periphery, 0.25; unknown, 0.50 and well within the area of distribution, 1.00.

Table 2 Scoring method of the reliability of biostratigraphic events as observed in samples (after Riedel 1981).

Factor \ Multiplier	0.10	0.25	0.50	0.75	1.00
a) Abundance of taxon (numbers of specimens observed in each sample)		1	2–5		>5
b) Average of relative abundances of ancestor and descendant above and below the event (applies only to evolutionary transitions)			<2x	2–5x	>5x
c) Type of event			cryptogenetic origin or extinction	evolutionary extinction, or morphotypic limit, or evolutionary offshoot	evolutionary transition
d) Ease of distinguishing the taxon from co-occurring fossils	difficult		moderately easy		very easy
e) Is the assemblage well preserved, relative to the preservation threshold of the taxon concerned, or are samples immediately above or below the event poorly preserved or highly diluted?	no		uncertain, or some difficulty due to preservation		yes
f) Constancy of occurrence in the sequence, above or below the event (not applicable to evolutionary transitions)				interrupted	constant
g) Relation of locality to periphery of area of distribution of taxon		near periphery	unknown		well within
h) Effects of reworking	the limit as here interpreted may be unreliable due to reworking			occurrences outside this range are certain to be due to reworking	no reworking

Figure 5 Artistic vision on 'Quantitative Biostratigraphic Analysis, created in 1972 by J.. Van Der Linden (Utrecht).

(8) Effects of reworking score in proportion to the confidence that assignment of an event position is below a reworked one. It is 0.75 when a natural event is thought to be below a reworked one, no reworking scores 1.00 and 0.10-0.50 means that the range limits as interpretated are unreliable due to reworking.

The original proponents of this detailed scheme did some research into the effects of this scoring system on their considerable Neogene data in Deep Sea Drilling Sites. It was generally found that the frequency of assigning specific values to the reliability score was more or less evenly spread over all the data and not particularly biased or non-sensitive.

That the index caters to the need for an objective system of morphological description is beyond dispute. A rating of the data according to such a scheme facilitates weighting of input. On the other hand, the index is a composite of different qualifiers, which means that similar indices for two different events may in fact be the result of a completely different set of qualifiers. Behind such equal score values a different series of unique qualities is found. Particularly when algorithms are used to analyze and interpret a weighted set of data, this dissimilarity of weight numbers may be circumvented. The data are manipulated with only one qualifier at a time and analysis is performed separately for each weight criterion. The finding of mutually exclusive results for each solution provides justification for non-addition of scores for each qualifier. Further research on this subject and practical applications are badly needed.

PALEONTOLOGICAL CORRELATION

Zonations are empirical schemes, which emphasize the temporal and spatial restriction of morphologically distinct fossil taxa. Good zonations have zonal units with well defined lower and upper limits, are easily recognizable in many sections, correlate well, and have been compared to other regional or extra-regional zonations. Some evolutionary considerations are desirable. Chronostratigraphic meaning may have been verified in the stratotype or some other classical sections.

Zonations are used to correlate and date sedimentary strata. Although it is possible to statistically express the similarity of two, presumably correlative fossil assemblages, this index does not necessarily bear on their degree of the assemblages being coeval. The famous Lyellian percentages for the increasing degree of mollusc fauna communality through time, with the Recent as standard, fare poorly in lateral correlation. Facies changes easily lead to misleading conclusions as they have in the past. Particularly for detailed stratigraphy, communality is as much or more a question of environment than of time.

A zone is a body of strata commonly characterized by the presence of certain fossil taxa. The most common type of zones are (after Hedberg ed 1976; Figure 4):

(1) Assemblage Zone ------- a group of strata characterized by a distinctive assemblage of fossil taxa;

(2) Range Zone ------ a group of strata corresponding to the stratigraphic range of a selected taxon in a fossil assemblage;

(3) Concurrent Range Zone ------ the overlapping part of the range zones of two or more selected taxa. The use of two or more taxa whose range zones overlap reinforces correlation.

(4) Phylo-Zone ------ a body of strata containing a taxon representing a segment of a morphological-evolutionary lineage, defined between the predecessor and the successor. The taxon is part of a lineage with morphologically well defined increments assumedly in stratigraphic order; and

(5) Interval Zone ----- the stratigraphic interval between two successive biostratigraphic events. Zones in drill cutting samples are interval zones.

The use of Acme-Zones, based on the presence of many specimens of a taxon, is tenuous and probably more of ecological significance than anything else. Partly under the influence of the development of a paleomagnetic reversal scale, which promises virtually isochronous correlations for specific horizons, efforts have been made to arrive at detailed sequences of evolutionary fossil datums. This effort has been particularly successful in the marine planktonic record of the last 150 Ma as preserved in Deep Sea Drilling sites. These datums are based on the First and Last Occurrences of calcareous and silicious microfossils as recorded in many ocean and also on land sections (FODs and LODs in biochronology; Berggren and Van Couvering 1978). In theory this allows for more or less reliable point correlation in time. We assume that correlation lines established through connecting zones correspond to time lines, but this remains a hypothesis (Drooger 1974). To equate biostratigraphy with chronostratigraphy and a-priori substitute biozone by chronozone is misleading. A biostratigraphically perfect correlation may be strongly diachronous, it may nevertheless be of value for basin analysis. The assumption of contemporaneity will have to be verified through other means, particularly multiple correlations.

Careful and innovative field and laboratory techniques lead to a more objective biostratigraphy, and a better understanding of its limits. Such techniques include:

(1) The application of multiple correlation criteria;

(2) Replicate sampling and replicate taxonomic and stratigraphic procedures by different observers;

(3) Quantification of conventional biostratigraphic analysis; and

(4) Quantitative taxonomy.

The remedies under 1. and 2. are commonplace in any type of geological research where the stratigraphic underpinning is critical, as for instance in the Deep Sea Drilling Project. Quantitative taxonomy was mentioned earlier, but is outside the scope of this book.

QUANTIFICATION OF CONVENTIONAL BIOSTRATIGRAPHIC ANALYSIS

Stratigraphers try to develop models that with a minimum of data will yield a maximum of predictive potency (Agterberg 1978). The properties of sedimentary strata, such as nature and order of fossil content are the input, and the output is, for example a matching or correlation of stratigraphic sections. The cartoon in Figure 5 is an artistic vision on quantitative biostratigraphic analysis, using paleontological input in a computer that produces literature-ready output. Critics of inroads of statistics in paleontology may be heard to say that 'the paleontological record defies statistical methods'. True as that may be in the traditional concept of stratigraphy, no one should lightly dismiss the geological record as artistic. The qualitative nature of conventional stratigraphic models is both their attraction and their weakness. The models are very flexible, but do not provide, for example, a rigorous stratigraphic correlation or a numerical match.

W.R. Riedel (1981) once wrote 'Biostratigraphy will continue to be regarded as an art rather than a science, until it is possible to attach confidence limits to suggested correlations.'

Schwarzacher (this volume) is of the opinion that correlations are not probabilistic and he argues that a correlation may only be right or wrong. Of course, it may be possible to asses the quality and reliability of the observed occurrences, like fossils or rock types, to be correlated. Events or zones may be poorly or well constrained in the actual rock sections, in relative-, or in linear time. Under certain conditions it is possible to attach a measure of uncertainty in time or in depth to the position of correlation lines.

It is in this spirit that recently attention has been focused on the theoretical, mathematical solution of correlation-type problems, which view biostratigraphic sequences as random deviations from a true solution. The solution faces four sources of uncertainty:

(1) The uncertainty that the optimum, or 'true,' sequence of fossil events has not been established. This ranking problem is discussed in detail by several authors (see under Sequencing Methods in Table 3). Under influence of Hay's (1972) original paper on "Probabilistic Stratigraphy," ranking of events in time to arrive at the stratigraphic order is often referred to as Probabilistic Stratigraphy. But, as Agterberg and Nel (1982a and Chapter II.5) have pointed out, there are no simple models to rank stratigraphic events according to a numerical probability. The problem is that order in time is based both on direct

and on indirect estimates. For example, the fact that A occurs above B in several sections ranks the same as that A in some sections occurs above events C, D, E, F and G, and that in some other sections C, D, E, F and G occur above B. Both cases lead to the conclusion that A occurs above B, although there is no simple test to extract this in a numerical probability.

(2) The uncertainty that the distance of fossil events along a relative time scale is not known (spacing or scaling problem). In conventional biostratigraphy extensive use is made of distances in time between events or (non) overlap of ranges to produce assemblage zones. In the simple, graphical technique of the composite standard (Shaw 1964), distance between two or more successive events is a function of the relative dispersion of each event in the sections considered; first occurrence levels are minimized and last occurrence levels are maximized, but no direct standard errors are available on the composite positions.

(3) The uncertainty that the geographic distribution of an event is not known. Drooger (1974) refers to this as traceability. Since, as pointed out earlier, proportionally less taxa occur in more sites, the majority of species is always rare and hence recovery will be strongly affected by the vagaries of lateral change in facies and rate of sedimentation. Nevertheless, given enough sampling points and counts, interpolations may be used, like confidence intervals in scatter ellipses, to predict the potential presence of each species.

(4) The error in fixation of biostratigraphic events at the scale in a well, or outcrop section. This is basically a technical error which calls for an understanding and mathematical expression of the errors in field and laboratory techniques.

In order to arrive at an optimum zonation and in order to attach some kind of confidence limits to correlations, considerable quantitative insight into these four parameters is required.

In recent years a number of quantitative methods have been proposed (Table 3), which deal with one or more of these four uncertainties. None is all-encompassing, but the insight gained in the stratigraphic properties of the data allows more objective inferences in general. Key zonation and correlation methods, each of which have been properly tested and used to solve specific geologic problems, include:

(a) Multivariate methods for zonation, including Q and R modes of covariance, clustering, and Seriation

(b) Sequencing methods for zonation, including Composite Standard technique and the methods of Unitary Association, Ranking and Scaling.

(c) Quantitative correlation based on ranking and scaling results.

I will summarize objectives and results for each of the key methods for making biozonations, and will give an example of the dual, ecologic and stratigraphic, value of the multivariate methods dealt with in detail in Chapter II.2. For details on the theory and the geological application of Unitary Association, Ranking and Scaling the reader is referred to Chapters II.4 through II.8. Chapter III.1 deals with quantitative biostratigraphic correlation, including confidence limits to tiepoints of correlation lines.

MULTIVARIATE METHODS

Multivariate methods of correlation, using sample by sample matrices of similarity, or distance coefficients, seek clustering of samples (Q-mode) as a function of comparative fossil content. In the final dendrogram, the level of clustering of samples may be selected according to a value which is a function of the degree of association of the original taxa observed. Biostratigraphic fidelity is a simple numerical expression of the preference of a species for a particular cluster (zonal) unit. Depending on the similarity coefficient and weighting procedure selected, multivariate cluster analysis and expression of biostratigraphic fidelity for taxa in the final dendrogram will define assemblage type zonations. Excellent recent reviews are in Brower et al. (1978) and Millendorf et al. (1978). Individual dendrogram clusters may be either of paleoecologic or stratigraphic significance, or both. The same is true for multivariate clustering on species by species matrices (R-mode). The latter may be insensitive to rare and scattered first and last occurrences of taxa, but such may be an advantage for robust correlation.

An advantage of R-mode clustering over either Ranking and Scaling or Unitary Association is its application to small data sets. It is probably true that the latter two, more rigorously stratigraphic methods work best using 10 or more stratigraphic sections (Gradstein and Agterberg 1982; Doeven et al. 1982; Baumgartner 1984). On the other hand R-mode clustering may be successfully applied to much smaller data sets.

An example of such an application with both paleoecologic and stratigraphic meaning is drawn from the vertical distribution of 66 taxa of deep sea Jurassic foraminifers in 119 core samples in Deep Sea Drilling Project (DSDP) Site 534, Atlantic Ocean (Gradstein, 1983). Clustering is based on the Dice coefficient of similarity between species i and j (S_{ij}),

$$S_{ij} = \frac{2\ C_{ij}}{N_i + N_j}$$

Table 3 Selected studies in quantitative biostratigraphy, discussed in this book.

HISTORICAL - PHILOSOPHICAL OVERVIEW

Hay and Southam 1978
Brower 1981
Edwards 1982

MULTIVARIATE ANALYSIS

Q AND R MODES	SERIATION	'COMPOSITE STANDARD'
Hazel 1970; 1977	Brower and Burroughs 1982	Hohn 1982
Brower et al 1978		
Hohn 1978		
Millendorf et al 1978		

SEQUENCING METHOD

COMPOSITE STANDARD	RANKING AND SCALING	UNITARY ASSOCIATION
Shaw 1964	Hay 1972	Rubel 1978; 1984
Miller 1977	Worsley and Jorgens 1977	Guex 1977
	Edwards and Beaver 1978	Davaud and Guex 1978
	Edwards 1979	Guex and Davaud 1982
	Blank 1979; 1984	Davaud 1982
	Harper 1981; 1984	Baumgartner 1984
	Gradstein and Agterberg 1982	
	Agterberg and Nel 1982 a, b	
	Doeven et al 1982	
	Blank and Ellis 1982	
	Hudson and Agterberg 1982	
	Heller et al 1983	
	Agterberg 1984	
	Gradstein 1984	
	Baumgartner 1984	
	Williamson in press	

QUANTITATIVE CORRELATION

Agterberg and Gradstein 1983
Williamson in press

where C_{ij} = number of samples in which taxon i and taxon j co-occur; N_i = number of samples with taxon i; N_j = number of samples with taxon j. To reduce the chance on trivial results, it was stipulated that participating taxa had to occur in at least 5 out of 66 samples; in this way weight was given to frequency of occurrence.

The dendrogram in Figure 6 shows a low level of clustering of taxa, but one cluster with **Lenticulina quenstedti** is made up of species restricted to a short stratigraphic interval of cores 104 to 99. The other "clusters" are probably more typical of abyssal assemblages in the Jurassic ocean. The **L. quenstedti** cluster is known to occur in three other DSDP Sites, the age is approximately Kimmeridgian, in agreement with multiple chronostratigraphic interpretations using also radiolarians, nannofossils, dinoflagellates and geomagnetic reversal interpretations. This preliminary, and relatively simple, objective, interpretation of the fossil distribution in Site 534 may serve as a working hypothesis for further research in the actual correlative value of this **L. quenstedti** cluster and its optimal composition.

UNITARY ASSOCIATIONS

The method of Unitary Association (U.A.) finds a stratigraphic succession of concurrent range zones, using the observed or virtual co-occurrences of taxa (Guex 1977; Davaud 1982). During execution of the computer program, complex ranking relationships have to be resolved, akin to cycle destruction in the Ranking and Scaling Program. The method computes a value of reproducibility, which is the ratio of the number of sites in which a U.A. (the approximate equivalent of a concurrent range zone) is positively identified and the number of sites in which it might occur. Higher reproducibility means better zonal body or better correlation value. Baumgartner (1984) also involves more subjective criteria to establish the final number of U.A.s, like number and chronostratigraphic usefulness of defining species, coincidence of zonal boundaries with significant geological boundaries, geographic distribution of the U.A. units and their stratigraphic extent.

RANKING AND SCALING

Ranking techniques produce optimum sequences of events in time, based on estimates of relative order of pairs of events. Such events are first or last occurrences. Depending on the method used, estimates of average and total stratigraphic range may result. The latter is more sensitive to outliers in the data. The optimum sequence shows an average order of fossil events. By combining the results based on lowest and highest occurrences, charts of optimum stratigraphic ranges (a type of range chart) may be constructed.

The scaling technique uses the degree of inconsistency in position of the events as a measure of interfossil distance. This inconsistency is the cross-over frequency in all pairs of events over all

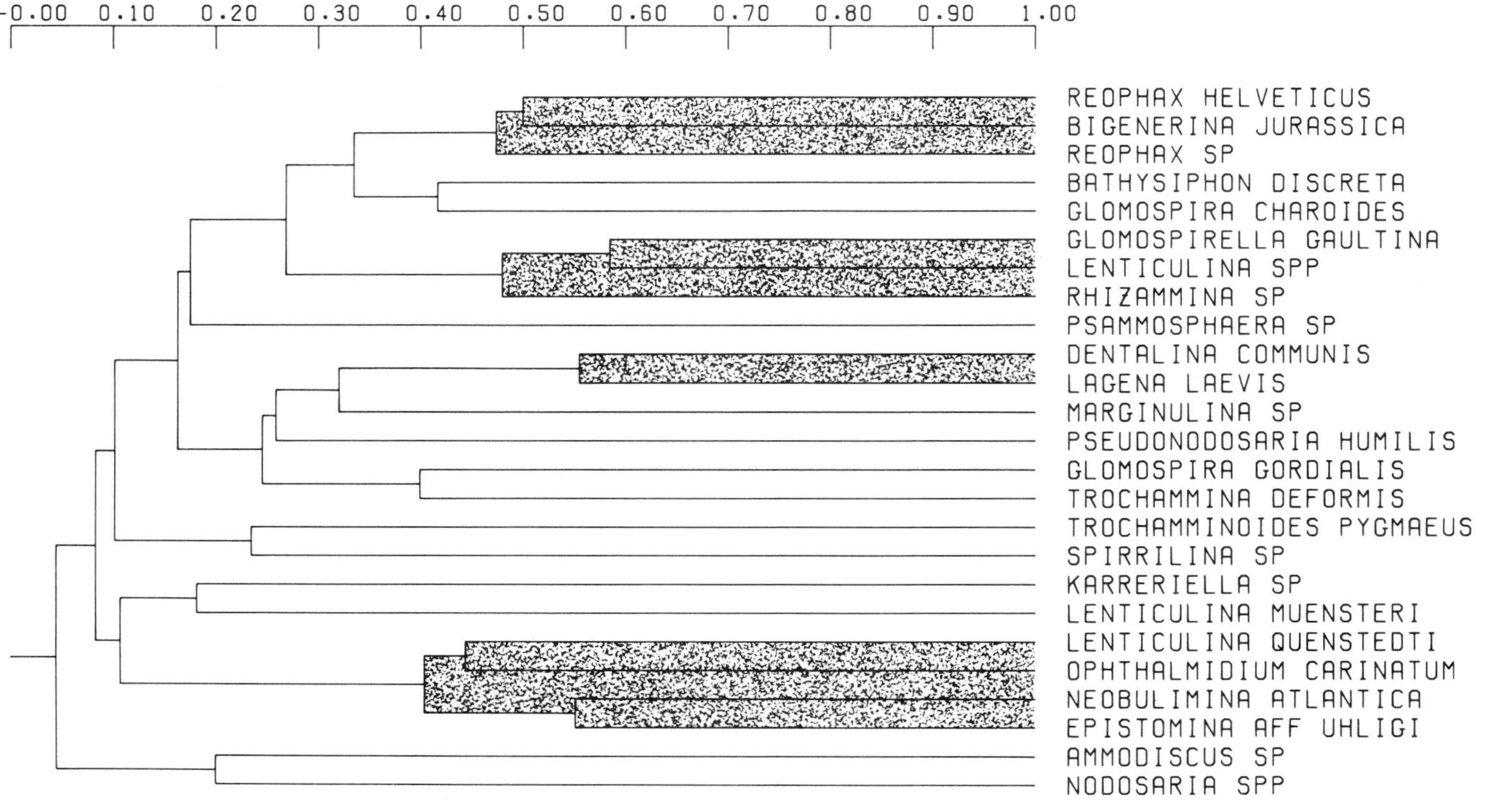

Figure 6 *Dendrogram using weighted R-mode clustering based on the Dice coefficient of similarity of those Jurassic deep sea foraminifers that occur in at least 5 out of 66 samples of Deep Sea Drilling Site 534 (after Gradstein 1983).*

sections. Scaled optimum sequences resemble assemblage zonations.

Normality testing evaluates the degree of correspondence of the individual stratigraphic record to the standard. A simple and effective tool is the scattergrams of the optimum sequence versus the individual ones. This way stratigraphically anomalous events, the result of reworking, sampling etc., may be more objectively accounted for. A cleaned set of stratigraphic sections, in which anomalous events have been eliminated, then produces a potentially improved zonation for detailed correlation and age interpretation. Provincialism or traceability in the fossil record may be dealt with through separate ranking and (or) scaling solutions for partial data sets. The data may be partitioned geographically in such a way that the difference between the quantitative zonations for each sub-set of data is maximized. Details on the ranking and scaling method and its applications in micropaleontology are in Chapters II.4 through III.1.

CONCLUSION

The properties of the fossil record are amenable to a variety of quantitative techniques. The attractions of quantitative stratigraphy may be summarized as follows:

(1) Standardization of the data and methods used gives non-specialists access to the data and the interpretations based on it;

(2) The methods and answers are more objective than in conventional stratigraphy;

(3) It optimizes stratigraphic quality and resolution, using most data;

(4) Unlike subjective biostratigraphy, it generally provides more than one possible solution, depending on input parameters which can be specified.

(5) It can handle large and complex data; and

(6) Under certain conditions, explained in Chapter III.1 it provides quantitative correlations, including confidence limits in time or in depth units of the lines that connect corresponding zones or events.

Successful marriage of the more flexible, subjective analysis with the more rigorous and objective quantitative approach will considerably enhance the quality of biostratigraphy.

REFERENCES

Adams, C.G., 1970, A reconsideration of the East Indian Letter Classification of the Tertiary. Bull. Brit. Museum (Nat. Hist.), 19 (3), 87-137.

Agterberg, F.P., 1978, Analysis of spatial patterns in the Earth Sciences. In Geomathematics: Past, Present and Prospects, ed. D.F. Merriam - Syracuse University, Geology Contr. 5, 7-19.

Agterberg, F.P. and Nel, L.D., 1982, Algorithms for the Ranking of Stratigraphic Events. Computers and Geosciences 8 (1), 69-90.

Baumgartner, P.O., 1984, Comparison of Unitairy Associations and probabilistic Ranking and Scaling results applied to Mesozoic radiolarians. Proc. 5th Int. Symp., IGCP Project 148, Geneva 1982. Computers and Geosciences 10 (1), 167-185.

Berggren, W.A., and van Couvering, J.A., 1978, Biochronology. In Cohee, G.V., Glaessner, M.F., and Hedberg, H. (eds.), Contributions to the Geological Time Scale, Am. Assoc. Petroleum Geol., Studies in Geology, No. 6, 39-57.

Blank, R.G., 1979, Applications of probabilistic biostratigraphy to chronostratigraphy. J. Geol., 87, 647-670.

Blank, R.G., 1984, Comparison of two binomial models in probabilistic biostratigraphy. Proc. 5th Int. Symp., IGCP Project 148, Geneva, 1982, Computers and Geosciences 10 (1), 59-69.

Blank, R.G., and Ellis, C.H., 1982, The probable range concept applied to the biostratigraphy of marine microfossils. J. Geol., 90 (4), 415-533.

Brower, J.C., Millendorf, S.A., and Dyman, T.S., 1978, Methods for the quantification of assemblage zones based on multivariate analysis of weighted and unweighted data. Computers and Geosciences 4, 221-227.

Buzas, M.A., Koch, C.F., Culver, S.J. and Sohl, N.F., 1982, On the distribution of species occurrence. Paleobiology 8 (2), 143-150.

Davaud, E., 1982, The automation of biochronological correlation. In Quantitative Stratigraphic Correlation ed. J.M. Cubitt and R.A. Reyment, J. Wiley & Sons Ltd., 85-99.

Doyle, P.S. and Riedel, W.R., 1979, Cretaceous to Neogene ichthyoliths in a giant piston core from the central North Pacific. Micropal. 25, 337-364.

Doeven, P.H., Gradstein, F.M., Jackson, A., Agterberg, F.P., and Nel, L.D.: 1982, A quantitative nannofossil range chart. Micropal. 28 (1), 84-92.

Drooger, C.W., 1974, The boundaries and limits of stratigraphy. Proc. K. Ned. Akad. Wet. B11 (17), 159-176.

Drooger, C.W., 1984, Evolutionary patterns in lineages of orbitoidal Foraminifera. Proc. K. Ned. Akad. Wet. B 87 (1), 103-130.

Drooger, M.M., Raju, D.S.N. and Doeven, P.H., 1979, Details of **Planorbulinella** evolution in two sections of the Oligocene of Crete. Utrecht Micropal. Bull. 21, 59-128.

Edwards, L., and Beaver, R., 1978, The use of a paired comparison model in ordering stratigraphic events. J. Math. Geol. 10, 261-272.

Edwards, L., 1982, Quantitative Biostratigraphy: The methods should suit the data. In Quantitative Stratigraphic Correlation, ed. J.M. Cubitt and R.A. Reyment, J. Wiley & Sons Ltd., 45-60.

Gorsel, J.T. van, 1975, Evolutionary trends and stratigraphic significance of the Late Cretaceous **Helicorbitoides - Lepidorbitoides** lineage Utrecht Micropal. Bull. 12, 99 pp.

Gradstein, F.M., 1984, On stratigraphic normality. Proc. 5th Int. Symp., IGCP, Project 148, Geneva 1982, Computers and Geosciences 10 (1), 43-57.

Gradstein, F.M., 1983, Paleoecology and stratigraphy of Jurassic abyssal Foraminifera in the Blake-Bahama Basin, Deep Sea Drilling Project Site 534. In Sheridan, R.S., Gradstein, F.M. et al. Int. Rept. Deep Sea Drilling Project, 76, Washington (U.S. Govt Printing Office), 537-559.

Gradstein, F.M., and Agterberg, F.P., 1982, Models of Cenozoic foraminiferal stratigraphy - Northwestern Atlantic margin. In Quantitative Stratigraphic Correlation, ed. J.M. Cubitt and R.A. Reyment, John Wiley & Sons Ltd., 119-173.

Guex, J. and Davaud, E., 1982, Récherche automatique des associations unitaires en biochronologie. Bull. Soc. Vaud. Sc. Nat. 361 (76), 53-69.

Harper, C.W., 1981, Inferring succession of fossils in time, the need for a quantitative and statistical approach. J. Paleont. 55 (2),442-452.

Harper, C.W., 1984, A Fortran IV Program for comparing ranking algorithms in quantitative biostratigraphy. Proc. 5th Int. Symp., IGCP Project 148, Geneva 1982, Computers and Geosciences 10 (1), 3-29.

Hay, W.W., 1972, Probabilistic stratigraphy. Eclog. geol. Helv. 65 (2), 255-266.

Hay, W.W., and Southam, J.R., 1978, Quantifying biostratigraphic correlation. Ann. Rev. Earth Sci. 6, 353-75.

Healy Williams, N., 1983, Fourier shape analysis of **Globorotalia truncatulinoides** from late Quaternary sediments in the southern Indian Ocean. Mar. Micropal. 8, 1-15.

Hedberg, H.D. (ed.), 1976, International Stratigraphic Guide. J. Wiley & Sons, Ltd., 200 pp.

Hinte, J.E. van, 1965, An approach to ***Orbitoides.*** *Proc. Kon. Ned. Akad. Wet. B, 68 (2), 57-71.*

Hudson, C.B. and Agterberg, F.P., 1982, Paired comparison models in biostratigraphy. J. Math. Geol. 14 (2), 141-159.

Millendorf, S.A., Brower, J.C., and Dyman, T.S., 1978, A comparison of methods for the quantification of assemblage zones. Computers and Geosciences 4, 229-242.

Riedel, W.R., 1981, DSDP Biostratigraphy in retrospect and prospect, In The Deep Sea Drilling Project: A decade of progress. Ed. J.E. Warme et al., SEPM Spec. Publ. Tulsa, 32, 253-261.

Riedel, W.R., and Westberg, M.J., 1982, Neogene radiolarians from the eastern tropical Pacific and Caribbean. Deep Sea Drilling Project Leg 68, in Prell, W.L., Gardner, J.V. et al. In Rep. Deep Sea Drilling Project 68, Washington (U.S. Govt. Printing Office), 289-300.

Scott, G.H., 1974, Biometry of the Foraminiferal shell. In Hedley, R.H., and Adams, C.G. (eds.), Foraminifera 1. Academic Press, 55-151.

Shaw, A.B., 1964, Time in Stratigraphy. McGraw-Hill Book Co., New York, 365 pp.

Westberg, M.J., and Riedel, W.R., 1982, Radiolarians from the Middle America trench off Guatemala. Deep Sea Drilling Project Leg 67. in Aubouin, J., van Huene, R., et al. 1982. Init. Rept. Deep Sea Drilling Project, 67, Washington (U.S. Govt. Printing Office), 401-424.

Williamson, M.A., in press, Quantitative biozonation of the late Jurassic and Early Cretaceous of the East Newfoundland Basin. Micropaleontology.

Worsley, T.R., and Jorgens, M.L., 1977, Automated biostratigraphy. In Ramsay, A.T.S. (ed), Oceanic Micropalaeontology 2, Academic Press, London, 1201-1229.

Zachariasse, W.J., Riedel, W.R., Sanfilippo, A., Schmidt, R.R., Brolsma, M.J., Schrader, H.J., Gersonde, R., Drooger, M.M., and Broekman, J.A., 1978, Micropaleontological counting methods and techniques - an exercise on an eight metres section of the Lower Pliocene of Capo Rossello, Sicily. Utrecht Micropal. Bull. 17, 1-265.

Part II

BIOSTRATIGRAPHIC ZONATIONS

Contents

II.1

THE INDEX FOSSIL CONCEPT AND ITS APPLICATION TO QUANTITATIVE BIOSTRATIGRAPHY

JAMES C. BROWER

INTRODUCTION

The biostratigraphic attributes and the ways in which to use fossils for correlation have been debated for many years. Most early discussion revolved around index versus non-index fossils. Although index fossils were employed for tracing and correlating strata long before the time of Darwin, the exact origin of the concept is obscure. The idea is often attributed to Albert Oppel (1856-1858) who named each of his zones after a characteristic fossil termed the 'index' (e.g., Eicher, 1976, Donovan, 1966). This is most likely erroneous because Oppel visualized the 'index' as only one species in the assemblage defining the zone (Hancock, 1977). The index fossil concept probably predates 1856. William Smith was definitely dealing with index fossils in 1817 when he stated (on p. iv and v of the Stratigraphical System of Organized Fossils):

> "By the tables it will be seen which Fossils are peculiar to any Stratum and which are repeated in others" (p. iv)

and

> "The organized Fossils which may be found, will enable him to identify the Strata of his own estate with those of others" (p. v).

The idea of an index fossil may have appeared earlier. For example, Mallory (1970) pointed out that Fuchsel noted the characteristic nature of the fossils in several rock units of different ages around 1760. Regardless of origin, William Smith probably developed the concept independently because his limited reading of previous authors dictated that he was largely uninfluenced by earlier ideas (Hancock, 1977). Smith's approach was empirical and not theoretical.

Correlation by index fossils flourished and was common practice by the late nineteenth century. Such correlation is still employed by many biostratigraphers as for example Arkel (1956) and Jeletzky (1956, 1965). Jeletzky (1965) criticized the application of quantitative methods to biostratigraphy, arguing (p. 135):

> "Any attempt at the quantification of biochronological correlation is, thus, precluded by the fundamentally qualitative and non-statistical nature of its most valuable data (index fossils)."

Subsequent discussion shows that this is not true.

The first practical attempt at measuring the biostratigraphic value of fossils was that of Hazel (1970; see also 1977) who advocated indices developed by ecologists. Initially Hazel defined assemblage zones by multivariate analysis of presence-absence data of the range-through type. Two measures were utilized. The Biostratigraphic Fidelity of a species is the percent of its total occurrences within a certain assemblage zone or biostratigraphic unit. The Biostratigraphic Constancy comprises the percentage of samples within a biostratigraphic unit containing that particular taxon. In conjunction, constancy and fidelity measure some attributes of an index fossil. I will not discuss Hazel's technique further because it is 'a posteriori' in that one must have prior knowledge of the zones. The methods of concern here are 'a priori' so they can aid in establishing the zones.

Cockbain (1965) applied the information function to estimating the utility of a species with the following equation:

$$H = -\Sigma \, p_i \log p_i$$

where H equals the information derived from knowing the range of a species and the probability of its occurrence within a given interval, and p_i is the probability of finding the taxon in the ith biostratigraphic unit. Cockbain noted problems in selecting probability categories and in associating probabilities with the different events. Also it is not clear as to how the H values are to be used. The main difficulty of this technique is that only the vertical range is considered; the geographical distribution and facies independence are conspicuously absent.The latter two parameters are clearly important. It is extremely difficult to estimate the probability of observing a given species in a particular biostratigraphic unit. Consequently I will not consider the entropy function further in this Chapter. The entropy function, like the fidelity and constancy of Hazel (1970, 1977), provides an 'a posteriori' rather than an 'a priori' measure.

McCammon (1970), Brower et al. (1978), Millendorf et al. (1978) and Brower (1981) published a threefold strategy for the quantification of index fossils. Initially, the biostratigraphic parameters of an index fossil were determined. Secondly, simple indices, called relative biostratigraphic value, were structured to tabulate the biostratigraphic information conveyed by recording the presence of a particular species. Third, the relative biostratigraphic values were employed in several ways to aid in correlating the data. A detailed discussion of the biostratigraphic attributes of fossils and their relative biostratigraphic values is available in Brower (1984).

THE NUMERICAL MODELS

Introduction

The biostratigraphic utility of different taxa varies greatly, ranging from the classic index fossil to species with little or no useful information. In addition the spectrum is relative. A guide fossil for large scale correlation between continents may have a stratigraphic range which is too long to be useful within a single region. The numerical data presented later show that the relative biostratigraphic value (hereafter termed RBV) of fossils follows a continuous probability distribution. This is also true for the biostratigraphic parameters such as vertical range. Some paleontologists (e.g. Jeletzky, 1965) consider fossils in two ways relative to their biostratigraphic properties. Index fossils are valued and retained whereas the other taxa are ignored. Thus a continuous spectrum is divided into two discontinuous parts. This retains some useful data but other pertinent information is discarded. I maintain that a continuous sequence of biostratigraphic properties should be treated continuously. It is also feasible to calculate the amount of biostratigraphic information conveyed by different taxa and to employ these as weights.

Measurement of Biostratigraphic Attributes

The primary biostratigraphic properties of fossils are their vertical range, geographic distribution, and amount of facies independence. For the examples discussed here, the data represent the presence or absence of numerous species in a series of samples taken from different stratigraphic sections. An individual section can be complete or composite, but all sections should be roughly the same age, for example the entire Eocene. Correlations are to be derived inside of the sampled time interval. Such data are necessary so that simple methods will be adequate to calculate the biostratigraphic attributes of the species. Given these data, the three attributes can be determined in the following way. More complex data sets require more complicated computations. Note that each attribute is standardized in that the characters must lie within a range of 0.0 to 1.0.

1. Vertical range (V) represents the most critical attribute. It is important to maximize V as much as possible so that the most conservative estimate of relative biostratigraphic value (RBV) be obtained. Thus the value of V for species i (V_i) is computed by the maximum figure of (thickness of sediment within section m occupied by species i)/(total thickness of section m)

for each stratigraphic section studied (Figure 1). This procedure has the advantages of simplicity and easy measurement. Although this measure of V ignores factors such as variation of rates of sedimentation within and between facies, local unconformities, etc., it has proven adequate for the case studies known to me. It is also practical to omit duplicate parts of sections which have been repeated by faulting; however, this problem is not present in the data analyzed by McCammon (1970) or myself.

V can be calculated in other ways. In one scheme, each section would be separated into a set of vertical intervals. The greatest proportion of intervals spanned by the taxon would yield another estimate of V. A composite maximum proportion of intervals can be pieced together for several stratigraphic sections. A third method is based on the vertical sequence of taxa. Begin by calculating the most likely order of the biostratigraphic events for all sections. The 'variance' of positions for each taxon relative to its average or median position would yield a measure of V, because in general greater 'variance' should be associated with larger V.

2. The facies independence (F) is given by (number of facies in which species i is known to occur)/(total number of facies present). This produces the most conservative estimate of F. Alternatively the facies independence could be calculated by (number of facies in which species i is known to occur)/(total number of facies within the biozone of species i). This index is intended to analyze the distribution in facies within the range zone of the taxon. An inherent difficulty of this method is that it requires an a priori knowledge of the range zone.

3. G indicates the geographical persistence or range. Like the other attributes, I prefer a conservative figure and G equals (number of sections or localities in which species i occurs)/(total number of sections or localities). This formula yields total geographic distribution. An obvious alternative comprises (number of localities or sections in which species i is found)/(number of sections or localities containing a sedimentary facies from which the taxon is known). The latter parameter denotes the geographical persistence within the stratigraphic sections with the appropriate sedimentary facies.

The three biostratigraphic attributes are constrained to range from 0.0 to 1.0. A text-book index fossil is geographically widespread, facies independent, and short ranged. This type of species exhibits large values of G and F in conjunction with small V. As the species become less widespread, more facies restricted, and longer ranging, the values of G and F decline whereas V increases. The nature of G, F, and

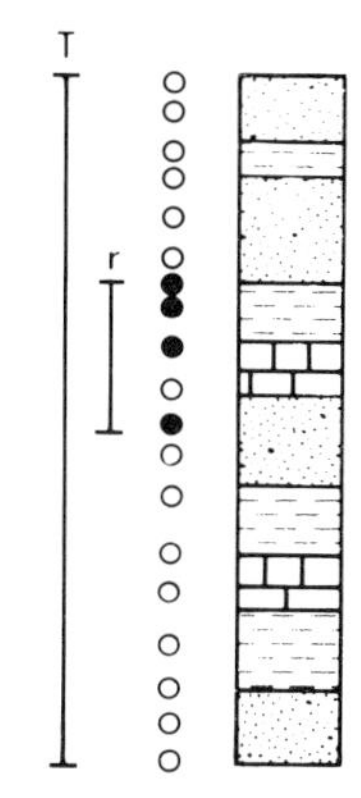

Figure 1 *Stratigraphic section showing local range zone (r) of a species and total thickness sampled (T). The solid dots indicate samples in which the species is present; Open circles point out samples where the species is not present. The value of V for this species is r/T.*

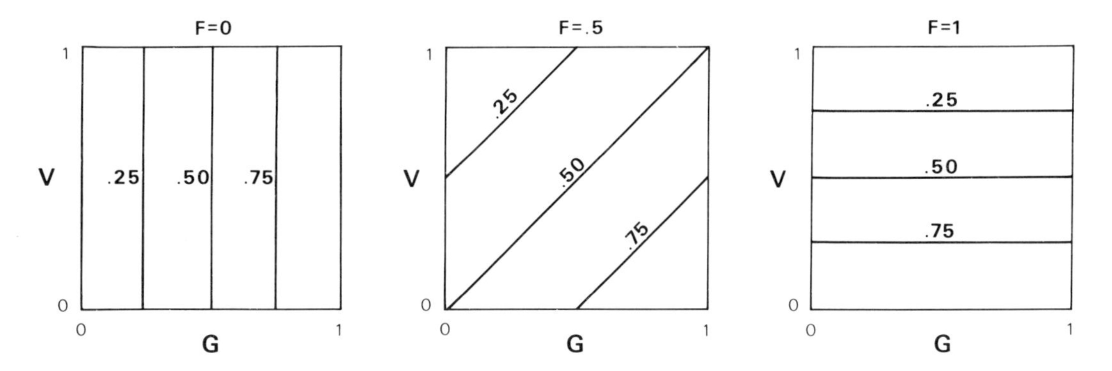

Figure 2. Properties of RBV1. Plots of V and G for fixed values of F. The straight contour lines are for constant RBV1.

Table 1 Relative biostratigraphic values for various types of taxa.

Type of species	G	F	V	RBV1	RBV2	RBV3
Classic index fossil	1.00	1.00	0.01	0.990	0.990	0.990
Near index fossil	0.90	0.95	0.10	0.900	0.832	0.900
Short-ranged facies tracer	1.00	0.10	0.10	0.990	0.495	0.900
Long-ranged facies tracer	1.00	0.10	0.90	0.910	0.055	0.100
Index fossil in one stratigraphic section	0.10	1.00	0.10	0.900	0.495	0.900
Non-occurring species	0.00	0.40	0.00	0.400	0.200	1.000
'Trash fossil'	0.05	0.10	0.90	0.055	0.008	0.100

V almost certainly dictates an inherent covariation between the attributes. In general, fossils with long ranges will occur in more sedimentary facies and will be more widespread geographically, a situation which is supported by the subsequent examples.

Relative Biostratigraphic Value

After calculating the attributes, the next step is to assign relative biostratigraphic values (RBV) to the various species or taxa. Observe that the numerical models for RBV are independent of the methods for estimating the biostratigraphic attributes. Any deficiencies in the calculations of G, F, and V should not argue against the basic statistical models. The RBV of Species i should indicate the amount of biostratigraphic information which is derived from observing its presence. Both McCammon (1970) and I elected to work with indices which range from 0.0 to 1.0 rather than deal directly with probability. An RBV of near zero identifies a species with little or no useful biostratigraphic information. A taxon which is useful for correlation is characterized by an RBV of almost 1.0. Our experience shows that these indices are superior to statistics such as the information function. A major advantage is that RBV's are easier to compare and use as weights.

Previous authors published three formulations of RBV. McCammon (1970) proposed the following for species i:

$$RBV1_i = F_i(1-V_i) + (1-F_i)\,G_i$$

where RBV1 denotes McCammon's model. This index exhibits three main characteristics. First, RBV1 is a linear model because if V and G are changed while keeping F constant, equal contours of RBV1 plot as straight lines (Figure 2). Secondly, RBV1 points out two different aspects of biostratigraphic correlation. Begin with an index fossil that is facies independent, geographically widespread, and short ranged, i.e., large G and F, and low V. This species will possess a large RBV1 of nearly 1.0 (Table 1). Each of the biostratigraphic parameters contributes to its large RBV1 and this form is valued for all three attributes. The large RBV1 of an index fossil reflects its time stratigraphic utility. A second kind of fossil is geographically widespread but restricted to a single biofacies or lithofacies. Thus G approaches 1.0 whereas F is near zero; note that large variation in V produces little change in RBV1. This taxon also exhibits a large RBV1 (Table 1). This second fossil is useful because it is able to trace a particular facies laterally and vertically. RBV1 is obviously designed to provide a compromise between time-stratigraphic correlation and establishing the persistence in time and space of a particular facies. It is important to observe that all three parameters are weighted equally. The third property postulates a critical value of RBV1 for a non-occurring species that is not found within the area studied. The G and V of this form are nil. According to McCammon (1970) its facies independence is F which is defined outside of the local area. Thus the RBV1 of a non-occurring species equals F (Table 1). McCammon (1970, p. 51, 52) presented the following reasoning:

> "Consider now a species which occurs within an interval and which has the same degree of facies independence as a species which is not found to occur in the same interval. Clearly, if the relative biostratigraphic value of the reported species is less than the value of the measure of its degree of facies independence, it is no more useful for biostratigraphic correlation than is the reported absence of the other species. For a given species to be considered useful for correlation, then, R.B.V. > α_c"

Expressed differently the observed RBV1 must exceed F if the presence of the species is to convey any meaningful biostratigraphic information. The empirical studies of Millendorf et al. (1978) demonstrate that this is not always the case. In some data sets, fossils with low RBV1 do provide useful similarities in the numerical analysis of assemblage zones. I believe that the problem resides in the ambiguity inherent in the absences of biostratigraphic data; this is because a species can be recorded as absent for several reasons; examples are sampling error, misidentification, the sample might be located outside of the range zone of the species, the sample may lie within the range zone of the species but in a facies not inhabited by the taxon (see Brower et al., 1978; Millendorf et al., 1978; Brower, 1981, 1984). Table 1 also illustrates RBV1 for various taxa with other values of G, F and V.

Brower et al. (1978) and Brower (1984) discussed a second index (RBV2):

$$RBV2_i = [F_i(1-V_i) + G_i(1-V_i)]/2.$$

The main characteristics of the second index are as follows. First, RBV2 represents a curvilinear statistic versus the linear model of RBV1. If F is kept constant, whereas V and G are varied, contours of equal RBV2 curve and converge as larger values of V and G are approached (Figure 3). Secondly, RBV2 is designed to detect the time-stratigraphic properties of fossils. A textbook index fossil will have G and F of almost 1.0 and a V near nil which produces an RBV2 of roughly 1.0 (Table 1). Consider a short-ranged and geographically-widespread species which is only found in a single facies (G near 1.0, F and V near 0). This taxon will exhibit an RBV2 of almost 0.5 which is reasonable because it can be used for correlation within one facies. Another example is an index fossil only recorded from several stratigraphic sections in the area studied; here F is large whereas G and V are small. This taxon

Table 2 Correlation matrix for biostratigraphic parameters.

G	F	V
1.00		
0.69	1.00	
0.49	0.71	1.00

Table 3 Principal components for biostratigraphic parameters.

	Eigenvector		
	I	II	III
G	0.83	0.52	-0.19
F	0.93	-0.02	0.37
V	0.84	-0.49	-0.22
Percent of variance	75.6	16.9	7.5

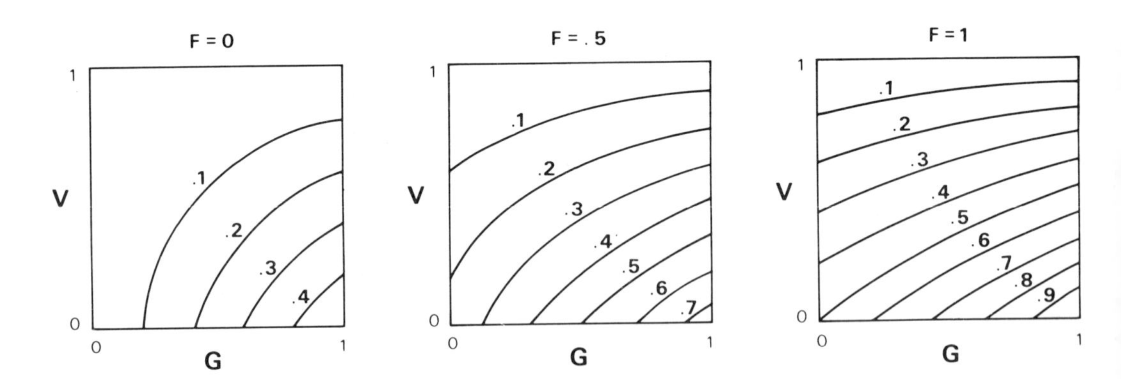

Figure 3. Properties of RBV2. Plots of V and G for fixed values of F. The curved contour lines denote constant RBV2.

could aid in correlating several of the sections and its RBV2 would equal an intermediate value of roughly 0.5. Both of the latter types of taxa have some biostratigraphic utility for time correlation but less than the classic index fossil. RBV2 places equal value on F and G but V is given double weight relative to either of the other attributes. Thus the weight for V equals the sum of those for G and F. This differs greatly from the equal weights of RBV1. Thirdly, if the reasoning of McCammon (1970) about non-occurring taxa is correct, then the minimum value of RBV2 equals F/2 if the presence of a fossil is to provide meaningful biostratigraphic information (Table 1).

RBV3 was suggested by Brower et al. (1978) and Brower (1984), as follows:

$$RBV3_i = (1-V_i)$$

This index has been applied to data such as Deep Sea Drilling Project sections where many taxa are geographically widespread and largely facies independent.

Numerous other measures of relative biostratigraphic value can be computed, one example being simply the average attribute, namely

$$[G_i + F_i + (1-V_i)]/3.$$

However, the unweighted average parameter is not particularly informative. In the search for index fossils, large values for G and F are only useful if V is small. Clearly, the biostratigraphic parameters must be combined or compounded to be useful. Although only V is directly involved in the computation of RBV3, the index is, in reality, compounded because it is only calculated for data where the taxa have large values of G and F. The Biostratigraphic Constancy and Biostratigraphic Fidelity of Hazel (1970, 1977) are also available. One can also specify that species with certain threshold values of G, F, and V be included in a particular analysis. In one case, only taxa with G and F exceeding 0.6 and V below 0.4 might be used. The critical values might be derived statistically so that better values, namely larger F and G and smaller V, are only shared by 32 percent of the fossils. The 32 percent point corresponds to one standard deviation above the mean for F and G and one standard deviation below the mean for V, assuming normally distributed data. Indices can be designed to incorporate the desired biostratigraphic attributes and statistical properties. The possibilities are ad infinitum.

APPLICATIONS OF BIOSTRATIGRAPHIC ATTRIBUTES AND RELATIVE BIOSTRATIGRAPHIC VALUES

Once the biostratigraphic attributes and the relative biostratigraphic values have been measured, the information can be used in different ways (see Brower, 1984, for details). Biostratigraphers have posed various hypotheses which can be tested with data on the biostratigraphic attributes and RBV's, as exemplified by: What relations exist between the different parameters and also between the RBV's and the biostratigraphic attributes? Have the biostratigraphic properties and/or the RBV's of a particular group of organisms changed over geologic time? Do pelagic and bottom dwelling taxa exhibit different biostratigraphic properties? Does the environment in which an organism lives dictate its biostratigraphic attributes? Do the biostratigraphic parameters of long-ranging taxa vary more than those with short range zones? Questions of this type are annotated extensively by Brower (1984).

A more direct issue is how the RBV's may be employed in quantitative correlations. The RBV's provide weights which serve two functions. First the RBV's can select a subset of species for analysis with other techniques, such as cluster analysis or various schemes for estimating sequences of biostratigraphic events (see review by Brower, 1981).In a different approach, Brower, Millendorf and Dyman (1978) and Millendorf, Brower and Dyman (1978) applied the RBV's to weight the presences of species. The object was to generate a weighted data matrix so that the presence of a taxon contributed an amount of similarity between samples which was directly proportional to its RBV. Multivariate analysis of the weighted similarity matrices yielded clusters and ordinations which were inspected to find the assemblage zones. Comparison of the weighted and unweighted similarity matrices reveals the costs and benefits of working with weighted versus unweighted data. Both of these applications are annotated subsequently in this Chapter and in Chapter II.2, which deals with multivariate analysis of assemblage zones (see also Brower, 1984).

CASE STUDY

Introduction

An example will illustrate some applications of the biostratigraphic attributes and RBV's to problems of correlation. A detailed discussion is available in Brower (1984). The data include 205 species of microfossils from the Eocene - Oligocene boundary strata of the Gulf Coast in the U.S.A. published by Deboo (1965). The 62 samples were collected from five stratigraphic sections, located roughly parallel to the depositional strike in Mississippi and Alabama. The typical lithologic sequence, from top to bottom, ranges from limestone to clay to marls (see Deboo, 1965; Hazel, 1970; Millendorf et al., 1978; and Chapter II.2, on 'Multivariate Analysis of Assemblage Zones', for correlation chart and locations of samples). The stratigraphy is basically simple.

Most lithologic units exhibit little geographic variation with the exception of the Mint Springs Marl and Forest Hill Sand which interfinger with the upper beds in the western section. Generally the clastic contents and thickness of the sediments decrease in an easterly direction as one moves further away from the ancestral Mississippi delta. The 205 microfossils consist of 55 ostracodes, 134 benthic foraminifers and 16 planktic foraminifers. Four lithofacies can be recognized, namely clay, limestone, marl, and glauconitic marl. The Deboo data are commonly used for testing quantitative methods in biostratigraphy, for example Brower (1984), Hazel (1970), McCammon (1970), Millendorf et al., (1978; refer to treatment of multivariate methods).

Relationships Between Parameters

It is critical to unravel the relations among the different biostratigraphic attributes in order to illustrate redundant information, some points about the distribution patterns of organisms, and the prediction of the RBV's. Figure 4 contains bivariate graphs for the 55 species of ostracodes; these results are representative of the whole data set. The correlation matrix for the 205 taxa and its principal components are listed in Tables 2 and 3, respectively.

The pattern of the correlation matrix is simple. The three attributes exhibit large positive correlations which are significantly larger than zero, subject to any reasonable risk level. This situation accords with evolutionary theory and it corroborates the empirical observations of many workers (see review in Stanley, 1979). Inasmuch as the three biostratigraphic attributes reflect the ecological strategies of organisms, they should covary (Jackson, 1977). The main ground rules are well known, at least for organisms inhabiting the continental shelves. The Eocene-Oligocene data represent shelf environments (Deboo, 1965). Eurytopic taxa are generalized, broadly adapted, commonly found in variable and unstable environments, have high ability to migrate, and are widespread geographically. Such species often have long stratigraphic ranges, probably because widespread environmental and geographic distribution cushions them from extinction and operates against speciation. On the other hand the more specialized stenotopic taxa are geographically and environmentally restricted. Their population structure is biased toward high rates of evolution and extinction; these, in turn, dictate short stratigraphic ranges (Jackson, 1977; Stanley, 1979). It is generally believed that this pattern does not apply to deep water faunas in the ocean basins. Such environments are more uniform and geographically widespread than the continental shelf habitats. There are few barriers to migration, so many deep sea organisms are very widespread (e.g. Taylor, 1977).

The first principal component extracts the positive correlations between the three parameters; observe that all loadings are positive and nearly equal (Table 3). Most of the variance, 76 percent, is explained

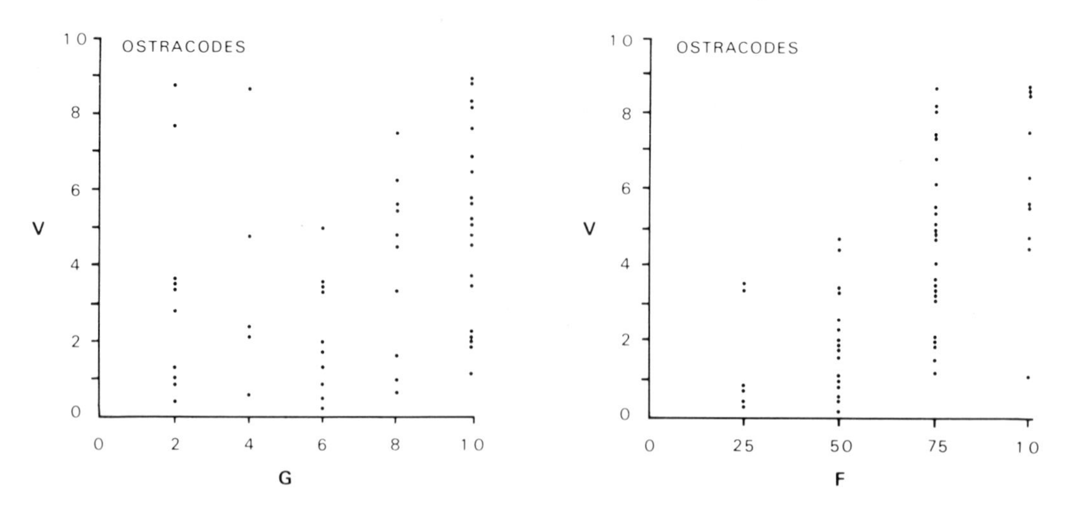

Figure 4. Bivariate scatterplots of biostratigraphic attributes for ostracodes.

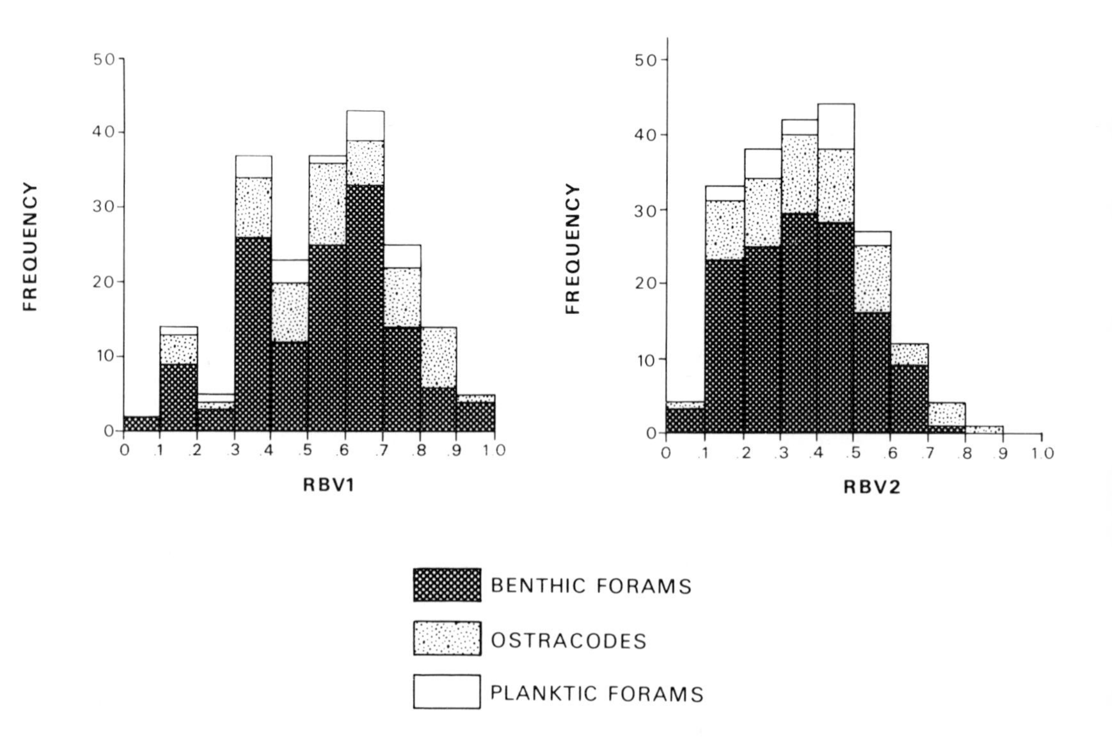

Figure 5. Histograms of RBV's for the Deboo data.

by Principal Component I. The principal component scores are continuously distributed. Taxa that are geographically widespread, occur in many facies, and have long stratigraphic ranges exhibit high scores. Low scores identify species that are provincial, facies restricted and short ranged with respect to time. In general such animals are rare and carry little useful biostratigraphic information. These results are typical of those obtained from other data (Brower, 1984).

Principal Component II accounts for almost 17 percent of the information in the matrix. G is compared inversely with V; the low loading for F indicates that it is not involved in this component. Organisms with high scores are short-ranged and geographically widespread, but have variable amounts of facies independence. Long- ranged taxa which are only found in one or two sections exhibit low scores.

The third principal component is linked with 7.5 percent of the variance. A large positive loading is present for F along with smaller negative loadings for V and G. Animals with high scores are facies independent, but short ranged and provincial. High scores indicate fossils with the opposite attributes. In this data set, none of the principal components identify classic index fossils (i.e., G and F contrasted with V) or the facies tracers of McCammon (namely G inverse with F in conjunction with a low loading for V).

The patterns of the principal components suggest several major conclusions:

1. The strong positive correlations between the three attributes are striking.

2. The loadings suggest that classic index fossils and facies tracers are rare animals, an inference which is supported by the literature (e.g., Kauffman and Hazel, 1977).

3. All principal component scores follow continuous probability distributions. Obviously the biostratigraphic properties are arrayed along a gradational spectrum.Neither index fossils nor facies tracers are discontinuously separated from any of the other species. These three themes are pervasive in all data examined by me.

4. The presence of an index fossil or facies tracer within the Gulf Coast data is a chance event. However, this is probably not true for the Middle Devonian fossils reported by Brower (1984).

The three RBV's are correlated. The largest correlation coefficient, 0.91, links RBV1 and RBV2. Although the two indices compound the three attributes differently, both mainly identify the same species. Lower correlations are observed when RBV3 is compared with RBV1 and RBV2 where the correlations are 0.64 and 0.47, listed in the same order. Obviously this is because only V affects RBV3 but the other two RBV's are controlled by all three attributes. If RBV3 is only calculated for the subset of taxa with large G and F as previously advocated, the correlations increase so that all three RBV's usually point out the same fossils.

It is important to unravel how the RBV's depend on the attributes, a subject that has been explored with stepwise multiple regression. The equations follow along with the squared multiple correlations (R^2):

$$RBV1 = 0.507 + 0.433\ G + 0.074\ F - 0.798\ V \quad (R^2 = 0.772),$$
$$RBV2 = 0.191 + 0.388\ G + 0.354\ F - 0.736\ V \quad (R^2 = 0.890).$$

The high R^2 values indicate that the three biostratigraphic attributes explain most of the information contained in the RBV's. The first variable incorporated into both equations is V which indicates that the vertical range represents the single best predictor of RBV. This clearly supports my previous argument that V is the most critical biostratigraphic attribute. The stepwise order for F and G is reversed in the two equations. For RBV1, G is entered in Step 2, followed by F in the last step. F is stepped into the equation for RBV2 prior to G. It is not known whether or not this order is typical of the two types of RBV. Observe that the R^2 value for RBV2 exceeds that for RBV1. This is rather surprising inasmuch as RBV1 is a linear model statistic whereas RBV2 is not. Perhaps the double weight placed on V in the computation of RBV2 simply overwhelms the contribution of the other attributes.

Relative Biostratigraphic Value and Correlation

Figure 5 shows histograms for RBV1 and RBV2 of the Deboo data. Both distributions are mainly continuous. Inasmuch as numerous species are involved in the tabulations, the graphs can be visualized as probability distributions. Note that the likelihoods of finding fossils with high RBV's (greater than 0.8 for RBV1 and over 0.6 for RBV2) are quite low. As mentioned earlier, index fossils are rare animals. RBV1 almost always exceeds RBV2 which reflects the different weights placed on the biostratigraphic attributes when the indices are calculated.

The critical issue for the RBV's is if they can identify taxa which are useful for correlation. This was studied by plotting range zones of species with various RBV's on a correlation chart in which the horizontal and vertical axes represent space and time, respectively. The columns of dots show locations of the 62 samples within the five stratigraphic sections. The chart is taken from Millendorf et al. (1978; see also Chapter II.2, on multivariate analysis; Deboo, 1965; Hazel, 1970). The graphs of Figure 6 illustrate representative species with high and low RBV's. Note that only fossils present in all five sections are pictured.

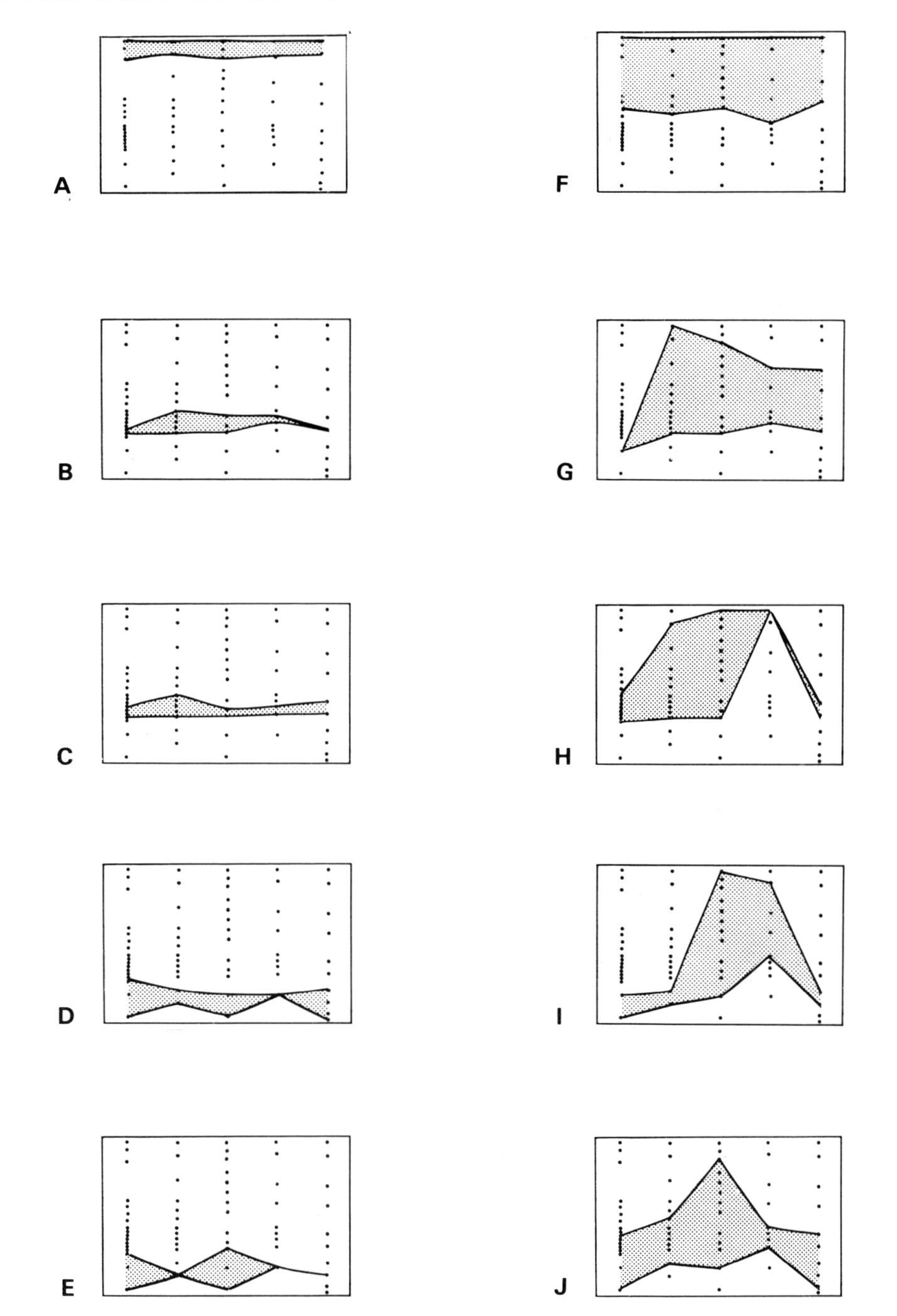

Figure 6. Plots of local range zones for selected species in five sections. A-E, fossils with high RBV's. Ranges of RBV1 and RBV2 are 0.775 to 0.905 and 0.608 to 0.89. F-J, fossils with low RBV's. Ranges of RBV1 and RBV2 equal 0.076 to 0.490 and 0.076 to 0.280. Names of taxa are listed by Brower (1984).

Data on the taxa with the highest RBV's, including 19 fossils with RBV1 over 0.8 (9.3 percent) and 17 species with RBV2 exceeding 0.6 (8.3 percent), are listed in Table 4 where the animals are arranged in order of increasing RBV2. In general the same fossils are recognized by both RBV's. Figures 6A-E illustrate the distributions of five of these select fossils. Inspection of the figures discloses two important points.

1. The local range zones are confined within narrow time intervals.

2. The highest and lowest occurrences are also located within small spans of time.

Deboo (1965) divided the strata into four assemblage zones and subzones, beginning with the **Floridina antiqua** zone at the base and ending with the **Lepidocyclina mantelli** zone at the top (see Table 4). Subsequent workers recognized three or four of these units (Hazel, 1970; McCammon, 1970; Millendorf, Brower and Dyman, 1978; see Chapter II.2 on multivariate analysis). Table 4 denotes that the species with the highest RBV's mostly occur within one zone or subzone and no taxon is found in more than two units. Not one species with a high RBV is restricted to the **'Cythereis' blanpiedi** subzone so this unit cannot be identified by index species. Deboo (1965) defined this subzone as an overlap extending from the lowest occurrence of **'Cythereis'** (actually **Trachyleberidea) blanpiedi** to the base of the **Lepidocyclina mantelli** zone. The fossils with the highest RBV's in the **'Cythereis' blanpiedi** subzone are also present in the overlying **Lepidocyclina mantelli** zone (e.g., **Trachyleberidea blanpiedi)** or in the underlying **Cribrohantkenia 'danvillensis'** subzone (e.g., **Siphonia danvillensis, Globigerina pseudoampliapertura).** Such zones obviously pose difficulties for this method of correlation. However if the fossils with the highest RBV's that occur in the 'C'. blanpiedi subzone are included, samples within this unit can be identified (McCammon, 1970; Millendorf et al., 1978; see multivariate analysis section, Chapter II.2).

Figures 6F-J picture data for five taxa with low RBV's which are present in all stratigraphic sections. Except for **Trachyleberidea blanpiedi,** fossils that consistently range through the highest and/or lowest samples are not illustrated. The RBV's of the five fossils range from 0.076 to 0.490 for RBV1 and from 0.076 to 0.280 for RBV2, figures that group with the lowest 25 percent of the RBV's for fossils found throughout the area. Typically the range zones of these taxa and their highest and lowest occurrences vary within wide limits between the different sections (Figure 6G-J). **Trachyleberidea blanpiedi** exhibits a different pattern (Figure 6F). Despite the fact that the RBV1 and RBV2 of this species are only 0.490 and 0.280, its lowest occurrence is quite uniform; the apparent straight-line highest occurrence is simply because this form also occurs in the overlying sediments. Deboo (1965) recognized this fact and employed **T. blanpiedi** to identify the base of his **'Cythereis' blanpiedi** subzone. Aside from **Trachyleberidea blanpiedi,** the range zones of the low-RBV species in Figure 6 are obviously too

Table 4 Biostratigraphic parameters and distribution of species with high RBV's.

Biostratigraphic data						Percent of range zone occurrences within listed biostratigraphic units			
Species	G	F	V	RBV1	RBV2	Floridina antiqua Zone	Cribohantkenina 'danvillensis' Subzone	'Cythereis' blanpiedi Subzone	Lepidocyclina mantelli Zone
Cytheretta jacksonensis	1.0	1.0	0.110	0.890	0.890	100			
Cibicides sp. 1	1.0	0.75	0.194	0.854	0.705	92.3	7.7		
N. gen. n. sp. 1	1.0	0.75	0.200	0.850	0.700	100			
Clithrocytheridea grigsbyi	1.0	0.75	0.200	0.850	0.700	100			
C. garretti	1.0	0.75	0.200	0.850	0.700	100			
Textularia sp. 2	1.0	0.50	0.076	0.962	0.693	85.7	14.3		
T. sp. 1	1.0	0.50	0.081	0.960	0.689	100			
Bolivina alazanensis	1.0	0.75	0.234	0.824	0.670		100		
Vulvulina advena	1.0	0.75	0.234	0.824	0.670		100		
"Cythereis" hysonensis	0.8	0.75	0.150	0.838	0.659	100			
Anomalina danvillensis	1.0	1.0	0.353	0.647	0.647		92.3	7.7	
Textularia adalta	0.8	0.75	0.171	0.822	0.642	85.7	14.3		
Digmocythere watervalleyensis	0.8	0.50	0.052	0.874	0.616	100			
Saracenaria ornatula	1.0	0.75	0.300	0.775	0.612		100		
Jugosocythereis vicksburgensis	1.0	0.50	0.190	0.905	0.608				100
Textularia dibollensis	1.0	0.50	0.194	0.903	0.604	100			
Sigmomorphina costifera	1.0	0.50	0.194	0.903	0.604	100			
Actinocythereis n. sp. 2	0.6	0.75	0.120	0.810	0.954	71.4	28.6		
Flabellina sp. 1	0.6	0.75	0.127	0.805	0.589		100		
Cyamocytheridea watervalleyensis	0.8	0.25	0.087	0.828	0.479	100			
Siphonia advena eocenica	0.8	0.25	0.166	0.808	0.438	100			

variable to be used for correlation. In fact the mean standard deviation for variation of the local range zone of microfossils with low RBV's is almost twice that of taxa with high RBV's (Brower, 1984).

Several subsets of species with different RBV's were used to zone the data and Figure 7 shows the results for the high-RBV fossils. The highest occurrences fall into three bands in which Groups 1, 2, and 3 contain two, eight and eight taxa which are found in the **Lepidocyclina mantelli** zone, **Cribrohantkenina 'danvillensis'** subzone, and the **Floridina antiqua** zone, respectively (Figure 7A). The highest occurrences of each group only span narrow intervals of time. There is no species present in all or most of the stratigraphic sections which has a highest occurrence restricted to the **'Cythereis' blanpiedi** subzone, so this part of the correlation chart is empty.

The lowest occurrences are similar to the highest ones (Figure 7B). Groups 4, 6 and 7 represent microfossils which come from the **Lepidocyclina mantelli** zone, **Cribrohantkenina 'danvillensis'** subzone, and the **Floridina antiqua** zone. All groups fall into narrow time zones. The lowest occurrences of three species which Deboo (1965) suggested for defining the base of the **'Cythereis' blanpiedi** subzone comprise Group 5. These fossils possess lower RBV's than those of the other groups; the applicable figures equal: RBV1, 0.490 to 0.670 versus 0.700 to 0.962; and RBV2, 0.280 to 0.493 versus 0.525 to 0.890. Consequently the events of Group 5 occupy a comparatively large interval on the correlation chart.

Figure 8 contains the highest and lowest occurrences of selected taxa with low RBV's. Events which generally extend above or below the Eocene-Oligocene boundary strata are omitted. The events are extremely variable and cross-cutting relationships are frequent. Clearly zonation of these and similar events is worthless for correlating the data. This example demonstrates that the RBV's can identify species which will provide the most reliable zonations of a series of biostratigraphic events.

The RBV's have been employed in two other ways, both of which are discussed in the chapter on multivariate analysis. First is the search for parsimony or how many species are necessary to correlate the data set of interest. This is certainly a function of the degree of stratigraphic resolution needed because it takes more taxa to produce more detailed zonations. One should also consider the probability of finding the species selected for correlation. Approximately 13 taxa with the highest value of RBV can identify three zones in the Deboo data namely the **Lepidocyclina mantelli** zone, the **Floridina antiqua** zone and one zone which includes the **Cribrohantkenia 'danvillensis'** and **'Cythereis' blanpiedi** subzones (Table 4). Adding three taxa with the highest possible RBV's allows recognition of the two subzones. Thus about 16 taxa (roughly eight percent of the total) can replicate the four biostratigraphic units of Deboo (1965). In fact, one can correlate most of the samples by using one fossil with high RBV from each of the four units. In a second approach the RBV's have been employed as weights in the multivariate analysis of assemblage zones.

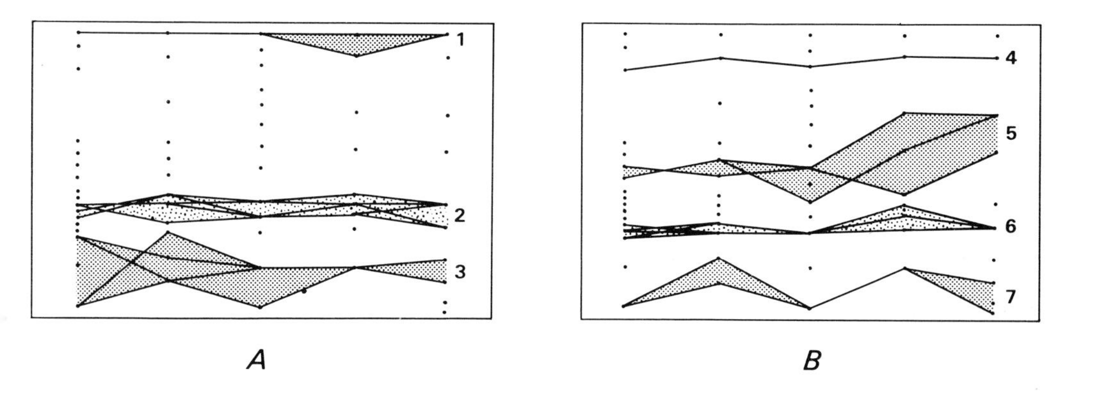

Figure 7. Plots of biostratigraphic events for species with high RBV's. A, highest occurrences. B, lowest occurrences. Ranges of RBV's for groups of species follow. Groups 1-4, 6, 7: RBV1 0.700 to 0.962; RBV2 0.525 to 0.890. Group 5: RBV1 0.490 to 0.670; RBV2 0.280 to 0.493. Species names are given by Brower (1984).

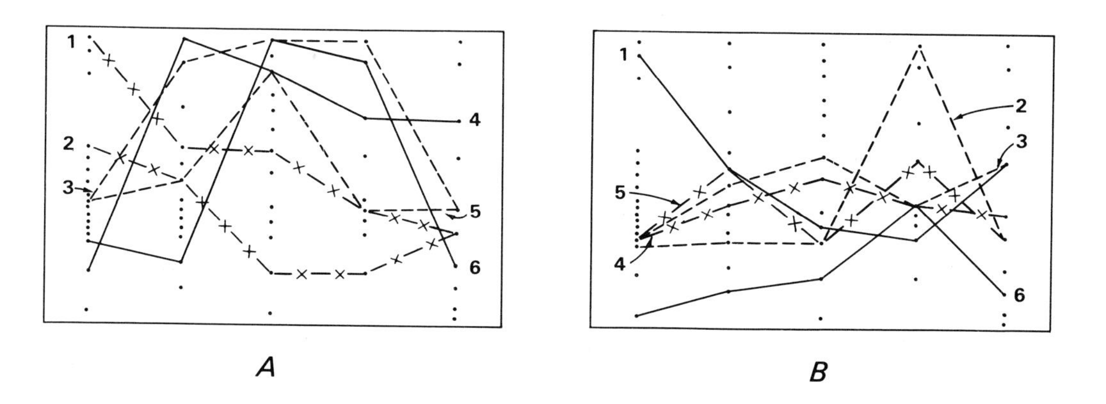

Figure 8. Plots of biostratigraphic events for species with low RBV's. A, highest occurrences, B, lowest occurrences. Ranges of RBV1 and RBV2 consist of 0.076 to 0.483 and 0.076 to 0.272. See Brower (1984) for species names.

SUMMARY AND CONCLUSIONS

The three main biostratigraphic attributes of a species are geographical range, facies distribution, and vertical range. These can be determined in several ways. Relative biostratigraphic value (RBV) represents the amount of information provided by the presence of a particular species. The three most useful measures are discussed here. RBV1 weights all three attributes equally and therefore yields a compromise between time-stratigraphic correlation and establishing the persistence of a particular facies. RBV2 places double weight on the vertical range compared to the other two parameters; this index is intended to point out species that are most useful for time correlation. In some cases a reasonable estimate, RBV3, can be calculated solely from the vertical range.

The biostratigraphic attributes and RBV's can be applied in several ways, such as to ascertain the relations between the parameters, to compare and contrast the biostratigraphic properties of various data sets and groups of taxa, to employ the RBV's as weighting functions, and to identify subsets of taxa with high RBV's for correlation. As far as correlation is concerned, my conclusions are straightforward and unambiguous. Species with high RBV's should be selected assuming that they can be consistently found when present. Most quantitative biostratigraphic methods are structured so they should work best for species with high RBV's.

REFERENCES

Arkell, W.J., 1956, Jurassic geology of the world. Hafner Pub. Co. Inc., New York, xv, 806 pp.

Brower, J.C., 1981, Quantitative biostratigraphy, 1830-1980. In Computer applications in the earth sciences, an update of the 70's, D.F. Merriam, ed., Plenum Press, New York and London, 63-103.

Brower, J.C., 1984, The relative biostratigraphic values of fossils. Computers & Geosciences, 10,(1), 111-132.

Brower, J.C., Millendorf, S.A., and Dyman, T.S., 1978, Quantification of assemblage zones based on multivariate analysis of weighted and unweighted data. Computers & Geosciences, 4, (3), 221-227.

Cockbain, A.E., 1966, An attempt to measure the relative biostratigraphic usefulness of fossils. J. Paleont., 40, (1), 206-207.

Deboo, P.B., 1965, Biostratigraphic correlation of the Type Shubuta Member of the Yazoo Clay and Red Bluff Clay with their equivalents in southwestern Alabama. Alabama Geol. Surv., Bull. 80, 84 pp.

Donovan, D.T., 1966, Stratigraphy: An introduction to principles. Rand McNally and Co., Chicago, 199 pp.

Eicher, D.L., 1976, Geologic time. (2nd ed.); Prentice-Hall Inc., New Jersey, 150 pp.

Hancock, J.M., 1977, The historic development of concepts of biostratigraphic correlation. In Concepts and methods of biostratigraphy, Kauffman, E.G., and Hazel, J.E., eds., Dowden, Hutchinson & Ross Inc., Stroudsburg, Pennsylvania, 3-22.

Hazel, J.E., 1970, Binary coefficients and clustering in biostratigraphy. Geol. Soc. America Bull., 81, (11), 3237-3252.

Hazel, J.E., 1977, Use of certain multivariate and other techniques in assemblage zonal biostratigraphy: examples utilizing Cambrian, Cretaceous, and Tertiary benthic invertebrates. In Concepts and methods of biostratigraphy, Kauffman, E.G., and Hazel, J.E. eds., Dowden, Hutchinson & Ross Inc., Stroudsburg, Pennsylvania, 187-212.

Jackson, J.B.C., 1977, Some relationships between habitat and biostratigraphic potential of marine benthos. In Concepts and methods of biostratigraphy, Kauffman, E.G., and Hazel, J.E. eds., Dowden, Hutchinson and Ross, Inc., Stroudsburg, Pennsylvania, 65-72.

Jeletzky, J.A., 1956, Biochronology, basis of practical geochronology. Am. Assoc. Petr. Geol. Bull., 40, 679-706.

Jeletzky, J.A., 1965, Is it possible to quantify biochronological correlation? J. Paleont., 39, (1), 135-140.

Kauffman, E.G., and Hazel, J.E., (editors), 1977, Concepts and methods of biostratigraphy. Dowden, Hutchinson and Ross, Inc., Stroudsburg, Pennsylvania, xiii, 658 pp.

Mallory, V.S., 1970, Biostratigraphy - a major basis for paleontologic correlation. Proc. North Am. Paleont. Conv., Chicago, 1969, F, 553-566.

McCammon, R.B., 1970, On estimating the relative biostratigraphic values of fossils. Bull. Geol. Inst. Univ. Uppsala (n.s.), 2, 49-57.

Millendorf, S.A., Brower, J.C., and Dyman, T.S., 1978, A comparison of methods for the quantification of assemblage zones. Computers & Geosciences, 4, (3), 229-242.

Oppel, A., 1856-1858, Die Jura Formation Englands, Frankreichs und des sudwestlichen Deutschlands, nach ihren einzelnen Gliedern eingeteilt und verglichen. Von Ebner and Seubert, Stuttgart (originally published in three parts in Abdruck der Wurttemb. Naturw. Jahreshefte 12-14, 1856, 1-438, 1857, 439 bis-594 + map, 1858, 695-857 + table).

Smith, W., 1817, Stratigraphical system of organized fossils, with reference to the specimens of the original geological collection in the British Museum: explaining their state of preservation and their use in identifying the British strata. E. Williams, London, 118 pp.

Stanley, S.M., 1979, Macroevolution, pattern and process. W.H. Freeman and Co., San Francisco, xi, 332 pp.

Taylor, M.E., 1977, Late Cambrian of western North America, trilobite biofacies, environmental significance and biostratigraphic implications. In Concepts and methods of biostratigraphy, Kauffman, E.G., and Hazel, J.E., eds., Dowden, Hutchinson and Ross, Inc., Stroudsburg, Pennsylvania, 397-425.

II.2

MULTIVARIATE ANALYSIS OF ASSEMBLAGE ZONES

JAMES C. BROWER

INTRODUCTION

Assemblage zones were the first kind of zone recognized by biostratigraphers. The zones proposed by William Smith (see 1817) and Charles Lyell (1830-1833) belong to this category. Lyell became the first quantitative biostratigrapher when he employed percentages of living molluscs to define several Tertiary epochs. Numerical biostratigraphy then languished for over 100 years. The renaissance was initiated by Simpson (1947, 1960) and Sorgenfrei (1958) who applied similarity coefficients to measure faunal similarity in biogeography and stratigraphy. Recently, statistical methods have become routine tools in the search for assemblage zones (e.g., Hazel, 1970, 1971, 1977; Brower et al., 1978; Millendorf et al., 1978; Brower, 1981; Brower and Burroughs, 1982).

An assemblage zone is characterized by a particular suite of taxa regardless of their ranges. Such zones typically combine ecological and stratigraphical information. The most popular techniques, such as cluster analysis and multidimensional scaling, are rooted in multivariate analysis and numerical taxonomy. As discussed in Chapter II.3, seriation of original data can also depict assemblage zones.

All methods outlined here are restricted to presence-absence data which contain origination and extinction events of taxa. Algorithms which normally function to analyze proportions, counts, and similar information in order to categorize ecological assemblages and gradients are excluded. The reader interested in these methods should consult Whittaker (1978a,b) and Gauch (1982) for ecological examples, and Cisne and Rabe (1978) and Scott and West (1976) for paleoecological applications. These techniques can identify ecological changes, such as eustatic shifts of sea level, which may yield time correlations. Cisne and Rabe (1978), Rabe and Cisne (1980) and Cisne et al. (1982) present a case study on the Ordovician of New York which involves correspondence analysis or reciprocal averaging, correlation of time series, and trend surfaces. The data are placed in a framework defined by time, space and reciprocal averaging score.

The presence-absence data matrix for biostratigraphy is rectangular with samples on one axis and taxa arrayed on the other. I assume that one is dealing with many taxa and samples which are derived from numerous stratigraphic sections. The algorithms for multivariate establishment of assemblage zones will be discussed in the following order: Q-mode, which analyzes samples, R-mode, which analyzes taxa, and dual space methods which handle taxa and samples in one sequence of operations.

Q-MODE ANALYSIS

Determination of the relations between samples proceeds in a sequence of steps, each of which presents the biostratigrapher with decisions to be made.

Step 1

Should the original data be used or converted to the range-through format? Range-through data list a species as present in all samples within its local range zone for each stratigraphic section (Figure 1 in Chapter II.1). This tends to minimize sampling problems and other factors that can cause a species to be absent from a particular sample. Absences pose problems in biostratigraphy because they cannot be completely evaluated. A species can be absent from any sample for the following reasons: the sample might lie outside of the range zone of the taxon; the fossil might be present but was not recovered or recognized in the sample; the species could be absent from a sample within its local range zone if the sample comes from a facies not inhabited by the organism; misidentification and reworking are also possible sources of error. The essential justification for range-through data is that the time duration of the species is of interest rather than the vagaries of its distribution. Most biostratigraphers advocate this type of data (e.g., Hazel, 1970, 1977; Brower et all, 1978; Millendorf et al., 1978; Brower, 1981; Brower and Burroughs, 1982; Millendorf and Millendorf, 1982). With range-through data similarities between samples are produced by similar patterns between range zones rather than the original presences and absences. Range-through data mean that the biostratigrapher is working with local range zones.

In some cases the range-through method can obliterate meaningful information such as where several biofacies intertongue repeatedly. These data are obviously not applicable to counts, proportions, etc.

When compiling data, it may be desirable to score a taxon as present only if it can be found consistently, say with 95 percent likelihood. Given a species which comprises a certain proportion of a population, the probability of encountering one or more specimens of that organism can be calculated from binomial or Poisson theory (Dennison and Hay, 1967). Admittedly such statistics are limited to chance distributions, rather than clumped or patchy patterns. Most biostratigraphic samples include intervals of strata and areas of bedding planes. To some extent, this practice homogenizes the samples so binomial and Poisson tests ought to provide satisfactory results.

Step 2

Are relative biostratigraphic values (RBV's) to be considered or not? As outlined in Chapter II.1 the RBV's can provide weights in several ways. A first strategy is to discard the taxa with low RBV's and to complete the analysis based on the species with the highest RBV's. This provides a parsimonious solution, i.e., a reasonable zonation with the smallest possible number of species. Secondly one can multiply the presences of a species by its RBV before moving on to the next step. An absence remains an absence regardless. The intent is to force the presence of Species k to contribute an amount of similarity between Samples i and j that is proportional to its RBV. Thirdly one can specify that only species with threshold values of geographical distribution (G), facies independence (F), and vertical range (V) be retained for a particular analysis. For example many biostratigraphers recommend that a species be present in half or more of the stratigraphic sections to be included in an analysis.

The purpose of the analysis dictates the course of action. If stratigraphical groupings are sought, the weights given by the RBV's can be useful because they commonly produce clusters or groupings of samples that are tighter and stratigraphically more homogeneous than those obtained from unweighted data. Conversely where ecological considerations are paramount, then straight presence-absence data are generally adequate.

Step 3

How should the similarity or dissimilarity matrix be computed for the samples? Given n samples, this matrix will have n rows and columns. The matrix contains coefficients which relate all possible pairs of samples. The literature on coefficients that are suitable for biofacies analysis is vast and recent reviews include Buzas (1970), Cheetham and Hazel (1969), Hazel (1970, 1977), Henderson and Heron (1977), Hohn (1976), Kaesler (1966), Raup and Crick (1979), Simpson (1960), Sepkoski (1974), Sneath and Sokal (1973), and Sokal and Sneath (1963). Most similarity and difference coefficients are symmetrical so that $S_{ij} = S_{ji}$ and $D_{ij} = D_{ji}$, where S and D denote similarity or dissimilarity, respectively, and i and j indicate the two samples being compared. Obviously the diagonal elements, S_{ii} and D_{ii}, equal 1.0 and 0.0, respectively, and therefore convey no useful information. Consequently one need only consider the upper or lower half of the matrix. Similarity and dissimilarity measures fall into two basic categories: those where only mutual presences (positive matches) contribute toward similarity and those where both mutual presences and absences (negative matches) produce similarity. Most workers have dealt with similarities, and similarity coefficients will be emphasized in subsequent discussion.

In calculating similarities between samples only coefficients based on mutual presences (positive matches) are recommended. A mutual presence is observed if the taxon is present in the two samples being compared, for example if species k is present in samples i and j. A mutual absence would be scored if the taxon is absent from the two samples, say if species l is absent from samples i and j. Mutual absences or negative matches of a taxon from the two samples being considered should not contribute directly toward similarity. This is because a species can be recorded as absent from a sample for several reasons; in other words the absences cannot be fully defined (see Step 1). The experience of biogeographers and at least some paleoecologists is similar (e.g., Rowell and McBride, 1972; Rowell et al., 1973). For presence/absence data, where presence is scored as 1.0 and absence equals 0.0, four widely used similarity measures are listed below.

Jaccard $$S_{ij} = \frac{C_{ij}}{N_i + N_j - C_{ij}}$$

Dice $$S_{ij} = \frac{2C_{ij}}{N_i + N_j}$$

Otsuka $$S_{ij} = \frac{C_{ij}}{N_i N_j}$$

Simpson $$S_{ij} = \frac{C_{ij}}{N_i} \qquad \text{where } N_i \leq N_j$$

S_{ij} = similarity between samples i and j,
C_{ij} = number of species present in both samples,
N_i = number of species present in sample i,
N_j = number of species present in sample j.

Table 1 gives an example for the Dice coefficient. The properties of these coefficients in biostratigraphy are discussed by Cheetham and Hazel (1969), Hazel (1970, 1977) and Simpson (1960). Typically the Simpson coefficient emphasizes similarity whereas the Jaccard coefficient stresses differences. The Dice and Otsuka coefficients are intermediate between the other two. The Simpson coefficient can provide misleading similarities, especially if N_i and N_j differ greatly; for example the presence of a few long-ranging taxa would inflate the similarity between the two samples. Consequently the Simpson index is probably the least satisfactory (e.g., Hazel, 1970; Rowell et al., 1973; Henderson and Heron, 1977). Several probabilistic similarity coefficients were formulated recently by Henderson and Heron (1977) and Raup and Crick (1979). Although not tested on biostratigraphic data, these coefficients should provide useful results. The Henderson and Heron (1977) equality and Equation 4 of Raup and Crick (1979) are derived directly from probability. Both assume equal odds of recovering all species. Raup and Crick (1979, p. 1218, 1219) also applied Monte Carlo methods to the problem. Despite the disadvantage of being expensive with respect to computer time, this last similarity index is more reasonable because it incorporates the probability of finding the various species.

All of these similarity coefficients employ unmodified presence/absence information. The data set can include all taxa or only those species with the highest RBV's or certain threshold biostratigraphic attributes.

As annotated in the previous step, one can weight the data by multiplying the presences by the RBV of the species involved. This is accomplished by substituting the RBV's for the presences of the taxa as shown below.

Jaccard $$S_{ij} = \frac{RBV_{ij}}{RBV_i + RBV_j - RBV_{ij}}$$

Dice $$S_{ij} = \frac{2\ RBV_{ij}}{RBV_i + RBV_j}$$

Simpson $$S_{ij} = \frac{RBV_{ij}}{RBV_i} \qquad \text{where } RBV_i \leq RBV_j$$

S_{ij} = similarity between samples i and j,
RBV_{ij} = total RBV's of species present in both samples,
RBV_i = total RBV's of species present in sample i,
RBV_j = total RBV's of species present in sample j.

Table 1 Hypothetical example showing calculation of Dice coefficient for unweighted and weighted data.

Unweighted Data

Species	Samples 1	Samples 2
A	0	1
B	1	1
C	1	1
D	1	0
E	1	0

1 = Present
0 = Absent
$C_{12} = 2$
$N_1 = 4$
$N_2 = 3$
$S_{12} = (2 \times 2)/(4+3) = 0.57$

Weighted Data

Species	Samples 1	Samples 2
A	0	.5
B	.8	.8
C	.2	.2
D	.7	0
E	.9	0

$RBV_{12} = 1.0$
$RBV_1 = 2.6$
$RBV_2 = 1.5$
$S_{12} = (2 \times 1.0)/(2.6+1.5) = 0.49$

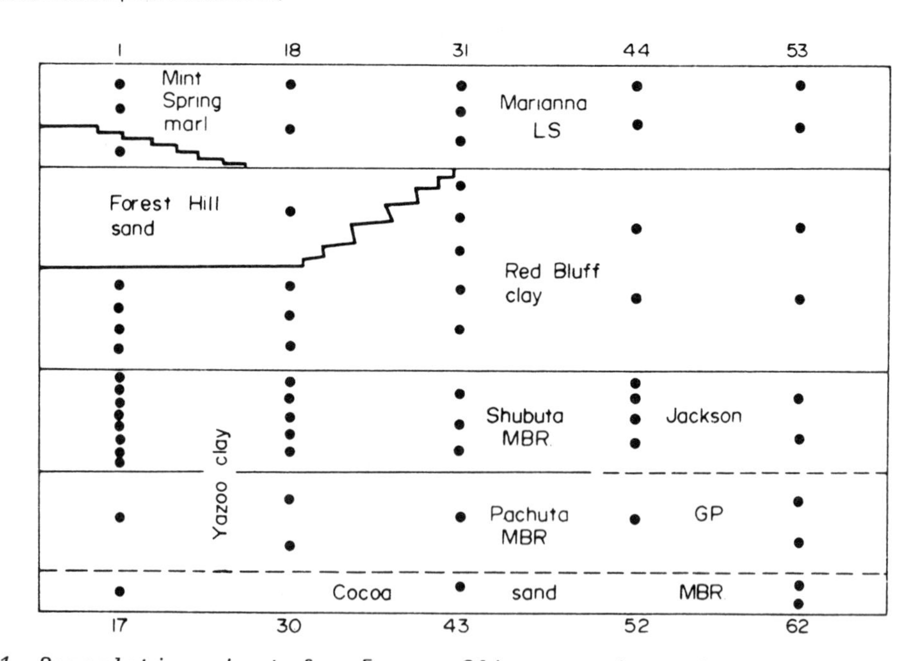

Figure 1 Correlation chart for Eocene-Oligocene data showing distribution of stratigraphic units in relation to 62 samples from 5 Gulf Coast sections. Sections are arranged from west to east. Each dot indicates the location of a sample; top and bottom samples in each section are numbered (Millendorf et al., 1978).

For the Otsuka coefficient, the standard cosine theta formula should be used inasmuch as the weighted data are continuous. The advantage of weighting is that it forces each species to contribute an amount of similarity that is proportional to its RBV. Consequently taxa with larger RBV's will result in more similarity between the samples than forms with smaller RBV's (example of Dice similarity in Table 1).

Step 4

The final decision is how should the main patterns be extracted from the similarity or difference matrix? A horde of algorithms is available for this purpose of which cluster analysis is probably used most widely. Our approach follows agglomerative cluster analysis with the flexible strategy of Lance and Williams (1967). Currently the unweighted-pair-group-method (UPGM) and the weighted-pair-group-method (WPGM) are most fashionable. UPGM minimizes the amount of distortion in the dendrogram relative to the original similarity matrix (Sneath and Sokal, 1973; Sokal and Rohlf, 1962; Hazel, 1970, 1977; Brower et al., 1978; Millendorf et al., 1978). However UPGM also tends to produce equal-sized clusters. If some of the clusters are much smaller than others, then the WPGM algorithm frequently gives superior results (Brower et al., 1978; Hazel, 1970, 1977). Several divisive clustering schemes can also be used, namely the polythetic technique dissimilarity analysis (Gill and Tipper, 1978; Tipper, 1979) and the monothetic method association analysis (Gill et al., 1976). I have no experience with either algorithm but they ought to yield similar results to agglomerative clustering. The agglomerative and divisive methods are probably not practical for similarity or difference matrices much larger than about 300 by 300.Within recent years, a number of non- hierarchical fast-clustering algorithms has been implemented for large data sets (e.g., Hartigan, 1975). Experimenting with various data sets convinces me that the partitions generated by fast-clustering are less optimum than those of the UPGM and WPGM agglomerative techniques.

Ordination techniques should be used in conjunction with clustering because these techniques are subject to differential distortion (Sneath and Sokal, 1973; Rohlf, 1970, 1972; Rowell et al., 1973). The reader should note that my comments on ordinations must be limited because I have only experimented with the eigenvector types and non-metric multidimensional scaling. Ordinations are standard practice in ecological work on classification and gradients as well as biogeography (see Whittaker, 1978 a,b for introduction to this literature). Cluster analysis faithfully preserves distances between similar items but the major groups are commonly arranged incorrectly. Conversely ordinations retain large scale patterns at the expense of small scale deformation. The most common ordinations are founded on eigenvectors such as principal components and principal coordinates (Hazel, 1977), factor analysis (Lynts, 1971, 1972; Symons and DeMeuter, 1974), and correspondence analysis or reciprocal averaging (Cisne and Rabe, 1978). Eigenvectors assume linear relations and are usually adequate for simple data sets (e.g. Hazel, 1977). Unfortunately many ecological, paleoecological, biogeographical and biostratigraphical data are strongly curvelinear and often fall along strange manifolds (see below and Whittaker, 1978 a,b for examples). In such situations, the eigenvectors provide only the most obvious and trivial information which is consistent with the views of Brower et al. (1978), Rohlf (1970, 1972), Rowell et al. (1973), and Sneath and Sokal (1973).

Some ordinations can handle nonlinear data. Two similar algorithms are nonmetric multidimensional scaling (Brower et al., 1978; Rowell et al., 1973) and nonlinear mapping (Howarth, 1973). Both methods are iterative and can be expensive in terms of computer time. The algorithms can also become trapped within a local region of the data so that an optimum global or overall solution for the entire data is never achieved. This situation can generally be avoided by rearranging the input data and checking the data with a global method such as principal components. Various ecologists and paleoecologists have applied polar ordination to curvilinear data although this scheme requires that the endpoints of the ordination axes be drawn from the samples (Ali, Lindemann and Feldhausen, 1976; Feldhausen, 1970; Park, 1974; Whittaker, 1978 a,b; Cisne and Rabe, 1978). Although I have not worked with nonlinear mapping or polar ordination, I find nonmetric multidimensional scaling superior to the eigenvector techniques for biostratigraphic data (Brower et al., 1978).

Several other ordinations should also be mentioned because they could prove useful. Shier (1978) outlined a simple technique termed sample ordering which produces a one-dimensional ordination of samples that can be used for paleoecological analysis. This strategy also could be applied to biostratigraphic data. A matrix of similarity coefficients derived from samples in two stratigraphic sections can be contoured (Reyre, 1974). Biostratigraphers might profitably borrow some of the seriation techniques developed by archaeologists. A list of summary papers is contained in the section on seriation of an original data matrix. A similarity matrix can also be seriated. These methods range from the exceedingly simple where one just rearranges the rows and columns of the matrix to concentrate the most similar samples along the diagonal to exceedingly complex methods which combine multidimensional scaling with principal components.

Lateral linkage represents a third category of methods of prime interest to biostratigraphers. Here the intent is to pair or merge samples over two or more stratigraphic sections. Lateral tracing (Sneath, 1975 a,b; Brower et al., 1978; Millendorf et al., 1978; Millendorf and Millendorf, 1982) is the simplest of these techniques. Samples from two adjacent stratigraphic sections are to be matched. The most similar pair of samples is selected subject to the constraint that they be located in different sections. The idea is to force the samples to join between the two stratigraphic sections rather than within them. After pairing, the two samples are deleted from further consideration. The scheme then finds the next most similar pair of samples that links across the two sections. Such pairings are determined until no more can be found and the computations terminate. The algorithm commonly produces crossovers which join stratrigraphically high samples in one section with lower ones in the other section. Crossovers are most frequent for the last samples to be joined. The crossover frequency measures the amount of structure in the data because crossovers are rare for tightly structured data. The vertical weighting factor denotes the degree of constraint required to eliminate crossovers. For the examples discussed here, this factor equals nil. Sneath's version of lateral tracing can be modified in various ways. For example, one can link each sample in the section with fewer samples to two samples in the other section, and so on.

Although the computations differ greatly, slotting is conceptually similar to lateral tracing (Gordon and Reyment, 1979; Gordon, 1980). The object is to merge the samples from two or more sections into a single sequence. A dissimilarity matrix is calculated for the samples in the sections. Dynamic programming recursively solves for the shortest path through the matrix which produces the slotted sequence.

For the examples presented by Millendorf et al. (1978), the best results were obtained with UPGM cluster analysis, nonmetric multidimensional scaling, and lateral tracing. The first two techniques display the major faunal zones and the discontinuities between them whereas lateral tracing increases the resolution within the zones.

A two-fold strategy is adopted. Initially the data are treated 'en masse' and, hopefully, the clusters and ordinations display the desired zones or correlations. This produces the overall pattern of the data and highlights major changes. Secondly pairs of adjacent sections are analyzed to develop a line of sections or a fence diagram. This usually obtains more detailed correlations. The nature of facies changes in most regions gives the rationale for this methodology. At any one time, environmental conditions and faunas will usually differ over an area. Such variation is generally continuous so the faunal elements are gradually replaced, both vertically and laterally. For an isochronous unit showing systematic geographical variation, samples collected from distant points might not be similar enough to cluster. Obviously the similarity between two synchronous samples declines with increasing rate of environmental change and/or distance. Working with adjacent sections allows lateral change without loss of stratigraphic resolution. A unit showing faunal variation along a transect can be followed by a pairwise comparison of the sections.

R-MODE ANALYSIS

An analysis of the taxa can be carried out with the same techniques with a few exceptions. Lateral tracing is clearly not applicable in this mode. The presence-absence data can be in either the original or range-through form. Weighting the presences by the RBV's serves no useful purpose.

R-mode results can be difficult to interpret relative to stratigraphic position (Hazel, 1970, 1977; see later discussion). Species restricted to a single assemblage zone will generally group together. Long-ranging species found in several zones obscure the results and introduce distortion into the dendrograms and ordinations. R-mode analyses are sometimes misleading because they mainly reveal information about the 'centroids' of the range zones. The endpoints of the range zones are often more useful than their 'centroids'.

SEQUENTIAL R- AND Q-MODE METHODS

Dual space techniques are available for biostratigraphic data. Two examples in the literature are principal components and factor analysis (Hazel, 1977; see later discussion). A matrix of similarities, typically phi coefficients, is derived for the taxa or samples, whichever is smaller. Usually the samples outnumber the species. Note that information in this matrix mainly reflects the 'centroids' of the range zones. The principal components or factors of this similarity matrix give the relations between the species whereas the principal component or factor scores array the samples. Other techniques, for example correspondence analysis, should yield similar results. The reader should remember that the eigenvector based algorithms are designed for linear data.

Although not a true dual space method, two-way cluster analysis aids greatly in data interpretation. First, separate Q- and R-mode clusters are done for the data matrix. Secondly, a rectangular array of the original data is printed with the taxa and samples in the same orders as on the dendrograms. Inspection of this matrix reveals blocks of species and samples with similar patterns of presences and absences. Such programs have been implemented in several statistical libraries, such as BMDP and the Ecological Analysis Package.

VERIFICATION

After establishing the assemblage zones, the biostratigrapher may wish to further explore the data to determine whether or not the differences between the zones are statistically significant, percentages of overlap between zones, etc. The algorithms fall into two main categories, namely discriminant analysis and significance tests on clusters (Sneath, 1977, 1979). The advantage of discriminant methods is that they find the maximum contrast or contrasts between the groups being compared. Unfortunately, these techniques are sensitive to assumptions such as multivariate normality which are rarely met by biostratigraphic information. If too many variables are entered into a discriminant function, artificial separation will often result. If discriminant functions are to be used, one should have access to a stepwise program. Discriminant analysis with presence-absence data presents both practical and theoretical problems; for example, it is difficult to obtain a unique answer. The cluster significance tests usually fail to detect the maximum separation between the groups being studied. However these methods are essentially independent of the number of variables and they are less sensitive to statistical assumptions (Sneath, 1977).

CASE STUDIES

Next two examples will be presented. The first illustrates the advantages of employing RBV's to weight the data. The second allows comparison of the results derived from clustering, lateral linkage, and several ordination algorithms.

Gulf Coast Data

The first example consists of the Eocene-Oligocene boundary strata of the American Gulf Coast reported by Deboo (1965; see Millendorf et al., 1978; Hazel, 1970 for more details). The data include 205 species of microfossils, namely 134 benthic foraminifers, 16 planktic foraminifers, and 55 ostracodes. The 62 samples were collected from five stratigraphic sections in Mississippi and Alabama as shown on the correlation chart of Figure 1. The sections roughly parallel the depositional strike so the facies changes are relatively simple.A generalized lithologic sequence, listed from oldest to youngest, equals marl, clay and limestone. The stratigraphic thickness and clastic content increase from east to west as the ancestral Mississippi delta is approached. The Forest Hill Sand and Mint Spring Marl form tongues in the westernmost section.

Figure 2 contains the dendrogram for 62 samples based on unweighted range-through data for the 71 species of ostracodes and planktic foraminifera. The sample numbers correspond to those of Figure 1. Three large and two small clusters are evident. All samples from the Marianna Limestone, the Forest Hill Sand and Mint Spring Marl, one upper Jackson Group sample, and numerous Red Bluff Clay collections form the large cluster at the top of the dendrogram. Note that samples from the Red Bluff and Marianna mainly cluster together. The lowermost cluster has samples from the Shubuta Member along with equivalent-aged samples from the Jackson Group. The other large cluster consists of samples from the Pachuta Marl and Cocoa Sand Members, an equivalent-aged sample from the Jackson Group and one Shubuta Member sample. Two Shubuta Member samples occur in the smallest cluster whereas six samples from the Red Bluff Clay are present in the other small cluster. The dendrogram reveals strong discontinuities between the major clusters and biozones. The general impression is that the clusters obtained from the unweighted data reflect stratigraphic horizons and units. However, the clusters are structured loosely and contain several stratigraphic units. For example, the large cluster at the top of the dendrogram includes samples from five of the higher units whereas the second cluster from the base of the dendrogram contains four of the lower units. I suggest

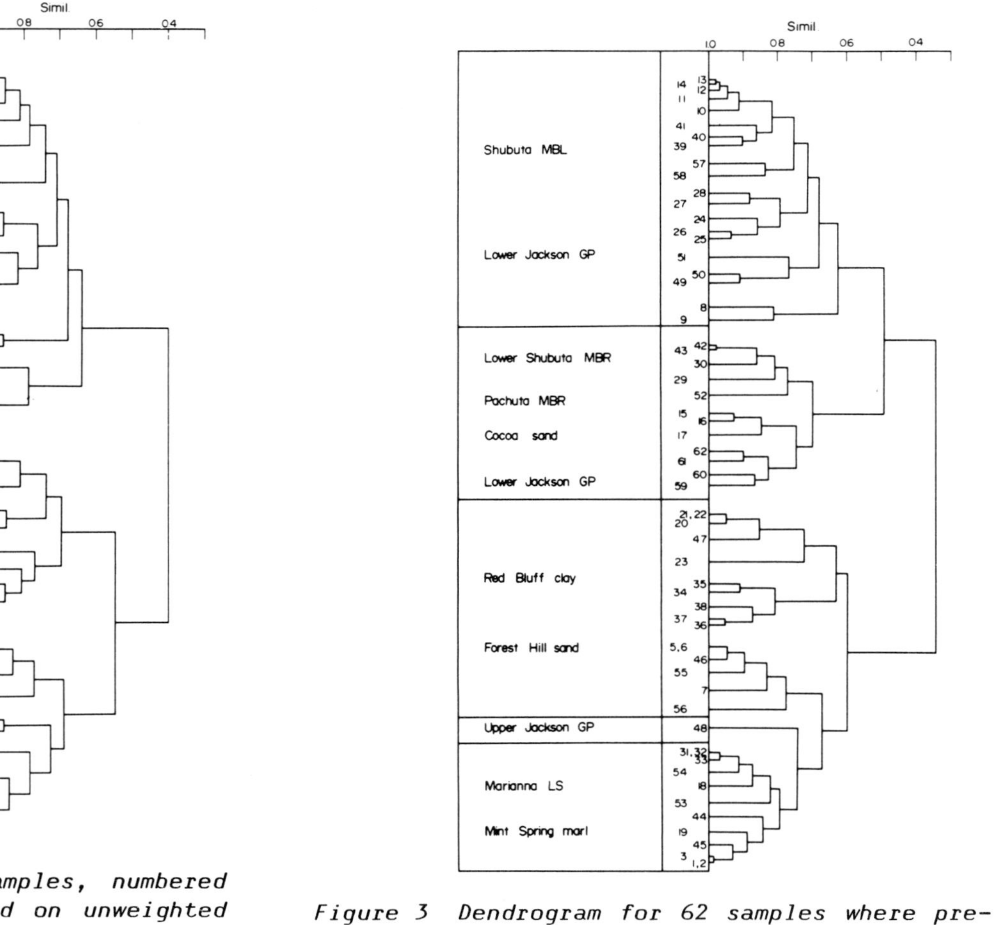

Figure 2 *Dendrogram for 62 samples, numbered as in Figure 1, based on unweighted data of 71 species. All dendrograms are based on range-through data, Dice or modified Dice similarities, and UPGM clustering. Stratigraphic units are in Figure 1.*

Figure 3 *Dendrogram for 62 samples where presences of 71 species are weighted by RBV1.*

that the reader check the data by plotting the clusters on the correlation chart.

As discussed previously, the data matrix can be weighted through multiplying the presences by the RBV's. Figures 3 and 4 illustrate cluster analyses where the weightings are obtained from RBV1 and RBV2. The groupings are tighter and more homogenous inasmuch as each cluster usually only includes one or two adjacent stratigraphic horizons.

The data for 205 species was reduced by eliminating forms that supposedly convey no useful information. For the data weighted by RBV1, each species whose RBV is lower than its facies independence (F) was removed which yielded a residue of 95 species. Where RBV2 is involved, each species deleted has an RBV of less than F/2 which left 137 taxa. The RBV1 analysis produced a dendrogram that is intermediate between those obtained from all species using unweighted and weighted data (compare Figures 2 and 3). One cluster includes the highest samples from the Marianna and Mint Spring units. The samples from the Shubuta and the two lower units, the Pachuta and Cocoa Members, mainly fall into two clusters. The relationships between the Red Bluff clay samples are considerably distorted. These samples occur in two clusters, one of which is linked with the oldest samples. These results are inconsistent with the stratigraphy and they suggest problems with the RBV1 model. Cluster analysis of 62 samples based upon unweighted distribution data for the 95 species used above yielded nearly identical results.

Another dendrogram was determined for 62 samples calculated from data weighted by RBV2 for 137 species which fulfilled the criterion RBV of greater than F/2. The groups for this run are almost identical to those of Figure 4 in which the presences of 71 species were multiplied by their RBV2 values. Only three samples, one from each of the Marianna, Forest Hill, and Shubuta units, move from one cluster to another. The clusters are tightly structured and clearly conform to the known stratigraphy.

Figure 5 gives the dendrogram calculated from unweighted data for the 16 species with the highest values of RBV1. Even though only about eight percent of the taxa are used, the four major clusters of the other figures for weighted data are apparent. Only about six samples are assigned to different clusters. Three of these are clearly anomalous, namely samples 7, 28, and 30, and these groupings are dictated by the occurrence of a single species. The very tight structure of the dendrogram is because many samples join at similarities of 1.0 which indicates identical faunal compositions.

These analyses follow a systematic pattern. For correlation, the best results are obtained from weighted data which can be created in two ways. First is to select a subset of species with high RBV's for later analysis. Second is to multiply the presences by the RBV's of the taxa. In this strategy, the best dendrograms are produced when all species are retained.

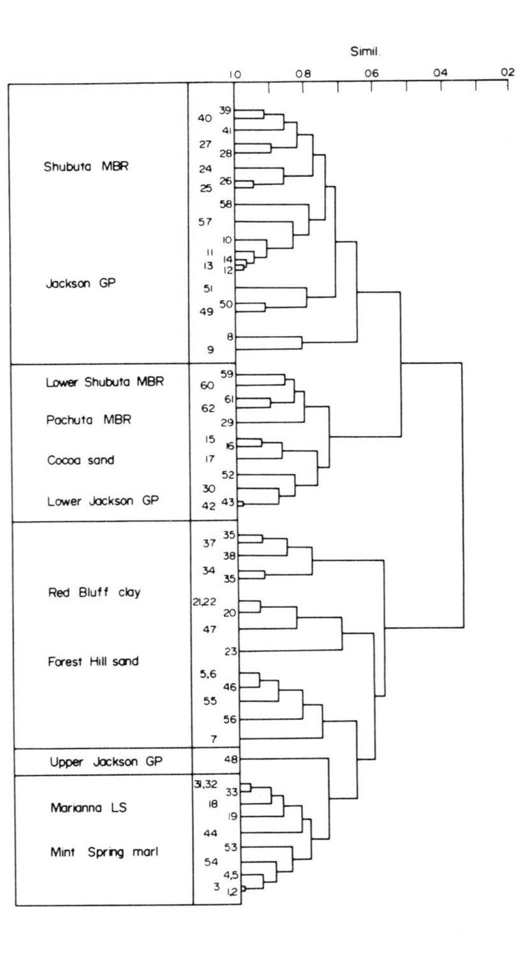

Figure 4 Dendrogram for 62 samples where presences of 71 species are weighted by RVB2.

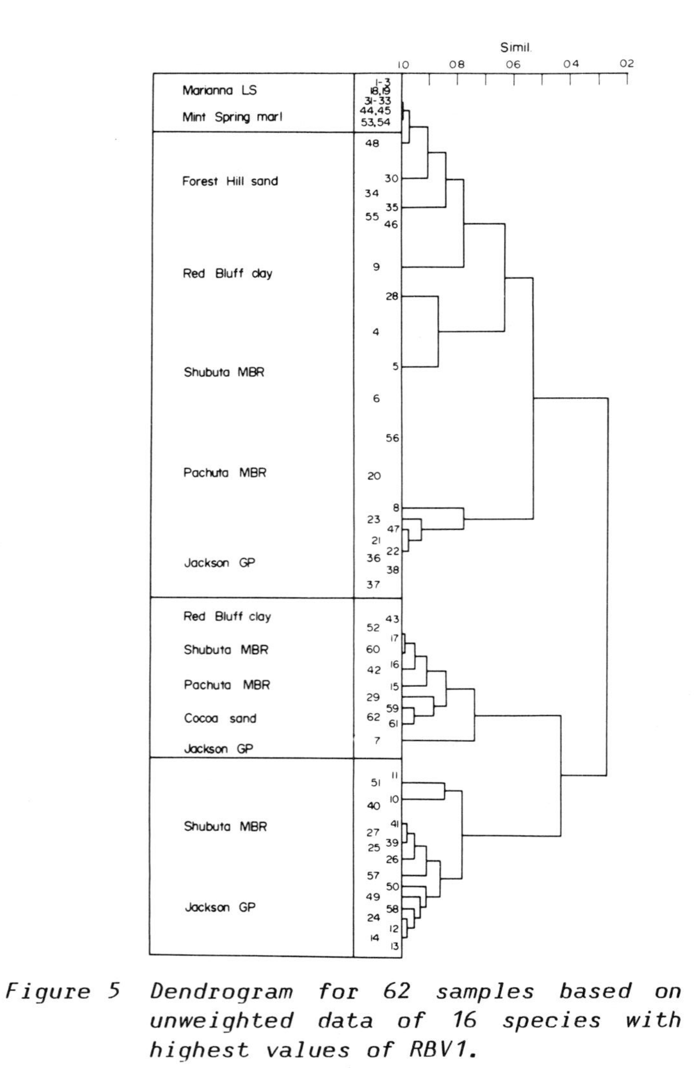

Figure 5 Dendrogram for 62 samples based on unweighted data of 16 species with highest values of RBV1.

The lateral tracings are consistent with the stratigraphy (Figure 6). Most linkages are roughly parallel to time lines. Although a number of crossovers occur in the lateral traces, most of these are confined within individual zones. Also most crossovers deal with the later pairings between samples. Essentially the lateral traces confirm the assemblage zones found by cluster analysis and, to some extent, provide correlations within the zones.

Upper Cretaceous Data

Introduction

The second study consists of the Upper Cretaceous foraminifers from the Western Interior Seaway of the United States (Eicher and Worstell, 1970). The 117 samples were obtained from eight stratigraphic sections ranging from Kansas and Colorado to South Dakota and Wyoming. The 50 species which occur in four or more sections are included. For subsequent discussion these species are numbered from 1 to 50 in the same order as listed in the systematic descriptions of Eicher and Worstell (1970). Species 1 to 20 and 21 to 50 comprise planktic and benthic taxa, respectively. The data are of the range-through form. Unfortunately RBV's cannot be estimated for the taxa because most of the samples are channel types which commonly contain several lithologies. Consequently all analyses are based on unweighted data.

In terms of the Kansas and Colorado nomenclature, the rock units range from the uppermost beds of the Graneros Formation, through the Greenhorn Formation, and into the Fairport and Blue Hill Members of the Carlile Formation (Figure 7). The correlations are well known and these are founded on bentonite beds in conjunction with time-stratigraphic zones of fossils (Eicher and Worstell, 1970). The dominant lithology of the Graneros and Blue Hill consists of dark, noncalcareous shale. The Greenhorn and Fairport are more calcareous and typical rocks are chalky and calcareous shales and limestones. Systematic facies changes exist and the clastic content and stratigraphic thickness generally increase from east to west and south to north. The sedimentary pattern represents a transgression followed by a regression. The water depthsare estimated at 30 to 90 m for the shallowest environments to 150 to 600 m for the deepest assemblages (Eicher, 1969; Eicher and Worstell, 1970; Hattin, 1975). The maximum depth of water is found in the upper beds of the Greenhorn.

Analysis of Samples

Four main clusters can be identified on the dendrogram for the 117 samples (Figure 8). The structure is reasonably tight and most items join at Dice similarities of 0.5 and higher. The clusters are plotted on the correlation chart in Figure 9, and they correlate closely with the stratigraphy. Although the boundaries of the clusters are

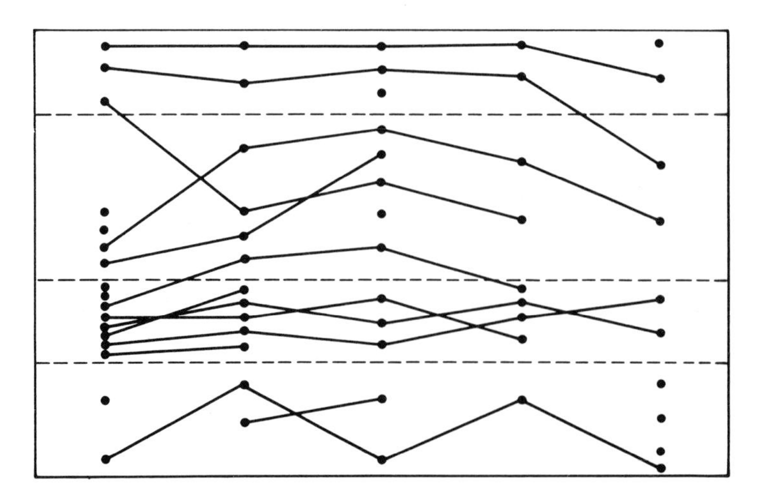

Figure 6 *Lateral tracing based on unweighted presences of 205 species plotted on correlation chart of Figure 1. The algorithm is that of Sneath (1975a,b).*

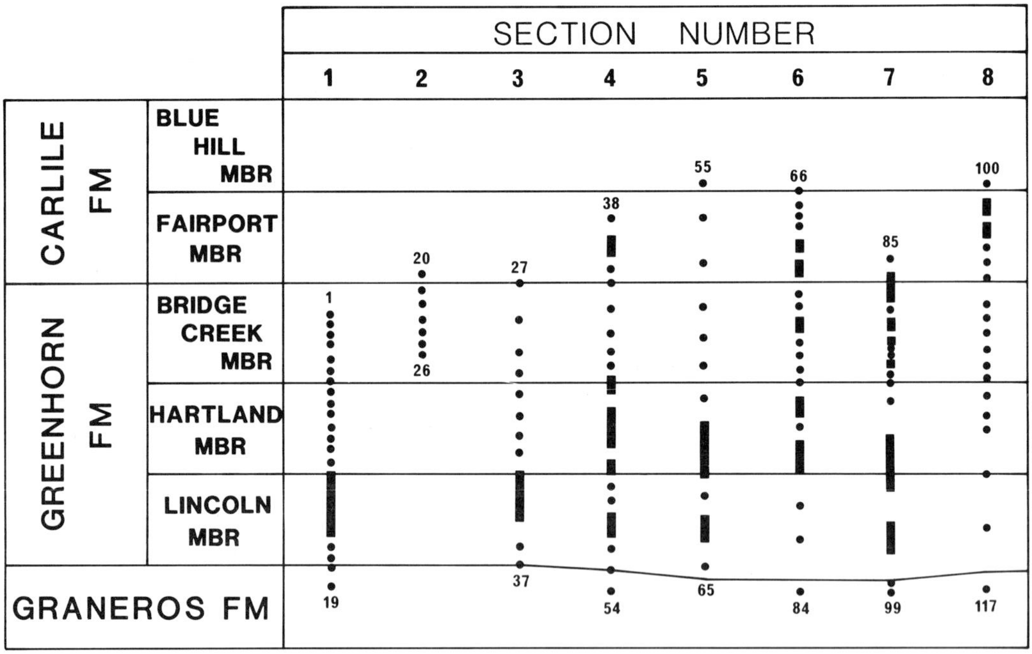

SECTION NUMBER

1 2 3 4 5 6 7 8

CARLILE FM: BLUE HILL MBR; FAIRPORT MBR

GREENHORN FM: BRIDGE CREEK MBR; HARTLAND MBR; LINCOLN MBR

GRANEROS FM

Section 1: 1 – 19
Section 2: 20 – 26
Section 3: 27 – 37
Section 4: 38 – 54
Section 5: 55 – 65
Section 6: 66 – 84
Section 7: 85 – 99
Section 8: 100 – 117

Figure 7 *Correlation chart for Upper Cretaceous data. Section numbers are those of Eicher and Worstell (1970). The sections are arranged approximately from north to south. The sample numbers indicate the highest and lowest samples in each stratigraphic section.*

diachronous to some extent, they approximately parallel time lines. Subdivisions of the main clusters typically yield biogeographic data as discussed later.

Cluster A exhibits a disjunct distribution and it is found in the base of most sections and in the highest beds at three localities. A low diversity fauna is present where only three long ranging taxa (species 2, 10, and 11) are common. In fact these taxa occur in all four clusters. Cluster A developed twice, once during the earlier phase of transgression and again in the later stages of regression. Apparently, these long-ranging taxa were hardy animals which appeared when conditions first became favorable for foraminifers and they survived after most other taxa became locally extinct.

Cluster C is mainly distributed within the lower part of the Greenhorn Formation. Faunal diversity exceeds that of the underlying strata and 36 species, 14 planktic and 22 benthic, have been identified. The core of this assemblage zone comprises the 11 planktic foraminifers found in over half of the samples. Although numerous benthic taxa are present, they only occur in a few samples and less than 10 percent of the individuals are benthic. The increased diversity demonstrates that environmental conditions were more favorable than those prevailing for Cluster A. The parameters involved represent increased depth and perhaps higher salinity (Eicher, 1969; Eicher and Worstell, 1970). The scarcity of benthic individuals reflects a deficiency of dissolved oxygen on the seabed.

Cluster D is confined to a narrow interval within the Greenhorn and it comprises the 'Benthonic zone' of Eicher and Worstell, (1970). Diversity attains a maximum here and 47 foraminifers, 17 planktic and 30 benthic types, are counted. Forty-one of these animals, 26 benthic and 15 planktic taxa, are common and are recorded from over half of the samples. Benthic specimens are more frequent than in the other zones and they range up to 20 percent of all foraminifers. Assemblage zone D marks the maximum transgression. The proliferation of benthic species and increased diversity of planktic forms evidences a much better environment (Eicher, 1969; Eicher and Worstell, 1970). The dissolved oxygen content was large enough to support many benthic taxa. The larger number of planktic organisms also implies more favorable conditions in the water column. Figure 9 reveals that the 'Benthonic zone' is diachronous and becomes younger toward the Black Hills. The largest subcluster isolates the Black Hills samples in Sections 1 to 4 from the others. This is due to a lower diversity of planktic taxa which suggests less open waters (Eicher and Worstell, 1970).

Cluster B includes samples in the adjacent beds of the Greenhorn and Carlile Formations and it represents the initiation of the regression. The number of species declines to 42. The overall composition of Cluster B mirrors that of C in that the fauna is dominated by planktic species and individuals. Of the species present in over half of the samples, 11 are planktics with only two being benthics. Like Cluster C,

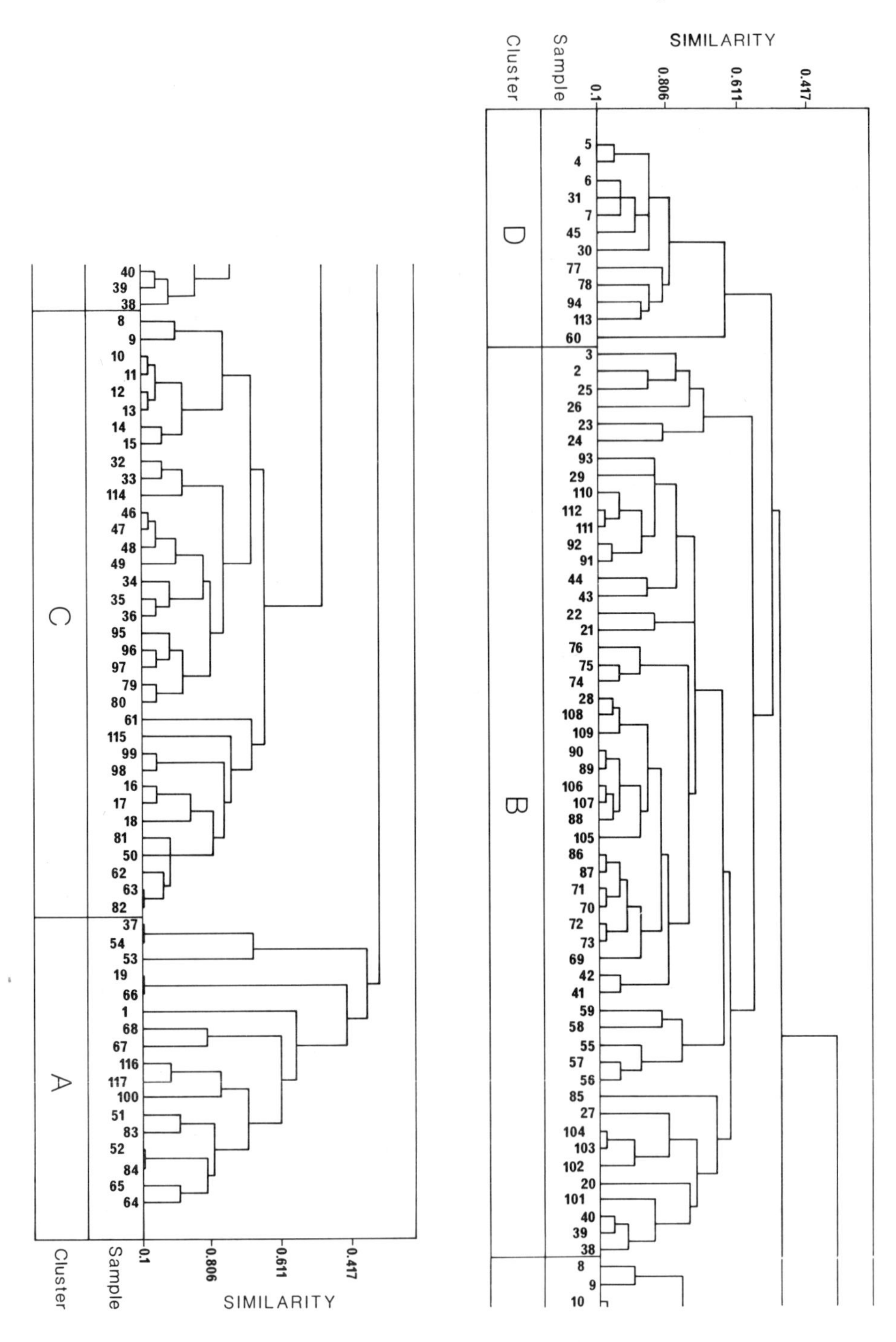

Figure 8 Dendrogram for 117 samples. Cophenetic correlation equals 0.778.

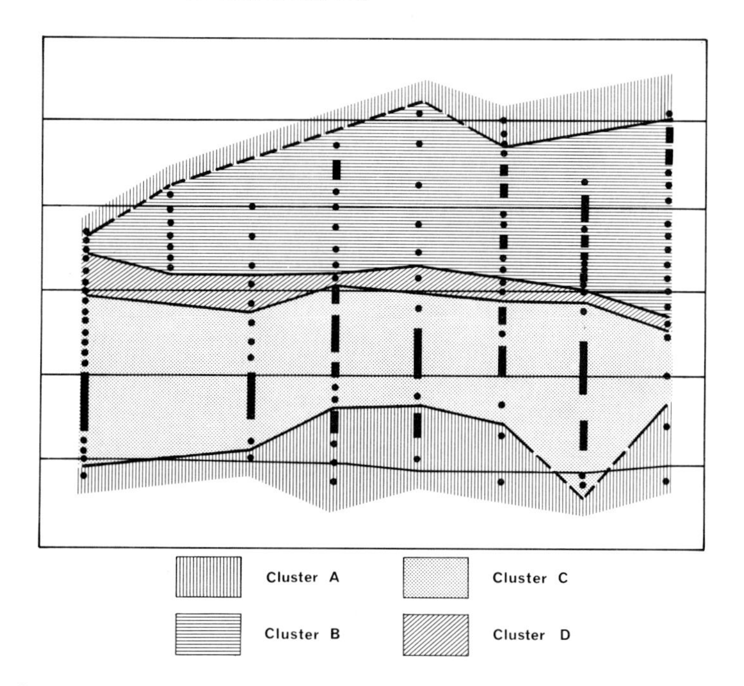

Figure 9 Graph showing distribution of clusters from Figure 8 plotted on the correlation chart of Figure 7.

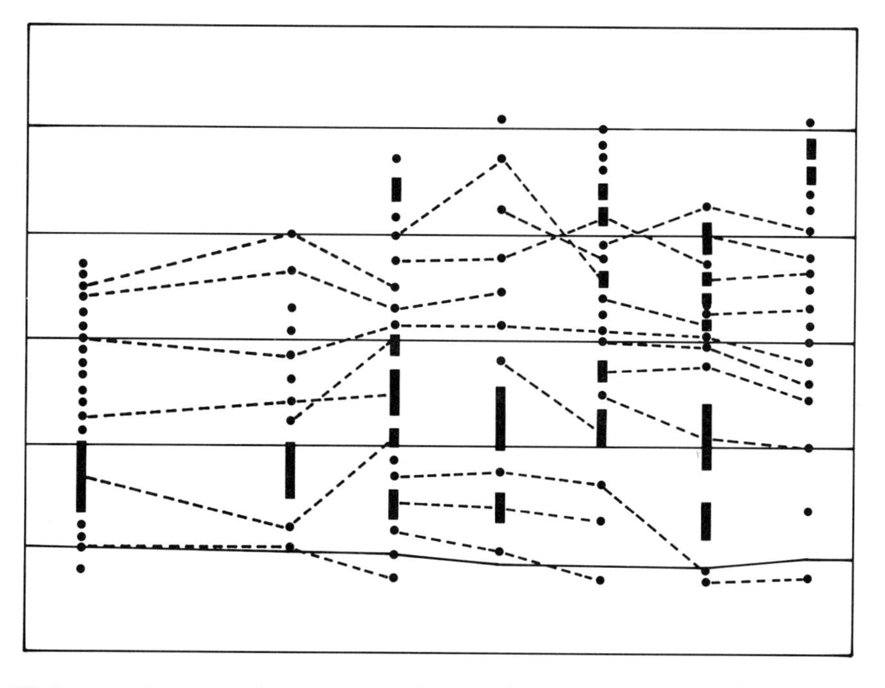

Figure 10 Lateral tracings superimposed on correlation chart of Figure 7. Section 2 was omitted.

benthic animals never make up more than 10 percent of all individuals. The falling sealevel restricted the dissolved oxygen on the seafloor which curtailed the benthic fauna (Eicher, 1960; Eicher and Worstell, 1970). Decreased water depth, a smaller depth range of oxygenated water, and possibly lower salinity may have affected the planktic foraminifers. Continued regression further reduced faunal diversity to produce the upper samples of Cluster A.

The lateral tracings are shown in Figure 10. The samples in section 2 were omitted because they lie within a narrow stratigraphic interval. Comparison of Figures 9 and 10 denotes that the lateral tracings provide somewhat more resolution than the cluster analysis of all samples. Most linkages roughly parallel time lines, and few crossovers are observed. Although several lateral traces cut across stratigraphically adjacent clusters, most join samples within single clusters.

Cluster analyses were done for pairs of adjacent sections (Figure 11); as before section 2 was not included. Placing the boundaries between the clusters on the correlation chart shows the stratigraphic information in the clusters. Observe that some cluster boundaries change between different pairs of sections whereas others can be traced through numerous localities. The results of pairwise clustering are intermediate between lateral tracing and cluster analysis of the entire data set (Figures 9-11). The 'en masse' cluster analysis displays the major patterns and faunal discontinuities whereas lateral tracing yields fine tuning.

The data were also treated with several ordination techniques because ordination and clustering are subject to different types of distortion. Figures 12 and 13 contain the two-dimensional solution by non-metric multidimensional scaling and the scores for the first two principal components. The data points are identified by the cluster designations. According to the dendrogram (Figure 8) Clusters D and B are most similar, followed by a join between C with D and B; A is only linked with the other clusters at low similarity. The relations interpreted from the ordinations differ somewhat. Similar patterns are observed for both ordinations. Clusters A and D are located at opposite ends of the graphs. Clusters B and C are separated and they lie between Clusters A and D. The overall arrangement of the multidimensional scaling follows a hollow ellipse. The principal component scores are similar but the base is flattened. Although the ordination is nonlinear, principal components performs reasonably well. Other eigenvector methods yield similar plots to those of the principal components. Multidimensional scaling better preserves the known relations between the clusters. Cluster C was derived from A by increasing diversity during transgression. The upper samples of Cluster A contain the protean species of Cluster B that survived the unfavorable environments developing during regression. Cluster D appeared at the maximum transgression. The diversified fauna of Cluster D contains many elements of Cluster B and C as well as A. These linkages are more accurately shown by the ordinations, especially multidimensional scaling, than by the clusters.

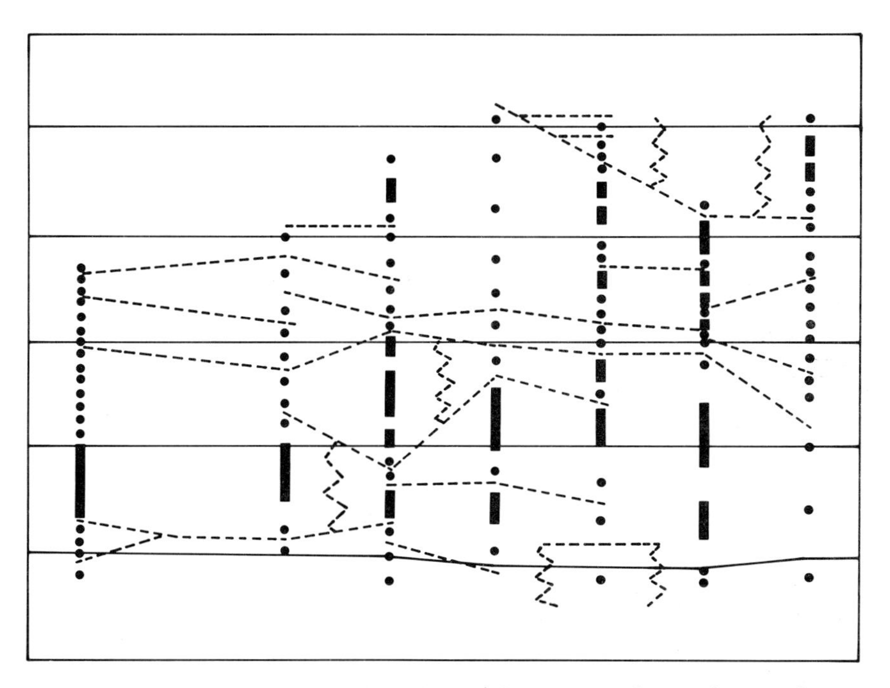

Figure 11 Boundaries between pairwise clusters plotted on the correlation chart of Figure 7. Section 2 was omitted.

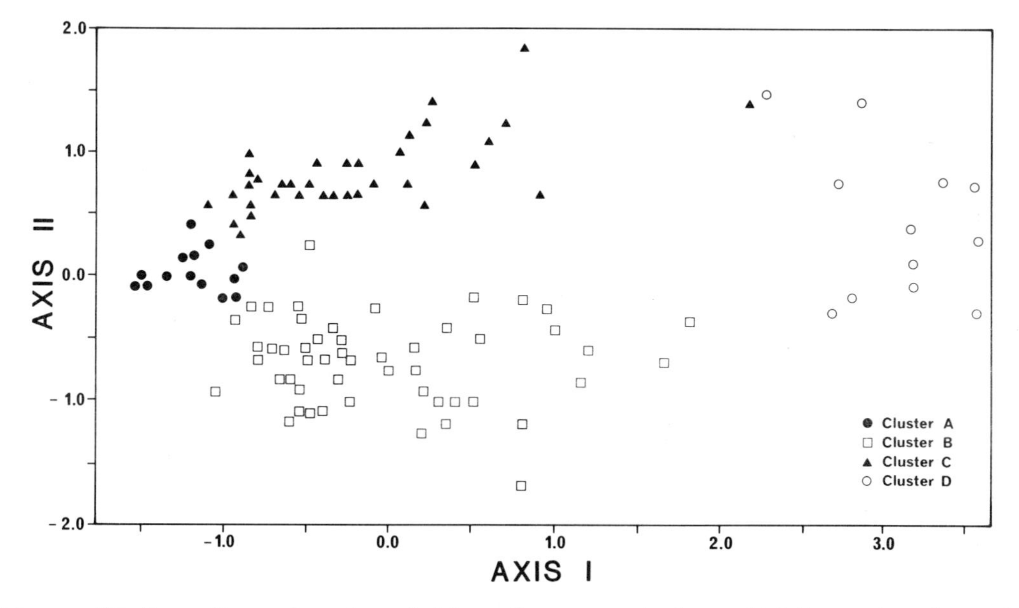

Figure 12 Two dimensional solution for nonmetric multidimensional scaling of the 117 samples. Euclidean distances were used. Stress and R values are 0.143 and 0.933, respectively. The samples are identified by the cluster designations of Figure 8.

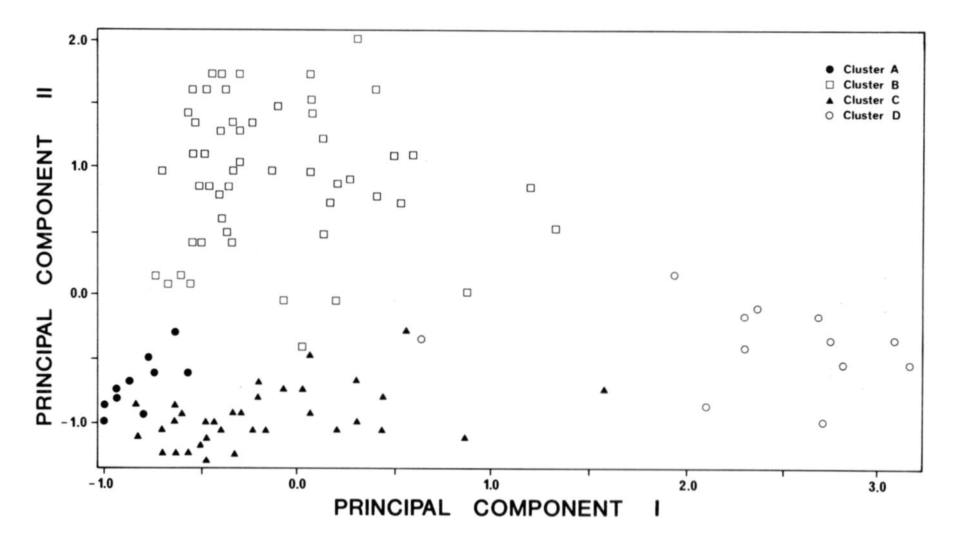

Figure 13 *Plot of scores of 117 samples for first two principal components. The eigenvectors were extracted from a matrix of phi coefficients for the 50 species. The first two eigenvalues account for 28.2 and 12.6 percent of the variance. The samples are identified by the cluster names in Figure 8.*

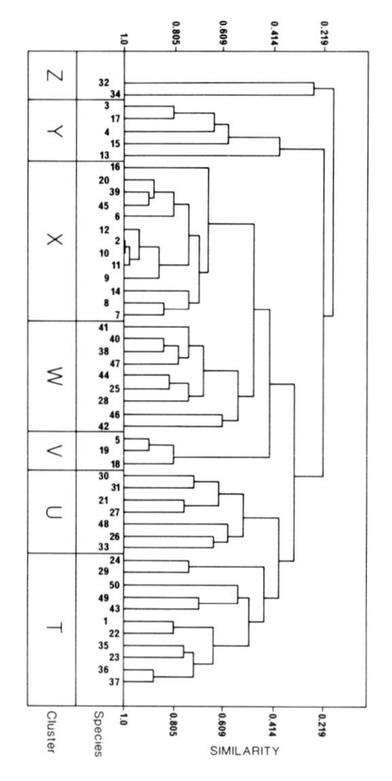

Figure 14 *Dendrogram for 50 species. Cophenetic correlation equals 0.791. Species numbers correspond to those in Table 2.*

Analysis of Species

The R-mode dendrogram exhibits seven groups, labeled from T to Z (Figure 14). Numerous clusters were selected to accommodate groups with slight differences in their distributions. Generally, the planktic (species 1 to 20) and the benthic (species 21 to 30) animals are concentrated in different clusters. As mentioned before the Biostratigraphic Constancy and Biostratigraphic Fidelity yield 'a posteriori' measures of the index fossil concept (Hazel, 1970, 1977; Brower, 1981, 1984). Constancy consists of the percent of samples within Cluster or Zone k that contain Species i. Fidelity represents the percent of occurrences of species i within Zone k. These data were compiled from a two-way cluster analysis for the 50 species (Table 2).

The clusters are summarized as follows:

Cluster T, mostly benthic animals with high constancy and fidelity for Zone D.

Cluster U, bottom dwellers with high constancy for Zone D and moderate fidelity for D and C.

Cluster V, planktic taxa with high constancy and fidelity for Zone C.

Cluster W, benthic organisms with moderately long ranges, high constancy for Zone D, moderate fidelities for B, D and C.

Cluster X, predominantly long ranged planktic taxa, many of which are found in all four zones, high constancies for all zones except A, moderate fidelities for B and C.

Cluster Y, planktic taxa with high fidelity and moderate constancy for Zone B.

Cluster Z, two rare benthic species.

The analysis can be 'turned around' because forms with high fidelity and constancy for a particular zone represent its indices. Study of Table 2 provides the following 'index fossils':

Cluster A, satisfactory 'indices' cannot be defined. The most common taxa, numbers 11, 10, and 2, possess constancies of about 80 percent but the highest fidelity is only 12.3 percent. This is because the taxa in Cluster A have long ranges.

Cluster B, species 3 and 17, to a lesser extent, taxa 16, 20, 39, 45, 6 and 12.

Cluster C, species 5, 19 and 18.

Cluster D, numbers 37, 36, 35, 43, 23, 1, 50 and 29.

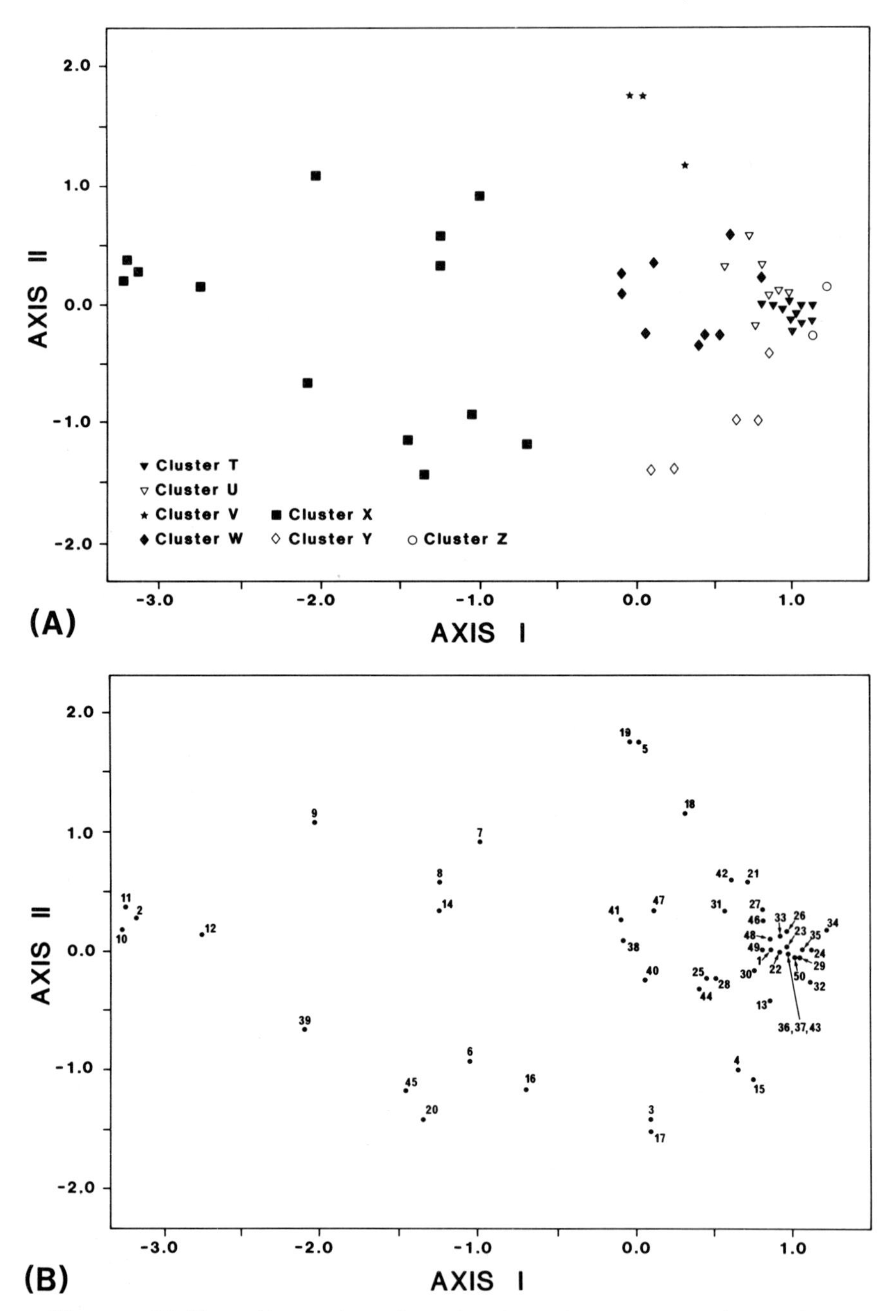

Figure 15 *Two dimensional solution for nonmetric multidimensional scaling of the 50 species. The distance matrix is based on Euclidean distances. Stress and R consist of 0.111 and 0.963. Species are identified by the cluster designations (A) of Figure 14 or the numbers in Table 2 (B).*

Table 2 Biostratigraphic Fidelities and Biostratigraphic Constancies for the Cretaceous data. Species identification numbers and ordering follow the dendrogram of Figure 14.

	Biostratigraphic Fidelity For Clusters				Biostratigraphic Constancy For Clusters			
Species	A	B	C	D	A	B	C	D
37				100				66.7
36				100				66.7
23		20.0		80.0		3.8		66.7
35				100				50.0
22		35.7	7.1	57.1		9.4	2.9	66.7
1	7.1	21.4		71.4	5.9	5.7		83.3
43				100				66.7
49		26.7	13.3	60.0		7.6	5.7	75.0
50		14.3		85.7		1.9		50.0
29		14.3		85.7		1.9		50.0
24				100				33.3
33		10.0	20.0	70.0		1.9	5.7	58.3
26		16.7	41.7	41.7		3.8	14.3	41.7
48		21.4	28.6	50.0		5.7	11.4	58.3
27	5.3	21.1	36.8	36.8	5.9	7.6	20.0	58.3
21		20.8	50.0	29.2		9.4	34.3	58.3
31		20.8	29.2	50.0		9.4	20.0	100
30		40.0	20.0	40.0		15.1	11.4	66.7
18			75.0	25.0			68.6	66.7
19		4.6	79.5	15.9		3.8	100	58.3
5	7.0		76.7	16.3	17.6		94.3	58.3
42		16.7	33.3	50.0		7.6	22.9	100
46		11.8	35.3	52.9		3.8	17.1	75.0
28		51.9	7.4	40.7		26.4	5.7	91.7
25		50.0	3.8	46.2		24.5	2.9	100
44		53.8	3.8	42.3		26.4	2.9	91.7
47		33.3	30.3	36.4		20.8	28.6	100
38		47.4	23.7	28.9		34.0	25.7	91.7
40		54.3	14.3	31.4		35.8	14.3	91.7
41		44.7	26.3	28.9		32.1	28.6	91.7
7		33.9	45.8	20.3		37.7	77.1	100
8	1.6	42.6	37.7	18.0	5.9	49.1	65.7	91.7
14		50.0	34.5	15.5		54.7	57.1	75.0
9	4.8	39.3	41.7	14.3	23.5	62.3	100	100
11	11.7	45.9	31.5	10.8	76.5	96.2	100	100
10	12.3	46.5	30.7	10.5	82.4	100	100	100
2	11.5	46.9	31.0	10.6	76.5	100	100	100
12	2.0	52.5	33.3	12.1	11.8	98.1	94.3	100
6		67.8	13.6	18.6		75.5	22.9	91.7
45	1.5	72.1	8.8	17.6	5.9	92.5	17.1	100
39	1.2	63.9	20.5	14.5	5.9	100	48.6	100
20	1.6	78.1	6.2	14.1	5.9	94.3	11.4	75.0
16		70.0	12.0	18.0		66.0	17.1	75.0
13		64.3		35.7		17.0		41.7
15		100				35.8		
4		100				34.0		
17		96.8		3.2		56.6		8.3
3		100				56.6		
34			80.0	20.0			11.4	8.3
32		58.3	25.0	16.7		13.2	8.6	16.7

Ordinations were also computed as exemplified by nonmetric multidimensional scaling (Figure 15). The taxa are identified by the cluster designations of Figure 14. The first axis represents a generalized measure of stratigraphic range. Long-ranging species occupy a wide spectrum on the left whereas the taxa of Cluster T, which are mostly restricted to the 'Benthonic zone', are located on the right side. The foraminifers most diagnostic of Clusters C and B are found in Cluster V and Y, and these animals have high and low projections on the second axis. The taxa with moderately-long ranges occupy intermediate regions on the plot. The nonmetric multidimensional scaling displays the relations between the major groups of taxa more clearly than the dendrogram.

SUMMARY

Multivariate methods are quite useful in Assemblage Zonal Biostratigraphy. To apply these algorithms, the biostratigrapher must answer questions at each step ranging from data preparation to final analysis. There are no right or wrong decisions! All depend on the nature of the data and the intent of the analysis (Edwards, 1982). The data matrix is a rectangular block of presences and absences with the species arrayed on one axis and samples on the other. Typically the samples derive from numerous stratigraphic sections. When compiling the information the data can be recorded as collected or converted to the range-through form. The probability of finding a species can be considered when tabulating the presences and absences. After assembling the data matrix one must decide whether or not to incorporate the RBV's and similar information. The RBV's or the biostratigraphic attributes can identify a select group of taxa for later treatment. One can weight the presences by the RBV's. The multivariate techniques belong to three main categories: clustering, ordination and lateral linkage. All are useful because they provide different insights into the data. Clustering illustrates the distribution of faunal discontinuities and retains small scale relationships. Ordinations tend to preserve major patterns but at the expense of distorting the distances between similar items. Clustering and ordination can be employed in two strategies, namely to treat all data together and to operate on two adjacent stratigraphic sections at a time in order to construct a line of sections or fence diagram. Lateral linkages yield detailed correlations within zones.

REFERENCES

Ali, S.A. Lindemann, R.H., and Feldhausen, P.H., 1976, A multivariate sedimentary environment analysis of Great South Bay and South Oyster Bay, New York. J. Math. Geology, 8 (3), 283-304.

Brower, J.C., 1981, Quantitative biostratigraphy, 1830-1980. In Computer applications in the earth sciences, an update of the 70's, D.F. Merriam, ed., Plenum Press, New York and London, 63-103.

Brower, J.C., 1984, The relative biostratigraphic values of fossils. Computers and Geosciences, 10 (1), 111-133.

Brower, J.C., and Burroughs, W.A., 1982, A simple method for quantitative biostratigraphy. In Quantitative stratigraphic correlation, J.M. Cubitt and R.A. Reyment, eds., John Wiley and Sons, Ltd., 61-83.

Brower, J.C., Millendorf, S.A., and Dyman, T.S., 1978, Quantification of assemblage zones based on multivariate analysis of weighted and unweighted data. Computers and Geosciences, 4 (3), 221-227.

Buzas, M.A., 1970, On the quantification of biofacies. Proc. North Am. Paleont. Conv., Chicago, 1969, pt. B, 101-116.

Cheetham, A.H., and Hazel, J.E., 1969, Binary (presence-absence) similarity coefficients. J. Paleont., 43 (5), 1130-1136.

Cisne, J.L., Karig, D.E., Rabe, B.D., and Hay, B.J., 1982, Topography and tectonics of the Taconic outer trench as revealed through gradient analysis of fossil assemblages. Lethaia, 15 (3), 229-246.

Cisne, J.L., and Rabe, B.D., 1978, Coenocorrelation, gradient analysis of fossil communities and its applications in stratigraphy. Lethaia, 11 (4), 341-364.

Deboo, P.B., 1965, Biostratigraphic correlation of the type Shubuta Member of the Yazoo Clay and the Red Bluff Clay with their equivalents in south-western Alabama. Alabama Geol. Survey Bull.80, 84 pp.

Dennison, J.M., and Hay, W.W., 1967, Estimating the needed sampling area for subaquatic ecologic studies. J. Paleont., 41, 706-708.

Edwards, L.E., 1982, Quantitative biostratigraphy: the methods should suit the data. In Quantitative stratigraphic correlation, J.M. Cubitt and R.A. Reyment, eds., John Wiley and Sons, Ltd., 45-60.

Eicher, D.L., 1969, Paleobathymetry of Cretaceous Greenhorn Sea in eastern Colorado. American Assoc. Petr. Geol. Bull., 53 (5), 1075-1090.

Eicher, D.L., and Worstell, P., 1970, Cenomanian and Turonian Foraminifera from the Great Plains, United States. Micropal., 16 (3), 269-324.

Feldhausen, P.H., 1970, Ordination of sediments from the Cape Hatteras continental margin. J. Math. Geology, 2 (2), 113-129.

Gauch, H.G., Jr., 1982, Multivariate analysis in community ecology. Cambridge Univ. Press, New York, x, 298 pp.

Gill, D., Boehm, S., and Erez, Y., 1976, ASSOCA: FORTRAN IV program for Williams and Lambert association analysis with printed dendrograms. Computers and Geosciences, 2 (2), 219-248.

Gill, D., and Tipper, J.C., 1978, The adequacy of non-metric data in geology: tests using a divisive-omnithetic clustering technique. J. Geology, 86 (2), 241-259.

Gordon, A.D., 1980, SLOTSEQ: a FORTRAN IV program for comparing two sequences of observations. Computers and Geosciences, 6 (1), 7-20.

Gordon, A.D., and Reyment, R.A., 1979, Slotting of borehole sequences. Math. Geol., 11, (3), 309-327.

Hartigan, J.A., 1975, Clustering algorithms. John Wiley & Sons, New York, xiii, 351 pp.

Hattin, D.E., 1975, Stratigraphy and depositional environment of Greenhorn Limestone (Upper Cretaceous) of Kansas. Kansas Geol. Surv., Bull. 209, 128 pp.

Hay, W.W., 1972, Probabilistic stratigrahy. Eclogae Geol. Helvetiae, 65 (2), 255-266.

Hazel, J.E., 1970, Binary coefficients and clustering in biostratigraphy. Geol. Soc. America Bull., 81, (11) 3237-3252.

Hazel, J.E., 1971, Ostracode biostratigraphy of the Yorktown Formation (Upper Miocene and Lower Pliocene) of Virginia and North Carolina. U.S. Geol. Survey Prof. Paper 204, 13 pp.

Hazel, J.E., 1977, Use of certain multivariate and other techniques in assemblage zonal biostratigraphy, examples utilizing Cambrian, Cretaceous, and Tertiary benthic invertebrates. In Concepts and methods of biostratigraphy, Kauffman, E.G., and Hazel, J.E., eds., Dowden, Hutchinson & Ross, Inc., Stroudsburg, Pennsylvania, 187-212.

Henderson, R.A., and Heron, M.L., 1977. A probabilistic method of paleobiogeographic analysis. Lethaia, 10 (1), 1-15.

Hohn, M.E., 1976, Binary coefficients, a theoretical and empirical study. J. Math. Geology, 8 (2), 137-150.

Howarth, R.J., 1973, Preliminary assessment of a nonlinear mapping algorithm in a geological context. Math. Geol., 5 (1), 39-57.

Kaesler, R.L., 1966, Quantitative re-evaluation of ecology and distribution of Recent Foraminifera and Ostracoda of Todos Santos Bay, Baja. California, Mexico, Kansas Univ. Paleont. Contr., Paper 10, 50 pp.

Lance, G.N., and Williams, W.T., 1967, A general theory of classification sorting strategies, I, Hierarchical systems. Computer J., 9, 373-380.

Lyell, C., 1830-33, Principles of geology. J. Murray, London, 3 vols.

Lynts, G.W., 1971, Analysis of the planktonic Foraminifera fauna of core 6275, Tongue of the ocean, Bahamas. Micropal., 17 (2), 152-166.

Lynts, G.W., 1972, Factor-vector analysis models in ecology and paleoecology. 21st Intern. Geol. Congress (Montreal) Sect. 7, Paleontology, 227-237.

Millendorf, S.A., Brower, J.C., and Dyman, T.S., 1978, A comparison of methods for the quantification of assemblage zones. Computers and Geosciences, 4 (3), 229-242.

Millendorf, S.A., and Millendorf, M.T., 1982, The conceptual basis for lateral tracing of biostratigraphic units. In Quantitative stratigraphic correlation, J.M. Cubitt and R.A. Reyment, eds., John Wiley and Sons, Ltd., 101-106.

Park, R.A., 1974, A multivariate analytical strategy for classifying paleoenvironments. J. Math. Geology, 6 (4), 333-352.

Rabe, B.D., and Cisne, J.L., 1980, Chronostratigraphic accuracy of Ordovician ecostratigraphic correlation. Lethaia, 13, 109-118.

Raup, D.M., and Crick, R.E., 1979, Measurement of faunal similarity in paleontology. J. Paleont., 53, (5) 1213-1227.

Reyre, Y., 1974, Les methodes quantitatives en palynologie. In Elements de palynologie, applications geologiques, Chateauneuf, J.-J., and Reyre, Y., eds., BRGM, Orleans, 271-312.

Rohlf, F.J., 1970, Adaptive hierarchical clustering schemes. Systematic Zoology, 19, 58-82.

Rohlf, F.J., 1972, An empirical comparison of three ordination techniques. Systematic Zoology, 21, 271-280.

Rowell, A.J., and McBride, D.J., 1972, Faunal variation in the **Elvinia** Zone of the Upper Cambrian of North America - a numerical approach. 21st Intern. Geol. Congress (Montreal), Sect. 7, Paleontology, 246-253.

Rowell, A.J., McBride, D.J., and Palmer, A.R., 1973, Quantitative study of Trempealeauian (Latest Cambrian) trilobite distribution in North America. Geol. Soc. America Bull., 84 (10), 3429-3442.

Scott, R.W., and West, R.R., eds., 1976, Structure and classification of paleocommunities. Dowden, Hutchinson & Ross, Inc., Stroudsburg, Pennsylvania, 291 pp.

Sepkoski, J.J., Jr., 1974, Quantified coefficients of association and measurement of similarity. J. Math. Geology, 6 (2), 135-152.

Shier, D.E., 1978, Sample ordering - a new statistical technique for paleoecological analysis. Trans. Gulf Coast Assoc. Geol. Soc., 28, 461-471.

Simpson, G.G., 1947, Holarctic mammalian faunas and continental relationships during the Cenozoic. Geol. Soc. America Bull., 58 (7), 613-687.

Simpson, S.G., 1960, Notes on the measurement of faunal resemblance. Am. J. Sci., 258a, 300-311.

Smith, W., 1817, Stratigraphical system of organized fossils, with reference to the specimens of the original geological collection in the British Museum, explaining their state of preservation and their use in identifying the British strata. E. Williams, London, 118 pp.

Sneath, P.H.A., 1975a, Quantitative method for lateral tracing of sedimentary units. Computers and Geosciences, 1 (3), 215-220.

Sneath, P.H.A., 1975b, Clarification on a quantitative stratigraphic correlation technique. Computers and Geosciences, 1 (4), 353-354.

Sneath, P.H.A., 1977, A method for testing the distinctness of clusters, a test of the disjunction of two clusters in Euclidean space as measured by their overlap. Math. Geol., 9 (2), 123-143.

Sneath, P.H.A., 1979, The sampling distribution of the W Statistic of Disjunction for the arbitrary division of a random rectangular distribution. Math. Geol., 11 (4), 423-429.

Sneath, P.H.A., and Sokal, R.R., 1973, Numerical taxonomy. W.H. Freeman & Co., San Francisco, 573 pp.

Sokal, R.R., and Rohlf, F.J., 1962, The comparison of dendrograms by objective methods. Taxon, 11, 33-40.

Sokal, R.R., and Sneath, P.H.A., 1963, Principles of numerical taxonomy. W.H. Freeman & Co., San Francisco, 359 pp.

Sorgenfrei, T., 1958, Molluscan assemblages from the marine Middle Miocene of South Jutland and their environments. Geol. Survey Denmark (2nd ser.), 79, 2 vols., 503 pp.

Symons, F., and De Meuter, F., 1974, Foraminiferal associations of the mid-Tertiary Edegem sands at Terhagen. Belgium, J. Math. Geology, 6 (1), 1-15.

Tipper, J.D., 1979, An ALGOL program for dissimilarity analysis, a divisive-omnithetic clustering technique. Computers and Geosciences, 5 (1), 1-13.

Whittaker, R.H., (editor), 1978a, Ordination of plant communities. Junk Publ., The Hague, 388 pp.

Whittaker, R.H., (editor), 1978b, Classification of plant communities. Junk Publ., The Hague, 408 pp.

II.3

ARCHAEOLOGICAL SERIATION OF AN ORIGINAL DATA MATRIX

JAMES C. BROWER

INTRODUCTION

Archaeologists have faced problems closely allied to those of biostratigraphy for many years. Samples containing various objects must be arranged or seriated into a one-dimensional sequence which represents 'time' or 'evolution'. Examples include artifacts in graves, Iron Age brooches, word and sentence structure in manuscripts, etc. Numerical solutions to archaeological seriation date back to 1899 and excellent reviews are given by Cowgill (1972), Doran and Hodson (1975), Gelfand (1971), Hodson et al. (1971), Johnson (1972), Kendall (1971), Marquardt (1978) and Wilkinson (1974).

One can seriate two kinds of data. In the first and most common type, a matrix of similarity or distance coefficients is calculated between the objects or samples. Then the desired variety of seriation is performed on this matrix. In effect the similarities or distances serve as 'middle-men' in the process, which raises questions such as: What sort of coefficient is appropriate? What is the metric nature of the data, etc.? Most studies deal with seriation in this form (see Chapter II.2, on multivariate analysis). The second variety of seriation directly manipulates an original data matrix with n rows and m columns. Although this involves working with a rectangular block of data which can be computationally awkward, it has the advantage of eliminating the 'middle-men' in the form of similarity or distance coefficients. Seriation of an original data matrix is the technique of interest here. A complete discussion and a computer program are available in Brower and Burroughs (1982) and Burroughs and Brower (1982).

Strangely enough, few seriation techniques have been noticed by paleontologists which is surprising inasmuch as the similarity between artifacts in graves and fossils in rocks is obvious. Scott (1974) first suggested that archaeological seriation of both types could be applied to biostratigraphy. Smith and Fewtrell (1979) adopted a non-quantitative seriation approach to several original data matrices of microfossils. Doveton et al. (1976) and Tipper (1977, 1980) seriated some ecological and paleoecological data.

THE ALGORITHM

Introduction

The data matrix consists of the presence/absence of m taxa taken from n samples in p stratigraphic sections. Biostratigraphers have one major advantage over the typical archaeologist because the sequence of the taxa is known in the individual stratigraphic sections; rarely do archaeologists possess analogous information. The presences are scored as 1.0 and absences as 0.0. The range-through method of data recording is generally employed for reasons discussed in Step 1 of multivariate analysis (Chapter II.2). Many biostratigraphers omit rare species which only occur in a few of the stratigraphic sections under study because such animals only provide information about several samples. Although not strictly necessary, elimination of taxa only known from a few sections and samples as well as samples with only a few species does improve the quality of the seriation because ambiguous information is removed.

The problem is to arrange the data into a range chart with the taxa in the columns and the samples in the rows. This is done by concentrating the presences along the diagonal of the matrix so the range zones of the taxa are minimized. For ideal data the seriated matrix will have a solid block of presences along the diagonal with the absences in the off-diagonal elements. The goodness of fit between the observed matrix and a perfectly seriated matrix can be measured.

Two types of seriations can be determined. In an unconstrainted solution, the data on superposition of the samples within any given stratigraphic section are ignored and the samples are free to group in any order. The constrained solution uses information on the stratigraphic position of the samples within the individual stratigraphic sections so that the samples are forced to remain in stratigraphic order within the final seriated matrix.

The Unconstrained Solution

Wilkinson (1974) proposed the simple method adopted for unconstrained seriation. This heuristic scheme alternately orders the rows and columns of the matrix based on the average position of the presences. In the computer program of Brower and Burroughs (1982), these positions are numbered from top to bottom for each column and from left to right for each row. The computations are performed in a series of iterations, each of which includes the following four steps:

(1) Calculate the mean position of the presences in the rows.

(2) Order the rows according to these means.

(3) Calculate the mean position of the presences in the columns.

(4) Order the columns according to these means.

Iterations continue until there is no change in the positions of the rows and columns.

This type of seriation represents a simple averaging process so the resulting zonation of the species and samples will be an average rather than a conservative zonation (see Brower, 1981 and Chapters I.2 and II.4).

The Constrained Solution

Data on the stratigraphic position for the samples within the individual sections are introduced into the analysis for a constrained solution. This forces the samples within each section to remain in stratigraphic order unlike unconstrained seriation where two samples in a single section can be seriated out of stratigraphic order.

The input data matrix has samples in the rows and the species or taxa in the columns. The data are read in order of the stratigraphic sections which are numbered from 1 to p; within each section the samples are numbered from 1 at the top to whatever the last consecutive sample number is at the bottom of that section. Each sample is thus identified by two numbers, one for the stratigraphic section and a second for the sample number within that stratigraphic section. Each sample is also given a number from 1 to n within the total sequence of all samples.

The program operates in a series of iterations, each consisting of five steps.

(1) Calculate the mean position of the presences in the rows (samples).

(2) Arrange the rows into order according to their means.

(3) The rows of this matrix are scanned to determine if the samples are seriated in stratigraphic order within the individual sections. If all samples are properly sequenced within their respective stratigraphic sections, the algorithm moves to the next step. If some samples are out of stratigraphic order within an individual section, these samples are interchanged within the seriated matrix to array them in proper stratigraphic order for that section. This is done for all samples within each stratigraphic section.

(4) The mean position of the presences is determined for each column (taxon or species).

(5) The columns are ordered according to these means.

Additional iterations are carried out until the data are stabilized.

The Test Criterion

A simple index reflects the amount of concentration of the presences along the diagonal of the seriated matrix. As discussed before, the basic idea is to minimize the number of embedded absences. (An embedded absence represents any absence between the lowest and highest presences in any one column.) The index consists of

$$1 - \left(\sum_{j=1}^{m} A_j \Big/ \sum_{j=1}^{m} R_i \right)$$

in which A_j is the number of embedded absences in column j and R_j is the total range of presences in column j, that is, the number of entries between the highest and lowest presences in each column, inclusive. A perfect seriation in which all of the presences lie along the diagonal of the matrix results in a test value of 1.0. Lower values are associated with more embedded absences and less perfect seriations.

The algorithm iterates until the seriated matrix reaches a stable state with the presences more or less clustered about the diagonal. Both the size and complexity of the original data interact to dictate the number of iterations necessary. Both types of seriation converge rapidly. The number of elements in the matrices examined by Brower and Burroughs (1982) ranges from about 40 to over 4000 and four to 42 iterations were required. The unconstrained seriation always ends with an optimum amount of concentration where the means of all rows and columns are arranged in ascending order. The stopping point and the final iteration are determined by the highest value attained by the test criterion. Conversely, the constrained solution does not invariably reach an absolutely stable configuration. Commonly, the final iterations in this latter situation will fluctuate about a particular degree of concentration of presences along the diagonal. Here, the problem usually stems from taxa that are poorly known stratigraphically (see Brower and Burroughs, 1982, for further discussion). These oscillating iterations always have very similar test values and nearly identical arrangements of the taxa and samples. The rather tortuous logic of the computer program of Burroughs and Brower (1982) traps the iteration with the largest test criterion and the maximum concentration of presences along the diagonal of the seriated matrix.

Comparison of Results from Unconstrained and Constrained Seriations

Study of the final test criteria and inspection of the orders for the row and column means displays the 'cost' of the stratigraphic constraint. On one hand the unconstrained solution relentlessly groups samples with similar faunal composition and taxa with similar patterns of occurrence until the best possible ordering of the position means for the rows and columns is found. Commonly this results in a better degree of concentration along the diagonal than seen in constrained seriation.

On the other hand, unconstrained seriation ignores the stratigraphic 'ground truth' inherent in biostratigraphic data, namely the relative position of taxa and samples in a stratigraphic section. The vagaries of the distribution of a taxon (for example due to facies control, nonpreservation, misidentification, etc.) can impose an apparent ordering of events in an unconstrained solution which may violate the actual order of occurrence of the samples or taxa in a given section while maximizing the degree of concentration of presences along the diagonal of the matrix. The constrained solution preserves the stratigraphic 'ground truth' but in doing so commonly decreases the amount of concentration somewhat. The unconstrained solution is also handicapped by a lack of linkage between groups of samples and taxa which may be found in faunal zones that do not overlap. Inasmuch as no stratigraphic information is included in an unconstrained seriation, the groups of samples and taxa are usually not in stratigraphic sequence. The overall results are similar to those obtained in cluster analysis and other multivariate methods. The clusters identify coherent groups of species and/or samples but these are usually not in stratigraphic order. To ascertain the stratigraphic distributions of the clusters, one must inspect the original data (see Chapter II.2, on multivariate analysis). Unconstrained seriation is more sensitive to the initial arrangement of the input data matrix than is constrained seriation because adjacent rows or columns with equal or tied means are left in their original order. The stratigraphically constrained seriation consistently produces range charts that are better for correlation.

CASE STUDIES

Several examples will be annotated to show how seriation works and to compare the results derived from the various techniques. Further details can be obtained from Brower and Burroughs (1982).

Hypothetical Data

This example deals with five species distributed among nine samples in three stratigraphic sections. The species are in the columns, numbered from one to five. The rows contain the samples and are numbered as follows. The first digit gives the section number whereas the subsequent digit or digits designate the location number for that sample within each stratigraphic section; these numbers increase from stratigraphiclly highest to lowest in each section. In the data matrix, the presences are black and the absences are white. This system is followed for all matrices in this paper. The original data have 18 embedded absences or blanks (Figure 1A).

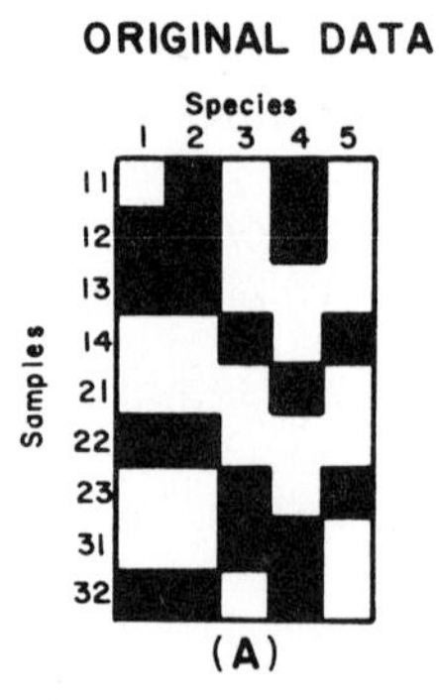

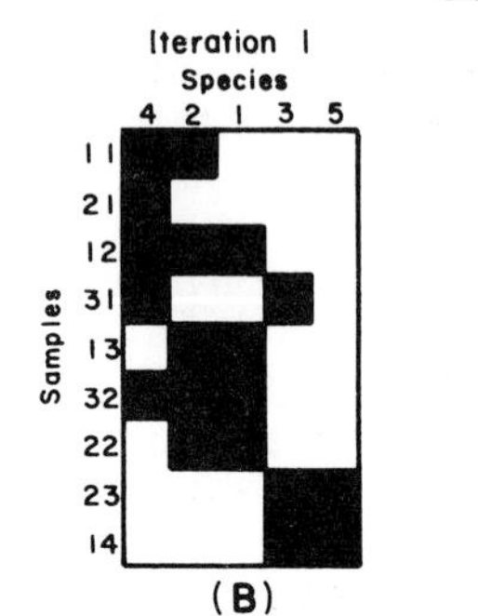

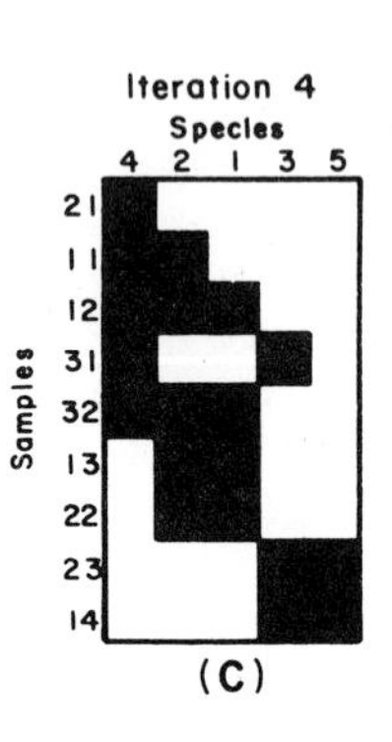

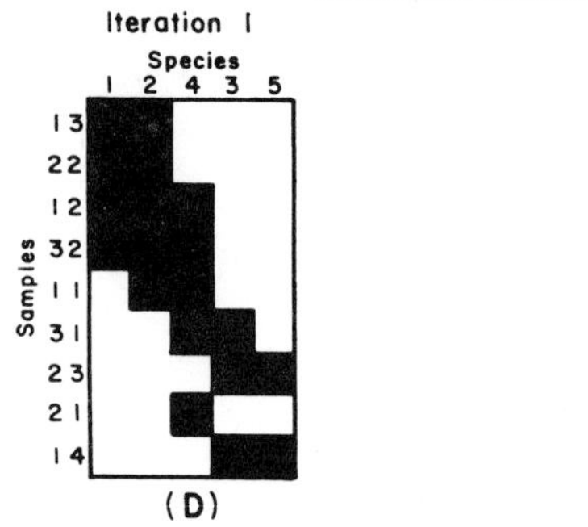

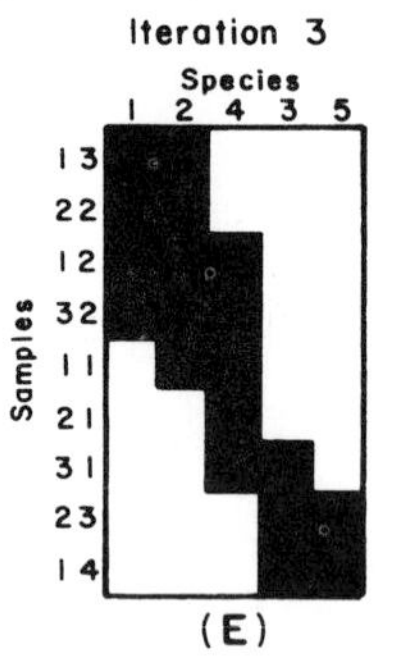

Figure 1 Results for hypothetical data set. Species are numbered 1 to 5, sample-numbers are explained in the text. Presences are black, absences are white.

A. Original data.
B. First iteration for constrained seriation.
C. Final iteration for constrained seriation.
D. First iteration for unconstrained seriation.
E. Final iteration for unconstrained seriation.

Both seriations converge rapidly. Only three iterations were required for unconstrained seriation and four iterations for the constrained solution. There are seven embedded absences after the first iteration of the constrained data which yields a test value of 0.73 and roughly achieves stratigraphic order (Figure 1B). The final iteration reduces the number of embedded blanks to five and raises the test criterion to 0.79 (Figure 1C).

Three embedded absences appear after the first iteration of the unconstrained seriation versus the 18 embedded absences of the original data (Figure 1D). The second and third iterations are identical perfect seriations with no embedded absences and test values of 1.0 (Figure 1E). Comparison of Figures 1C and 1E illustrates the differences between the two solutions. Although the adjacent rows and columns of the unconstrained seriation are characterized by similar patterns of presences and absences, neither the species nor samples are in any stratigraphic order. Despite the fact that the unconstrained solution violates the known information about the stratigraphic distribution of the samples and taxa, it still gives a test criterion of 1.0 versus the lower figure of 0.79 for the constrained solution. Here, the stratigraphic constraint 'costs' five absences.

Triassic Data

This example treats 11 species of Lower Triassic ammonites from the Salt Ranges of Pakistan (data from Guex, 1978 and Kummel, 1966). The 31 samples were taken from four stratigraphic sections distributed over about 180 km. Only species known from more than one section are included. Figure 2A illustrates a time-correlation chart abstracted from Guex (1978) and Kummel (1966). The main purposes of this rather simple case study are to compare the results of the two seriations with each other and with the time-correlation chart. More details and a list of the species are available in Brower and Burroughs (1982).

The test value for the stratigraphically constrained seriation equals 0.64 which is an intermediate value. The samples are seriated from youngest at the top of the matrix to oldest at the base (Figure 2B). The upper eight samples (11-42) all belong to Zone VII on the correlation chart. These samples typically contain some or all of the species 1-5. The only form present in three of the next four samples (43, 24, 14) is species 5 which is a long-ranging ammonite. These samples lie in Zone VI or at the base of Zone VII. The samples at the top of Zone V include 44-47 and 31; the last of these, containing only species 10, is slightly out of position in the seriation. The typical fossils of these samples are species 4, 10, 11, all of which are short-ranged, and the ubiquitous species 5. The next group of samples is found in the middle of Zone V and is characterized by the common genera Meekoceras and Anasibrites (species 6 and 7); these samples include 48-410, 32, 33, 15 and 16. Species 5 is the only taxon listed in the oldest three samples of Zone V (411-413). These are misplaced by the seriation which blocks these samples within a group of collections with

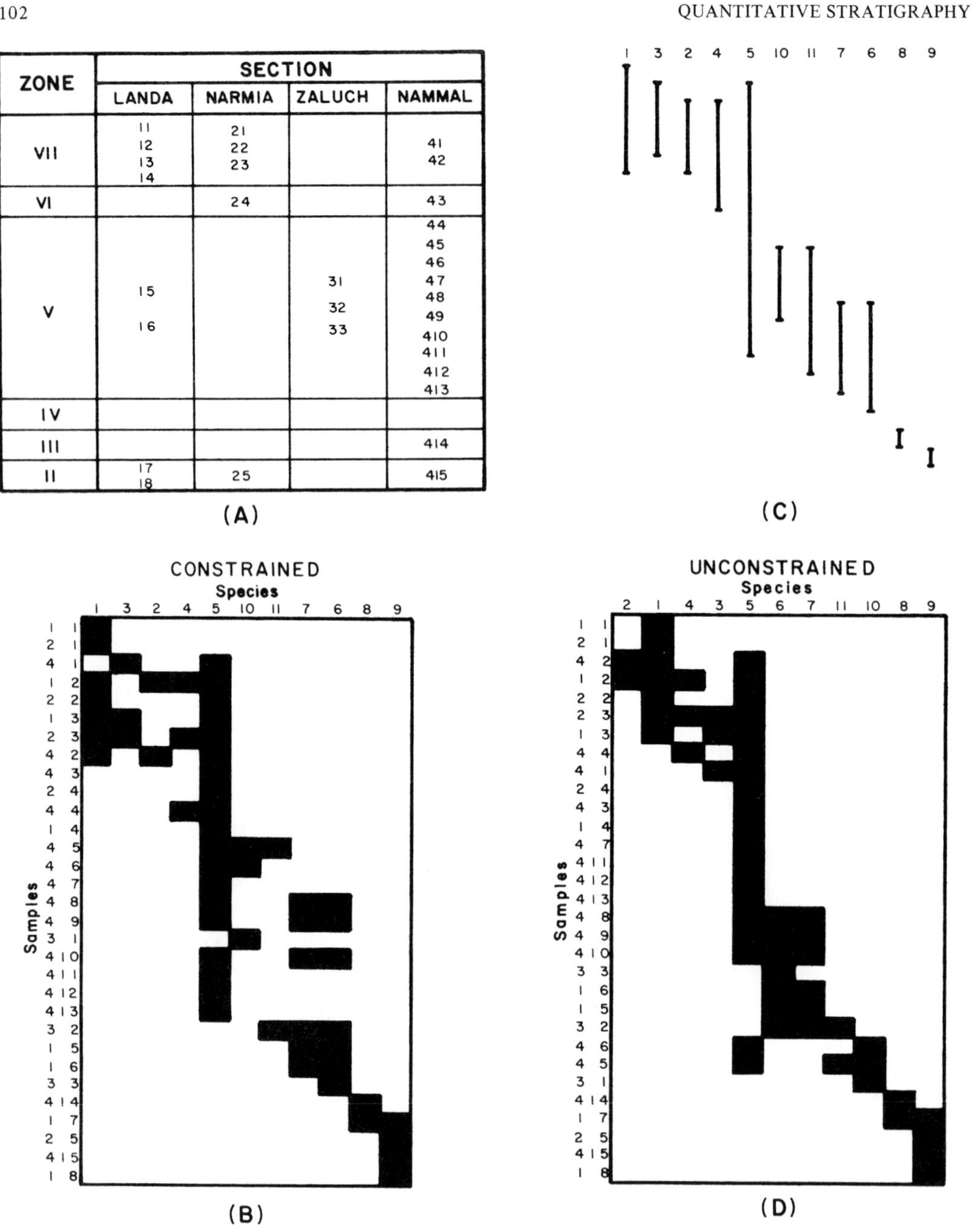

ZONE	SECTION			
	LANDA	NARMIA	ZALUCH	NAMMAL
VII	11 12 13 14	21 22 23		41 42
VI		24		43
V	15 16		31 32 33	44 45 46 47 48 49 410 411 412 413
IV				
III				414
II	17 18	25		415

Figure 2 Data for Triassic ammonities.

A. Correlation chart compiled from Kummel (1966) and Guex (1978).

B. Constrained seriation.

C. Range chart of taxa (see Brower and Burroughs, 1982 for list of species).

D. Unconstrained seriation.

species 6 and 7. Because of the averaging nature of the technique, this is common with samples having only a few species which are either long-ranged and/or show erratic distributions. The oldest samples are in Zones II and III where species 8 and/or 9 are always present. These samples (414, 415, 25, 17 and 18) are assigned correctly by the seriation. Clearly, a high degree of agreement exists between the correlation chart and the stratigraphically constrained seriation.

A range chart for the species can be read from the seriation (Figure 2C). Construction of the chart assumes that adjacent samples with equal means and the same faunal composition are of equal age; consequently the range chart differs slightly from a direct overlay of the constrained seriation. Species 1, 2 and 3 are the youngest whereas species 8 and 9 represent the oldest. The long range of species 5 is quite apparent. The only possible inconsistency between the range chart and the actual data involves species 10. According to the seriation, the base of the range zone of species 10 lies below the upper limits of species 6 and 7. Although not impossible, this relationship is not present in the original data and does not seem probable. As expected, the confusion is produced by a rare taxon, species 10, which only occurs in three samples.

The unconstrained seriation produces a test value of 0.87 compared to 0.64 for the constrained solution. Eliminating the stratigraphic restriction reduces the number of embedded absences from 32 to 9. The assemblages of taxa reconstructed from the seriation are only partially in stratigraphic order. The youngest ammonites, species 1-4, are in the left columns. The wide distribution of species 5 is evident. Proceeding from left to right the next two groups comprise species 6 and 7 on one hand versus species 10 and 11 on the other which are reversed relative to their stratigraphic distribution. Although the unconstrained seriation does place taxa with similar distribution patterns together, these groupings are not arranged in exact stratigraphic order. Study of the samples in the seriated matrix for the unconstrained seriation reveals the same situation where many samples are not in stratigraphic order. For example, most of the samples with species 10 and 11 are seriated below those with species 6 and 7. Sample 44 is over 41 in the rearranged matrix but their stratigraphic order is the opposite. All samples containing only species 5 are blocked together although these are found at widely separated horizons. Obviously the unconstrained seriation reduces the number of embedded absences, but only at the expense of known information about the stratigraphic relationships of the data. Brower and Burroughs (1982) demonstrate that the unconstrained seriation is much more similar to cluster analyses done on the same data than is the constrained solution.

Gulf Coast Data

This example, which deals with the data on the Eocene-Oligocene boundary strata published by Deboo (1965), was discussed earlier in

Chapter II.2 on multivariate analysis. The seriated data include 71 species of ostracodes and planktic foraminifers and 62 samples. The unconstrained and constrained seriations yield nearly identical test criteria of 0.65 and 0.64. The results are illustrated differently than for previous cases. Here, the row positions of the samples in the final seriations are contoured on the correlation chart (Figure 3A, B). The stratigraphically constrained solution arranges the samples largely in stratigraphic order but the unconstrained seriation almost completely inverts the stratigraphy. Unlike many data sets, the two solutions are highly correlated where the correlation coefficient between the two sets or row positions equals -0.9995. The contours of equal seriation position nearly parallel the time lines in the middle of the chart. Along the upper and lower margins the contours roughly follow time lines although some pronounced bends appear in the lines. When reading these comments, bear in mind that either the seriations and/or the correlation chart could be erroneous. In fact the results of the seriations are similar to the patterns obtained by plotting the clusters for unweighted data on the correlation chart.

This was researched further with linear regression analysis for the constrained seriation using three variables for each of the 62 samples: the vertical location on the correlation chart measured with a ruler, V; mileage location on each stratigraphic section along a linear transect, D; and the row position in the constrained seriation, C. Observe that the data do not fit the assumptions of the analysis because both V and C are subject to error. Consequently the results can only be interpreted in a general fashion. The correlation matrix is listed below (see graphs in Figure 4).

C	D	V	
1.0	0.157	0.944	C
	1.0	-0.009	D
		1.0	V

C and V are highly correlated whereas the correlations with D are much lower. The first regression equals

$$V = 5.06 + 1.53\ C$$

with a multiple correlation coefficient (R^2) of 0.892 which is significant at any reasonable risk level. Thus C explains slightly over 89 percent of the variance of V. Incorporating D into the analysis produces

$$V = 13.1 + 1.57\ C - 0.183\ D$$

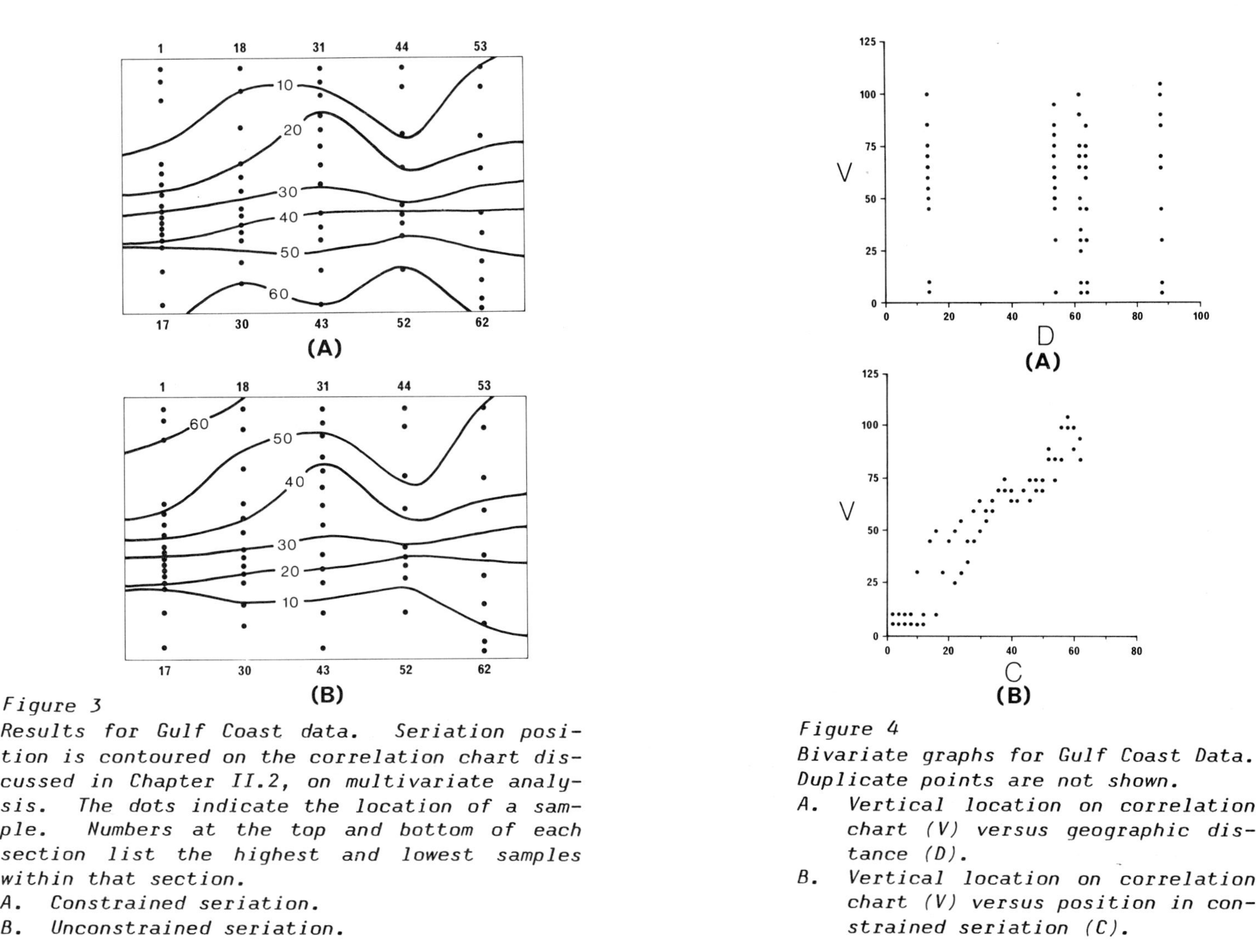

Figure 3
Results for Gulf Coast data. Seriation position is contoured on the correlation chart discussed in Chapter II.2, on multivariate analysis. The dots indicate the location of a sample. Numbers at the top and bottom of each section list the highest and lowest samples within that section.
A. Constrained seriation.
B. Unconstrained seriation.

Figure 4
Bivariate graphs for Gulf Coast Data. Duplicate points are not shown.
A. Vertical location on correlation chart (V) versus geographic distance (D).
B. Vertical location on correlation chart (V) versus position in constrained seriation (C).

where R is 0.918. The coefficient for D is significantly different from nil at a risk level of 0.05. This evidences a small systematic change in the data as one goes toward the east. Inspection of Figure 3A discloses that the average seriation position rises from left to right. Similar results are obtained from an analysis of the unconstrained seriation.

The only similar application known to me is the study of Cisne et al. (1982). These authors calculated reciprocal averaging scores for approximately 200 samples based on a variety of Ordovician taxa in New York. Their data are basically ecological because they represent relative abundances of taxa and few, if any, origination and extinction points are involved. The scores were regressed on geographic location and position in a relative time framework. The resulting R equals 0.51 which is much lower than the equivalent figure of 0.918 for the Eocene-Oligocene example. Study of their Figure 7 reveals a strong geographic trend in the data. Clearly the time-stratigraphic signal within the Gulf Coast example is much larger than in the Ordovician data.

Other Examples

Seriation has been applied to a variety of data sets ranging from Cambrian macrofossils to Tertiary microfossils. Discussion of the results and comparisons with other techniques are available in Brower and Burroughs (1982).

SUMMARY

Archaeological seriation of an original data matrix provides a simple and powerful tool for numerical biostratigraphy. The basic idea is to array the data in a range chart with the taxa in the columns and the samples in the rows. This is accomplished by grouping the presences along the diagonal of the matrix so that the number of embedded abences on and near the diagonal is minimized. Seriation simultaneously generates the range chart for the taxa and groups the samples with the most similar faunal composition in adjacent rows of the seriated matrix. A test criterion measures the degree of concentration of the presences along the diagonal. Two types of solutions can be determined. In an unconstrained solution, the data on superposition of the samples within the stratigraphic sections are ignored and the samples can group in any order. This generally produces suites of species and samples that are similar to those obtained from multivariate analysis of assemblage zones. The constrained solution uses information on the stratigraphic position of the samples within the individual stratigraphic sections so that the samples are forced to remain in stratigraphic order within the final seriated matrix. This usually yields a better range chart for correlation. Study of numerous examples demonstrates that the results from seriation compare closely with those of correlation charts and output derived from various other quantitative techniques commonly employed in biostratigraphy (Brower and Burroughs, 1982).

REFERENCES

Brower, J.C., 1981, Quantitative biostratigraphy, 1830-1980. In Computer applications in the earth sciences, an update of the 70's. D.F. Merriam, ed., Plenum Press, New York and London, 63-103.

Brower, J.C., and Burroughs, W.A., 1982, A simple method for quantitative biostratigraphy. In Quantitative stratigraphic correlation, J.M. Cubitt and R.A. Reyment, eds., John Wiley and Sons Ltd., 61-83.

Burroughs, W.A., and Brower, J.C., 1982, SER, a FORTRAN program for the seriation of biostratigraphic data. Computers and Geosciences, 8, (2), 137-148.

Cisne, J.L., Karig, D.E., Rabe, B.D., and Hay, B.J., 1982, Topography and tectonics of the Taconic outer trench as revealed through gradient analysis of fossil assemblages. Lethaia, 15, (3), 229-246.

Cowgill, G.L., 1972, Models, methods and techniques for seriation. In Models in archaeology, D.L. Clarke, ed., Methuen & Co. Ltd., London, 381-424.

Deboo, P.B., 1965, Biostratrigraphic correlation of the type Shubuta Member of the Yazoo Clay and Red Bluff Clay with their equivalents in south-western Alabama. Alabama Geol. Surv. Bull. 80, 84 pp.

Doran, J.E., and Hodson, F.R., 1975, Mathematics and computers in archaeology. Harvard Univ. Press, Cambridge, xi, 381 p.

Doveton, J.H., Gill, D., and Tipper, J.C., 1976, Conodont distributions in the Upper Pennsylvanian of eastern Kansas, binary pattern analysis and their paleoecological implications. Geol. Soc. America, Abstracts with Programs, 8, (7), 842.

Gelfand, A.E., 1971, Rapid seriation methods with archaeological applications. In Mathematics in the archaeological and historical sciences, F.R. Hodson, D.G. Kendall, and P. Tautu, eds., Edinburgh Univ. Press, Edinburgh, 186-201.

Guex, J., 1978, Le Trias inferieur des Salt Ranges (Pakistan), problèmes biochronologiques. Eclogae geol. Helv., 71, (1), 105-141.

Hodson, F.R., Kendall, D.G., and Tautu, P., (editors), 1971, Mathematics in the archaeological and historical sciences. Edinburgh Univ. Press, Edinburgh, vii, 565 pp.

Johnson, L., 1972, Introduction to imaginary models for archaeological scaling and clustering. In Models in archaeology, D.L. Clarke, ed., Methuen & Co., Ltd., London, 309-379.

Kendall, D.G., 1971, Seriation from abundance matrices. In Mathematics in the archaeological and historical sciences, F.R. Hodson, D.G. Kendall and P. Tautu, eds., Edinburgh Univ. Press, Edinburgh, 215-252.

Kummel, B., 1966, The Lower Triassic Formations of the Salt Range and Trans-Indus Ranges. West Pakistan, Mus. Comp. Zoology, Bull. 134, (10), 361-429.

Marquardt, W.H., 1978, Advances in archaeological seriation. In Advances in archaeological method and theory, I, M.B. Schiffer, ed., Academic Press, 257-314.

Scott, G.H., 1974, Essay review: stratigraphy and seriation. Newsl. Stratigr. 3, 93-100.

Smith, D.G., and Fewtrell, M.D., 1979, A use of network diagrams in depicting stratigraphic time-correlation. J. Geol. Soc. London, 136, 21-28.

Tipper, J.C., 1977, Some distributional models for fossil marine animals. Geol. Soc. America, Abstracts with Programs, 9, 1202.

Tipper, J.C., 1980, Some distributional models for fossil animals. Paleobiology, 6, (1), 77-95.

Wilkinson, E.M., 1974, Techniques of data analysis-seriation theory. Archaeo-Physika, 5, 1-142.

II.4

RANKING AND SCALING IN EXPLORATION MICROPALEONTOLOGY

F.M. GRADSTEIN

"... Statistical techniques are essential to exploit the vast amounts of micropaleontological data that are being accumulated during routine petroleum exploration drilling..."
In, "Principles of Sedimentary Basin Analysis", A.D. Miall, 1984

INTRODUCTION

The microfossil record of appearances and disappearances, as preserved in samples from stratigraphically long and thick well sequences is often complex and 'noisy'. Frequent inconsistencies in relative order must be accounted for, the result of sampling or geological factors like dissolution and reworking. An account of factors that affect the use of fossils for the purpose of geological basins analysis is presented in Chapter I.2.

To cope in an objective manner with a large amount of biostratigraphic data, and with stratigraphically 'noisy' information, we set out in 1977 to develop and apply quantitative biostratigraphic techniques (Gradstein and Agterberg, 1982). The models should use as much information as possible, organize the information in a novel way, and provide good zonations, detailed correlations and reliable geological interpretations. Varying thickness of the sections studied called for a relative scale, in which depth per se would not initially be considered. Only relative order and spacing in time of fossil events would be modelled.

In this Chapter I present a non-technical discussion of the Ranking and Scaling (RASC) method and two case histories. The viewpoint is as much as possible geological rather than statistical. The latter is treated by F.P. Agterberg in Chapters II.5 and II.6.

The case histories are on Cenozoic foraminiferal and Upper Cretaceous nannofossil distributions of the western North Atlantic margin. A third application of RASC, using Mesozoic Tethyan radiolarians, is treated in the context of Unitary Associations, presented in Chapters II.7 and II.8.

SEQUENCING METHODS

Quantitative stratigraphic methods suited to deal with stratigraphic events, as opposed to stratigraphic ranges, largely fall in the categories of certain multivariate methods (J. Brower, Chapter II.2), seriation (J. Brower, Chapter II.3) and of ranking and scaling. As outlined in more detail in Chapter I.2, a paleontological event is the presence of a taxon in its time context, derived from its position in a rock sequence. In this Chapter the types of events used are first and last occurrences. Ranking is a pairwise comparison technique which tries to determine the most likely sequence of biostratigraphic events as recorded in different stratigraphic sections. Scaling determines spacing of these events in relative time. The relative position of the events in the optimum sequence is more or less an average of all the relative positions encountered.

In the context of this study, ranking is used with biostratigraphic events, but the basic method is valid for average ranking of any type of elements along a relative scale. For example, ranking can list the average order in which twelve yachts of a local sailing fraternity passed the finish line in 20 regattas in which they participated last summer. The data should have the different orders as observed in trials. In geological applications, stratigraphic sections are the trials, fossils are the yachts, and biostratigraphic events are the positions of the yachts in all the trials.

There are several statistical variants to the basic method of pairwise comparison which lead to the construction of optimum sequences. Discussions on the philosophy, methods and geological applications may be found in several recent publications as shown in Table 1 of Chapter I.2.

The resulting most likely sequences are generally of the average type, as opposed to conservative types. In general, the most likely first (or last) occurrence of a taxon is slightly younger (older) than the very first (last) observed occurrence of this taxon. Conservative sequences derive from sequencing methods which rank events based on the highest (or lowest) positions encountered. Examples are Shaw's (1964) composite standard and the method developed by Edwards & Beaver (1978). In this sense these conservative methods are not probabilistic and results are particularly vulnerable to outliers in the data. The inclusion of observations on badly reworked or otherwise wayward specimens will influence the final results much more than in averaging methods.

Ideally, there might be little difference in relative event position when comparing average and conservative approaches. An example is the situation where the last common or consistent occurrence and the very last occurrence of events as sampled in many sections, are stratigraphically close. For this reason Doeven et al. (1982) found an excellent match between 'probabilistic' and subjective range charts using Late Cretaceous nannofossils.

RANKING AND SCALING (RASC)

The RASC - Ranking and Scaling method deals with most likely sequences and most likely distances (spacing) of events in relative time. The method is applied under the assumption that the probable order of events does not differ significantly from section to section. RASC tests of stratigraphic normality evaluate the validity of this assumption. Deletion of badly-placed events improves the zonation. When a coherent group of sections all show a deviating sequence, two zonations (= optimum sequences) may be required, one for this coherent subgroup and for the other sections.

From an input consisting of the sequence of fossil events within each of several sections, RASC will produce the following answers:

(1) Optimum (or average) sequence of paleontologic events along a relative time scale
= RANKING;

(2) Scaled optimum sequence, which calculates the average relative distance in time for the events in the optimum sequence
= SCALING;

(3) Tests of the order of the events in each individual section for deviations in rank when compared to the scaled optimum sequence
= NORMALITY TESTING; and

(4) The position of unique events (like rare index fossils) or marker horizons (like seismic or mechanical log events) in the optimum sequence
= UNIQUE EVENT OPTION.

Optimum sequences for both first and last occurrences produce average range charts. Scaled optimum sequences resemble assemblage zonations. Tests on stratigraphic normality are of three types: Bivariate scatter plots of each individual sequence versus the optimum sequence, the step-model and normality testing.

Data Preparation

Preparation of the data for ranking and scaling in the RASC program is a simple procedure (Heller et al., 1983), which can be accomplished in four or five steps:

(1) List the raw data. The list should contain the name of each section and for each section a stratigraphic sequence of samples. Each sample is listed with the events observed;

(2) Produce an index of all participating fossil taxa. Each taxon has to be assigned a unique index number. Numbering is consecutive starting at 1;

(3) Code the observed event sequences in the individual sections, using the numbers as assigned in the index. Both the index and coded sequences have to be keypunched on cards or stored on magnetic tape or disc;

(4) Identify all unique events or marker horizons; and

(5) Establish the threshold parameters k_c and m_c (see below).

Raw Data

The preparation of data starts with defining which events are to be used. Commonly this will be first occurrence, first consistent occurrence, last consistent occurrence, or last occurrence of the species identified (Figure 1 in Chapter I.2), where first and last correspond to oldest and youngest, respectively.

Next comes a listing of the biostratigraphic events, according to observed depth, in each stratigraphic section studied. This tabulation generally will take the format of a succession of samples per section. For each sample is shown which fossils appear and/or disappear.

Table 1 shows the downhole succession of foraminifers as observed in the Egret K-36 well, Grand Banks. This listing is part of the last occurrence (tops) of Cenozoic foraminifers in 24 wells discussed under Cenozoic Foraminiferal Zonation (see below).

Index

The purpose of the index (also referred to as dictionary) is to number (code) each event in a manner compatible with the RASC program. Each species that contributes to the list of event occurrences is assigned a unique index number. Numbering is consecutive, starting at 1. The index to be digitized must list its assigned numbers consecutively and continuously. This means that subsequent deletions of index numbers or accidental numerical gaps, have to be replaced by dummies (listed as "dummy" in the input). Table 1 shows the sequence of foraminiferal events in the Egret K-36 well, coded according to the regional Cenozoic foraminiferal dictionary.

If appearance and disappearance of the same fossil are being coded, it helps with the interpretation to number appearances for example from 1 through 100 (if less than 100 taxa) and disappearances from 101 through 200. The appearance of fossil A may then have code number 49 and its disappearance 149. If there are more observed appearances than disappearances (or vice versa), dummy numbers fill the gaps. For error finding and further extensions to the data, the RASC program prints both the numerical and a derived alphabetical index.

Table 1 Downhole succession of Cenozoic foraminiferal extinctions as observed in Egret K-36 well, Grand Banks, with index code numbers assigned.

DEPTH IN FEET	SPECIES NAME	DICTIONARY NUMBER
860	Asterigerina gurichi	17
1040	Uvigerina dumblei	26
1340	Ceratobulimina contraria	16
1520	Gyroidina girardana	20
-	Guttulina problema	21
-	Spiroplectammina carinata	18
-	Epistomina elegans	71
-	Globigerina praebulloides	15
1580	Turrilina alsatica	24
1610	Eponides umbonatus	27
-	Cibicidoides alleni	42
1950	Nodosaria sp. 8	69
2240	Globigerina linaperta	82

Table 2 Coded sequence data for the Egret K-36 well of Table 2. This data is digitized for RASC treatment. Number 202 is seismic event 2, treated as a marker horizon.

EGRET K-36

17 26 16 20 -21 -18 -71 -15

24 27 -42 202 69 82

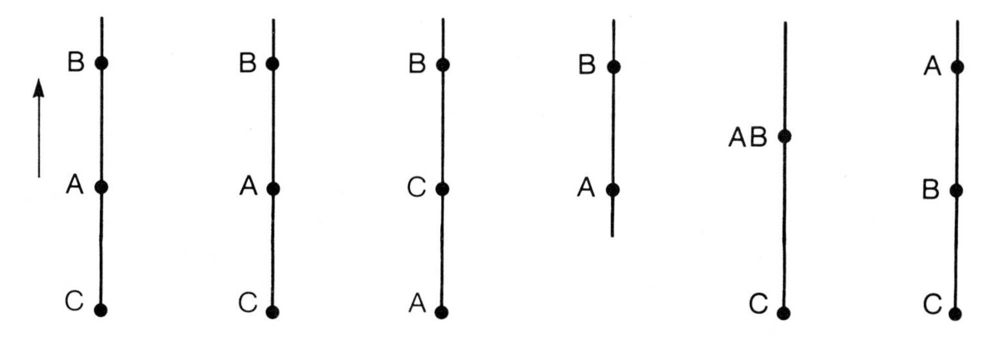

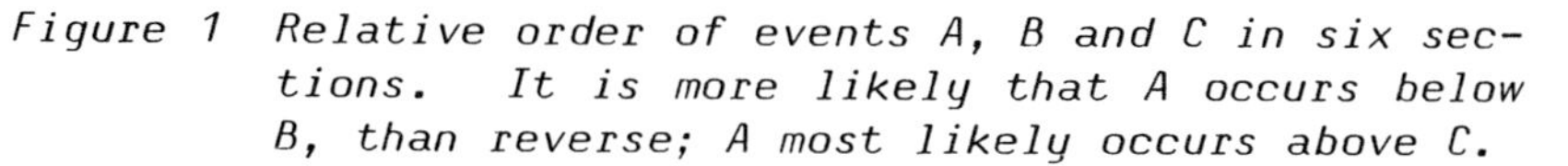

Figure 1 Relative order of events A, B and C in six sections. It is more likely that A occurs below B, than reverse; A most likely occurs above C.

Coded Sequences

Table 1 shows the sequence of foraminiferal events in the Egret K-36 well, coded according to the regional Cenozoic foraminiferal index. Events which appear in the same geological sample are considered to occur simultaneously. Simultaneous events are identified by a hyphen sign between them. For example in the Egret K-36 well of Table 1 the following order of events has been observed (Table 2).

The events are ordered from left to right, representing a stratigraphically downward sequence. Events 20, 21, 18, 71 and 15 occur simultaneously, and so do 27 and 42.

Optimum Sequence

In ranking, the frequency that each event occurs above, simultaneous with, or below all other events is recorded. The final arrangement of the events in the optimum sequence is based on rank. The rankings are not necessarily significant in a statistical sense, but the derived sequence is the most likely one.

In the original method as proposed by Hay (1972) simultaneous events were deleted from the original sequence data, but this produces a loss of information. In RASC, simultaneous events receive a score of 0.5 (see Chapter II.5). If certain events only occur simultaneously, they compete for each others positions in the optimum sequence. This situation is covered under Range (see below). In order to briefly demonstrate the principle of the method, let us take a look at a theoretical example.

Three stratigraphic events - in this case the last occurrences of species A, B and C - are found in 6 stratigraphic sections. Their relative order is shown in Figure 1.

RASC lists the participating taxa A, B and C in the margins of a so-called cumulative order matrix, in the rather arbitrary sequence in which they have actually been identified. It is the objective to reorder these margin - taxa until their sequence "averages" all sequences found. The elements of the matrix are scores, indicating in how many stratigraphic sections one event (B) occurs above (or below) another event (A). In our example of Figure 1, A occurs above B in 1 out of 6 sections, and B occurs above A in 4 out of 6 sections. Also, A and B occur simultaneously in one section, thus AB and BA each score an additional 0.5 points. In the order matrix of Figure 2, A over B gives a score of (1x1) + (1x0.5) = 1.5 and B over A gives a score of (4x1) + (1x0.5) = 4.5. Similarly, A occurs over C 4 times, and under C once, and B occurs over C 5 times, while C does not occur above B. Thus, these scores are A over C = 4, B over C = 5 and C over B = 0.

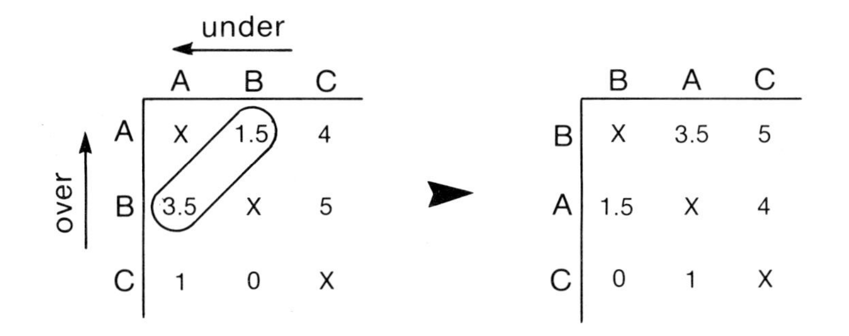

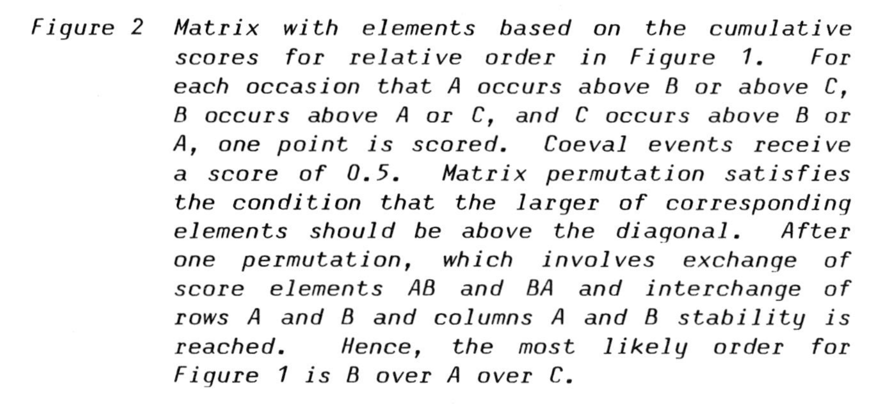

Figure 2 *Matrix with elements based on the cumulative scores for relative order in Figure 1. For each occasion that A occurs above B or above C, B occurs above A or C, and C occurs above B or A, one point is scored. Coeval events receive a score of 0.5. Matrix permutation satisfies the condition that the larger of corresponding elements should be above the diagonal. After one permutation, which involves exchange of score elements AB and BA and interchange of rows A and B and columns A and B stability is reached. Hence, the most likely order for Figure 1 is B over A over C.*

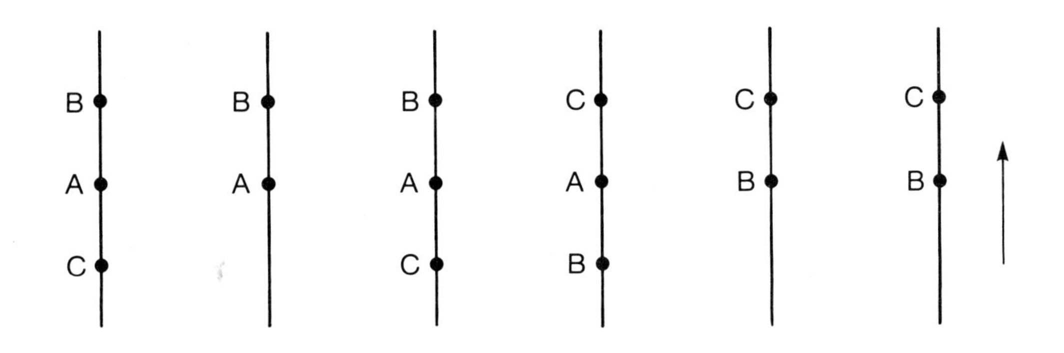

Figure 3 *Relative order of events A, B and C in six sections, such that the order is ABCA, which is not possible. Such a 3-event-cycle is relatively common when dealing with patchy data.*

SEQUENCE POSITION	FOSSIL NUMBER	RANGE	FOSSIL NAME
1	10	0-2	Uvigerina canariensis
2	17	1-3	Asterigerina gurichi
3	16	2-4	Ceratobulimina contraria
4	67	3-5	Scaphopod sp. 1
5	21	4-6	Guttulina problema
6	18	5-7	Spiroplectammina carinata
7	71	6-8	Epistomina elegans
8	20	7-9	Gyroidina girardana
9	26	8-12	Uvigerina dumblei
10	15	8-11	Globigerina praebulloides

Figure 4 *Example of optimum sequence, as taken from the application on Cenozoic foraminifers, discussed below;* $k_c \geq 7$, $m_{c1} \geq 2$.

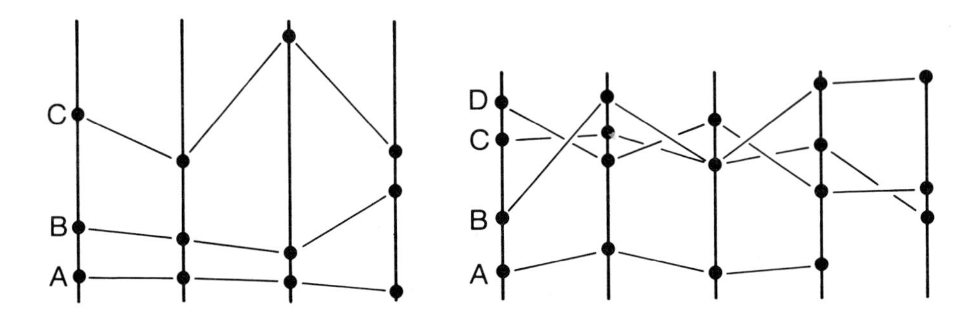

Figure 5 *(left) The fossil events A, B and C occur in the same order in all four stratigraphic sections examined.*
(right) Fossil events B, C and D have a high cross-over frequency, which means that these events have stratigraphically indistinct positions and will cluster well, using scaling.

Table 3 *Frequency tabulation of event occurrences as printed in the output of the RASC program. The second row shows that 59 taxa only occur in one well, 30 other taxa only occur in two wells, etc. The third row is the cumulative total. It shows that all the taxa (168) occur in at least one well, 109 occur in at least two wells, etc. Each (first or last) occurrence of a taxon is an event.*

NUMBER OF WELLS	1	2	3	4	5
NUMBER OF EVENTS	59	30	9	13	10
CUMULATIVE NUMBER OF EVENTS	168	109	79	70	57

When a matrix element (score value) in the column above the diagonal - the upper triangle - is greater than - or equal to - a corresponding element in the lower triangle, its position is left unchanged. If this is not the case, as in the example of Figures 1 and 2, the elements are exchanged on each side of the diagonal, and the corresponding rows and columns are interchanged. In this example, the order of A and B was exchanged, to conform to the geological situation in which A occurs more often below B than above B. The final order is B over A over C.

The size of the matrix is dependent on the number of taxa used. The larger the number of taxa along the matrix margins, the more corresponding matrix elements may have to be exchanged, to satisfy the condition that the larger of the two (AB versus BA, AC versus CA, BC versus CB, etc.), is above the diagonal. For each of these socalled matrix permutations, the corresponding rows and columns are interchanged, until there is stability. Insufficient observations may lead to complex order relationships, and special conditions do arise.

Cycles

In some cases the matrix permutation and reordering of events will not produce a solution. Cycles will occur and may be due to:

(1) A, B, C (etc.) occur simultaneously;

(2) The data used is patchy and there are insufficient observations on relative order, leading to conflicting situations in rank. Reworking or otherwise wayward events also contribute, as discussed under Cenozoic Foraminiferal Zonation (see below).

For example, a fairly common situation is that in some sections species A occurs above species B and above species C, and in several different sections species C occurs above species A. Or, in more formal terms, if the relative order of events A, B and C, shown in Figure 3, is such that:

A occurs more often before B, than B before A;

B occurs more often before C, than C before B;

C occurs more often before A, than A before C;

then the order of events A, B and C is CABC etc., and is non-determined. The algorithm developed for RASC would result in cycling of the three events and no solution would be reached, unless special steps are undertaken. Rather than ignoring cycling events (Hay, 1972) which leads to a loss of information, RASC tries to break the cycles (Agterberg and Nel, 1982a). For example in the cycle CABC, the links CA, AB, BC are tested to see if the scores of CA and AC are smaller than those of AB and BA and of BC and CB. If so (in this case yes), this pair of events (CA,AC) is replaced by zeros in the order matrix, and so on, until there is a ranking of events (in this case ABC). In our experience, up to 6 events may participate in a cycle. In the output of the RASC program each cycle is printed with its participating taxa, and may guide the biostratigrapher to scrutinize those taxon events in each section.

Threshold Conditions and Range

As mentioned, the use of few stratigraphic sections, or the use of data with many missing pair values or with many coeval events, causes uncertainty in the optimum sequence and subsequent results. Successive events in the optimum sequence in reality may not have been found together in any individual section. The optimum sequence is thus an ideal composite sequence.

In order to provide at least a minimum number of observations on relative order of each pair of events, it is useful to require that each event occurs in at least k_c sections and that each pair of events in the optimum sequence occurs in at least m_c sections ($k_c \geq m_c$). The latter means that two events have to be recorded in a certain order (including coeval situations) at least m_c times. The threshold parameters make sense from a geological correlation point of view; the values of k_c and m_c are to be selected as desired.

RASC output provides a table showing in how many sections each taxon occurs, and then derives a table of the cumulative frequencies of events which occur at least 1, 2, 3 .... times. For example in the application discussed below on Cenozoic Foraminiferal Zonation, all 168 taxa occur in at least one well, 109 of those 168 taxa occur in at least two wells, 79 in at least three wells, etc. (Table 3).

This (Table 3; see also Figure 3 in Chapter I.2) will help to decide on a RASC solution with a sufficient number of events. The middle row - number of events - shows that exactly 59 taxa occur in one well only, 30 other taxa in two wells only, etc.

The Range (i, j) (see Figure 4) of an event in the optimum sequence indicates that the event occurs somewhere between events i and j (i < event < j). It is not to be confused with stratigraphic range, and arrives from insufficient information on relative order. Three criteria are involved (Agterberg and Nel, 1982a; Heller et al., 1983) in determining whether the relative position of two successive events i and j remains undecided in the optimum sequence: If the order score S_{ij} in the upper triangle of the final order matrix is

(a) ignored, because $S_{ij} < m_c$;

(b) zeroed, because S_{ij} and S_{ji} were pairs destroyed in re-ordering cycles; and

(c) 'tied', because S_{ij} = S_{ji} and events i and j are simultaneous in the optimum sequence.

In these situations, the ranges of events i and j involve the same positions in the optimum sequence. Strictly speaking, ranking of an event and its calculated range values should use a small m_c value, but a large m_c value is desirable for scaling, where a minimum number of pairs is necessary for good results. For this reason, recent versions of the RASC computer program allow a smaller (or zero) m_c value for ranking than for scaling. The threshold m_{c1} pertains to ranking, m_{c2} to scaling.

In the RASC output the optimum sequence prints sequence position, fossil number, range and fossil name (Figure 4).

At the top and at the bottom of each optimum sequence, event positions are less certain than in the middle part. This is simply the result of the fact that there is less information on the relative order of events. The end members cannot be compared to events respectively above position 1 or below position n (n = number of events in the optimum sequence).

Scaled Optimum Sequence

Biostratigraphic events cluster along a relative timescale when they change position frequently with respect to one another in stratigraphic sections. This feature is used in the RASC program to calculate cross-over frequencies for pairs of events. The number of times that the relative positions of two events in a number of secions is exchanged - i.e. the number of times the correlation lines cross over - provides a measure of inconsistency. This inconsistency is taken as a measure of stratigraphic distance between events.

A simple example will serve to clarify the theory behing this technique which uses the same events as occur in the optimum sequence. Consider the extreme case that all fossil events occur in the same order in all sections examined (Figure 5, left). There are no cross-overs of events. In this case the optimum sequence of events is easily determined. Since the scaling of the optimum sequence is a function of the cross-over frequency of events, no distances between events can be determined and there will not be clustering. In this case, distances are set to some constant value (Chapter II.6).

The real situation, particularly with 'noisy' records, is that some events show a more regular, consistent order than others in the section studied (Figure 5, right). The greater the tendency for adjacent events to cross-over, the better they cluster in the scaled optimum sequence diagram; conversely, the tighter events cluster in this diagram, the less their order is determined. Large breaks in the clustering or scaled optimum sequence correspond to stratigraphic horizons separating assemblages of events (see Figures 12, 14, and 15 in paragraph on Cenozoic Foraminiferal Zonation), thus achieving the same goals as a subjective zonation. Since the objective zonation based on scaling only uses the events common enough to occur in the optimum sequence, it has a high practical value.

In the statistical technique developed by F.P. Agterberg and tested for statistical ruggedness and computer efficiency by C.B. Hudson (Agterberg and Nel, 1982b; Hudson and Agterberg, 1982), the position of each event along the relative timescale is modelled by a normal (Gaussian) probability distribution with equal variance (σ^2). The cross-over frequency is also normally distributed with unit variance. The crossover frequency of events is tabulated, e.g. if in 10 stratigraphic sections A occurs five times above B and five times below B, than $P_{AB} = 0.5$, or if A occurs eight times above B and two times below B, than $P_{AB} = 0.8$ (Figure 6). In the first case A and B are probably closer in time than in the second, and we consider the cross-over frequency P_{AB} to be an approximation of the distance between the mean positions of A and that of B. In order to increase the number of observations, the distance d_{AB} between events A and B is also measured from their cross-over frequency with event C, such that

$$d_{AB} = d_{AC} - d_{BC}$$

The stability of these estimates of distance between fossil events clearly depends on sample size, i.e. the number of pairs of events used to calculate the interfossil distance. To adjust for this a weight-corrected equation was formulated, which emphasizes distance as a function of the number of observations. Successive distances also have a standard deviation calculated (Chapter II.6).

Scaling calculates interfossil distances for the events as ranked in the optimum sequence; the distances are expressed in dendrogram format (see further on in Figures 12, 14 and 15). The stratigraphically successive dendrogram clusters have nothing in common with results based on so-called cluster analysis. Some calculated interfossil distances initially may come out as negative values, in which case the optimum sequence events are re-ordered and new (generally positive) distances calculated. This explains why the order in the ranked and scaled optimum sequences may be slightly different.

STRATIGRAPHIC NORMALITY

Stratigraphic normality refers to the degree of correspondence between the individual stratigraphic record and the standard record. The standard summarizes the consistency in the order, the duration, and the interfossil distance or spacing of events in time. Typically, (bio)stratigraphic normality is judged by qualitative, section-by-section comparison of the order and the spacing of the events. Badly-placed events in sections, the result of reworking, misidentification,

A B A B A B A B A B

B A B A B A B A B A

P_{AB} =0.5

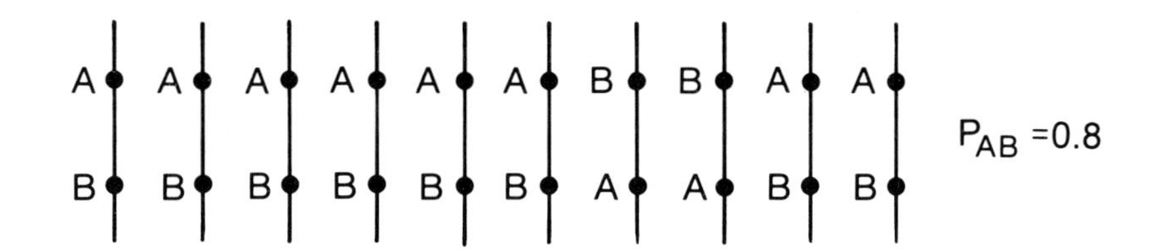

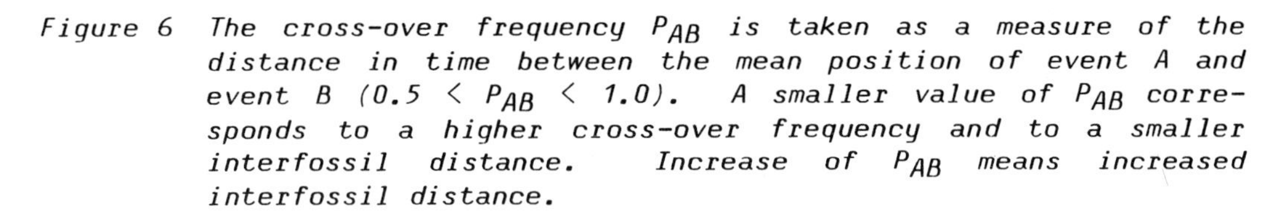

Figure 6 *The cross-over frequency P_{AB} is taken as a measure of the distance in time between the mean position of event A and event B ($0.5 < P_{AB} < 1.0$). A smaller value of P_{AB} corresponds to a higher cross-over frequency and to a smaller interfossil distance. Increase of P_{AB} means increased interfossil distance.*

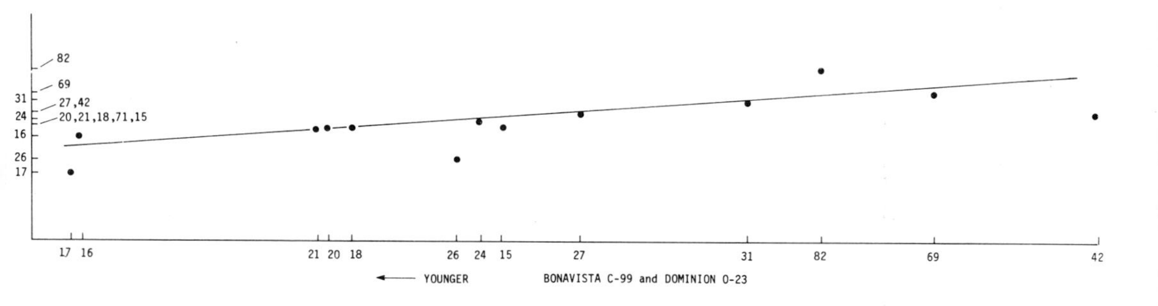

Figure 7 *Comparison in a scatter diagram of the sequence of foraminifers in the Egret K-36 well and of the same events in the combined sequences in the nearby Bonavista and Dominion wells. Events 26 (**Uvigerina dumblei**) and 42 (**Cibicidoides alleni**) in Egret are stratigraphically too high; 82 is too low. Scale is in depth units observed in the wells.*

hiatuses, non-observation or sampling contaminants, if recognized may be deleted. Yet, such sources of noise, which make the record imperfect, and conventional biostratigraphic decision-making often difficult, cannot be easily eliminated.

In quantitative biostratigraphy there are several strategies to recognize and delete 'noise' and increase stratigraphic normality prior to final analysis:

(1) In methods which express clustering of species in time (and, unfortunately, generally also in space) and define time zones and ecozones for correlation, data may be weighted by the degree of 'biostratigraphic fidelity' of taxa (see Brower et al., 1978, for a review). Measures of relative biostratigraphic value (RBV) give an expression of geographic persistence, facies independence, and stratigraphic range. Species with favourable RBV values (the ideal for index fossils) are used as input for further analysis.

(2) In methods concerned with the succession of fossils in time, events that are weak links, due to varying degrees of uncertainty in rank, should be deleted. Typically, such methods compare the relative order of time-successive pairs of events in many stratigraphic sections. Hay (1972) and Worsley and Jorgens (1977) employ tests to find out if two events occur in any order. First, occurrences of coeval events are deleted from the data. Also, if two events occur separated in any order in N sections and occur in a particular order in n sections, only those pairs may be used, where $1/2\ N < n < N$. Then a binomial test is applied to find significance of departure from randomness for the order in time of two events, A and B. Blank (1979) and Harper (1981) allow the test to remain undecided when insufficient pairs of events A and B are available. However, if event A in some sections occurs above many events, which in turn occur (in some other sections) above B, then the relative position of A and B is well determined, although the binomial test might not suggest so.

RASC (Ranking and Scaling) tries to reduce weak links by means of the threshold values of k_c and m_c. Only those events are allowed to participate that occur in at least k_c wells; each pair of events in the optimum sequence occurs in at least m_c wells; $k_c \geq m_c$ and both thresholds typically vary from 3 to 10.

Analysis and testing of stratigraphic normality in the data are performed by RASC following initial ranking and scaling. Three methods are used:

(i) A scatter diagram is used to compare two sequences of events, one of which is the (scaled) optimum sequence, the optimum standard. It shows the offset of the scatter points to a line of best fit. The offset is a function of stratigraphic mismatch;

(ii) The RASC step-model scores penalty points for reversals in order, when comparing each individual sequence of events to the RASC average sequence; and

(iii) The RASC normality test calculates if the distance between events, which are neighbors in an individual sequence, in the scaled average sequence exceeds a normalized limit.

Each method will be reviewed below.

Scatter Diagram

Scatter diagrams, which compare sequences of paleontological events in strata as observed in two or more stratigraphic sections, were vigorously advocated by Shaw (1964). The sequence of entries and exits of taxa in Section A is plotted along one axis and that of Section B along the other axis of a conventional two-axis graph. Scale units are the depth or thickness units of the wells or outcrop sections compared, but in a simplified procedure, rank order only can be used. The best fit of the resulting scatter of homologous points is called line of correlation. In the original method regression analysis is used as line fitting technique, although x and y are probably both subject to uncertainty. There are two regression lines, one for x over y and one for y over x. Spline-fitting techniques, as employed in Chapter III for this purpose, may be more suitable, particularly if no straight-line fit is possible due to changes in sedimentation rate.

One might choose to hand-fit the scatter of points to arrive at some kind of composite standard, particularly so if individual deviations from a standard are limited (Edwards, 1984). The actual standard from two sequences is a projection of the individual scatter points on a line of best fit. When a third stratigraphic section is added, the standard for the first two sections becomes one of the axes of the graph. This way, sections are successively added and eventually a composite standard results, which expresses the variation in all sections examined. The scale, in the meantime, has become a composite of all spacing values between successive events.

As an example of how this technique displays stratigraphic normality I evaluate the sequence of Cenozoic foraminiferal exits in 3 Canadian Atlantic margin wells. Depth in the well(s) is quoted in Appendix I.

The x-axis of the graph (Figure 7) is the best fit of the scatter of Cenozoic foraminiferal exits with depth in the upper part of two adjacent wells, Bonavista and Dominion. These wells were chosen as 'standard', because the record is relatively 'regular' and complete. Homologous scatter points were projected perpendicularly on a visualized best-fit line. Event 42, only present in the Bonavista well, was added to the standard using the Bonavista depth and its estimated depth in the

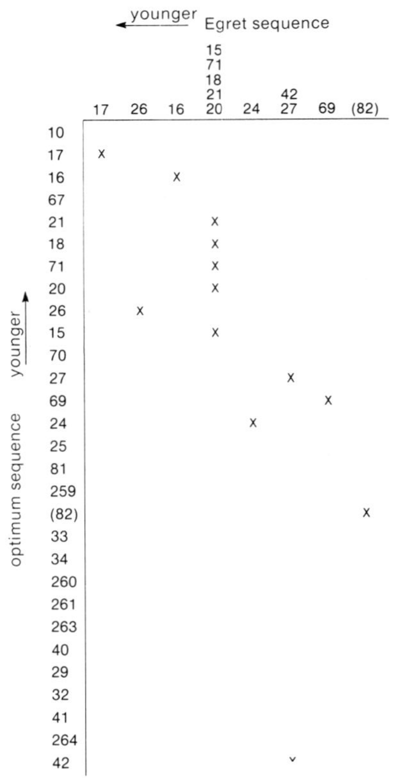

Figure 8 *Scatter plot of partial list of the (scaled) optimum sequence of foraminifer events in twenty-one wells ($k_c \geq 7$, $m_{c1} \geq 2$, $m_{c2} \geq 4$) and the observed sequence of foraminifers in the Egret K-36 well. Events 26* ***Uvigerina dumblei)*** *and 42 (**Cibicidoides alleni)** in Egret occur stratigraphically too high.*

Table 4 *The RASC step-model compares the relative order of every pair of events in the Egret K-36 well with that of the relevant part of the (scaled) optimum sequence of the same events. If the order is different one penalty point is scored; coeval events each receive 0.5 points.*

EGRET WELL SEQUENCE	SCALED OPTIMUM SEQUENCE	EGRET PENALTY POINTS
	10	
17	17	
26	16	1.0
16	67	
20	18	3.0
-21	71	3.0
-18	21	3.0
-71	20	3.0
-15	15	3.0
24	26	6.0
27	70	
-42	24	
69	25	
(82)	27	0.5
	69	1.0
	81	
	259	
	(82)	
	33	
	34	
	260	
	261	
	263	
	40	
	32	
	29	
	264	
	42	1.5(2.5)

Dominion well. This position in Dominion was estimated from comparison to the optimum sequence of events using RASC.

Along the y-axis is the sequence with depth in the Egret well. If a straight-line best fit is reasonable, one may conclude that Events 26 **(Uvigerina dumblei)** and 42 **(Cibicidoides alleni)** in Egret are stratigraphically too high (young), and less so 17 **(Asterigerina guerichi).** Event 82 **(Globigerina linaperta)** occurs stratigraphically slightly too low (old).

A convenient and simple way to identify stratigraphic normality consists of scatter plots, using RASC data. The scaled optimum sequence of events, calculated by the RASC method, is plotted along one axis and an individual (well) sequence is plotted along the other axis of the scatter diagram. The bivariate score, if the sequences are perfectly matched, should define a straight line. Strong deviations point to misplaced occurrences in the individual sequence.

Figure 8 is a scatter plot of the younger part of the optimum sequence in 21 wells (including Egret K-36) on the Labrador Shelf and Grand Banks, ($k_c \geq 7$; $m_{c1} \geq 2$; $m_{c2} \geq 4$) and of the sequence of events in the Egret well. Note that Events 21, 18, 20, 71 and 15 and also 27 and 42 in Egret occur simultaneously (same sample). Events 26 **(Uvigerina dumblei)** and 42 **(Cibicidoides alleni)** are more out of line than any of the other Egret events. Since they fall below a best fit trend, it is reasonable to conclude that both events 42 and 26 are too high in Egret. This may be due either to misidentification or to reworking. This result agrees with the conclusion using the composite standard.

Step-model

In the step-model, which is also a standard feature of the RASC computer program, the relative order of every successive pair of events in the individual sequence of events is compared to their order in the scaled version of the optimum or average sequence. If the order is different, one penalty point is scored. Coeval events in individual sections each receive 0.5 point.

The example of Table 4 shows the step-model as applied to the Egret well in the $k_c \geq 7$, $m_{c2} \geq 4$ run for the 21 wells, used earlier. Event 16 receives in Egret one penalty point because it occurs below 26, whereas in the average sequence it occurs above it. Event 18 receives in Egret two penalty points (4 x 0.5) because it is coeval with 71, 21, 20 and 15, and one penalty point because of its reversed position with 26. The same reasoning applies to events 71, 21, 20 and 15. Event 26 has more penalty points than the other ones in this well section, in agreement with the scatter of 26 in Figures 7 and 8, discussed previously. In runs with a lower k_c value more events as found in the Egret sequence, like 82 (in brackets shown in Table 4) participate in ranking and scaling. These events in Egret occur below 42, whereas in the average sequence their order is reversed. As a result 42 also receives above average penalty points, as suggested in the bivariate plots.

A disadvantage of the step-model is that clusters score high penalty points. A cluster may be due to many events occurring in one sample or to an indeterminate order from stratigraphic section-to-section comparison of several adjacent events. An example of a cluster due to sampling was shown in Egret (Events 18, 71, 21, 20 and 15). Such a cluster is a stratigraphic inconsistency, but at least one of the events in a cluster may be correctly placed.

A slight convenience in the step-model results as printed by the RASC computer program would be to list the contribution of clustering, due to coeval events, in the final penalty score.

Normality Test

The normality test, which is also a standard feature of the RASC program, provides a test to express the mismatch of the order in each stratigraphic section, relative to the scaled standard sequence. Normality testing follows scaling.

In order to test an event position B in a section, the event's stratigraphic neighbors A and C in that section are identified. All three of these events have a position in the scaled optimum sequence, which means that their relative distance is known and their cumulative distance from the origin (stratigraphically highest point in the scaled sequence). When the three position values are close, the first-order difference between the distance values of B and A and of B and C, and the second-order differences between the caluculated first-order differences, will be close to zero. When the three position values are far apart, when the events should not be neighbors in the tested section, the second-order difference is much different from zero.

The principle is illustrated in Figure 9, where events 1 through 7 form the scaled optimum sequence and bars depict the interfossil distances and cumulative distances (a-g) from origin, (top of sequence). In the stratigraphic sections A and B to be tested, events 3, 4, and 5 and 1, 4 and 7 respectively are neighbours. Events 3, 4 and 5 are also neighbours in the scaled optimum sequence and the difference between the cumulative distances of events 5 and 4, minus the cumulative distances of 4 and 3 is small, $\left((d-c)-(e-d)\right)$. On the other hand, the differences between the cumulative distances of events 7 and 4, minus the cumulative distances between events 4 and 1, which are not neighbours in the scaled optimum sequences, is much larger, $\left((d-a)-(g-d)\right)$.

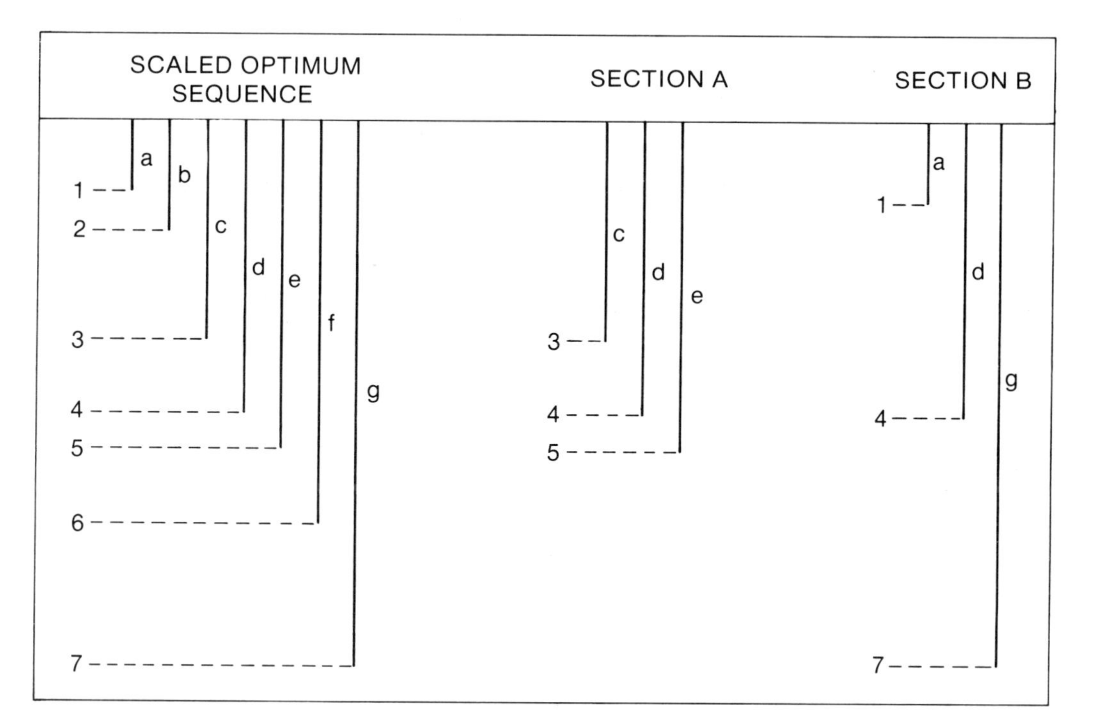

Figure 9 Principle of normality testing. Events 1-7 form a scaled optimum sequence, with bars depicting the interfossil distances and cumulative distances (a-g) from origin, (top of sequence). In the stratigraphic sections A and B to be tested, events 3, 4, and 5 and events 1, 4, and 7 respectively are neighbours. Events 3, 4 and 5 are also neighbours in the scaled optimum sequence and the difference between the cumulative distances of events 5 and 4, minus the cumulative distances of 4 and 3, (d-c)-(e-d), is small. On the other hand, the difference between the cumulative distances of events 7 and 4, minus the cumulative distances between events 4 and 1, (d-a)-(g-d), which are not neighbours in the scaled optimum sequence, is much larger. These second-order differences can be tested against expected values.

The first- and second-order differences can be regarded as realizations of independent normal random variables. Second-order differences that exceed 95% and 99% probability limits of a normal distribution, respectively, receive one or two asterisks (Chapter II.7).

Table 5 and 6 show the normality test for the Egret well succession as part of the $k_c \geq 7$, $m_{c2} \geq 4$ run of 21 wells used earlier.The top- and bottom most events have only one neighbour and are always omitted from testing. Note that both events 26 and 42 are calculated to be out of place with probabilities of 99% or more. This result agrees with that obtained in the step-model or scatter diagram, but such agreement is not necessarily to be expected for each stratigraphic section tested by the different normality methods.

A note of warning is in order before results of the normality test are used for further study. First, in the scaled optimum sequence, distances between successive events may have a large standard deviation. Hence, our events A, B and C or 1, 2, 3, 4, 5, 6 and 7 may have poorly defined cumulative distances. No expression of this original uncertainty is provided in the normality test.

Second, and more seriously, second-order differences are interdependent. The method involves testing of positions B to A and of B to C, and an anomalous position of B will often implicate A and C as well. This artifact of the method implies that, in the test result for each section, single and double asterisk events often occur in pairs, more so with higher second-order difference values. This explains the one asterisk event number 16 in Table 6, adjacent to 26. In reality, only one event - in this case 26 - of a cluster of two or three may be out of place, which may be verified in the original record, in an RASC scatter diagram or in the step-model.

Testing of individual sections with large hiatuses will also give rise to 'significant' second-order differences for the events on each side of the hiatus. This does not mean that these event positions in that section are necessarily anomalous. The step-model score and RASC scatter diagram deviation both should be low.

An interesting feature of the normality test is that, since the second-order differences may be expected to satisfy a normal distribution, a dataset can be checked to see whether it has more mismatch than expected (Agterberg and Nel, 1982b). One could reasonably expect 5% of the events with one asterisk and 1% with two. In the data file for the 24 wells discussed below, there are 845 events. Not counting the first and last events of each well, which lack one neighbor and cannot be tested for distance, leaves 797 events. Of these, 31 have one asterisk and 14 have two, which is 4% and 1%, respectively. If one allows that an anomalous position of one event may implicate another event to be out of place, as explained earlier, these percentage values are low. This low mismatch in the data is the result of anomalous events filtering, treated further on in the paragraph on "Subjective and Objective Zonations".

Table 5 Principle of normality testing; events 15, 71, 18 and 21 occur in the same sample as events 20, as indicated with hyphens. For explanation see text.

CUMULATIVE DISTANCES IN SCALED OPTIMUM SEQUENCE		EGRET WELL SEQUENCE		FIRST ORDER DIFFERENCE	SECOND ORDER DIFFERENCE
18	1.6474	-21	1.9185		
71	1.7985	-18	1.6474	0.2711	0.1200
21	1.9185	-71	1.7985	-0.1511	
		-15			
		24			

Table 6 Normality test for the succession of foraminiferal exits in the Egret K-36 well. Both events 26 and 42 are calculated to be out of place with a probability of 99% or more, in agreement with results of the step-model and scatter plots.

NORMALITY TEST

EGRET K36

		CUM. DIST.	2ND ORDER DIFF.
ASTERIGERINA GURICHI	17	.3905	
UVIGERINA DUMBLEI	26	2.4864	-3.6704 **
CERATOBULIMINA CONTRARIA	16	.9120	2.7707 *
GYROIDINA GIRARDANA	20	2.1082	-.8653
GUTTULINA PROBLEMA	-21	1.9185	-.0814
SPIROPLECTAMMINA CARINATA	-18	1.6474	.4222
EPISTOMINA ELEGANS	-71	1.7985	.4925
GLOBIGERINA PRAEBULLOIDES	-15	2.4421	-.4018
TURRILINA ALSATICA	24	3.2045	-.5345
EPONIDES UMBONATUS	27	3.4325	2.0776
CIBICIDOIDES ALLENI	-42	5.2175	-4.0087 **
NODOSARIA SP8	69	3.5144	

* -GREATER THAN 95% PROBABILITY THAT EVENT IS OUT OF POSITION
** -GREATER THAN 99% PROBABILITY THAT EVENT IS OUT OF POSITION

As explained by Agterberg and Nel (1982b), Heller et al. (1983), and in Chapter II.7, the RASC program prints a table which compares the observed and expected frequencies of the second-order difference values, calculated for each of 10 classes of equal probability of the standard normal distribution.

The Perfect Data

As mentioned in the previous paragraph, it is reasonable with scaling and normality testing to expect at least 5% of events with one asterisk and 1% of events with two asterisks. Removal of all single and double asterisks events in the sections, followed by renewed application of RASC, does not mean that no further single and double asterisks events will appear. As a result of filtering the interfossil distances in the final scaled dendrogram will have smaller numerical values, but there will always be anomalously-placed events in each solution. Paradoxically, the perfect data do not exist, at least not in a numerical sense. Nevertheless, in the 'cleaned' data, interfossil distances have decreased somewhat, and tighter and stratigraphically often more distinct dendrogram clusters may result.

To summarize, three different methods provide insight in the stratigraphic normality of the fossil record through time in different geological sections. Of these three, the scatter diagrams are simple, flexible, and intuitively attractive. The RASC normality test is more sophisticated and the only one to provide objective degrees of mismatch of the individual record relative to the (scaled) standard sequence of events.

All tests are designed to identify potential outliers in the data. Removal of these outliers may result in better zonations.

MARKER HORIZONS AND UNIQUE EVENTS

For the purpose of chronostratigraphic interpretation of the dendrogram clusters in a scaled optimum sequence, it is useful to incorporate age-diagnostic index fossils. Unfortunately, the occurrence of such taxa may be a relatively rare event, below the threshold k_c. For this reason, the RASC data can incorporate so-called <u>Unique Events</u>. Unique events are defined as events that occur in less than k_c sections, and would not normally participate in ranking or scaling.

For the purpose of interrelation of a biostratigraphic zonation with seismic -, well log-, or lithological data, it is useful to define <u>Marker Horizons</u>. A good example of marker horizons are bentonite beds. Marker horizons must participate in more than k_c sections, and are treated as stratigraphic events with zero variance with regards to their position.

Both types of events are assigned a dictionary number and entered in the original sequence data like any other of the events. When specifying the input parameters (like k_c and m_c) for a RASC program run, both unique events and marker horizons have to be designated as such with their dictionary numbers.

The average relative stratigraphic position of a marker horizon is shown both in the optimum sequence and in the scaled optimum sequence. It is recognized in the final results by a single asterisk. Unique events are simply fitted in the scaled optimum sequence (following termination of scaling and normality testing), using their (average) position in the few (or one) section(s) where they occur. Little can be said of the reliability of the position of unique events, identified by means of double asterisks in the final scaling solution. Details on the statistical and computer technical treatment are in Agterberg and Nel (1982b), Heller et al. (1983), and in Chapters II.6 and II.7.

THE STRATIGRAPHICAL MEANING OF THE MODELS

In order to appreciate the statistical and paleontological meaning of the optimum sequence and the scaled optimum sequence, we will briefly discuss their characteristics.

The optimum sequence shows the most likely order of fossil events. An event generally is defined as the stratigraphically lowest or highest occurrence of a taxon. It is required that the events occur in at least k_c sections and each pair of events in at least m_c sections ($k_c \geq m_c$). The relative position of the events in the optimum sequence is an 'average' of all the relative postions encountered. By combining the results based on the lowest and highest stratigraphic occurrences, charts of average stratigraphic ranges can be constructed.

As discussed in the next paragraphs, the method provides information on regional markers. A comparison of the optimum sequence to the sequence of events in the individual sections may provide insight into reworking, misidentifications, and other sources of mismatch. Rigorous solutions based on high k_c and high m_c values serve in point correlations. This is promising and simple for geological studies which require local, detailed well-to-well correlations.

Scaling of the optimum sequence in relative time, is based on interfossil distances calculated from cross-over frequencies of each pair of events in the optimum sequence. The stratigraphically successive dendrogram clusters resemble assemblage zones. There is one important advantage of a zonation based on dendrogram clusters over a conventional assemblage zonation. A conventional, subjective zonation frequently makes no distinction between species that occur in few or many sections. It will give weight to index forms, however rare, whose stratigraphic value is understood. Such a procedure will lead to an idealistic zonation which reflects the best stratigraphic resolution to be obtained in the area studied, but may be impractical for correlation. The zonation based on scaling principally uses events frequent enough to occur in the optimum sequence. Less weight is given to unique events. In this respect, the objective zonation obtained by RASC is superior for stratigraphic correlation, without loss of resolution in relative time.

CENOZOIC FORAMINIFERAL ZONATION

Introduction

In 1982 the first biozonation was published, using the RASC - Ranking and Scaling method (Gradstein and Agterberg, 1982). A nine-fold Cenozoic zonation was proposed, based on 555 last occurrences of 157 benthic and planktonic foraminiferal taxa in 16 Grand Banks and Labrador Shelf wells. Seven plates with scanning electron micrographs documented the morphology of the stratigraphically important taxa.

Since that time, the RASC zonation has been routinely applied to zone newly drilled wells and further taxonomic studies were performed. As a result, several wells and many new taxa were added to the original data. Application of the stratigraphic normality test procedures in RASC have shown that several dozen events, out of almost 900, were badly placed in the wells. Scrutiny of these events in many cases has led to their redefinition in terms of depth of occurrence or taxonomic assignment. The result is a cleaner, less noisy data file for more detailed biostratigraphy and less uncertainty in correlation.

In this Chapter I will briefly discuss the filtering or cleaning of the data file and improvements to and stratigraphic quality of the RASC zonation. At this point it is instructive to start with the regional geological rationale for development and application of the RASC method. As I will show, the search for a more objective and practical biostratigraphy in general, is coupled to a specific paleontological reason for using sequencing methodology.

Why should 'Probabilistic' Biostratigraphy be Applied

It is well known that the sequence of first and last occurrences of planktonic foraminiferal species in open marine Cenozoic sediments in the low-latitude regions of the world is closely spaced and shows an astonishingly regular order. As a result, standard, non-probabilistic planktonic zonations provide a stratigraphic resolution of 30 to 45 zones over a time span of 65 x 10 y (Stainforth et al., 1975). Although several Cenozoic taxa are indigneous to mid-latitudes, the absence of many essentially lower-latitude forms and the longer stratigraphic ranges of mid-latitude taxa causes stratigraphic resolution to decrease away from the lower-latitude belt. In high latitudes (50°N and S) the virtual absence of planktonic foraminiferal taxa makes the standard zonations inapplicable.

The northwestern Atlantic margin, offshore eastern Canada, spans the mid- to high-latitudinal realm (north of 42°), and although there were temporal northward incursions of lower latitude taxa in early or middle Eocene times, there is a drastic overall diminution of the number of biostratigraphically useful Cenozoic planktonic species from the Scotian Shelf (75 + taxa) to the Grand Banks to the Labrador Shelf (15 + taxa). A change from a generally deeper, open marine facies in the Paleogene to more nearshore, shallower conditions in the Oligocene to Neogene (Gradstein and Srivastava, 1980) also curtails the number of taxa present in the younger Cenozoic section.

As a consequence, the construction of a planktonic zonation is mainly applicable to the southern Grand Banks and Scotian Shelf where 12 zones have been recognized based on species of standard zonations which are not too rare locally to be of practical value in correlation. Similarly, on the northern Grand Banks and Labrador Shelf a sevenfold planktonic subdivision of the Cenozoic sedimentary strata is possible (see below). The regional application is limited but the zonal markers and associated planktonic species improve chronostratigraphic calibration for the benthonic zones.

Independently, the Cenozoic benthonic foraminiferal record also shows temporal and spatial trends in taxonomic diversity and number of specimens. Calcareous benthonic species diversity and number of specimens decreases northward from the Scotian Shelf to the Grand Banks to the Labrador Shelf, whereas the early Cenozoic agglutinated species diversity and numbers of specimens drastically increases on the Labrador Shelf. This benthonic provincialism is complicated by incoherent geographic distribution of some taxa, which at least in part is due to sampling.

Few of the agglutinated taxa (only 20 or so out of more than 60 determined so far) are of biostratigraphic value (Gradstein and Berggren, 1981), but among the hundreds of calcareous benthonic forms determined, many more potentially locally useful or widely-known index species occur. As a consequence of the ecological sensitivity of many of these bottom dwellers, and because of their relatively long stratigraphic ranges, facies changes can be expected to somewhat modify stratigraphic ranges. This is known as the problem of total versus local stratigraphic range. As a result the benthonic stratigraphic correlation framework based on exits forms a kind of weaving pattern with many slight and some large scale cross correlations. Considerable mismatch in correlation is the result of misidentification and reworking, or due to large differences between local stratigraphic ranges of a taxon. Some correlation lines only go part of the way over the combined shelves area.

The above summary provides insight into some of the constraints on a traditional regional foraminiferal zonation, but there are others as well. The most important constraint is the method of sampling. Usually, only samples of cuttings, taken mostly over 30 ft intervals, are available from the wells. This means that instead of entry, relative range, peak occurrence, and exit, only the exit of taxa is known. Downhole contamination in the cuttings hinders recognition of stratigraphically separate benthonic- or planktonic homeomorphs. Other limiting factors are that many species often occur in low numbers and that reworking of tests is common in the younger Neogene section of the Labrador Shelf. Finally, there is always uncertainty regarding consistent taxonomic identifications during the regional investigation.

In summary, the data show the following properties, ranked according to their importance toward stratigraphic resolution:

(1) The samples are almost exclusively cuttings, which forces use of the highest part of stratigraphic ranges or of the highest occurrences (tops, exits), and restricts the number of stratigraphically useful taxa;

(2) The application of standard planktonic zonations is limited due to the mid to high latitude setting of the study area and the general tendency for shallower water environments of deposition in the younger sediments;

(3) There are many minor and few major inconsistencies in the relative extinction levels of many benthonic taxa;

(4) Most of the samples are relatively small, which limits the detection of species represented by few specimens; this contributes to factor (3) and to the erratic, incoherent geographic distribution pattern of some taxa; and

(5) There is geographic and stratigraphic provincialism in the benthonic record from the Labrador to the Scotian Shelves which makes detail in a general zonation difficult.

Despite the limiting factors it is quite possible to use traditional methods to erect a zonation when it is based on a partial data file that uses few taxa. Gradstein and Williams (1976) used four Labrador Shelf-northern Grand Banks wells to produce an eightfold (benthonics) subdivision of the Cenozoic section. Similar stratigraphic resolution and better zone delineation was obtained by Gradstein (unpublished) using nine wells on the Labrador Shelf and northern Grand Banks. However, some of the zones were tentative and their relative ages not well determined.

Since the attempt at a zonation using a few wells only, study of many more wells has clarified the broader correlation pattern and increased the number of chronostratigraphic calibration points based on occurrences of planktonic foraminifera. It has also increased the noise in the stratigraphic signal (our factors 3 and 4) due to more stratigraphic inconsistencies and more geographic incoherences of exits. Therefore, in an attempt to optimize stratigraphic resolution based on as

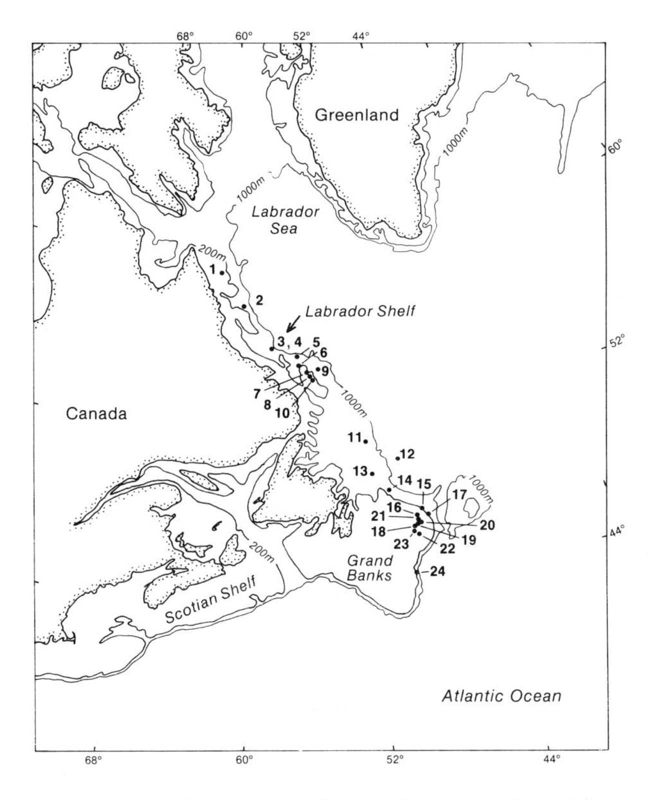

Figure 10 Location of twenty-four deep exploration wells sampling Cenozoic strata on the Labrador Shelf and northern Grand Banks, which contain the foraminiferal sequences used in Ranking and Scaling.

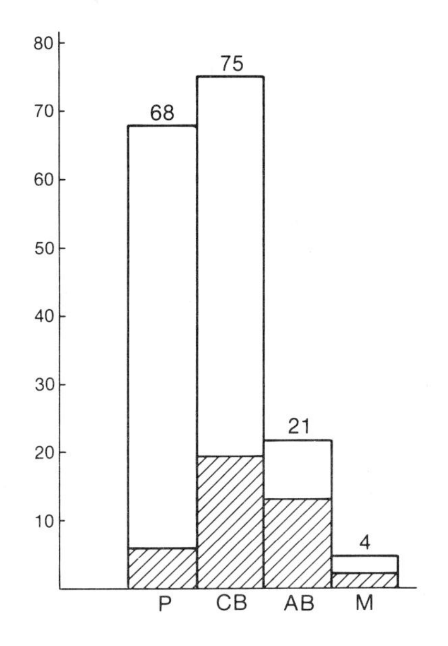

Figure 11 Cenozoic strata sampled in the 24 wells contain 168 dictionary taxa, including 68 planktonic foraminifers (P), 75 calcareous benthic (CB) and 21 arenaceous benthic (AB) ones. There are 4 miscellaneous morphotypes (M), including a megaspore, and a scaphopod-type microskeleton. The preferred $K_c \geq 7$, $m_{c2} \geq 4$ RASC zonation in 21 wells contains 40 taxa, represented in the hatched part of the histogram.

many of the observations as could be used for a zonation, the RASC Ranking and Scaling method was developed.

The Data

The main concern in wells selection has been broad regional coverage. The wells now span latitudes from 44° to 60° N. The wells are numbered from north to south and include (Figure 10):

LABRADOR SHELF		Sample depths below rotary table
1.	Karlsefni H-13	1760 - 12960'
2.	Snorri J-90	1260 - 9820'
3.	Herjolf M-92	3750 - 6450'
4.	Bjarni H-81	2860 - 6590'
5.	Gudrid H-55	1660 - 8550'
6.	Cartier D-70	2220 - 4960'
7.	Leif E-38	1210 - 3520'
8.	Leif M-48	1300 - 5620'
9.	Indian Harbour M-52	1740 - 10230'
10.	Freydis B-87	1000 - 5230'
GRAND BANKS		
11.	Hare Bay H-31	393 - 1910 m
12.	Blue H-28	2090 - 4760 m
13.	Bonavista C-99	1860 - 11910'
14.	Cumberland B-55	920 - 10930'
15.	Dominion O-23	1380 - 10230'
16.	Flying Foam I-13	990 - 5650'
17.	Adolphus D-50	1140 - 8726'
18.	Hibernia O-35	520 - 1295 m
19.	Hibernia P-15	255 - 1400 m
20.	Hibernia B-08	540 - 1745 m
21.	Hibernia K-18	700 - 1580 m
22.	Egret K-36	860 - 2240'
23.	Egret N-46	1080 - 1950'
24.	Osprey H-84	1190 - 2450'

Depths are expressed in the original units of measurements, also used on the microfossil slides. The wells numbers 5, 9, 13, 14, 15, 16 and 17 are more fossiliferous than others. At some sites Paleocene, Oligocene and late Neogene strata are missing or poorly represented.

In each well penetrating Cenozoic strata, the foraminiferal record was artifically truncated at the Cretaceous-Tertiary boundary, as determined from the highest occurrences of Cretaceous taxa. The Cenozoic index or dictionary, listed in Appendix I, now consists of 232 foraminifers and a few miscellaneous microfossils, numbered from 1 to 275. There are 43 blanks (dummy taxon numbers) due to subsequent deletion of taxa.

The 24 wells contain 168 dictionary taxa, consisting of 68 planktonic, 75 calcarous benthic, 21 arenaceous benthic, and 4 non-foraminiferal morphotypes (Figure 11). Ostracods are too rare to be of use for Cenozoic stratigraphic correlations.

The microfossil record in each well was coded, using the index numbers and three simple rules:

(a) a local identification with the qualifier aff. (affinis) is the same as the nominate taxon;

(b) a local identification with the qualifier cf (confer) for a regular dictionary taxon has not been taken into account;

(c) obviously reworked highest occurrences of taxa have gradually been eliminated, either from initial observation or through normality testing, discussed below. Reworking is apparent from anomalous, poor preservation of tests relative to the rest of an assemblage at that particular depth, and from erratic stratigraphic positions.

RASC computer runs to evaluate the quality of the data file, using the normality test procedures, used the thresholds $k_c \geq 5$; $m_{c2} \geq 4$ and $k_c \geq 7$, $m_{c2} \geq 4$ in 18-24 wells. Normality was evaluated from (a) high scores in the step-model, (b) a high probability that an event is out of place using normality testing, and (c) markedly deviating scores in the bivariate scatter plots.

The well file originally also included Gabriel C-60, a well located in 1134 m of water, outside the present continental shelf area. Normality testing of the sequence of events in this well showed an above average number of badly placed events, and (deeper water) taxa unique to this location. The sequence probably is complicated by a larger than usual reworking component. This well was omitted from the file.

Normality testing of the observed species sequences in the remaining 24 wells led to re-examination of one or more occurrences of 34 taxa, including:

Name	Index Number	Name	Index Number
Fursenkoina gracilis	9	**Spiroplectammina navarroana** (was **Textularia plummerae**)	54
Nonionella pizarrensis	11	**Nodosaria** sp. 11	63
Textularia agglutinans	14	**Cassidulina islandica**	64
Asterigerina guerichi	17	**Coscinodiscus** sp. 1	65
Coscinodiscus sp. 3	22	**Nodosaria** sp. 8	69
Coscinodiscus sp. 4	23	**Alabamina wolterstorffi**	70
Turrilina alsatica	24	**Globigerina venezuelana**	81
Coarse arenaceous spp	25	**Acarinina** aff. **broedermanni**	93
Eponides umbonatus	27	**Cibicidoides tenellus**	131
Pteropod sp 1	31		

Name	Index Number	Name	Index Number
Ammosphaeroidina sp. 1 (was **Quadrimorphinella incauta**)	32	**Alabamina takanagayakii** (was **Alabamina** sp. 3)	169
		Anomalina sp. 1	173
Spiroplectammina dentata	35	**Allomorphina** sp. 1	176
Bulimina midwayensis	43	**Plectofrondicularia** sp. 3	182
Bulimina trigonalis	45	**Lenticulina** sp. 8	231
Anomalina sp 5	48	**Eponides** sp. 4	236
Osangularia expansa	49	**Anomalina** sp. 4	251
Uvigerina batjesi	53		

Most of these taxa were re-assigned a lower or higher depth, where specimens occur that correspond better to the dictionary morphotype or are not likely reworked. Taxa numbers 22, 23, 31, 48, 65, 131, 176, 182, 231, 236, and 251 were deleted. Their last occurrences fluctuated erratically over several RASC zones, or the unique morphological identity of the taxa could not well be established due to lack of specimens.

The last appearances of another taxon, **Spiroplectammina spectabilis** were differentiated into a last consistent occurrence (number 57) and a last occurrence (number 68). Also, the occurrences of **Asterigerina guerichi** were split into a top occurrence (number 17) and a peak occurrence (number 265), to lessen the chance of including reworked highest occurrences from above a regional Upper Miocene disconformity. Obviously reworked presences in glacial drift deposits in Indian Harbour, Gudrid and Leif wells were deleted. **Cyclammina amplectens** (number 29) was redefined to only include specimens with a sharp periphery, rather than those with a somewhat more rounded periphery. This means that in almost all wells the last occurrence of **Cyclammina amplectens** is stratigraphically lower than previously assigned.

Relative stratigraphic positions were added of 4 seismic markers as observed in northern Grand Banks wells (pers. communication, A. Grant, Dartmouth, N.S.), numbered 201, 202, 203, and 204.

The coded data file and index are listed in Appendix I.

Objective and Subjective Zonations

The original zonation for the Grand Banks - Labrador was based on 555 events, being the stratigraphic distribution of last occurrences of 157 taxa in 16 wells. The preferred RASC solutions were based on $k_c \geq 5$ and $k_c \geq 3$ with $m_c \geq 3$. Figure 12 shows the $k_c \geq 3$, $m_c \geq 3$ scaling for this well file, which retains the optimum relative position and relative distance in time of 68 taxa. The taxa group in 9 distinct and progressively younger clusters, resembling assemblage zones.

Within each cluster relative positions also have stratigraphic meaning, depending on the standard error of the relative distances. Another way to "test" the stratigraphic meaning of the model is to directly compare optimum clustering to subjective zonations. For this purpose a comparative species score was made of the optimum sequence and the subjective zonations mentioned earlier, based on respectively 4 and 9 of the same wells.

The subjective biostratigraphic solutions produced an 8- to 9-fold subdivision of the Cenozoic section (Figure 12) which already lead to much improved regional geological correlation. Some of the zones were tentative, based on few taxa only, and not well defined.

The sequence of species in the RASC - assemblages shows a similar sequential arrangement as in the two conventional assemblage zonations. There is a match between species in the corresponding successive assemblages, with deviating scores being a minority. This is also obvious from the order of key markers, which are in common between the probabilistic and the conventional (T) zonations and are indicated with black dots. Each group of key markers in the succession of T zones occurs in successively younger and stratigraphically corresponding probabilistic ones, with the exception of **Asterigerina guerichi.** This species, which is thought to occur below a latest Miocene regional disconformity, is incorporated at the base of the **Cassidulina teretis** zone of Pliocene/Pleistocene age. This may have been caused by reworked highest occurrences of this species in some wells, and shows that it is difficult to separate younger Neogene strata. The comparison establishes that there is no noticeable effect from the fact that the statistical method produces average positions, whereas a subjective zonation would not tend to average, but prefers the highest possible occurrence.

The filtered and most recent Cenozoic data file in 24 wells is based on 845 last occurrences of 168 taxa. Under the hypothesis that one zonation provides the best stratigraphic resolution for this large region, one RASC solution might suffice. However, as explained earlier, the northern Grand Banks and Labrador Shelf differ paleontologically. In a nutshell, the two areas, each over 1000 km^2 large, are two sedimentary basins separated by a southwestern-northeastern trending ridge (Gradstein and Berggren, 1981). The southern region is more fossiliferous and contains more planktonic taxa, but the wells there did not sample the youngest part of the stratigraphic section, the **Cassidulina teretis** zone of Pliocene to Pleistocene age. The **Gavelinella beccariiformis** zone of (late) Paleocene age also is infrequently encountered, as apparently most or all of Paleocene beds are missing in the southern wells. More important than stratigraphic differences is that several benthic taxa show a preference for one of the basins only. An argument against the 24 wells file also is that it contains at least 4 (Hibernia) wells, which are all closely spaced on the same geological structure. This provides a regional bias.

OPTIMUM CLUSTERING OF CENOZOIC FORAMINIFERAL EXITS
(16 Wells; $k_c \geq 3 / m_c \geq 3$)

Pliocene - Pleistocene

Late Miocene

Early - Middle Miocene

Oligocene

Middle - Late Eocene

(Early) Middle Eocene

Early Eocene

Early Eocene

Paleocene

0.6564 0.5437 0.4310 0.3183 0.2056 0.0929 -0.0198

INTERFOSSIL DISTANCES

			GRADSTEIN (Unpublished; 9 Wells)								GRADSTEIN & WILLIAMS (1976; 4 Wells)							
			NEOGENE			PALEOGENE					NEOGENE				PALEOGENE			
			T1	T2	T3	T4	T5	T6	T 7a	T 7b	C. TERETIS	A. GURICHI	C. CONTRARIA	S. CARINATA	T. ALSATICA	PTEROPOD SP1	S. SPECTABILIS	G. CORONA
9	0.1645	FURSENKOINA GRACILIS																
11	0.1388	NONIONELLA PIZARRENSE																
77	0.1958	ELPHIDIUM SP •	●								X							
10	0.1627	UVIGERINA CANARIENSIS		X														
65	0.2507	COSCINODISCUS SP1		X									X					
228	0.0461	CASSIDULINA TERETIS •	●								X							
17	0.6924	ASTERIGERINA GURICHI		●								X						
22	0.0603	COSCINODISCUS SPP		X									X					
16	0.0813	CERATOBULIMINA CONTRARIA											X					
67	0.3474	SCAPHOPOD SP1		●														
18	0.0820	SPIROPLECTAMMINA CARINATA •		?	●									X				
14	0.1331	TEXTULARIA AGGLUTINANS																
71	0.2209	EPISTOMINA ELEGANS			X									X				
181	0.1242	CYCLOGYRA INVOLVENS		X									?					
21	0.0329	GUTTULINA PROBLEMA		?	●									X				
20	0.2321	GYROIDINA GIRARDANA •			●									X				
15	0.1968	GLOBIGERINA PRAEBULLOIDES •			●									X				
26	0.3877	UVIGERINA DUMBLEI				X												
70	0.4157	ALABAMINA WOLTERSTORFFI			X		?							X				
24	0.0375	TURRILINA ALSATICA •				●									X			
25	0.1457	COARSE ARENACEOUS SPP.				X												
27	0.1675	EPONIDES UMBONATUS			?													
69	0.2848	NODOSARIA SP8			X									X				
81	0.0157	GLOBIGERINA VENEZUELANA				X									?	X		
33	0.0835	TURBOROTALIA POMEROLI •					●									X		
82	0.0425	SUBBOTINA LINAPERTA •					●									X		
31	0.0957	PTEROPOD SP1 •					●		?							X		
29	0.0908	CYCLAMMINA AMPLECTENS •					●									X		
84	0.0988	SUBBOTINA YEGUAENSIS					●											
173	0.0159	ANOMALINA SP1																
85	0.0010	PSEUDOHASTIGERINA MICRA					●											
34	0.1274	MARGINULINA DECORATA				?	●									X		
40	0.0129	BULIMINA ALAZANENSIS					X											
118	0.0261	EPISTOMINA SP5						X									X	
75	0.0528	LENTICULINA ULATISENSIS															X	
41	0.1572	PLECTOFRONDICULARIA SP1					●	X										
74	0.0622	EPONIDES SP5					X										X	
42	0.1450	CIBICIDOIDES ALLENI					X										X	
30	0.0301	CIBICIDOIDES BLANPIEDI						X									X	
35	0.0234	SPIROPLECTAMMINA DENTATA					X											
32	0.0524	QUADRIMORPHINELLA INCAUTA																
36	0.0007	PSEUDOHASTIGERINA WILCOXENSIS						●										
86	0.0184	TURRILINA BREVISPIRA					X	X										
57	0.0735	SPIROPLECTAMMINA SPECTABILIS •						●									X	
90	0.0757	ACARININA DENSA •						●										
88	0.0735	SIPHOGENEROIDES ELEGANTA						X										
49	0.0100	OSANGULARIA EXPANSA																
190	0.0580	ANOMALINOIDES ACUTA					X									X		
53	0.0232	UVIGERINA BATJESI						X										
44	0.2613	CIBICIDOIDES AFF WESTI																
45	0.0791	BULIMINA TRIGONALIS					X											
63	0.1353	NODOSARIA SP11						X										
43	0.0438	BULIMINA MIDWAYENSIS							X									
50	0.1167	SUBBOTINA PATAGONICA •							●									
46	0.0333	MEGASPORE SP1 •							●									X
96	0.0031	ACARININA INTERMEDIA WILCOXENSIS							●									
93	0.0291	ACARININA AFF. BROEDERMANNI						X	●									
230	0.0666	BULIMINA OVATA							X									
176	0.3711	ALLOMORPHINA SP1						X										
54	0.0049	TEXTULARIA PLUMMERAE								X								X
52	0.2143	ACARININA SOLDADOENSIS																
164	0.2189	NUTTALIDES TRUMPYI																
62	0.0820	GAVELINELLA DANICA																
47	0.0333	PLANOROTALITES PLANOCONICUS •							●									
159	0.4381	MOROZOVELLA ARAGONENSIS							●									
56	0.5972	GLOMOSPIRA CORONA							X	X								X
59	0.0342	RZEHAKINA EPIGONA								X								
55		GAVELINELLA BECCARIIFORMIS •																

Based on this geological and paleontological information, the data was partitioned in four separate files:

(a) 24 wells - all data;

(b) 21 wells - 3 out of 4 Hibernia wells omitted;

(c) 14 wells - Grand Banks only;

(d) 10 wells - Labrador Shelf only.

Table 7 contains the cumulative frequency distribution of the events in each of the four files. Over half of the taxa occur in three or less than three wells and are of limited value for practical correlation, unless lateral substitutes can be traced. In general RASC solutions are preferred that include taxa which occur in at least one-third of the wells, but such a threshold may be different for other geological basins.

Next, different RASC computer runs were performed for the purpose of finding:

(a) the most suitable zonations for the northern Grand Banks and Labrador treated separately;

(b) the best overall zonation;

(c) if the new zonation increased stratigraphic resolution when compared to the original (1982) version that used 16 wells and minimal noise cleaning.

Table 8 lists different input files used to perform the RASC runs. Variables include the number of wells, the thresholds k_c, m_{c1} and m_{c2}, INEG, the unique events and the marker horizons. For comparison I show the specification of the original 16 well RASC zonation,

*Figure 12 Comparison of $k_c \geq 3$, $m_c \geq 3$ scaled optimum sequence based on sixteen Grand Banks and Labrador wells with conventional zonations of Gradstein and Williams (1976) based on four of the wells, and of Gradstein, (unpublished), based on nine of the same wells. Both four and nine wells data produce an eight-fold biostratigraphic scheme. The scaled optimum sequence retains sixty-eight events (exits of taxa) grouped in nine distinct and progressively younger dendrogram clusters, resembling assemblage zones. A shading pattern was used to enhance stratigraphically useful part of individual clusters. The species that are marked with **x** or **o** and are common to the scaled optimum sequence and conventional zonations, show stratigraphically comparable sequential arrangements (after Gradstein and Agterberg, 1982).*

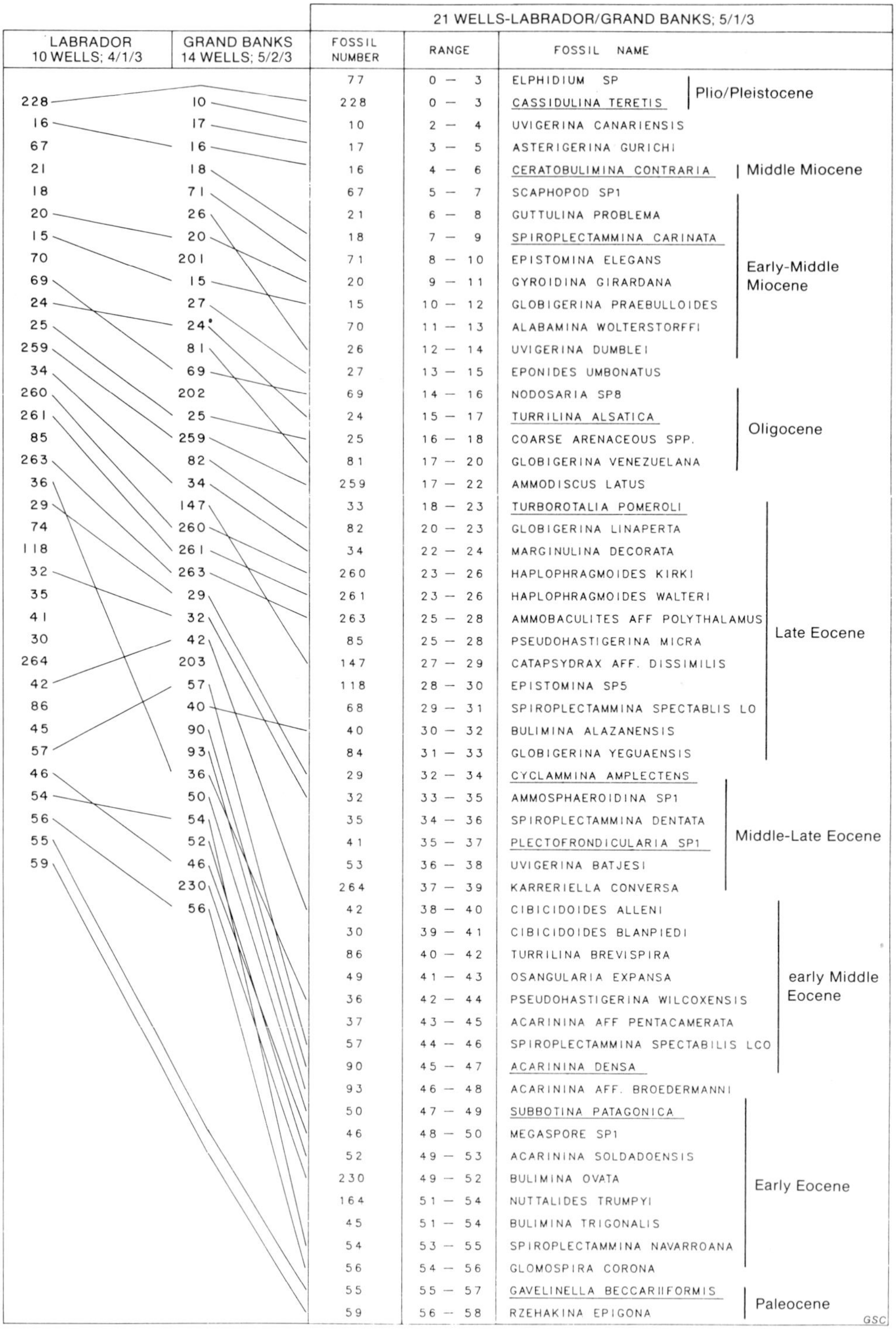

LABRADOR 10 WELLS; 4/1/3
228
16
67
21
18
20
15
70
69
24
25
259
34
260
261
85
263
36
29
74
118
32
35
41
30
264
42
86
45
57
46
54
56
55
59

GRAND BANKS 14 WELLS; 5/2/3
10
17
16
18
71
26
20
201
15
27
24*
81
69
202
25
259
82
34
147
260
261
263
29
32
42
203
57
40
90
93
36
50
54
52
46
230
56

21 WELLS-LABRADOR/GRAND BANKS; 5/1/3			
FOSSIL NUMBER	RANGE	FOSSIL NAME	
77	0 — 3	ELPHIDIUM SP	Plio/Pleistocene
228	0 — 3	CASSIDULINA TERETIS	Plio/Pleistocene
10	2 — 4	UVIGERINA CANARIENSIS	
17	3 — 5	ASTERIGERINA GURICHI	
16	4 — 6	CERATOBULIMINA CONTRARIA	Middle Miocene
67	5 — 7	SCAPHOPOD SP1	Early-Middle Miocene
21	6 — 8	GUTTULINA PROBLEMA	Early-Middle Miocene
18	7 — 9	SPIROPLECTAMMINA CARINATA	Early-Middle Miocene
71	8 — 10	EPISTOMINA ELEGANS	Early-Middle Miocene
20	9 — 11	GYROIDINA GIRARDANA	Early-Middle Miocene
15	10 — 12	GLOBIGERINA PRAEBULLOIDES	Early-Middle Miocene
70	11 — 13	ALABAMINA WOLTERSTORFFI	Early-Middle Miocene
26	12 — 14	UVIGERINA DUMBLEI	Early-Middle Miocene
27	13 — 15	EPONIDES UMBONATUS	
69	14 — 16	NODOSARIA SP8	Oligocene
24	15 — 17	TURRILINA ALSATICA	Oligocene
25	16 — 18	COARSE ARENACEOUS SPP.	Oligocene
81	17 — 20	GLOBIGERINA VENEZUELANA	Oligocene
259	17 — 22	AMMODISCUS LATUS	
33	18 — 23	TURBOROTALIA POMEROLI	Late Eocene
82	20 — 23	GLOBIGERINA LINAPERTA	Late Eocene
34	22 — 24	MARGINULINA DECORATA	Late Eocene
260	23 — 26	HAPLOPHRAGMOIDES KIRKI	Late Eocene
261	23 — 26	HAPLOPHRAGMOIDES WALTERI	Late Eocene
263	25 — 28	AMMOBACULITES AFF POLYTHALAMUS	Late Eocene
85	25 — 28	PSEUDOHASTIGERINA MICRA	Late Eocene
147	27 — 29	CATAPSYDRAX AFF. DISSIMILIS	Late Eocene
118	28 — 30	EPISTOMINA SP5	Late Eocene
68	29 — 31	SPIROPLECTAMMINA SPECTABLIS LO	Late Eocene
40	30 — 32	BULIMINA ALAZANENSIS	Late Eocene
84	31 — 33	GLOBIGERINA YEGUAENSIS	Late Eocene
29	32 — 34	CYCLAMMINA AMPLECTENS	Middle-Late Eocene
32	33 — 35	AMMOSPHAEROIDINA SP1	Middle-Late Eocene
35	34 — 36	SPIROPLECTAMMINA DENTATA	Middle-Late Eocene
41	35 — 37	PLECTOFRONDICULARIA SP1	Middle-Late Eocene
53	36 — 38	UVIGERINA BATJESI	Middle-Late Eocene
264	37 — 39	KARRERIELLA CONVERSA	Middle-Late Eocene
42	38 — 40	CIBICIDOIDES ALLENI	early Middle Eocene
30	39 — 41	CIBICIDOIDES BLANPIEDI	early Middle Eocene
86	40 — 42	TURRILINA BREVISPIRA	early Middle Eocene
49	41 — 43	OSANGULARIA EXPANSA	early Middle Eocene
36	42 — 44	PSEUDOHASTIGERINA WILCOXENSIS	early Middle Eocene
37	43 — 45	ACARININA AFF PENTACAMERATA	early Middle Eocene
57	44 — 46	SPIROPLECTAMMINA SPECTABILIS LCO	early Middle Eocene
90	45 — 47	ACARININA DENSA	early Middle Eocene
93	46 — 48	ACARININA AFF. BROEDERMANNI	
50	47 — 49	SUBBOTINA PATAGONICA	Early Eocene
46	48 — 50	MEGASPORE SP1	Early Eocene
52	49 — 53	ACARININA SOLDADOENSIS	Early Eocene
230	49 — 52	BULIMINA OVATA	Early Eocene
164	51 — 54	NUTTALIDES TRUMPYI	Early Eocene
45	51 — 54	BULIMINA TRIGONALIS	Early Eocene
54	53 — 55	SPIROPLECTAMMINA NAVARROANA	Early Eocene
56	54 — 56	GLOMOSPIRA CORONA	Early Eocene
55	55 — 57	GAVELINELLA BECCARIIFORMIS	Paleocene
59	56 — 58	RZEHAKINA EPIGONA	Paleocene

GSC

published in 1982, when the method had just been developed. INEG is an option that forces the scaled optimum sequence to maintain the same order of events as in the ranked only optimum sequence. In our data that order is rather stable, and the difference between the ranked and scaled optimum sequences would not be significant (Gradstein, 1984). Such is not true for the ranked and scaled optimum sequences using Mesozoic radiolarians (Baumgartner, 1984, and Chapter II.8). This author noted that extreme sampling problems, in part due to the dissolution of tests and variations in specimen frequency for most species, causes many taxa to have high cross-over frequencies with many other taxa. Also, coeval events are frequent. As a result there is considerable uncertainty in rank. Scaling led to considerable, arbitrary reshuffling in the ranked sequence. If scaling is deemed necessary, the INEG option preserves the order in the ranked sequence.

Under the assumption that the southern and northern part of the region each may require one zonation, Figure 13 was constructed. It shows that there is little difference in rank for the optimum sequences in 10 Labrador wells ($k_c \geq 4$, $m_{c1} \geq 1$) and in 14 Grand Banks wells ($k_c \geq 5$, $m_{c1} \geq 2$). The majority of events is shared. Statistical ranges for all average event positions are small and cycles during matrix re-ordering were few. The only event that occurs in a stratigraphically rather different position in the two optimum sequences is **Pseudohastigerina wilcoxensis.** In the Labrador wells it disappears in younger Eocene strata. The species may need further taxonomic investigation to distinguish it properly from the younger **P. micra.** Unfortunately, the latter is invariably represented by few specimens, as is P. wilcoxensis in the younger part of its range.

For comparison the rankings are also traced to the optimum sequence for the combined data file, using the file with 21 Grand Banks and Labrador wells. Lenient thresholds of $k_c \geq 5$, $m_{c1} \geq 2$ leave 57 ranked events. Reshuffling of events from the 14 to 21 wells files is subordinate, if one bears in mind that it largely involves slight adjustments within (bio)stratigraphical clusters as determined from scaling.

The exception is **Bulimina alazanensis,** which is ranked five positions higher in the Eocene part of the 21 wells optimum sequence. Tentatively, we conclude that one regional Cenozoic biozonation sufficiently explains the stratigraphic order of events for the entire region. Of course, the real order is an elusive phenomenon, as it can be modelled only from the available observations. Stability in rank is to some extent a function of the number of observations, but there is a practi-

Figure 13 Reversals in ranks for the three ranking solutions in 10, 14 and 21 wells. ***Bulimina alazanensis*** *(40) and* ***Pseudohastigerina wilcoxenis*** *(36) show considerable position re-ordering, but otherwise reshuffling is within, biostratigraphically meaningful, clusters as determined from scaling.*

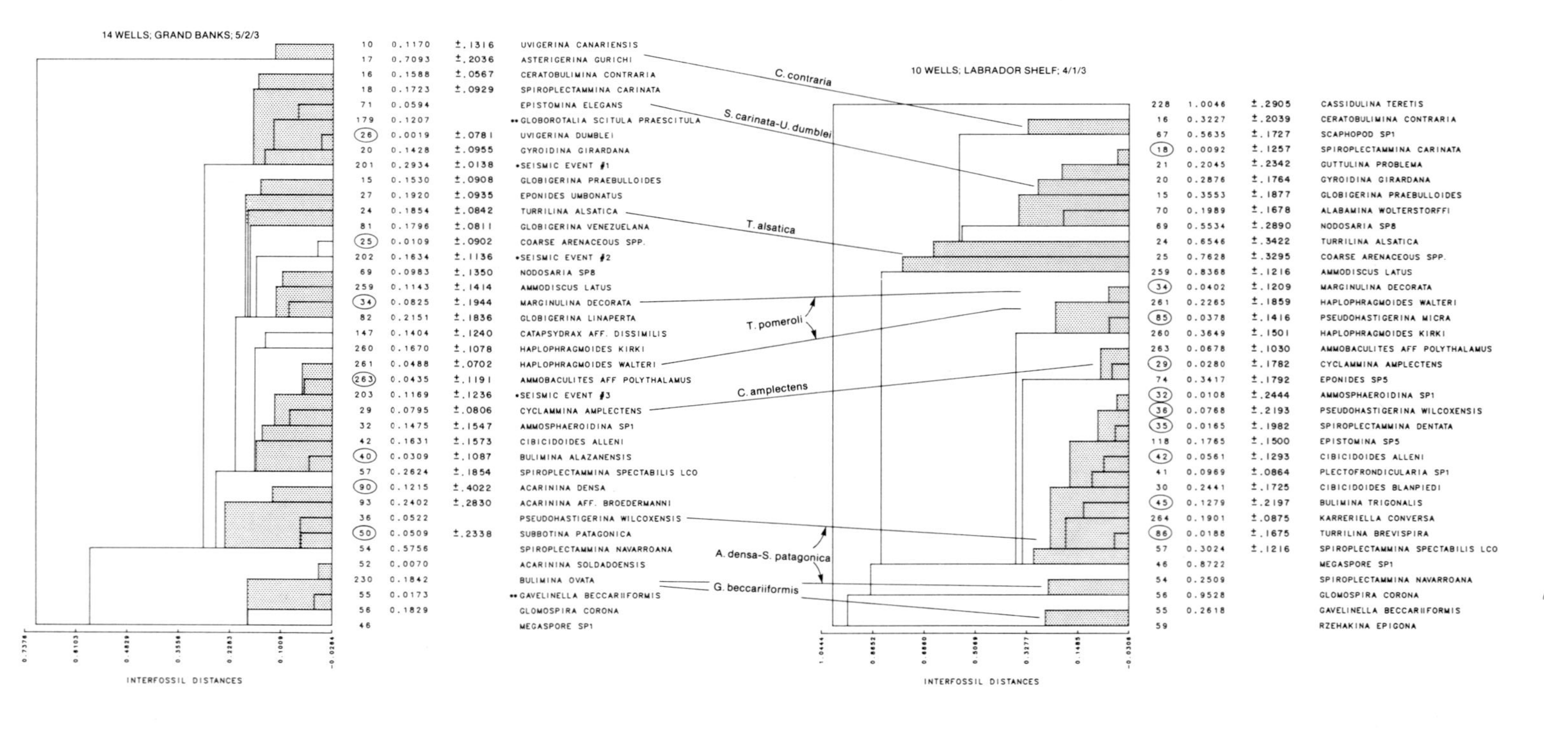

Figure 14 Scaled optimum sequences for the 10 (Labrador) and 14 (Grand Banks) well files. The regional zonations show differences discussed in the text. Circled events have interfossil distances with large standard deviations.

cal limit to the size of a data file. After all, stratigraphers try to develop models that with a minimum of data will yield a maximum of predictive potency.

If stratigraphic order is comparable in both regions, and the majority of common events is shared, scaling might still be rather different. Figure 14 shows the companion scaling solutions to the ranked optimum sequences discussed for the Grand Banks and Labrador. Both for the 10 Labrador wells and for the 14 Grand Banks wells, each pair of events must occur in at least 3 wells ($m_{c2} \geq 3$). The observed number of pairs, for each pair of events in the scaled optimum sequence varies from 3 to 24, with the higher (10-25) values in the Eocene - Oligocene part of the sequences. Ten of the events (all but one Eocene) in the Labrador file and 7 of the Eocene-Miocene events in the Grand Banks wells have standard deviation (SD) several times the average interfossil distance values. Despite this spread, the uncertainty in interfossil distances does not affect the make-up of the shaded part of the dendrograms. It is these shaded parts that either contain distinct index fossils or compare favourably to the new, single, standard zonation, proposed below for the region as a whole.

Table 7 Number of qualifying events, under different minimum occurrence (k_c) conditions, in each of the 4 Cenozoic foraminiferal files; a = 24 wells, b = 21 wells, c = 14 wells (Grand Banks only), d = 10 wells (Labrador only).

k_c →		1	2	3	4	5	6	7	8	9	10	11	12	13	14	15	16	17	18	19	20	21	22	23	24
NUMBER OF QUALIFYING EVENTS	a	169	115	87	70	61	51	44	36	33	31	27	25	21	15	12	9	9	8	7	4	3	3	0	0
	b	168	109	80	70	57	47	40	33	30	27	22	17	14	10	8	7	5	3	3	0	0			
	c	147	94	66	51	37	29	24	18	15	14	11	6	4	1										
	d	94	60	44	35	28	21	19	14	7	0														

GSC

Table 8 Different input files (runs) used to perform RASC.

RUNS	DATAFILE	WELLS	DICTIONARY TAXA	EVENTS	THRESHOLDS	EVENTS IN OPTIMUM SEQUENCE	INEG	UNIQUE EVENTS	MARKER HORIZONS
-	LAB	16	157	555	5/3	41			
-					3/3	68			
1*	LAB 11	24	169	841	5/2/3	61	0	61, 265, 270, 1	201, 202, 203
2	"	"	"	"	"	"	1	"	"
3	"	"	"	"	7/2/4	44	0		
4	"	"	"	"	"	"			202
5	LAB 10	21	168	745	5/1/3	57	0	265, 228, 1, 270, 61	
6	"	"	"	"	6/4	47			
7	"	"	"	"	7/2/4	40	0	265, 228, 1, 270, 61	
8*	"	"	"	"	"	"	1	1, 228, 265, 270, 4, 5, 61, 85, 93, 137, 179, 194 ,253, 269	
9	LAB 12	14(G.B.)	147	522	4/2/3	51	0		
10	"	"	"	"	"	"	1		201, 202, 203
11	"	"	"	"	5/2/3	37	0	55, 179	201, 202, 203
12*	"	"	"	"	"	"	1	"	"
13*	LAB 13	10 (LAB.)	94	319	4/1/3	35	1		
14	"	"	"	"	5/2/3	28	1		

GSC

Despite the fact that scaling tends not to produce good results for low numbers of pairs of events, as generally found in 10 or less wells, both scaled optimum sequences show stratigraphically and statistically reasonable cluster definition. An exception is the **Turrilina alsatica** Zone (Oligocene) in Labrador wells. This zone is better defined to the south and stands out in the standard zonation in 21 wells, discussed below. Other differences are:

(a) the Grand Banks zonation suggests a separate zone between the **Gavelinella beccariiformis** Zone (Paleocene) and the **Subbotina patagonica** Zone (Early Eocene);

(b) the **Subbotina patagonica** and **Acarinina densa** Zones (Early-Middle Eocene) are not well developed on the Labrador Shelf;

(c) the **Cyclammina amplectens** Zone (Middle-Late Eocene) is more distinct on the Labrador Shelf;

(d) the Grand Banks data suggests a separate zone with **Ammodiscus latus** between the **Turrilina alsatica** (Oligocene) and **Turborotalia pomeroli** (late Eocene) Zones;

(e) on the Labrador Shelf there is a clear break between Eocene and Oligocene strata as defined by the preliminary zones. This reflects the shallowing of the basin (Gradstein and Berggren, 1981), not so apparent in many Grand Banks wells.

Addition of 10 to 20 well sequences to each of the two data files will provide more distinct Grand Banks and Labrador zonations. The differences presumably will reflect less on the order and more on the relative stratigraphic distances between events, as far as shared. Each region also has its distinctive taxa.

The Cenozoic Foraminiferal Zonation

Despite the paleontological differences cited before, the combined data file on the wells for both Labrador and Grand Banks regions provides a practical and easily-interpretable biostratigraphic sequence. It also preserves aspects of both sub-zonations. In order to avoid much bias towards the Grand Banks, the 21-well file, rather than the 24-well file, was chosen as representative for the region. The $k_c \geq 7$, $m_{c1} \geq 2$, $m_{c2} \geq 4$ solution retains 40 events, out of a total of 168 (Figure 15). Only 3 cyles had to be broken to achieve ranking, whereas this number exceeded 35 for the ranking of the, non-filtered, 16 wells file. The stratigraphic normality, as expected is excellent, with 8 double asterisks and 15 single asterisk events in the normality test and only 10 events with more than 5 penalty points in the step model. The Snorri well has some of the highest scatter and the Adolphus well the least. This is graphically depicted in Figure 16, where the observed sequences of events in these two wells are plotted against the order in the 7-2-4 optimum sequence in the 21 wells. The scatter in Snorri at least in part is due to a high degree of stratigraphic condensation in the Eocene.

The scaled optimum sequence for the 21 Grand Banks and Labrador wells (Figure 15) contains ten, sharply defined, stratigraphically successive dendrogram clusters. These clusters can be compared to successive assemblage zones, internally also organised according to the 'arrow of time'. Improvements relative to the original 16 wells zonation of Figure 12 are as follows:

Firstly, the successive dendrogram clusters are more clearly separated at twice to three times higher interfossil distance values. The data filtering and expansion has tightened the successive clusters and at the same time expanded the cumulative distance scale from 8.9394 for 68 events in the 16 wells, to 8.0326 for 40 events in the 21 wells.

Secondly, the Eocene clusters in particular are defined more satisfactory and overall Cenozoic resolution is 9 or 10 rather than 8 or 9 zones, depending on the thresholds considered.

Thirdly, only 10 of the (Eocene-Oligocene) events have standard deviations exceeding the interfossil distances to their neighbours, whereas this number was far greater in the original scaled optimum sequence. Interesting is that despite the filtering and increased data, the RASC method is sufficiently robust that the basic zonation already stood out well in the original RASC solution for 16 wells (Figure 12). Figure 17 graphically illustrates the order and preferred clusters in the old and new data files. The main stratigraphic improvement involves distinction of the **Uvigerina dumblei** Zone, Miocene and the **Plectofrondicularia** sp. 1 Zone, Eocene. The re-ordering of average positions is within zones, and the drop in average position of 29, **Cyclammina amplectens,** is due to a redefinition of the morphotype, as explained earlier.

Shifts in average positions and events additions and deletions is highest in the Eocene **(S. patagonica - T. pomeroli** Zones). Eocene was the time of maximum transgression, generally middle to upper continental slope conditions and fine-grained sediments. Assemblages are rich in species and specimens with many taxa showing more differences between average and local ranges than in the pre- and post Eocene. Data filtering and taxonomic addition were concentrated on the Eocene.

For the purpose of better chronostratigraphic interpretation of the successive RASC zones, 11 rare planktonic taxa were designated unique events. Three benthic events, i.e. **Asterigerina guerichi** (peak; 265), **Cassidulina teretis** (228) and **Cibicidoides grossa** (270), that fell below the thresholds of k_c and m_c, were also included. These fourteen unique events are shown with double asterisks in the scaled optimum sequence. Their relative stratigraphic position is less well determined than that of the neighbouring taxa.

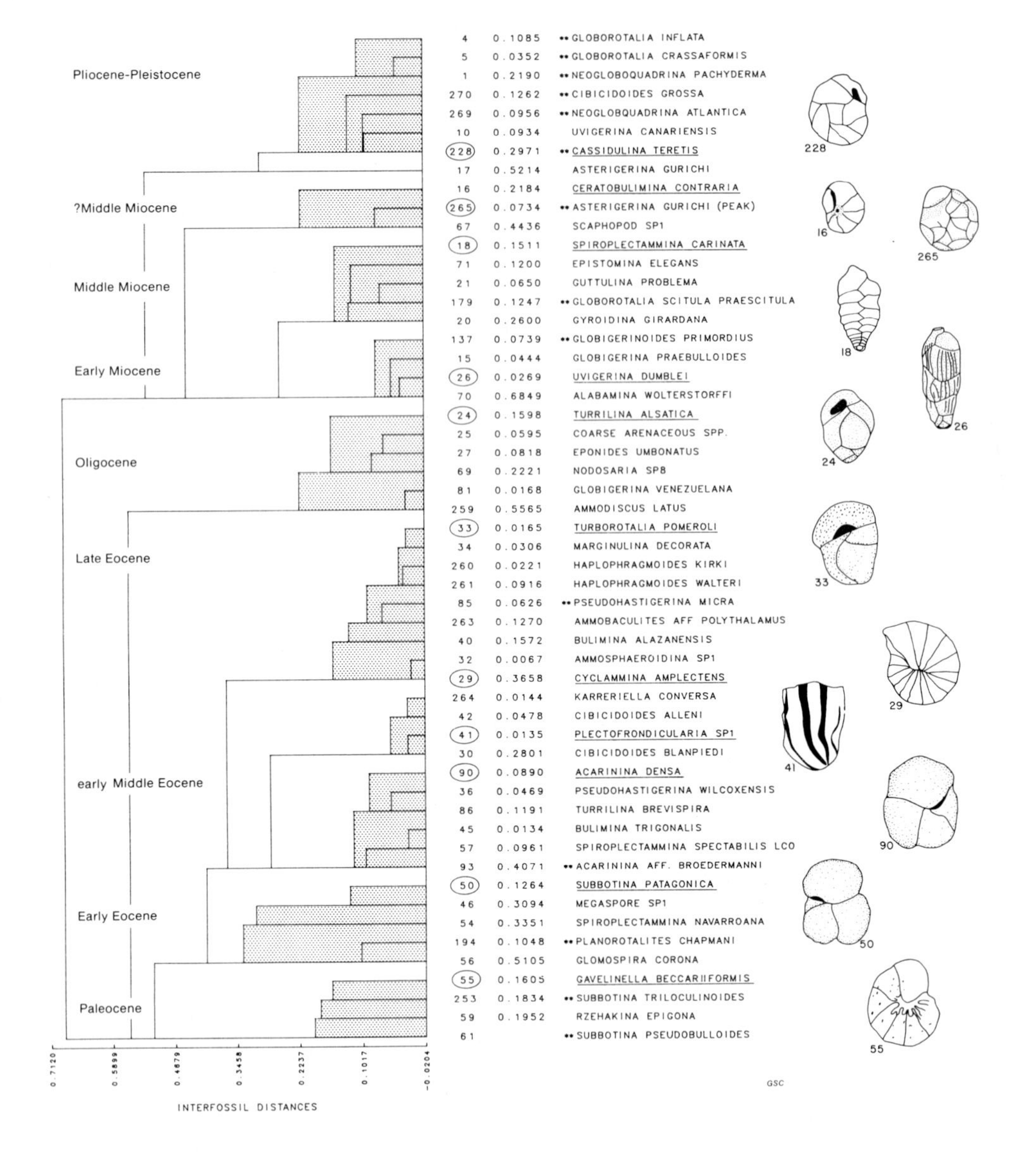

*Figure 15 Scaled optimum sequence for 21 wells, Labrador-Grand Banks, $k_c \geq 7$, $m_{c1} \geq 2$, $m_{c2} \geq 4$. Dendrogram values along x-axis are interfossil distances. The values along the y-axis are distances between an event and its successor. Scaling is stratigraphically downward, in line with the study of exploration wells. The tenfold zonation is representative for the regional Cenozoic stratigraphy. There are eleven unique events, shown with **. A shading pattern was used to enhance the stratigrahically most useful part of the dendrogram. The large distances on either side of the Eocene, Oligocene and Miocene assemblages are sedimentary cycle boundaries.*

The ten successive zones, with the characteristic species listed in ascending stratigraphic order and the chronostratigraphic interpretation are:

1. **Gavelinella beccariiformis** Zone - Paleocene.

   Subbotina pseudobulloides, Rzehakina epigona, Subbotina triloculinoides and **Gavelinella beccariiformis.**
   This zone is rarely encountered in Grand Banks wells.

2. **Subbotina patagonica** Zone - Early Eocene.

   Glomospira corona, Planorotalites chapmani, Spiroplectammina navarroana, Megaspore sp 1 and **Subbotina patagonica. Nuttalides trumpyi, Bulimina ovata** and **Acarinina soldadoensis** occur in 5 of the wells.

3. **Acarinina densa** Zone - early Middle Eocene.

   Spiroplectammina spectabilis (last common occurrence), **Bulimina trigonalis, Turrilina brevispira, Pseudohastigerina wilcoxensis** and **Acarinina densa. Acarinina pentacamerata** occurs in 5 wells only.

4. **Plectofrondicularia** sp 1 Zone - Middle-Late Eocene.

   Cibicidoides blanpiedi, Plectofrondicularia sp 1, **Cibicidoides alleni** and **Karreriella conversa. Uvigerina batjesi** on average occurs within or just above this zone.

5. **Turborotalia pomeroli - Cyclammina amplectens** Zone - Late Eocene.

   Cyclammina amplectens, Ammosphaeroidina sp 1, **Bulimina alazanensis, Ammobaculites** aff. **polythalamus, Pseudohastigerina micra, Haplopragmoides walteri, H. kirki, Marginulina decorata, Turborotalia pomeroli. Globigerina linaperta** and **Catapsydrax** aff. **dissimilis** occur in at least 5 Grand Banks wells. **Epistomina** sp 5 is a Labrador Shelf marker.

6. **Turrilina alsatica** Zone - Oligocene.

   Ammodiscus latus, Globigerina venezuelana, Nodosaria sp 8, **Eponides umbonatus,** coarse arenaceous foraminifers, **Turrilina alsatica. Cyclogyra** sp 3 (probably **C. decorata)** occurs in some Hibernia wells.

   The disappearance of coarsely built arenaceous foraminifers is a distinct event in the North Sea and Labrador Sea, (Gradstein and Berggren, 1981). It is not synchronous, but on the Labrador Shelf is a good indicator for the shallowing upward following the Eocene transgression.

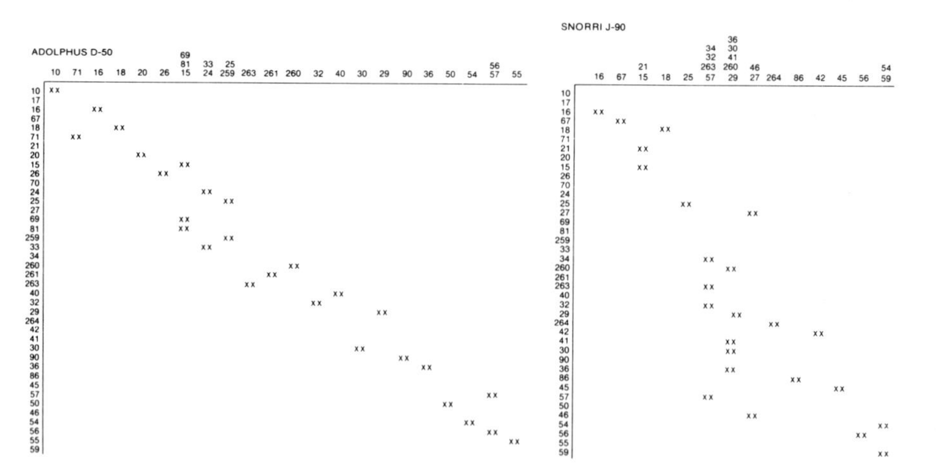

Figure 16 *Scatter diagrams of the observed sequences of events in the Adolphus and Snorri wells versus the optimum sequence of events in 21 wells.*

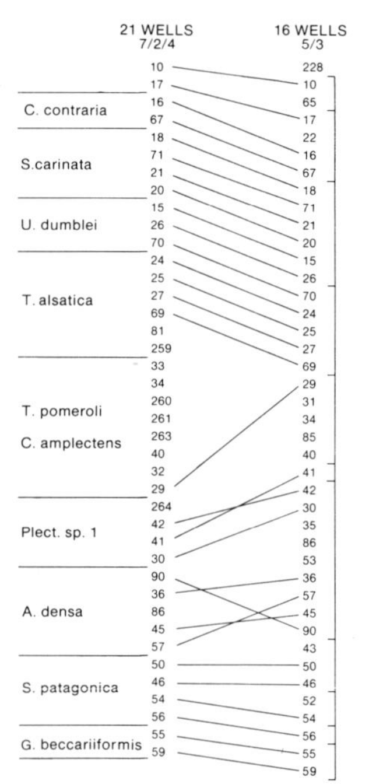

Figure 17 *Comparison of the scaled optimum sequences in 16 wells (Gradstein and Agterberg, 1982) with the new scaled optimum sequence in 21 wells of Figure 15. The preferred dendrogram clusters (zones) are shown as rectangular blocks. Species 22 and 31 in the 16 well solution have since been deleted. Mismatch is minimal and within zones (see text).*

7. **Uvigerina dumblei** Zone - Early Miocene.

   Alabamina wolterstorffi, Uvigerina dumblei, Globigerina praebulloides, Globigerinoides primordius. Best recognized on the Grand Banks.

8. **Spiroplectammina carinata** Zone - Middle Miocene.

   Gyroidina girardana, Globorotalia praescitula, Guttulina problema, Epistomina elegans, Spiroplectammina carinata.

9. **Ceratobulimina contraria** Zone - ? Middle Miocene.

   Scaphopod sp 1, **Asterigerina guerichi** (peak), **Ceratobulimina contraria.** This is a Labrador Shelf zone.

10. **Cassidulina teretis** Zone - Pliocene/Pleistocene.

   Cassidulina teretis, Uvigerina canariensis, Neogloboquadrina atlantica, N. pachyderma, Cibicidoides grossa, Globorotalia crassaformis, G. inflata. Sampled on the Labrador Shelf, where the planktonic species of **Neogloboquadrina** and **Globorotalia** are rare; **Elphidium** spp is present in 6 wells.

Details regarding the chronostratigraphy and an illustrated taxonomy are in Gradstein and Agterberg (1982).

LATE CRETACEOUS NANNOFOSSIL ZONATION

Introduction

A regional study of Late Cretaceous nannofossils on the Canadian Atlantic margin has led to a sixteenfold zonation and a detailed subsurface chronostratigraphy (Doeven et al., 1982; Doeven, 1983). The author used the RASC Ranking and Scaling program to aid in the definition of this detailed zonation.

The Data

The data file comprises the observed sequence of Late Cretaceous nannofossil events in side-wall cores in ten exploratory wells: Triumph P-50, Onondaga E-84, Primrose A-41, Missisauga H-54, MicMac H-86, Wyandot E-53, Heron H-73, Osprey H-84, Adolphus D-50 and Indian Harbour M-52, on the Scotian Shelf (6 wells), Grand Banks (3 wells) and Labrador Shelf. The original sequence data are in Appendix 1.

In all, 119 nannofossil events were selected, representing 64 taxa.Each event was given a unique code number, which was recorded in a dictionary or index. The computer input consisted of the observed

WELL I

SAMPLES	SPECIES A			SPECIES B		
	OCCURRENCE	TYPE OF EVENT	CODE NUMBER	OCCURRENCE	TYPE OF EVENT	CODE NUMBER
1						
2	•	HO	98			
3				X	HO	99
4				X		
5	•	ST	298	X		
6	•			XX	ST	299
7	•			XX		
8	•	SB	398	XX		
9				XX		
10				XX	LO,	199
11	•	LO	198			
12						

WELL II

SAMPLES	SPECIES A			SPECIES B		
	OCCURRENCE	TYPE OF EVENT	CODE	OCCURRENCE	TYPE OF EVENT	CODE
13				X	HO	99
14				X		
15	•	HO, ST	98-298	X		
16	•			XX	ST	299
17	•			XX		
18	•			XX		
19	•			XX		
20	•	SB, LO	398-198	XX	LO	199
21						
22						

HO = HIGHEST OCCURRENCE
ST = SUBTOP
SB = SUBBOTTOM
LO = LOWEST OCCURRENCE
X = RARE
XX = COMMON

Figure 18 Example of the subtop and subbottom concept to differentiate and weigh stratigraphic ranges of nannofossils. For an explanation see text.

Table 9 The number of late Cretaceous nannofossil events that occur in at least 1, 2, 3, ... or all ten wells, satisfying conditions of minimum occurrence (k_c).

NUMBER OF QUALIFYING EVENTS	119	117	115	110	106	91	77	68	52	1
$k_c \geqslant$	1	2	3	4	5	6	7	8	9	10

sequences of (coded) events, accompanied by the index (Appendix I). Simultaneous events were connected by hyphens, which means the computer reads those as occurring in the same sample.

The events consisted of the lowest occurrences, highest occurrences, 'subbottoms' (lowest regular or common occurrences), and 'subtops' (highest regular or common occurrences) of nannofossil species. In a subjective approach, stratigraphically isolated lowest or highest occurrences might be considered due respectively to contamination or reworking and consequently ignored. This data file is not subject to such bias.

The principle of the subtop-subbottom concept is illustrated in Figure 18. This figure shows the stratigraphical distribution of species A and B in wells I and II. In well I species A was consistently found in samples 5 through 8 and occurred isolated in samples 2 and 11. The highest occurrence of species A (code number 98) was in sample 2, the subtop (code number 298) in sample 5, the subbottom (code number 398) in sample 8, and the lowest occurrence (code number 198) in sample 11. In well II species A ranged continuously through samples 15 to 20. The highest occurrence and the subtop coincided as did the lowest occurrence and the subbottom. The top of species A was coded 98-298, the base 398-198. In an extreme case,where species A occurs in only one sample in a well, all four types of events would coincide and the occurrence would be entered as 98-298-398-198.

The distribution of species B illustrates the other application of the subbottom-subtop concept. In both wells species B shows a significant change in abundance in the top part of its range. The highest common occurrence of species B (sample 6 in well I and sample 16 in well II) was recorded as a separate event (subtop). Of the 119 entered events 23 are subtops and subbottoms.

Most nannofossil taxa occur in all the wells, leading to few missing data. The coherence of the data is illustrated in Table 9, which shows the number of events that occur in at least 1, 2, 3, ... or all 10 wells. Even under the severe condition that an event must occur in 8 of the 10 wells, still more than half of the entered events qualified.

Ranking Results

After several runs of the RASC program under varying k_c and m_c conditions the optimum sequence obtained under $k_c \geq 3$, $m_c \geq 3$ conditions was chosen as a basis for further evaluation. Out of the 119 entered events 115 qualified, including virtually all marker events used in the previously established conventional (subjective) zonation. Two analyses have been conducted, each with a slightly different data file. In the first one the basic set of data from the 10 wells was entered. In the second one, wells with the best sample coverage and a relatively good score in the normality test, namely Adolphus D-50, Primrose A-41, and Triumph P-50 were entered twice, increasing the number of wells from 10 to 13. Thus weighting of good records was introduced. A range chart was constructed from the obtained optimum sequences of events.

Figure 19 shows the obtained optimum sequences and stratigraphic ranges; the results of the first analysis (10 wells) are in thick lines, those of the second analysis (13 wells) are in thin lines. The average range of a species is the interval between its lowest and its highest occurrence position as determined in the optimum sequence. The intervals between the lowest occurrence and the subbottom and between the highest occurrence and the subtop of a species are represented by dotted lines. For species with a known entry level below the range of the samples only highest occurrences were used; for most of the species with a known extinction level at the end of the Cretaceous only lowest occurrences were entered. To produce subjectively truncated ranges of these species their lowest or highest occurrence positions in the optimum sequence have been connected with the bottom or top of the chart.

The two highest-placed events in the optimum sequence, the highest occurrence of **Arkhangelskiella cymbiformis** (code number 3) and the highest occurrence of **Nephrolithus frequens** (code number 58), occur simultaneously. Their ordering in the optimum sequence merely reflects the order in which their code numbers were recorded.

The positions in the optimum sequence of the events used in the subjective zonation are at the horizontal lines dividing the range chart of Figure 19 in successive zonal units. This zonation and the chronostratigraphic interpretations are shown in the left-hand columns of the chart.

The optimum sequence (in both analyses) maintains the order of the marker events as established in the subjective approach. The only anomaly is the highest occurrence of **Podorabdus albianus** (code number 63). In the subjective zonation the latter, as a subzone marker, was placed above the lowest occurrence of **Quadrum gartneri.** In the optimum sequence the order of these two events is reversed, reflecting the situation in the majority of the studied wells. Between the zone markers in the optimum sequence there are several other events, which are potentially useful for further biostratigraphic refinement. A number of these are incorporated in the biostratigraphic framework, discussed in Doeven (1983). The choice of marker events has to be considered thoughtfully, because the positions of closely spaced events may not always be well defined. The dendrogram displaying the clustering of events along a relative time axis, one of the options of the RASC program, may serve as a guide in selecting marker events.

The optimum sequence of the second analysis, in which data from the Adolphus D-50, Primrose A-41, and Triumph P-50 wells were entered twice, compares favourably to the optimum sequence of the first analysis. There is some minor reshuffling of positions, mainly in the Cenomanian and Lower Maastrichtian intervals. As to the zone markers, the lowest occurrence of **Ceratolithoides aculeus** has dropped some places in the sequence and the lowest occurrence of **Quadrum gothicum** is situated some positions higher. An improvement is that the highest occurrence of **Watznaueria britannica,** used to delineate the Albian-Cenomanian boundary, now satifies the imposed $k_c \geq 3$, $m_c \geq 3$ conditions.

One of the most detailed zonations of the Upper Cretaceous is by Sissingh (1977). In the conventional approach a number of Sissingh's index species did not seem to be useful. Some of these species could not be consistently identified (owing to poor preservation or to morphological transition to other species), others showed irregular occurrences that did not conform to this author's scheme. Surprisingly, Sissingh's zonation is almost exactly duplicated in the optimum sequence, if the subtop-subbottom concept is utilized. The right-hand column of Figure 19 shows Sissingh's Upper Cretaceous marker events, which are numbered stratigraphically downward from 1 to 19. With the exception of number 14, these events are positioned in the same order in the optimum sequence. The anomalous position of event 14, the lowest occurrence of **Micula** ex gr. **staurophora** (in which Sissingh included **M. decussata** and **M. concava)** may be due to a different concept of this species but also the frequency of occurrence may be a factor. On the Canadian Atlantic margin **M. decussata** usually occurs in low numbers in the Turonian and becomes more frequent in the Coniacian (a distinct subbottom, however, could not be recognized). Sissingh put its lowest occurrence in the Upper Coniacian. The only other deviations are two of Sissingh's subzone markers, the highest occurrence of **Broinsonia parca** (b) and the lowest occurrence of **Reinhardtites levis** (c). In Sissingh's scheme, event b is situated between events 4 and 5 (in the optimum sequence between 3 and 4) and event c between 5 and 6 (in the optimum sequence between 6 and 7). The base of **Reinhardtites levis** (c) is difficult to pinpoint, because it appears to have evolved gradually from **Reinhardtites anthophorus.**

Analyzing the ranges of some species with regard to the subtop-subbottom concept, one faces a number of different situations. In some cases the subbottom and lowest occurrence events have adjacent positions in the optimum sequence (e.g. **Marthasterites furcatus,** code number 211, 148 and **Quadrum gartneri,** code number 219, 168). Here the (rare) 'anomalous' lowest occurrences in some wells (presumably due to drilling mud contamination) do not affect the relative position of the event and the use of subbottoms appears not to have been meaningful.

In other cases, however, there is a substantial difference in position between the subtops-subbottoms and the corresponding highest-lowest occurrences. Irregular and relatively rare occurrences below or above the regular or common range of a species may be attributed to drilling mud contamination or reworking, but other factors can also be involved. For example, **Eiffellithus eximius** has been found only sporadically above the Middle Campanian; its (average) highest occurrence level is near the Middle Maastrichtian. In literature the extinction datum of **Eiffellithus eximius** is often placed at the end of the Campanian. Perhaps the observed decrease in abundance of **Eiffellithus eximius**

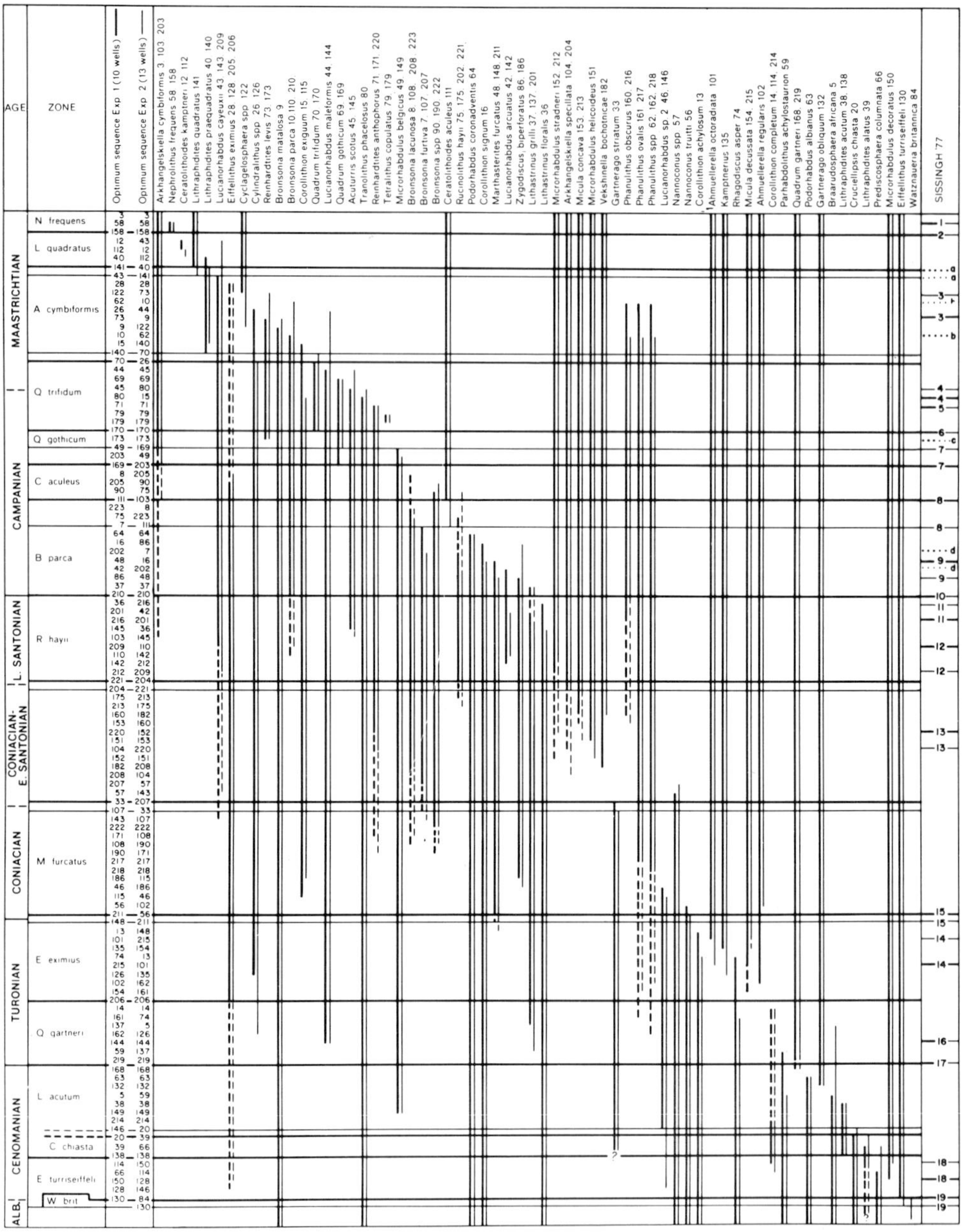

AGE
ZONE
Optimum sequence Exp 1 (10 wells)
Optimum sequence Exp 2 (13 wells)

Arkhangelskiella cymbiformis 3, 103, 203
Nephrolithus frequens 58, 158
Ceratolithoides kamptneri 12, 112
Lithraphidites quadratus 141
Lithraphidites praequadratus 40, 140
Lucianorhabdus cayeuxii 43, 143, 209
Eiffellithus eximius 28, 128, 205, 206
Cyclagelosphaera spp 122
Cylindralithus spp 26, 126
Reinhardtites levis 73, 173
Broinsonia matalosa 9
Broinsonia parca 10, 110, 210
Corollithion exiguum 15, 115
Quadrum trifidum 70, 170
Lucianorhabdus maleformis 44, 144
Quadrum gothicum 69, 169
Acuturris scotus 45, 145
Tranolithus phacelosus 80
Reinhardtites anthophorus 71, 171, 220
Tetralithus copulatus 79, 179
Microrhabdulus belgicus 49, 149
Broinsonia lacunosa 8, 108, 208, 223
Broinsonia furtiva 7, 107, 207
Broinsonia spp 90, 190, 222
Ceratolithoides aculeus 111
Rucinolithus hayii 75, 175, 202, 221
Podorhabdus coronadventis 64
Corollithion signum 16
Marthasterites furcatus 48, 148, 211
Lucianorhabdus arcuatus 42, 142
Zygodiscus, biperforatus 86, 186
Lithastrinus grilli 37, 137, 201
Lithastrinus floralis 36
Microrhabdulus stradneri 152, 212
Arkhangelskiella specillata 104, 204
Micula concava 153, 213
Microrhabdulus helicoideus 151
Vekshinella bochotnicae 182
Gartnerago striatum 33
Phanulithus obscurus 160, 216
Phanulithus ovalis 161, 217
Phanulithus spp 62, 162, 218
Lucianorhabdus sp. 2 46, 146
Nannoconus spp. 57
Nannoconus truitti 56
Corollithion achylosum 13
Ahmuellerella octoradiata 101
Kamptnerius 135
Rhagodiscus asper 74
Micula decussata 154, 215
Ahmuellerella regularis 102
Corollithion completum 14, 114, 214
Parhabdolithus achylostaurion 59
Quadrum gartneri 168, 219
Podorhabdus albianus 63
Gartnerago obliquum 132
Braarudosphaera africana 5
Lithraphidites acutum 38, 138
Cruciellipsis chiasta 20
Lithraphidites alatus 39
Prediscosphaera columnata 66
Microrhabdulus decoratus 150
Eiffellithus turriseiffeli 130
Watznaueria britannica 84
SISSINGH '77

AGE	ZONE	Exp 1	Exp 2
MAASTRICHTIAN	N frequens	3	3
		58	58
		158	158
	L quadratus	12	43
		112	12
		40	112
		141	40
		43	141
	A cymbiformis	28	28
		122	73
		62	10
		26	44
		73	9
		9	122
		10	62
		15	140
		140	70
		70	26
	Q trifidum	44	45
		69	69
		45	80
		80	15
		71	71
		79	79
		179	179
		170	170
CAMPANIAN	Q gothicum	173	173
		49	169
		203	49
		169	203
	C aculeus	8	205
		205	90
		90	75
		111	103
		223	8
		75	223
		7	111
	B parca	64	64
		16	86
		202	7
		48	16
		42	202
		86	48
		37	37
		210	210
L. SANTONIAN	R hayii	36	216
		201	42
		216	201
		145	36
		103	145
		209	110
		110	142
		142	212
		212	209
		221	204
CONIACIAN-E. SANTONIAN		204	221
		175	213
		213	175
		160	182
		153	160
		220	152
		151	153
		104	220
		152	151
		182	208
		208	104
		207	57
		57	143
		33	207
CONIACIAN	M furcatus	107	33
		143	107
		222	222
		171	108
		108	190
		190	171
		217	217
		218	218
		186	115
		46	186
		115	46
		56	102
		211	56
TURONIAN	E eximius	148	211
		13	148
		101	215
		135	154
		74	13
		215	101
		126	135
		102	162
		154	161
		206	206
	Q gartneri	14	14
		161	74
		137	5
		162	126
		144	144
		59	137
		219	219
CENOMANIAN	L acutum	168	168
		63	63
		132	132
		5	59
		38	38
		149	149
		214	214
		146	20
	C chiasta	20	39
		39	66
		138	138
	E turriseiffeli	114	150
		66	114
		150	128
		128	146
	W brit	130	84
ALB.			130

1
2
a
a
3
3
b
4
4
5
6
c
7
7
8
8
d
9
d
9
10
11
11
12
12
13
13
?
15
15
14
14
16
17
18
18
19
19

Figure 19 Quantitative Late Cretaceous nannofossil range chart using the $k_c \geq 3$, $m_c \geq 3$ ranking solution for the distribution of 64 taxa in ten wells, North Atlantic margin. The range chart was constructed by connecting average first and last occurrences and average consistent first (subbottom) and consistent last (subtop) occurrences in the optimum sequence (after Doeven, 1983).

in the Upper Campanian resulted from a slow extinction process, accelerated by climatic deterioration in Late Campanian-Maastrichtian time. In any case, the subtop of **Eiffellithus eximius** (code number 205) seems to be a consistent and useful marker on the Canadian Atlantic margin and is more reliable than the highest occurrence event (code number 28).

The position of the lowest occurrence of **Eiffellithus eximius** considerably below the subbottom presumably is due in part to some contaminated samples, but on the other hand its supposed evolution from **Eiffellithus turriseiffeli** (Verbeek, 1977) may be a factor. Transitional types do occur and an overall '**Eiffellithus turriseiffeli** population' may contain rare early froms of the '**Eiffellithus eximius** type'. The lowest regular occurrence (subbottom) of **Eiffellithus eximius** usually can be readily located, and its stratigraphic position agrees with the position of the entry of **Eiffellithus eximius** as reported by Manivit et al. (1977) and Verbeek (1977).

Corollithion completum also has a substantial interval between its highest occurrence (code number 14) and its subtop (code number 214). The position of its highest occurrence in the optimal sequence (high up in the Turonian) is deceptive, because it is recorded only in the Upper Cenomanian-lowest Turonian interval (coinciding with the subtop), or, in a few wells, as isolated occurrences in the Upper Campanian and Maastrichtian. The resulting average top just below the subbottom of **Eiffellithus eximius** is meaningless, while the (average) subtop of this distinctive species in the Upper Cenomanian has some biostratigraphic value. **Corollithion completum** has seldom been reported in literature. It has been regularly recorded in the Cenomanian-lowest Turonian interval of the El Kef section in Tunisia and isolated occurrences are known in the Upper Campanian in southern Spain (Verbeek, 1977) and Maastrichtian on Madagascar (Perch-Nielsen, 1973; original description). In view of this stratigraphic distribution, reworking may not be satisfactory to explain the disjunct stratigraphic range of **Corollithion completum** on the Canadian Atlantic margin.

Lucianorhabdus cayeuxii, Phanulithus obscurus, Phanulithus ovalis and nannoconids are usually considered indicative of nearshore environments and their occurrences are influenced by environmental factors. Nevertheless, the tops and bottoms of their regular occurrences (the subtops and subbottoms) can be regionally useful markers.

The application of the RASC quantitative stratigraphic method to the Cretaceous nannofossil record of the Canadian Atlantic margin results in a zonation that closely matches the most detailed subjective zonation. It provides readily available guidelines for further biostratigraphic refinement. The subtop-subbottom concept, where applicable, gives a better insight into the stratigraphic distribution of a species than the mere use of lowest and highest occurrences. In many cases subtop and subbottom events are more reliable markers than the highest or lowest occurrences of species.

Scaling Results

Scaling of the ranking solution in the 10 wells shows 10-13 successively younger clusters (assemblages). A preliminary analysis suggests an approximate correlation to the zones proposed in Figure 19 based on range end points. The range chart shows that there is a regular staggering of range end points, without dramatic breaks or an "explosion" of new taxa. This fact, together with the observation that nannofossil events cross-over little from well to well and the relatively low number of wells (10), may explain that scaling does not produce such clear cut dendrogram clusters, as for example in the Cenozoic foraminifers file in 21 wells. The detailed range chart and zonation derived from ranking of (2 types of) first and last occurrences makes scaling less necessary.

CONCLUSIONS

The RASC-ranking and scaling computer program provides the paleontologist with an objective and stratigraphically effective method to obtain a practical and high resolution biozonation in sedimentary basins. Ranking averages event positions, like first or last occurrences of taxa, and is not sensitive to outliers in the data. It produces optimum sequences that are used for detailed (point) correlation and average range charts.

A minimum of 10 wells or outcrop sections is desirable, particularly when scaling of the events is required, to obtain a measure of distance between the events along a relative time-axis. Scaling of the optimum sequence is a pair-wise comparison technique that works when positions of events show some cross-over from well to well. The cross-over frequencies are inversely proportional to the distances calculated. The method produces assemblage zones, organised with the arrow of time. Large distances between time-successive assemblages coincide with sedimentary cycle boundaries.

Normality testing compares the sequence of events in one section with the (scaled) optimum one. Normality testing allows objective data filtering, like removal of misplaced events, and helps to assess geographic trends.

The RASC method effectively reduces large and complex paleontological files to practical zonations. A detailed analysis of the stratigraphic and geographic distribution of 168 taxa of Cenozoic foraminifers in 24 Grand Banks and Labrador Shelf wells leads to a ten-fold RASC zonation ($k_c \geq 7$; $m_{c1} \geq 2$; $m_{c2} \geq 4$) with 40 common taxa and 14 unique events. The zones span the Paleocene through Pliocene/Pleistocene. Each shelf area has its distinctive taxa for local correlation.

Ranking ($k_c \geq 3$) of the first occurrences, first common occurrences, last common occurrences and last occurrences of 64 taxa of Late Cretaceous nannofossils in 10 northwestern Atlantic margin wells (Doeven, 1983) produces an average range chart that matches a standard Tethyan zonation.

REFERENCES

Agterberg, F.P. and Nel, L.D., 1982a, Algorithms for the ranking of stratigraphic events. Computers and Geosciences, 8 (1), 69-90.

Agterberg, F.P. and Nel, L.D., 1982b, Algorithms for the scaling of stratigraphic events. Computers and Geosciences, 8 (1), 163-189.

Baumgartner, P.O., 1984, Comparison of unitary associations and probabilistic ranking and scaling as applied to Mesozoic radiolarians. Computers and Geosciences, 10 (1), 162-183.,

Blank, R.C., 1979, Applications of probabilistic biostratigraphy to chronostratigraphy. J. Geol., 87, 647-670.

Brower, J.C., Millendorf, S.A. and Dyman, T.S., 1978, Methods for the quantification of assemblage zones based on multivariate analysis of weighted and unweighted data. Computers and Geosciences, 4 (3),221-227.

Doeven, P.H., 1983, Cretaceous nannofossil statigraphy and paleoecology of the Canadian Atlantic margin. Geol. Surv. Canada, Bull. 356, 69 pp.

Doeven, P.H., Gradstein, F.M., Jackson, A., Agterberg, F.P. and Nel, L.D., 1982, A quantitative nannofossil range chart. Micropaleontology, 28 (1), 85-92.

Edwards, L.E., 1984, Insights on why graphic correlation (Shaw's method) works. J. Geol., 92, 583-597.

Edwards, L.E. and Beaver, R., 1978, The use of a paired comparision model in ordering stratigraphic events. Math. Geol., 10 (3), 261-272.

Gradstein, F.M., 1984, On stratigraphic normality. Proc. 5th Int. Symp., IGCP, Project 148, Geneva 1982, Computers and Geosciences, 10 (1), 43-57.

Gradstein, F.M. and Agterberg, F.P., 1982, Models of Cenozoic foraminiferal stratigraphy - northwestern Atlantic margin. In Quantitative Stratigraphic Correlation, Editors J.M. Cubitt and R.A. Reyment, John Wiley and Sons Ltd., 119-173.

Gradstein, F.M. and Berggren, W.A., 1981, Flysch-type agglutinated foraminifera and the Maestrichtian to Paleogene history of the Labrador and North Seas. Marine Micropal. 6, 211-268.

Gradstein, F.M. and Srivastava, S.P., 1980, Aspects of Cenozoic stratigraphy and paleoceanography of the Labrador Sea and Baffin Bay. Palaeogeogr. Palaeocl., Palaeoecol., 30, 261-295.

Gradstein, F.M. and Williams, G.L., 1976, Biostratigraphy of the Labrador Shelf. Geol. Surv. Canada Rept. 349, 40 pp.

Harper, C.W., 1981, Inferring succession of fossils in time: the need for a quantitative and stratistical approach. J. Paleont., 55 (2), 42-452.

Hay, W.W., 1972, Probabilistic stratigraphy. Eclog. Geol. Helv., 65 (2), 255-266.

Heller, M., Gradstein, W.S., Gradstein, F.M. and Agterberg, F.P., 1983, RASC - Fortran IV Computer Program for ranking and scaling of biostratigraphic events. Geol. Surv. Canada, Rept. 922, 54 pp.

Hudson, C.B. and Agterberg, F.P., 1982, Paired comparison models in biostratigraphy. Math. Geol., 14 (2), 141-159.

Manivit, H., Perch-Nielsen, K., Prins, B. and Verbeek, J.W., 1977, Mid Cretaceous calcareous nannofossil biostratigraphy. Proc. Kon. Nederl. Akad. Wetensch., Series B, 3 (80), 169-181.

Perch-Nielsen, K., 1973, Neue Coccolithen aus dem Maastrichtien von Dänemark, Madagaskar und Agypten. Bull. Geol. Soc. Denmark, 22, 306-333.

Shaw, A., 1964, Time in stratigraphy. McGraw-Hill Book Co., New York, 365 pp.

Sissingh, W., 1977, Biostratigraphy of Cretaceous calcareous nannoplankton. Geologie en Mijnbouw, 56 (1), 37-65.

Stainforth, R.M., Lamb, J.L., Luterbacher, H., Beard, J.H. and Jeffords, R.M., 1975, Cenozoic planktonic foraminiferal zonation and characteristics of index forms. Univ. Kansas Paleont. Contr. 62, 1-162 + Appendix.

Verbeek, J.W., 1977, Calcareous nannoplankton biostratigraphy of Middle and Upper Cretaceous deposits in Tunisia, southern Spain and France. Utrecht Micropal. Bull. No. 16, 1-157.

Worsley, T.R. and Jorgens, M.L., 1977, Automated biostratigraphy. In, Oceanic Micropalaeontology, Editor A.T.S. Ramsay, Academic Press, 2, 1201-1229.

II.5

METHODS OF RANKING BIOSTRATIGRAPHIC EVENTS

FREDERIK P. AGTERBERG

INTRODUCTION

Ranking

According to Kendall (1975, p. 1), a number of individuals are ranked when they are arranged in order according to some quality which they all possess to a varying degree. The arrangement as a whole is called a ranking in which each member has a rank. Stratigraphic events such as the highest or lowest occurrences of fossil taxa are ranked on the basis of geologic time. The result of a ranking in stratigraphy is also called an optimum sequence. Observed sequences of stratigraphic events are frequently inconsistent with respect to one another. For example, event A is observed to occur above event B in some geological sections or wells but A occurs below B in others. Several methods have been developed to eliminate inconsistencies of this type in a systematic manner in order construct the optimum sequence (Shaw, 1964; Edwards and Beaver, 1978; Agterberg and Gradstein, 1981; Brower and Burroughs, 1982; Rubel and Pak, 1984; Guex and Davaud, 1984). Much of this work has also been reviewed by Brower (1981), Edwards (1982), Agterberg (1984), Gradstein (1984) and Harper (1984).

Worsley and Jorgens (1977) have pointed out that it may be difficult to determine the optimum sequence when many fossils are missing from the observed sequences. The generally irregular distribution of stratigraphic events among the natural sections observed in a region prevents the design of controlled experiments as generally are used for the ranking of individuals in other fields of science. Another characteristic feature of sequences of stratigraphic events is that the inconsistencies are restricted to relatively short segments of a much longer sequence. Any group of events with internal inconsistencies among them generally is clearly above or below other such groups of events along the axis of relative time. The effects of missing data and the continuity of geologic time have to be considered when stratigraphic ranking techniques are employed. Although the number of observations (sample size) used for the comparison of two events is generally small, the total number of comparisons can be very large in biostratigraphic applications and this may allow the precise ranking of stratigraphic events.

The present Chapter contains a discussion of the RASC method for ranking of stratigraphic events (Agterberg and Nel, 1982a), followed by an introduction to other approaches involving ranking.

Sedimentation Rate, Patchiness and Sampling Problems

One of the problems of quantitative stratigraphic correlation of sections or wells with one another consists of the fact that stratigraphic units deposited in different places almost always have different thicknesses. The data base on which RASC was developed covers an area that spans about 18° of latitude and differences in thickness as well as lithological composition of the stratigraphic units are considerable. The problem of variable thickness of stratigraphic units can be circumvented to a large degree by considering only the relative order of the stratigraphic events in each well and by constructing the optimum sequence on this basis. Only extreme condensation of sections, which produces many seemingly coeval exits, precludes meaningful ranking and scaling.

Figure 1 shows a hypothetical relationship between relative abundance, observed highest occurrence and relative time for two taxa. The purpose of Figure 1 is to clarify the assumptions to be made for the statistical model underlying the RASC program. The frequency curves of Figure 1 are relative in that their frequency density values add to 100 per cent. The concept of relative abundance in time can be illustrated as follows. Suppose that a cylinder is constructed around an axis which is approximately perpendicular to all time planes. Any horizontal slice of this cylinder then would provide a sample of a geological environment for a relatively short time interval. The total number of specimens (n_t) for a fossil taxon in the entire cylinder depends on the total area of the geological environment sampled. A relative abundance curve as shown in Figure 1 represents a ratio n/n_t where n is the total number of specimens in a segment of the cylinder of constant width which is slid through the cylinder along the time axis. The number n depends on the width of the segment used. An abundance curve for n/n_t might become continuous and positive within its entire range if this width were increased sufficiently. In practice, it may not be feasible to measure the relative abundance n/n_t in a precise manner. However, this model for a broader geological environment can be regarded as representing a 'population' of which a single well drilled approximately perpendicular to the time planes provides a 'sample'.

Because of variable thickness of stratigraphic units, the time scale is arbitrary at any one location. Even if the relative abundance pattern for fossils of a taxon were identical in two places which are relatively far apart in distance, the relative abundance curve for this taxon would assume different shapes for the two locations because of differences in the rate of sedimentation. This problem, however, was circumvented to a large extent by considering only the relative order of stratigraphic events at any one place as already pointed out. The relative time scale of Figure 1 represents a standardized measure of this type. It should, however, be kept in mind that the unit along the time axis in Figure 1 is not a measure of absolute time.

Suppose that a stratigraphic event such as highest occurrence is observed for a taxon. Normally, only a very small portion of the larger geological environment is sampled. Hence the probability that the truly highest occurrence of the taxon in this environment is determined is very small. Disregarding special cases such as rare taxa which may not be detected at all, the position along the relative time scale at which the highest occurrence is observed can be regarded as a random variable in which relative abundance of the taxon in the broader geological environment is combined with a probability of detection. The latter probability depends on such factors as the patchiness of the spatial distribution of the fossils and the sampling method used. The possibilities of reworking, poor preservation, slow dispersion, early extinction, facies control, incomplete sampling and misidentification also have to be considered. Hypothetical frequency functions for positions of observed exits are shown by solid lines in Figure 1 for taxa A and B. Note that the exits as observed for the taxa A and B can be coeval (Figure 1), although their frequency curves are not. This can be due to sampling or due to abundance of both taxa in 'patches'. The disappearance of groups of microfossils at many places in wells on the Canadian Atlantic Continental margin is partly caused by the fact that cuttings for 30-ft intervals along the wells were studied. It is noted that different taxa do not necessarily have the same type of frequency distribution for their highest occurrence. For example, planktonic species are likely to have narrower relative abundance curves and exit location curves than benthonic species.

RASC

Introduction

Two ranking algorithms, which were first described by Agterberg and Nel (1982a), form the first part of the RASC program for ranking and scaling of stratigraphic events. This FORTRAN IV computer program was originally developed to erect a zonation of the Cenozoic benthonic and planktonic foraminiferal record (highest occurrences of 206 taxa) in 22 wells on the Canadian Atlantic continental margin between latitudes 42° and 60° (Gradstein and Agterberg, 1982). The first ranking technique (presorting option) consists of the calculation of ranks based on a comparison of each event to all other events. The second technique (modified Hay method) finds permutations in which each event is located before the events with which it shows inconsistencies, by using the relative order frequencies. The position of each event in the stratigraphic ranking, or optimum sequence, obtained by these techniques is an 'average' of all positions encountered for it. See below and Chapter II.5 for in-depth discussions of the RASC ranking techniques.

The ranking algorithms can be employed without the scaling algorithms contained in the second part of RASC. For example, ranking without scaling was used by Doeven et al. (1982) in the construction of a quantitative nannofossil range chart for the Upper Cretaceous strata along the Canadian Atlantic margin.

However, the ranking solutions obtained by the techniques described in this Chapter can also be used as the starting point for the application of the scaling algorithms of the RASC program which require a good initial ranking. The purpose of the scaling techniques is to calculate distances between successive stratigraphic events along a relative time scale. Further discussion of this topic follows later in this chapter and in Chapter II.6. Application of both ranking and scaling algorithms have also been described in Gradstein and Agterberg (1982). Different types of ranking techniques have also been compared by Hudson and Agterberg (1982).

A CDC CYBER 74 computer was used for the original program development, while other versions of RASC have been implemented and tested on different types of IBM, Univac, and CDC mainframe computers, on DEC-System 10, and on several microcomputers. It has been attempted to make the program essentially machine-independent.

Practical Examples

I. Lower Tertiary Nannoplankton from the California Coast Range

In order to explain the practical application of the ranking techniques, they will be applied to two examples. The first example, which was already discussed in Chapter I.1, is relatively simple in that there are only 10 stratigraphic events in 9 sections. It was used by Hay (1972) who extracted the information from publications by Sullivan (1965, first eight sections) and Bramlette and Sullivan (1961, Section I).

The columns of symbols at right in Figure 1 of Chapter I.1 show (1) a hypothetical sequence of events suggested by inspection of the more complete sections, and (2) the 'most probable' sequence of events obtained by Hay (1972, Figure 4 on p. 263).

The second sequence (2) was obtained from the first (1) by (a) counting how many times each event occurs above or below each of the other events, and (b) reordering the events in such a manner that each event occurs more frequently above than below (or as frequently as) the events below it in the 'most probable' sequence. A modified version of this method will be discussed in more detail later in this Chapter, because the ranking method used in RASC is based on it.

Stratigraphic events to be considered for ranking and scaling must be uniquely identified. A dictionary of events forms part of the

Table 1 RASC program output for number of times that successive events occur in a well; e.g. event 1 occurs in 2 wells and event 2 in 1 well.

TABULATION OF EVENT OCCURRENCES:
DICTIONARY CODE NUMBER VERSUS FREQUENCY OF OCCURRENCE

1- 2	53- 5	105-0	157-2	209-0
2- 1	54- 9	106-0	158-1	210-1
3- 1	55- 6	107-0	159-4	211-1
4- 1	56-12	108-0	160-0	212-0
5- 1	57-12	109-2	161-2	213-1
6- 1	58- 1	110-0	162-2	214-0
7- 1	59- 6	111-0	163-0	215-0
8- 1	60- 2	112-1	164-3	216-0
9- 3	61- 2	113-0	164-0	217-0
10- 5	62- 3	114-0	166-1	218-0
11- 4	63- 3	115-0	167-0	219-1
12- 1	64- 2	116-0	168-0	220-0
13- 1	65- 5	117-2	169-0	221-0
14- 3	66- 0	118-4	170-0	222-0
15-14	67- 8	119-1	171-0	223-0
16-15	68- 0	120-0	172-0	224-0
17- 7	69-10	121-0	173-4	225-0
18-15	70- 7	122-1	174-0	226-1
19- 1	71- 6	123-1	175-1	227-0
20-13	72- 0	124-0	176-4	228-5
21-11	73- 2	125-0	177-1	229-0
22- 6	74- 4	126-0	178-0	230-3
23- 1	75- 3	127-0	179-1	231-0
24- 9	76- 2	128-0	180-1	232-0
25-12	77- 4	129-0	181-4	233-0
26- 7	78- 2	130-0	182-2	234-1
27- 8	79- 1	131-1	183-0	235-0
28- 1	80- 1	132-1	184-1	236-1
29-12	81- 4	133-0	185-0	237-1
30-10	82- 4	134-0	186-0	238-1
31-13	83- 2	135-0	187-1	239-0
32- 4	84- 4	136-1	188-1	240-0
33- 4	85- 5	137-1	189-0	241-0
34-11	86- 5	138-0	190-3	242-0
35- 5	87- 1	139-0	191-1	243-0
36- 7	88- 3	140-2	192-0	244-1
37- 2	89- 2	141-0	193-0	245-0
38- 2	90- 5	142-1	194-2	246-0
39- 2	91- 0	143-0	195-0	247-0
40- 5	92- 1	144-1	196-1	248-0
41-11	93- 3	145-1	197-0	249-0
42-11	94- 2	146-1	198-0	250-1
43- 5	95- 0	147-2	199-0	251-0
44- 3	96- 3	148-1	200-0	252-0
45- 7	97- 0	149-0	201-0	253-1
46-12	98- 0	150-0	202-0	254-1
47- 4	99- 0	151-2	203-0	255-0
48- 1	100- 0	152-0	204-0	256-0
49- 3	101- 0	153-0	205-0	257-0
50-10	102- 0	154-0	206-2	258-0
51- 2	103- 0	155-0	207-0	259-0
52- 5	104- 0	156-1	208-0	260-0

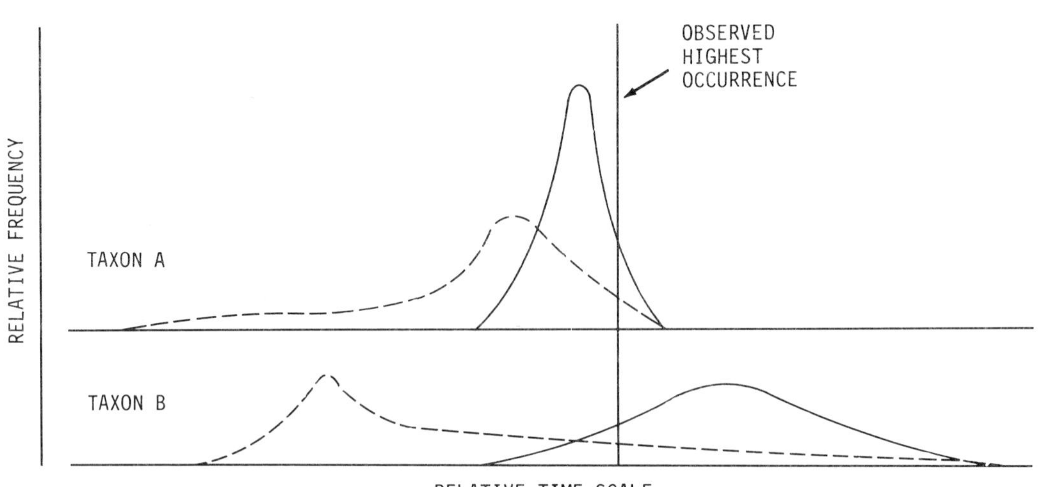

Figure 1 *Schematic diagram representing frequency distributions for relative abundance (broken lines) and location of observed highest occurrence (solid lines) for two taxa. Vertical line illustrates that observed highest occurrences of two taxa can be coeval even when the frequency distributions of these two taxa are different.*

Table 2 *Frequency distribution of numbers tabulated in Table 1; e.g. the number 1 occurs 56 times for number of wells in Table 1 and 2 occurs 26 times.*

TABULATION OF EVENT OCCURRENCES:

NUMBER OF WELLS	1	2	3	4	5	6	7	8	9	10	11	12	13	14	15	16
NUMBER OF EVENTS	56	26	13	14	11	4	5	2	2	3	4	5	2	1	2	0
CUMULATIVE NUMBER	150	94	68	55	41	30	26	21	19	17	14	10	5	3	2	0

RASC input. The example with ten events from Figure 1 in Chapter I.1 was shown in the top part of Table 1 of that chapter. It consists of nine lowest occurrences and one highest occurrence of the Eocene calcareous nannofossils considered. The order of the events in a dictionary for the RASC program is arbitrary. It can be alphabetical or approximately follow the order in one or more of the observed sequences.

The method used for coding the sequences observed in wells or in geological sections was also shown in Table 1 of Chapter I.1. The nine sections (A to I) each have between four and nine events. The events were ordered from left to right, moving in the stratigraphically upward direction. The downward direction also could have been used for ordering the sequences from left to right. Coeval events are separated by a hyphen, coded as a minus sign. If a number is preceded by a minus sign it will be stored in the memory of the computer as having been observed at the same level as the event immediately preceding it in the series. The remainder of the RASC procedures and the results will be discussed later, using this example as well as the following, larger, data set.

II. Cenozoic Foraminifera from the Northwestern Atlantic Continental Margin

The larger data base of Gradstein and Agterberg (1982) will be used for a second example. The events in this example are highest occurrences of Cenozoic foraminifera and other microfossils in 16 wells on the Labrador Shelf and northern Grand Banks.

It is required that the numbers assigned to the events in the dictionary are consecutive and, of course, they should correspond exactly to the numbers in the coded sequences. In practical applications involving large data bases, both the dictionary and the set of coded sequences may have to be subjected to extensive editing. A separate computer file must be created for the dictionary when RASC is used. The coded sequences are to be read in directly either from cards or from disc file.

Because of the patchy spatial distribution of most types of fossils it is desirable to subject the data base to a preliminary statistical analysis. Three tables with frequencies are constructed by the RASC program for this purpose.

(1) The number of sections in which a given event occurs is counted for each event.

(2) The counted number of sections is listed against the frequency by which this count occurs in the first tabulation.

(3) The cumulative frequency F_i for the number of events occurring in i or more sections is printed out.

Tables 1 and 2 show these three tabulations for the example dealing with Cenozoic continental margin stratigraphy.

From the results in Table 2 it is clear that the number of events occurring in n wells may decrease rather rapidly when n increases. In order to work with sample sizes that are not too small it is useful to define a threshold parameter k_c for the minimum number of wells or sections containing an event. Only 26 events occur in 7 or more wells (k_c = 7) and these will be used for the example in this Chapter. Gradstein and Agterberg (1982) used 41 and 68 events for k_c equal to 5 and 3, respectively. The total number of events retained after setting k_c equal to a specific value and used for analysis will be written as N.

The two ranking techniques to be discussed are

(1) A presorting option based on a simple method described by Kendall (1975, p. 151), and

(2) A modification of Hay's (1972) method of permutation of stratigraphic events.

In practice, these methods can be used separately or consecutively. In the latter case, (1) is followed by (2).

In general, optimum sequences obtained by the RASC program are independent of the order in which the events are entered for the sections. This means that the events can be ordered either stratigraphically upwards or stratigraphically downwards starting at the left side of each row. However, the ranking resulting from the presorting option depends to some extent on whether upward or downward sequences are used in the input. The optimum sequence obtained by the modified Hay method may be influenced by this order if there is extensive cycling in the data set with inconsistencies involving more than two events (see 'Three-event cycles,' later in this Chapter).

Presorting Option

For this technique as well as for the modified Hay method, the starting point is a table of frequencies representing the number of times each event occurs below (or above), or coeval to any one of the other events. Suppose that these two frequencies are written as f_{ij} and t_{ji} for events labelled i and j, respectively. Then the score $s_{ij} = f_{ij} + (1/2)t_{ij}$ can be calculated. This means that coeval events are weighted half as much as superimposed events. Because $t_{ij} = t_{ji}$, f_{ij} and f_{ji} are increased by equal amounts when s_{ij} and s_{ji} are calculated for mixtures of coeval and superimposed events. The total number of times that the events i and j occur in the same section can be written as $r_{ij} = s_{ij} + s_{ji}$.

The frequencies f_{ij}, t_{ij}, s_{ij} and r_{ij} may be regarded as the elements of the (NxN) matrices F, T, S and R, respectively. These matrices all have N columns and N rows with N representing the total number of events considered. Table 3 shows the four matrices for the first example. The matrices T and R are symmetrical.

The presorting option of the RASC program functions as follows. A simplified matrix A with elements a_{ij} is derived from the matrix S by comparing each score s_{ij} with its counterpart s_{ji}. If $s_{ij} > s_{ji}$, $a_{ij} = 1$; if $s_{ij} = s_{ji} \neq 0$, $a_{ij} = 0.5$; if $s_{ij} = s_{ji} = 0$, $a_{ij} = 0$; and if $s_{ij} < s_{ji}$, $a_{ij} = 0$. The matrix A and its row totals a_i are shown in Table 4. The events can be ranked on the basis of the values a_i in this table.

In practice, the matrix S may contain pairs of zero elements with $s_{ij} = s_{ji} = 0$ (and $a_{ij} = 0$) because of missing data. It is desirable to distinguish between pairs of zero scores and pairs of scores which are equal to some positive value. This can be accomplished by determining the frequency b_i of zeros in the i-th row of the S-matrix corresponding to pairs of events with $s_{ij} = s_{ji} = 0$. The count b_i can be combined with the row total a_i to give a ranking number

$$A_i = (N-1)a_i\ (N-1-b_i)^{-1}$$

In practice, the numbers A_i are used for ranking instead of the row totals a_i. For the example of Table 4, $A_i = a_i$ $(i=1,\ldots,N)$. The relative order assigned to events with the same ranking number A_i (e.g. events 2 and 3 in Table 4) is arbitrary.

The ranking numbers A_i for the example of Table 1 are shown in Table 5. The results of the presorting option can be used either as input for the modified Hay method or as input for the scaling algorithms. In data bases with relatively large values of b_i, due to missing data, the ranks resulting from the presorting option may not be satisfactory because of the uncertainties introduced. In the method presented in the next section, the relative order of the events is determined by comparing each event to all other events. Then the immediate neighbours of an event will have more influence on the position of this event in the optimum sequence than in the preliminary ranking resulting from the presorting option.

Modified Hay Method

This method consists of attempting to obtain a permutation of the events for which $s_{ij} \geq s_{ji}$ $(i=1,\ldots,N-1;\ j=i+1,\ldots,N)$.

Table 3 *The four matrices F, T, R and S for example of Table 1 in Chapter I.1. Events are compared in the sequence of the dictionary of Table 1 in Chapter I.1. (Event-numbers are not given in this table).*

F-matrix

x	1	1	4	3	2	4	5	8	4
4	x	2	2	3	2	4	5	8	4
1	2	x	3	3	2	3	4	6	3
1	2	0	x	1	2	2	3	6	3
2	1	1	3	x	2	3	5	9	4
0	0	0	0	0	x	1	1	4	1
2	1	1	3	3	1	x	4	7	4
0	0	0	1	0	1	0	x	5	2
0	0	0	1	0	0	0	0	x	3
0	0	0	1	0	0	0	0	3	x

T-matrix

x	2	3	1	3	1	0	0	0	1
2	x	2	2	4	2	1	0	0	1
3	2	x	3	2	2	1	0	0	1
1	2	3	x	3	2	1	0	0	1
3	4	2	3	x	2	1	0	0	2
1	2	2	2	2	x	1	0	0	1
0	1	1	1	1	1	x	1	0	1
0	0	0	0	0	0	1	x	0	1
0	0	0	0	0	0	0	0	x	0
1	1	1	1	2	1	1	1	0	x

R-matrix

x	7	5	6	8	3	6	5	8	5
7	x	6	6	8	4	6	5	8	5
5	6	x	6	6	4	5	4	6	4
6	6	6	x	7	4	6	4	7	5
8	8	6	7	x	4	7	5	9	6
3	4	4	4	4	x	3	2	4	2
6	6	5	6	7	3	x	5	7	5
5	5	4	4	5	2	5	x	5	3
8	8	6	7	9	4	7	5	x	6
5	5	4	5	6	2	5	3	6	x

S-matrix

x	2	2½	4½	4½	2½	4	5	8	4½
5	x	3	3	5	3	4½	5	8	4½
2½	3	x	4½	4	3	3½	4	6	3½
1½	3	1½	x	2½	3	1½	3	6	3½
3½	3	2	4½	x	3	3½	5	9	5
½	1	1	1	1	x	1½	1	4	1½
2	1½	1½	3½	3½	1½	x	4½	7	4½
0	0	0	1	0	1	½	x	5	2½
0	0	0	1	0	0	0	0	x	3
½	½	½	1½	1	½	½	½	3	x

Table 4 *A-matrix derived from S-matrix in Table 3. Row sums (a_i) are used for ranking in presorting option.*

i	j=1	2	3	4	5	6	7	8	9	10	a_i
1	X	0	½	1	1	1	1	1	1	1	7½
2	1	X	½	½	1	1	1	1	1	1	8
3	½	½	X	1	1	1	1	1	1	1	8
4	0	½	0	X	0	1	0	1	1	1	4½
5	0	0	0	1	X	1	½	1	1	1	5½
6	0	0	0	0	0	X	½	½	1	1	3
7	0	0	0	1	½	½	X	1	1	1	5
8	0	0	0	0	0	½	0	X	1	1	2½
9	0	0	0	0	0	0	0	0	X	½	½
10	0	0	0	0	0	0	0	0	½	X	½

Table 5 *Ranking numbers A_i obtained by means of presorting option applied to 26 events of Table 1 which occur in k_c = 7 or more wells. Original event numbers are shown in column 1. The new ranks obtained from the ranking numbers A_i are shown in the last column.*

Event	i	A_i	Rank
15	1	19,5	7
16	2	24.0	2
17	3	25.0	1
18	4	21.5	4
20	5	20.0	6
21	6	20.5	5
24	7	15.5	10
25	8	15.0	11
26	9	18.2	8
27	10	14.0	13
29	11	11.5	15
30	12	7.0	19
31	13	12.0	14
34	14	10.0	16
36	15	5.5	20
41	16	9.0	17
42	17	8.0	18
45	18	4.5	22
46	19	3.0	23
50	20	2.5	24
54	21	1.0	25
56	22	0.0	26
57	23	4.5	2
67	24	23.9	3
69	25	14.0	12
70	26	17.0	9

Use is made of a second threshold parameter $m_{c1} \leq k_c$ which controls the minimum sample size $r_{ij} = s_{ij} + s_{ji}$ permitted. Scores for samples with $r_{ij} < m_{c1}$ will be ignored. This means that if two events, labeled i and j, occur together in less than m_{c1} stratigraphic sections, their scores will not be used for construction of the optimum sequence. It is desirable to set m_{c1} smaller than k_c because, otherwise, some events may end up in undetermined positions in the optimum sequence. Ideally m_{c1} should be equal to one. However, it may then not be possible to obtain an optimum sequence when cycles are numerous (see next section). Until 1983, only two threshold parameters, k_c and m_c, had to be chosen for running the RASC computer program. At present, there are three such parameters, k_c, m_{c1} and m_{c2}, because it is desirable to define separate parameters for numbers of pairs in ranking by modified Hay method and scaling. If the notation m_c is used in the text, this implies that a pre-1983 RASC run is discussed with $m_{c1} = m_{c2} = m_c$.

A permutation with $s_{ij} \geq s_{ji}$ (i=1,...,N-1; j=i+1,...,N) is shown in Table 6 for the example of Table 1 in Chapter I.1. This solution, which is not unique because successive events with $s_{ij} = s_{ji}$ are interchangeable, was obtained by using the following algorithm. Initially the first event is compared to the second event. If $s_{12} \geq s_{21}$, which implied that event 1 occurs more frequently before event 2 than 2 before 1 or that the events are tied in the observed sequences, then the relative order of events 1 and 2 is left unchanged. However, if $s_{21} > s_{12}$, the order of events 1 and 2 is reversed. After a reversal, the values of i and j for the events are interchanged. Consequently, the situation in which $s_{21} > s_{12}$ is eliminated. When $s_{12} \geq s_{21}$, event 1 is compared to event 3. Again the order of the events is reversed only if $s_{13} < s_{31}$. This process is continued until event 1 has been compared to all events in the first row. After each interchange of events, the first position is occupied by a new event which has not yet been compared to the events (j=2,3,...,N) preceding it in the sequence before the interchange. Hence the process of comparing s_{1j} to s_{j1} must be repeated beginning with the second event of the first row after each interchange.

The first event of the optimum sequence has been found only when $s_{1j} \geq s_{j1}$, (j=2,...,N). Then the second event (i=2) can be compared to all subsequent events in the second row until $s_{2j} \geq s_{j2}$ (j=3, ...,N).

In general, s_{ij} is compared to s_{ji} for i=1,...,N-1 and j=i+1,...,N. The final ranking with $s_{ij} \geq s_{ji}$ (i=i,...,N-1; j=i+1, ...,N) is the optimum sequence.

Three Event Cycles

Worsley and Jorgens (1977) have found that the preceding algorithm does not necessarily yield an optimum sequence because cyclical

Table 6 Optimum sequence obtained from S-matrix in Table 3 without use of presorting option. The ranked event-numbers are shown in both margins.

	2	1	3	5	7	4	6	8	9	10
2	x	5	3	5	$4\frac{1}{2}$	3	3	5	8	$4\frac{1}{2}$
1	2	x	$2\frac{1}{2}$	$4\frac{1}{2}$	4	$4\frac{1}{2}$	$2\frac{1}{2}$	5	8	$4\frac{1}{2}$
3	3	$2\frac{1}{2}$	x	4	$3\frac{1}{2}$	$4\frac{1}{2}$	3	4	6	$3\frac{1}{2}$
5	3	$3\frac{1}{2}$	2	x	$3\frac{1}{2}$	$4\frac{1}{2}$	3	5	9	5
7	$1\frac{1}{2}$	2	$1\frac{1}{2}$	$3\frac{1}{2}$	x	$3\frac{1}{2}$	$1\frac{1}{2}$	$4\frac{1}{2}$	7	$4\frac{1}{2}$
4	3	$1\frac{1}{2}$	$1\frac{1}{2}$	$2\frac{1}{2}$	$2\frac{1}{2}$	x	3	3	6	$3\frac{1}{2}$
6	1	$\frac{1}{2}$	1	1	$1\frac{1}{2}$	1	x	1	4	$1\frac{1}{2}$
8	0	0	0	0	$\frac{1}{2}$	1	1	x	5	$2\frac{1}{2}$
9	0	0	0	0	0	1	0	0	x	3
10	$\frac{1}{2}$	$\frac{1}{2}$	$\frac{1}{2}$	1	$\frac{1}{2}$	$1\frac{1}{2}$	$1\frac{1}{2}$	$1\frac{1}{2}$	3	x

Table 7 Example of cycling events A,B, C and D from Worsley and Jorgens (1977). The algorithm for interchanging events, which is explained in the previous section, results in a return of the matrix (a) after six steps (b-g).

(a)

A	B	C	D
x	2	3	2
1	x	5	1
4	2	x	3
0	7	4	x

(b)

C	B	A	D
x	2	4	3
5	x	1	1
3	2	x	2
4	7	0	x

(c)

B	C	A	D
x	5	1	1
2	x	4	3
2	3	x	2
7	4	0	x

(d)

A	C	B	D
x	3	2	2
4	x	2	3
1	5	x	1
0	4	7	x

(e)

C	A	B	D
x	4	2	3
3	x	2	2
5	1	x	1
4	0	7	x

(f)

B	A	C	D
x	1	5	1
2	x	3	2
2	4	x	3
7	0	4	x

(g)

A	B	C	D
x	2	3	2
1	x	5	1
4	2	x	3
0	7	4	x

Figure 2

Graph for example of Table 7. The three-event cycle involving events A, B and C is characterized by successive arrows pointing in the same direction at both sides of each corner point of the triangle ABC. An arrow pointing from i to j between two events i and j indicates that j follows i more frequently than i follows j in the observed sequences.

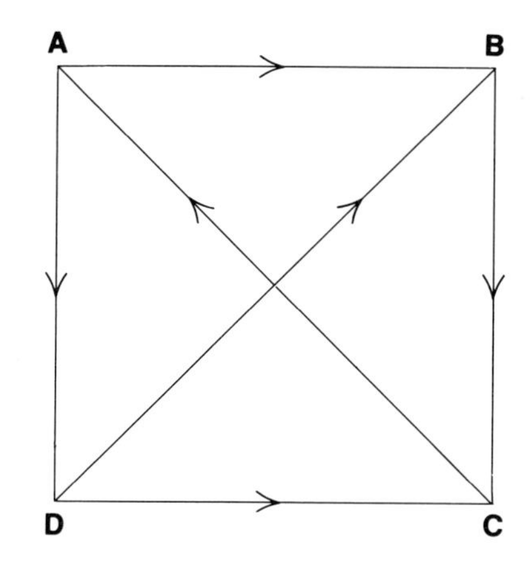

Figure 3

Graphical illustration of algorithm developed to locate a three event cycle which is hidden in a large order-relation matrix S. The elements in successive rows of the upper triangle are tested proceeding from left to right. Row and column interchanges only take place when an element is less than its counterpart in the lower triangle. In the example, the element circled in the margin (C) will then be replaced by K which, in turn, will be followed by F. The cycle CKF would repeat itself indefinitely.

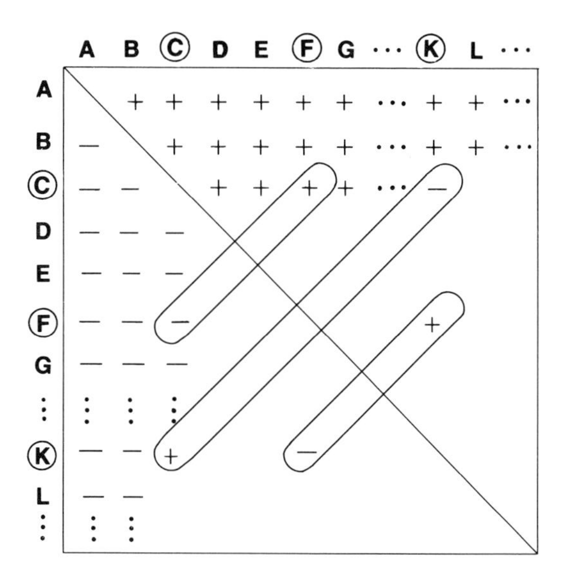

inconsistencies may occur in which more than two events are involved. Their original example of cycling events is shown as the first matrix of Table 7. When the algorithm is applied, the original S-matrix reoccurs after every set of six consecutive iterations. Hence an optimum sequence could never be determined by means of the preceding algorithm.

In the example of Table 7, A occurs more frequently before B ($s_{AB}>s_{BA}$), B before C ($s_{BC}>s_{CB}$), and C before A ($s_{CA}>s_{AC}$). The three events A, B and C are involved in a cyclical inconsistency and are said to form a three-event cycle. It is useful to represent this type of situation by means of a graph. The relationships of Table 7 are represented by arrows in the graph shown in Figure 2. The three-event cycle involving A, B and C is immediately apparent in Figure 2 because the arrows in the triangle ABC point in the same direction at both sides of each of the corner points of this triangle. It should be noted that D does not participate in a cycle.

If there are no cycles, all inconsistencies can be eliminated by disregarding situations in which $s_{ij} < s_{ji}$. Suppose that each situation $s_{ij} \geq s_{ji}$ is indicated by a + sign for s_{ij} in the upper triangle above the diagonal of the S-matrix where $j > i$, and a - sign for the corresponding element in the lower triangle where $j < i$. Then the S-matrix of Table 6 would be replaced by a matrix with exclusively + signs in the upper triangle and - signs in the lower triangle. If a 3-event cycle occurs it is not possible to achieve a clear subdivision of this nature as is illustrated in Figure 3 for an artificial example. The events of Figure 3 are indicated by means of letters. C, F and K form a 3-event cycle. The elements in the first two rows could be tested by means of the previous algorithm. However, iterations would continue indefinitely for the elements in the third row which is for one of the cycling events (C). The event in the margin of the third column of Figure 3 can be scanned by putting a 'window' on it in the computer algorithm. For the 3-event cycle of C, F and K, this window will begin showing the sequence CKFCKF... which can be readily detected. Once the events involved in a cycle have been identified, the + sign corresponding to the pair of scores with the smallest difference $|s_{ij} - s_{ji}|$ can be allowed to remain in the lower triangle. In the algorithm, this is accomplished by temporary replacement of its scores by zeros.

It is possible that two pairs of scores for events involved in a 3-event cycle have equal smallest difference values, or that all three pairs have equal differences. In those situations only the first pair encountered will be ignored.

For example, the data of Table 2 were run setting the threshold parameters equal to $k_c = 7$ and $m_c = 5$, respectively. For N = 26 events, it is possible to make $(1/2)N(N-1) = 325$ comparisons. However, because of $m_c = 5$, forty pairs were not used. The presorting option was used (see Table 5) and the 26 events were reordered using the ranks in the last column of Table 5. A three-event cycle involving events 25, 27 and 69 was identified with the corresponding output shown in Table 8.

The event positions printed below the cycling events are temporary and can be used to identify which pair of events (11 and 12) was ignored in order to break the cycle.

In the original input the three cycling events were encountered together in four wells: Freydis (69, -27, 25), Gudrid (69, 25, 27), Bonavista (25, 27, 69) and Dominion (27, 25, 69). In these expressions, relative order is indicated by means of a comma and coeval events are separated by a comma followed by a hyphen (e.g., in Freydis, 69 and 27 are coeval and both precede 25). For abbreviation, the four expressions can be rewritten as (2-31, 213, 132, 312) where 25, 69 and 27 have been replaced by 1, 2 and 3, respectively. Two of the three events were encountered together in seven wells with relative orders (21, 21, 13, 12, 21, 13, 32). The scores of Table 8 can be obtained by counting subsequences for two events (e.g., 21 occurs five times while 12 occurs three times). All three events participate in a cycle because the preferred subsequences 21, 13 and 32 cannot all hold true simultaneously.

In this application, the optimum sequence is almost equal to the result obtained by means of the presorting option. In addition to a change in order corresponding to the 3-event cycle, only the events with ranks 21 and 22 change places in the sequence. Every cycle is allowed to run 100 times before it is broken. Hence the total number of iterations in Table 8 is 102 instead of 2. Extra iterations may be needed to eliminate possible pseudo-cycles which can develop initially before a truly periodic cycle appears. This subject will be explained in a later section of this chapter which also contains a discussion of situations in which cycles involving more than three events can develop.

Cycles tend to occur frequently if one or both of the following two conditions are satisfied:

(1) Many small samples are used (e.g. $r_{ij} < 3$), and

(2) The expected values of many of the frequencies $f_{ij} = s_{ij} r_{ij}^{-1}$ are close to 0.5.

A tolerance parameter (TOL) can be used in the RASC program to reduce the number of cycles. If TOL is equal to a positive value (e.g. 0.5 or 1.0), scores with $s_{ij} + \text{TOL} \geq s_{ji} > s_{ij}$ will be allowed to occur in the lower triangle ($j < i$) in addition to the values $s_{ji} < s_{ij}$.

A print-out of the optimum sequences with (uncertainty) range values and the original numbers and names of the events also is part of the output (see Table 9 and below under 'Ranges of Events in Optimum Sequence').

Ranges of Events in Optimum Sequence

The position of an event in the optimum sequence can be uncertain because of three reasons (see also Chapter II.4):

Table 8 Selected output from RASC program including information on a single 3-event cycle encountered when data of Table 1 are run with k_c = 7 and m_c = 5. See text for explanations.

```
RUN FOR 7 OR MORE OCCURRENCES AND 5 OR MORE PAIRS.

        CYCLING EVENTS:           27      25      69

        EVENT POSITIONS:          11      13      12

        MATRIX ELEMENTS:

        0.0     2.0     3.5

        4.0     0.0     3.0

        1.5     5.0     0.0

C(11, 13) AND C(13, 11) ZEROED

RANKING SOLUTION OBTAINED WITH:

        102 ITERATIONS OUT OF MAXIMUM 9000

        TOLERANCE OF 0.0
```

Table 9 RASC program output of optimum sequence of data of Table 1 with k_c = 7 and m_c = 5.

Sequence Position	Fossil Number	Range	Fossil Name
1	17	0- 2	Asterigerina gurichi
2	16	1- 3	Ceratobulimina contraria
3	67	2- 4	Scaphopod spl
4	18	3- 6	Spiroplectammina carinata
5	21	3- 6	Guttulina problema
6	20	5- 7	Gyroidina girardana
7	15	6- 8	Globigerina praebulloides
8	26	7-10	Uvigerina dumblei
9	70	7-12	Alabamina wolterstorffi
10	24	8-11	Turrilina alsatica
11	27	10-12	Eponides umbonatus
12	69	11-13	Nodosaria sp8
13	25	12-14	Coarse arenaceous spp.
14	31	13-16	Pteropod spl
15	29	13-16	Cyclammina amplectens
16	34	15-17	Marginulina decorata
17	41	16-18	Plectofrondicularia spl
18	42	17-19	Cibicidoides alleni
19	30	18-20	Cibicidoides blanpiedi
20	36	19-23	Pseudohastigerina wilcoxensis
21	45	19-22	Bulimina trigonalis
22	57	21-23	Spiroplectammina spectabilis
23	46	22-25	Megaspore spl
24	50	22-25	Subbotina patagonica
25	54	24-26	Textularia plummerae
26	56	25-27	Glomospira corona

(1) The event does not occur together with one or more of its neighbours in the optimum sequence; this condition usually is introduced artifically by ignoring pairs of scores s_{ij} and s_{ji} with $r_{ij} < m_{c1}$;

(2) The event belongs to a pair of events ignored because they have the smallest difference in a cycle; and

(3) The event is tied with one or more of its neighbours; this situation arises when $s_{ij} = s_{ji}$.

Only the first two of these three conditions were considered when the ranges for the optimum sequences in Gradstein and Agterberg (1982) were determined. Although each pair with the smallest difference in a cycle was ignored previously and at present only the first pair encountered is ignored, it is likely that uncertainty ranges calculated by the algorithm discussed here are wider than those calculated before.

The uncertainty range (h, k) for an event i in the optimum sequence indicates that i occurs somewhere between events h and k ($h < i < k$). If $h=i-1$ and $k=i+1$, as for most events in Table 9, the position of event i in the optimum sequence is fully determined by its two neighbours. Otherwise, it can occur anywhere between the events h and k. If $h < i-1$, there occur $(i-h-1)$ scores above the diagonal in the i-th column which either belong to a pair of ignored scores or which satisfy the condition $s_{ij} = s_{ji}$. Likewise, if $k > i+1$, there occur $(k-i-1)$ scores of these types to the right of the diagonal in the i-th row of the S-matrix for the optimum sequence. This matrix is called the final order relation matrix in the RASC output and can be printed out if required.

If its uncertainty range is wider than $(i-1, i+1)$ the fact that an event occupies a specific position within its range is caused by its position along the margin of the original S-matrix which, in turn, merely reflects the ranking number initially and randomly assigned to the event in the dictionary, or obtained after application of the presorting option.

For example, the uncertainty ranges of Table 9 can be interpreted as follows: As mentioned before, the positions of most events are fully determined by their immediate neighbours because $h-i=k-i=1$. Events 4 and 5 have the same range (3-6) indicating that they are interchangeable. The same applies to events 23 and 24. The relationship between events 8 (range 7-10), 9(7-11) and 10(8-11) is somewhat more complicated. Event 8 can be interchanged with event 9, and event 9 with event 10. However, events 8 and 10 are not interchangeable and event 8 should remain before event 10. The events 13, 14 and 15 have similar interrelationship.

Finally, event 20 is interchangeable with both 21 and 22, but event 21 must remain before 22 in the optimum sequence. Thus the only three possible arrangements of these three events in the optimum sequence are (20, 21, 22), (21, 20, 22) and (21, 22,20).

It is noted that the range of an event in the optimum sequence is determined by constructing all possible permutations of the events for which the condition $s_{ij} \geq s_{ji}$ $(i=1,\ldots,N-1;\ j=i+1,\ldots,N)$ is satisfied. It does not mean that the event occurs within this range with certainty when the optimum sequence is interpreted as the statistical average for a set of observed sequences. This interpretation arises if it is attempted to estimate the expected ranks of the events in an infinitely large population of observed sequences. In order to solve the latter type of problem, a statistical theory is needed as developed in Gradstein and Agterberg (1982) where each event is positioned along a scale. This allows the visualization of the distances between all events and the estimation of their standard deviations. For further discussion of this theory the reader is referred to Chapters II.4 and II.6.

Higher-order Cycles and Pseudo-Cycles

Suppose that four events (A, B, C and D) with $s_{ij} = s_{ji}$ $(i=A,B,C,D \neq j=A,B,C,D)$ are subject to the relationships $s_{AB} > s_{BA}$, $s_{BC} > s_{CB}$, $s_{CD} > s_{DC}$, and $s_{DA} > s_{AD}$.

This situation was in fact shown in Table 7. Worsley and Jorgens (1977) assumed that all four events participated in the inconsistency. However, when the algorithm of this paper is applied, only the events A, B and C are involved in what is called a 3-event cycle.

In general, it can be shown that, if $s_{ij} \neq s_{ji}$ $(i \neq j)$ for four events, then there must be two 3-event cycles in the system for the situation defined at the beginning of this section. The scores for A in comparison to C satisfy either $s_{AC} > s_{CA}$ or $s_{CA} > s_{AC}$. If $s_{AC} > s_{CA}$, A, C and D form a 3-event cycle; if $s_{CA} > s_{AC}$, A, B and C form a cycle. Likewise, either A, B and D or B, C and D form a 3-event cycle. If the algorithm is applied, a 3-event cycle will be identified (cf. Table 8). When this cycle is broken, the other cycle either remains in the system and would be identified next, or it is broken at the same time as the first cycle. Whether or not two cycles will be identified depends on relative magnitudes of the differences $|s_{ij}-s_{ji}|$.

A 4-event cycle for $s_{AB} > s_{BA}$, $s_{BC} > s_{CB}$, $s_{CD} > s_{DC}$, and $s_{DA} > s_{AD}$ arises only if $s_{AC} = s_{CA}$ and $s_{BD} = s_{DB}$ as illustrated in Figure 4.

Higher-order cycles, including the 5-event and 6-event cycles which are also shown in Figure 4, only occur if all arrows on the circumference of the graph point in the same direction, while all indirect connections between nodes are undirected with $s_{ij} = s_{ji}$ $(i \neq j; j \neq i+1)$.

Higher-order cycles are identified and eliminated in the same manner as 3-event cycles. It is noted that in Gradstein and Agterberg (1982) all pairs of scores with equal minimum differences were ignored, whereas, in the algorithm described here, only the first pair encountered will be ignored.

Four-event cycles frequently occur in practice but 5-event cycles are rare. In numerous runs of RASC, the author has encountered a 6-event cycle only twice. The RASC program would identify and break cycles of up to nine events.

The concept of a pseudo-cycle is illustrated in Figure 5. The initial order ABCD is changed into ACDB after four iterations. The sequence ACDB contains a single 3-event cycle (ACD) and reappears with a periodicity of six iterations. When a window is placed on the first event, the observed sequence is ADCBADCADCA... This initially would suggest a 4-event cycle involving all four events. However, this pseudo-cycle is unstable and is automatically replaced by the 3-event cycle for A, C and D.

Presorting and Ranking by Harper

In a recent study evaluating various ranking techniques, Harper (1984) found that the presorting option provided slightly better rankings than the modifed Hay method. Harper was interested in comparing competing ranking algorithms in stratigraphic paleontology on the basis of computer-simulated sections. By means of a computer program he

(1) generated a hypothetical, and thus known, succcession of taxa in time, and

(2) simulated their succession in strata at several local sample sites.

If desired, (1) and (2) may be repeated for many (50 or 100, for example) iterations and the local site data for each iteration sent to user routines for inferred rankings (inferred succession of events in time). First, data for first and last occurrences (entries and exits) taken together, then data for exits-only, then data for entries-only is sent. For each submision of data to a user routine, Kendall rank correlation coefficients and Spearman coefficients (for equations, see "Computer Simulation Experiments" in Chapter II.6) are computed, comparing the inferred rankings generated by the user routine with the known succession of events in time. The performance of two competing ranking algoriths may be compared by

(1) obtaining for each submitted dataset the differences between corresponding Kendall and Spearman coefficients computed for the two algoriths, and

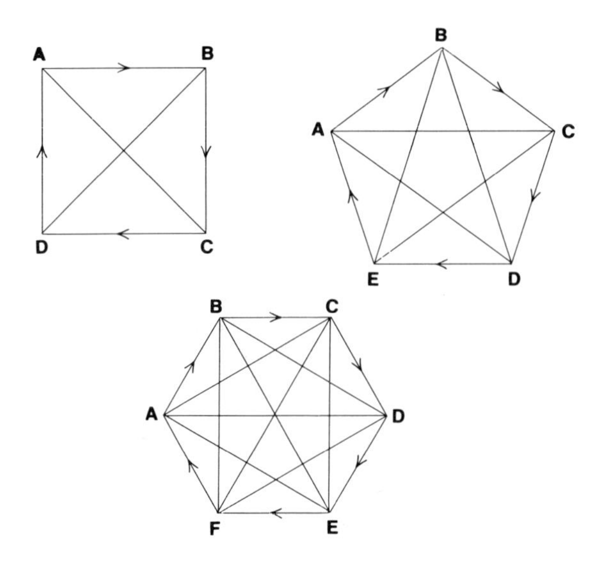

Figure 4 *Cycles involving more than three events occur only when all events, except those participating in the cycle, are pairwise simultaneous ($s_{ij} = s_{ji}$). The latter situation may also be the result of missing data or pairs of scores ignored because $r_{ij} < m_{c1}$. Pairs that are simultaneous on the average have connecting lines without arrows. Examples shown are for four-, five-, and six-event cycles, respectively.*

Figure 5

Illustration of pseudo-cycle (ADCB) which initially develops when the algorithm is applied but is automatically replaced by the three-event cycle (ADC). Events with hats are being observed at a 'window' and checked for periodicity in the algorithm.

	$\hat{A}$	B	C	D
A	x	+	+	-
B	-	x	+	o
C	-	-	x	+
D	+	o	-	x

	$\hat{D}$	B	C	A
D	x	o	-	+
B	o	x	+	-
C	+	-	x	-
A	-	+	+	x

	$\hat{C}$	B	D	A
C	x	-	+	-
B	+	x	o	-
D	-	o	x	+
A	+	+	-	x

	$\hat{B}$	C	D	A
B	x	+	o	-
C	-	x	+	-
D	o	-	x	+
A	+	+	-	x

	$\hat{A}$	C	D	B
A	x	+	-	+
C	-	x	+	-
D	+	-	x	o
B	-	+	o	x

	$\hat{D}$	C	A	B
D	x	-	+	o
C	+	x	-	-
A	-	+	x	+
B	o	+	-	x

	$\hat{C}$	D	A	B
C	x	+	-	-
D	-	x	+	o
A	+	-	x	+
B	+	o	-	x

	$\hat{A}$	D	C	B
A	x	-	+	+
D	+	x	-	o
C	-	+	x	-
B	-	o	+	x

	$\hat{D}$	A	C	B
D	x	+	-	o
A	-	x	+	+
C	+	-	x	-
B	o	-	+	x

	$\hat{C}$	A	D	B
C	x	-	+	-
A	+	x	-	+
D	-	+	x	o
B	+	-	o	x

	$\hat{A}$	C	D	B
A	x	+	-	+
C	-	x	+	-
D	+	-	x	o
B	-	+	o	x

(2) testing the observed differences for statistical significance.

Harper (1984) used the computer program to compare three ranking algorithms (presorting, ranking and scaling) provided by Agterberg and Nel (1982a,b) as well as to determine whether the algorithms work as well for datasets combining exits and entries versus datasets for exits-only or entries-only. He concluded from a series of experiments that Agterberg and Nel's presorting algorithm performed slightly better than their ranking and scaling algorithms. All three performed slightly but significantly better on data for exits-only or entries-only as opposed to combined data. The reader is referred to Harper (1984) for a full discussion of his approach and complete results for all experiments performed. Only a few examples will be given here with emphasis on how Harper's approach can be used in practice; e.g., for choosing the threshold parameters k_c and m_c.

The computer program begins by generating ranges for 50 taxa over 80 time intervals. A random number generator is used for determining 'true' entries and exits of each taxon in a range chart. Next stratigraphic succession data for n_s sample sites are generated by random sampling of the range chart. This sampling is controlled by choosing a value for

(1) the probability (P_1) that a given taxon is present at a local site;

(2) the probability (P_2) that a taxon is sampled at a given horizon at a sample site given that it occurs in the time interval represented by the horizon; and

(3) the probability (P_3) that two adjacent horizons correspond to the same time interval.

For each sample site, n_t sets of stratigraphic succession data are obtained, with n_t being the number of iterations. Harper conducted 3 experiments (A, B and C) with the following parameters:

Run	n_s	P_1	P_2	P_3	n_t
A	22	0.20	0.55	0.10	50 (or 100)
B	16	0.20	0.80	0.10	50 (or 100)
C	6	0.10	0.85	0.20	50

Run, sample site, and sequence data were sent to RASC to obtain three types of optimum sequences (a) Presorting only; (b) Modified Hay method only; and (c) Scaling sequence as derived from (b). The following threshold parameters were employed:

Run	k_c	m_c	k_c/m_c
A	5	4	1.25
B	7	5	1.40
C	3	3	1.00

Harper (1984, Figures 4-6) compared experimentally obtained optimum sequences with the 'true' optimum sequence on the range chart by using Kendall's rank correlation coefficients. The latter are also called tau-values and range from -1 to +1 like product-moment correlation coefficients. In total, 1950 tau-values were calculated, one for each comparison; all turned out to be relatively close to +1, and significantly greater than zero. This signifies that all rankings were good. However, by comparing methods with one another, and looking at small differences between average tau-values, it can be determined which one of a pair of techniques is better (see Harper 1984, Tables 2-7). Average differences between tau-values for comparing presorting with modified Hay method were as follows:

Run	k_c	k_c/m_c	exits	entries	both	both(100)
B	7	1.40	-0.003	-0.003	-0.001	-0.000
A	5	1.25	-0.013	-0.014	-0.004	-0.005
C	3	1.00	-0.022	-0.020	-0.007	n.o.

In this tabulation, the runs were re-ordered according to decrease in k_c and k_c/m_c ratio. Each of the values shown is the average of 50 differences between tau-values, except the two values in the last column which were based on 100 differences; n.o. indicates that an average for 100 runs was not obtained for Run C. A negative value signifies that modified Hay method gave poorer rankings than presorting. Except for Run B (first run), the negative values are significantly different from zero as determined by Student's t-test (Harper, 1984, Tables 2-7). The results for exits and entries are similar as can be expected, and the first two values in the last two columns also duplicate one another.

It may be concluded that, for the experiments performed, presorting gives better results than use of the modified Hay method only, when k_c is relatively small. When k_c is large, the two methods probably give rankings that are equally good. The results of the experiments also suggest the possibility that, by increasing the ratio k_c/m_c, the performance of the modified Hay method can be improved. The presorting option was introduced in Agterberg and Nel (1983a) and has been used in all RASC runs performed by us after 1980. The results of presorting are independent of the choice of the threshold parameter m_c which applies to modified Hay method and scaling only. As a result of Harper's experiments, a modification in the RASC program was introduced in 1983 to allow the choice of two m_c values: m_{c1} for modified Hay method and m_{c2} for scaling. Before then, all runs including those performed by Harper had $m_{c1} = m_{c2} = m_c$. The optimum sequences for the data bases in Appendix 1 were obtained by presorting followed by modified Hay method, except for Baumgartner's data base (No. 4) and Rubel's combined data base (No. 6D). In the latter two instances, presorting was applied only, because of numerous cycles occurring when it was attempted to use the modified Hay method.

Application of the modified Hay method after presorting can be regarded as a fine-tuning operation. The presorting option may yield poor results when many frequencies are undetermined because of missing data. It should then be useful to compare the ranking of each event with all others in order to find the permutation of the optimum sequence as is done in the modified Hay method. Ideally, the threshold parameter m_{c1} should be set equal to 1 so that all frequencies are considered. However, as pointed out before, a decrease in m_{c1} frequently corresponds to an increase in number of cycles. It then is necessary to use a value greater than 1 in order to reduce the number of iterations.

Harper (1984) also found negative differences between tau-values when the modified optimum sequence resulting from scaling was compared to the optimum sequence resulting from the modified Hay method only. However, the lower tau-values in this instance may have been caused by the fact that Harper (1984, p. 16) regarded as tied successive events which were less than 0.05 apart along the RASC scale. A modified formula for estimating Kendall's rank correlation coefficient was used to accommodate tied events. On average, events preceding other events along the RASC scale, probably occur before those other events on the range chart as well, even when distances between successive events are small. Scoring them as tied, therefore, results in a somewhat smaller tau-value. This may explain why the optimum sequence from the modified Hay method, in which no ties were allowed, yielded somewhat higher tau-values.

Finally, Harper (1984) showed that exits and entries, run separately, gave somewhat higher tau-values than when both were mixed together. This was to be expected (also see Edwards and Beaver, 1978) because, on the average, exits will be moved downward, and entries upward, with respect to their relative positions on the range chart when stratigraphic succession data for sample sites are generated using probabilities of occurrence (P_1, P_2 and P_3). If exits or entries are considered on their own, this bias will not show up. However, if they are mixed, some exits will probably assume final positions, in any type of optimum sequence, below entries of other taxa which occur above them on the range chart. Although smaller tau-values are to be expected for sequences of mixed entries and exits, these differences were almost negligibly small in the results of Harper's experiments.

DIFFERENT METHODS FOR RANKING

The ranking methods described in the previous sections are those programmed by Agterberg and Nel (1982a) for the ranking part of the RASC program. In this section several alternative methods of approach will be briefly discussed.

The ranking method recently described by Blank and Ellis (1982) is very similar to the procedure described in this Chapter. Presorting is followed by an application of the Hay method (see Figure 1). The ranking numbers of Blank and Ellis differ from those shown in Tables 5 and 6. Their method consists of counting how many times a given event occurs above any event. This frequency is then divided by the total number of pairs of events used for comparison. The resulting relative frequency provides the ranking number of the event considered. Blank and Ellis (1982) use the original Hay method for further ranking. Cyclicities may present a problem but are eliminated by judging each situation on its own merits.

In the original Hay method, coeval events are ignored. On the other hand. Davaud and Guex (1978) and Rubel (1978) in their methods assign more weight to ties (coeval events) than is done in the modified Hay method. In this respect, we have adopted the practice of authors including Kendall (1975), and Bunk (1960) who scored ties as 0.5 above and below the principal diagonal of the matrix for frequencies. However, arguments that ties should be ignored in some situations have been presented by Hemelrijk (1952) and Tocher (1950). It is also noted that, in the absence of cycling, the modified Hay method produces the same optimum sequence as the original Hay method. In the methods of Davaud and Guex (1978) and Rubel (1978), occurrences of fossil species are considered to be coeval if they are observed to be coeval at least once. For example, even if fossil A is observed to occur above fossil B in several sections, their coexistence in a single section results in the two fossils to co-occur in the standard constructed on the basis of all sections. Clearly, then more weight is assigned to ties than in either the Hay method or modified Hay method. Guex and Davaud (1984) have made extensive use of graph theory in developing their technique. This allowed them to construct an optimum sequence of multiple events which may be subdivided into parts called "unitary associations" which can be identified in the original sections and used for correlation. Further discussion of Guex and Davaud's methods can be found later in this Chapter and in Chapters II.7 and II.8.

BINOMIAL TEST FOR RANDOMNESS

Finally, the binomial test for randomness will be briefly discussed (cf. Hay, 1972; Southam et al.. 1975; Blank and Ellis, 1982). If the sequence of a pair of biostratigraphic events is random, the probability of one event preceding the other is $p = 1/2$. Suppose that two events (A and B) both occur in M sections. Then the probability that A occurs above B k times satisfies

$$P(k) = \binom{M}{k} 2^{-M}$$

with the binomial coefficient being

$$\binom{M}{k} = M! \, [k! \, (M-k)!]^{-1}$$

For example, if M = 5, P(0)= P(5)= 1/32; P(1) = P(4)= 5/32; and P(2)= P(3)= 10/32. These probabilities add to one. It is also possible to write P(0 or 5) = 1/16, P(1 or 4) = 5/16 and P(2 or 3) = 10/16. In practice, the observation that A occurs k times above B generally can not be distinguished from B occurring k times above A when the hypothesis $p = E(k/M) = 1/2$ is being tested. In this expression, E(...) denotes expected value. The test hypothesis can obviously not be rejected if $k/M = 1/2$, a situation which may be observed when M is even. For $k > M/2$, the probability

$$P_c(k) = 2 \sum_{r=k}^{M} \binom{M}{r} 2^{-M}$$

may be computed where the subscript c denotes that this probability is cumulative. For the preceding example, $P_c(5) = 1/16$, $P_c(4) = 1/16 + 5/16 = 6/16$, and $P_c(3) = 6/16 + 10/16 = 1$. This probability was tabulated by Hay (1972, Table 1 on p. 264). Next a level of significance (e.g. $\alpha = 0.05$) can be selected. Then the hypothesis $p = 1/2$ will be rejected only if $P_c(k) < \alpha$.

The binomial test is useful when only two events are being compared to each other. If many events are to be considered simultaneously while most values of M are small, this approach is less useful. For example, in Figure 2 of Chapter 1, event 10 occurs 4 times above event 1. According to the binomial test $P_c(4) = 1/8 = 0.125$ for M = 4. This exceeds $\alpha = 0.05$ and the hypothesis that events 1 and 10 are coeval ($p = 1/2$) may therefore not be rejected. Strictly speaking, it would have to be accepted. On the other hand, event 10 is separated from event 1 by four intermediate levels with other events in three of the four sections considered. This would suggest that event 10 probably occurs above event 1.

A multivariate statistical approach would be needed to test whether or not two events are coeval when observations on many other events are also available. In Chapter II.6 a multivariate approach (scaling method) will be developed that permits the use of significance tests in which all events can be considered simultaneously.

OTHER APPLICATIONS OF GRAPH THEORY

Earlier in this Chapter, graphs (Figures 2 and 4) have been used for representing relationships between biostratigraphic events. The applications in this section will be to co-occurrences and superpositional relationships of fossil taxa. Graph theory is a branch of applied mathematics in which properties of graphs are established and used to solve specific problems. Roberts (1976, 1978) has provided an excellent introduction to the topic (also see Berge, 1973; and Carré, 1979).

Guex (1980) has made an important contribution to quantitative stratigraphy by adopting a graph theoretical approach. The Guex approach differs from the probabilistic one underlying the algorithms discussed earlier in this Chapter in that co-occurrences of fossils are used as the basic building stones for constructing 'unitary associations' of fossils which can be used for correlation. Guex and Davaud (1984, p. 71) stated that "observed co-occurrences between species must be accepted as true unless the contrary is demonstrated. No deterministic analysis of the problem can be performed otherwise". Later in this volume, results obtained by the RASC computer program will be compared with results obtained by the unitary association method for several examples (Chapter II.7). The purpose of this section is to introduce the additional concepts of graph theory needed for this. Figure 6 (from Guex, 1980) will be used for illustration. For a further application of the graph method of 'unitary associations', the reader is referred to Chapter II.8.

Graphs consist of vertices and arcs or edges. An arc is an edge with an arrow indicating the direction for an ordered pair of vertices. Hypothetical space-time domains of eight fossil species are shown in Figure 6. Observations were made in four stratigraphic sections (heavy black lines in Figure 6b). All observed relationships of co-occurrence or superposition are shown in the graph G of Figure 6f which can be decomposed into an undirected graph (Figure 6g, G_e with edges only) and a directed graph (Figure 6h, G_a with arcs only). The same information is contained in the so-called adjacency matrix of Figure 6a. Each of the fossils has a row and a column in Figure 6a. If two species are observed to co-occur, this is shown by a pair of ones in the adjacency matrix (e.g., 1 and 2). An ordered pair (e.g., 4 and 1) is coded by means of a one in the column for 4 (and row for 1, above the diagonal of zeros in Figure 6a) and a zero in the row for 4 and column for 1 (below the diagonal). If a fossil is observed above another fossil in one or more sections and below it elsewhere, this pair of fossils will be scored as a pair of ones in the adjacency matrix.

An undirected graph G_e is called complete if it contains all possible edges. A complete subgraph of an undirected graph is called a clique. A clique is maximal if it is not contained in a larger clique. Figure 6g has six maximal cliques labelled I to VI in Figures 6c-e. For example, the subgraph (4,8) is complete in Figure 6g. It is referred to

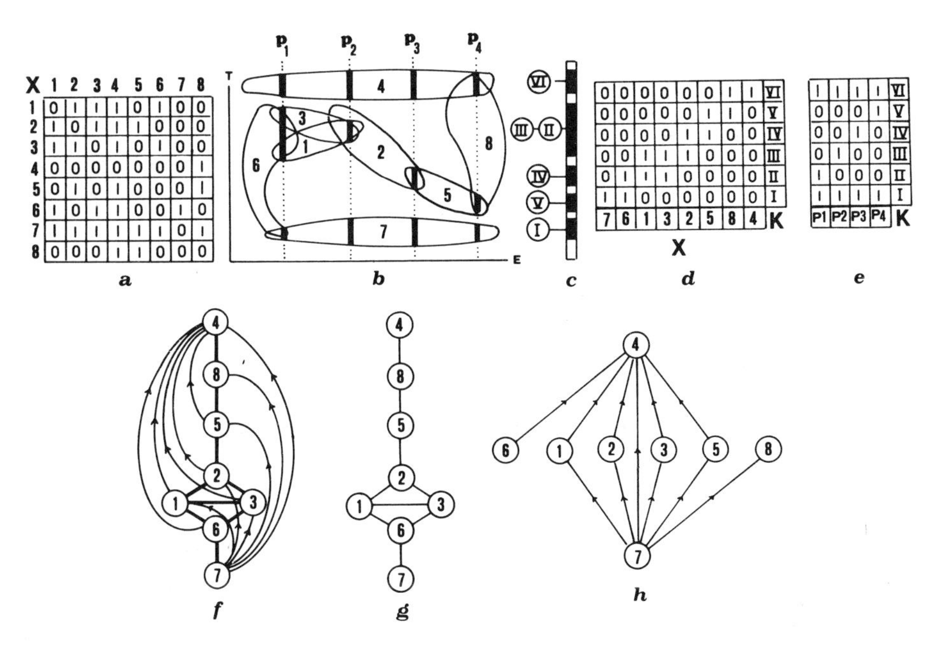

Figure 6 *Example of concepts of graph theory applied in biostratigraphy (from Guex, 1980).*

(a) Adjacency matrix containing same information as Figure 6f for sections in Figure 6b.

(b) Space-time relationship of 8 species numbered 1 to 8; heavy black vertical lines represent stratigraphic sections with observations on domains of existence (closed regions) of the eight species; T = time, E = space.

(c) Relative chronological position of the intervals I to VI for maximal cliques representing 'unitary associations' derived from Figures 6d and 6g.

(d) Matrix relating maximal cliques (K) of Figure 6g to the eight species (X).

(e) Maximal cliques (K) identified in four sections (p_1-p_2) of Figure 6b.

(f) Biostratigraphical graph G representing co-occurrences and superpositional relationships between the 8 species as observed in the four sections.

(g) Undirected graph G_e representing co-occurrences of Figure 6f only.

(h) Directed graph G_a with arcs for superpositional relationships.

It is noted that the original purpose of this diagram was to illustrate, for a simple example, that construction of an interval graph (see Figure 7) normally does not result in a chronological ordering. Only 'reproducible unitary associations' are chronologically ordered as shown in Figure 6e (Guex, 1980).

as maximal clique VI with two consecutive ones in the matrix of Figure 6d. Another example of a maximal clique is III (for fossils 1, 2 and 3) with three consecutive ones in Figure 6d. In the example of Figure 6, the maximal cliques are 'unitary associations' which can be recognized in individual sections without ambiguity (see Figure 6e) and used for correlation. In general, the situation is more complex than that shown in the example of Figure 6 and additional concepts and methods of graph theory are needed.

In general, a set of intervals on the real line can be represented by means of a so-called interval graph. Only graphs with an interval assignment (Figure 7, graphs 1, 2, 3 from Roberts, 1976) are interval graphs. The interval J(i) of a vertex i of an interval graph overlaps at least in part with the intervals of vertices to which i is connected by an edge.

The special graph Z_4 (Figure 7) is not an interval graph because it is not possible to assign intervals to it. The vertices of Z_4 are labelled u, v, w and x in Figure 7. According to the preceding definition of an interval assignment, the intervals J(u) and J(v) would have to overlap because u and v are connected by an edge. J(v) extends to the right of J(u) in Figure 7 because it cannot completely lie within J(u). Otherwise, J(w) could not be overlapping J(v) without overlapping J(u) as required. According to the relationships drawn in Z_4, J(w) overlaps J(v) but not J(u) and must be depicted in the interval assignment as shown. It is not possible now to draw the interval for J(x) which should overlap with J(w) and J(u) but not J(v). This completes the proof that Z_4 does not have an interval assignment and is not an interval graph.

A graph G_e with vertices V and edges E can be written as G_e = (V,E). A graph H_e = (W,F) is a subgraph of G_e = (V,E) if W is a subset of V and F a subset of E. H_e is called a generated subgraph if F consists of all edges from E joining vertices in W. It can be seen that if G_e is an interval graph, then every generated subgraph (but not every subgraph) must also be an interval graph.

Any graph G_e representing associations of fossil species should be an interval graph because pairs of fossils coexisted during specific time intervals with or without overlap. The question of when a graph is an interval graph can be answered in several ways. Fulkerson and Gross (1965) have proved the theorem that a graph G_e is an interval graph if and only if there is a ranking of the maximal cliques of G_e which is consecutive. A ranking K_1, K_2, ..., K_p of the maximal cliques of G_e is called consecutive if whenever a vertex u is in K_i and K_j for $i < j$, then for all $i < r < j$, u is in K_r. It is easy to see that the maximal cliques of G_e in Figure 6 are consecutive. Consequently, G_e of Figure 6 is an interval graph.

Gilmore and Hoffman (1964) proved the following theorem: A graph G_e is an interval graph if and only if it satisfies the following conditions:

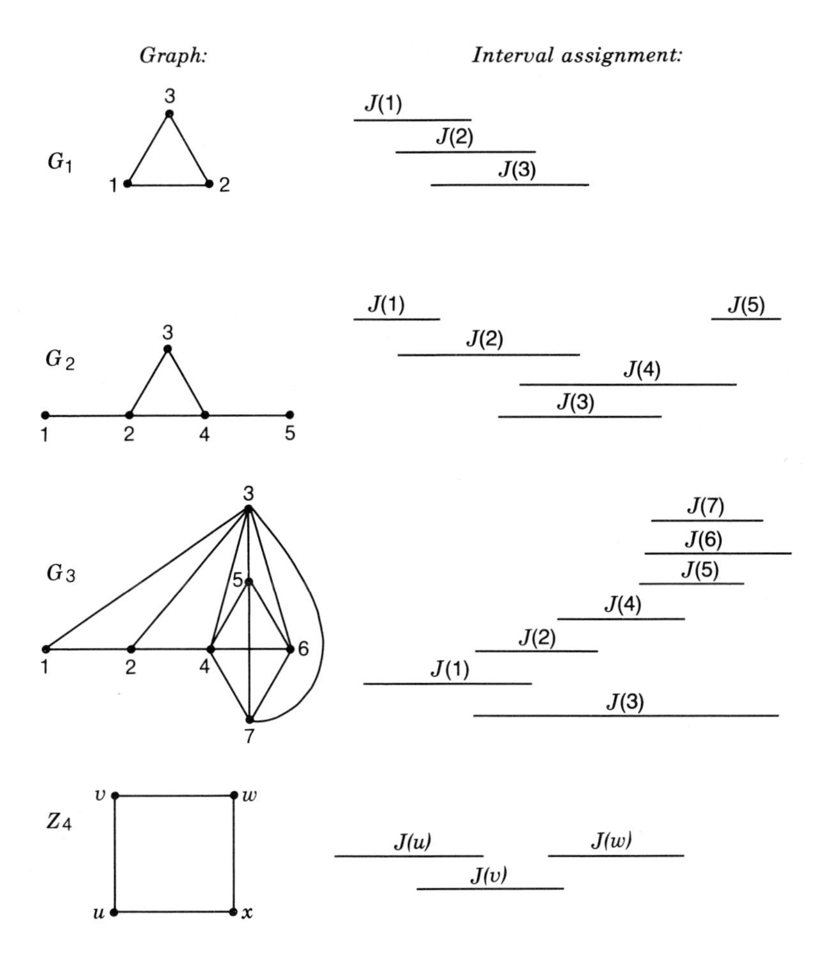

Figure 7 Examples of interval assignments $J(i)$, $i = 1,2, \ldots$ for undirected graphs. An interval assignment for Z_4 with vertices u, v, w and x does not exist (after Roberts, 1976).

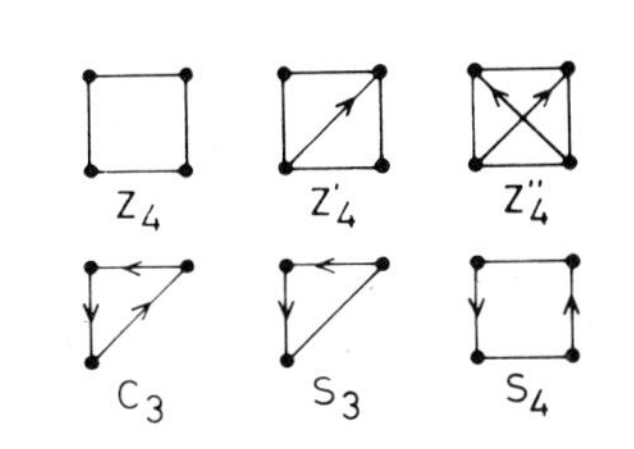

Figure 8 The most frequent structures (subgraphs and generated subgraphs) in G which are 'forbidden' in an interval graph (from Guex and Davaud, 1984).

(a) Z_4 is not a generated subgraph of G_e, and

(b) G_e^c is able to be transitively oriented.

G_e^c is the complementary graph of G_e. It has the same vertices as G_e but edges only between those vertices which are not connected by edges in G_e. If G_e is an interval graph, G_e^c has edges connecting vertices representing nonoverlapping intervals only. Suppose that arrows are assigned to these edges, thus changing them into arcs either pointing in the direction for 'before' or 'after'. It is easy to see that, if G_e is an interval graph, these arrows all point either in the forward or in the backward direction of the real line. Conversely, if G_e^c has the preceding property, then G_e (without Z_4's) is an interval graph according to the theorem of Gilmore and Hoffman. The formal definition of a transitively oriented graph G_a is that, if (travelling in the directions of the arrows) a vertex v can be reached from another vertex u, and a vertex w from v, then w can be reached from u.

A graph G representing stratigraphic relationships (e.g., Figure 6f) generally is a mixture of an undirected graph G_e and a directed graph G_a. From the preceding two theorems, it can be seen that the complement of G_e for the example (Figure 6g) is transitively orientable. The directed graph G_a (Figure 6h) for observed superpositional relationships is a subgraph of the oriented complement of G_e. In a situation that the relationships between all possible pairs of fossils are fully known, the biostratigraphic graph G would be the union of G_e and its transitively oriented complement. If G_e is an interval graph, G cannot contain any of a number of 'forbidden' generated subgraphs. The most frequent forbidden structures are shown in Figure 8 (after Guex and Davaud, 1984). For example, the cycle C_3 with vertices (u,v, and w) shows u before v, v before w and w before u. This is comparable with the 3-event cycle introduced earlier in this Chapter (e.g., cycle ABC in Figure 2). In a biostratigraphical graph G, C_3 is not a possible generated subgraph because it would mean that G_e^c is not transitively orientable and G_e is not an interval graph.

C_3 constitutes the most frequently encountered cycle in biostratigraphical graphs G. C_3 are likely to occur in the strong component of G if it exists. The strong component of a graph is defined as the generated subgraph which is strongly connected and has the maximum number of vertices. A directed graph is called strongly connected if for every pair of its vertices u and v, v is reachable from u and u from v. Guex and Davaud (1984) introduced a special coefficient s = c/r for each arc (e.g., u to v) where c represents the number of times this arc occurs in a C_3 within the strong component while r is the total number of times it occurs in the same strong component. If the coefficient of an arc is high, this may indicate reworking or contamination. If reworking is suspected, v is omitted in beds where it was observed to occur above u. For contamination, u would be removed from below v.

Guex and Davaud (1984) have developed further rules for interactive or automated elimination of other forbidden structures from G. For example, Z_4 is removed by assuming 'virtual' co-occurrence for either a pair of two or all four of the fossils involved. Two fossil species are said to co-occur 'virtually' if their co-occurrence was not observed but inferred. After elimination of all inconsistencies, the biostratigraphic graph G yields an interval graph G_e of which the maximal cliques can be determined. These are the initial unitary associations. (I.U.A.). They are called 'initial' because Guex and Davaud (1984) have recently added the following method for combining some of the I.U.A.s with one another in order to form the U.A.s. The I.U.A.s are identified in sections as previously illustrated for the 'unitary associations' in Figure 6e. A complete I.U.A. may not be observed in a section. However, a given I.U.A. is fully characterized by any one of its unique species or pairs of species. I.U.A.s characterized by 'virtual' (inferred, not observed) co-occurrences of fossils only cannot be identified in sections. Guex and Davaud (1984) then proceed by constructing the directed graph G_k of superpositional relations between the I.U.A.s as identified in the sections. The construction of G_k with the I.U.A.s as vertices is identical to the extraction of G_a from the original biostratigraphical graph G. Next they find the I.U.A.s with the longest path in G_k. In general, a vertex in a directed graph G_a is connected to another vertex by means of a 'path' if the arrows on the arcs between these two vertices point in the same direction. Each I.U.A. not on the longest path is combined with the I.U.A. on the path with which it has an interval in common. This gathering process yields the final unitary associations (U.A.s) which are identified in the sections as the I.U.A.s were before. If the new I.U.A.-U.A. method is applied to the example of Figure 6, the initial unitary associations II and III would be combined with one another.

SUMMARY AND CONCLUDING REMARKS

The two RASC methods (presorting option; modified Hay method) for ranking biostratigraphic events were presented in the first part of this Chapter. The presorting option of the RASC computer program is simple, does not require the selection of a threshold parameter, and takes little computing time. In a recent study evaluating various ranking techniques, Harper (1984) found in his computer simulation experiments on artificial series that the presorting option provided slightly better ranks than the modified Hay method only. The recommended procedure therefore is to use presorting first, followed by modified Hay method for fine-tuning. For the latter method, a relatively large value of m_{c1} has to be chosen if the number of cycles becomes too large to obtain an optimum sequence. The optimum sequences for nine data bases in Appendix I were obtained by presorting followed by the modified Hay method except for Baumgartner's data base (No. 4) and Rubel's combined data base (No. 6D). In the latter two instances, only presorting was applied because of numerous cycles occurring when it was attempted to use the modified Hay method.

It is noted that the original method as presented by Hay (1972) can only be used when cycles involving 3 or more stratigraphic events are missing. In that situation, the original and modified Hay methods yield the same optimum sequence which then is independent of the assignment of scores from the T-matrix to events observed to be coeval. The conditions for this special situation were not fulfilled in any of the data bases listed in Appendix I.

The binomial test developed by Hay (1972) and Southam et al. (1975) for testing the null hypothesis that two stratigraphic events are coeval on the average, was discussed later in this Chapter. This test should be used only when the two events co-occur relatively frequently in the same sections. Otherwise, the estimated confidence belts tend to become so wide that the null hypothesis is too readily accepted. The binomial test was not used in our applications. The uncertainty range for positions of events in the optimum sequence reflects missing data and events observed to be coeval on average. A multivariate approach for testing the hypothesis that two events are coeval on the average, based on scaling, will be presented in the next Chapter.

It has been pointed out that the probabilistic ranking methods are not 'conservative' in that average positions of events in the optimum sequence are biased estimates of true first or last occurrences. The average last occurrence of an event may fall (considerably) below its last occurrence in sections where this event was observed close to its true extinction point in time. As pointed out by Edwards (1982), range charts based on probabilistic methods generally have shorter stratigraphic ranges than those based on conservative methods, in which it is attempted to approximate true origins and extinctions in time as closely as possible. In conservative methods, also more weight is given to all observed co-occurrences of fossil taxa, which is true for the graph theory approach as well.

Graph theory can be fruitfully applied in biostratigraphy. The last part of this Chapter contained an introduction to the method of unitary associations (Guex and Davaud, 1984) based on interval graphs. This method yields zonations with somewhat longer ranges of taxa than when probabilistic methods are used. More detailed comparisons between the unitary association method and probabilistic ranking and scaling will be made in Chapters II.7 and II.8.

REFERENCES

Agterberg, F.P., 1984, Binomial and trinomial models in quantitative biostratigraphy. Computers and Geosciences 10, (1), 31-41.

Agterberg, F.P., and Gradstein, F.M., 1981, Workshop on quantitative stratigraphic correlation techniques. Ottawa, February, 1980, J. Math. Geol. 13, (1), 81-91.

Agterberg, F.P., and Nel, L.D., 1982a, Algorithms for the ranking of stratigraphic events. Computers and Geosciences 8, (1), 69-90.

Agterberg, F.P., and Nel, L.D., 1982b, Algorithms for the scaling of stratigraphic events. Computers and Geosciences 8, (2), 163-189.

Berge, C., 1973, Graphes et hypergraphes. Dunod, Paris, 516 pp.

Blank, R.G., and Ellis, C.H., 1982, The probable range concept applied to the biostratigraphy of marine microfossils. J. Geol. 90, (4), 415-433.

Bramlette, M.N., and Sullivan, F.R., 1961, Coccolithophorids and related nannoplankton of the Early Tertiary in California. Micropal. 7, 129-188.

Brower, J.C., 1981, Quantitative biostratigraphy, 1830-1980. In Merriam, D.F., ed., Computer Applications in the Earth Sciences, an Update for the 70's, Plenum Press, New York, 63-103.

Brower, J.C., and Burroughs, W.A., 1982, Simple methods for quantitative biostratigraphy. In J.M. Cubitt and R.A. Reyment, eds., Quantitative Stratigraphic Correlation, Wiley, Chichester, 61-83.

Bunk, H.D., 1960, Mathematical models for ranking from paired comparisons. J. Am. Stat. Assoc. 55, 503-520.

Carré, B., 1979, Graphs and networks. Clarendon Press, Oxford, 277 pp.

Davaud, E., and Guex, J., 1978, Traitement analytique 'manuel' et algorithmique de problèmes complexes de correlations biochronologiques. Eclogae Geol. Helv, 71, (3), 581-610.

Doeven, P.H., Gradstein, F.M., Jackson, A., Agterberg,F.P., and Nel, D., 1982, A quantitative nannofossil range chart. Micropal. 28, (1), 85-92.

Edwards, L.E., 1982, Numerical and semi-objective biostratigraphy: Review and predictions. Third N. Am. Paleont. Convention, Montreal, Canada, August 1982, Proc. 1, 147-152.

Edwards, L.E., and Beaver, R.J., 1978, The use of paired comparison models in ordering stratigraphic events. J. Math. Geol. 10, (3), 261-272.

Fulkerson, D.R., and Gross, O.A., 1965, Incidence matrices and interval graphs. Pacific J. Math. 15, 835-855.

Gilmore, P.C., and Hoffman, A.J., 1964, A characterization of comparability graphs and interval graphs. Can. J. Math. 6, 539-548.

Gradstein, F.M., 1984, On stratigraphic normality. Computers and Geosciences 10, (1), 43-57.

Gradstein, F.M., and Agterberg, F.P., 1982, Models of Cenozoic foraminiferal stratigraphy - Northwestern Atlantic Margin. In Cubitt, J.M., and Reyment, R.A., eds., Quantitative Stratigraphic Correlation, John Wiley, Chichester, p. 119-173.

Guex, J., 1980, Calcul, caractérisation et identification des associations unitaires en biochronologie. Bull. Soc. Vaud. Sci. Nat. 75, (358), 111-126.

Guex, J., and Davaud, E., 1984, Unitary associations method: Use of graph theory and computer algorithm. Computers and Geosciences 10, (1), 69-96.

Harper, C.W., Jr., 1984, A Fortran IV program for comparing ranking algorithms in quantitative biostratigraphy. Computers and Geosciences 10, (1), 3-29.

Hay, W.W., 1972, Probabilistic stratigraphy. Eclogae Geol. Helv. 65, (2), 255-266.

Hemelrijk, J., 1952, A theorem on the sign test when ties are present. Kon. Nederl. Akad. Wetensch., Proc., 55, 322.

Hudson, C.B., and Agterberg, F.P., 1982, Paired comparison models in biostratigraphy. J. Math. Geol. 14, (2), 141-159.

Kendall, M.G., 1975, Rank Correlation Methods. Griffin, London, 202 pp.

Roberts, F., 1976, Discrete Mathematical Models. Prentice-Hall, Inc., Englewood Cliffs, N.J., 559 p.

Roberts, F., 1978, Graph theory and its applications to problems of society. Regional Conference Series in Applied Mathematics 29, SIAM, Philadelphia, Penn., 122 pp.

Rubel, M., 1978, Principles of construction and use of biostratigraphical scales for correlation. Computers and Geosciences 4, (3), 243-246.

Rubel, M., and Pak, D.N., 1984, Theory of stratigraphic correlationby means of ordinal scales. Computers and Geosciences 10, (1), 97-105.

Shaw, A.B., 1964, Time in Stratigraphy. McGraw-Hill Book Co., New York, 365 pp.

Southam, J.R., Hay, W.W., and Worsley, T.R., 1975, Quantitative formulation of reliability in stratigraphic correlation. Science 188, 357-359.

Sullivan, F.R., 1965, Lower Tertiary nannoplankton from the California Coast Ranges; II. Eocene. Univ. Calif. Publ. Geol. Sc. 53, 1-52.

Tocher, K.D., 1950, Extension of the Neyman-Pearson theory of tests to discontinuous variates. Biometrika 37, 130.

Worsley, T.R., and Jorgens, M.L., 1977, Automated biostratigraphy. In: A.T.S. Ramsay, ed., Oceanic Micropaleontology, Academic Press, London, 2, 1201-1229.

II.6

METHODS OF SCALING BIOSTRATIGRAPHIC EVENTS

FREDERIK P. AGTERBERG

INTRODUCTION

The scaling algorithms presented in this Chapter form the second part of the RASC program for ranking and scaling of biostratigraphic events and other events which can be uniquely identified. An optimum sequence constructed by means of a ranking algorithm provides the starting point for estimating average 'distances' (in time) between successive events. The frequency of cross-over (mismatch) of the events in the sections is used for this purpose. These distances are then clustered by constructing a dendrogram which can be used as a standard and permits the establishment of assemblage zones. This Chapter also contains artificial examples in which the theory of scaling is illustrated and tested by applying it to sets of random normal numbers in computer simulation experiments.

The techniques described in this Chapter have in common that numbers are generated that can serve as 'distance'- values between successive events in the optimum sequence which was obtained by the ranking algorithms described in the previous Chapter. The RASC computer program has originally been developed to erect a zonation for the Cenozoic benthonic and planktonic foraminiferal record (highest occurrences of 206 taxa) in wells on the Canadian Atlantic Continental Margin (Gradstein and Agterberg, 1982). In a ranking, the successive events follow each other and no allowance can be made for the situation that some events should be closer together than others along a relative time scale. It can be useful to position the events along such a scale with variable intervals between them if the record seems to indicate such variations. For example, suppose that two microfossils have observed extinction points (A and B) in 10 sections with A occurring 5 times above B, and 5 times below B. If a diagram were constructed, in which each event is connected to itself in other sections, the lines connecting event A would cross those connecting the event B in a number of places. It can be said that the relative cross-over (mismatch) frequency is p_{AB} = 0.5. If the number of sections is not too small, this frequency can be regarded as an estimate of the probability that A occurs above B. The corresponding distance between A and B along the relative time scale then should be close to zero.

The preceding case can be distinguished from the case that A occurs, for example, 9 times above B and only once below B. Then A and B should be separated by a longer distance along the time scale, corresponding to $p_{AB} = 0.9$.

The purpose of the scaling techniques of this Chapter is to establish 'distances' in time between successive events, not only from the cross-over frequencies between successive events, but also by using the cross-over frequencies between all events.

Figure 1 from Gradstein and Agterberg (1982) provides an example of output from a scaling algorithm. The number codes of the events (exits of microfossils) and the microfossil names are shown on the right side. Each code is followed by the estimated distance from its event to the event below it. These distances have also been plotted in the horizontal direction toward the left. They were clustered as follows during a sequence of linking steps. The two successive events (36 and 86) in the optimum sequence with the shortest distance (0.0007) between them were linked first. After scanning the set of unused inter-event distances, single events or clusters of events were linked pairwise, at each linking step, by using the shortest distances between them until the longest inter-event distance (between 17 and 22) was reached. The resulting cluster, based on inter-event distances in time, resemble assemblage zones.

The solution of Figure 1 for 68 taxon exits in 16 wells on the Labrador Shelf and northern Grand Banks shows a number of distinct and progressively younger clusters. A shading pattern was used to enhance the stratigraphically most useful parts of individual clusters. In total, nine preferred assemblages are shown. The input data for Figure 1 were processed by using the modified Hay (see Chapter II.5) method with threshold parameters $k_c = 3$ and $m_c = 3$. The resulting optimum sequence was used as a starting point for scaling. It was reordered during the application of the scaling algorithm as will be discussed later in this chapter. Consequently, the scaled optimum sequence shown on the right side of Figure 1 differs from the optimum sequence originally obtained by the ranking algorithm. The distances between successive events shown in Figure 1 can be added in order to obtain the distance of each event from a common origin that coincides with the first event (No. 9 in Figure 1). The resulting RASC distances can be related to geological time on the basis of those events for which the age is relatively well known (see Figure 2).

In order to construct Figure 1, the output of the RASC program listed in Agterberg and Nel (1982) was combined with a DISSPLA graphics package (copyrighted in 1975 by Integrated Software System Corporation). A version of this DISSPLA program called DENO was published by Jackson et al. (1984). DENO was used to construct all optimum sequences and dendrograms for the nine data bases in Appendix I.

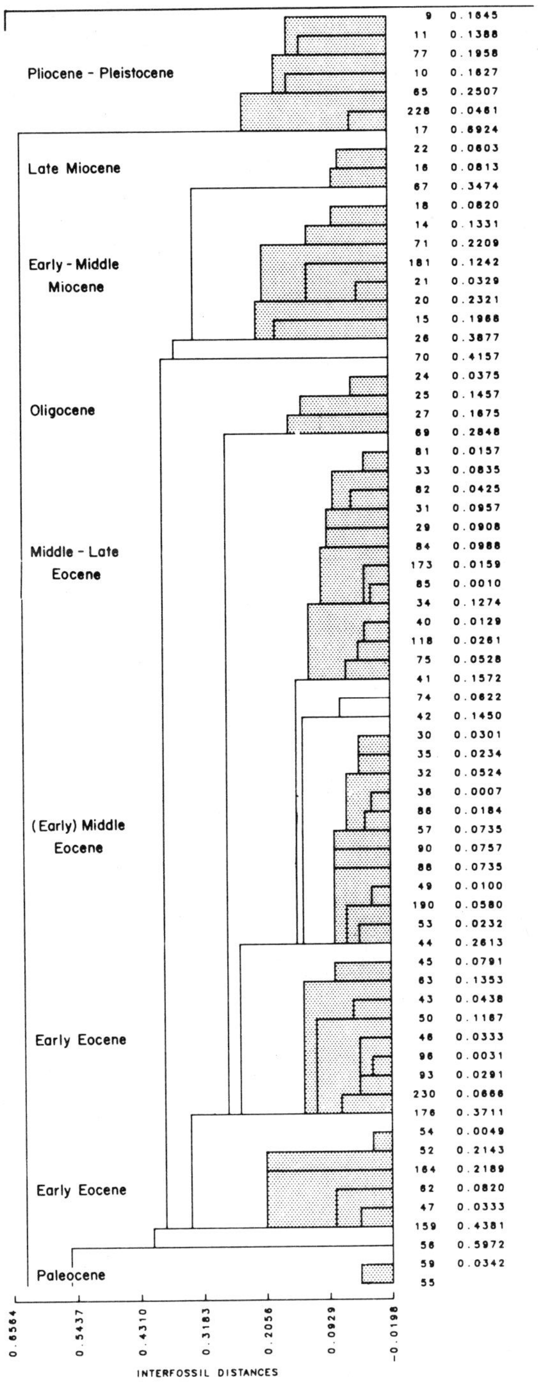

Figure 1

Example of result obtained by the scaling of biostratigraphic events. Numbers represent coded events as in Gradstein and Agterberg (1982). Exits of 206 taxa were run with $k_c = 3$ and $m_c = 3$. Code numbers are followed by distances between successive events which also have been plotted and clustered.

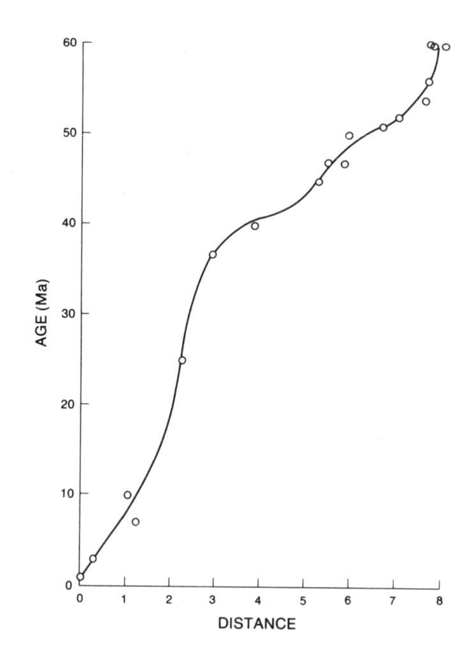

Figure 2 Relation between geological time in Ma and RASC distances for Cenozoic foraminifera in 18 wells, NW Atlantic continental margin (from Agterberg and Gradstein, 1983). A spline-curve was fitted to circles representing selected biostratigraphic events.

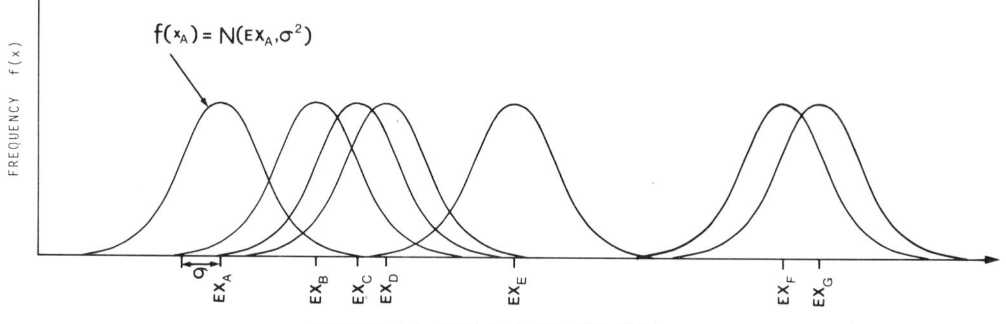

Figure 3 Probabilistic model for the clustering of biostratigraphic events (A, B, C, ...) along a relative time scale (x-axis). The relative position of each event (e.g., A) in a section or well is a random variable (X_A) which is normally distributed around its average location (EX_A) with standard deviation σ. Random variables are denoted by capital letters (e.g., X_A) and values assumed by random variables in lower case (e.g., x_A).

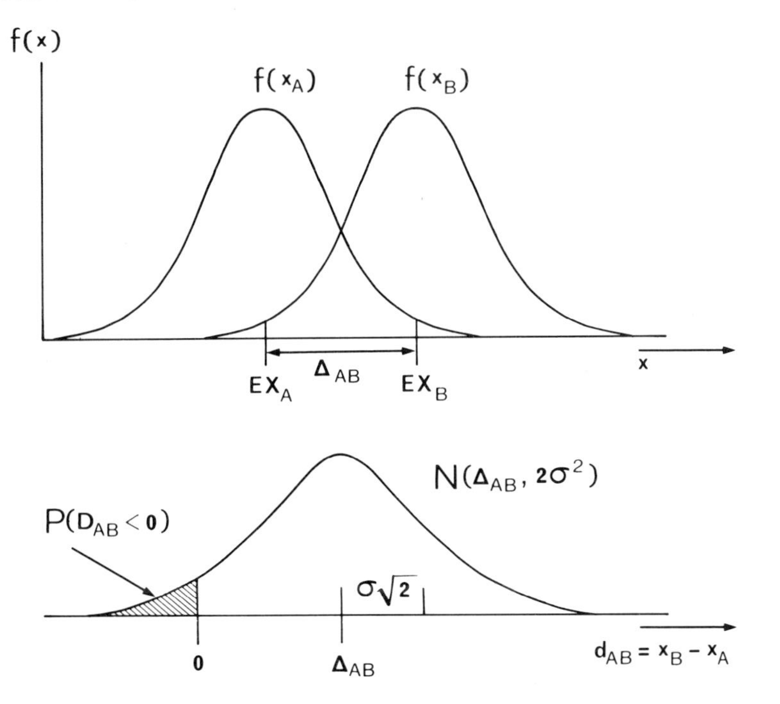

Figure 4 *Direct estimation of distance Δ_{AB} between events A and B from their relative cross-over frequency p_{AB} which is assumed to be equal to $P(D_{AB} < 0)$. The random variable D_{AB} $(=X_B - X_A)$ is negative only when the order of A and B in a section is the reverse of the order of EX_A and EX_B. The variance of D_{AB} is twice as large as the variance σ^2 of the individual events A and B.*

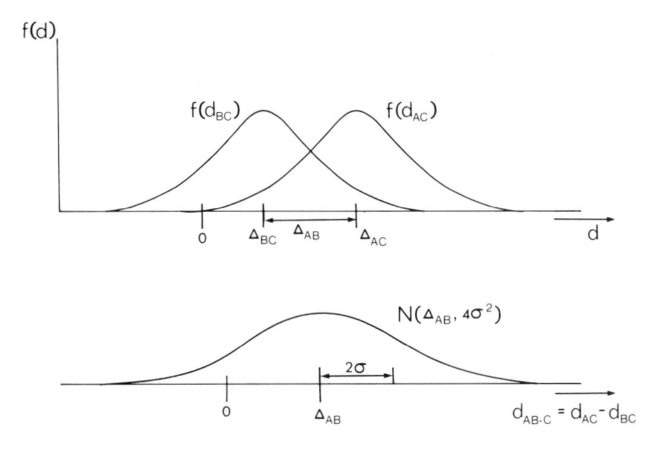

Figure 5 *Indirect estimation of distance Δ_{AB} from relative cross-over frequencies p_{AC} and p_{BC}. The indirect distance $D_{AB.C} = D_{AC} - D_{BC}$ has a variance which is four times as large as the variance of the individual event-positions A, B, and C.*

SCALING OF STRATIGRAPHIC EVENTS

Statistical Model

The situation of events that interchange places with one another in different sections can be explained by assuming that the place of each event along the 'time-line' is described by a different probability distribution. As pointed out before, the exact probability distributions of the events are not known. However, it can be assumed that all distributions are of a specific type. It will be shown that this allows the estimation of the parameters of the model. The usefulness of this statistical approach is that, later, the fitted model can be tested against the observed data. This final testing either verifies or negates the results obtained by means of the statistical model.

Figure 3 shows the basic model initially adopted for the scaling algorithms. Each event (e.g. A) would assume a position x_{Ai} in section i where x_{Ai} is the distance to A from an origin with arbitrary location along the relative time scale. The distance x_{Ai} is assumed to be the realization of a random variable X_A whose probability distribution is shown in Figure 4. Similar random variables are defined for the other events B, C,...

The random variable X_A satisfies the normal (Gaussian) probability distribution $N(EX_A, \sigma^2)$ with expected (or mean) value EX_A and variance σ^2. The mean values of the event-positions differ from one another but the standard deviations of all event-positions are assumed to be equal to σ in the model of Figure 3.

The normal distribution curves for events A and B are shown separately in Figure 5. Because the time scale is relative, it will not be possible to estimate σ. (In the RASC program σ^2 is arbitrarily set equal to 0.5.) However, it is possible to estimate the ratio $\Delta_{AB}/(\sigma\sqrt{2})$ for the distance between the population means $\Delta_{AB} = EX_B - EX_A$ from the relative cross-over frequency p_{AB}. For this purpose, p_{AB} is considered to provide a good estimate of the probability $P(X_B - X_A > 0) = P(D_{AB} > 0)$ which satisfies

$$P(D_{AB}>0) = \frac{1}{\sqrt{(2\pi)}} \cdot \int_{-\Delta_{AB}/\sigma\sqrt{2}}^{\infty} e^{-z^2/2} dz$$

This formula follows from the fact that the difference $D_{AB} = X_B - X_A$ has a normal distribution $N(\Delta_{AB}, 2\sigma^2)$ which is shown in the bottom part of Figure 4. The distance between events A and B for a specific section can be written as $d_{AB} = x_B - x_A$. A precise estimate of p_{AB} which would allow the determination of Δ_{AB} is seldom available in practical applications because this would require a very large number of sections containing both A and B. However, it is generally possible to estimate Δ_{AB} indirectly by using pairs of cross-over frequencies linking A and B to other events, for example, by using the pair p_{AC} and p_{BC}. A distance of this type will be written as $D_{AB.C}$. As illustrated in Figure 5, $D_{AB.C} = D_{AC} - D_{BC}$ is normally distributed with $N(\Delta_{AB}, 4\sigma^2)$. Because σ^2 is arbitrary, the variance of the normal distribution was set equal to a constant ($\sigma^2 = 0.5$).

It can be argued that the model of Figure 3 is not realistic because it is unlikely that all events would have the same normal curve with variance equal to σ^2 for their exit location distributions. However, in practice, each of the indirect distances $D_{AB.C}$ is based on two independent distances (D_{AC} and D_{BC}) and, in turn, each of these two variables is based on two independent distances from an arbitrary origin, although X_C is used twice. Hence $D_{AB.C}$ is based on three random variables that can not be estimated separately. Because of the central-limit theorem of statistical theory, $D_{AB.C}$ tends to be normally distributed even if the frequency curves of events A, B and C were not normal and had unequal variances.

Table 1 shows the relative cross-over frequencies for the example of Table 3 in Chapter II.5. The order of the events is that of the optimum sequence obtained by means of the modified Hay method for ranking. Each of the frequencies of Table 1A was changed into a fractile of the standard normal distribution or z-value (= probit - 5) (see matrix in Table 1B).

Table 2 shows z-values for selected relative cross-over frequencies. Because $p_{ji} = 1 - p_{ij}$, it follows that $z_{ji} = -z_{ij}$. When the optimum sequence is used as a starting point, all or most of the z-values in the upper triangle of the Z-matrix are positive. Negative values occur in the upper triangle only for elements with $p_{ij}<0.5$ corresponding to events whose scores were temporarily ignored in order to break a cycle in which these events were participating. It is noted that scores ignored for constructing the optimum sequence are restored to their positions before use of the scaling algorithms of RASC is initiated.

Clearly, a relative frequency p_{ij} for a small sample will be subject to a considerable uncertainty and this error is propagated into the z_{ij}-value derived from it. For this reason, it is desirable to define a minimum sample size m_{c2}. It means that z_{ij}-values based on fewer than m_{c2} pairs of occurrences will not be used. In the original RASC program no distinction was made between m_{c1} and m_{c2}. However, later work has shown that better results can be obtained by setting $m_{c2}>m_{c1}$. For the example of Table 1, $m_{c1} = m_{c2} = m_c = 3$. Then two values will not be used (see Table 1). When an average distance is estimated in practice, for example between events 3 and 5, it would generally be based on nine estimates of the distance. The direct estimate of the distance between events i and j follows from z_{ij} and the indirect estimates from the differences $z_{ik} - z_{jk}$ ($k\neq i,j$) where

Table 1 (a) *Matrix of relative cross-over frequencies p_{ij} for example of 10 events. The order of events is according to the optimum sequence.*
(b) *Corresponding z-values. Note that two values are not determined because $m_c = 3$ and the value $q = 1.645$ was used for all cross-over frequencies equal to one.*

A.	(2)	(1)	(3)	(5)	(7)	(4)	(6)	(8)	(9)	(10)
(2)	x	5/7	3/6	5/8	4½/6	3/6	3/4	5/5	8/8	4½/5
(1)		x	2½/5	4½/8	4/6	4½/6	2½/3	5/5	8/8	4½/5
(3)			x	4/6	3½/5	4½/6	3/4	4/4	6/6	3½/4
(5)				x	3½/7	4½/7	3/4	5/5	9/9	5/6
(7)					x	3½/6	1½/3	4½/5	7/7	4½/5
(4)						x	3/4	3/4	6/7	3½/5
(6)							x	1/2	4/4	1½/2
(8)								x	5/5	2½/3
(9)									x	3/6
(10)										x

B.	(2)	(1)	(3)	(5)	(7)	(4)	(6)	(8)	(9)	(10)
(2)	x	0.566	0.000	0.319	0.674	0.000	0.674	1.645	1.645	1.282
(1)		x	0.000	0.157	0.431	0.674	0.967	1.645	1.645	1.282
(3)			x	0.431	0.524	0.674	0.674	1.645	1.645	1.150
(5)				x	0.000	0.366	0.674	1.645	1.645	0.967
(7)					x	0.210	0.000	1.282	1.645	1.282
(4)						x	0.674	0.674	1.068	0.524
(6)							x	ND*	1.645	ND*
(8)								x	1.645	0.967
(9)									x	0.000
(10)										x

*ND, not determined because m_c=3 for this example

Table 2 *Example of z-values for selected relative cross-over frequencies p.*

p	z
0.05	-1.645
0.10	-1.282
0.20	-0.842
0.30	-0.524
0.40	-0.253
0.50	0.000
0.60	0.253
0.70	0.524
0.80	0.842
0.90	1.282
0.95	1.645

i and j = i+1 are successive rows. However, because $z_{ij} = -z_{ji}$, the differences $z_{kj} - z_{ki}$ ($k \neq i,j$), where i and j = i+1 are successive columns, also can be used. For the example, the direct estimate of distance satisfies $d_{35} = z_{35}(\sigma\sqrt{2}) = z_{35} = 0.431$ because $\sigma^2 = 0.5$. The indirect estimates are $z_{25} - z_{23} = 0.319 - 0.000 = 0.319$, $z_{15} - z_{13} = 0.157 - 0.000 = 0.157$, and six other, similar differences between z-values in adjacent columns or rows. In the RASC program, z-values in the upper triangle are used only. The lower triangle is used to retain information on sample sizes. Addition of indirect and direct estimates yields the sum $S_o = (0.319-0.000) + (0.157-0.000) + 0.431 + (0.524-0.000) + (0.674-0.366) + (0.674-0.674) + (1.645-1.645) + (1.645-1.645) + (1.150-0.967) = 1.922$. The average of all nine estimates of the distance between events 3 and 5 amounts to $S_o/9 = 0.214$. This is called an unweighted estimate of distance in the output of the RASC program. Because of missing values (see Table 1) or pairs of cross-over frequencies which both are equal to one (see later), distance estimates may be based on fewer than N-1 (=9 for the example) pairs of events.

The relative cross-over frequencies p_{ij} are calculated from scores (s_{ij}) on samples of different sizes (r_{ij}). For this reason, it is preferable to compute weighted mean distances $\overline{\Delta_{ij}}$ in which the weights assigned to the direct and indirect estimates of distance are primarily determined by the sizes of the samples used to obtain the z-values. The weight-corrected equation for estimating the distance between events i and j is:

$$\overline{\Delta_{ij}} = \frac{w_{ij} z_{ij} + \sum_{k \neq i,j} w_{ij.k}(z_{ik} - z_{jk})}{w_{ij} + \sum_{k \neq i,j} w_{ij.k}}$$

where the weights w_{ij} and $w_{ij.k}$ are

$$w_{ij} = \frac{r_{ij} e^{-z_{ij}^2}}{2\pi p_{ij}(1-p_{ij})} \qquad w_{ij.k} = \frac{(w_{jk})(w_{ik})}{w_{ik} + w_{jk}} \qquad k \neq i,j$$

Weights were derived in the following manner. The observed proportion p_{ij} is assumed to be the realization of a random variable P which is related to a standard normal variable Z such that

$$P = \frac{1}{\sqrt{2\pi}} \int_{-\infty}^{Z} e^{-s^2/2}\, ds$$

where s denotes position along the linear scale used.

The proportion P can be assumed to originate from a binomial random variable with expected value $E(P) = p_{ij}$ and variance

$$\sigma^2(P) = \{p_{ij}(1-p_{ij})\}/r_{ij}$$

where r_{ij}, as before, is the number of times E_i and E_j occurred in the same section.

It is known that

$$\sigma^2(P) = \left(\frac{\partial p}{\partial z}\right)^2 \sigma^2(Z)$$

where p and z represent the density functions of P and Z. These equations can be combined into

$$\sigma^2(P) = \frac{1}{2\pi} e^{(-z_{ij}^2)} \sigma^2(Z)$$

Each weight w_{ij} is obtained as

$$w_{ij} = \frac{1}{\sigma^2(Z)} = \frac{r_{ij}e^{-z_{ij}^2}}{2\pi p_{ij}(1-p_{ij})}$$

Weights $w_{ij.k}$ are obtained by addition of similar variances $\sigma^2(Z)$ of the values z_{ik} and z_{jk}.

Table 3 shows weighted distances $\bar{\Delta}_{i,i+1}$ (i = 1, ..., N-1) estimated for successive events in the optimum sequence. For example, the weighted distance between events 3 and 5 is calculated as follows. From Table 1 it follows that $r_{35} = 6$, $p_{35} = 4/6$ and $z_{35} = 0.931$. Consequently, $w_{35} = 3.569$. Likewise, for example, $w_{25} = 4.907$ and $w_{23} = 3.820$. Hence $w_{35.2} = 2.148$. The sum of the 9 weights is W = 2.148 + 1.952 + 3.569 + 1.749 + 1.835 + 1.078 + 0.497 + 0.806 + 0.985 = 14.619. The corresponding sum satisfies S = 2.148(0.319-0.000) + 1.952(0.157-0.000) + 3.569 x 0.431 + 1.749(0.524-0.000) + 1.835(0.674-0.366) + 0.985(1.150-0.967) = 4.19. The weighted distance is therefore $\bar{\Delta}$ = S/W = 0.287. This value is among those listed in Table 3.

For simplification, the preceding formula can be rewritten as:

$$\bar{x} = W^{-1} \sum_{i=1}^{N*} w_i x_i$$

with $\bar{x} = \overline{\Delta_{AB}}$

$$W = \sum_{i=1}^{N*} w_i$$

where N* represents the number of pairs of z-values, while

$$x_1 = z_{AB}, \qquad w_1 = w_{AB}$$

$$x_2 = z_{ZC} - z_{BC}, \; w_2 = w_{AB.C}$$

$$x_3 = z_{AD} - z_{BD}, \; w_3 = w_{AB.D}$$

$$x_4 = \ldots, \text{ etc.}$$

where A and B denote successive events, and other events are written as C, D, ... The weight W and sum $\Sigma\, w_i\, x_i$ are given in Table 3. Note that N* includes the z-value for the direct estimate. The standard deviation s(x) shown in Table 3 is the positive square root of

$$s^2(\bar{x}) = W^{-1}(N^*-1)^{-1} \sum_{i=1}^{N*} w_i (x_i - \bar{x})^2$$

The standard deviation for the distance between events 3 and 5 amounts to 0.060 (see Table 3). This is considerably less than the distance itself (=0.287). It indicates that the distance is significantly different from zero. A rapid test of this hypothesis consists of multiplying the standard deviation by 2 and subtracting the result from the estimated distance. If the difference is negative, the distance could well be zero.

When all possible comparisons can be made, N* = N-1. However, N* may be less than N-1 for the following two reasons:

(1) The total number of comparisons is reduced by one for each value x_i that cannot be computed because one of the z-values needed is missing (this includes the case that both z-values are missing);

(2) If $s_{ij} = r_{ij}$, $p_{ij} = 1$ and the corresponding z-value is set equal to a value q (=1.645 in Table 1).

Pairs of z-values both equal to q, and with zero-difference, are not used for estimating the average distance Δ_{ij} unless a pair of this type is contained within a cluster of mutally inconsistent events. For this reason, pairs of values (z_{ik}, z_{jk}) in successive rows are tested by letting k increase from k = i+2. Suppose that, for a given value of k, $z_{ik} = z_{jk} = q$. This pair is not used for the distance estimation unless a pair of z-values, which are not both equal to q, is found for a larger value of k. In the RASC program, it is assumed that this situation is encountered as soon as five pairs of z-values equal to q have been identified for increasing k. Likewise, pairs of values (z_{ki}, z_{kj}) in successive columns are tested by letting k decrease from k = i-1.

The preceding situation occurs in Table 1 for estimation of the distance between events 8 and 9. Because four pairs of values both are equal to q = 1.645, and because this example also has a non-determined value, N* = 9 - 5 = 4. The corresponding weight (W) in Table 3 is only 4.2, and the standard deviation (=0.306) for this distance (= 0.861) is greater than any other standard deviation estimated.

When a large number of events for a long time interval is used, N* is likely to be much smaller than N-1 in all distance calculations, because events belonging to relatively young assemblages (e.g. Late Miocene in Figure 1) normally all occur above events in older assemblages (e.g. Early Eocene in Figure 1). Distance estimates based on few pairs of z-values are relatively imprecise. In the RASC program there is an option that distances based on N* less than m_{c2} are replaced by zeros.

The choice of a value of q is usually not critical, because most pairs of q-values will not be used for distance estimation. As a 'default', q is set equal to 1.645 in the RASC program. This corresponds to a cross-over frequency of p = 0.95 (see Table 2). The user can replace the default value by any other value. In general, q should be greater than 1 and less than 2. A value of q is selected because, theoretically, a cross-over frequency of 1 corresponds to an infinitely large z-value and distance estimation would not be possible. It can be assumed that the scores from which cross-over frequencies are calculated satisfy binomial frequency distributions. For small samples, the probability that a cross-over frequency is equal to 1(or 0), is then relatively large even when a minimum sample size (m_{c2}) has been defined.

Table 3 (a) *Calculation of distances between successive events in optimum sequence with each z-value weighted according to its sample size (N*).*
(b) *The events have been re-ordered on the basis of their cumulative distance from the first event. Results of final re-ordering are shown in the last three columns.*

A	Pair of events	Distance	Standard deviation	Cumulative distance	Sum and weight (W)	N*
	2-1	0.111	0.134	0.111	1.63(14.8)	9
	1-3	-0.110	0.082	0.000	-1.47(13.3)	9
	3-5	0.287	0.060	0.287	4.19(14.6)	9
	5-7	0.148	0.079	0.435	2.32(15.7)	9
	7-4	0.157	0.153	0.592	2.34(15.0)	9
	4-6	0.266	0.163	0.858	2.44(9.2)	7
	6-8	0.770	0.203	1.628	3.62(4.7)	6
	8-9	0.861	0.306	2.488	3.59(4.2)	4
	9-10	-0.317	0.100	2.171	-3.27(10.3)	8

B	Reordered sequence	Distance from origin	Pair of events	Recalculated distance	Standard deviation
	3	0.000	3-2	0.006	0.124
	2	0.006	2-1	0.117	0.147
	1	0.123	1-5	0.195	0.082
	5	0.318	5-7	0.157	0.085
	7	0.475	7-4	0.157	0.153
	4	0.632	4-6	0.266	0.163
	6	0.898	6-8	0.770	0.203
	8	1.668	8-10	0.176	0.289
	10	1.844	10-9	0.317	0.100
	9	2.161			

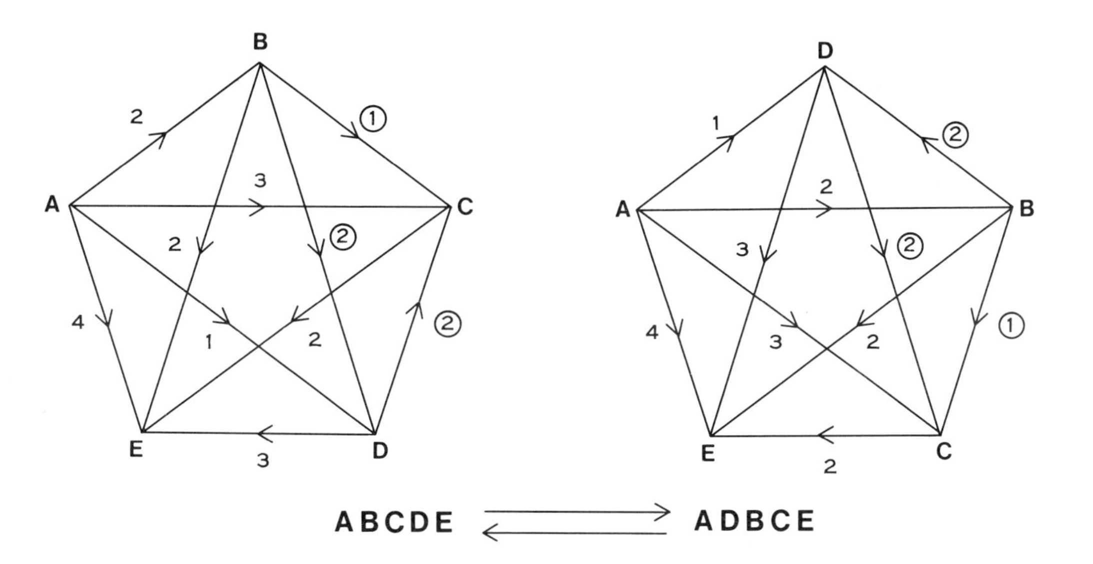

Figure 6 Artificial example for demonstrating that the final re-ordering option of the RASC program does not necessarily converge to unique solution. See text for further explanations.

This problem is restricted to the tails of the normal frequency curve and can be solved by choosing a q-value which, effectively, changes the range of the normal curve from $(-\infty, \infty)$ to $(-q, q)$.

Two of the distances estimated in Table 3A are negative. For this reason, it is desirable to reorder the events before a dendrogram of successive inter-event distances is constructed. The cumulative distance from the first event in the original optimum sequence obtained by the modified Hay method (event 2 in Table 3) can be calculated for each event. The events may be reordered on the basis of this cumulative distance. This allows the clustering of successive distances as shown in Figure 1.

The standard deviations of the distances between successive events cannot be recalculated readily after a reordering which removes negative distances. This is because successive distance estimates are not statistically independent. In order to obtain the new standard deviations, it is necessary to repeat all calculations taking the reordered optimum sequence as the starting point. Because different z-values then are used for estimation, the distance estimates will change as is illustrated in Table 3B. New negative distances may be computed at this stage and the procedure would have to be repeated again. These new calculations can be performed by using the final reordering option of the RASC program. The objective of final reordering is to obtain a set of distances between successive events which are all positive so that the corresponding standard deviations are also known. This result could be achieved for the examples discussed in this Chapter. However, when the data base is large, and when k_c and m_{c2} are small, it may not be possible to obtain a single set of consecutive distances which are all positive. This is because the iterative process does not necessarily converge to a single solution, and at most four complete reorderings are allowed in the RASC program. If convergence to a situation of positive distances is not obtained in four steps, either the results without final re-ordering can be accepted, or the result obtained after four reorderings. In the latter solution, the number of negative distances probably will have been reduced considerably. In practice, the improvements gained by employing the final reordering option may be small as can be evaluated by comparing standard deviations of distances before and after reordering.

Figure 6 illustrates that the preceding iterative process for final reordering does not necessarily converge to a single solution. Suppose that the numbers in Figure 6 represent estimated distances between pairs of events A, B, C, D and E. A positive distance from one event to another is indicated by an arrow pointing from the one event to the other. For example, the distance from A to B is +2 and that from C to D is -2. Let the optimum sequence ABCDE have only one negative distance (between C and D). Because this distance is greater than that between B and C (= +1), the reordered sequence becomes ADBCE. The distances for this artificial exmaple have been chosen in such a way that this new sequence has again only one negative distance (between D and B)

Table 4 *Array of random normal numbers with $E(X) = 2$ and $\sigma = 1$ taken from Table A-23 of Dixon and Massey (1957, p. 452-453). Headings of columns indicate subsample number and event letter (A, B, or C).*

1A	1B	1C	2A	2B	2C	3A	3B	3C	4A	4B	4C	5A	5B	5C
2.422	0.130	2.232	1.700	1.903	0.725	2.031	0.515	–0.684	1.911	0.626	2.289	1.628	1.638	2.676
0.694	2.556	1.868	1.263	2.115	1.516	1.972	3.627	1.482	3.196	2.979	2.447	2.099	1.273	2.733
1.875	2.273	0.655	2.299	0.055	1.955	–0.147	2.168	2.193	0.398	2.304	1.019	0.363	1.286	2.428
1.017	0.757	1.288	1.322	2.080	2.170	1.502	2.953	0.171	1.228	2.134	0.300	1.785	2.547	1.566
2.453	4.199	1.403	2.017	3.496	0.165	2.556	1.003	1.973	1.190	3.020	0.954	2.907	2.916	1.279
2.274	1.767	1.564	2.412	2.207	0.475	2.656	1.579	0.394	0.953	2.127	1.723	2.302	1.474	0.826
3.000	1.618	1.530	2.224	2.881	2.715	3.103	1.941	2.179	1.479	1.956	1.280	1.722	0.938	0.922
2.510	2.256	1.146	5.177	1.931	1.693	1.021	3.337	2.137	1.509	0.952	1.258	–0.864	1.620	1.789
1.233	2.085	2.251	1.578	3.796	3.017	2.863	2.514	1.615	0.627	2.404	0.571	2.940	2.705	1.709
3.075	1.730	2.427	2.990	1.680	3.250	3.050	3.243	1.846	1.923	2.765	2.422	1.725	1.009	2.372
1.344	–0.095	2.166	4.116	2.500	1.939	1.567	3.047	1.385	2.760	2.633	3.011	2.277	1.539	0.873
1.246	3.860	1.253	1.876	4.373	1.993	1.262	2.319	2.488	2.009	3.204	1.114	2.269	0.912	0.831
0.889	2.299	2.458	1.790	1.048	2.302	0.138	2.383	1.170	0.876	1.124	2.137	1.448	1.236	1.699
1.154	1.401	1.935	3.106	1.548	–0.096	2.153	2.333	1.761	1.430	1.920	2.969	1.518	1.543	1.509
3.031	1.048	0.719	1.474	2.779	0.292	2.341	2.707	1.741	3.422	2.307	2.919	1.833	1.792	3.090
0.534	1.155	1.705	1.662	0.457	0.602	1.365	2.663	3.755	3.304	1.292	1.863	2.785	1.666	0.323
2.230	3.096	0.045	3.639	0.580	0.970	1.593	2.117	2.395	2.329	2.671	3.353	1.166	1.016	3.036
2.355	1.761	1.816	1.822	1.434	2.259	3.788	3.280	1.317	1.402	1.964	1.505	1.746	1.912	1.202
1.461	0.947	0.717	2.923	2.133	2.526	2.687	2.144	1.692	2.274	1.209	1.450	2.241	1.678	1.565
3,034	1.778	2.122	2.025	3.008	1.447	–0.305	2.452	1.726	1.205	0.531	2.975	3.024	3.357	2.558
2.761	0.473	3.726	1.893	2.455	1.633	1.654	3.006	3.523	2.462	1.328	1.301	3.312	1.959	2.010
1.961	0.965	1.481	1.402	2.106	2.214	1.727	3.670	3.795	0.227	3.166	1.989	2.976	2.188	1.399
2.639	4.010	1.915	1.713	1.484	1.443	1.444	2.394	1.688	2.142	2.926	1.634	1.940	0.875	2.331
1.349	2.225	0.644	1.404	2.583	2.149	2.359	2.274	1.432	2.558	0.903	0.082	1.299	2.366	2.554
2.959	2.797	4.635	3.268	2.889	2.349	0.933	3.403	2.206	0.818	3.174	0.123	–1.149	1.606	2.118
2.440	2.919	1.455	0.695	1.466	1.124	1.257	1.265	0.096	3.083	2.287	2.379	2.909	2.520	0.708
3.078	3.279	0.352	2.583	1.690	0.729	2.072	1.332	1.158	2.517	1.470	2.621	0.880	1.931	1.495
1.736	1.968	0.011	2.418	1.026	1.342	2.103	1.792	2.175	2.594	1.571	1.218	2.346	2.267	0.946
3.275	3.147	2.800	2.172	0.004	1.763	3.801	2.510	2.517	0.411	0.760	1.114	1.842	1.756	3.951
2.579	2.297	2.030	2.725	3.721	2.545	1.631	–0.346	–0.011	2.853	3.054	2.421	2.418	1.542	2.070

Table 5

Sequences of A, B, and C generated from random normal numbers of Table 4. For explanations, see text.

Subsample	1	2	3	4	5
	BAC	ACB	CBA	BAC	ABC
	ACB	ACB	ACB	ACB	ABC
	ACB	BAC	ABC	ACB	ABC
	ABC	ABC	ACB	ACB	ACB
	ACB	CAB	BAC	ACB	CAB
	ABC	CAB	CBA	ABC	ACB
	BAC	ABC	BAC	ACB	ABC
	ACB	BCA	ACB	ABC	ABC
	ABC	ACB	ACB	ACB	ACB
	BAC	BAC	ACB	ABC	ABC
	BAC	CBA	ACB	ABC	ACB
	ACB	ACB	ABC	ACB	BAC
	ABC	ABC	ACB	ABC	ABC
	ABC	CBA	ACB	ABC	ABC
	BCA	ACB	ACB	BAC	ABC
	ABC	BAC	ABC	BAC	CBA
	CAB	BCA	ABC	ABC	ABC
	ABC	ABC	CAB	ABC	ACB
	ABC	ABC	ABC	BAC	ABC
	BAC	ACB	ACB	ABC	ACB
	BAC	ACB	ABC	BAC	BAC
	ABC	ABC	ABC	ACB	CAB
	ACB	ABC	ACB	ACB	BAC
	ACB	ABC	ACB	CBA	ABC
	ABC	CAB	ACB	ACB	ABC
	ACB	ABC	ACB	ABC	CAB
	CAB	CAB	ABC	BAC	ABC
	CAB	BAC	ABC	BAC	ACB
	ABC	BAC	BAC	ABC	ABC
	ABC	ACB	BCA	ACB	ABC

and reordering ADBCE gives the original sequence ABCDE. Consequently, a unique solution with positive distances between successive events does not exist. Situations similar to the one illustrated in Figure 6 can occur in practice, especially where the estimated distances are not very precise.

Artificial Example

The purpose of this section is to illustrate the theory of scaling developed in the previous sections by using an artificial example based on random normal numbers. Although the theory leads to valid results for large samples, small-sample fluctuations may be considerable. This aspect will be evaluated in this section. In general, the understanding of statistical models applied to observed data can be helped considerably by simulation experiment models. Nevertheless, it should be kept in mind that numbers are used, of which it may be known beforehand that they should fit well, while in practical applications to real data the conditions artifically created for a simulation experiment may not be satisfied. The following artificial example clearly demonstrates some features of the theory outlined in the previous sections, but differs from natural situations by

(1) small number of events,

(2) large number of sequences,

(3) all events are observed in all sequences, and

(4) the positions of the events satisfy normal distributions with equal variances.

Table 4 shows an array of random normal numbers with E(X) = 2 and σ = 1 taken from Table A-23 of Dixon and Massey (1957, pp. 452 and 453). There are 15 columns and 30 rows. Table 5 was created from the random normal numbers in Table 4 according to the following device. Suppose that the relative order of three events (A, B and C) has been observed in 150 sections, and that $E(X_A) = 1$, $E(X_B) = 2$, $E(X_C) = 2.5$, and $\sigma(X_A) = \sigma(X_B) = \sigma(X_C) = 1$. Values of X_A and X_C can be obtained by, respectively, subtracting 1 from and adding 0.5 to random normal numbers with E(X) = 2 and σ = 1. For example, the first three numbers in row 1 of Table 4 (2.422, 0.130 and 2.232) can be transformed into $x_A = 1.422$, $x_B = 0.130$, and $x_C = 2.732$. The relative order of the three events in the first hypothetical section then is BAC (see Table 5). The sample of 150 sequences of Table 5 has been divided into 5 columns for subsamples each consisting of 30 sequences.

By counting, it was determined that A is followed by B in 116 of the 150 sequences. (This implies that B precedes A in 34 sequences.) Likewise, A precedes C in 130 sequences and B comes before C in 85 sequences. These three numbers (n) are shown in the first column of

Table 6 RASC method of scaling applied to data of artificial example. For meaning of column headings, see text.

	n	f	z	D (direct)	D (indirect)	D_o (Ave)	D (Ave)	E (D)
AB	116	0.7733	0.750	1.061	1.335	1.198	1.152	1.000
AC	130	0.8667	1.111	1.571	1.297	1.434	1.480	1.500
BC	85	0.5667	0.167	0.236	0.510	0.373	0.327	0.500
SSD				0.079	0.152	0.060	0.053	

Table 7 Statistical analysis of Table 6 repeated for the five sub-samples.

Subsample		D (direct)	D (indirect)	D_o (Ave)	D (Ave)
1 ...	AB	1.030	1.089	1.060	1.050
	AC	1.571	1.512	1.542	1.551
	BC	0.482	0.541	0.512	0.502
	SSD	0.006	0.010	0.006	0.006
2 ...	AB	0.741	0.911	0.826	0.798
	AC	1.030	0.860	0.945	0.973
	BC	0.119	0.289	0.289	0.176
	SSD	0.459	0.462	0.426	0.424
3 ...	AB	1.191	1.809	1.500	1.112
	AC	1.571	0.953	1.262	1.365
	BC	−0.238	0.380	0.071	−0.032
	SSD	0.586	0.968	0.491	0.314
4 ...	AB	0.881	2.236	1.559	1.333
	AC	2.594	1.239	1.917	2.142
	BC	0.358	1.713	1.036	0.810
	SSD	1.231	3.067	0.774	0.619
5 ...	AB	1.571	1.089	1.330	1.410
	AC	1.571	2.053	1.812	1.732
	BC	0.482	0.000	0.241	0.321
	SSD	0.331	0.564	0.273	0.254

Table 6. They were transformed into relative frequencies (f) by dividing them by 150 (see column 2 of Table 6). By consulting a table of cumulative frequencies for the normal distribution in standard form, the f-values were converted into z-values. Multiplication by $\sqrt{2}$ then yields direct estimates of the distances between the events. For example, $D_{AB} = 0.750 \times \sqrt{2} = 1.061$. Only one indirect estimate of the distance D_{AB} can be obtained. It is equal to 1.335 which represents the difference between 1.571 (direct estimate of D_{AC}) and 0.236 (direct estimate of D_{BC}). The arithmetic average of the direct and indirect estimates of distance is shown in column 6 of Table 6. This is followed by a weighted distance estimate which satisfies D(Ave) = (1.061 + 1.335/2)/1.5 = 1.152. Finally, the expected value E(D) is shown in the last column.

Comparison of the three estimates of distance is facilitated by computing the sum of squared deviations (SSD) from the expected value of each estimate. For example, for the direct estimate of D in Table 6,

$$SSD = (1.061-1.000)^2 + (1.571-1.500)^2 + (0.236-0.500)^2 = 0.079$$

The SSD values are also shown in Table 6. The results suggest that the variance of the indirect estimate which is proportional to its SSD value is about twice as large as that of the direct estimate. The weighted average distances are most precise because they have the smallest variance.

The analysis shown in Table 6 was repeated for the 5 smaller subsamples. In all instances, the weighted mean distance provided the best estimate (see Table 7). It can also be seen, however, that in small samples, the estimated distance may differ considerably from its expected value.

In the preceding statistical analysis of which the results are reported in Tables 6 and 7, the weighted distance estimate D(Ave) was obtained by assigning twice as much weight to the direct estimate as to the indirect estimate. Because weights are inversely proportional to variances, this simply reflects the fact that, on the average, the variance of indirect estimates is twice as large as that of direct estimates (see Figure 5). Suppose, however, that the equation for estimating weighted distances $\Delta_{i,i+1}$ as in the RASC program is used. From the values in the columns for f and z in Table 6, it is readily computed that $w_{AB} = 77.593$, $w_{AC} = 60.139$, and $w_{BC} = 94.549$. On the other hand, $w_{AB.C} = 36.758$, $w_{AC.B} = 42.618$, and $w_{BC.A} = 33.880$. The latter three weights are not exactly half as large as the first three weights. The reason for this discrepancy is that the values of f and z are approximations only. They were estimated from samples and used instead of the population values in the RASC method. The RASC weighted distances become 1.149, 1.457 and 0.308, instead of 1.152, 1.480 and 0.327 shown in Table 6 for D(Ave). Their corresponding SSD value becomes 0.061 indicating that the D(Ave) values (with SSD = 0.053) of Table 6 are better in this artificial example.

Computer Simulation Experiments

The type of experiment described in the previous section can also be performed on a computer using a pseudo-random number generator. Computer simulation experiments on biostratigraphic events have recently been performed by Edwards (1982) and Harper (1984).

Edwards (1982) dealt with the problem graphically represented in Figure 3 earlier in this chapter. She assumed that a taxon has a true extinction point in time which was randomly displaced upwards or downwards in a section due to sediment mixing. The aim of experiments of this type is to model the distribution of exits (or entries).

Harper (1984) has performed computer simulation to create artificial successions of taxa in sections. Three types of optimum sequences were obtained for exits and entries of these taxa by means of the RASC program:

(a) Presorting option only;

(b) Modified Hay method only;

(c) Scaling the sequence derived from (b).

As already discussed in more detail in the previous chapter, Harper has demonstrated that, for his successions, (a) consistently gave results that were better than those of (b) and recommended that (a) instead of (b) be used as the input sequence for (c). However, results of (b) could be improved by making m_{c1} as small as possible. A recent revision made in the RASC program partly on the basis of Harper's results is that it is now possible to use different threshold values for number of pairs of events (m_{c1} and m_{c2}; see before).

Our own computer simulation experiments were as follows. Twenty artificial stratigraphic events were studied in each of 50 sections whereby the interval between the expected positions of the events was kept constant in each run. For comparison, in the experiment of Table 4, 3 events were studied in each of 150 sections and, in total, 3 x 150 = 450 random normal numbers were used. For each experiment to be described now, 20 x 50 = 1000 random normal numbers with σ = 1 were used. Every experiment was performed twice using a different set of 1000 random normal numbers but the same set of 1000 random normal numbers was used in each of two sets of experiments where the interval between expected positions E(D) was put equal to 1.0, 0.5, 0.3, 0.2, 0.1 and 0.0, respectively. Table 8 shows the artificial sequences of events created for the first two sections for all 6 intervals in one of the two sets of experiments. Table 8 illustrates that for E(D) = 1.0 (σ=1.0),

Table 8 First two artificial sequences used in set of computer simulation experiments (20 events in 50 sections) with E(D) equal to 1.0, 0.5, 0.3, 0.2, 0.1, and 0.0, respectively.

Series No.	E(D)	Expected Sequence Numbers																			
		1	2	3	4	5	6	7	8	9	10	11	12	13	14	15	16	17	18	19	20
1	1.0	1	2	4	5	3	6	8	10	9	7	11	12	13	14	15	17	16	18	19	20
2	1.0	1	4	3	2	7	6	5	8	9	11	10	12	13	15	14	16	18	17	19	20
1	0.5	1	2	5	4	3	6	10	8	9	11	13	14	12	15	7	17	16	18	19	20
2	0.5	1	4	3	2	7	8	9	6	11	5	12	13	10	15	18	19	16	14	17	20
1	0.3	5	1	4	2	10	3	6	8	11	9	15	13	14	17	12	16	7	19	18	20
2	0.3	1	4	3	7	2	8	9	11	6	12	13	18	15	5	10	19	16	20	17	14
1	0.2	5	4	1	2	10	11	3	6	8	9	15	17	14	13	12	16	19	18	20	7
2	0.2	1	4	7	3	11	8	9	13	18	12	2	15	6	19	20	10	16	5	17	14
1	0.1	5	10	4	2	1	11	17	15	14	8	9	13	6	3	16	12	19	20	18	7
2	0.1	1	4	7	18	11	19	9	13	8	12	3	15	20	2	6	16	17	10	14	5
1	0.0	5	10	17	15	11	4	14	13	16	19	2	9	20	8	1	18	6	12	3	7
2	0.0	1	18	19	4	20	11	13	15	7	12	9	8	17	16	3	10	6	14	2	5

there are relatively few inconsistencies in the observed sequences. On the other hand, for E(D) = 0.1 it is difficult to recognize from the sequences that the expected sequence is 1,2,...,20. The sequences for E(D) = 0.0 are, of course, completely random. Table 9 shows optimum sequences obtained by:

(a) Presorting option;

(b) ditto, followed by modified Hay method;

(c) Scaling (unweighted differences);

(d) Scaling (weighted differences);

(e) ditto, after final reordering.

For scaling, the constant q (see discussion of Table 3) was set equal to 2.326, which corresponds to $p = 0.99$. Ranking correlation coefficients (Kendall's $\hat{\tau}$ and Spearman's $\hat{\rho}$) for comparison to the expected sequence 1,2,...,20 are also shown in Table 9. Except for E(D) = 0.0, these coefficients are very large and highly significant. From the results in Table 9, it cannot be decided which optimum sequence is best.

The ranking correlation coefficients in Table 9 were estimated by using the equations (cf. Kendall, 1975):

$$\hat{\tau} = \frac{2S}{n(n-1)} \qquad \hat{\rho} = 1 - \frac{6\ SSD}{n(n^2-1)}$$

where S is a total score of +1 for pairs of elements having the same order in both series and -1 otherwise. The total number of elements is written as n. Spearman's ρ is based on the sum of squared differences (SSD) of rankings of the elements in the two series compared to one another.

Table 10 shows estimated distances between successive events and their standard deviations, as obtained after final reordering for E(D) = 0.1 in set 1. Because $\sigma = 1$, the expected distance between two successive events in the expected sequence 1,2,...,20 amounts to $0.1\sqrt{2} = 0.0707$. The order of the events in Table 10 differs from that of the expected sequence. Nevertheless, there is general agreement between the estimated distances and the expected value of 0.0707. An exception is the value of 0.4008 between events 19 and 20 which is larger than expected (probably due to the artificial truncation with q = 2.326). The total expected distance between events 1 and 20 amounts to 19 x 0.07071 = 1.3435 and is only slightly less than the observed total distance of 1.5501.

Table 9 Various optimum sequences obtained in computer simulation experiments and their ranking coefficients with the 'true' sequence 1, 2, ..., 20.

Set 1	E(D) = 1.0					0.5					0.3				
	a	b	c	d	e	a	b	c	d	e	a	b	c	d	e
1	*	*	*	*	*	1	*	*	*	*	1	*	*	*	*
2						3					3				
3						2					2				
4						4					4				
5						5					5				
6						6					6				
7						7					7				
8						8					8				
9						9					9				
10						10					11				
11						11					10				
12						12					12				
13						13					13				
14						14					14				
15						15					15				
16						16					16				
17						17					17				
18						18					18				
19						19					19				
20						20					20				
Kendall's $\hat{\tau}$	1	1	1	1	1	0.990	0.990	0.990	0.990	0.990	0.979	0.979	0.979	0.979	0.979
Spearman's $\hat{\rho}$	1	1	1	1	1	0.999	0.999	0.999	0.999	0.999	0.997	0.997	0.997	0.997	0.997

Set 2	E(D) = 1.0					0.5					0.3				
	a	b	c	d	e	a	b	c	d	e	a	b	c	d	e
1	*	*	*	*	*	*	*	*	*	*	*	*	1	*	*
2													2		
3													3		
4													4		
5													5		
6													6		
7													7		
8													8		
9													9		
10													10		
11													12		
12													11		
13													13		
14													14		
15													15		
16													16		
17													17		
18													19		
19													18		
20													20		
Kendall's $\hat{\tau}$	1	1	1	1	1	1	1	1	1	1	1	1	0.979	0.979	0.979
Spearman's $\hat{\rho}$	1	1	1	1	1	1	1	1	1	1	1	1	0.997	0.997	0.997

* – no changes with respect to column to the left
a – presorting option
b – ditto, followed by modified Hay method
c – scaling (unweighted differences)
d – scaling (weighted differences)
e – ditto, after final reordering

		0.2					0.1					0.0		
a	b	c	d	e	a	b	c	d	e	a	b	c	d	e
1	*	1	1	*	3	*	1	1	*	15	15	15	15	*
3		3	3		4		3	3		14	11	14	14	
2		2	2		5		4	4		11	4	3	3	
4		4	4		1		2	5		5	5	19	11	
5		6	5		2		5	2		3	14	6	19	
6		5	6		6		6	6		4	17	11	4	
7		7	7		8		8	8		19	18	18	6	
8		8	8		11		7	7		18	3	16	18	
11		11	11		7		11	11		16	16	4	16	
9		9	9		9		9	9		6	6	17	5	
10		10	10		10		10	10		8	19	5	17	
12		12	12		15		15	15		17	8	8	8	
15		13	13		13		14	14		1	1	13	13	
13		14	14		12		12	12		13	9	10	1	
14		15	15		14		13	13		12	7	1	10	
16		16	16		16		16	16		10	20	12	12	
17		18	18		18		18	18		20	13	7	7	
18		17	17		17		17	17		7	12	2	9	
19		19	19		19		19	19		9	10	9	2	
20		20	20		20		20	20		2	2	20	20	
0.947	0.947	0.947	0.968	0.968	0.853	0.853	0.884	0.874	0.874	-0.116	-0.042	-0.137	-0.095	-0.095
0.990	0.990	0.991	0.994	0.994	0.955	0.955	0.974	0.970	0.970	-0.128	-0.111	-0.193	-0.125	-0.125

		0.2					0.1					0.0		
a	b	c	d	e	a	b	c	d	e	a	b	c	d	e
1	*	1	1	*	1	1	1	1	*	19	19	20	20	*
2		2	2		2	2	2	2		16	16	5	5	
3		3	3		3	3	5	5		5	10	1	19	
4		4	4		4	4	3	3		20	5	17	16	
5		5	5		5	5	4	4		17	20	19	17	
6		6	6		6	6	7	7		10	17	16	10	
7		7	7		8	8	6	6		1	3	2	1	
8		8	8		7	7	9	8		3	1	8	2	
10		10	10		10	10	8	9		4	14	10	3	
9		9	9		9	9	10	10		2	7	4	8	
11		12	12		12	12	12	12		8	9	9	4	
12		11	11		14	14	14	14		7	2	18	9	
14		14	14		11	11	16	16		18	4	7	7	
13		13	13		16	16	11	11		14	12	3	18	
15		15	15		13	13	13	13		9	8	12	14	
16		17	16		17	19	17	17		12	18	14	12	
17		16	17		19	17	19	19		6	6	15	6	
19		19	19		15	15	15	15		15	13	6	15	
18		18	18		18	18	18	20		13	11	13	13	
20		20	20		20	20	20	18		11	15	11	11	
0.979	0.979	0.958	0.968	0.968	0.895	0.884	0.863	0.863	0.863	-0.032	-0.063	-0.021	-0.021	-0.021
0.997	0.997	0.994	0.996	0.996	0.979	0.970	0.961	0.959	0.959	-0.089	-0.122	-0.035	-0.065	-0.065

Table 10 Computer simulation experiment with E(D) = 0.1. Estimated distances between successive events (weighted difference method) after final re-ordering, and their standard deviations. Total distance between events 1 and 20 amounts to 1.5501. Expected total distance is equal to 19 x 0.1 $\sqrt{2}$ = 1.3435.

DISTANCE ANALYSIS REPEATED WITH WEIGHTED DIFFERENCES

POSITION	FOSSIL PAIRS	FOSSIL DISTANCE	CUMULATIVE DISTANCE	SUM DIFF Z VALUES	WEIGHT	S.D.
1	1- 3	.0055	.0055	1.4314	258.9	.0395
2	3- 4	.0235	.0290	6.1615	262.7	.0467
3	4- 5	.1293	.1582	35.2188	272.5	.0394
4	5- 2	.0114	.1697	3.1952	279.4	.0354
5	2- 6	.0337	.2034	9.2798	275.5	.0337
6	6- 8	.2177	.4210	61.6319	283.2	.0415
7	8- 7	.0360	.4570	10.6016	294.7	.0282
8	7- 11	.1019	.5589	30.3554	297.8	.0405
9	11- 9	.1303	.6892	38.8631	298.2	.0431
10	9- 10	.0359	.7251	10.6802	297.7	.0379
11	10- 15	.0073	.7324	2.1837	298.0	.0309
12	15- 14	.0753	.8078	22.0608	292.9	.0340
13	14- 12	.0247	.8325	7.1684	289.7	.0261
14	12- 13	.0230	.8555	6.7007	291.2	.0372
15	13- 16	.1359	.9915	38.7111	284.8	.0409
16	16- 18	.0486	1.0400	13.6101	280.3	.0402
17	18- 17	.0605	1.1006	16.6032	274.3	.0424
18	17- 19	.0487	1.1493	13.0438	267.7	.0417
19	19- 20	.4008	1.5501	97.4825	243.2	.0512

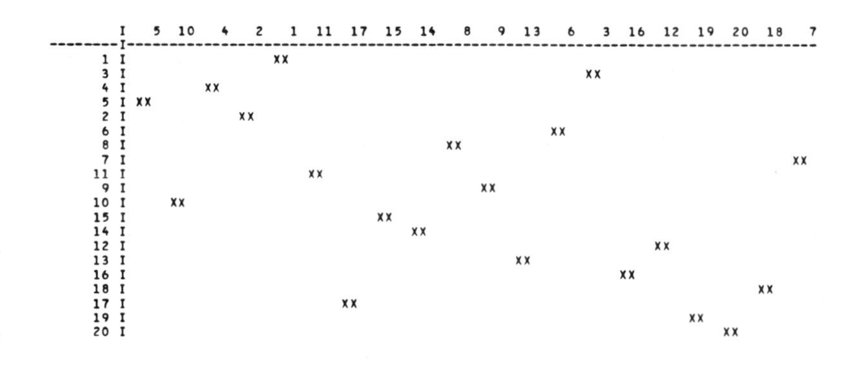

Figure 7 Example of scattergram. Sequence for section (first sequence in Table 8 for E(D) = 0.1) plotted in horizontal direction against scaled optimum sequence after re-ordering (see Table 9).

NORMALITY TEST

The normality test, which follows scaling in the RASC computer program, is an important tool to spot anomalous events which occur either much higher or much lower than expected in a section when individual sections or sequences are compared with the optimum sequence obtained through scaling. The RASC program also provides three optional outputs immediately preceding the normality test. These are:

(1) An occurrence table with the final ranking in the vertical direction. Sections are in this table represented by columns. If an event occurs in a section its presence is indicated by an X.

(2) Each section can be compared to the optimum sequence by using a system of scoring penalty points when an event is out of place. This procedure is called the Step Model. The relative order of every pair of events is checked against their order in the optimum sequence. If the order is different, one penalty point is scored. Coeval events each receive half a point. Obviously, an event with many penalty points is likely to be either too high or too low in a section. A drawback of the step model is that events which belong to clusters of events with many internal inconsistencies are likely to accumulate high total scores even if they occur in normal positions. It may not be easy to distinguish between anomalous events which are out of place and events which are part of a cluster. For this reason one should use the step model in conjunction with other ways to test normality.

(3) Scattergrams: The order of events in the sections is plotted against the optimum sequence. An example is shown in Figure 7.

The preceding three outputs can be obtained before as well as after scaling. The normality test can only be performed after scaling. Two outputs for the normality test are shown in Table 11 for the computer simulation experiment with E(D) = 0.1 in set 1. After scaling, each event occurs at a fixed distance from the origin, which was arbitrarily set at the position of the first event. Hence, the distance-score of each event, with the exception of the first and last events, can be compared with the distance-scores of its two neighbours in a particular section. The amount by which the event is out of place can be evaluated statistically by calculating 'second order difference values' for each event in relation to its neighbours.

If the second-order difference value of an event in a particular section indicates that the event is out of place with a probability of 99 percent, the event receives two asterisks in the output, while one asterisk signifies an event that occurs too high or too low with a probability of 95 percent. It is to be expected that, on the average, one and five percent of all events tested will be assigned two asterisks and one asterisk, respectively, when there are no anomalies.

Table 11 Two kinds of RASC output for normality test. Computer simulation experiment for E(D) = 0.1 was used. The output is explained in the text.

	CUM. DIST.	2ND ORDER DIFF.
5	.1582	
10	.7251	-1.2630
4	.0290	.8368
2	.1697	-.3104
1	0.0000	.7286
11	.5589	-.0173
17	1.1006	-.9097
15	.7324	.4434
14	.8078	-.4621
8	.4210	.6550
9	.6892	-.1020
13	.8555	-.8184
6	.2034	.4543
3	.0055	1.1838
16	.9915	-1.1449
12	.8325	.4757
19	1.1493	.0840
20	1.5501	-.9109
18	1.0400	-.0730
7	.4570	

COMPARISON OF OBSERVED AND EXPECTED OCCURRENCES OF SECOND ORDER DIFFERENCE VALUES

CLASS NO.	OBSERVED	EXPECTED	DIFFERENCE	DELTA
1	85	90.000	-5.000	.068
2	88	90.000	-2.000	.011
3	100	90.000	10.000	.270
4	79	90.000	-11.000	.327
5	84	90.000	-6.000	.097
6	86	90.000	-4.000	.043
7	103	90.000	13.000	.457
8	107	90.000	17.000	.781
9	97	90.000	7.000	.132
10	71	90.000	-19.000	.976

CHI SQUARED = 3.163

* -GREATER THAN 95% PROBABILITY THAT DIFFERENCE IS NOT ZERO
** -GREATER THAN 99% PROBABILITY THAT DIFFERENCE IS NOT ZERO

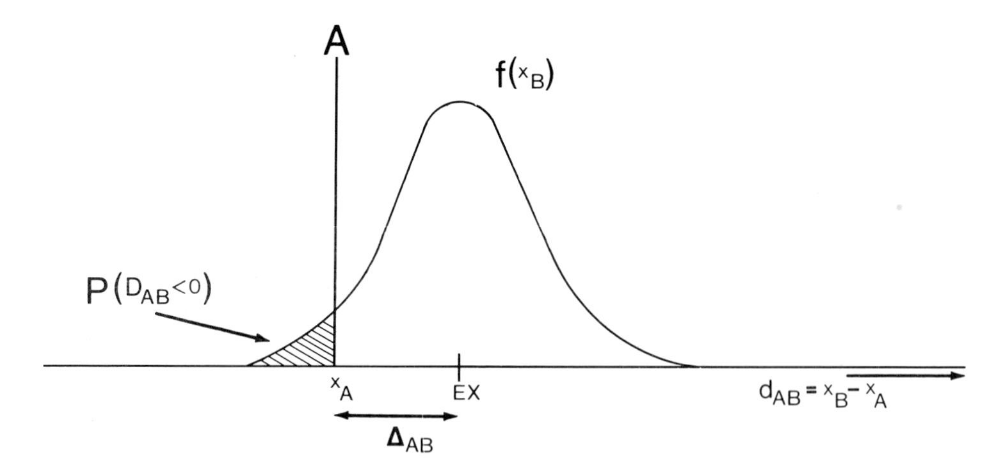

Figure 8 Direct estimation of distance Δ_{AB} between events A and B from relative cross-over frequency when A is a marker horizon with zero variance. The variance of D_{AB} is equal to the variance of the event B.

The model which assumes that each event position has a normal probability curve with the same variance, can be used to estimate a set of expected frequencies for all observed event locations. The 'second-order differences' are then also normally distributed. For convenience, this theoretical distribution is divided into 10 classes with equal expected frequencies. In total, 10 expected frequencies for 900 values are compared to the corresponding observed frequencies in the bottom part of Table 11.

In the last column, the squared difference, divided by expected value corrected for autocorrelation (see later) is shown. Each of these values is approximately distributed as χ^2 with a single degree of freedom, and the total for 10 classes has seven degrees of freedom. Probable χ^2 departures from normality are marked by asterisks. A more detailed explanation of the procedures followed in the normality test will be given in the next chapter (Chapter II.7).

MARKER HORIZON OPTION OF THE RASC METHOD

This section deals with an option of the RASC computer program in which the location of chronostratigraphic (marker) horizons such as seismic markers or bentonite beds resulting from volcanic ash fall can be considered. If events of this type can be correlated with certainty between sections, they should be assigned zero variance along the relative time scale when they are considered in conjunction with other stratigraphic events. In practice, this means that marker horizons will be given more weight than other events in the calculations. Marker horizons are entered like other stratigraphic events in the data block for the RASC program. However, they are identified as such in the input specifications.

The underlying statistical theory for marker horizons is explained in Figure 8 which can be compared to Figure 4 earlier in this Chapter. The event A has been replaced by a marker horizon with zero variance in Figure 8. This means that D_{AB} is normally distributed with mean Δ_{AB} and variance σ^2. The z-value for relative cross-over frequency p_{AB} has to be divided by $\sqrt{2}$ before it can be used as a direct estimate of the distance between events A and B which is compatible with the direct estimate of Figure 5 and the indirect estimates of Figure 6. Its weight also has to be adjusted accordingly.

When indirect estimates such as $\Delta_{AB.C}$ involving a marker horizon are obtained, there are two possibilities:

(1) The event used for the indirect comparison (C) is a marker horizon; and

(2) either A or B is a marker horizon.

Table 12 RASC method of scaling applied to data of artificial example. Event B is marker horizon. These results should be compared to those shown in Table 7 where B, like A and C, had unit variance.

	n	f	z	D(direct)	D(indirect)	D_0(Ave)	D(Ave)	E(D)
AB	129.5	0.8633	1.095	1.095(1)	1.283(3)	1.189	1.142	1.000
AC	130	0.8667	1.111	1.571(2)	1.383(2)	1.477	1.477	1.500
BC	92	0.6133	0.288	0.288(1)	0.476(3)	0.382	0.335	0.500
SSD				0.059	0.094	0.050	0.048	

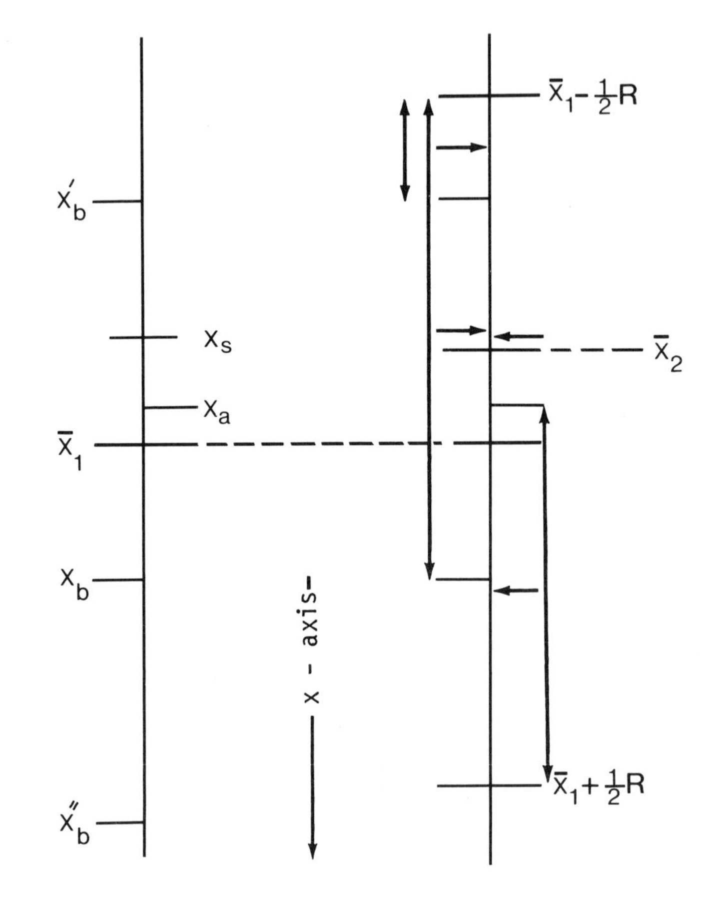

Figure 9 Simple example of application of unique event option. A unique event U is observed in a single section with ...A U-S B B' B"... Positions of events, A, S, B, and B' in the standard are used to obtain successive approximations ($\bar{x}_1$ and $\bar{x}_2$) for position of U in the standard.

In the first case, D_{AC} and D_{BC} are normal with variance σ^2 (as D_{AB} in Figure 8). Consequently, their difference can be assigned variance equal to $2\sigma^2$. The second case results in difference $D_{AB.C}$ with variance $3\sigma^2$.

The preceding theory of marker horizons can further be illustrated by using a modified version of the artificial example of Tables 4 to 7. Table 12 shows results comparable to those of Table 6 under the assumption that B is a marker horizon. As before, a table of sequences for A, B and C was derived from Table 4. However, this time the random normal numbers for B were all replaced by 2.000. For example, the first three numbers of Table 4 previously (see Table 5) gave 1.422, 0.130, 2.732 and, therefore, BAC. This time they result in 1.422, 2.000, 2.732 and, therefore, ABC. (For one sequence, the locations of A and B became both equal to 2.000 and 0.5 was added to the tallies of both AB and BA.)

Comparison of Table 12 to Table 6 shows the following differences. The cross-over frequencies for AB and BC have increased significantly. This reflects the fact that the variance of B was set equal to zero. The direct estimates of distance for AB and BC were not multiplied by $\sqrt{2}$ as in Table 6. However, D (indirect) was estimated from D (direct) by simple addition or subtraction as before. The relative variances of D (direct) and D (indirect) are shown in brackets in Table 12. As before, D_0(Ave) is the arithmetic average and D(Ave) represents the weighted average. The weights are inversely proportional to the variances. For example, D(Ave) of AB is equal to (1.095+1.283/3)(1+1/3) = 1.142. The SSD values in Table 12 are less than those in Table 6.

UNIQUE EVENT OPTION OF THE RASC PROGRAM

The purpose of this option is that a rare (unique) event can be entered into a regional standard by comparing its position in a single or few sections to those of the more abundant taxa used for constructing the RASC standard sequence. The unique event option is useful when an index fossil is observed in one or a few sections only. Because of its rarity, the index fossil cannot be used for construction of the optimum sequence. However, it can be fitted in with the other events afterwards and this may help to define assemblage zones from the dendrogram. Like marker horizons, the unique events have to be identified in the input specifications of the RASC program.

The unique event option can also be used to solve this following hypothetical problem. A feature is observed in two sections and it is of interest to determine whether this feature represents a single stratigraphic event. The feature, then, can be entered by codes as a different unique event to each section. The resulting positions of these two unique events in the standard can be compared to each other and this may be helpful for deciding whether the same event is present in both sections.

Figure 9 shows the technique used for treating unique events. The event A with distance x_a from the origin is observed immediately above the unique event in a section. Event S(with x_s) is simultaneous to the unique event and event B (with x_b) occurs below it. Also shown on the l.h.s. of Figure 9 are the locations x'_b and x''_b for two other events B' and B" observed below b. As a first approximation, the unique event is assigned the position $\bar{x}_1$ representing the arithmetic average of x_a, x_s, and x_b. In practical applications, S may be missing. The special situation that A or B is missing would occur only if the unique event were to occupy the first or last position in a section.

More than a single event may be observed in the positions immediately above, simultaneous to, or below the unique event. Then, x_a, x_s, or x_b will be computed as averages for these events which, in turn, will be averaged to estimate $\bar{x}_1$.

A range of $\bar{x}_1 \pm (1/2)R$ can be defined for all events encountered within the vicinity of $\bar{x}_1$ with a probability greater than 5 percent. Because $\sigma^2 = 0.5$, $(1/2)R = 1.96\sigma = 1.386$. The events in the standard with locations in the interval $\bar{x}_1 \pm (1/2)R$ can be identified. For the simplified example of Figure 9, these are the events A, S, B, and B' (but not B"). For each event above the unique event in the section (A), a value is computed which is the average of its location (x_a) and the value $\bar{x}_1 + (1/2)R$. Similarly, for each event below the unique event (B or B'), a value is computed which is the average of its location (x_b or x'_b) and the value $\bar{x}_1 - (1/2)R$. These average values which are shown as arrows in the diagram on the r.h.s. of Figure 9 are averaged together with the values (x_s) for events observed to be simultaneous with the unique event. This gives the second approximation $\bar{x}_2$. If the unique event occurs in more than one section, the preceding calculation is performed for each section and the resulting values $\bar{x}_2$ are averaged.

The choice of a range R in the method of Figure 9 is arbitrary. However, the parameter $\bar{x}_2$ is independent of R when the number of events within the interval $\bar{x}_1 \pm (1/2)R$ remains constant.

BINOMIAL AND TRINOMIAL MODELS FOR SCALING BIOSTRATIGRAPHIC EVENTS

Review

The final sections of this Chapter are based on a review paper by Agterberg (1984). First the RASC model for scaling, discussed in this Chapter, is evaluated in terms of observed probabilities by which the events succeed one another in the wells or were observed to be coeval. In addition, several other binomial and trinomial models will be considered.

Suppose that two stratigraphic events (either entries or exits or one of each) for different taxa are expressed as E_i and E_j. If E_i and E_j occurred relatively close in geologic time it may be that

E_i is observed to occur above E_j in some outcrop sections or wells, and E_j above E_i in others. The observation may also be that E_i and E_j are locally coeval. In order to avoid confusion, events for pairwise comparison will here be denoted by using the letter A instead of E. Binomial models provide estimates of the probability of A_1 or that E_i occurs above E_j in a section. If the probability is written as p_1, and the probability of A_2 (E_j occurs above E_i) as p_2, then $p_2 = 1-p_1$. As originally pointed out by Edwards and Beaver (1978), only a trinomial model can result in an estimate of the probability of occurrence of A_3 or that E_i and E_j are observed to be coeval in a section in addition to the probabilities that A_1 and A_2 occur. The development of trinomial models is of importance because biostratigraphic events are frequently observed to be coeval and this possibility should be considered in the statistical models.

Statistical theory of the binomial and trinomial distributions can be found in standard reference volumes such as Johnson and Kotz (1969, Chapters 3 and 11). Consider a series of independent trials, in each of which just one of 3 mutually independent events A_1, A_2 and A_3 must be observed, and in which the probability of occurrence of event A_k (k = 1, 2, or 3) is equal to p_k for each trial with

$$\sum_{k=1}^{3} p_k = 1.$$

The trinomial distribution then is the joint distribution of the random variables n_1, n_2 and n_3 representing the numbers of occurrences of the events A_1, A_2 and A_3, respectively, in r trials with

$$\sum_{k=1}^{3} n_k = r.$$

It is defined by

$$P(n_1,n_2,n_3) = r! \prod_{k=1}^{3} (p_k^{n_k}/n_k!)$$

in which

$$n_k > 0, \qquad \sum_{k=1}^{3} n_k = r$$

The distribution of n_1, n_2 or n_3 considered separately is binomial with

$$P(n_k) = \binom{r}{n_k} p_k^{n_k} (1-p_k)^{r-n_k}$$

in which

$$k = 1, 2, \text{ or } 3, \qquad n_k = 0, 1, \ldots, r$$

Also, if one of the events, say A_3, is ignored, then the other two satisfy this equation provided that r is replaced by $r' = (r-n_3)$ and p_k by $p_k' = p_k r/(r-n_3)$. The maximum likelihood estimator of p_k (k = 1, 2, or 3) is n_k/r; that of p_k' (k = 1, or 2) is $n_k/(r-n_3)$.

The preceding theory can be illustrated by means of the following simple example. For a set of 18 wells on the Canadian Atlantic margin, **Uvigerina canariensis** (Fossil no. 10) and **Asterigerina gurichi** (Fossil no. 17) were both observed in r=5 wells. The exit of no. 10 occurred twice (n_1=2) above and once (n_2=1) below that of no. 17, respectively. In the remaining two wells (n_3=2), the events were observed to be coeval. Consequently, it can be estimated that no. 10 occurs above, below or coeval to no. 17 with probabilities of 40, 20, and 40 percent, respectively. If coeval events are ignored, then the estimates of the probabilities that no. 10 occurs above and below no. 17 are 67 and 33 percent, respectively.

Of course the uncertainties of the preceding estimated percentages are considerable. For the observed relative frequency 2/5, the 95 percent confidence limits for p_1 are 5.3 and 85.3 percent respectively. For 2/3, the 95 percent confidence limits are 9.4 and 99.2 percent. These confidence limits were looked up in Hald's (1952) statistical tables. They can also be computed by using various approximation formulas (see Johnson and Kotz, 1969, Chapter 3).

The preceding practical example illustrates clearly that simple binomial or trinomial theory results in imprecise estimates of the probabilities if the sample size r is small. For large samples, the theory is, however, satisfactory. It has been used extensively in biostratigraphy, e.g., by Southam et al. (1975) and Blank and Ellis (1982). These authors used the binomial model which arises from the trinomial one by ignoring coeval events.

A Binomial Model Based on Multiple Pairwise Comparisons

Gradstein and Agterberg (1982) have developed the scaling technique mainly to cope with the problem that nearly all of their samples were small. Each stratigraphic event E_i is assumed to occupy a position along a linear scale L. The positions asumed by E_i in individual sections fluctuate at random about an average value along L. The 'distance' d_{ij} between average positions of two events E_i and E_j then can be converted into the probability that E_i occurs above E_j. Alternately, the probability can be converted into the distance. The advantage of this method is that the distance need not only be estimated from the relative positions of E_i and E_j in the sections but (N-1) double pairs (E_i,E_k) and (E_j,E_k) with $k \neq i$, j can also be used. Even if all samples are small, N may be large and precise estimates of the average positions of the events along the linear scale L can be obtained. Standard deviations of the distances between successive events form part of the output of Agterberg and Nel's (1982) RASC computer program for ranking and scaling of stratigraphic events.

Another advantage of using multiple pairwise comparisons is as follows. Let E_i be observed to occur r times above E_j for a small sample of r pairs. Suppose that r=5. The estimated probability that E_i occurs above E_j then is equal to $p_1 = 1.0$ with 95 percent confidence interval ranging from 0.478 to 1.0. This would admit the possibility that $p_1 = 0.5$, or that E_i and E_j are coeval on the average. The latter conclusion would be drawn regardless of whether or not E_i occurs above many other events which, in turn, occur above E_j. In a multiple paired comparison model, E_i can be placed above E_j with certainty ($p_1 = 1.0$) even if r is small.

Ties were treated as follows by Gradstein and Agterberg (1982). If $f(=n_1)$ represents the observed frequency of A_1 (event E_i occurs above E_j), and $t(=n_3)$ the number of ties for these two events, then the score $s = n_1 + n_3/2$ can be used for estimating the probability of occurrence of A_1 with $p_1 = s/r$. This implies that A_2 is observed to occur $n_2 + n_3/2 = r-s$ times, and its probability of occurrence is $p_2 = 1-s/r$. In this approach, an observed tie receives the same weight as either one of the direct observations of the events A_1 or A_2. However, it is recognized that no preference can be given to A_1 or A_2 if A_3 is observed. Although the average positions of E_i and E_j along the linear scale could concide (with $p_1 = p_2 = 0.5$), and observed ties will tend to decrease the distance between the average positions, the scaling model does not allow for explicit estimation that two events are observed to be coeval in a section. Instead of this, a tie is interpreted as a coincidence due to sampling method (e.g. use of well cuttings) or occurrence of sudden events at the time of deposition which favoured fossilization of several taxa in 'patches'.

It is noted that a distinction should be made between the frequency curve for relative abundance of occurrence of a fossil taxon through time at a given place and the probability curves for its entry and exit (cf. Agterberg and Nel, 1982). Brower (1981) and Edwards (1982) have pointed out that methods by which frequencies are averaged may give range zones which are shorter than those resulting from 'conservative' methods in which more weight is assigned to places where events occur relatively high or low in the stratigraphic column in relation to other events. For example, if exit E_1 is observed above exits E_2 and E_3 in one section but below E_2 and E_3 in many other sections, then a 'conservative' method would place the upper limit of the range for taxon 1 above those of taxa 2 and 3. On the other hand, this point would fall below those of taxa 2 and 3 when the average location of E_1 is determined.

From a statistical point of view, the estimation of an average exit is more satisfactory because the position of the endpoint is more susceptible to random fluctuations. Moreover, the average value is more robust if events are locally out of place due to anomalous circumstances such as sediment mixing or misidentification. As discussed before, the RASC computer program can also compare individual sections to the standard. The standard is the scaled optimum sequence, which consists of a set of average distance values along the linear scale L. This procedure is called normality test (see 'Normality Test' in this Chapter and also Gradstein, 1984).

Trinomial Models Based on Multiple Pairwise Comparisons

In a recent review paper, Hudson and Agterberg (1982) have listed three trinomial models by means of which the probabilities p_1, p_2 and p_3 (for occurrence of A_1, A_2 or A_3) can be estimated using all possible pairwise comparisons. These are:

(1) Glenn and David's (1960) model,

(2) Rao and Kupper's (1967) model,

(3) Davidson's (1970) model.

Davidson's model was successfully applied by Edwards and Beaver (1978) and later by Hudson and Agterberg (1982) to several data sets. Drawbacks, pointed out in the latter publication, were that this method, because of many iterations required, becomes time-consuming even for digital computers when the number of events exceeds 40. Also, the model is not able to handle the situation that many events in the upper parts of a large stratigraphic column occur with certainty above many events in its lower parts. It will be shown in this section that a modification of Glenn and David's model is not subject to these constraints and can be used in situations where Davidson's model is definitely not applicable.

Possible applications of Rao and Kupper's model in biostratigraphy have not been investigated in detail, mainly because of the great similarity between this model and that of Davidson.

Glenn and David's model is an extension of the Thurstone-Mosteller model which uses Gaussian curves for the distribution of positions of events along a linear scale L as is done in the RASC model of

Agterberg and Gradstein (1981). It is noted that the Thurstone-Mosteller model does not permit ties. As a first step for calculating average distances between events along this scale, the observed 'cross-over' frequencies ($p_{ij} = s_{ij}/r_{ij}$ in the RASC model) are converted to z-values according to the transformation $\Phi^{-1}(p) = z$. This is the inverse of $p = \Phi(z)$ where Φ denotes the fractile (cumulative frequency) of a normal distribution in standard form. Mosteller (1951) has shown that, under certain conditions, the best position of an event along the scale L is obtained by averaging all z-values for pairwise comparisons of this event to all other events. The resulting position is 'best' in a least squares sense. If the RASC model is used in a situation that none of the frequencies p_{ij} are missing or equal to one, then the unweighted method (simple averaging of z-values regardless of sample size) yields results identical to those of the Thurstone-Mosteller model. Modifications were made in the RASC model to avoid missing values and frequencies equal to one or zero. These modifications were discussed in detail earlier in this Chapter. In this section, they will also be applied to Glenn and David's model.

Suppose that the random variable D represents 'distance' along the linear scale L between two events in a single section. D is assumed to have unit variance and its average value is d. Glenn and David (1960) have introduced a threshold parameter τ. A tie of the two events is assumed to occur when D is less than τ and greater than $-\tau$. The probability of a tie (p_3) then depends on both τ and the mean distance (d) between the two events considered. This relationship is illustrated in Figure 10 for $\tau = 0.2$ and $\tau = 0.4$.

It is readily shown that Glenn and David's model results in the following three probabilities for A_1, A_2 and A_3:

$$p_1 = \Phi(d-\tau)$$

$$p_2 = 1 - \Phi(d+\tau)$$

$$p_3 = \Phi(d+\tau) - \Phi(d-\tau)$$

Consequently,

$$p_1 + p_3 = \Phi(d+\tau)$$

$$p_2 + p_3 = \Phi(-d+\tau)$$

This indicates that d and τ can be estimated from p_1, p_2 and p_3. A set of observed frequencies using the format (f,t/r) are shown in Table 13. This example was previously used for Table 3 in Chapter II.5 and Table 1 earlier in this Chapter. It is convenient to define

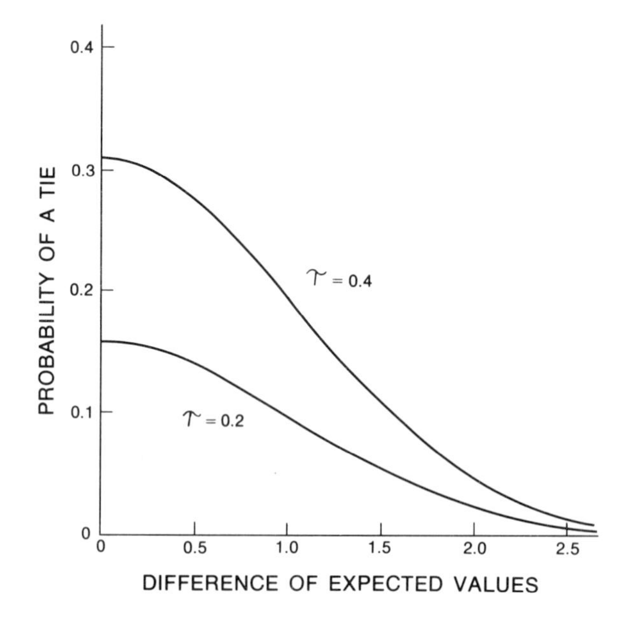

Figure 10 Probability of a tie as a function of 'distance' (d) between mean positions of events along linear distance scale (after Glenn and David, 1960).

Table 13 Example of 10 biostratigraphic events forming 'optimum' sequence as in Agterberg and Nel (1982b, Table 7, p. 74). Numbers f,t/r are for pairwise comparison using a trinomial model. If rows are labelled by the index i and columns by j, then f denotes the number of times E_j follows E_i in the sequences, t represents number of ties, and r is number of times E_i and E_j were observed in the same section. Example: the first entry of the second column (4, 2/7) indicates that event 1 follows event 2 four times while the two events were observed to be coeval in two sections. Because r = 7, this implies that event 1 precedes event 2 in one section.

	2	1	3	5	7	4	6	8	9	10
2	x	4,2/7	2,2/6	3,4/8	4,1/6	2,2/6	2,2/4	5,0/5	8,0/8	4,1/5
1	1,2/7	x	1,3/5	3,3/8	4,0/6	4,1/6	2,1/3	5,0/5	8,0/8	4,1/5
3	2,2/6	1,3/5	x	3,2/6	3,1/5	3,3/6	2,2/4	4,0/4	6,0/6	3,1/4
5	1,4/8	2,3/8	1,2/6	x	3,1/7	3,3/7	2,2/4	5,0/5	9,0/9	4,2/6
7	1,1/6	2,0/6	1,1/5	3,1/7	x	3,1/6	1,1/3	4,1/5	7,0/7	4,1/5
4	2,2/6	1,1/6	0,3/6	1,3/7	2,1/6	x	2,2/4	3,0/4	6,0/7	3,1/5
6	0,2/4	0,1/3	0,2/4	0,2/4	1,1/3	0,2/4	x	1,0/2	4,0/4	1,1/2
8	0,0/5	0,0/5	0,0/4	0,0/5	0,1/5	1,0/4	1,0/2	x	5,0/5	2,1/3
9	0,0/8	0,0/8	0,0/6	0,0/9	0,0/7	1,0/7	0,0/4	0,0/5	x	3,0/6
10	0,1/5	0,1/5	0,1/4	0,2/6	0,1/5	1,1/5	0,1/2	0,1/3	3,0/6	x

$$a_{ij} = (f_{ij} + t_{ij})/r_{ij}$$

$$a_{ji} = (f_{ji} + t_{ij})/r_{ij}$$

These values are shown in matrix form in Table 14, with a_{ij} in the upper triangle and a_{ji} in the lower triangle. The transformation $\Phi^{-1}(a_{ij})$ was made with the result shown in Table 15. Finally, separate estimates of d and τ were obtained as

$$d' = (1/2)\left[\Phi^{-1}(a_{ij}) - \Phi^{-1}(a_{ji})\right]$$

$$\tau' = (1/2)\left[\Phi^{-1}(a_{ij}) + \Phi^{-1}(a_{ji})\right]$$

The d'-values computed from the values of Table 15 are shown in the upper triangle of Table 16 and the τ'-values in its lower triangle. The d'-values can be treated in exactly the same way as the z-values are treated in the RASC computer program for obtaining average distances between events along the linear scale L. Each of the τ'-values can be regarded as an estimate of τ. A frequency distribution of the 32 observed τ'-values of Table 16 is shown in Table 19. Their average amounts to $\bar{\tau}$ = 0.4520 which seems to be a fairly precise estimate of τ. (The standard deviation of $\bar{\tau}$ is 0.046).

Glenn and David (1960) have shown that the preceding simple averaging method does not result in a least squares solution of τ and d_{ij}. They proposed a modified model replacing the Gaussian curves along the distance scale L by cosine curves. Then the preceding expressions for d' and τ' represent the least squares solution when $\Phi^{-1}(a_{ij})$ is replaced by arcsin $(2a_{ij} - 1)$. Application of the arcsin transformation to the values of Table 14 yields Table 17 instead of Table 15. Table 18 was derived from Table 17 in the same way as Table 16 from Table 15 and can also be used for estimating τ and the distances d_{ij}. The modified average value now amounts to $\bar{\tau}$ = 0.4080 as shown in Table 19.

A more elaborate test of the preceding version of Glenn and David's model consisted of its application to 48 events each occurring at least 5 times in the set of 18 wells used by Gradstein (1984). First an optimum sequence was obtained (with presorting option, and m_{c1} = 1, c.f. Agterberg and Nel, 1982). This sequence was arbitrarily split into two segments consisting of 21 and 27 events, respectively. τ was estimated separately by the two methods (Gaussian Model and Cosine Model) for these groups which contain 75 (Group 1 in Table 20) and 173 (Group 3)

Table 14 Matrix consisting of elements a = (f+t)/r corresponding to Table 13.

	2	1	3	5	7	4	6	8	9	10
2	x	6/7	4/6	7/8	5/6	4/6	4/4	5/5	8/8	5/5
1	3/7	x	4/5	6/8	4/6	5/6	3/3	5/5	8/8	5/5
3	4/7	4/5	x	5/6	4/5	6/6	4/4	4/4	6/6	4/4
5	5/8	5/8	3/6	x	4/7	6/7	4/4	5/5	9/9	6/6
7	2/6	2/6	2/5	4/7	x	4/6	2/3	5/5	7/7	5/5
4	4/6	2/6	3/6	4/7	3/6	x	4/4	3/4	6/7	4/5
6	2/4	1/3	2/4	2/4	2/3	2/4	x	1/2	4/4	2/2
8	0/5	0/5	0/4	0/5	1/5	1/4	1/2	x	5/5	3/3
9	0/8	0/8	0/6	0/9	0/7	1/7	0/4	0/5	x	3/6
10	1/5	1/5	1/4	2/6	1/5	2/5	1/2	1/3	3/6	x

Table 15 Values $\Phi^{-1}(a)$ corresponding to Table 14. Values for samples with r = 2 were not used and are written as x. Values corresponding to 1 and 0 are written as a and -a, respectively. For some subsequent calculations a was set equal to 1.645 (cf. Agterberg and Nel, 1982).

	2	1	3	5	7	4	6	8	9	10
2	x	1.068	0.430	1.150	0.967	0.430	a	a	a	a
1	-0.180	x	0.841	0.674	0.430	0.967	a	a	a	a
3	0.430	0.891	x	0.967	0.841	a	a	a	a	a
5	0.318	0.318	0.000	x	0.180	1.068	a	a	a	a
7	-0.430	-0.430	-0.253	0.180	x	0.430	0.430	a	a	a
4	0.430	-0.430	0.000	0.180	0.000	x	a	0.674	1.068	0.841
6	0.000	-0.430	0.000	0.000	0.430	0.000	x	x	a	x
8	-a	-a	-a	-a	-0.841	-0.674	x	x	a	a
9	-a	-a	-a	-a	-a	-1.068	-a	-a	x	0.000
10	-0.841	-0.841	-0.674	-0.930	-0.841	-0.253	a	-0.430	0.000	x

Table 16 Values d' (in upper triangle) and r' (in lower triangle) obtained by Eq.(5). The values aa and aaa are undetermined.

	2	1	3	5	7	4	6	8	9	10
2	x	0.624	0.000	0.416	0.699	0.000	0.823	aaa	aaa	1.243
1	0.444	x	0.000	0.178	0.430	0.699	1.038	aaa	aaa	1.243
3	0.430	0.841	x	0.484	0.547	0.823	0.823	aaa	aaa	1.160
5	0.734	0.496	0.484	x	0.000	0.444	0.823	aaa	aaa	1.038
7	0.269	0.000	0.294	0.180	x	0.215	0.000	1.243	aaa	1.243
9	0.430	0.269	0.823	0.624	0.215	x	0.823	0.674	1.068	0.547
6	0.823	0.607	0.823	0.823	0.430	0.823	x	x	aaa	x
8	aa	aa	aa	aa	0.402	0.000	x	x	aaa	1.038
9	aa	aa	aa	aa	aa	0.000	aa	aa	x	0.000
10	0.402	0.402	0.486	0.607	0.402	0.294	x	0.607	0.000	x

individual τ'-values, respectively. Group 2 in Table 20 is for 39 τ'-values arising from comparison of events in Group 1 to events in Group 3. The average values of τ' (Gaussian Model) are 0.2419, 0.1914 and 0.2179, respectively. These values are not significantly different from each other at the 5 percent level of significance when analysis of variance is applied. This demonstrates that Glenn and David's trinomial model can indeed be used for describing the frequencies of observed ties.

The d'-values were treated as z-values in the RASC computer program (now setting m_{c2} = 3 and using the unweighted method; cf. Agterberg and Nel, 1982). The resulting dendrograms are shown in Figure 11 (Gaussian Model) and Figure 12 (Cosine Model). It may be concluded that the differences between results obtained by these two models are minimal. On average, successive distances in Figures 11 and 12 are shorter than those in dendrograms resulting from runs with the RASC program (Agterberg and Nel, 1982). All successive distances in Figure 11 are less than 0.5. Because τ is approximately equal to 0.2, most probabilities of a tie between successive events are about 15 percent (see Figure 10).

A detailed comparison of estimated trinomial and binomial probabilities with observed frequencies is shown in Table 21 for Group 1 only. The distances d_f are as in Figure 11. For example, the distance 0.32 between Events 10 and 17 is equal to the sum of three successive differences (0.0643, 0.1760 and 0.0814) in Figure 11. According to the original equations for the Glenn-David model, the estimate of τ(=0.2419) should be subtracted from these distances and the fractile of the normal distribution in standard form determined for estimation of p_1. In order to distinguish it from another estimate of p_1 (see later), this estimate will be written as P_f. For example, the distance d_f = 0.3217 gives $d_f-\overline{\tau}$ = 0.3217-0.2419 = 0.0798 from which P_f = 0.53 was derived. Multiplication of the estimated probability P_f by r=5 resulted in the estimated frequency f_e = 2.7 for number of times event no. 10 occurs above 17. This estimated frequency can now be compared to the observed frequency f=2 in the second column of Table 21.

It is also possible to estimate the parameters p_2 and p_3. Because $p_2 = 1-p_1-p_3$, the probability of a tie, written as P_t, is shown only, followed by the corresponding estimated frequency t_e. For the previous example, P_t = 0.18 and t_e = 2.8 (to be compared to t=2).

The 75 pairs of events are divided into 9 groups in Table 21. The estimated frequencies t_e and f_e were added for these groups, with the totals shown as T_e and F_e in Table 22 for comparison to corresponding sums of observed frequencies written as T_o and F_o. The quantities $U_t = (T_o - T_e)^2/T_e$ and $U_f = (F_o - F_e)^2/F_e$ are also shown in Table 22.

If the model provides a good fit to the observations, each of the quantities U_t and U_f is approximately distributed as χ^2 with a

Table 17 Same as Table 15 except that the transformation arcsin (2a - 1) was used. For some subsequent calculations a was set equal to 1.571 (instead of 1.645 used for some a's in Table 15).

	2	1	3	5	7	4	6	8	9	10
2	x	0.796	0.340	0.848	0.730	0.340	a	a	a	a
1	-0.143	x	0.644	0.524	0.340	0.730	a	a	a	a
3	0.340	0.644	x	0.730	0.644	a	a	a	a	a
5	0.253	0.253	0.000	x	0.143	0.796	a	a	a	a
7	-0.340	-0.340	-0.201	0.143	x	0.340	0.340	a	a	a
4	0.340	-0.340	0.000	0.143	0.000	x	1.571	0.524	0.796	0.644
6	0.000	-0.340	0.000	0.000	0.340	0.000	x	x	a	x
8	-a	-a	-a	-a	-0.644	-0.524	x	x	a	a
9	-a	-a	-a	-a	-a	-0.796	-a	-a	x	0.000
10	-0.644	-0.644	-0.524	-0.340	-0.644	-0.201	x	-0.340	0.000	x

Table 18 Same as Table 16 except that the transformation arcsin (2a-1) was used.

	2	1	3	5	7	4	6	8	9	10
2	x	0.469	0.000	0.298	0.535	0.000	0.785	aaa	aaa	1.107
1	0.326	x	0.000	0.135	0.340	0.535	0.955	aaa	aaa	1.107
3	0.340	0.644	x	0.365	0.422	0.785	0.785	aaa	aaa	1.047
5	0.550	0.388	0.365	x	0.000	0.326	0.785	aaa	aaa	0.955
7	0.195	0.000	0.221	0.143	x	0.170	0.000	1.107	aaa	1.107
4	0.340	0.195	0.785	0.469	0.170	x	0.785	0.524	0.796	0.422
6	0.785	0.615	0.785	0.785	0.340	0.785	x	x	aaa	x
8	aa	aa	aa	aa	0.464	0.000	x	x	aaa	0.955
9	aa	aa	aa	aa	aa	0.000	aa	aa	x	0.000
10	0.464	0.464	0.524	0.615	0.464	0.221	x	0.615	0.000	x

single degree of freedom. The totals ΣU_t and ΣU_f would be distributed as χ^2 with approximately 9 degrees of freedom. The 95 percent confidence limit for this distribution amounts to 16.9. This suggests that the observed frequencies f are well described by Glenn and David's model. On the other hand, the discrepancy that the T_e-values are less than the T_o-values in the upper part of Table 22 and greater in its lower part may be statistically significant. The number of degrees of freedom is not known exactly for this test. It is, however, probably less than 9 and this would increase the 95 percent confidence limit to below ΣU_t = 16.8.

CONCLUDING REMARKS

The method of scaling was presented and initially illustrated by using the two examples in the previous Chapter on ranking of Lower Tertiary nannoplankton from the California Coast Range and Cenozoic

Table 19 Frequency distribution of τ'-values shown in lower triangles of Tables 16 and 18. G.M. denotes Gaussian Model; C.M. - Cosine Model; N - sample size; S.D. - Standard Deviation.

Class Limits	G.M.	C.M.
0.000	4	4
0.001 - 0.200	1	4
0.201 - 0.400	5	8
0.101 - 0.600	11	7
0.601 - 0.800	5	9
0.801 - 1.000	6	0
N	32	32
Mean	0.4520	0.4080
S.D.	0.2603	0.2489
S.D./$N^{\frac{1}{2}}$	0.0460	0.0440

Table 20 Glenn and David's trinomial model applied to 48 exits of Cenozoic foraminifera observed in 18 wells on Northwestern Atlantic Margin. Abbreviations as in Table 19. Groups resulted from splitting the optimum sequence after 21 events. Group 1 (see Table 21 for original data) is for pairwise comparisons of events belonging to first 21 events, Group 3 is same for last 27 events, and Group 2 is for comparison of events of Group 1 to events of Group 3.

	Group 1		Group 2		Group 3	
Class Limits	G.M.	C.M.	G.M.	C.M.	G.M.	C.M.
0.000	28	28	20	20	73	73
0.001 - 0.200	5	14	1	2	13	21
0.201 - 0.400	22	14	14	8	47	43
0.401 - 0.600	11	11	8	6	27	25
0.601 - 0.800	6	6	3	3	10	10
0.801 - 1.000	1	1	0	0	2	2
1.001 - 1.200	1	0	0	0	1	0
N	75	75	39	39	173	173
Mean	0.2419	0.2242	0.1914	0.1854	0.2179	0.2008
S.D.	0.2402	0.2321	0.2196	0.2249	0.2288	0.2204
S.D./$N^{\frac{1}{2}}$	0.0277	0.0268	0.0352	0.0360	0.0174	0.0168

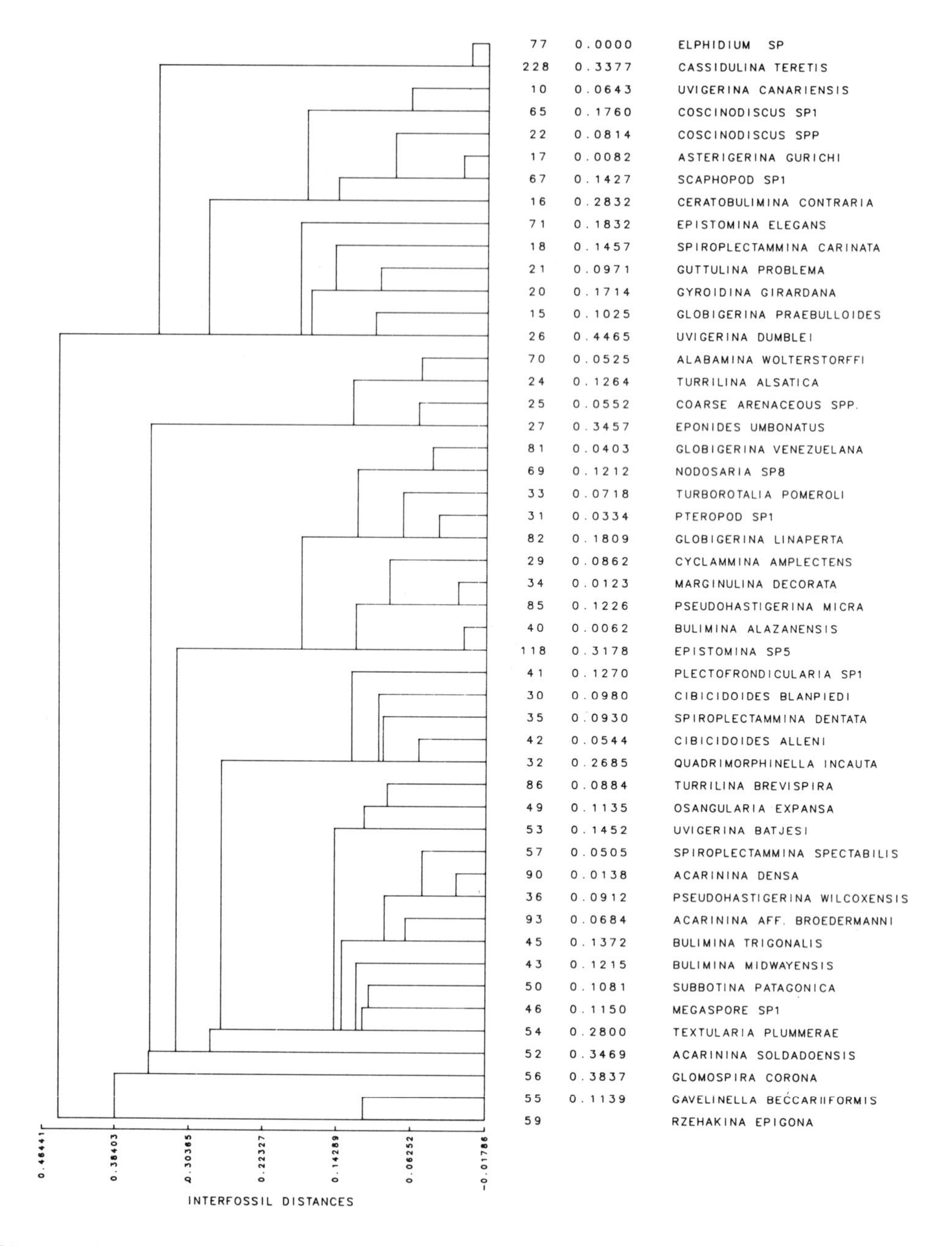

Figure 11 *Dendrogram for distances between successive events estimated by Glenn and David's trinomial model assuming Gaussian probability curves for events. Each event (except the last one) is followed by estimate of ditance connecting it to the event immediately below it. These distances were plotted toward the left and clustered.*

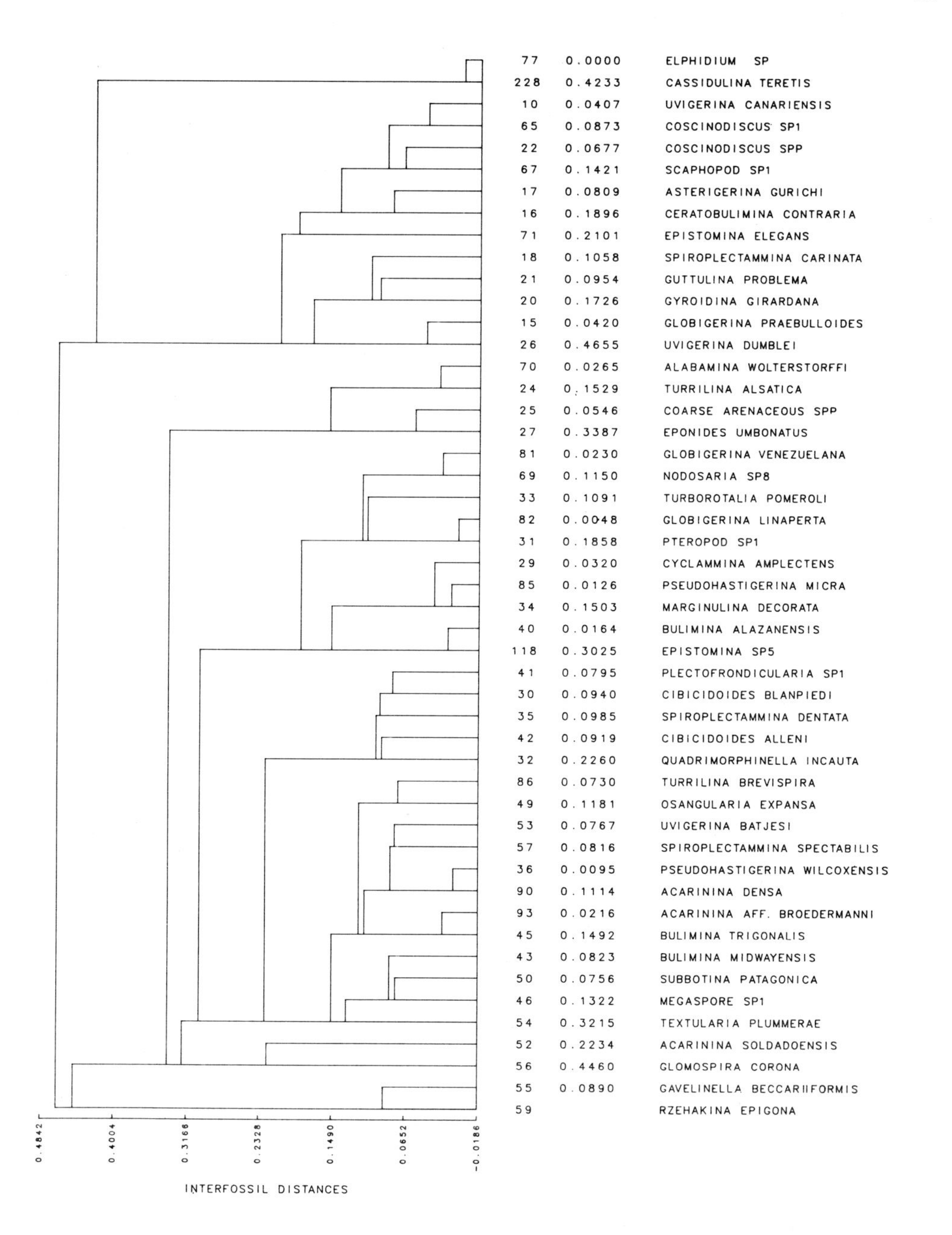

Figure 12 Same as Figure 11 except that cosine-shaped probability curves (instead of Gaussian curves) were assumed for events. Note that differences between patterns of Figures 11 and 12 are small, indicating that choice of shape of probability curves for events is probably not of critical importance.

Table 21 *Estimation of probabilities (P_f and P_t) and frequencies (f_e and t_e) corresponding to observed successions (f) and ties (t). Trinomial model was applied to first 21 events (Group 1) of optimum sequence for 48 exits of Cenozoic foraminifera also used in Table 20 and Figure 11. Last columns show estimated values for scores (s) based on modified binomial model. See text for explanations of column headings. Event numbers of column 1 are explained in Figure 11.*

E_i–E_j	f,t/r	d_f	P_f	f_e	P_t	t_e	s	d_s	P_s	s_e
10–17	2,2/5	0.32	0.53	2.7	0.18	0.9	3.0	0.39	0.65	3.3
10–16	5,1/6	0.47	0.59	3.5	0.17	1.0	5.5	0.94	0.83	5.0
10–17	2,1/3	0.76	0.70	2.1	0.15	0.4	2.5	1.43	0.92	2.8
17–16	3,3/6	0.15	0.46	2.8	0.19	1.1	4.5	0.54	0.71	4.2
17–71	2,0/3	0.43	0.58	1.7	0.18	0.5	2.0	1.03	0.85	2.5
17–18	6,1/7	0.62	0.65	4.5	0.16	1.1	6.5	1.19	0.88	6.2
17–20	6,1/7	0.86	0.73	5.1	0.13	0.9	6.5	1.49	0.93	6.5
17–15	6,1/7	1.03	0.79	5.5	0.11	0.8	6.5	1.73	0.96	6.7
65–16	4,1/5	-0.09	0.37	1.9	0.19	1.0	4.5	0.84	0.80	4.0
77–228	1,2/4	0.00	0.40	1.6	0.19	0.8	2.0	0.00	0.50	2.0
228–22	2,0/3	0.58	0.63	1.9	0.16	0.5	2.0	1.21	0.89	2.7
16–22	3,0/5	0.23	0.50	2.5	0.19	0.9	3.0	0.01	0.51	2.5
16–67	3,2/6	0.14	0.46	2.8	0.19	1.1	2.0	-0.32	0.49	2.9
16–71	3,1/6	0.28	0.52	3.1	0.18	1.1	3.5	0.49	0.69	4.1
16–18	11,3/16	0.47	0.59	9.4	0.17	2.8	12.5	0.64	0.74	11.8
16–20	11,2/14	0.71	0.68	9.5	0.15	2.1	12.0	0.95	0.83	11.6
16–15	11,4/15	0.88	0.74	11.1	0.13	2.0	13.0	1.18	0.88	13.2
16–26	7,0/8	0.98	0.77	6.2	0.12	1.0	7.0	1.33	0.91	7.3
22–71	2,1/3	0.52	0.61	1.8	0.17	0.5	2.5	0.48	0.68	2.0
22–21	2,1/3	0.84	0.73	2.2	0.13	0.4	2.5	0.80	0.79	2.4
22–18	3,0/5	0.70	0.68	3.4	0.15	0.8	3.0	0.63	0.74	3.7
22–20	4,1/5	0.94	0.76	3.8	0.12	0.6	4.5	0.93	0.82	4.1
22–15	4,0/5	1.11	0.81	4.0	0.10	0.5	4.0	1.17	0.88	4.4
67–21	4,1/5	0.75	0.70	3.5	0.15	0.7	4.5	0.84	0.80	4.0
67–18	5,1/6	0.61	0.64	3.9	0.16	1.0	5.5	0.68	0.75	4.5
71–21	1,1/3	0.33	0.54	1.6	0.18	0.5	1.5	0.32	0.63	1.9
71–18	2,2/6	0.18	0.48	2.9	0.19	1.1	3.0	0.15	0.56	3.4
71–20	4,1/6	0.43	0.57	3.4	0.18	1.1	4.5	0.46	0.68	4.1
71–15	4,2/6	0.60	0.64	3.8	0.16	1.0	5.0	0.70	0.76	4.5
71–26	4,0/5	0.70	0.68	3.4	0.15	0.8	4.0	0.84	0.80	4.0
71–27	2,1/3	1.38	0.87	2.6	0.08	0.2	2.5	1.58	0.94	2.8
21–18	3,5/11	-0.15	0.35	3.8	0.19	2.1	5.5	-0.17	0.43	4.8
21–20	3,6/9	0.10	0.44	4.0	0.19	1.7	6.0	0.14	0.55	5.0
21–15	4,2/10	0.27	0.51	5.1	0.18	1.8	5.0	0.38	0.65	6.5
21–26	2,0/4	0.37	0.55	2.2	0.18	0.7	2.0	0.52	0.70	2.8
21–70	5,1/6	0.82	0.72	4.3	0.14	0.8	5.5	0.85	0.80	4.8
21–24	7,1/8	0.87	0.74	5.9	0.13	1.1	7.5	1.00	0.84	6.7
21–27	5,0/6	1.05	0.79	4.7	0.11	0.7	5.0	1.26	0.90	5.4

E_i–E_j	f,t/r	d_f	P_f	f_e	P_t	t_e	s	d_s	P_s	s_e
18–20	7,5/15	0.24	0.50	7.5	0.19	2.8	9.5	0.30	0.62	9.3
18–15	10,3/16	0.41	0.57	9.1	0.18	2.8	11.5	0.54	0.71	11.3
18–26	7,0/8	0.52	0.01	4.9	0.16	1.3	7.0	0.68	0.75	6.0
18–70	6,1/8	0.96	0.77	6.1	0.12	1.0	6.5	1.01	0.84	6.8
18–24	10,1/12	1.02	0.78	9.4	0.12	1.4	10.5	1.16	0.88	10.5
18–25	11,0/12	1.14	0.82	9.8	0.10	1.2	11.0	1.43	0.92	11.1
20–15	8,2/14	0.17	0.47	6.6	0.19	2.6	9.0	0.24	0.60	8.3
20–26	5,1/8	0.27	0.51	4.1	0.18	1.5	5.5	0.38	0.65	5.2
20–70	6,0/7	0.72	0.68	4.8	0.15	1.0	6.0	0.71	0.76	5.3
20–24	9,1/11	0.77	0.70	7.7	0.14	1.6	9.5	0.86	0.81	8.9
20–25	9,1/10	0.90	0.74	7.4	0.13	1.3	9.5	1.13	0.87	8.7
20–27	5,0/7	0.95	0.76	5.3	0.12	0.9	5.0	1.13	0.87	6.1
15–26	4,0/8	0.10	0.45	3.6	0.19	1.5	4.0	0.14	0.56	4.4
15–70	4,3/8	0.55	0.62	5.0	0.16	1.3	5.5	0.47	0.68	5.5
15–24	9,1/12	0.60	0.64	7.7	0.18	2.1	9.5	0.62	0.73	8.8
15–25	10,1/12	0.73	0.69	8.2	0.15	1.8	10.5	0.89	0.81	9.8
15–27	5,0/7	0.78	0.71	4.9	0.14	1.0	5.0	0.89	0.81	5.7
15–81	4,1/6	1.13	0.81	4.9	0.10	0.6	4.5	1.30	0.90	5.4
26–24	5,1/7	0.50	0.60	4.2	0.17	1.2	5.5	0.48	0.69	4.8
26–25	1,0/4	0.63	0.65	2.6	0.16	0.6	1.0	0.75	0.77	3.1
26–27	3,1/5	0.68	0.67	3.4	0.15	0.8	3.5	0.75	0.77	3.9
70–24	4,0/6	0.05	0.43	2.6	0.19	1.1	4.0	0.15	0.56	3.4
70–25	6,1/8	0.18	0.48	3.8	0.19	1.5	6.5	0.42	0.66	5.3
70–27	2,0/3	0.23	0.50	1.5	0.19	0.6	2.0	0.42	0.66	2.0
70–81	2,0/3	0.58	0.63	1.9	0.16	0.5	2.0	0.83	0.80	2.4
70–31	3,0/4	0.81	0.72	2.9	0.14	0.6	3.0	1.02	0.85	3.4
24–25	6,1/9	0.12	0.45	4.1	0.19	1.7	6.5	0.26	0.66	5.4
24–27	4,0/6	0.18	0.48	2.9	0.19	1.1	4.0	0.27	0.61	3.6
24–81	3,0/4	0.53	0.61	2.4	0.17	0.7	3.0	0.83	0.80	3.2
25–27	3,0/5	0.06	0.43	2.1	0.19	1.0	3.0	0.00	0.50	2.5
25–31	7,1/9	0.63	0.65	5.9	0.16	1.4	7.5	0.60	0.73	6.5
27–81	2,0/3	0.35	0.54	1.6	0.18	0.5	2.0	0.41	0.66	2.0
27–31	5,0/7	0.58	0.63	4.4	0.16	1.1	5.0	0.60	0.73	5.1
27–82	2,0/3	0.61	0.65	1.9	0.16	0.5	2.0	0.68	0.75	2.3
81–31	3,0/5	0.23	0.50	2.5	0.19	0.9	3.0	0.19	0.58	2.9
81–82	3,1/4	0.27	0.51	2.0	0.19	0.7	3.5	0.27	0.61	2.4
31–82	4,0/5	0.03	0.42	2.1	0.19	1.0	4.0	0.08	0.53	2.7

Table 22 Comparison of observed and estimated frequencies for 75 pair-wise comparisons of Table 21. If model provides good fit, the U-values are approximately distributed as chi-squared with single degree of freedom. Totals are shown in bottom line.

T_e	T_o	U_t	F_e	F_o	U_f	S_e	S_o	U_s
9.09	13	1.69	33.31	39	0.97	45.85	45.5	0.00
10.91	12	0.11	44.53	49	0.45	53.49	53	0.00
9.14	12	0.89	40.30	41	0.01	45.77	47	0.03
8.90	15	4.18	30.07	29	0.04	35.93	36.5	0.01
10.52	10	0.02	46.72	51	0.39	54.95	56	0.02
8.86	5	1.68	36.00	42	1.00	42.51	44.5	0.09
8.35	6	0.66	34.28	36	0.09	39.58	39	0.01
10.34	4	3.88	32.14	39	1.46	40.39	41	0.01
7.15	2	3.71	22.60	29	1.81	26.29	30	0.52
83.25	79	16.83	319.95	355	6.22	384.76	392.5	0.70

foraminifera from the northwestern Atlantic continental margin. The basic assumptions of the scaling approach were tested by using artificial data sets consisting of ranking normal numbers and computer simulation experiments. Important options of the RASC computer program introduced in this Chapter are normality test, marker horizon option and unique event option.

In the last part of the Chapter it was shown that a modified version of the trinomial model of Glenn and David (1960) can be used for description of observed frequencies of coeval biostratigraphic events. The stratigraphic significance of the threshold parameter τ is not immediately obvious. It can be said that a new distribution for ties (see Figure 13) has been introduced in addition to the probability distributions for events along the linear scale L. The height of the new distribution for ties is roughly proportional to the value of τ. In general, τ therefore expresses the likelihood that events are coeval.

In the RASC model, observed ties are not ignored but each tie of two events E_i and E_j is scored as a 50 percent probability that E_i occurs above E_j and a 50 percent probability that E_j occurs above E_i. The last four columns of Table 21 shows observed scores s in comparison with estimated frequencies $s_e = P_s \times r$. The estimated probabilities P_s (for E_i occurring above E_j) satisfy $P_s = \Phi(d_s)$ where d_s was estimated by means of the weighted scaling option of the RASC computer program in which variations of sample size r are considered. The agreement between observed and estimated scores is excellent (see also Table 22 for comparisons of group totals).

The RASC method has other options (e.g., normality test, marker horizon and unique event option) in addition to the weighted scaling option, which are not available for the modified Glenn-David model. For this reason, this trinomial model should only be used when it is necessary to model observed frequencies of coeval events. In the second part of this Chapter, it was shown that the RASC model results in estimated probabilities which fit the observed data in the example used regardless of the assumption that event-positions are normally distributed.

REFERENCES

Agterberg, F.P., 1974, Geomathematics. Elsevier, Amsterdam, 596 pp.

Agterberg, F.P., 1984, Binomial and trinomial models in quantitative biostratigraphy. Computers and Geosciences 10, (1), 31-41.

Agterberg, F.P., and Gradstein, F.M., 1981, Workshop on quantitative stratigraphic correlation techniques. Ottawa, February, 1980, J. Math. Geol. 13, (1), 81-91.

Agterberg, F.P., and Gradstein, F.M., 1983, Interactive system of computer programs for stratigraphic correlation. Current Research, Geol. Survey of Canada, Paper 83-1A, 83-87.

Agterberg, F.P., and Nel, L.D., 1982a, Algorithms for the ranking of stratigraphic events. Computers and Geosciences 8, (1), 69-90.

Agterberg, F.P., and Nel, L.D., 1982b, Algorithms for the scaling of stratigraphic events. Computers and Geosciences 8, (2), 163-189.

Blank, R.G., 1984, Comparison of two binomial models in probabilistic biostratigraphy. Computers and Geosciences 10, (1), 59-67.

Blank, R.G., and Ellis, C.H., 1982, The probable range concept applied to the biostratigraphy of marine microfossils. J. Geol. 90, (4), 415-433.

Brower, J.C., 1981, Quantitative biostratigraphy, 1830-1980. In Merriam, D.F., ed., Computer Applications in the Earth Sciences, an Update for the 70's, Plenum Press, New York, p. 63-103.

Davidson, R.R., 1970, On extending the Bradley-Terry model to accommodate ties in paired comparison experiments. J. Amer. Stat. Assoc. 65, (329), 317-328.

Dixon, W.J., and Massey, F.J., 1957, Introduction to Statistical Analysis. McGraw-Hill, New York, N.Y., 488 pp.

Edwards, L.E., 1982, Numerical and semi-objective biostratigraphy; review and predictions. Third N. Amer. Paleont. Convention, Montreal, Canada, August 1982, Proc. 1, 147-152.

Edwards, L.E., and Beaver, R.J., 1978, The use of paired comparison models in ordering stratigraphic events. J. Math. Geol. 10 (3), 261-272.

Glenn, W.A. and David, H.A., 1960, Ties in paired-comparison experiments using a modified Thurtone-Mosteller model. Biometrics 16, (1), 86-109.

Gradstein, F.M., 1984, On stratigraphic normality. Computers and Geosci ences 10, (1), 43-57.

Gradstein, F.M., and Agterberg, F.P., 1982, Models of Cenozoic foraminiferal stratigraphy - Northwestern Atlantic Margin. In Cubitt, J.M., and Reyment, R.A., eds., Quantitative Stratigraphic Correlation, John Wiley, Chichester, 119-173.

Hald, A., 1952, Statistical tables and formulas. (Wiley Publications in Statistics), Wiley, New York, 97 pp.

Harper, C.W., Jr., 1984, A Fortran IV program for comparing ranking algorithms in quantitative biostratigraphy. Computers and Geosciences 10, (1), 3-29.

Hudson, C.B., and Agterberg, F.P., 1982, Paired comparison models in biostratigraphy. J. Math. Geol. 14, (2), 141-159.

Jackson, A., Lew, S.N., and Agterberg, F.P., 1984, DISSPLA program for display of dendrograms from RASC output. Computers and Geosciences 10, (1), 59-165.

Johnson, N.L., and Kotz, S., 1969, Distributions in statistics: Discrete distributions. Houghton-Mifflin, Boston, Mass., 328 pp.

Kendall, M.G., 1975, Rank Correlation Methods. Griffin, London, 202 pp.

Mosteller, F., 1951, Remarks on the method of paired comparisons, I, The least squares solution assuming equal standard deviations and equal correlations. Psychometrika 16, (1), 3-9.

Rao, P.V., and Kupper, L.L., 1967, Ties in paired-comparison experiments: A generalization of the Bradley-Terry model. J. Amer. Stat. Assoc. 62, (1), 194-204. Corrigenda, ibidem 63, (4), 1550.

Southam, J.R., Hay, W.W., and Worsley, T.R., 1975, Quantitative formulation of reliability in stratigraphic correlation. Science 188, 357-359.

II.7

NORMALITY TESTING AND COMPARISON OF RASC TO UNITARY ASSOCIATIONS METHOD

FREDERIK P. AGTERBERG

INTRODUCTION

In this Chapter, the RASC normality test will be explained in more detail. For comparison with results obtained by Guex and Davaud (1984), who used the unitary association method on a reworked bed of alveolinids from Yugoslavia (see Drobne, 1977), the RASC normality test will be applied to this same data-set as well. For this purpose use was made of a revised version of the RASC program originally published by Agterberg and Nel (1982a,b) and Heller et al. (1983). Several new options were added and the normality test subroutine was revised.

Appendix I contains nine data sets created by participants in IGCP Project No. 148. The data sets have different input formats and input conversion programs were used to convert unitary association input format to RASC input format and vice versa. Some stratigraphers commence coding at the top of a well and proceed stratigraphically downwards, whereas others move in the stratigraphically upward direction. For this reason, an option has been included in the conversion programs to reverse the sequences of a data base before processing by means of RASC. These nine data sets will be used for illustrating the use of the normality test in the next section. Appendix I also contains results obtained by applying RASC to the data.

REVISED NORMALITY TEST

The normality test option of the RASC program was briefly introduced in the previous Chapter. A more detailed explanation of the procedures followed will be given in this section. The normality test was developed for two reasons:

(a) to determine anomalous events which in a specified section occur much higher or lower than (at their average locations) in a regional standard developed on the basis of a number of sections in a region; and

(b) to test the assumption of normality, which was used to transform cross-over frequencies into z-values during scaling.

The normality test contributes useful information with respect to both these objectives.

In the original RASC program (Agterberg and Nel, 1982; Heller et al., 1983), the simplifying assumption was made that the second-order differences between positions of events in specific sections, on the one hand, and their positions in the average scaled sequence, on the other hand, would be approximately normally distributed with a standard deviation equal to 2σ. The original events were assumed to be normally distributed along the RASC scale with standard deviation equal to σ. It was realized that this simple model yields results which were at best approximately true. In the original applications, which were mainly variants of the data bases No. 1 and 2 in Appendix I, the final histogram of the normality test showed observed frequencies that were, on the average, equal to the expected frequencies, indicating that the simple model could be used. Three sets of frequencies for the original normality test are shown in Table 1.

Anomalous events would cause observed frequencies of the highest and lowest class (O_1 and O_{10}) to be greater than the expected frequency E_i (i = 1, 2, .., 10) which is equal for all classes of i. During 1982 and 1983 when the RASC program was applied to other data sets, several of which are listed in Appendix I, it turned out that the original normality test provided poor results in some situations because the number of anomalous events was either much higher or lower than expected. For example, too many anomalous events were found in data file No. 4 (Jurassic Tethyan radiolarians) and too few in No. 7 (Paleogene Californian nannofossils). It became difficult or even impossible in these situations to distinguish between anomalous events and departures from the normal distribution model originally assumed to hold approximately true for the second-order differences. It was decided to assess the problem systematically by means of computer simulation experiments.

Table 2 shows observed frequencies obtained by the original RASC program for 10 classes of 900 second-order differences. These second order differences were created in 10 of the computer simulation experiments described in the previous chapter. The 10 experiments were designed for different expected 'distances' E(D). The expected frequency is 90 for all 100 observed frequencies (O_1, O_2 ..., O_{10}) in Table 2. Clearly, the observed frequencies in the tails of these distributions for the second-order difference are too large when E(D) is greater than 0.5 and they are too small when E(D) is less than 0.2. It is noted that the runs for E(D) = 1.0 have a single greater-than-expected frequency near the centre of their distributions. This phenomenon is related to the use of pairs of z-values arbitrarily set equal to q (=2.326) and with zero difference between them (see previous chapter). It constitutes another problem which is not related to the problem at hand and does not arise for smaller values of E(D) in the experiments.

The examples of Table 1 may be compared to the experiments on artificial data sets, with E(D) between 0.2 and 0.5, for which the

Table 1 Normality test output from the original RASC program: Comparison of the observed frequencies (O_i) of second order difference-values in each of the ten classes i = 1, 2, ..., 10, with the expected frequencies (E_i) which are constant for each of the ten classes.

Source	E_i	O_1	O_2	O_3	O_4	O_5	O_6	O_7	O_8	O_9	O_{10}
Agterberg and Nel (1982, Table 6)	24.1	27	23	26	20	27	24	28	22	21	23
Heller et al. (1983, Table 6)	21.5	30	20	21	15	22	22	18	23	13	31
Gradstein (1984, Table 3)	39.8	50	36	32	41	43	31	39	42	38	46

Table 2 Normality test output for ten computer simulation experiments. Observed frequencies O_i are compared to the expected frequency (=90) for each of the ten classes i = 1, 2, ..., 10. E(D) represents the expected interval (or RASC 'distance') between event-positions along the RASC-scale in these experiments.

Original RASC	O_1	O_2	O_3	O_4	O_5	O_6	O_7	O_8	O_9	O_{10}
E(D) = 1.0, Set 1	156	55	32	69	127	145	54	39	48	175
E(D) = 1.0, Set 2	162	69	44	82	88	140	64	28	60	163
E(D) = 0.5, Set 1	119	98	77	52	78	117	55	80	104	120
E(D) = 0.5, Set 2	119	95	89	62	79	94	84	59	88	131
E(D) = 0.3, Set 1	89	111	75	84	87	88	85	77	107	97
E(D) = 0.3, Set 2	102	114	89	80	72	80	69	78	102	114
E(D) = 0.2, Set 1	84	101	83	76	80	98	106	100	88	84
E(D) = 0.2, Set 2	62	118	107	97	89	75	81	87	91	93
E(D) = 0.1, Set 1	18	77	91	135	115	123	153	111	62	15
E(D) = 0.1, Set 2	10	76	106	129	139	134	112	103	75	16

Table 3 Revised normality test output for the nine data sets of Appendix I using RASC program.

Data Base	E_i	O_1	O_2	O_3	O_4	O_5	O_6	O_7	O_8	O_9	O_{10}	$\hat{\chi}^2(7)$
1. Gradstein-Thomas	50.3	70	42	38	49	55	52	49	45	46	57	5.93
2. Gradstein	21.1	20	21	30	13	23	17	29	18	18	22	7.55
3. Doeven	64.1	78	53	65	68	53	64	70	67	53	70	3.36
4. Baumgartner	149.6	127	175	142	143	158	155	149	140	176	131	15.80
5. Blank	172.2	235	139	145	139	210	173	179	147	118	235	53.72
6A. Rubel, brachiopods	62.3	61	59	65	73	52	59	66	69	63	56	2.35
6B. Rubel, ostracods	36.8	43	37	21	36	41	46	33	30	39	42	5.07
6C. Rubel, thelodonts	35.9	39	37	39	29	37	40	36	32	27	42	2.31
6D. Rubel, combined	57.6	50	75	45	62	62	51	54	69	52	51	12.52
7. Sullivan	47.4	55	40	40	49	66	37	44	46	42	55	2.45
8A. Corliss, tops	1.8	1	1	3	1	2	2	2	1	3	2	2.94
8B. Corliss, bottoms	5.0	6	2	6	10	5	3	2	7	2	7	9.24
9A. Agterberg-Lew, E(D)=0.5	45.0	44	41	56	35	34	35	42	54	48	41	8.76
9B. Agterberg-Lew, E(D)=0.3	45.0	43	45	45	44	38	57	36	56	50	36	6.20
9C. Agterberg-Lew, E(D)=0.1	45.0	62	29	34	46	43	46	53	47	50	40	3.51

Table 4 *Normality test output for six computer simulation experiments. See text for further explanation.*

A. Revised RASC (Set 1 only)	0_1	0_2	0_3	0_4	0_5	0_6	0_7	0_8	0_9	0_{10}	$\hat{\chi}^2(7)$
E(D) = 0.5	70	117	93	58	86	132	66	107	95	76	52.6
E(D) = 0.3	81	91	90	90	94	90	93	94	96	81	1.9
E(D) = 0.2	84	102	82	78	78	98	106	100	87	85	6.3
E(D) = 0.1	85	88	100	79	84	86	103	107	97	71	3.2

B. Revised RASC	0_1	0_2	0_3	0_4	0_5	0_6	0_7	0_8	0_9	0_{10}	$\hat{\chi}^2(7)$
E(D) = 0.0, Set 1	98	90	73	86	94	98	83	120	85	73	0.5
E(D) = 0.0, Set 2	86	81	98	90	86	94	76	108	106	75	0.2

Table 5 *Some statistics for RASC results for 9 data sets of Appendix I. The equivalent number (n') of statistically independent values was derived from number of second-order differences (n), standard devation $\hat{\sigma}_2$ of Gaussian curve fitted to second-order differences (large values were not used, see text), and autocorrelation coefficient ($\hat{\rho}$).*

	Data Base	k_c	No. of Events	No. of Sections	n	$\hat{\sigma}_2$	$\hat{\rho}$	n'
1.	Gradstein-Thomas	7	44	24	503	1.223	0.420	206
2.	Gradstein	5	31	20	211	1.471	0.222	135
3.	Doeven	7	77	10	641	1.108	0.508	210
4.	Baumgartner	13	86	43	1496	1.701	0.027	1419
5.	Blank	15	80	81	1722	1.419	0.264	1003
6A.	Rubel, brachiopods	8	54	20	632	1.234	0.412	260
6B.	Rubel, ostracods	8	40	12	368	1.192	0.444	142
6C.	Rubel, thelodonts	8	34	20	359	1.188	0.447	137
6D.	Rubel, combined	13	43	35	576	1.659	0.063	507
7.	Sullivan	9	52	10	474	0.791	0.725	76
8A.	Corliss, tops	3	9	6	18	1.686	0.040	17
8B.	Corliss, bottoms	4	15	6	50	1.516	0.184	35
9A.	Agterberg-Lew, E(D)=0.5	25	20	25	450	1.512	0.187	309
9B.	Agterberg-Lew, E(D)=0.3	25	20	25	450	1.388	0.289	248
9C.	Agterberg-Lew, E(D)=0.1	25	20	25	450	0.881	0.668	90

Table 6 *Autocorrelation statistics for RASC runs of five computer simulation experiments. If the original values along the RASC-scale were stochastically independent, the ratio $\hat{\sigma}_2/\sigma$ would be equal to 1. Note extreme reduction from n to n' for E(D) = 0.0. The negative autocorrelation coefficients $\hat{\rho}_1$ apply to second-order differences (see text).*

E(D)	n	$\hat{\sigma}_2$	$\hat{\sigma}_2/\sigma$	$\hat{\rho}$	n'	$\hat{\rho}_1$
0.5	900	1.698	0.98	0.030	848	-0.658
0.3	900	1.528	0.88	0.173	634	-0.621
0.2	900	1.408	0.87	0.273	514	-0.597
0.1	900	0.966	0.56	0.609	219	-0.532
0.0	900	0.327	0.19	0.948	25	-0.501

observed frequencies, on the average, are equal to the expected frequency E_i (=90) in Table 2.

The revised normality test in the RASC program of Appendix II consists of fitting a doubly-truncated normal distribution to the second-order differences belonging to the classes with observed frequencies O_3 to O_8. If present, anomalous events are most likely to occur in the tails of an observed frequency distribution. Values in the classes of frequencies O_1, O_2, O_9, and O_{10} were therefore not used for estimating a theoretical normal distribution.

Each second-order difference value in the normality test is computed as follows. First, the difference of two successive values is calculated. If an event precedes the next event, their difference is corrected by subtracting a small amount. This correction is made because a gradual increase in distance from the origin is to be expected for successive events in each section. The small amount was set equal to the difference between the highest and lowest location values in the observed sequence divided by the total number of times an event precedes the next event without being coeval to it. No correction is made for pairs of coeval events. Next, the successive difference of two resulting values is determined. This procedure resembles the calculation of a second derivative with respect to location for every event except those in the first or last positions of an observed sequence. The second-order difference calculated is minus the difference between twice the distance of an event, on the one hand, and the sum of the distances of its two neighbouring events, on the other. If successive differences could be regarded as realizations of independent normal random variables with variances equal to $2\sigma^2$, the variance of the second-order difference would amount to $6\sigma^2$. This can be seen as follows. Suppose that three successive distance estimates x_{k-1}, x_k, and x_{k+1} were normally distributed with a mean of zero and variance σ^2. The second-order difference $-[2x_k - (x_{k-1} + x_{k+1})]$ would then be normal with a variance of $6\sigma^2$, because the variance $\sigma^2(2X_k) = 4\sigma^2$ and $\sigma^2(X_{k-1}) = \sigma^2(X_{k+1}) = \sigma^2$. However, the successive distance estimates have become autocorrelated because of the various manipulations to which the data were subjected during ranking and scaling. Suppose that the autocorrelation coefficient of successive distances X_k and X_{k+1} is written as ρ with $\rho = Cov(X_k,X_{k+1})/\sigma^2$. The variance of the second-order difference satisfies

$$\sigma_2{}^2 = E[(X_{k-1} - X_k) - (X_k - X_{k+1})]^2$$

if $$E[(X_{k-1} - X_k) - (X_k - X_{k+1})] = 0$$

It follows that

$$\sigma_2{}^2 = 2\sigma^2(\rho^2 - 4\rho + 3)$$

if $\quad Cov(X_{k-1}, X_{k+1}) = \rho^2\sigma^2$

The procedure followed in the RASC program consists of ordering the second-order differences from all sections from the smallest value to the largest one. The standard deviation of the central 60% of the ordered values is estimated and assumed to represent a truncated normal distribution. The relationship between standard deviations of truncated normal and normal distributions is given in statistical tables. Their ratio amounts to 0.463 if 20% is truncated from each tail. Division by 0.463 yields the estimate $\hat{\sigma}_2$. Not all second-order differences were used for this estimation because, if anomalous values are present, these are more likely to occur in the tails of the distribution. From $\sigma^2 = 1/2$, it follows that $\hat{\rho}$ can be estimated from $\hat{\sigma}_2$ by

$$\hat{\rho} = 2 - \sqrt{1 + \hat{\sigma}_2^{\,2}}$$

In general (cf. Agterberg, 1974, p. 302), it can be assumed that n autocorrelated values are equivalent to n' stochastically independent values with

$$1/n' = 1/n + 2\hat{\rho}[n/(1 - \hat{\rho}) - 1/(1 - \hat{\rho})^2]/n^2$$

This allows us to estimate n' which is part of the output of the RASC program. In the chi-squared test for goodness of fit, expected frequencies E_i of stochastically independent data in p classes are related to the corresponding observed frequencies by

$$\sum_{i=1}^{p} (E_i - O_i)^2/E_i = \hat{\chi}^2(p - 3)$$

if two parameters of the fitted distribution were estimated. For autocorrelated data, the sum on the left-hand side of this equation may be multiplied by n'/n in order to obtain an approximate estimate of chi-squared.

The 10 classes of the normality test in the RASC program (lower part of Table 11 in Chapter II.6) were constructed by dividing the expected ordered sequence of second-order differences into 10 equal parts in order to obtain 10 equal expected frequencies for comparison to the corresponding observed frequencies. The class limits are given by the z-values of the relative frequencies 0.1, 0.2, ..., 0.9 multiplied by $\hat{\sigma}_2$. This procedure provides a convenient normality test. The individual second-order differences (top part of normality test output) were compared to the 95% and 99% confidence intervals $\pm$ 1.960 $\hat{\sigma}_2$ and $\pm$ 2.576 $\hat{\sigma}_2$, respectively.

The preceding method generally yields sets of observed frequencies O_i (i = 2, 3, ..., 9) which are equal to one another (and to E_i) except for random fluctuations. The frequencies (O_1 and O_{10}) in the tails of the distribution may be too high when anomalous events occur in some of the sections.

Results of applying the revised normality test for the 9 data bases are shown in Table 3 and for 6 computer simulation experiments in Table 4. Other statistics for most of these computer runs are given in Tables 5 and 6.

The normal distribution model provides a good fit for 13 of the 15 tests in Table 3 according to the approximate chi-squared test (see last column of Table 3). The 95% and 99% confidence limits of $\chi^2(7)$ which should not be exceeded if the normality assumption holds true (with level of significance equal to 5% and 1%), amount to 14.1 and 18.5, respectively. Only $\hat{\chi}^2(7)$ = 53.7 of data base No. 5 clearly exceeds both confidence limits. According to Blank (1984, p. 65) a number of events in this data base were determined to be anomalous because of four main reasons:

(1) Taxonomic problems with Mesozoic events,

(2) Short sections that were truncated artificially at coring gaps,

(3) Contamination due to reworking and

(4) Provinciality because of the large latitudinal spread of control sites.

The chi-squared value for data base No. 4 exceeds the 95% confidence limit but is below the 99% confidence limit. There is the possibility that the tail frequencies O_1(=127) and O_{10}(=131) are slightly too low (in comparison with E_i = 149.6). Further research would be needed to demonstrate whether or not this new null hypothesis holds true.

The run for E(D) = 0.5 in Table 4 gave $\hat{\chi}^2(7)$ = 52.6 indicating non-normality. It is likely that the central frequency O_6 (=132) is significantly greater than its expected value (E_i = 90) for the same reason that O_6 was too high in the computer simulation experiment with E(D) = 1.0 (see Table 2).

In Part B of Table 4, the values of $\hat{\chi}^2(7)$ are equal to 0.5 and 0.2, respectively. The 1% and 5% confidence limits of $\chi^2(7)$ amount to 0.6 and 1.6, respectively. This suggests a degree of fit which is too good to be true. The approximate chi-squared test is based on the assumption that n autocorrelated values are equivalent to n' independent values (see before). As shown in Table 6, this reduction becomes very large (from n = 900 to n' = 25) when E(D) = 0. There are no definite trends in the two sets of O_i-values in Table 6. It may therefore be assumed that the procedure used for estimating the observed and expected frequencies remains valid when E(D) approaches 0 but that the reduction from n to n' has become too large.

Finally, it is noted that the autocorrelation coefficient $\hat{\rho}$ estimated from $\hat{\sigma}_2/\sigma$ applies to the successive distances X_k and not to the second-order differences $(X_{k-1}-X_k) - (X_k-X_{k+1})$. Suppose that the autocorrelation coefficient of the second-order differences is called ρ_1. Then,

$$\rho_1 = E[\{(X_{k-1}-X_k)-(X_k-X_{k+1})\} \cdot \{(X_k-X_{k+1})(X_{k+1}-X_{k+2})\}]/\sigma_2{}^2$$

It follows that

$$\rho_1 = \frac{\rho^3 - 4\rho^2 + 7\rho - 4}{2\rho^2 - 8\rho + 6}$$

if $\quad Cov(X_{k+i},X_k) = \rho^i\sigma^2;\ i = 1,2,3$

The latter condition would imply that the X_k satisfy a first-order Markov process (Agterberg, 1974). The autocorrelation coefficient ρ_1 of the second-order differences is negative and ranges from -0.6667 for $\rho = 0$ to -0.5 in the limit for $\rho \rightarrow 1$. Its values in five computer simulation experiments are shown in Table 6. It is also noted that the estimation of the autocorrelation coefficients ρ and ρ_1 has no bearing on the calculation of the observed and expected frequencies of the normality test. The theory of autocorrelation was only used to provide an approximate chi-squared test for comparing the observed and expected frequencies with one another.

UNITARY ASSOCIATIONS METHOD APPLIED TO DROBNE'S ALVEOLINIDS

Guex (1981) has coded biostratigraphic information on alveolinids collected by Drobne (1977) and applied his unitary association method to these data. The original information on 15 species in 11 sections as used by Guex (1981) is shown in Figure 1 and Table 7. The corresponding graph and its adjacency matrix are shown in Figure 2 and Table 8. Forbidden structures (see Figure 8, Chapter II.5) have to be identified and eliminated before an interval graph can be constructed from the observed co-occurrences in Figure 2 and Table 8. The computer program of Guex and Davaud (1984) initially gives the output shown in

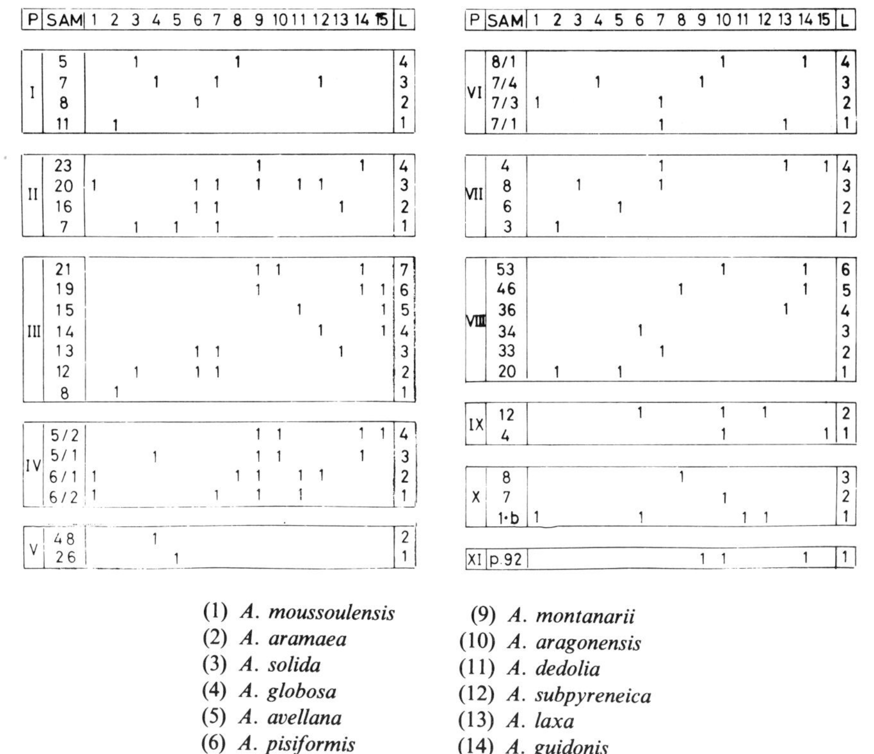

P	SAM	1	2	3	4	5	6	7	8	9	10	11	12	13	14	15	L
I	5			1					1								4
	7				1			1					1				3
	8						1										2
	11		1														1
II	23									1					1		4
	20	1					1	1		1		1	1				3
	16						1	1						1			2
	7			1		1		1									1
III	21									1	1				1		7
	19									1					1	1	6
	15											1				1	5
	14												1			1	4
	13						1	1						1			3
	12			1			1	1									2
	8		1														1
IV	5/2									1	1				1	1	4
	5/1				1					1	1				1		3
	6/1	1							1	1		1	1				2
	6/2	1						1		1		1					1
V	48				1												2
	26					1											1

P	SAM	1	2	3	4	5	6	7	8	9	10	11	12	13	14	15	L
VI	8/1										1				1		4
	7/4				1					1							3
	7/3	1						1									2
	7/1							1						1			1
VII	4							1						1		1	4
	8			1				1									3
	6					1											2
	3		1														1
VIII	53										1				1		6
	46								1						1		5
	36													1			4
	34						1										3
	33							1									2
	20		1			1											1
IX	12						1				1		1				2
	4										1					1	1
X	8								1								3
	7										1						2
	1·b	1					1					1	1				1
XI	p.92									1	1				1		1

(1) *A. moussoulensis*
(2) *A. aramaea*
(3) *A. solida*
(4) *A. globosa*
(5) *A. avellana*
(6) *A. pisiformis*
(7) *A. pasticillata*
(8) *A. leupoldi*
(9) *A. montanarii*
(10) *A. aragonensis*
(11) *A. dedolia*
(12) *A. subpyreneica*
(13) *A. laxa*
(14) *A. guidonis*
(15) *A. decipiens*

Figure 1 Occurrence of 15 alveolinids (1 to 15) from Yugoslavia (data from Drobne, 1977) in 11 sections (I to XI). SAM: Sample numbers originally used by Drobne. Successive beds are numbered in the stratigraphically upward direction for each section (see last column).

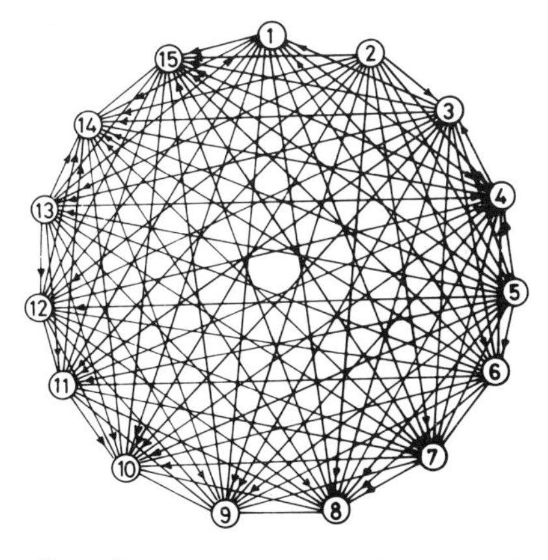

Figure 2 Graph representing the stratigraphic relationships of Figure 1 and Table 7.

Table 7 *Data of Figure 1 coded using U.A. input format. Each species number is followed by codes for its lowest and highest occurrence, respectively.*

```
  ALVEOLINIDS (DROBNE)
11 15
7
2 1 1 3 4 4 4 3 3 6 2 2 7 3 3 8 4 4 12 3 3
10
1 3 3 3 1 1 5 1 1 6 2 3 7 1 3 9 3 4 11 3 3 12 3 3 13 2 2 14 4 4
11
2 1 1 3 2 2 6 2 3 7 2 3 9 6 7 10 7 7 11 5 5 12 4 4 13 3 3 14 6 7 15 4 6
10
1 1 2 4 3 3 7 1 1 8 2 2 9 1 4 10 3 4 11 1 2 12 2 2 14 3 4 15 4 4
2
4 2 2 5 1 1
7
1 2 2 4 3 3 7 1 2 9 3 3 10 4 4 13 1 1 14 4 4
6
2 1 1 3 3 3 5 2 2 7 3 4 13 4 4 15 4 4
8
2 1 1 5 1 1 6 3 3 7 2 2 8 5 5 10 6 6 13 4 4 14 5 6
4
6 2 2 10 1 2 12 2 2 15 1 1
6
1 1 1 6 1 1 8 3 3 10 2 2 11 1 1 12 1 1
3
9 1 1 10 1 1 14 1 1
```

Table 8 *Adjacency matrix for data of Figure 1.*

	1	2	3	4	5	6	7	8	9	10	11	12	13	14	15
1	0	0	0	1	0	1	1	1	1	1	1	1	0	1	1
2	0	0	1	1	1	1	1	1	1	1	1	1	1	1	1
3	1	0	0	0	1	1	1	1	1	1	1	1	1	1	1
4	0	0	1	0	0	0	1	1	1	1	0	1	0	1	1
5	1	1	1	1	0	1	1	1	1	1	1	1	1	1	1
6	1	0	1	1	0	0	1	1	1	1	1	1	1	1	1
7	1	0	1	1	1	1	0	1	1	1	1	1	1	1	1
8	1	0	1	1	0	0	0	0	1	1	1	1	0	1	1
9	1	0	0	1	0	1	1	1	0	1	1	1	0	1	1
10	0	0	0	1	0	1	0	1	1	0	0	1	0	1	1
11	1	0	0	1	0	1	1	1	1	1	0	1	0	1	1
12	1	0	1	1	0	1	1	1	1	1	1	0	0	1	1
13	1	0	0	1	0	1	1	1	1	1	1	1	0	1	1
14	0	0	0	1	0	0	0	1	1	1	0	0	0	0	1
15	0	0	0	0	0	1	1	0	1	1	1	1	1	1	0

Table 9. If a strong component of a graph can be detected in the biostratigraphical graph G this may provide useful information on biostratigraphical inconsistencies. For the example, the strong component in G_e involves fossils 1, 3, 4, 11, and 13. The frequencies of arcs of the strong component belonging to any cycle C_n ($n = 3, 4, ...$) and to cycle C_3 only are tabulated next in Table 9. The s-ratio (see Chapter II.5) is determined. In Table 9, the arc from 4 to 3, which occurs only in Section 1, has the highest s-ratio (=3.00). The next tabulation in the output from Guex and Davaud's (1984) computer program displays frequencies of edges and arcs participating in the forbidden structures S_3 and S_4. Finally, the number of times that each of the fossils participates in an S_3 or S_4 are shown at the bottom of Table 9. The arrows a-d in Table 9 were entered after visual inspection of the tabulations. They indicate that an abnormally large proportion of the inconsistencies is due to the occurrence of fossils 3, 4, and 8 in Section 1. In the original plot for individual sections (Figure 1) it can be seen that species 3 occurs higher in Section 1 than in the other sections where it was observed. Drobne (1977, p. 83) specifically stated that bed No. 4 in Section 1, which contains fossils 3 and 8, is reworked. For this reason, Guex and Davaud (1984) decided to delete fossil 3 from bed No. 4 in Section 1 and to repeat their analysis. Final results for the modified computer run (without species 3 in Section 1) are shown in Table 10. The method followed to obtain the unitary associations and the resulting 'range chart' was as described previously. These U.A.s which result from the union of some I.U.A.s correspond closely to the original definition of 'zones' introduced by Oppel in 1856.

RASC PROGRAM APPLIED TO DROBNE'S ALVEOLINIDS

The information of Table 1 was converted into RASC input by replacing each fossil number i (=1,2,...5) by two numbers: (2i-1) for highest occurrences and 2i for lowest occurrences, respectively. RASC was run on the resulting data set with $k_c = 4$, $m_{c1} = 0$, and $m_{c2} = 2$. Setting $k_c = 4$ ensured that no events were eliminated as in the U.A. computer program. However, it became immediately apparent that 7 of the 15 species were observed in only one bed in the sections containing them. Because the highest and lowest occurrences of these 7 species coincided everywhere, it was decided to maintain a single number for these species indicating occurrence only (instead of two numbers indicating coinciding highest and lowest occurrences). The RASC input for the new 4-0-2 run performed after this modification is shown in Table 11. The presorting option was applied and followed by the modified Hay method. Three cycles occurred (Table 12) and each of these involved the species 3 and 4. The optimum sequence is shown in Table 13. Based on $m_{c2} = 2$, 42 out of 253 pairs of matrix elements were zeroed for scaling. The results of weighted distance analysis (after final reordering) are shown in Table 14 and in the dendrogram (original RASC output listing) of Figure 3. The results of the normality test are shown in Table 15. As in the previous section, it may be concluded that species 3 occurs too high in Section 1 (because of reworking). Figure 4

```
28   .0259   HI A. GUIDONIS
20   .2304   HI A. ARAGONENSIS
18   .2258   HI A. MONTANARII
19   .1443   LO A. ARAGONENSIS
15   .0155   LO A. LEUPOLDI
27   .1471   LO A. GUIDONIS
30   .0823   HI A. DECIPIENS
 7   .2479   LO A. GLOBOSA
29   .1285   LO A. DECIPIENS
17   .2225   LO A. MONTANARII
22   .0801   HI A. DEDOLIA
23   .0657   LO A. SUBPYRENEICA
 2   .2134   HI A. MOUSSOULENSIS
21   .1475   LO A. DEDOLIA
 1   .0667   LO A. MOUSSOULENSIS
12   .1282   HI A. PISIFORMIS
14   .1643   HI A. PASTICILLATA
25   .1259   LO A. LAXA
11   .1979   LO A. PISIFORMIS
 5   .1562   LO A. SOLIDA
13   .8140   LO A. PASTICILLATA
 9   .7353   LO A. AVELLANA
 3           LO A. ARAMAEA

.8460  .7022  .5585  .4148  .2710  .1273  -.0164
   .7741  .6304  .4866  .3429  .1992  .0554
```

Figure 3 Dendrogram in RASC output for run on Drobne's alveolinids.

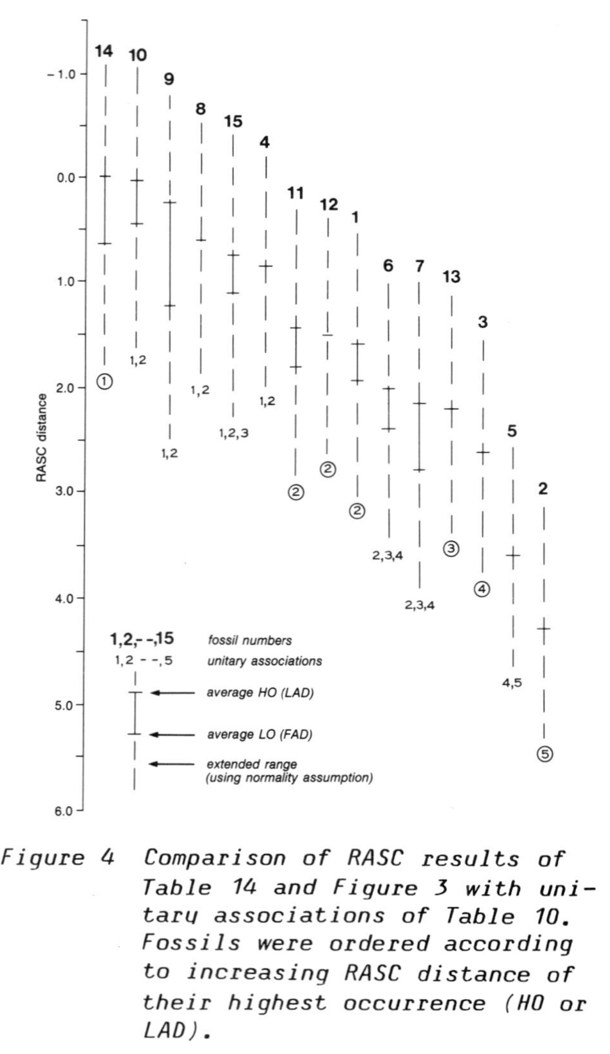

Figure 4 Comparison of RASC results of Table 14 and Figure 3 with unitary associations of Table 10. Fossils were ordered according to increasing RASC distance of their highest occurrence (HO or LAD).

shows a comparison of the five Unitary Associations of Table 10 with the scaled optimum sequence of Table 14 and Figure 3. The highest occurrences of the 15 fossils were ordered in Figure 4 according to their RASC distance in Table 14. Because ranking and scaling estimate average highest and average lowest occurrences, the distances between these two types of events on the RASC scale are less than their true stratigraphic ranges. According to the original scaling model, events in sections are normally distributed about their average position with standard deviations equal to $\sigma = 0.7071$. Consequently, the observed highest occurrence of a fossil in a section would occur with a probability of 95% below its RASC value decreased by $1.645 \times \sigma = 1.16$. This value provides a more reasonable estimate of the true highest occurrence or last appearance datum (LAD) than the original RASC value. Likewise 1.16 can be added to the RASC distance estimated for a lowest occurrence or first appearance datum (FAD) along the RASC scale. The resulting enlargements of the RASC ranges are shown as dashed lines in Figure 4. According to the probabilistic range chart of Figure 4, fossil 14 probably co-occurred with 3 and probably not with 2. The dashed lines are based on the assumption that all events satisfy a normal distribution with the same standard deviation along the RASC scale. This assumption may not hold true in reality and care should be taken in interpreting the ranges of Figure 4. For example, Guex (personal information, 1984) has pointed out that fossil 5 probably never co-existed with 11 although their ranges overlap in Figure 4.

The U.A. numbers of the fossils are also shown in Figure 4 and circled if a fossil belongs to a single U.A. only. The order of the overlapping U.A.s is very similar to that of the sequence of RASC ranges for the fossils. The only discrepancy is that fossil 15, which belongs to U.A. 3, occurs in fifth position in Figure 4, while the other fossils of U.A. 3 (6, 7, and 130) occupy positions 10, 11, and 12, respectively.

CONCLUDING REMARKS

The preceding comparison using Drobne's alveolinids is interesting in that similar results for ranking as well as stratigraphic 'normality' were obtained by means of two methods (U.A. and RASC), which are built upon different premises. In the U.A. method, observed co-occurrences of fossils are augmented by virtual occurrences partly to resolve inconsistencies (forbidden structures) in order to obtain assemblage zones. In the RASC model, the observed highest and lowest occurrences of fossils in sections are considered to be realizations of random variables with fixed average positions along a linear scale. The two methods have in common that each provides a way of eliminating inconsistencies and filling in the gaps due to missing data. In the U.A. method, this is done by adopting rules based on graph theory, whereas in the RASC method the observed data are considered to belong to small samples derived from (infinitely large) statistical populations, of which the parameters (rankings, means, and standard deviations) can be estimated. The 'zones' resulting from the U.A. method are primarily based on

Table 9 Output from Guex and Davaud's (1984) computer program for unitarity associations. See text for explanation.

STRONG COMPONENT DETECTED 1 3 4 11 13

SUBMATRIX : STRONG COMPONENT ALVEOLINIDS (DROBNE)

		TAXA				
		1	3	4	11	13
	1:	1	1	0	1	1
	3:	0	1	1	0	0
TAXA	4:	1	0	1	1	1
	11:	1	1	0	1	1
	13:	0	1	0	0	1

ARC	CYCLE OCC. FREQ.	STRATIGR. OCC. FREQ.	RATIO	CYCLE 3 OCC. FREQ.	S RATIO	SECTION NUMBER	
1 -> 4	2	2	1.00	1	.50		
3 -> 1	1	1	1.00	1	1.00	2	
3 -> 11	1	2	.50	1	.50		
3 -> 13	3	3	1.00	1	.33		
4 -> 3	5	1	5.00	3	3.00	1	← (a)
11 -> 4	2	1	2.00	1	1.00	4	
13 -> 1	1	2	.50	0	.00		
13 -> 4	1	1	1.00	1	1.00	6	
13 -> 11	1	2	.50	0	.00		

ITERATION : 1 30 S3, S4 HAVE BEEN DETECTED

EDGES OCCURRENCE FREQUENCIES IN SEMI-ORIENTED CYCLES

EDGES	SEM. OR. CY. OCC. FREQ.	STRATIGR. OCC. FREQ.	SECTION NUMBER	
1 - 6	1	2		
3 - 8	8	1	1	← (b)
1 - 7	1	3		
1 - 8	2	1	4	
6 - 10	4	1	9	
4 - 7	3	1	1	
3 - 5	1	1	2	
3 - 6	1	1	3	
3 - 7	1	3		
4 - 8	1	0		
6 - 9	1	1	2	
6 - 11	1	2		
7 - 9	1	2		
7 - 11	1	2		
3 - 12	1	0		
8 - 11	2	1	4	
4 - 9	1	2		
4 - 10	2	1	4	
7 - 15	3	1	7	
4 - 14	3	1	4	
12 - 15	2	1	3	
6 - 15	2	0		
8 - 10	1	0		
11 - 15	3	1	3	
8 - 14	2	1	8	
13 - 15	3	1	7	

ARCS OCCURRENCE FREQUENCIES IN SEMI ORIENTED CYCLES

ARC	SEM. OR. CY. OCC. FREQ	STRATIGR. OCC. FREQ.	RATIO	
8 -> 15	6	1 (4)	6.00	
3 <- 4	6	1 (1)	6.00	← (c)
7 -> 8	7	3 ( )	2.33	
4 <- 11	2	1 (4)	2.00	
8 <- 13	2	1 (8)	2.00	
3 -> 9	3	2 ( )	1.50	
1 -> 15	1	1 (4)	1.00	
4 <- 5	1	1 (5)	1.00	
12 -> 14	2	3 ( )	.67	
1 <- 13	1	2 ( )	.50	
7 -> 10	2	4 ( )	.50	
1 -> 4	1	2 ( )	.50	
1 -> 10	1	3 ( )	.33	
4 -> 15	6	1 (4)	6.00	
4 <- 6	3	1 (1)	3.00	
6 -> 8	7	3 ( )	2.33	
3 -> 10	2	1 (3)	2.00	
1 <- 3	2	1 (2)	2.00	
4 <- 13	1	1 (6)	1.00	
3 -> 11	2	2 ( )	1.00	
10 <- 11	2	3 ( )	.67	
3 -> 13	2	3 ( )	.67	
3 -> 14	1	2 ( )	.50	
12 <- 13	1	2 ( )	.50	
11 -> 14	1	3 ( )	.33	
7 -> 14	1	5 ( )	.20	

TAXA OCCURRENCE FREQUENCY IN ORIENTED AND SEMI-ORIENTED CYCLES

TAXON	OCC. FREQ.	TAXON	OCC. FREQ.	TAXON	OCC. FREQ.	TAXON	OCC. FREQ.	TAXON	OCC. FREQ.	
8	38	4	30	3	30	15	26	7	20	← (d)
6	20	11	14	10	14	14	10	13	10	
1	10	12	6	9	6	5	2	2	0	

Table 10 Final unitary associations for Drobne's alveolinids and their occurrence in the sections.

RANGE CHART 1 ALVEOLINIDS (DROBNE)

U.A.	5	2	6	7	3	15	13	4	8	9	10	1	11	12	14
1:	0	0	0	0	0	1	0	1	1	1	1	0	0	0	1
2:	0	0	1	1	0	1	0	1	1	1	1	1	1	1	0
3:	0	0	1	1	0	1	1	0	0	0	0	0	0	0	0
4:	1	0	1	1	1	0	0	0	0	0	0	0	0	0	0
5:	1	1	0	0	0	0	0	0	0	0	0	0	0	0	0

RANGE CHART 2 ALVEOLINIDS (DROBNE)

U.A.	5	2	6	7	3	15	13	4	8	9	10	1	11	12	14
1:	0	0	0	0	0	1	0	1	1	1	1	0	0	0	1
2:	0	0	1	1	0	1	0	1	1	1	1	1	1	1	0
3:	0	0	1	1	0	1	1	0	0	0	0	0	0	0	0
4:	1	0	1	1	1	0	0	0	0	0	0	0	0	0	0
5:	1	1	0	0	0	0	0	0	0	0	0	0	0	0	0

(a)

ALVEOLINIDS (DROBNE)

SECTION 1

LEVEL	U.A.	
3	-	2
2	2	4
1	-	5

SECTION 2

LEVEL	U.A.	
4	-	1
3	-	2
2	-	3
1	-	4

SECTION 3

LEVEL	U.A.	
7	-	1
6	-	1
5	-	2
4	-	2
3	-	3
2	-	4
1	-	5

SECTION 4

LEVEL	U.A.	
4	-	1
3	-	1
2	-	2
1	-	2

SECTION 5

LEVEL	U.A.	
2	1	2
1	4	5

SECTION 6

LEVEL	U.A.	
4	-	1
3	1	2
2	-	2
1	-	3

SECTION 7

LEVEL	U.A.	
4	-	3
3	-	4
2	4	5
1	-	5

SECTION 8

LEVEL	U.A.	
6	-	1
5	-	1
4	-	3
3	2	4
2	2	4
1	-	5

ALVEOLINIDS (DROBNE)

SECTION 9

LEVEL	U.A.	
2	-	2
1	1	2

SECTION 10

LEVEL	U.A.	
3	1	2
2	1	2
1	-	2

SECTION 11

LEVEL	U.A.	
1	-	1

IDENTIFIED U.A. IN STRAT. SECTIONS ALVEOLINIDS (DROBNE)

U.A. \ SECT	1	2	3	4	5	6	7	8	9	10	11
1:	0	1	1	1	0	1	0	1	0	0	1
2:	1	1	1	1	0	1	0	0	1	1	0
3:	0	1	1	0	0	1	1	1	0	0	0
4:	0	1	1	0	0	0	1	0	0	0	0
5:	1	0	1	0	0	0	1	1	0	0	0

Table 11 Data of Figure 1 coded using RASC input format.

```
SECTION 1
15  -5  23 -14 -13  -7  12 -11   3

SECTION 2
28 -27 -18  23 -22 -21 -17 -14 -12  -2  -1  25 -11   9  -5 -13

SECTION 3
28 -20 -19 -18  30 -27 -17  22 -21  29 -23  25 -14 -12  13 -11  -5

SECTION 4
30 -29 -28 -20 -18  27 -19  -7  23 -22 -15  -2  21 -17 -14 -13  -1

SECTION 5
 7   9

SECTION 6
28 -27 -20 -19  18 -17  -7  14  -2  -1  25 -13

SECTION 7
30 -29 -25 -14  13  -5   9   3

SECTION 8
28 -20 -19  27 -15  25  12 -11  14 -13   9  -3

SECTION 9
23 -20 -12 -11  30 -29 -19

SECTION 10
15  20 -19  23 -22 -21 -12 -11  -2  -1

SECTION 11
28 -27 -20 -19 -18 -17
```

Table 12 Cycles in 4-0-2 RASC run on Drobne's alveolinids.

```
 CYCLING EVENTS:     17    7    5

EVENT POSITIONS:      9   10   11

MATRIX ELEMENTS:
  0.0  0.5  2.0
  1.5  0.0  0.0
  0.0  1.0  0.0

C(9, 10) AND C(10, 9) ZEROED


 CYCLING EVENTS:     23    7    5

EVENT POSITIONS:     10   12   11

MATRIX ELEMENTS:
  0.0  0.5  2.0
  1.5  0.0  0.0
  1.0  1.0  0.0

C(10, 12) AND C(12,10) ZEROED


 CYCLING EVENTS:      7    5   22

EVENT POSITIONS:     10   12   11

MATRIX ELEMENTS:
  0.0  0.0  1.0
  1.0  0.0  0.0
  0.0  2.0  0.0

C(10, 12) AND C(12,10) ZEROED
```

Table 13 Optimum sequence for Drobne's alveolinids.

SEQUENCE POSITION	FOSSIL NUMBER	RANGE	FOSSIL NAME
1	20	0- 3	HI A. ARAGONENSIS
2	28	0- 3	HI A. GUIDONIS
3	18	2- 5	HI A. MONTANARII
4	19	2- 6	LO A. ARAGONENSIS
5	30	3- 6	HI A. DECIPIENS
6	27	5- 8	LO A. GUIDONIS
7	29	5- 8	LO A. DECIPIENS
8	15	7- 9	LO A. LEUPOLDI
9	17	8- 13	LO A. MONTANARII
10	7	7- 11	LO A. GLOBOSA
11	22	10- 13	HI A. DEDOLIA
12	2	10- 13	HI A. MOUSSOULENSIS
13	21	12- 15	LO A. DEDOLIA
14	23	11- 15	LO A. SUBPYRENEICA
15	14	14- 17	HI A. PASTICILLATA
16	1	14- 17	LO A. MOUSSOULENSIS
17	25	16- 19	LO A. LAXA
18	12	14- 19	HI A. PISIFORMIS
19	5	18- 21	LO A. SOLIDA
20	11	18- 21	LO A. PISIFORMIS
21	13	20- 22	LO A. PASTICILLATA
22	9	21- 23	LO A. AVELLANA
23	3	22- 24	LO A. ARAMAEA

Table 14 Results of weighted distance analysis of Drobne's alveolinids.

DISTANCE ANALYSIS REPEATED WITH WEIGHTED DIFFERENCES

POSITION	FOSSIL PAIRS	FOSSIL DISTANCE	CUMULATIVE DISTANCE	SUM DIFF Z VALUES	WEIGHT	S.D.
1	28- 20	.0259	.0259	.3609	14.0	.0665
2	20- 18	.2304	.2563	2.9342	12.7	.0746
3	18- 27	.3856	.6419	3.9125	10.1	.1134
4	27- 19	-.1598	.4821	-2.3629	14.8	.1097
5	19- 30	.3070	.7890	3.3904	11.0	.1025
6	30- 15	-.1627	.6264	-.8425	5.2	.3154
7	15- 29	.4929	1.1193	2.7709	5.6	.2838
8	29- 17	.1285	1.2478	1.4337	11.2	.1507
9	17- 7	-.3765	.8714	-2.5858	6.9	.1597
10	7- 22	.5990	1.4703	3.0570	5.1	.1941
11	22- 23	.0801	1.5504	1.0990	13.7	.0971
12	23- 2	.0657	1.6161	.7805	11.9	.0961
13	2- 21	.2134	1.8295	1.9549	9.2	.1038
14	21- 1	.1475	1.9770	1.3385	9.1	.0681
15	1- 12	.0667	2.0437	.7104	10.7	.1145
16	12- 14	.1282	2.1719	2.0582	16.1	.0921
17	14- 25	.1643	2.3361	1.6730	10.2	.1784
18	25- 11	.1259	2.4621	1.2744	10.1	.1944
19	11- 5	.1979	2.6600	1.7322	8.8	.1592
20	5- 13	.1562	2.8162	1.3585	8.7	.1133
21	13- 9	.8140	3.6302	3.5662	4.4	.1247
22	9- 3	.7353	4.3655	1.4909	2.0	.0848

NEW SEQUENCE	DISTANCE FROM 1ST POSITION	FOSSIL PAIRS	INTER FOSSIL DISTANCE
28	0.0000	28- 20	.0259
20	.0259	20- 18	.2304
18	.2563	18- 19	.2258
19	.4821	19- 15	.1443
15	.6264	15- 27	.0155
27	.6419	27- 30	.1471
30	.7890	30- 7	.0823
7	.8714	7- 29	.2479
29	1.1193	29- 17	.1285
17	1.2478	17- 22	.2225
22	1.4703	22- 23	.0801
23	1.5504	23- 2	.0657
2	1.6161	2- 21	.2134
21	1.8295	21- 1	.1475
1	1.9770	1- 12	.0667
12	2.0437	12- 14	.1282
14	2.1719	14- 25	.1643
25	2.3361	25- 11	.1259
11	2.4621	11- 5	.1979
5	2.6600	5- 13	.1562
13	2.8162	13- 9	.8140
9	3.6302	9- 3	.7353
3	4.3655		

Table 15 *RASC output for normality test. Second column lists dictionary numbers of events named in the first column. Single and double asterisks in last columns indicate departures from normality for 95 and 99% confidence limits, respectively. For this run, $\hat{\sigma}_2 = 1.268$, $\hat{\rho} = 0.385$, and $n' = 43$ ($n = 96$). A double asterisk was assigned when a second-order difference exceeded $2.576\ \hat{\sigma}_2 = 3.266$.*

SECTION 1

		CUM. DIST.	2ND ORDER DIFF.
LO A. LEUPOLDI	15	.6264	
LO A. SOLIDA	-5	2.6600	-4.3895 **
LO A. SUBPYRENEICA	23	1.5504	2.9773 *
HI A. PASTICILLATA	-14	2.1719	.0229
LO A. PASTICILLATA	-13	2.8162	-2.5891 *
LO A. GLOBOSA	-7	.8714	1.8708
HI A. PISIFORMIS	12	2.0437	.4924
LO A. PISIFORMIS	-11	2.4621	.2387
LO A. ARAMAEA	3	4.3655	

SECTION 2

		CUM. DIST.	2ND ORDER DIFF.
HI A. GUIDONIS	28	0.0000	
LO A. GUIDONIS	-27	.6419	-1.0275
HI A. MONTANARII	-18	.2563	.7411
LO A. SUBPYRENEICA	23	1.5504	-.4356
HI A. DEDOLIA	-22	1.4703	.4393
LO A. DEDOLIA	-21	1.8295	-.9409
LO A. MONTANARII	-17	1.2478	1.5057
HI A. PASTICILLATA	-14	2.1719	-1.0522
HI A. PISIFORMIS	-12	2.0437	-.2994
HI A. MOUSSOULENSIS	-2	1.6161	.7885
LO A. MOUSSOULENSIS	-1	1.9770	-.9405
LO A. LAXA	25	2.3361	.7056
LO A. PISIFORMIS	-11	2.4621	.1034
LO A. AVELLANA	9	3.6302	-1.1996
LO A. SOLIDA	-5	2.6600	1.1264
LO A. PASTICILLATA	-13	2.8162	

SECTION 3

		CUM. DIST.	2ND ORDER DIFF.
HI A. GUIDONIS	28	0.0000	
HI A. ARAGONENSIS	-20	.0259	.4304
LO A. ARAGONENSIS	-19	.4821	-.6820
HI A. MONTANARII	-18	.2563	.0310
HI A. DECIPIENS	30	.7890	.0477
LO A. GUIDONIS	-27	.6419	.7531
LO A. MONTANARII	-17	1.2478	-1.1110
HI A. DEDOLIA	22	1.4703	.8642
LO A. DEDOLIA	-21	1.8295	-1.7970
LO A. DECIPIENS	29	1.1193	1.8690
LO A. SUBPYRENEICA	-23	1.5504	-.3731
LO A. LAXA	25	2.3361	-.2224
HI A. PASTICILLATA	-14	2.1719	.0361
HI A. PISIFORMIS	-12	2.0437	.1730
LO A. PASTICILLATA	13	2.8162	-.3990
LO A. PISIFORMIS	-11	2.4621	.5520
LO A. SOLIDA	-5	2.6600	.7800
LO A. ARAMAEA	3	4.3655	

SECTION 4

		CUM. DIST.	2ND ORDER DIFF.
HI A. DECIPIENS	30	.7890	
LO A. DECIPIENS	-29	1.1193	-1.4495
HI A. GUIDONIS	-28	0.0000	1.1451
HI A. ARAGONENSIS	-20	.0259	.2046
HI A. MONTANARII	-18	.2563	-.2408
LO A. GUIDONIS	27	.6419	-.1495
LO A. ARAGONENSIS	-19	.4821	.5491
LO A. GLOBOSA	-7	.8714	-.1062
LO A. SUBPYRENEICA	23	1.5504	-.3632
HI A. DEDOLIA	-22	1.4703	-.7639
LO A. LEUPOLDI	-15	.6264	1.8337
HI A. MOUSSOULENSIS	-2	1.6161	-1.1724
LO A. DEDOLIA	21	1.8295	-.3991
LO A. MONTANARII	-17	1.2478	1.5057
HI A. PASTICILLATA	-14	2.1719	-.2797
LO A. PASTICILLATA	-13	2.8162	-1.4834
LO A. MOUSSOULENSIS	-1	1.9770	

SECTION 6

		CUM. DIST.	2ND ORDER DIFF.
HI A. GUIDONIS	28	0.0000	
LO A. GUIDONIS	-27	.6419	-1.2579
HI A. ARAGONENSIS	-20	.0259	1.0723
LO A. ARAGONENSIS	-19	.4821	-1.6207
HI A. MONTANARII	18	.2563	2.1560
LO A. MONTANARII	-17	1.2478	-1.3680
LO A. GLOBOSA	-7	.8714	.7382
HI A. PASTICILLATA	14	2.1719	-.9175
HI A. MOUSSOULENSIS	-2	1.6161	.9166
LO A. MOUSSOULENSIS	-1	1.9770	-.9405
LO A. LAXA	25	2.3361	1.0597
LO A. PASTICILLATA	-13	2.8162	

SECTION 7

		CUM. DIST.	2ND ORDER DIFF.
HI A. DECIPIENS	30	.7890	
LO A. DECIPIENS	-29	1.1193	.8866
LO A. LAXA	-25	2.3361	-1.3811
HI A. PASTICILLATA	-14	2.1719	-.3836
LO A. PASTICILLATA	13	2.8162	.3916
LO A. SOLIDA	-5	2.6600	-.0657
LO A. AVELLANA	9	3.6302	-.2349
LO A. ARAMAEA	3	4.3655	

SECTION 8

		CUM. DIST.	2ND ORDER DIFF.
HI A. GUIDONIS	28	0.0000	
HI A. ARAGONENSIS	-20	.0259	.4304
LO A. ARAGONENSIS	-19	.4821	-1.1695
LO A. GUIDONIS	27	.6419	.6977
LO A. LEUPOLDI	-15	.6264	.8522
LO A. LAXA	25	2.3361	-2.0022
HI A. PISIFORMIS	12	2.0437	1.5839
LO A. PISIFORMIS	-11	2.4621	-1.5817
HI A. PASTICILLATA	14	2.1719	1.8076
LO A. PASTICILLATA	-13	2.8162	-.7034
LO A. AVELLANA	9	3.6302	.7943
LO A. ARAMAEA	-3	4.3655	

SECTION 9

		CUM. DIST.	2ND ORDER DIFF.
LO A. SUBPYRENEICA	23	1.5504	
HI A. ARAGONENSIS	-20	.0259	3.5424 **
HI A. PISIFORMIS	-12	2.0437	-1.5995
LO A. PISIFORMIS	-11	2.4621	-1.0230
HI A. DECIPIENS	30	.7890	.9349
LO A. DECIPIENS	-29	1.1193	-.9675
LO A. ARAGONENSIS	-19	.4821	

SECTION 10

		CUM. DIST.	2ND ORDER DIFF.
LO A. LEUPOLDI	15	.6264	
HI A. ARAGONENSIS	20	.0259	1.7321
LO A. ARAGONENSIS	-19	.4821	-.0631
LO A. SUBPYRENEICA	23	1.5504	-.4732
HI A. DEDOLIA	-22	1.4703	.4393
LO A. DEDOLIA	-21	1.8295	-.1450
HI A. PISIFORMIS	-12	2.0437	.2042
LO A. PISIFORMIS	-11	2.4621	-1.2643
HI A. MOUSSOULENSIS	-2	1.6161	1.2069
LO A. MOUSSOULENSIS	-1	1.9770	

SECTION 11

		CUM. DIST.	2ND ORDER DIFF.
HI A. GUIDONIS	28	0.0000	
LO A. GUIDONIS	-27	.6419	-1.2579
HI A. ARAGONENSIS	-20	.0259	1.0723
LO A. ARAGONENSIS	-19	.4821	-.6820
HI A. MONTANARII	-18	.2563	1.2173
LO A. MONTANARII	-17	1.2478	

COMPARISON OF OBSERVED AND EXPECTED OCCURRENCES OF SECOND ORDER DIFFERENCE VALUES

CLASS NO.	OBSERVED	EXPECTED	DIFFERENCE	DELTA
1	4	9.600	-5.600	1.463
2	14	9.600	4.400	.903
3	12	9.600	2.400	.269
4	7	9.600	-2.600	.315
5	10	9.600	.400	.007
6	9	9.600	-.600	.017
7	8	9.600	-1.600	.119
8	14	9.600	4.400	.903
9	9	9.600	-.600	.017
10	8	9.600	-1.600	.119

CHI-SQUARED = 4.134

observed and inferred co-occurrences of fossil species, while the 'zones' resulting from the RASC method are primarily based on estimated proximity of stratigraphic events in time. As shown in this Chapter, the two approaches can yield similar results for anomalous occurrences.

REFERENCES

Agterberg, F.P., 1974, Geomathematics. Elsevier, Amsterdam, 596 pp.

Agterberg, F.P. and Nel, L.D., 1982a, Algorithms for the ranking of stratigraphic events. Computers and Geosciences 8 (1), 69-90.

Agterberg, F.P. and Nel, L.D., 1982b, Algorithms for the scaling of stratigraphic events. Computers and Geosciences 8 (1), 163-189.

Blank, R.G., 1984, Comparison of two binomial models in probabilistic biostratigraphy. Computers and Geosciences 10 (1), 59-67.

Drobne, K., 1977, Alvéolines Paléogènes de la Slovênie et de l'Istrie. Mém. Suisses Paleont. 99, 175 pp.

Gradstein, F.M., 1984, On stratigraphic normality. Computers and Geosciences 10 (1), 43-57.

Guex, J., 1981, Associations virtuelles et discontinuités dans la distribution des espèces fossiles: un exemple intéressants. Bull., Soc. Vaud. Sci. Nat. 75 (359), 179-197.

Guex, J., and Davaud, E., 1984, Unitary assocations method: Use of graph theory and computer algorithm. Computers and Geosciences 10 (1), 69-96.

Heller, M., Gradstein, W.S., Gradstein, F.M., and Agterberg, F.P., 1983, RASC-Fortran IV Computer Program for ranking and scaling of biostratigraphic events. Geol. Survey of Canada, Rept. 922, 54 pp.

II.8

UNITARY ASSOCIATIONS AND RANKING OF JURASSIC RADIOLARIANS

F.M. GRADSTEIN

"Can you do additions?", the White Queen asked. "What is one and one and one and one and one and one and one and one and one and one?" "I don't know", said Alice. "I lost count."
In, *'Through the Looking-Glass, and what Alice found there', Lewis Caroll, 1896 ed.*

INTRODUCTION

In Chapter I.2 on 'Stratigraphy and the Fossil Record' a number of factors were listed that affect the presence or absence of a fossil taxon in a sample. Included are cases where a taxon is absent as a result of inadequate sampling, or stratigraphically abnormal presences due to contamination. The uncertainties attached to observations of single taxon events, particularly non-evolutionary end-points of ranges, have led to the development of a method that stresses the actual ranges and not the end-points.

A well-known, rigorous technique for reliable biostratigraphic correlation makes use of the overlaps of successive ranges of fossils as zonal units. This is the concept of the concurrent range zone (Figure 4 in Chapter I.2), which emphasizes the body of a zone as correlative unit, rather than its end-points. A rigorous adherence to this concept for subdivision of the fossil record may lead to a certain loss of stratigraphic resolution (Figure 1). Resolution is sacrificed to regional reproducibility. The following excerpt and example of a quantitative stratigraphic method based on concurrent range zones, largely follows Guex and Davaud (1982), Davaud (1982) and the impressive application by Baumgartner (1984a, b).

THE UNITARY ASSOCIATION METHOD

The rigorous, quantitative, stratigraphic method based on Unitary Associations (UA), as originally developed by Guex (1977) and Rubel (1976) starts with the premise that the only definitive chronological 'datums' which may be locally recognized are those based on intervals of co-occurring taxa. I would prefer to substitute chronological for stratigraphical, because time correlations (chronozones) are based on hypotheses, whereas any correlation using fossil events is a biostratigraphic act. A perfectly valid geological correlation may not necessarily be synchronous everywhere.

The succession of intervals with co-occurring taxa defines the most reliable biostratigraphic scale. Such intervals resemble concurrent range zones or Oppel- or assemblage zones. Each interval corresponds to the "ensemble maximum d'espèces compatibles", and stratigraphically successive intervals lead to "une succession irreversible d'associations unitaires" (Guex, 1977); (freely translated as: all assemblages of real or virtual co-occurring taxa, which lead to an irreversible succession of unitary associations). Virtual co-occurrence is defined as:

(a) two taxa exchange order in lateral sections, without actually co-occurring;

(b) two taxa co-exist with a third taxon, without actually co-occurring.

Given enough sections with real or virtual co-occurring taxa, a conservative, but reliable biozonation will result. The actual method searches for all possible unitary associations (UA) with the maximum number of (virtual) co-occurring species and then tries to arrange those stratigraphically, using the stratigraphic order of the zonal species in them. The result is a range chart showing maximum ranges of each species, relative to all other species, no matter whether this species has a patchy or continuous stratigraphic distribution (Baumgartner, 1984a).

Principal Steps

The following main steps are involved in the analysis of unitary associations:

1. Express the stratigraphic relation between the taxa in the sections.

2. Extract the UA, based on real or virtual co-occurrence of taxa.

3. Order the UA stratigraphically.

4. Determine the correlative value of the UA.

The actual computer program follows a step by step interactive approach to the problem of extracting and correlating meaningful UA.

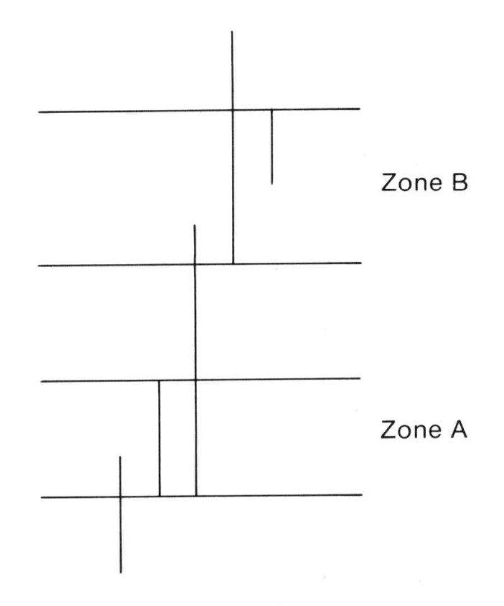

Figure 1 *Zones A and B are concurrent range zones, which may have a good reproducibility for regional correlation. Individual range segments, outside the zonal body of co-occurring taxa are ignored.*

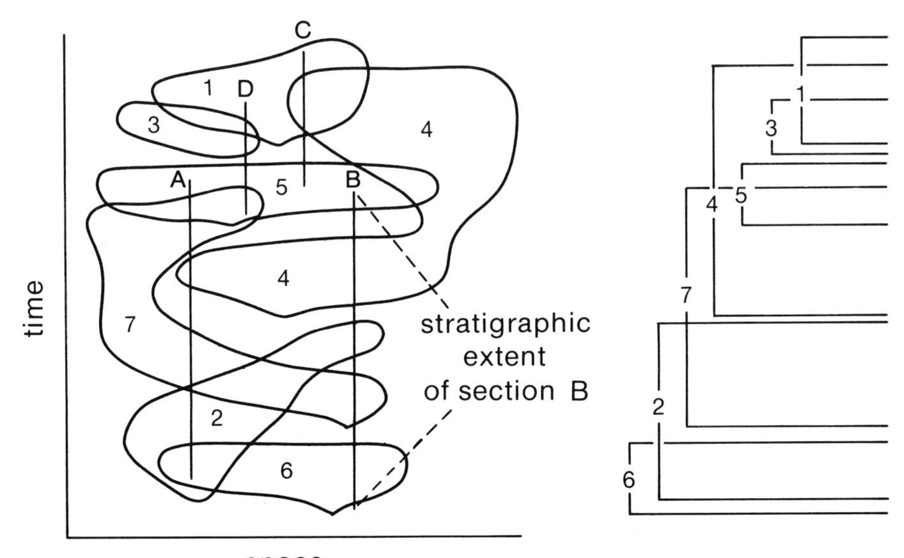

Figure 2 *Temporal and spatial distribution of 7 fossil taxa and their record in 4 stratigraphic A, B, C and D sections. To the right is the succession of ranges (after Davaud, 1982).*

Table 1 *Indirect estimate in 4 stratigraphic sections that taxon j occurs above taxon 1.*

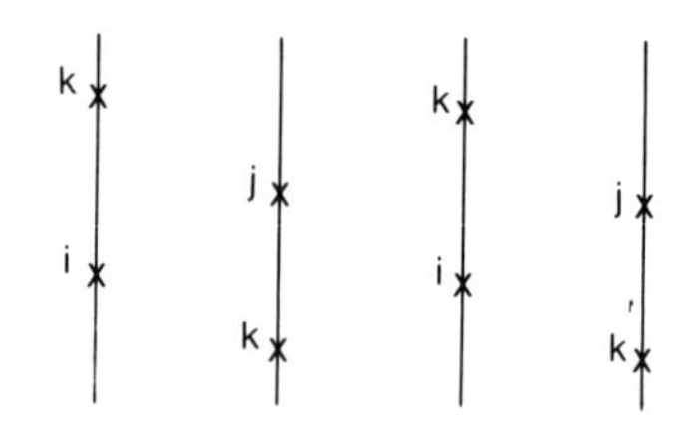

Table 2 *Matrix of the order relationship of the taxa in the four sections of Figures 2 and 3. For explanation, see text.*

	1	2	3	4	5	6	7
1	1	+	1	1	+	+	+
2		1	-	-	-	1	1
3			1	0	+	+	+
4				1	1	+	1
5					1	+	1
6						1	-
7							1

Table 3 *Fully determined matrix after elimination of taxon 3 in the matrix of Table 2.*

	1	2	4	5	6	7
1	1	+	1	+	+	+
2		1	-	-	1	1
4			1	1	+	1
5				1	+	1
6					1	-
7						1

Table 4 *Score of the relative stratigraphic order of the Unitary Associations in Table 3.*

	1,4	2,6	2,7	4,5,7
1,4	1	+	+	+
2,6		1	-	-
2,7			1	-
4,5,7				1

The following example illustrates the principle and visualizes the concept. Figure 2 shows a possible spatial and temporal distribution of 7 hypothetical fossil taxa and their resulting record in 4 stratigraphic sections. To the right is the 'real' succession of ranges.

We will now proceed to extract the stratigraphic record of the 7 taxa as ideally observed (Figure 3) in four wells or sections. The first step is to find all real or virtual co-occurrences of taxa through a tabulation of their relative positions.

In a matrix R the taxa observed are listed along each axis; if i and j are two taxa then score:

i above j	$r_{ij} = +$
i below j	$r_{ij} = -$
i together with j at least once	$r_{ij} = 1$
i and j non-determined	$r_{ij} = 0$

The following special cases have to be considered:

1. If the relative position of i and j alternates from section to section, then r_{ij} is also considered 1.

2. If the relative positions of i and j are non-determined in the sections considered, but if there is a taxon k which occurs always above (or below) i and below (or above) j than $r_{ij} = -$ or $r_{ij} = +$. This situation is depicted (Table 1) for 4 stratigraphic sections where taxon k occurs above taxon i and below taxon j. Hence, j occurs above i and $r_{ij} = -$.

Armed with this information the matrix R of the fossil record in the four sections looks as follows, (Table 2);

The following special relations were resolved;

$r_{12} = +$, since $r_{25} = -$ and $r_{15} = +$;

$r_{16} = +$, since $r_{15} = +$ and $r_{56} = +$;

$r_{23} = -$, since $r_{25} = -$ and $r_{35} = +$;

$r_{36} = -$, since $r_{37} = +$ and $r_{67} = -$;

r_{34} is indeterminate, and this is resolved by eliminating taxon 3 which only occurs in one section, versus taxon 4 in three sections;

$r_{45} = 1$, since the scores in sections A,B,C and D are -,-, + and 0 respectively.

The new, fully determined matrix reads as shown in Table 3. The UA are (1,4) (2,6) (2,7) (4,5,7); 2,6,7 is not an UA because of lack of unity between 6 and 7, but 2 occurs with both 6 and 7, reason for the UA (2,6) and (2,7).

We now score in Figure 3 which of the UA occurs above (+) or below (-) the other ones at least once. The rule is that if uncertainty arrives, the least determined taxon is eliminated and the tabulation repeated (Table 4). The stratigraphic order is 1,4 above 4,5,7, above 2,7 above 2,6.

As a final step we want to determine if the UA are reproducible in many sections. This can be done by simple inspection of the stratigraphic sections used.

An important step in the actual algorithms is the destruction of cycles, briefly discussed in Chapter II.7. These cycles are akin to the contradictory order of events in patchy data, a problem encountered also with RASC ranking, (Chapters II.4 and II.5).

Application of the Unitary Association Method

The procedure of UA is attractive to paleontologists working with core or outcrop sample material. One reason is that the method is relatively non-sensitive to discontinuous ranges of taxa and has excellent correlation potential. The fact that the ranges of many taxa may be discontinuous in individual sections, the result of local dissolution or sampling does not affect UA. Weak links, the result of aberrant, individual occurrences are also eliminated. The method emphasizes (potential) co-occurrence of as many taxa as may be found in the data.

A large scale application and model verification, using Mesozoic radiolarians, has been successfully attempted by Baumgartner et al. (1980) and Baumgartner (1984b). The data are from 226 samples in 43 Jurassic and Early Cretaceous localities in the Mediterranean and Atlantic regions of the ancient Tethys Ocean. Each locality has between 1 and 28 stratigraphically successive samples. In all 110 taxa are involved, the maximum allowed for in the standard computer program. Excluded were rare, long ranging and poorly defined morphotypes. The data, which are shown in the Appendix I, are characterized by a great many co-occurring forms in certain samples, alternating with samples with few (dissolution resistant) taxa.

The computer input consists of the lowest and highest occurrence, further on called 'events', of each morphotype for each locality. No record needs to be kept of the (patchy) occurrences in between. The following discussion closely follows the study by Baumgartner (1984a).

Cycles

A large number of three or four event cycles were found, which reflect on the patchiness of the original data. Suppose there are three

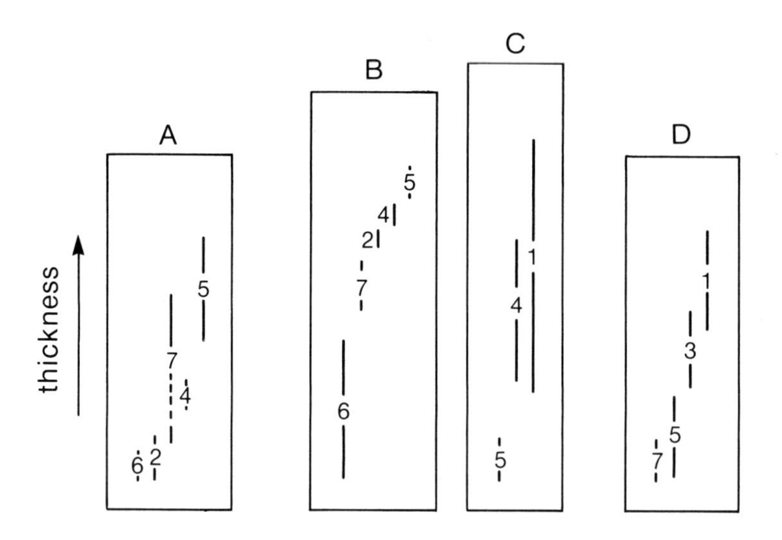

Figure 3 Record of the 7 taxa in Figure 2, as observed in the 4 stratigraphic sections A, B, C and D (after Davaud, 1982).

Table 5 Twenty-eight Initial Unitary Associations (IUA), as identified according to their vertical sequence in 25 of the 42 sections. The longest chain of vertically superimposed IUA (arrows) is determined using this information (after Baumgartner, 1984).

2	5	7	8	14	16	17	19	20	21	22	23	24	25	26	28	29	30	31	32	33	35	37	38	40
7	18	15	16	11	24	18	1	18	9	8	26	22	20	21	18	21	24	27	26	22	22	6	22	10
																↑		↑						
		1			25				18	9		18	4	17		14	16	28		26	26			
																↑		↑						
										18		5	19			18	12	23						
										↑		↑	↑					↑						
										11		3	17				13	21						
										↑		↑												
										5		15					20							
												↑												
												2					2							
																	↑							
												17					19							

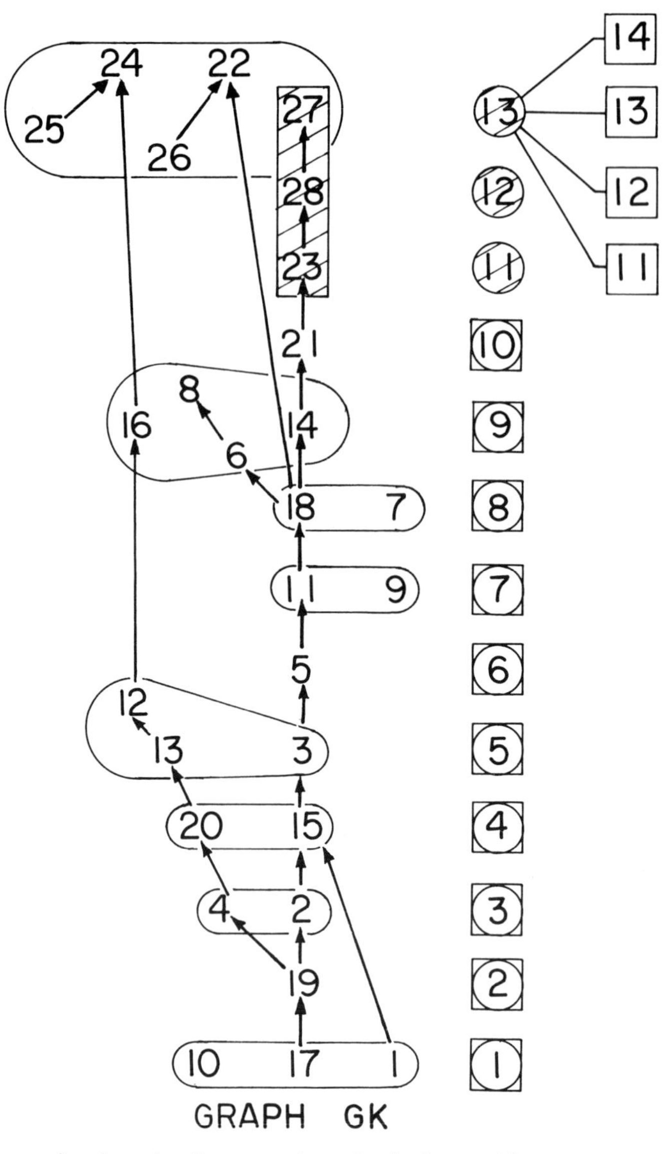

Figure 4 *Graph G_k constructed from the superpositional data of Initial Unitary Associations 1-28, given in Table 5. The vertical column of numbers and arrows is the longest chain, the ovals illustrate the gathering of IUA not belonging to the longest chain. This produces Unitary Associations 1-13 (in circles). UA 11 and 12 are disregarded and UA 13 is split in UA 11-14 (in squares). Numbers in squares correspond to final UA used in Figures 3-5 (after Baumgartner, 1984).*

sections. In section 1 (the range of) taxon A is above (the range of) taxon C, in section 2 taxon A occurs with B, and in section 3 taxon C occurs above B. This conflicting information is resolved by making A, B and C coeval. Obviously, there is a lack of information on the order, if any. Artificial co-occurrences may have occasionally resulted, particularly where many stratigraphically short and discontinuous sections were used, as in the Lower Cretaceous. The method works best if at least one or several species-rich sections span the entire stratigraphic range involved.

Construction of an Interval-Graph

The destruction of all cyclic structures has led to a modification of the species/species matrix because undefined stratigraphic relations are replaced by defined ones. From this updated species/species matrix a set of Initial Unitary Associations (IUA) can be calculated. Each IUA is defined by one of several pairs of species, which are unique to that IUA. In our example, 28 IUA have been calculated. In order to find the stratigraphic sequence of the IUA these are identified in the initial data of each locality. Table 5 gives the IUA and their vertical sequence, as identified in 25 out of 42 sections. From this information, the longest chain of vertically superposed IUA (the strong component) is determined, which is marked by arrows in Table 5. Obviously, this longest chain is composed from several sections, each of which has the most complete set of superposed IUA for a certain interval. Many IUA do not belong to the longest chain, but form shorter chains of superimposed IUA. In order to achieve correlation of these slightly different vertical sequences of IUA, a gathering process is applied, to obtain the final UA 1-13. If there is one IUA of the longest chain, which has a maximum of common species with a given IUA not belonging to the longest chain, this IUA is then gathered with the one of the longest chain (Figure 4). IUA 10 and IUA 1 for example are gathered with IUA 17, which belongs to the longest chain, to form UA 1. IUA 16, 8 and 6 are all gathered with IUA 14, which belongs to the longest chain, to form UA 9. This longest chain in Figure 4 is referred to as an Interval Graph. A comparison of the interval graph in Figure 4 and the sequential occurrence of IUA (Table 5) shows that isolated IUA or short chains of IUA result from localities with single samples or short stratigraphic sections which show slightly different species spectra and thus result in slightly different IUA.

A drastic case of gathering takes place at the top of the column (Early Cretaceous): chain 25 → 24 and chain 26 → 22 are gathered with IUA 27 to form UA 13. The two small chains occur in sections 16, 33 and 35, of Early Cretaceous age, whose data are in no way related to the Late Jurassic data body. This fact has resulted from cycles that were destroyed by assuming virtual associations (see above), which in turn resulted in partial destruction of the vertical separation of IUA. The computer program has selected section 31 for linking the Late Jurassic and the Early Cretaceous data bodies, because it represents the most detailed vertical sequence of IUA (21 → 23 → 28 → 27). However, IUA 23 = UA 11 and IUA 28 = UA 12 have no biostratigraphic significance as they occur in section 31 only. Furthermore, UA 13, gathered from the small chains which show biostratigraphic potential, covers the entire Early Cretaceous and thus does not seem useful.

Inspection of the original data resulted in a manual version which splits UA 13 into 4 UA (Figure 4, UA 11-14 in squares). All samples originally assigned to UA 13 could be assigned to UA 11, 12, 13 or 14, or, when lacking the species pairs of a single UA, they could be assigned to UA 11-14 or UA 13-14 etc. The superpositional control is given in sections 16, 30, 31, 32, 35 and 43, (Figure 5). UA 11, 13 and 14 would form the longest chain, whereas 12 is clearly above 11 but nowhere found below 13. The vertical sequence 12-13 is inferred from the original samples and dating by other fossil groups, (calpionellids). Thus, UA 12 does not belong to the longest chain; moreover, it only occurs in section 43 and therefore has, at present, no biostratigraphic value. UA 12 is, however, maintained in Figure 5 and Figure 6 (range chart) because its biostratigraphic potential may be demonstrated by further sampling of Early Cretaceous sections.

Biostratigraphic Interpretation

The biostratigraphic interpretation of UA, resulting in the definition of biozones is based on the principle of lateral reproducibility. The fundamental idea of this principle is to compare the amount of positively identified UA (black areas in Figure 5) with the amount of potentially identifiable UA, which are defined by bracketing identified UA in the vertical sequences (hatched areas in Figure 5). Two relative values of reproducibility are given: R_1, is a value which is dependent on the number of sections in which a given UA has been identified. For example, UA 1 is identified in all sections which reach down to the base of the studied interval but has a lower value R_1 than UA 8, which is identified in only 11 out of 22 possible sections. R_2 is a value of reproducibility which is independent on the number of sections and simply gives the proportion of identified UA versus potentially identified ones.

The values of reproducibility R_1 and R_2 are, of course, only guidelines for establishing biozones. Subjective criteria, such as number and usefulness of the defining species or species pairs, coincidence of zonal boundaries with major litho- and/or biostratigraphic units, geographic distribution of identified zones and the overall superpositional control were considered for the definition of the proposed biozones. Figure 5 clearly shows that most of the studied sections cover only part of the entire assessed time interval. The values of R_1 and R_2 may therefore be low, simply due to a low number of sections covering a certain interval. If for any given UA, R_1 falls below 3 and R_2 falls below 0.3, it is considered insufficient for the definition of a biozone or subzone. In the proposed zonation the proportion of identified versus potentially identifiable zones ranges between 0.5 (Zone A2) and 1 (Zone A0).

COMPARISON OF RANGE CHARTS USING UNITARY ASSOCIATIONS AND RANKING (RASC)

The same data as used in the UA analysis were also treated with the RASC program for ranking and scaling (Baumgartner, 1984a). At present only an incomplete ranking has been attempted, (based on pre-sorting only). Also, the large number of coeval events creates uncertainty in rank and leads to many small interfossil distances. For this reason, scaling should not be undertaken.

The ranking results are based on $k_c \geq 8$; $m_c \geq 6$ (for explanations of k_c and m_c see Chapter II.5). For this reason 32 out of 110 taxa were eliminated from the solution, mainly those that occur in the upper and lower sections only. In the final ranking 156 first and last occurrence events occur, [(110 species - 32 species) x 2 = 156 events]. As a result of uncertainty in rank, the uncertainty range is quite large, as reflected in the plot of Figure 4 (open part of bars). Three rare taxa (83, 96 and 107) have touching or overlapping bottom and top events and should be eliminated from further consideration.

Because of the fundamental difference between ranges defined by mutual association or exclusion of species and average ranges produced by ranking of first and last occurrences, interpretational differences should be expected for each solution. Average RASC ranges may turn out shorter in relative time, than maximal ranges produced by UA. This is exactly as shown in Figure 6.

The following comparison is an attempt to locate in the optimum sequence the zonal boundaries established with UA (Baumgartner, 1984a). This is accomplished by searching for those defining species that show maximal average ranges in the ranking solution.

- The limit between zones D and E cannot be located due to a lack of relevant species in the ranking solution.

- The limit between zones C and D is a major faunal break that occurs near the Jurassic/Cretaceous boundary (Late Tithonian). In UA it is defined by the appearance of over a dozen new taxa, (Figure 6).The same break, although less well defined, can be observed in the ranking solution. It must occur between the top of species 67 (event range 19-27) and the base of species 94 (event range 24-31),hence within positions 24 and 27.

- The limit between zones B and C is defined in UA by the exclusion of 12 older species with 4 incoming species, which co-occur with 17 outgoing species in UA 9. In the ranking solution, only species 86 (event range 28-34) of the incoming species occurs

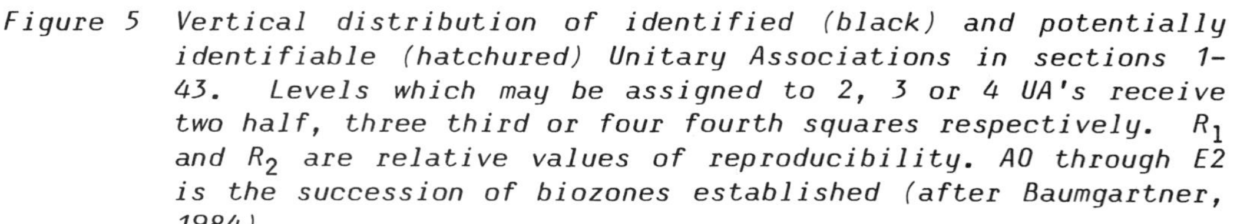

Figure 5 *Vertical distribution of identified (black) and potentially identifiable (hatchured) Unitary Associations in sections 1-43. Levels which may be assigned to 2, 3 or 4 UA's receive two half, three third or four fourth squares respectively.* R_1 *and* R_2 *are relative values of reproducibility. AO through E2 is the succession of biozones established (after Baumgartner, 1984).*

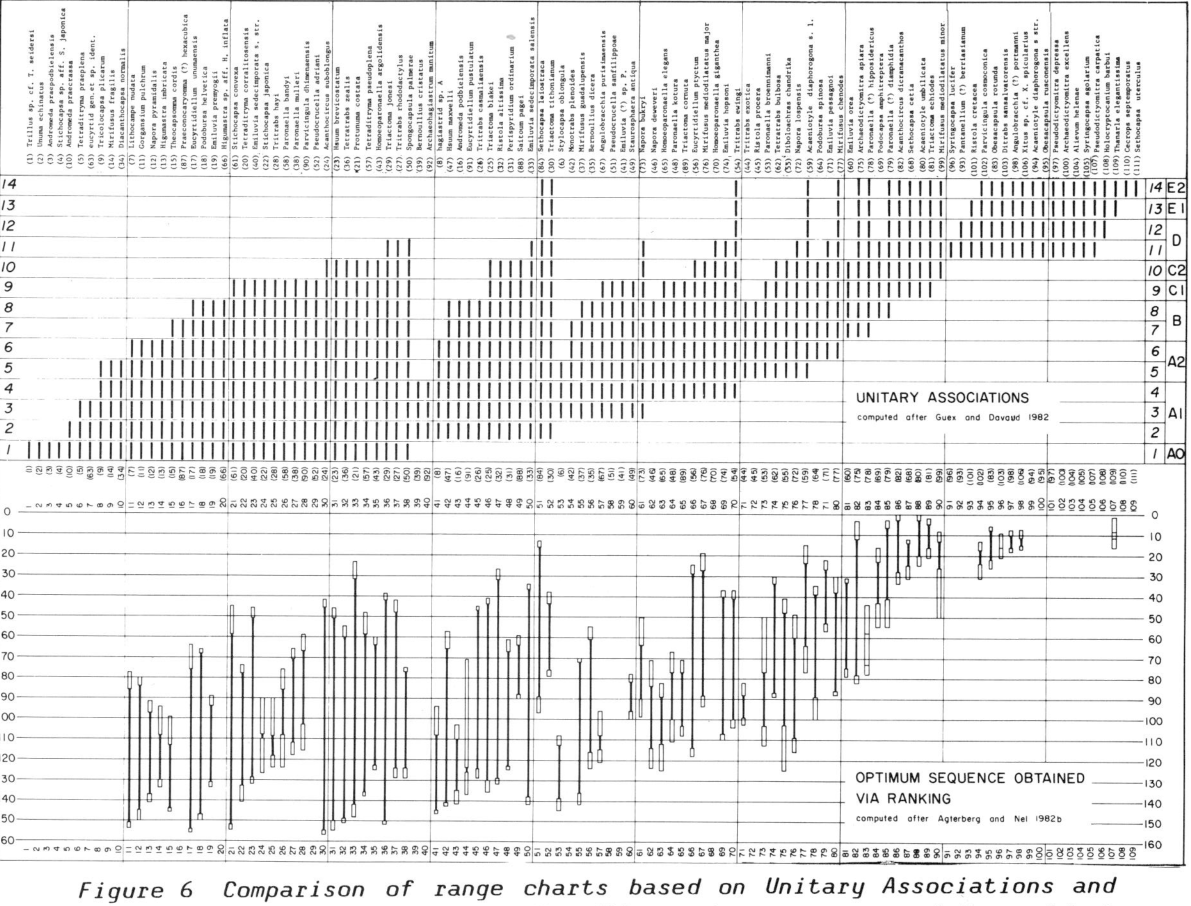

Figure 6 Comparison of range charts based on Unitary Associations and Ranking using RASC. The 109 species are arranged from oldest to youngest, according to UA results; Dictionary numbers are in parenthesis. The boxes on the end points of the average ranges using Ranking, show the uncertainty Range of the end points. Species lacking in the Ranking solution, but present in the UA zonation, did not meet the $k_c - m_c$ threshold values (after Baumgartner, 1984).

below the C/B limit (event range 24-27). The limit B/C thus may range between the base of species 86 and the top of species 45 (event range 45-47), hence within 28 and 47.

- The limit between zones A and B must occur in the ranking solution between the top of species 12 (event range 80-84) and the base of species 82 (event range 78-82), hence within 80 and 82.

- The limit between zones A1 and A2 has to be located below position 109 to be below the base of species 76 (event range 109-116), which has the lowest range of all the species that first occur in UA 5. No lower limit can be given for the boundary A1/A2, as all species with possible tops below 109 were eliminated from the ranking solution by the k_c-m_c limits.

- The limit between zones A0 and A1 has to be placed below position 145 to be below the base of species 41 (event range 145-147).

- It is evident that most species show a much shorter range relative to each other in the ranking solution than in UA. The zonal limits inferred from the UA-solution thus are defined only by the few species which show similar ranges in both solutions. A different zonation would result when using the average ranges produced by ranking for the definition of zonal boundaries.

This points to a problem in applying the average range chart concept to Mesozoic radiolarians. Almost every well preserved radiolarian sample contains a high, above average, number of species including rare and/or easily soluble ones, defined as high diversity species. These high diversity species are characterized by a patchy vertical distribution or, in other words, an inconsistent range. This adds to the problem of occurrences of many coeval events in such well preserved samples, which result in long UA-ranges and short average taxon-ranges in probabilistic ranking. Well preserved radiolarian samples thus may contradict the average range chart by showing species associations falling outside the average ranges. The question may be asked, if in this case one should work only with samples that show average preservation? An answer to this problem of patchiness of all but the most common taxa, is to erect suitable thresholds when applying U.A. Such threshold can be defined in the same manner as in the RASC program. For example, only taxa that occur more than n_1 times in the same section (and or samples) with other taxa, and are present in a minimum of n_2 sections ($n_2 \leq n_1$) might participate in U.A. analysis. Suitable selection of n_1 and n_2 during test runs will then rapidly provide insight in the stratigraphic properties of the data.

It seems that the comparison of maximum and average stratigraphic ranges is a valuable tool to estimate the consistency of occurrence and the resistence to dissolution of the species. Evidently, those species that show similar ranges in both solutions and have short event-ranges must be the most reliable ones for zonation. Only 27 out of 110 species meet these requirements. Some 10 more may result when lowering the k_c and m_c threshold parameters.

CONCLUSION

In the Unitary Association method the succession of intervals with co-occurring taxa defines the most reliable biostratigraphic scale. The final result is a range chart showing maximum ranges of each species, relative to all other species, no matter whether this species has a patchy or a continuous stratigraphic distribution. Baumgartner (1984a) has successfully used this method and Ranking to extract a zonation from the Jurassic radiolarians.

In order to compare Unitary Associations and Ranking, first and final occurrences of 110 species of Middle Jurassic-Early Cretaceous radiolarians in 226 samples from 43 localities were used to compute zonations.

A low-latitude radiolarian biostratigraphy is based on 14 Unitary Associations which define 7 biozones (including 4 subzones). The ages of these zones are:

Zone E_2	-	Late Valanginian to Hauterivian
Zone E_1	-	Early to Middle Valanginian
Zone D	-	Latest Tithonian to Berriasian
Zone C_2	-	Middle to Late Tithonian (=Volgian)
Zone C_1	-	Kimmeridgian to Early Tithonian =(Early Volgian)
Zone B	-	Late Oxfordian to earliest Kimmeridgian
Zone A_2	-	Latest Callovian - Middle Oxfordian
Zone A_1	-	Callovian
Zone A_0	-	Bathonian - Early Callovian

Each zone is defined by the co-occurrence of a number of pairs of species.

Since the Mesozoic radiolarian fossil record is mainly dissolution controlled, the sequence of events differs greatly from section to section. Unitary Associations produce maximum ranges of species relative to each other by stacking co-occurrence data from all sections and therefore compensate for local dissolution effects. Probabilistic methods, like Ranking and Scaling based on the assumption of a normal random distribution of events, produce average ranges which are shorter than the maximum ranges.

There are approximately 27 species with similar ranges in both solutions. These are the most dissolution resistant, 'low diversity' species, which provide a consistent occurrence throughout their range, even in sequences with average preservation. The conclusion is that these are the most useful species for biostratigraphic zonation. The comparison of UA maximum ranges and RASC average ranges, therefore, aids in the identification of taxa that will provide the most reliable correlation of stratigraphic units.

REFERENCES

Baumgartner, P.O., 1983, Summary of Middle Jurassic - Early Cretaceous radiolarian biostratigraphy of DSDP Site 534 (Blake Bahama Basin) and correlation to Tethyan sections. In Sheridan, R.S., Gradstein, F.M. et al. (eds.), Initial Reports Deep Sea Drilling Project, Washington (Govern. Printing Office), 76, 569-571.

Baumgartner, P.O., 1984a, Comparison of Unitary Associations and probabilistic Ranking and Scaling as applied to Mesozoic Radiolaria. Computers and Geosciences, 10 (1), 167-183.

Baumgartner, P.O., 1984b, A Middle Jurassic - Early Cretaceous low-latitude radiolarian zonation based on Unitary Associations and age of Tethyan radiolarites. Eclog. Geol. Helv. 77 (3), 729-837.

Baumgartner, P.O., De Wever, P. and Kocher, R., 1980, Correlation of Tethyan Late Jurassic-Early Cretaceous radiolarian events. Cah. Micropal. C.R.N.S. 2, 23-72.

Davaud, E., 1982, The automation of biochronological correlation. In Cubitt, J. and Reyment, R. (eds.), Quantitative Stratigraphic Correlation, J. Wiley, New York, 85-99.

Guex, J., 1977, Une nouvelle méthode d'analyse biochronologique. Bull. Soc. Vaud; Sc. Nat. 73/3 (351) et Bull. Géol. Univ. Lausanne, 224.

Guex, J. and Davaud, E., 1982, Recherche automatique des associations unitaires en biochronologie. Bull. Soc. Vaud. Sc. Nat. 76 (361), 53-69.

Rubel, M., 1976, On biological construction of time in geology. Eesti NSV Tead. Akad. Toimetised Keemia Geologia, 25 (2), 136-144.

II.9

A COMPARISON OF FIVE QUANTITATIVE TECHNIQUES FOR BIOSTRATIGRAPHY

JAMES C. BROWER
AND
DONALD T. BUSSEY

INTRODUCTION

Two approches are available for the comparison and evaluation of quantitative biostratigraphic methods (Harper, 1984). Generally, a real data set is analyzed with several techniques and the results are compared between the methods and also with "known correlations" which have been compiled with classical subjective biostratigraphy and time markers such as bentonites. Techniques which yield divergent answers and/or poor comparisons with the "known correlations" are presumed to perform poorly. The obvious flaw in this scheme is that the "known correlations" are subject to unknown errors. The "odd technique out" might provide the right answer. As Harper (1984) dryly observes, "Several techniques may agree on the wrong answer". Examples of this approach are Hazel (1977) and Millendorf et al. (1978) on different methods for making assemblage zones, Brower and Burroughs (1982) comparing seriation with numerous other methods, and Edwards (1982b), Hudson and Agterberg (1982), Agterberg (1984), Blank (1984) and Gradstein (1984) on various sequencing schemes.

A second approach is based on hypothetical data (Edwards, 1982a, 1984; Harper, 1984; see Agterberg's "computer simulation experiments" in Chapter II.6). Simulated or hypothetical data are subject to mixed advantages and disadvantages. The obvious advantage is the existence of an "answer". The inherent problem with simulation is to make it right. One can create a data set that is too difficult. This might be caused by an excessive amount of random noise which would prevent any technique from reaching the "answer". Alternatively a data set that is far too easy can also be structured.

Harper (1984) generates a range chart by random roll. At present, this procedure does not allow for systematic facies changes or geographic variation such as faunal provinces. These data are then randomly sampled over a selected number of localities for 50 to 100 iterations. Three algorithms formulated from binomial probability by Agterberg and Nel (1982a,b) determined the sequence of events for each iteration. The technique which most faithfully reproduces the original sequence of events is judged to be the best. According to Harper (op. cit.) the presorting option was marginally superior to the ranking and scaling algorithms. However, the reader should note that presorting fares poorly if there are many missing data (Chapter II.5).

Edwards (1982a, 1984) offered data sets for examination. These were published as correlation charts showing the range zones of 16 to 20 species. The range zones were subjectively designed to illustrate certain aspects of biostratigraphic data. Edwards tested her no-space graph technique (1982a) and graphical correlation (1984) on these data and sampled subsets thereof to test the ability of these methods to recover the true zonation of events.

Several authors advocated rank correlation coefficients for accessing the results (Brower and Burroughs, 1982; Edwards, 1982a; Harper, 1984; Gradstein, 1984). Generally, the various techniques yield similar results, most of which closely replicate the presumed or known answers.

This paper will work with the Edwards (1982a) data which hopefully lie somewhere in the middle ground of simulated data, namely neither too difficult nor excessively easy. As mentioned subsequently, this data set does contain some features often encountered in practical biostratigraphy. The data will test some of the most widely used biostratigraphic techniques, namely a slightly modified version of presort ranking (Chapter II.5), graphical correlation (Chapters II.4 and III.1), constrained seriation (Chapter II.3), lateral tracing and cluster analysis (Chapter II.2).

THE DATA SET

The hypothetical data were developed by Edwards (1982a) to provide a test for quantitative biostratigraphic methods. The 20 species are distributed in eight stratigraphic sections, numbered 1-8 of various ages (Figure 1). The vertical axis represents time whereas the horizontal axis gives space. Although artificial, the data set mimics some of the characteristics of real data. The time-value of the individual sections varies considerably. Twenty species provide low diversity compared to typical data sets. The number of species per sample ranges from zero to six with a mean of three to four taxa (Figure 2A, compiled for samples denoted by the tic marks shown for the sections of Figure 1). In general, low diversity creates ambiquity for biostratigraphic data. Edwards identified five species (A-E) as index fossils. These are geographically widespread with highest and lowest occurrences concentrated in narrow time intervals. Except for A all species exhibit short range zones in terms of relative time. However, several other taxa also possess reasonably good time-stratigraphic properties, namely H, O and T. Some species are highly provincial and the data can be divided into "faunal and/or floristic provinces." No offense is intended to those who work with fossil plants when we subsequently term

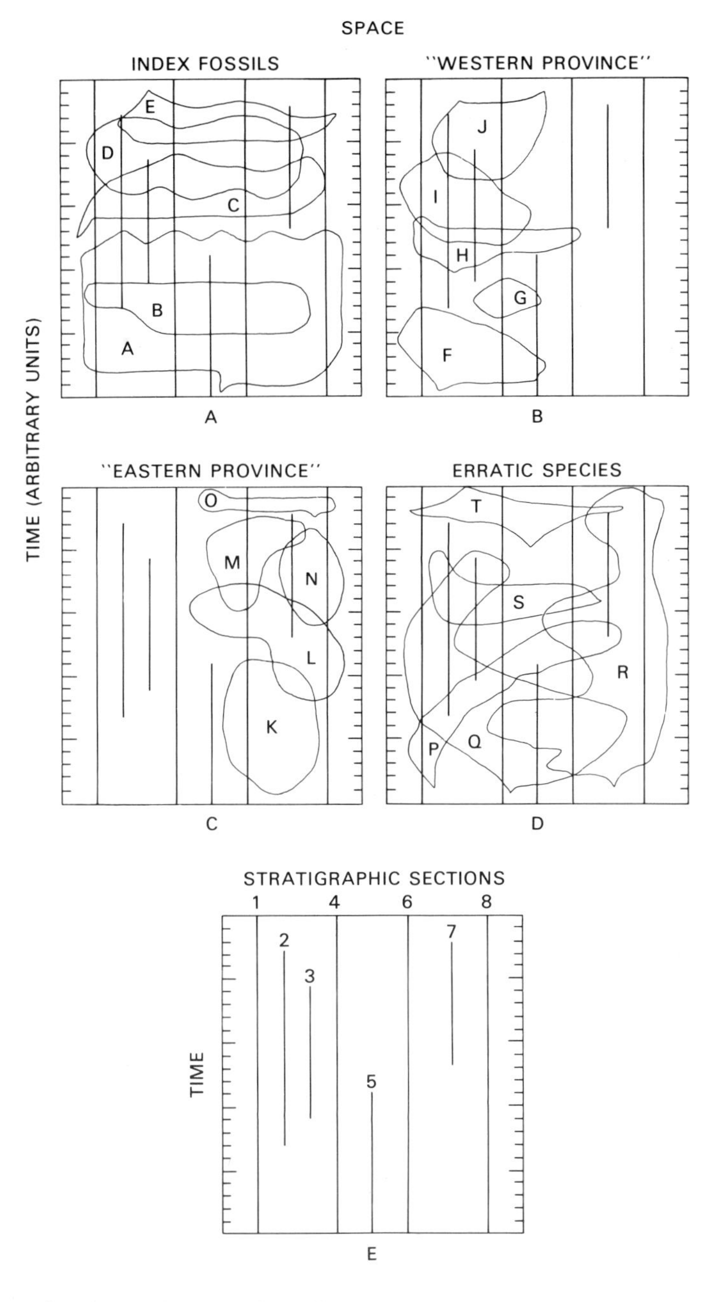

Figure 1 Correlation charts for Edwards (1982a) data set. Horizontal and vertical axes give time and space, respectively. Species are designated from A-T. (A) Index fossils. (B) Western "faunal province." (C) Eastern "faunal province." (D) Erratic species. (E) Distribution of stratigraphic sections 1 through 8.

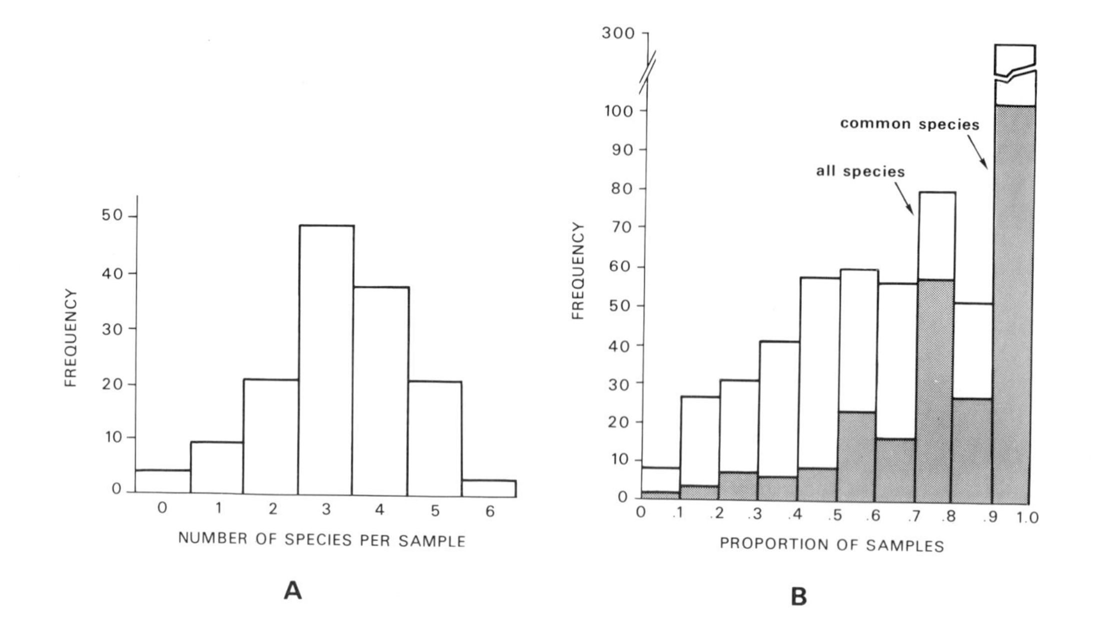

Figure 2 Graphs of data for sampling. (A) Frequency distribution of number of taxa in samples. Mean, median and standard deviation are 3.26, 3.0 and 1.27. (B) Frquency distribution for proportion of samples within range zone in which a species is known to be present. Data for 705 species; mean median and standard deviation equal 0.736, 0.800 and 0.237; 25th, 50th and 75th percentiles consist of 0.521, 0.793 and 0.941. Data for 250 common taxa; mean, median and standard deviation are 0.790, 0.819 and 0.216; 25th, 50th and 75th percentiles comprise 0.691, 0.815 and 0.939.

these as "faunal provinces." F-J are restricted to the "western" sections whereas K-O only occur in the "eastern" sections. The only overlap between the two "faunal provinces" is observed in section 6. In general the faunal diversity decreases from the stratigraphic sections in the middle of the correlation chart towards the eastern and western margins. The biostratigraphic properties of fossils P-R are far from ideal. P is strongly time transgressive although its local range zones are relatively narrow. Q and R are most kindly described as erratic. The range zones of S and to a lesser extent T vary somewhat in different sections and both are absent from the most easterly and westerly sections.

A variety of sampling strategies were applied to the data and the following will be reported:

Run 1. Completed recovery so that each species is found in all samples within its local range zone in all stratigraphic sections.

Run 2. As previously, this run assumes that a taxon is present in all samples within the local range zones. A perusal of the journal "Marine Micropaleontology" for the past six years indicates that many workers count about 300 specimens per sample. The observed figures range from a minimum of about 200 to an estimated maximum of 10,000 individuals. The reported percentages of "biostratigraphically useful" forms (not necessarily index fossils) vary from 0.7 to 100 percent, but most values hover around one to three percent. Binomial or Poisson probability can calculate the probability of recovering at least one specimen of a species representing a specified proportion of the population in an outcrop or subsurface sample of a selected size (e.g. Hay, 1972). Binomial and Poisson probabilities assume random distributions; this is not unreasonable because many paleontological samples tend to be homogenized by preservation and sampling factors. Based on the above figures, we reasoned that the index fossils A-E would be less common than the other taxa. We will follow a worst case approach and specify that index fossils comprise 0.5 percent of each "outcrop" sample whereas the more plebeian forms, F-T, make up five or more percent of each sample; the latter figures seem appropriate in view of the low diversity. For the index fossils, there is a 80 percent chance of finding a single specimen in a sample size of 300. The equivalent probability for the other species equals one for all practical purposes. These probabilities were applied to all samples using the random number operator in the Sharp version of the APL language.

Run 3. The supposition that a taxon occurs in every sample inside its local range zone is obviously unrealistic, and it is critical to "know" the probability that a species is present in any sample within its local range zone. Unfortunately it is not possible to measure this probability. However an estimate can be obtained from the observed distributions of taxa within their local range zones as defined by their highest and lowest occurrences. Obviously these data can only provide approximations because the true ranges are probably not seen and owing to sampling errors within their estimated ranges. We compiled data for 705 species of Cretaceous and Cenozoic microfossils from the journal "Marine Micropaleontology" over the past six years and from Eicher and Worstell (1970). Two distributions are pictured, one for all species and another for the common taxa which comprise one or more percent per sample (Figure 2B). Inasmush as all samples are based on counting 300 or more specimens, these species should be present with at least 95 percent confidence. Both distributions are highly asymmetrical and have median and mean probabilities; these values are 0.8 and 0.736 for all 705 species and 0.819 and 0.790 for the 250 common taxa. We adopted the lower figure and rounded it off to 0.75 which is easier to model. Harper (1984) selected similar values of 0.55, 0.80 and 0.85 for his simulations. As in run 2, the probabilities of finding at least one specimen within a sample where it is present equal 0.8 for the index species, A-E, and almost 1.0 for F-T. Hence the aggregate probability of scoring a species as present in a single sample outside of its range zone is 0.6 for species A-E and 0.8 for taxa F-T.

Run 4. This run is intended to represent poor recovery from all samples. Here we lowered the probability that a species is present in a sample inside of its range zone to 0.5. This value roughly corresponds to the 25th percentile (acutally 0.521) of the distribution of all taxa in Figure 2B. As previously the probability of finding one or more specimens of a fossil in a sample where it is present remain at 0.8 for species A-E and nearly 1.0 for fossils F-T. Thus the final probabilities consist of 0.4 for the "index fossils" and almost 0.5 for the others.

It is important to note that the only source of error in the design is due to sampling. Errors such as reworking, misidentification, unconformities, etc. are not included.

Our conventions for the data matrices obviously affect the results. The data for seriation, lateral tracing and cluster analysis consist of presences and absences in the range-through format for the species in each sample denoted by the tic marks on Figure 1E (see Chapters II.2 and II.3).

The data set used for graphical correlation and the ranking model is a rectangular array showing the distribution of the events in the various stratigraphic sections. A zero is recorded if the event is not observed in that stratigraphic section. The other elements represent the sample positions of the events in the various stratigraphic sections. The sample numbers increase from the base to the top of the stratigraphic sections; the basal sample in each section is designated as 1.0; higher samples are recorded 2.0, 3.0 and so forth. If several events are observed at the same level, there are two possibilities. If the events are of contrasting types, that is both highest and lowest occurrences, we assume that the observed data underestimate the range zones of the taxa and place the highest occurrences slightly above the lowest ones in the data (Harper, 1981). This adjustment is made by displacing the highest and lowest occurrences 0.3 sample units above and below the sample where they are located in the original data. If the events are of the same type, i.e. either highest or lowest occurrences, the ties are retained in the data. Ranking models are usually only calculated for events which are reasonably widespread in terms of geography (e.g., Agterberg and Nel, 1982; Hay, 1972). Consequently only events known from three or more of the eight sections are included in the data.

ALGORITHMS

The following common techniques were selected for comparison:

A. A Ranking model which is slightly modified from the presorting option of Agterberg and Nel (1982a; see Chapter II.5). In our computer program, the ranking numbers in the composite sequence of events consists of the average probability of finding the row event, I, over all of the column events for which data are available. The original presorting algorithm is discussed in Chapter II.5.

B. Graphical correlation (Shaw, 1964; Miller, 1977). Four rounds of compositing are performed for each sampling trial. As recorded, the data assume a constant relationship between sample position and time. Bivariate scatterplots denote that terraces and dog-legs are absent and linear regression lines are adequate for calculating the lines of correlation. The regression lines are fitted to all data points present in the composite section (X) and the one being incorporated into the composite (Y). X is regressed on Y and the data are updated as illustrated by Miller (1977).

C. The stratigraphically constrained version of seriation of an original data matrix (Brower and Burroughs, 1982; Burroughs and Brower, 1982; Chapter II.3).

D. Lateral tracing as performed by Millendorf et al. (1978; see Chapter II.2).

E. Cluster analysis as outlined in Chapter II.2.

Correlations for the stratigraphic sections are obtained as follows:

A. For the ranking model, all events present in the calculated composite sequence and in that particular stratigraphic section, for example M, are listed, and the rank order of both sets of events is tabulated. The absolute value for the difference between the two ranks for event I measures the difference in the order of event I in the composite sequence of events and section M. Agterberg and Nel (1982b) term these "penalty points." Any event with 2.0 or more penalty points is ignored when correlating section M. The remaining events can be used to correlate the sections in any desired fashion. To access the results, we plotted the ranking numbers in the composite sequence against the known locations of the events in each stratigraphic section. Composite sequence position was then regressed on the known locations of the events and these lines were used to calculate the predicted composite sequence position of each sample. The basic idea is shown later in Figure 4. Contouring the predicted composite sequence position of all samples on the correlation chart of Figure 1E gives a visual picture of the correlations with respect to the composite sequence of the events. Perfect correlations or time lines are shown by horizontal contours. The deflections of contours from horizontal lines yield a measure of "miscorrelation" relative to the correlation chart.

B. Graphical correlation follows a similar procedure. First, bivariate scatterplots were compiled for section M (Y axis) versus the composite (X axis). Inspection of these graphs shows which points should be deleted. Obviously if event I is not present in section M, it must be ignored. Also highest and lowest occurrences that are underestimated in section M compared to the composite should be omitted. These are lowest occurrences which fall to the left of the line of correlation and highest occurrences located to the right of the line (Miller, 1977). These points were indentified subjectively and we accepted a "reasonable amount" of scatter with respect to the line (see later examples in Figure 4). The remaining points serve to correlate section M with the composite zonation. The correlations were computed and contoured on the correlation chart (Figure 1E) as done for the ranking model.

Previous authors have usually only evaluated the techniques relative to either the sequence of biostratigraphic events or the samples. Whenever possible both events and samples should be considered. After all, the end product of the exercise equals the correlations for the samples!

For sequences of events, two answers were compiled from the correlations chart. One answer, hereafter termed "truth" gives the true zonation of the biostratigraphic events as seen in the eight stratigraphic sections. The second is the true average sequence, termed "average truth", which is determined from the mean position of each event in all sections as measured on the correlation chart. The resulting zonations from the various techniques were compared with the two answers by means of the absolute values of the Spearman's rank correlations coefficients (e.g. Brower and Burroughs, 1982; Edwards, 1982a). The two "truths" diverge somewhat inasmuch as the rank correlation coefficient is only 0.940.

The accuracy of the correlations is harder to measure for the samples:

A and B. For the two sequencing methods, the ranking model and graphical correlation, we calculated composite sequence position as a least squares function of the known position of the events in each stratigraphic section. Plotting and contouring the predicted composite sequence position of each sample on the correlation chart provides a graphic view as mentioned before. Spearmen's rank correlations between the predicted and observed location of the samples quantify the relationship.

C. Seriation. This evaluation is simple. We rank-correlated the row position of each sample in the final seriated matrix with the known position of the sample on the correlation chart.

D. Lateral tracing. For this method, horizontal links or lateral traces fall along time lines. The left sample of each lateral trace is assumed to represent the time value for the entire trace. Here, the rank correlations were determined for the time value of the entire lateral trace versus the observed sample location on the correlation chart for all samples in the trace except for the left-most one.

E. Cluster analysis. Plotting the groups of samples on the correlation chart shows how well the clusters parallel the time lines.

RESULTS

Four sampling runs will be analyzed. Following Edwards (1982a), we designate the highest and lowest occurrences of each species with lower and upper case letters, respectively, throughout the subsequent text.

Run 1

This run samples all species at the one-sample intervals on the correlation chart. In effect, this run shows how effective each method is at recovering the answer given optimum sampling. The true sequences of events and those determined by the several methods are shown in Figure 3 and Table 1 whereas Table 2 contains the rank correlations coefficients. All 40 events, highest and lowest occurrences for 20 species, were treated by graphical correlation and seriation but the ranking model was only applied to the 28 events found in three or more sections. In general all techniques recover similar sequences of events and the magnitudes of the correlations range from 0.916 to 0.993. The graphical correlation sequence is highly correlated with "truth" but less so with "average truth". The applicable rank correlations consist of 0.986 and 0.918 (Figure 3A, B). Conversely, the ranking model produces a zonation which is very similar to "average truth" but this solution diverges more from the "truth" and the correlations are 0.993 and 0.916 (Figures 3C,D). This situation was expected inasmuch as the ranking model and graphical correlation are designed to determine average and conservative sequences, respectively. Basically, the two methods perform as predicted and both are efficient at calculating the type of event sequence for which they are intended.

The contrasts between the results can be inspected in the graphs of the event sequences versus "truth" and "average truth" in conjunction with the correlation chart (Figures 1 and 3, Table 1). Species P, Q and R provide examples because all vary greatly in the different stratigraphic sections. Species P trangresses time and becomes younger towards the east. Events P and p are underestimated by the ranking model relative to "truth", so that the highest occurrence is too low in the final zonation whereas the lowest occurrence is too high. However, these events are zoned correctly compared to "average truth". It is important to note that this does not detract from the ability of the technique to correlate the samples because events are deleted from correlating sections were their observed rank order differs greatly from that in the composite sequence of events. Consequently, P and p were only used to correlate section 4 and sections 2 and 3, respectively. An analogous situation is observed for graphical correlation. Compared to "average truth", this methods places P too low and p too high in the composite zonation although the two events are assigned correctly relative to "truth". As with the ranking model this does not affect correlating because underestimated events are omitted when the individual sections are matched with the composite. Events P and p only contribute to correlate section 1 and sections 3, 4, 6 and 8, listed in the same order. The diachronous pattern of species P is shown by the correlation of each section relative to either the ranking or graphical correlation sequence of events. Both P and p become older toward the west when plotted versus either zonation.

Events q and r are erratically distributed. The ranking model calculates q correctly in terms of "average truth" but it is underestimated and too old compared to "truth". This event only passed the tolerance for the correlation of sections 1 and 5 and it was ignored in sections 2, 4 and 6. Graphical correlation rightly groups q with the higher events. Event q aids in correlating sections 2 and 4 where its placement is adequate but not sections 1, 5 and 6 in which it is underestimated with respect to "truth".

The ranking model roughly averages all sections and consequently, r is too old in the zonation compared to "truth" although it fits well with "average truth". This event is misplaced in all of the known sections and it is not involved in any of the correlations. Graphical correlation assigns r properly relative to "truth" but it is excessively young relative to "average truth". Although present in sections 4, 5, 6 and 8, r only enters into the correlation of the last section where it is reasonably placed.

Both the ranking model and graphical correlation yield similar results for the stratigraphically useful species A-E, H, O, and T (Figure 3, Table 1). Although the zonations of the events differ, both techniques can deal with the two "faunal provinces" with no difficulty.

Seriation generates somewhat intermediate sequences between those of the ranking model and graphical correlation where the magnitudes of the correlations equal 0.975 and 0.944 with the "truth" and

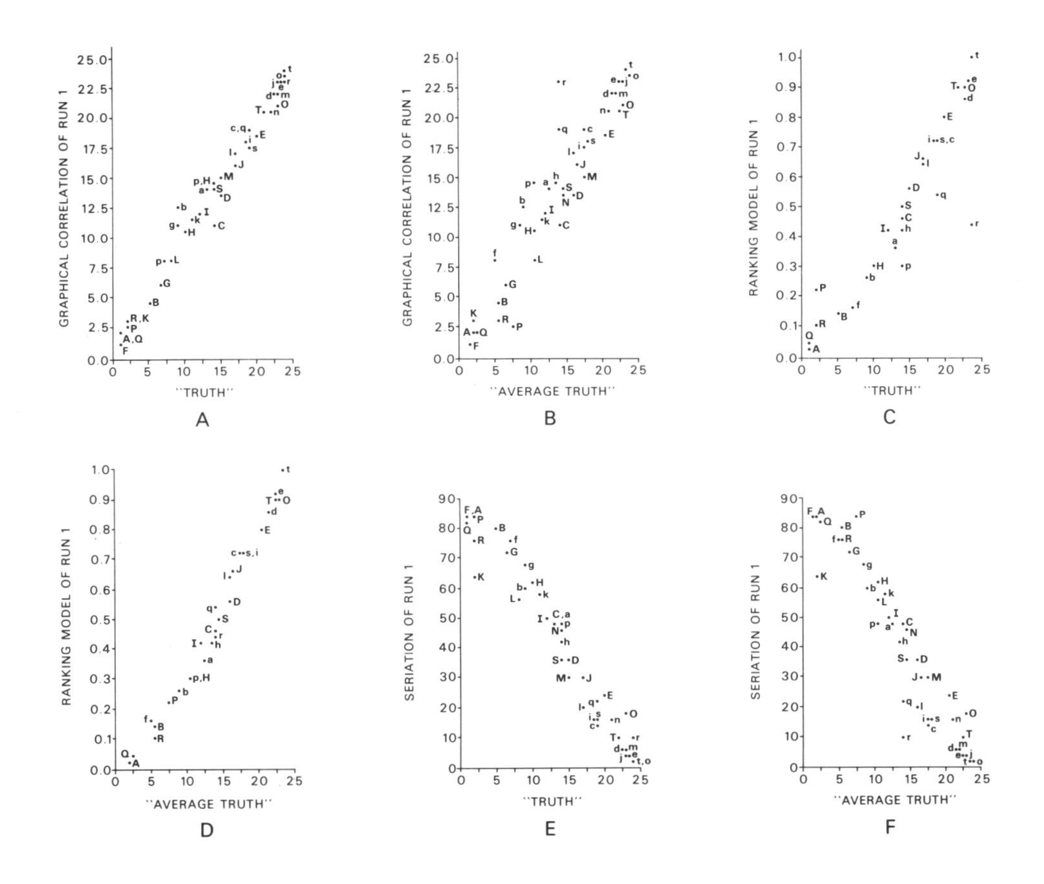

Figure 3 *Plots of computed sequences of events versus true zonations for Run 1.*

A. Graphical correlation versus "truth".
B. Graphical correlation versus "average truth".
C. Ranking model versus "truth".
D. Ranking model versus "average truth".
E. Seriation versus "truth".
F. Seriation versus "average truth".

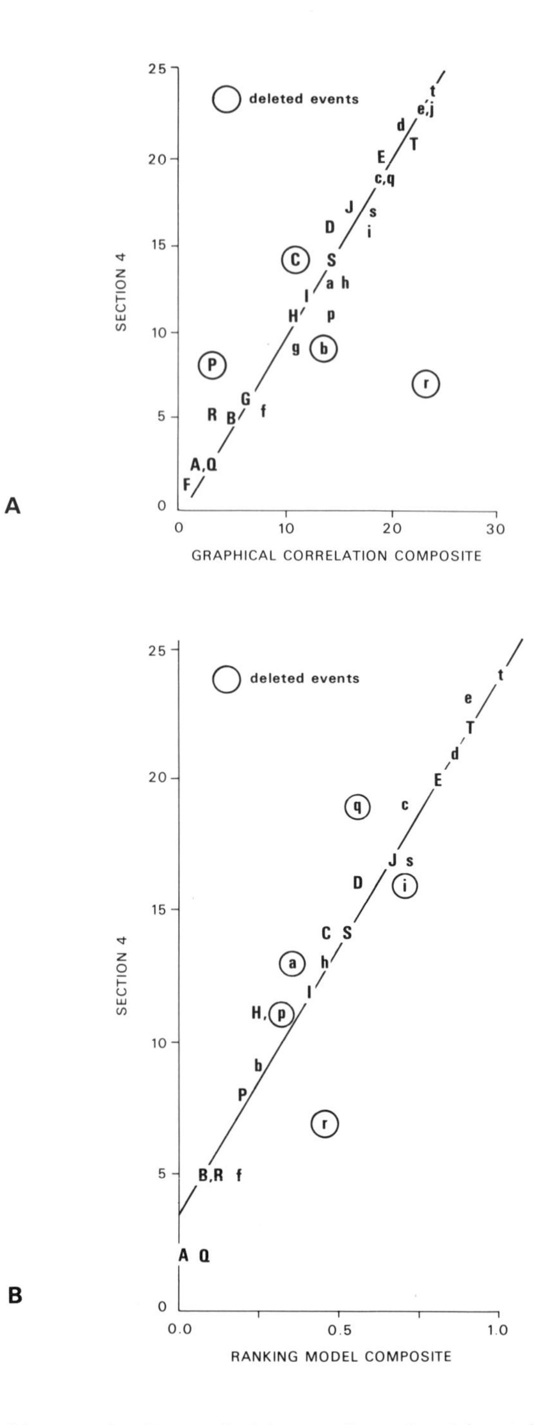

Figure 4 Correlations for Section 4 with the composite zonations for Graphical correlation and the Ranking model of Run 1.

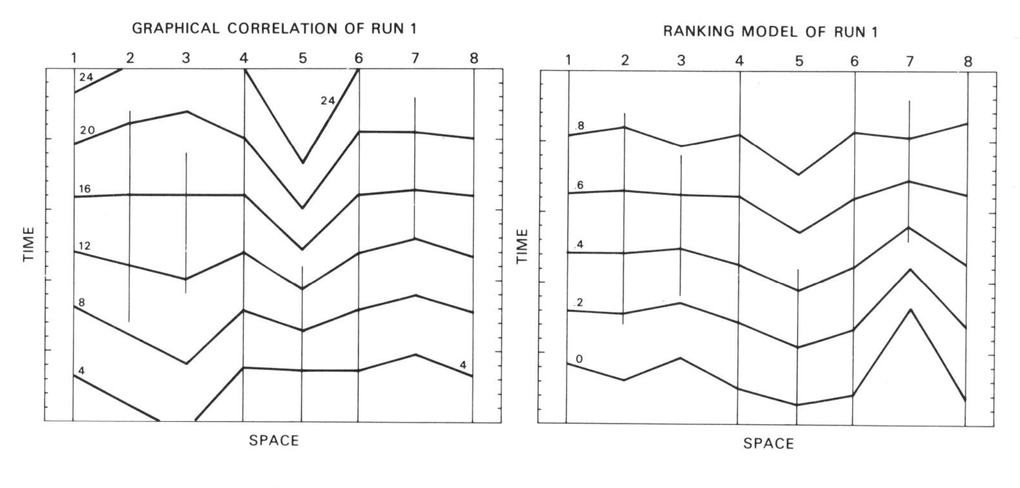

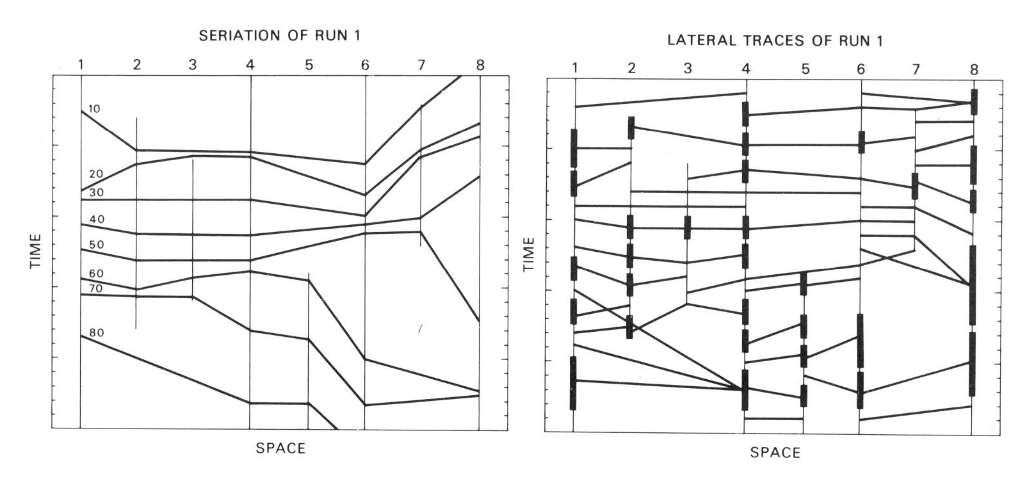

Figure 5 Plot of correlations for samples of Run 1.
A. Graphical correlation.
B. Ranking model.
C. Seriation.
D. Lateral tracing.

*Table 1 Rank order of events for "truth", "average truth", and calculated sequences. Youngest events are at top. Tied events are denoted by * or #.*

"Truth"	Graphical Correlation				Seriation				Ranking Model				"Average Truth"
	Run 1	Run 2	Run 3	Run 4	Run 1	Run 2	Run 3	Run 4	Run 1	Run 2	Run 3	Run 4	
*o	t	o	t	o	t	t	t	*o	–	–	–	–	o
*r	o	r	j	*e	o	*e	o	*O	t	t	–	t	t
*t	r	t	o	*j	*e	*j	j	t	–	–	–	e	j
e	*e	*e	e	*t	*j	o	e	e	e	O	–	–	O
*d	*j	*j	r	O	#d	d	T	j	O	T	e	j	e
*j	#d	#d	m	d	#m	m	#d	d	T	e	–	T	T
*O	#m	#m	*d	m	*r	*i	*m	m	–	–	–	–	m
m	O	O	*O	T	*T	*s	c	s	d	d	d	E	d
T	n	n	T	r	c	#r	s	i	–	–	–	–	n
n	T	T	n	q	*i	#T	r	*r	E	E	E	d	E
E	*c	*c	q	s	*s	n	i	*T	i	i	s	J	i
*c	*q	*q	s	i	n	*c	M	#c	c	s	i	s	s
*q	E	E	E	E	O	*l	q	#l	s	c	c	i	c
*s	i	*i	i	n	l	q	O	E	–	–	–	–	M
i	s	*s	c	c	q	J	*E	M	J	J	D	c	J
*J	l	l	J	J	E	M	*n	q	l	l	j	r	D
*l	J	J	M	l	J	O	J	J	D	D	l	–	l
#D	M	M	l	M	M	D	h	n	–	–	–	–	N
#M	h	a	h	h	D	S	l	C	q	q	q	D	S
*C	p	p	p	D	S	E	S	D	S	S	S	q	C
*h	a	h	a	p	h	h	D	h	C	C	C	C	*q
*N	S	*D	D	a	N	*C	N	S	r	r	I	S	*r
*p	D	*S	S	S	*C	*p	p	*L	h	h	h	I	h
*S	N	N	N	N	*p	#a	C	*p	I	I	r	h	a
a	b	b	b	C	a	#I	*a	N	a	a	a	p	I
I	I	I	C	I	I	N	*I	a	–	–	–	–	k
k	k	g	I	L	L	L	k	I	H	H	p	H	p
H	*C	k	k	b	k	H	b	H	p	p	H	a	H
*b	*g	C	g	g	b	k	H	*b	–	–	–	–	L
*g	H	H	H	k	H	b	K	*k	b	b	b	b	b
L	L	f	f	f	K	K	L	g	–	–	–	–	g
f	f	L	L	B	g	g	g	K	P	P	P	P	P
G	G	G	K	G	G	G	G	G	–	–	–	–	G
B	B	B	G	H	*f	f	*f	R	f	f	f	B	R
*K	R	P	B	K	*R	R	*R	f	B	B	B	R	B
*P	K	*K	R	F	B	B	#B	A	R	R	R	–	f
*R	P	*R	P	*P	Q	Q	#Q	*B	Q	Q	A	A	Q
#A	A	#F	A	#A	*A	*A	A	*Q	–	–	–	–	K
#F	Q	#Q	Q	#Q	*P	*P	P	#F	A	A	Q	Q	A
#Q	F	A	F	R	F	F	F	#P	–	–	F	–	F

Table 2 Spearman's rank correlation coefficients for sequences of events. Data tabulated are correlation coefficient, number of events, and Student's t values for calculated sequences of events versus "truth" and "average truth". The t values compare the observed correlations with a population value of nil.

	Run 1	Run 2	Run 3	Run 4
Graphical correlation versus "average truth"	0.918, 40, 14.3	0.913, 40, 13.8	0.927, 40, 15.2	0.935, 40, 16.3
Graphical correlation versus "truth"	0.986, 40, 36.1	0.984, 40, 33.9	0.985, 40, 35.2	0.978, 40, 28.6
Ranking model versus "average truth"	0.993, 28, 42.1	0.993, 28, 43.8	0.987, 26, 29.9	0.976, 26, 22.1
Ranking model versus "truth"	0.916, 28, 11.6	0.914, 28, 11.4	0.889, 26, 10.1	0.938, 26, 13.2
Seriation versus "average truth"	-0.944, 40, -17.7	-0.917, 40, -15.1	-0.938, 40, -16.7	-0.945, 40, -17.9
Seriation versus "truth"	-0.975, 40, -27.1	-0.953, 40, -19.4	-0.960, 40, -21.2	-0.960, 40, -21.1

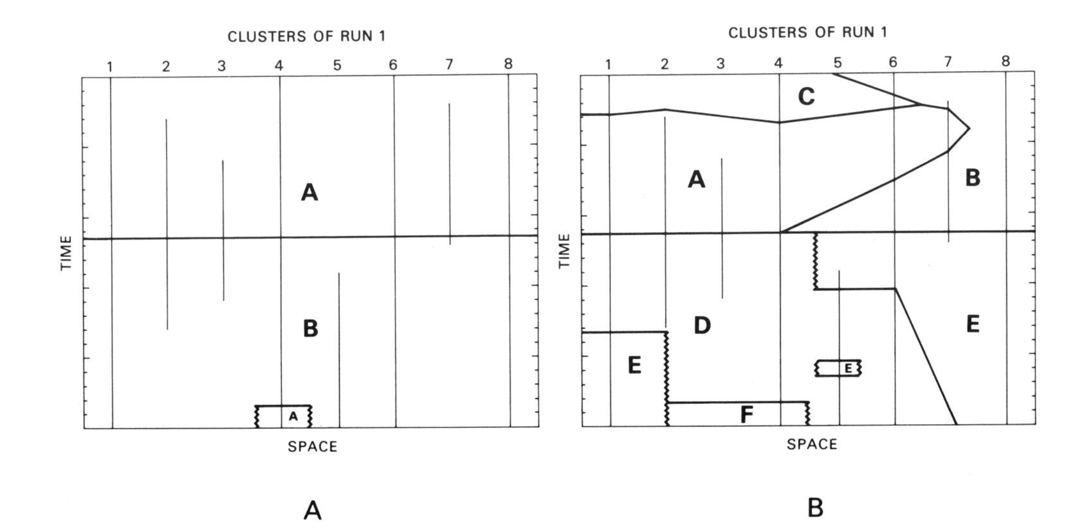

Figure 6 Clusters superimposed on correlation chart for Run 1.
A. Two clusters.
B. Six clusters.

Table 3 List of events which are poorly estimated by the techniques in the various sampling runs. Events listed in parentheses are considered too young by the calculated sequences.

	Run 1	Run 2	Run 3	Run 4
Graphical correlation versus "truth"	C (b)	C (a, b)	C, F (b, K)	H, k, n, R, r (B, L)
Ranking model versus "average truth"	R (f, I, q)	R (f, I, q)	R, r (f, I, q)	a, d (I, J, j, Q)
Seriation versus "truth"	E, O, r (K, L)	E, O, r (K, L)	E, n, O, r (c, K, M)	n, r (K, L, M)
Seriation versus "average truth"	E, O, P (K, q, r)	E, O, P (K, q, r)	E, n, O, P (K, q, r)	P, n, k (K, L, p, q, r)

Table 4 Spearman's rank correlation coefficients for the samples. Data listed are correlation coefficient, number of samples or links and Student's t values. The t figures compare the calculated correlation with a parametric value of zero.

Graphical Correlation			
Run 1	0.992,	142,	91.8
Run 2	0.993,	142,	98.9
Run 3	0.990,	142,	84.9
Run 4	0.997,	142,	54.7
Ranking model			
Run 1	0.992,	142,	93.5
Run 2	0.995,	142,	123.0
Run 3	0.992,	142,	90.4
Run 4	0.957,	142,	39.1
Seriation			
Run 1	-0.940	142,	-32.7
Run 2	-0.939	142,	-32.2
Run 3	-0.934,	142,	-30.8
Run 4	-0.906,	134,	-24.6
Lateral tracing			
Run 1	0.982,	59,	38.7
Run 2	0.979,	63,	37.6
Run 3	0.962,	54,	25.3
Run 4	0.832,	51,	10.5

"average truth" (Figures 3E,F). This can be explained by the algorithm. When operating on the samples or rows of the data, seriation first ascertains the average position of the presences in each row; the rows are then sorted according to these means; next the rows or samples that are out of stratigraphic order in each section are interchanged to correspond with their known stratigraphic positions. Thus the stratigraphic constraint forces the seriation solution to converge on that given by "truth" and graphical correlation.

Each method frequently provides poor estimates for the location of certain events when compared with the appropriate type of "truth" (Table 3). At least some of these events are consistently misplaced in several of the sampling trials, particularly runs 1 to 3.

Figure 4 illustrates graphs for correlating the stratigraphic sections. With the ranking model, events were dropped if their observed order differs by two or more ranks from that in the composite zonation. In section 4, five of 26 events were omitted. The data were treated in similar fashion for graphical correlation; events which were clearly underestimated in the stratigraphic section were ignored when matching it with the composite. Of the 30 events present in section 4, only 26 were employed for correlation. Over all sections, the percentages of unused events are almost identical, namely 19.2 for the ranking model and 18.8 for graphical correlation.

Figure 5 contains contour diagrams of the correlations and lateral traces for all methods. If the sections were matched perfectly, these contours and lateral traces would be horizontal and follow time lines. Deviations of the lines from the horizontal reflect the amount of error or miscorrelation. For graphical correlation and the ranking model, the composite zonation position is regressed on the observed location of the events in each stratigraphic section (Figure 4). These regressions then serve to compute the predicted composite sequence position for all samples which is contoured in Figures 5A and B. The contours approximately parallel time lines. Observe that most of the major deflections of the contours are concentrated in regions where no data exist. It is important to realize that such contours do not provide correlations for any of the samples. These areas fall below section 7 and above section 5 for the ranking model, and they lie below sections 2 and 3 and above section 5 in the case of graphical correlation. A downward displacement means the predicted position of the sample is too young whereas an upbend denotes the reverse. The rank correlations for the predicted and observed sample locations of the ranking model and graphical correlation methods constitute identical values of 0.922 for 142 samples (Table 4).

The row positions of the samples in the final seriated matrix represent the expected correlations and the map of these can be seen in Figure 5C. Although many of the contours have major bends, the absolute value of the rank correlation is 0.94 which is a large figure. Some deflections parallel the boundaries between clusters as mentioned later.

It is important to observe that many of the curves in the seriation contours are present in areas on the correlation chart which have data. This contrasts with the situation for graphical correlation and the ranking model. The figure demonstrates that the correlations derived from seriation are less precise than those given by the sequencing schemes. Many of the seriation errors can be traced to ambiguous samples in which few species are present (see Brower and Burroughs, 1982). The order of events given by seriation can function to correlate the sections by means of bivariate scatterplots as done with graphical correlation and the ranking model. Here, points that deviate markedly from the main trend are removed. This generates more accurate correlations.

Despite the low diversity of the fauna, lateral tracing yields excellent results (Figure 5D). The rank correlation between the observed and predicted locations of the lateral trace links comprises 0.982 although only 59 examples can be tabulated (Table 4). Most lateral traces roughly follow time lines. Compared to the sequencing techniques, lateral tracing is subject to several limitations. The lateral traces have a relatively small number of links and some areas cannot be joined such as the base of sections 6 and 8. The lateral traces of this paper provide less stratigraphic resolution than that given by the sequencing schemes.

The clusters of the multivariate cluster analysis were superimposed on the correlation chart of Figure 1E. The two clusters of Figure 6A are time-parallel. However, the results deteriorate if more clusters are considered and the boundaries between the adjacent clusters become diachronous (Figure 6B). The analysis is strongly influenced by the two "faunal provinces" and species P, Q and R which are distributed erratically. The poor performance of the cluster analysis is attributed to two factors, namely the poor biostratigraphic properties of 13 of the 20 species and the low faunal diversity of the data (see Millendorf, et al., 1978).

Run 2

Data were collected from all samples in this instance. The probabilities of scoring a species as present in a sample within its range zone are set at 80 percent for the index fossils, species A-E, and almost 100 percent for the other 15 taxa. The outputs obtained for this run are surprisingly good. Omitting some occurrences of the index fossils does not materially affect the composite zonation of events or the correlations of the samples. The rank correlations for graphical correlation against "truth" and the ranking model against "average truth" equal 0.984 and 0.993 which are nearly the same as those previously annotated for Run 1 (Tables 1, 2). Aside from minor changes of some events, the composites for the ranking model and graphical correlation are stable over both runs and the two methods are still effective at recovering the type of zonations for which they were designed. The absolute values of the rank correlations for seriation with respect to "truth" and "average truth" constitute 0.953 and 0.927 and both are lower than the equivalent run 1 figures of 0.975 and 0.944. Comparison of the seriation sequence of events with the two "truths" discloses more numerous and larger displacements of some events. Events E and O changed locations by 14 or more column positions in the final seriated matrices whereas events H, J, r, R and T move from 4 to 6 units. The reason why these particular events migrate so far remains unknown. All events are recorded from three to five sections. Except for R and r, the position of each event in the several sections is reasonably constant (Figure 1). As expected each algorithm produces similar zonations in both sampling runs and these rank correlations range from 0.998 to 0.989.

Turning to correlating the samples, the percentages of ignored events are slightly higher for this run, i.e. 20.9 for the ranking model and 19.9 for graphical correlation as compared to the run 1 numbers of 18.8 and 19.2. The rank correlation coefficients between the calculated and observed locations of the samples are extremely high and quite similar to those of run 1 (Table 4); the applicable figures for run 2 equal 0.993 for graphical correlation, 0.995 for the ranking model, -0.939 for seriation, and 0.979 for lateral tracing. Unexpectedly, the run 1 values for graphical correlation and the ranking model are slightly less than those of this run although the constrasts are not statistically significant. Deleting several more events marginally improved the results. The most striking change for the ranking model is found in the region with no data above section 5. The contours of predicted position in the composite zonation are deflected downward in run 1 but they bend up in run 2. We attribute the change to the omission of events a and q from the correlations of section 5 in the latter run. These two events are slightly underestimated in run 1 and their inclusion in the regression equation decreases the predicted correlations (see Figure 1). Comparison of the two outputs for graphical correlation reveals only minor differences, mostly located in areas not reached by the stratigraphic sections. Likewise the seriation solutions and the lateral traces have nearly identical patterns for the two sampling runs. The same technique produces similar correlations for both runs and the Spearman's coefficients only vary from 0.911 to 0.997.

The cluster analyses of run 2 are even worse than the ones previously mentioned for run 1. In the three-cluster plot most of the older samples are grouped in a single large cluster. The younger samples fall into two groups which interfinger laterally and vertically. Graphing six clusters displays a complex mosaic which is dictated by a combination of time relations, the two "faunal provinces", and the nasty characteristics of the infamous species P, Q, and R. In short, the clusters generate little information of stratigraphic value.

Run 3

Although the probabilities of observing the various species within their local range zones have declined to 0.6 for the index fossils and 0.75 for the other taxa, the zonations and correlations are still quite accurate for all techniques except the cluster analyses. The rank correlations for the sequences of events derived from the ranking model and graphical correlation versus the appropriate kinds of "truth" are essentially like those of run 1 where the sample recovery was complete (Tables 1, 2). The applicable correlations of run 3 consist of 0.985 for graphical correlation versus "truth" and 0.987 for the ranking model against the "average truth". Thus these algorithms continue to be effective at calculating their target sequences of events. All 40 events were included in the graphical correlation but only the 26 events present in three or more sections were utilized for the ranking computations. Inspection of Table 1 reveals that the main changes for the ranking model are seen for events A and Q, r relative to I and h, and a switch of l and D. The most striking displacements for the graphical correlation involve K, l and T.

Compared to run 1, the absolute values of the rank correlations for seriation decline from 0.975 to 0.960 relative to the "truth" but these values are roughly the same with respect to the "average truth" (0.944 versus 0.938, Tables 1, 2). This demonstrates that lowering the probabilities typically shortens the range zones of the taxa. The same phenomena affects graphical correlation where the Spearman's rank correlations with the "average truth" rise slightly from 0.918 for run 1 to 0.927 for this one. Numerous events have shifted positions in the seriation zonation of run 3 compared to run 1, and seven events, n, E, l, D, M, h and L, moved seven or more positions in the seriated matrices over the two runs. Highest and lowest occurrences can migrate in either direction so that an individual range zone can be extended or compressed. As previously, the reasons why these events are displaced farther than others remain obscure. The sequences of events generated by each quantitative technique are highly similar and these correlations range from 0.977 to 0.998.

Correlations for the samples closely parallel those of the previous runs. As expected, lower probabilities of recovering the fossils increase the number of events that must be omitted to correlate the stratigraphic sections. These percentages constitute 23.3 and 24.0 for the ranking model and graphical correlation which are more than the corresponding run 1 numbers of 19.2 and 18.8. The correlations calculated by the two sequencing methods largely follow time lines in areas of the correlation chart which have samples. Most major curves of the contours occur in parts of the chart lacking data. Some deflections of the contours can be traced through all three sampling runs. Examples include the large downward bend in the contours above section 5 for graphical correlation and the upward bulges in the ranking model contours below section 7. The contours for the ranking model above section

5 may bend up as in runs 2 and 3 or down like in run 1. The rank correlations denote that the predicted values continue to account for most of the variance in the known positions of the samples. The Spearman's coefficients equal 0.990 and 0.992 for graphical correlation and the ranking model, respectively. These values are almost identical to those annotated before for run 1 (Table 4).

The row positions of the samples in the final seriated matrix were superimposed on the correlation chart. The correlation for the seriation rows versus the sample locations has an absolute value of 0.934 which is lower than the 0.99 values for the ranking model and graphical correlation (Table 4). As with the other techniques, the seriation correlations are relatively stable and these only change from -0.940 to -0.934 during the three runs.

The seriation plots contain some large bends which persist through all sampling trials. Notable examples are the depressed contours in the tops of sections 2 through 7 and the diving lines near the bases of sections 1 to 3 and 6 to 8. Although the rank correlations between the expected and observed sample locations are high, the seriation contours give less stratigraphic correlation. As mentioned earlier, the sequence of events computed by seriation can be used in regressions to determine more accurate correlations similar to those of the other methods.

The multivariate scheme lateral tracing continues to furnish good correlations. The Spearman's rank correlation for lateral trace position with known sample level comprises 0.962 for this run which is only marginally smaller than the run 1 value of 0.982. Most linkages do not deviate widely from time lines although there are some exceptions, such as the nearly vertical, diachronous lines near the bases of sections 1 to 3.

The solutions for five and seven clusters were graphed and two features must be noted. First is the strong influence of the two "faunal provinces" which dominate the data, especially for the seven clusters. Secondly, the clusters only afford poor stratigraphic resolution, even within the eastern and western regions of the correlation chart. Despite the fact that the clusters do overlap and interfinger, a biostratigrapher would experience considerable difficulty in reconstructing any reasonable time stratigraphy from these data. Given the low faunal diversity and other tribulations within the data set, cluster analysis simply cannot unravel enough stratigraphic information to produce adequate correlations.

Run 4

The probabilities of recording the species as present within their range zones were decreased to 0.4 for the most useful taxa, A-E, and to 0.5 for the more plebeian fossils, F-T. The absolute values of the rank correlations for the various zonations versus the applicable types of "truth" range from 0.978 to 0.945 which indicate that the sequencing methods remain capable of recognizing the main features for which they are designed.

The correlations for the ranking model zonation equal 0.976 and 0.938 with the "average truth" and "truth", respectively (Tables 1, 2). An unexpected feature of the sampling is that this run yields the highest correlation between "truth" and the ranking model. Exactly how these results were achieved remains uncertain. Despite the vagaries of sampling, the ranking model continues to place the time-transgressive and erratically distributed events properly. Table 1 shows that the ranking model understates the range zone of the time-transgressive fossil P relative to the "truth". Thus, this technique considers events P and p as too young and too old, listed in the same order, because the algorithm places the events approximately in their average positions. Similarly, the ranking model evaluates r and q as too old with respect to the true situation. Due to the irregular sampling, the ranking model of Run 4 yields a poor zonation for somewhat different events than previously in terms of the "average truth" (Tables 1, 2).

The rank correlation coefficients for the sequence of the events by graphical correlation versus the "truth" and "average truth" equal 0.978 and 0.935, given respectively (Tables 1, 2). Obviously, the technique still renders its target zonation accurately. The Student's t values for graphical correlation exceed those of the ranking model (Table 2). The correlations and t figures for the two methods versus their intended sequences of events comprise 0.978 and 28.6 for graphical correlation and "truth" and 0.976 and 22.1 for the ranking model against the "average truth". The different t values mainly reflect the fact that graphical correlation uses all 40 events whereas the ranking model only deals with the 26 events found in more than two stratigraphic sections. Hence, the graphical correlation of this run is marginally superior to the ranking scheme although the contrasts are not statistically significant. As before with the ranking model and "truth", the correlation between graphical correlation and "average truth" increased somewhat in this sampling run (0.935) over previous ones. This implies that a declining caliber of sampling forces the two types of sequences, i.e. conservative and average or probabilistic, to converge on each other.

Events P, p, q and r are appropriately located by graphical correlation relative to the "truth". However, they are overextended when scaled against the "average truth" so the highest and lowest occurrences are too young and too old, listed in the same order (Table 1). Graphical correlation sequences the following events incorrectly relative to "truth" (Table 3); the computed bases of species B and L are too young (underestimated) whereas events H, R, k, n and r are assigned positions that are too old. Observe that this artificially lowers the two bases which could extend the range zones of species H and R. Strangely enough, the earlier sampling runs gave reasonable results for these events (Tables 1, 2).

As with the other runs, the magnitudes of the rank correlations for seriation fall between those obtained from the ranking model and graphical correlation; the pertinent numbers equal 0.960 and 0.945 for seriation against "truth" and "average truth" (Tables 1, 2). Despite the changes in sampling regime, the method continues to generate a compromise sequence of events between the conservative or average zonations. Seriation groups the "ringer events" adequately with respect to "truth", but these events deviate more widely compared to the "average truth". According to seriation, p, q and r are too young whereas P is too old. The seriation zonation derives poor results for numerous events in terms of either answer (Table 3). The current seriation solution is less stable than before, and eight events, namely C, E, h, J, L, M, O and p moved 10 or more column positions relative to the matrix of run 1.

The ranking model and graphical correlation produce correlations that resemble those of previous sampling trials (Figures 7A, B). Owing to the increasing uncertainties of sampling, numerous events must be ignored when computing the locations of the samples, i.e. 29.7 for the ranking model and 28.9 percent for graphical correlation.

The Spearman's correlation coefficient for the known sample locations versus those determined by graphical correlation constitutes 0.977 which indicates that the method can still yield useful correlations (Table 4). As before, most of the big bends in the contours occur in parts of the time-stratigraphic diagram which contain no samples, namely the large upward bulge above section 5 and the downward deflections below sections 2, 3 and 7. The main errors of correlation within the data consist of the bases of sections 2, 3 and 5 and, to a smaller degree, the upper samples in sections 2 and 3. As in earlier sampling runs, the complete sections are generally matched correctly.

Only 26 events are treated by the ranking model sequence. This along with the low probabilities of finding the species, drastically reduces the number of events that are present in and can serve to correlate the stratigraphic sections. In spite of the vagaries of the run 4 sampling trial, the rank correlation between the observed and predicted locations of the samples remains quite respectable at 0.957 (Table 4). In fact, this number is only marginally lower than the 0.977 value for graphical correlation. The ranking model correlates most sections reasonably well. The two notable exceptions are sections 5 and 8, each of which only has four useful events (Figure 7B). The diverging contours of section 8 denote that the ranking model underestimates the ages of the upper and lowermost samples. Likewise the base of section 5 appears excessively young. Comparison of Figures 7A and B demonstrates that graphical correlation tends to generate errors in almost all stratigraphic sections whereas the ranking model concentrates most mistakes in several sections. This is believed to be caused by the sampling of run 4 rather than any inherent properties of the techniques. Some of the major bends in the ranking model contours noted in earlier runs are more pronounced here, such as the large upbends above section 5 and

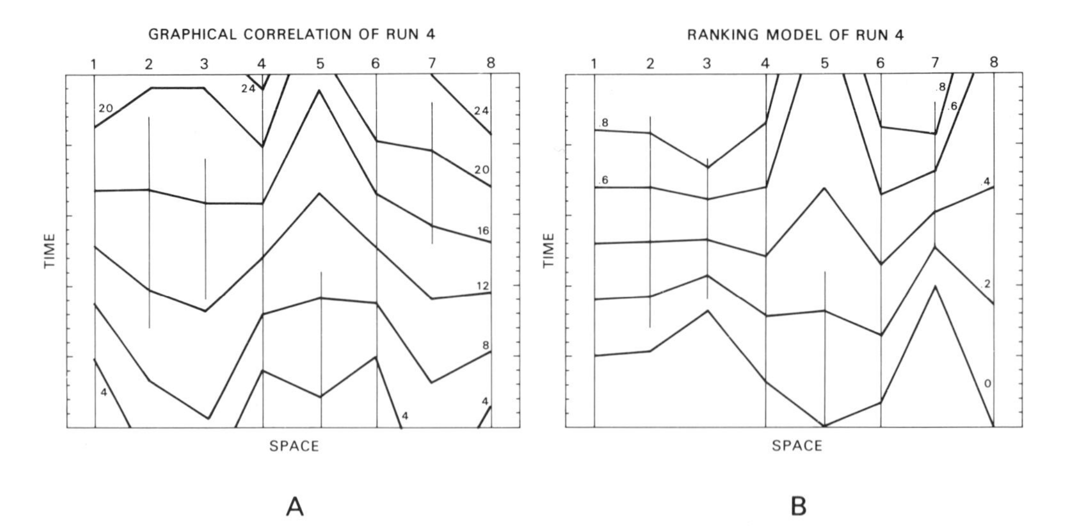

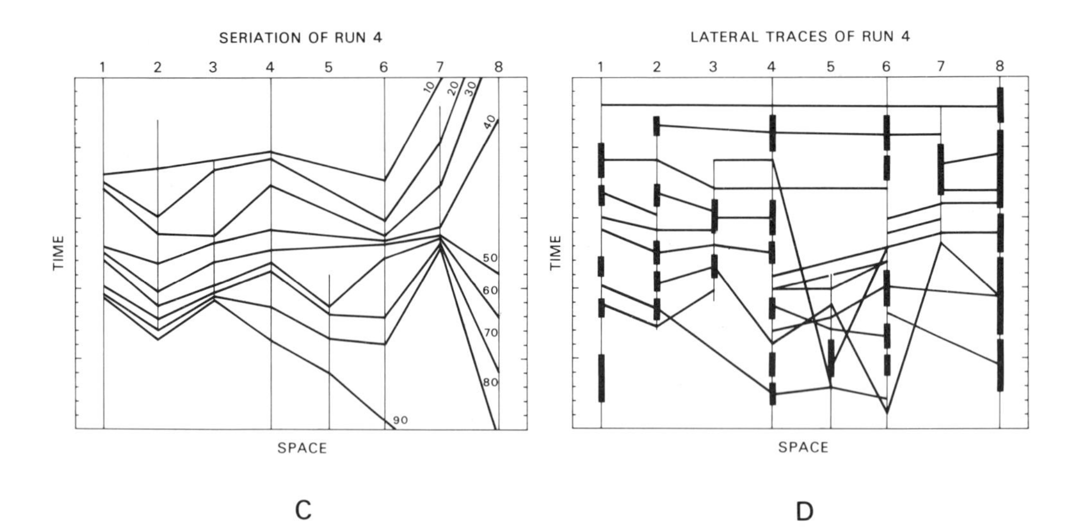

Figure 7 Plot of correlations for samples of Run 4.
A. Graphical correlation.
B. Ranking model.
C. Seriation.
D. Lateral tracing.

below section 7 and the smaller bulge under section 3 (see Figures 5B, 7B). Despite these minor features, the ranking model of run 4 continues to furnish the biostratigrapher with an adequate working hypothesis for the ages of the samples.

The correlations derived from the seriation are less sanguine. The rank correlation between the sample elevation on the correlation chart and the row position of the samples in the seriated matrix of run 4 is -0.906, whereas the values for previous runs vary from -0.940 to -0.934 (Table 4). The lower correlation profoundly affects the contour map of Figure 7C, and the technique generates poor results in several areas of the data. The upper and lower samples of section 8 are considered excessively old and young, respectively. Also the strongly clustered contours at the base of section 7 suggest the existence of an "unconformity" which does not exist. Less serious difficulties are seen in the top of section 7, the base of section 6 and in other parts of the graph. All in all, the performance of the seriation of run 4 is distinctly inferior to that of graphical correlation and the ranking model.

The rank correlation between the link positions of the lateral traces and the known locations of the samples is only 0.832 in contrast to the figures of previous runs which range from 0.982 to 0.962 (Table 4). Clearly the output of this method has begun to deteriorate. Inspection of the lateral traces in Figure 7D discloses links which markedly cut across the time lines. The most extreme example connects sample 19 in section 4 with number 3 in section 5. All other crossovers are concentrated in the lower half of the correlation chart. Most lateral traces of the younger samples closely parallel time lines. One difficulty of the lateral tracing algorithm is that links are conspicuously absent from certain regions of the time-stratigraphic diagram. A biostratigrapher could employ the lateral traces of the younger beds to correlate the samples. However, the linkages cannot help to unravel most of the older samples. Like most multivariate techniques, lateral tracing is sensitive to diversity effects. Essentially, sampling reduces the number of taxa present in strategically placed samples which dictates the relatively poor performance of the technique in run 4.

The cluster analysis of Figure 8 is most generously characterized as chaotic. It seems unlikely that any reasonable chronostratigraphy could be inferred from these data. Clearly, the clustering scheme is overwhelmed by a combination of low diversity, the two "faunal provinces", some erratically distributed species and the rigors of the sampling.

SUMMARY

The hypothetical data set of Edwards (1982a) affords a test of biostratigraphic methods. The main characteristics of the data represent low diversity, five stratigraphically useful species, 10 fossils in

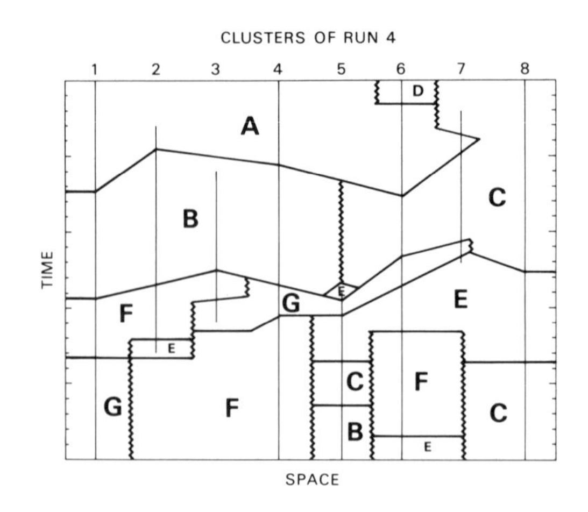

Figure 8 Clusters superimposed on correlation chart for Run 4.

two "faunal provinces" and five more or less erratically distributed taxa. The techniques evaluated comprise graphical correlation, a modified version of the ranking model developed by Agterberg and coworkers, the stratigraphically constrained type of seriation, lateral tracing and cluster analysis. The data are sampled in four trials during which the probabilities of scoring a species as present in a sample within its local range zone decline from 100 to 40 percent. These probabilities seem reasonable for the large samples typically collected by micropaleontologists. The techniques are accessed by Spearman's rank correlation coefficients. For sequences of events, these compare the calculated zonations with the known average and true range charts of the taxa. Similarly, the computed sample positions are scaled against their known elevations on the correlation chart.

As expected, the results of all techniques deteriorate as the probabilities of finding the species decrease. Likewise, the more complete stratigraphic sections can be correlated more accurately than incomplete ones due to their higher information content. In terms of overall effectiveness, the algorithms fall into three main categories, listed from best to worst: first, graphical correlation and the ranking model; second, seriation and lateral tracing; and a distant third, cluster analysis. The sequences of biostratigraphic events derived from graphical correlation and the ranking model are most highly correlated with the true and average range charts, respectively. Thus each method produces the type of zonation for which it is designed. Seriation generates a compromise sequence which is intermediate between the average and conservative types of event sequences. All three techniques can cope with fossils which are time-transgressive, erratically distributed or confined to a single "faunal province". For both samples and events, the graphical correlation and ranking model computations hold up relatively well over all sampling runs. Edwards (1982a) reported similar results based on comparisons of her no-space graph technique, which is basically a non-parametric version of graphical correlation, with "truth". Although seriation and lateral tracing provide the biostratigrapher with workable hypotheses in the first three sampling runs, the results decline in the last trial where the probabilities of locating the taxa were minimized. Plotting the clusters from multivariate cluster analysis on the correlation chart reveals little information of stratigraphic value. Cluster analysis is simply unable to handle a combination of low diversity, the "faunal provinces", and the several taxa with poor biostratigraphic properties.

REFERENCES

Agterberg, F.P., 1984, Binomial and trinomial methods in quantitative biostratigraphy. Computers and Geosciences, 10, 31-41.

Agterberg, F.P., and Nel, L.D., 1982a, Algorithms for the ranking of stratigraphic events. Computers and Geosciences, 8, 69-90.

Agterberg, F.P., and Nel, L.D., 1982a, Algorithms for the scaling of stratigraphic events. Computers and Geosciences, 8, 163-189.

Blank, R.G., 1984, Comparison of two binomial models in probabilistic biostratigraphy. Computers and Geosciences, 10, 59-67.

Brower, J.C., 1981, Quantitative biostratigraphy, 1830-1980. In Computer applications in the earth sciences, an update of the 70's, D.F. Merriam, ed., Plenum Press, New York and London, p. 63-103.

Brower, J.C., and Burroughs, W.A., 1982, A simple method for quantitative biostratigraphy. In Quantitative stratigraphic correlation, J.M. Cubitt and R.A. Reyment, eds., John Wiley and Sons, Ltd., p. 61-83.

Brower, J.C., and Millendorf, S.A., 1978, Biostratigraphic correlation within IGCP project 148. Computers and Geosciences, 4, 217-220.

Burroughs, W.A., and Brower, J.C., 1982, SER, a FORTRAN program for the seriation of biostratigraphic data. Computers and Geosciences, 8, 137-148.

Edwards, L.E., 1982a, Quantitative biostratigraphy: the methods should suit the data. In Quantitative stratigraphic correlation, J.M. Cubitt and R.A. Reyment, eds., John Wiley and Sons, Ltd. p. 45-60.

Edwards, L.E., 1982b, Numerical and semi-objective biostratigraphy: review and predictions. Third North American Paleont. Conv., Proc., 1, 147-152.

Edwards, L.E., 1984, Insights on why Graphic Correlation (Shaw's method) works. J. Geol., 92, 583-597.

Eicher, D.L., and Worstell, P., 1970, Cenomanian and Turonian Foraminifera from the Great Plains, United States. Micropaleontology, 16, 269-324.

Gradstein, F.M., 1984, On stratigraphic normality. Computers and Geosciences, 10, 43-57.

Harper, C.W., Jr., 1981, Inferring succession of fossils in time, the need for a quantitative and statistical approach. J. Paleont., 55, 442-452.

Harper, C.W., Jr., 1984, A FORTRAN IV program for comparing ranking algorithms in quantitative biostratigraphy. Computers and Geosciences, 10, 3-29.

Hay, W.W., 1972, Probabilistic stratigraphy. Eclogae Geol. Helvetiae, 65, 255-266.

Hay, W.W., and Southam, J.R., 1978, Quantifying biostratigraphic correlation. Ann. Rev. Earth and Planet. Sci., 6, 353-375.

Hazel, J.E., 1977, Use of certain multivariate and other techniques in assemblage zonal biostratigraphy: examples utilizing Cambrian, Cretaceous, and Tertiary benthic invertebrates. In Concepts and methods of biostratigraphy, Kauffman, E.G., and Hazel, J.E., eds., Dowden, Hutchinson & Ross, Inc., Stroudsburg, Pennsylvania, p. 187-212.

Hudson, C.B., and Agterberg, F.P., 1982, Paired comparison models in biostratigraphy. Math. Geol., 14, 141-159.

Millendorf, S.A., Brower, J.C. and Dyman, T.S., 1978, A comparison of methods for the quantification of assemblage zones. Computers and Geosciences, 4, 229-242.

Miller, F.X., 1977, The graphic correlation method in biostratigraphy. In Concepts and methods of biostratigraphy, Kauffman, E.G., and Hazel, J.E., eds., Dowden, Hutchinson & Ross, Inc., Stroudsburg, Pennsylvania, p. 165-186.

Shaw, A.B., 1964, Time in stratigraphy. McGraw-Hill Book Co., New York, 365 p.

Part III

BIOSTRATIGRAPHIC CORRELATION

Contents

III.1

QUANTITATIVE CORRELATION IN EXPLORATION MICROPALEONTOLOGY

F.M. GRADSTEIN
and
F.P. AGTERBERG

INTRODUCTION

This Chapter describes the theory and application in geological basin analysis of the computer program CASC. CASC stands for Correlation And SCaling in time. The method provides a precise, semi-automated and semi-objective means of correlation of rock sections for which an optimum sequence or a scaled optimum sequence of biostratigraphic events have been determined using the zonation method and computer program RASC. RASC is described in Chapters II.4 through II.7.

The use of RASC and CASC provides the stratigrapher with an integrated biostratigraphic method, particularly suitable to exploit the considerable amount of micropaleontological data that accumulates during sedimentary basin analysis. The method starts with a data file of the original observations on the distribution in time and in space in wells or outcrop sections of all taxa identified. Next, it will reduce this data file to biozonations that best explain the regional and temporal trends. Finally, it will calculate geologically reasonable correlations of the sections. Segmentation and correlation of the original sections can be achieved by means of fossil events and RASC biozones. Interpolation of the scaled optimum sequence in linear time makes it feasible to also correlate isochrons. Each correlation line carries an uncertainty limit, which is a conservative estimate of various original uncertainties in the data.

The CASC method of quantitative correlation draws on the RASC method and on the philosophical reviews and statistical methodology of several geologists and (geo)mathematicians, for example: Shaw (1964), Hay (1972), Drooger (1974), Millendorf and Heffner (1978), Blank (1979), Reinsch (1967) and De Boor (1978). A description of CASC from an operational point of view and the interactive computer program listing in Fortran are found in Agterberg et al. (1985).

Correlation is one of the most widespread, abstract undertakings of the mind and refers to causal linkage of present or past processes and events. Such events can be inorganic, organic or abstract. Correlation of geological attributes generally expresses the hypothesis that a mutual relation exists between stratigraphic units. In a more narrow stratigraphic sense it means that samples (or imaginary samples) from two separate rock sections occupy the same level in the known sequence of stratigraphic events. Without correlation, successions of strata or events in time derived in one area, are unique and contribute nothing to understand earth history elsewhere (McLaren, 1978).

Geological correlation traditionally is expressed in terms of:

1. rock units, like formations or well log intervals = lithostratigraphic correlation;

2. fossil units, like zones = biostratigraphic correlation;

3. relative age units or stages = chronostratigraphic correlation;

4. linear time units or ages = geochronologic correlations.

Rather than units with a certain thickness or a duration in time, it is events that are frequently correlated. Events or datum planes refer to the fossilized physical or organic occurrences of supposedly irreducible resolution along the geological scale.

An important contention of geological correlation is that once the events or rock-, fossil-, or relative age units have been properly determined and defined, these units can indeed be used for correlation. This statement, which might seem to be trivial and redundant, is made here because existing stratigraphic codes show how to define stratigraphic units, but they do not define how to correlate. The latter is the subjective domain of experts. Procedures for correlation or stratigraphic equivalence depend on subjective judgement of the unique relationship of each individual record to the derived and accepted standard.

It follows that correlation as practised in geology cannot be readily verified without a detailed review of all the underlying facts. It also follows that traditionally there is no expression of the uncertainty in fixation of the individual record to the standard. Or, as W.R. Riedel (1981) wryly wrote: "Biostratigraphy will be continued to be regarded as an art rather than a science, until it is possible to attach confidence limits to suggested correlations". An improvement in definition of the zonation through increased number of observations and taxa may increase the number of correlation tiepoints, but still leaves the question of uncertainty unanswered. Such an uncertainty will generally be couched in qualitative terms only. For many geological investigations such a subjective procedure yields satisfactory results, correlation being only a part of the scientific objectives. Situations do arise, however, where the quality of correlations determines the outcome of the study. This is particularly true in the field of operational biostratigraphy, where large and complex data sets may have to be reduced before they can be of assistance in basin history.

Let us assume that the stratigraphic distribution of hundreds of fossils has been sampled in dozens of cored wells or outcrop sections. Following a detailed analysis, which lasted many months or even several years, a range chart is proposed. The range of each fossil taxon divides geological time into three segments by its entry (first occurrence) and exit (last occurrence). The range chart synthesizes the information on all observed ranges to arrive at average or total (maximum) ranges for each species. The range chart is then segmented, using co-existence of taxa and discrete taxon events, to establish time-successive intervals. Each segment is called a zone. When only last occurrences of fossils are known, such a chart portrays a succession of events or partial ranges.

The critical and least understood step in the practice of correlation is to actually tie the zones back to the individual (well) sections. This may be a difficult undertaking when the individual stratigraphic record shows frequent inconsistencies, due to sampling problems, reworking, unfilled ranges due to facies changes, etc., (Chapter I.2).

Ideally, the individual fossil record as observed in each rock section, should be normalized prior to actual correlation. With this is meant that insight should be gained in the probability that the observed events occur where the standard suggests they should be found. In our hypothetical example, the paleontologist will make a judgement on the outliers, of events to be rejected or moved up and down the section. Such normalization is based upon geological or paleontological experience. Next, the paleontologist will in each rock section define the successive zones in such a manner that a minimum number of (key) taxa for each of the zones are observed outside the zonal limit. Mismatch of the zones and the individual record is explained by poor data or unfilled ranges due to facies changes.

The problem with the hypothetical case history is not so much that it leads to right or wrong stratigraphy, but that a single solution is proposed. What is a reasonable criterion for successful correlation, if there is no insight in the actual uncertainty in correlation, either in millions of years or in depth in meters? In the regional correlations there frequently is limited or no understanding of how much (in depth or in relative time units) the solution differs from alternate solutions, using the same data. In all likelihood it is difficult to propose or compare two alternative correlations, without a major review or analysis of all underlying facts.

The interactive computer program CASC was developed in an effort to provide a relatively straightforward and simple method of semi-objective biostratigraphic correlation. The method calculates error bars in depth, or in (linear) time units, for each desired correlation line through all sections examined. We will try to demonstrate that CASC produces geologically meaningful results, but first we will introduce two existing techniques for geological correlations, i.e. matching and composite standard.

Matching

A quantitative stratigraphic procedure that is particularly useful for well log or detailed lithological correlation is based on matching. A chosen parameter, like fossil content or lithological or physical properties is noted in two geographically separate rock sections. The two sections are slid by one another to locate the maximum of the ratio of matches to the number of comparisons.

Lateral tracing (Chapter II.9) is a special type of matching, suitable for a series of rock sections, each with a slightly different succession of highly fossiliferous samples. The hypothesis is that the faunal composition shows both stratigraphic and geographic gradients. For such type of data 'en masse' data reduction obscures the lateral faunal gradation. A matrix of dissimilarity coefficients (for example 1 - Dice similarity coefficient) is calculated between the samples from two adjacent sections. The procedure briefly runs as follows (Millendorf and Heffner, 1978):

1. The smallest dissimilarity (greatest similarity) is extracted from the matrix and its corresponding samples are matched. These two samples are then discounted in subsequent steps of analysis;

2. Step 1 is repeated, matching pairs of samples from the matrix, until all samples from the two sections are matched. The samples that remain unmatched, from the section with the greater number of samples, are disregarded;

3. Dissimilarity measures are then calculated between pairs of vertically adjacent samples within each section.

The combined lateral and vertical dissimilarity analysis may portray how successive zones can be traced. After analysis of the first two sections, the procedure is repeated for the second and third ones, and so on. Matching can be a useful technique to portray lateral correspondence of different units, but no estimate of confidence is available.

Composite Standard

Biostratigraphic correlation depends on the chances that:

(a) in each rock section the events defining a biostratigraphic increment have been detected and properly taxonomically determined; and

(b) the true (or natural) sequence of events is known.

This principle was succinctly stated by Hay (1972), who then went on to propose the principle of matrix permutation for construction of the most likely sequence of (nannofossil) events in time, (see Chapter II.5). In the ranked sequence each event position is an average of all the relative positions, but no direct insight is available in the uncertainty of rank.

As early as 1964, Shaw not only proposed a simple ranking method for biostratigraphic events, but also a correlation method of the sections in which the events occur. The original method runs as follows: From a number of individual geological sections (A, B, C, D, etc.), one (for example A) is selected that shows a more complete and reasonable 'normal' order and spacing of events. This particular sequence of entries and exits of taxa is plotted along one axis and that of a reasonable comparable sequence B along the other axis of a conventional two-axis graph. Scale units are in feet or meters, as found in each section, but in a simplified procedure use can be made of order only. The best fit of the resulting scattergram is called line of correlation. Shaw (1964) advocates regression analysis as a linear trend-fitting technique, although A and B are probably both subject to uncertainty. There are two regression lines, one for A over B and one for B over A. Spline fitting, as now available (see below) may be more suitable, particularly if no straight-line fit is possible due to changes in sedimentation rate.

The order and spacing of first and last occurrence events along the A-axis is now updated through projection of the homologous B-axis events, via the best fit line, on the A-axis. If the first occurrence of an event in B occurs relatively lower than in A, the range of this event in A is appropriately extended downward. If a last occurrence of an event in B occurs relatively higher than in A, the range of the event appropriately extended upward. It is generally tried to maximize the stratigraphic ranges. Next, the updated A-axis (composite section) is compared to section C, etc. In the final composite section the scale of the successive events has become a composite of all spacing values between successive events.

Actual correlation of events, clusters of events, etc. is achieved by making new bivariate plots for each individual section as a function of the final composite one. For each bivariate scattergram a new best fit line or best fit channel is calculated which serves to project the composite events on the individual section scale. In a mathematical sense, each value in the composite standard can be expressed as a function of its correlative (depth) value in the individual sections. Probably the best description of use of the composite standard method is by Miller (1977).

One might also choose to hand-fit the scatter of points to arrive at some kind of composite standard (Shaw, 1964; Edwards, 1984), particularly so if scatter is limited. In a variation on the basic technique, the best fit line of the scattergram can become the standard section, to arrive at some kind of average event positions. When a third section is added this standard section becomes one of the axis of the scattergram.

An important difference between Hay (1972) and Shaw (1964) is that the latter tries to find the highest and lowest possible occurrences out of all individual records. It maximizes possible stratigraphic ranges, but as a result no uncertainty limits can be attached to the range end points. In the Hay and RASC methods, average ranges are calculated (Chapters I.2 and II.4-7), which allows an expression of sampling error. The CASC method of quantitative correlation combines average sequence methodology with bivariate correlation technique.

CORRELATION AND SCALING IN TIME (CASC)

The Principle

The CASC method of correlation calculates the most likely depths in well sections of events, zones or ages as defined in the RASC (scaled) optimum sequence. Each depth has an error attached to it, which reflects uncertainty in the original data. Input for CASC is the RASC input file that shows the original sequence of events in each of the sections. In addition, the program requires a depth file, that shows the observed depth in feet or in metres for all the events in the original sequences file (for example, see the Depth File for Labrador Shelf and Grand Banks Cenozoic Foraminifera in Appendix I).

The correlation and scaling in time (CASC) program first computes the RASC optimum sequence and RASC scaled optimum sequence of events. Using the three normality testing techniques in the RASC method, bivariate graphs, stepmodel and normality testing, outliers in the individual sections may be eliminated. Based on the filtered data file, a new optimum sequence may be calculated, after which each individual sequence of events is compared to the (scaled) optimum one, and best fit curves (splines, see below) are calculated. A spline fit yields a function such that, for each optimum sequence position, the most likely stratigraphic equivalent position can be found in the individual sequences. These normalized tiepoints are then correlated.

Figure 1 graphically depicts the principal steps, executed for the correlation of event 29 (top of **Cyclammina amplectens**) in one Grand Banks well, Adolphus D-50. The y-axis is the optimum sequence in 21 of the wells ($k_c \geq 7$, $m_{c1} > 2$; see Chapter II.4). Instead of the optimum sequence, the scaled optimum sequence can be used. The x-axis is the observed sequence of events, whereas the z-axis is the depth scale of the well.

The lower scattergram expresses mismatch of the individual sequence with the optimum one. The best fit line for the graph (here visually estimated) is the line of correlation. Working with event

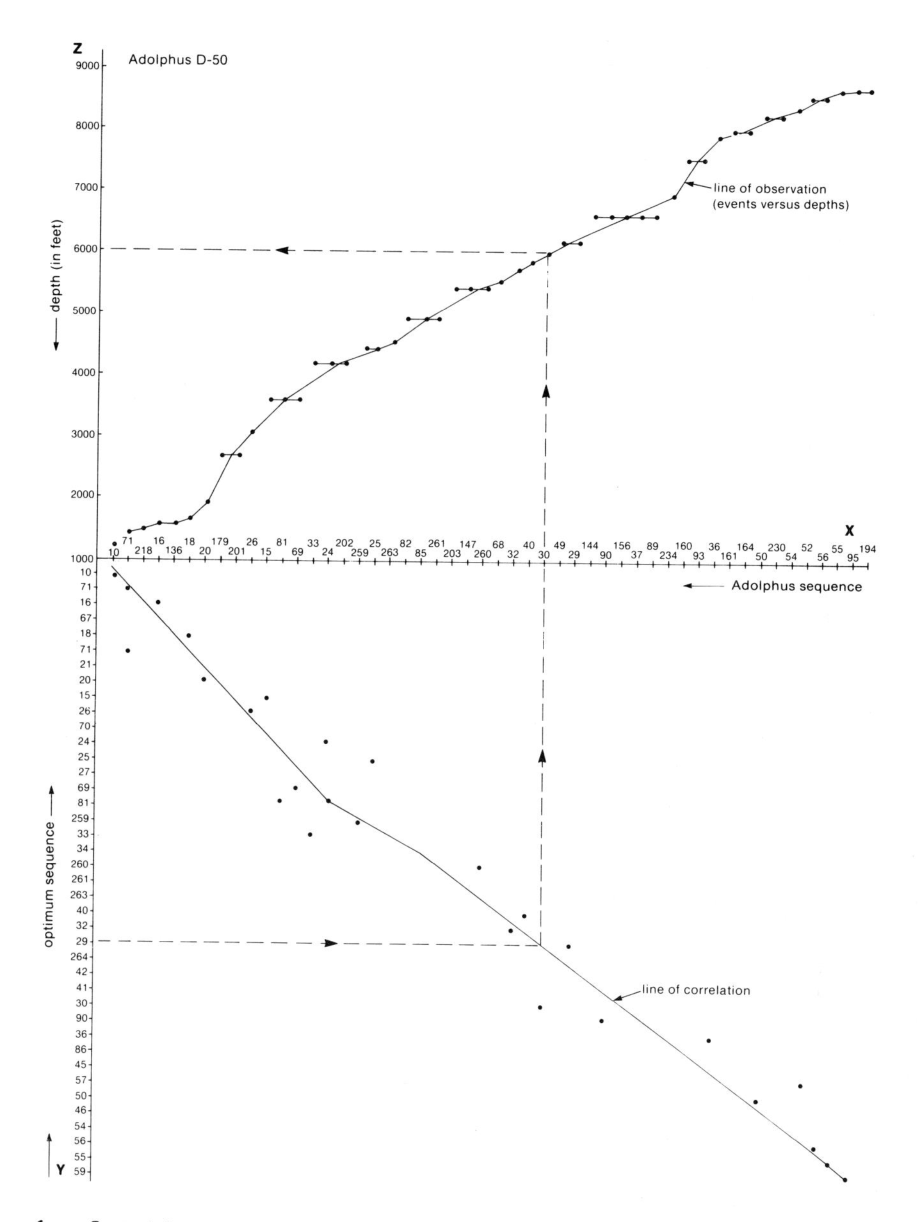

Figure 1 Graphical illustration of correlation, using the computer program CASC (Correlation and Scaling in time); x - observed sequence of events in Adolphus well; y - optimum sequence of events in 21 wells (7-2-4 condition); z - depth scale. The most likely position of selected optimum sequence events in the well is found by projecting these events via the two best fit lines (line of correlation and line of observation) onto the (well) depth scale. The most likely position of event 29 (top of ***Cyclammina amplectens***) in Adolphus well is at 6050' (observed 6200').

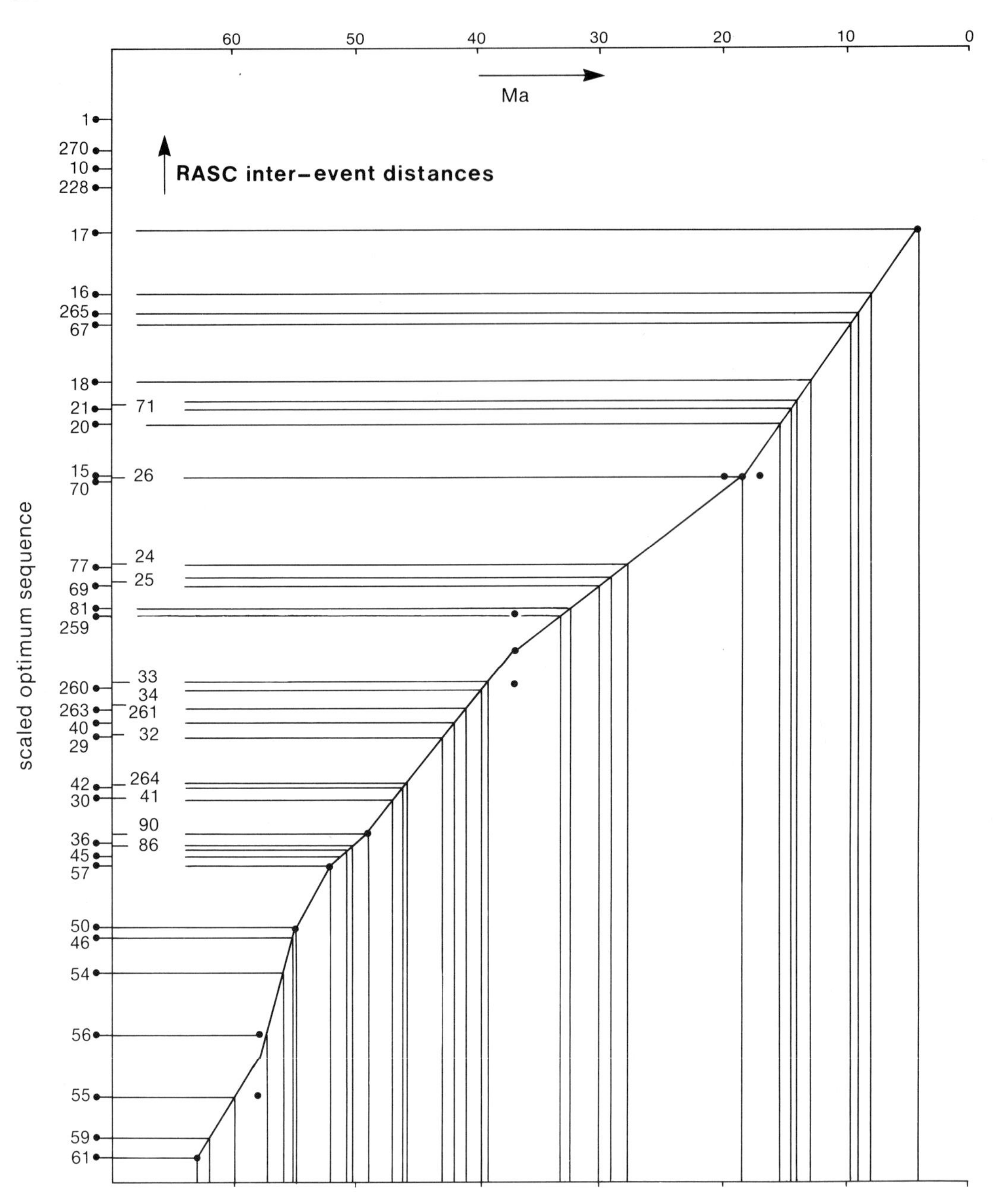

Figure 2 Plot of the Cenozoic scaled optimum sequence (21 wells; 7-2-4 run) versus linear time in Ma. The inter-event distances are plotted cumulatively. For some selected events in the scaled optimum sequence the numerical age is known (dot's), which allows to scale the whole fossil sequence in linear time.

scales initially has the advantage that complications due to different rates of sedimentation in different places which may be hundreds of km apart are avoided. Moreover, equal spacing of values for the independent (x-axis) variable in spline-curve fitting has the considerable advantage that the possibility of unrealistic oscillations of the fitted curve between irregularly spaced control points is avoided. However, the number of levels in the event scale differs from section to section in a 'random' manner. For correlation between wells it is required to replace the levels of the event scales by depths (in km). This replacement is shown in the upper part of the scattergram of Figure 1. The individual sequence x is a function of the depth z at which the events were observed. This function is shown here as the 'line of observation'. The most likely position of event 29 in the Adolphus well can be found by projecting its optimum position via the line of correlation in the individual sequence and from there via the line of observation in the depth scale. All optimum sequence events can thus be scaled in (well) depth. In a multiwell comparison, the most likely depth value (z-axis) in each well is calculated for selected y-axis values (event positions) in the optimum sequence.

In our example, event 29 in Adolphus should occur at at 6050' (observed 6200'). This depth then is the most likely correlation tie-point.

The standard deviation (SD) of the events relative to the line of correlation (and the line of observation) in the y direction (and parallel z direction, which is the depth-axis in the well) provides an estimate of the mismatch of each event. This is further on called the local error. When it is geologically unreasonable to expect a continuous sedimentation rate in the vicinity of a certain depth, the local SD can be modified such that it is restricted to shorter time increments.

The same procedure as shown in Figure 1, using the RASC optimum sequence, is applied when the scaled optimum sequence is chosen instead. The inter-event distances in the scaled optimum sequence reflect the average relative distance of the events in relative time. If it is possible to estimate the numerical geological ages of some of the events in the scaled optimum sequence, the relative distance estimates can be used to stretch the scaled sequence in linear time. This way the scaled optimum sequence becomes a (local) biochronology and hence isochrons can be traced through the wells or outcrop sections. For paleontologists this is a valuable method to find the numerical age of the most likely position in each well of principal zone boundaries. Such boundaries, as argued in Chapter II. 4, can delineate sedimentary cycles. The original standard deviations for the interfossil distances in the scaled optimum sequence now reflect the uncertainty in linear time between the events. This uncertainty is named the global error bar.

As a first test of the validity and use of this numerical time interpolation for geological analysis, Figures 2 and 3 were constructed. Along the y-axis of Figure 2 are plotted the interfossil distances for

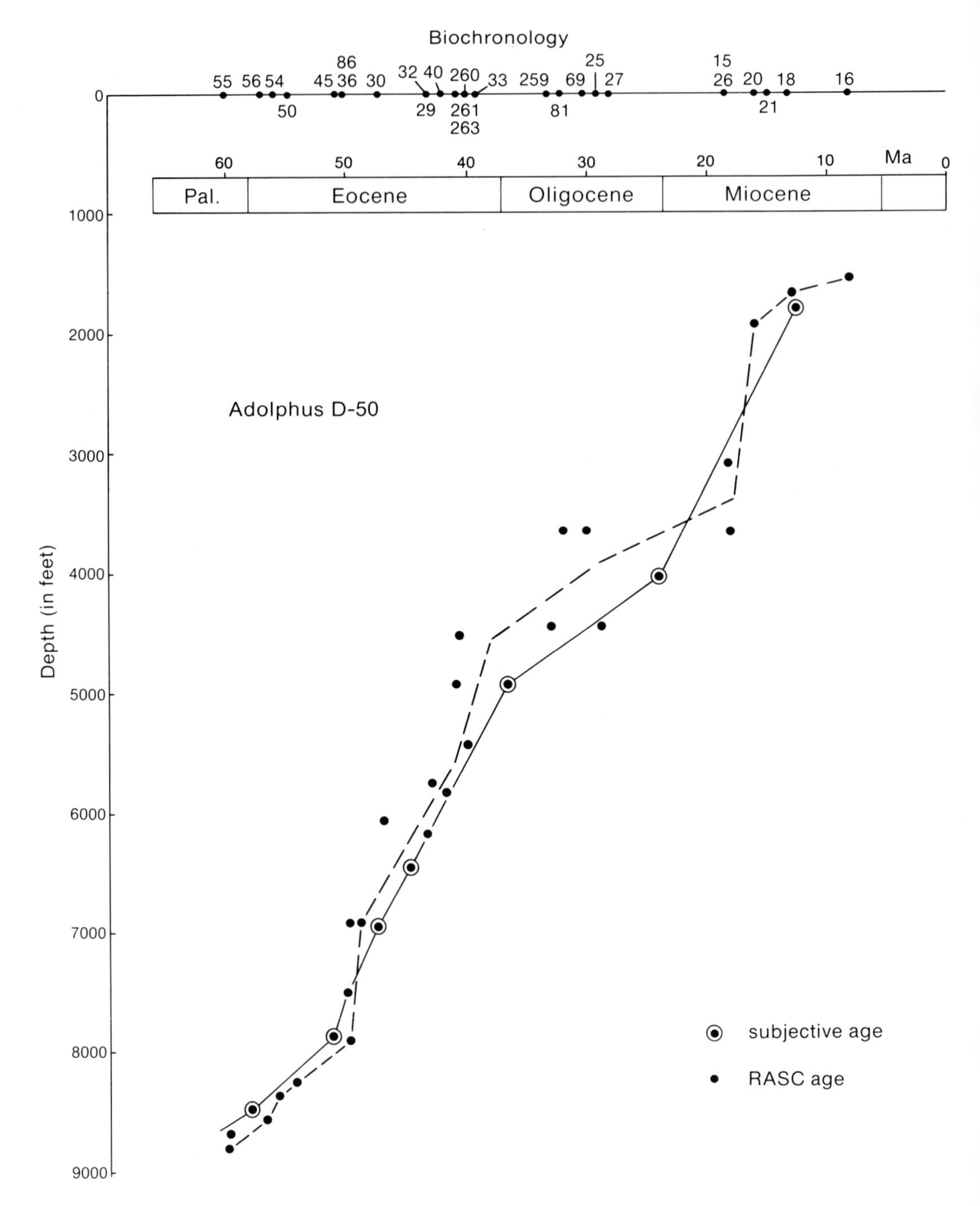

Figure 3 *The RASC biochronology of Figure 2 is used to estimate rate of sedimentation (dashed line) in the Adolphus well. The solid line (subjective) shows approximately the same trend, using independent well history data.*

the Cenozoic foraminiferal events in the scaled optimum sequence in 21 wells. For some events, listed in the paragraph on 'Application of CASC,' which are the (regionally averaged) last occurrences of key planktonic and a few benthic foraminifers, numerical ages can be estimated. This involves comparison of the regional to standard Atlantic zonations, details of which follow later. The horizontal scale is linear time in Ma, for the Cenozoic period.

The calibrated events are used to form a nomogram or line of correlation, such that all events in the scaled optimum sequence can be dated. In Figure 3 this new RASC biochronology (horizontal axis) is used to estimate the rate of sediment accumulation (dashed line) in one of the wells - Adolphus D50.

Several years earlier, prior to development of RASC and CASC an approximate chronostratigraphy of this well section was given, in system units, Paleocene, Eocene, etc. As shown in Figure 3, there is a close approximation of the two, independently arrived at, sediment accumulations. The earlier interpretation obscures a possible late Oligocene - early Miocene hiatus. Scaling in time of the scaled optimum sequence is a pragmatic way of erecting a local time scale.

In summary the CASC method of correlation is founded on three conditions:

1. each individual stratigraphic sequence of events is a sample of the optimum sequence;

2. the observed depths of the events in a stratigraphic section are estimates of the true depths;

3. the calculated relative interfossil distances of events in the scaled optimum sequence can be used to stretch this sequence along the numerical geological scale; known ages for index fossils in this sequence provide the necessary tiepoints.

Input for (semi-)automated correlation of fossil events (or zones) with confidence limits in depth - or in time units are:

a. depth in feet or in meters for all stratigraphic events in all wells or outcrop sections. The events are the same as used in the RASC methods;

b. ages of index fossils to stretch the scaled optimum sequence in linear time;

c. events, clusters of events (zones) or ages to be correlated;

d. wells or outcrop sections to be correlated.

We will now proceed with an explanation of the statistical aspects of the method and the interactive computer program.

The Method

The CASC program was developed using a CDC Cyber 730 computer with a Tektronix 4014 terminal and is coded in Fortran Extended Version 4. Two computer libraries are required to use CASC: IMSL Library and Tektronix Advanced Graphing Library. Also, mass storage facilities are used (Agterberg et al., 1985). To obtain the geologically most satisfactory bivariate fits, the program is best used interactively.

One of two different routes may be selected at the beginning of an interactive CASC session. The first route uses as starting point the RASC optimum sequence, plotted against the so-called event scale. The latter is the original sequence data for each stratigraphic section. Instead of the optimum sequence, the RASC scaled optimum sequence may be used. The latter combines average order and relative 'distance' in time. The optimum sequence option is simpler than the so-called distance option, based on the scaled optimum sequence, but not principally different. For further illustration the distance method is applied to results of RASC on the distribution of Cenozoic foraminifers in the 21 and 24 well files, Grand Banks and Labrador Shelf, discussed in Chapter II.4. The same data file is used in the final section of this Chapter that deals with 'Application of CASC.'

If RASC distances are used as input, it is necessary to replace them by ages (in Ma). The procedure for this is schematically shown in Figure 4. Suppose that approximate ages are known for a subgroup of events in the scaled optimum sequence shown in Figure 4a. Our objective is to fit a curve to these data in order to be able to replace any RASC distance by its age (see Figure 4d). First, RASC distances with the same age are averaged (see Figure 4a). Then a cubic spline curve is fitted to the age-distance pairs minimizing the sum of squares of deviations between points and curve in the vertical (age-) direction of Figure 4b. The smoothing factor SF can be chosen beforehand by the user of the interactive CASC computer program. It is equal to the square root of the mean square deviation between points and curve. Because this standard error is normally not known beforehand, the user can determine it by trial and error while experimenting with different plots on the screen of the monitor. In Figure 4b a curve was fitted to 5 original values (o) and 2 averages of two values (+). The standard deviation of the original data in relation to this curve is also shown on the screen (SD in Figure 4c). The fitted curve does not extrapolate outside the range of the RASC distances used for the curve fitting. Consequently, the circle with the highest RASC value is not considered for estimating SD in this example. It is noted that a curve could also be fitted directly through the 8 circles in Figures 4a and 4c. Then SD would be equal to SF.

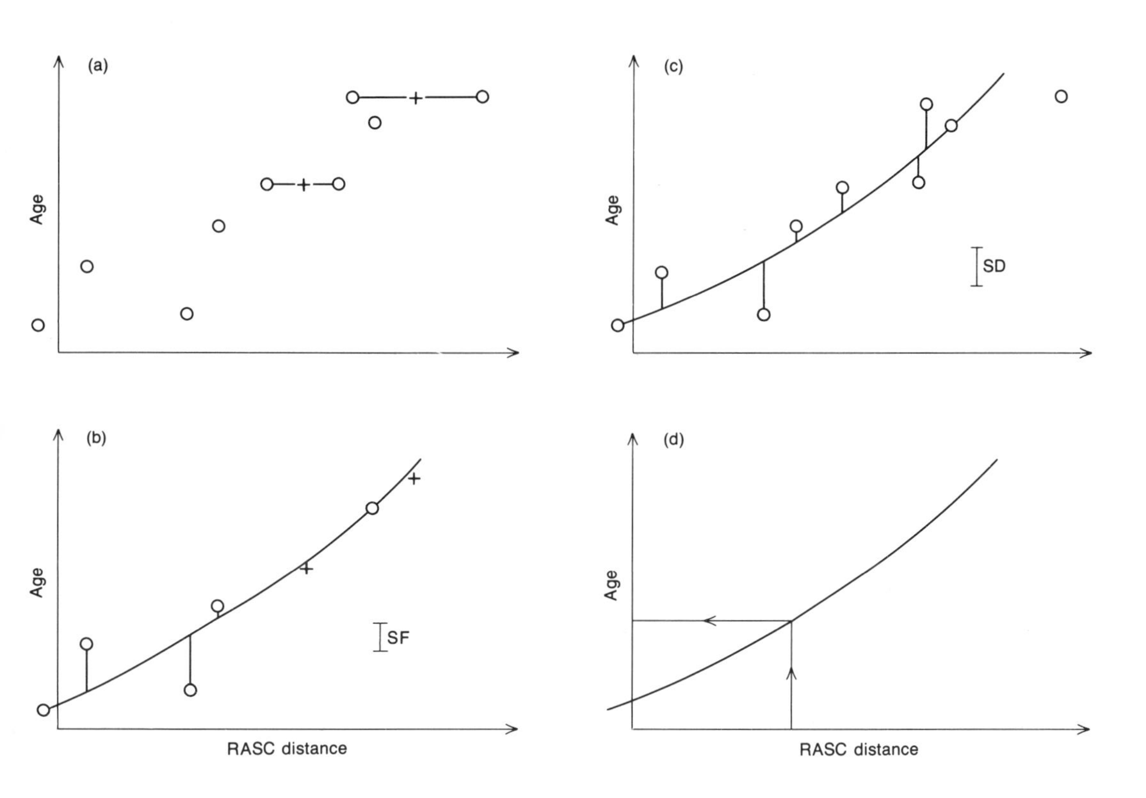

Figure 4 *Schematic illustration of method followed in CASC computer program to establish relation between RASC distance and age;*
(a) Two (or more) RASC distances for the same age are averaged;
(b) Cubic spline curve is fitted using age as the dependent variable; smoothing factor (SF) representing standard deviation of differences between event ages and curve is chosen in advance, before curve fitting;
(c) Standard deviation (SD) for differences between original values and curve is computed after curve fitting; and
(d) Fitted curve is used to convert any RASC distance into corresponding age.

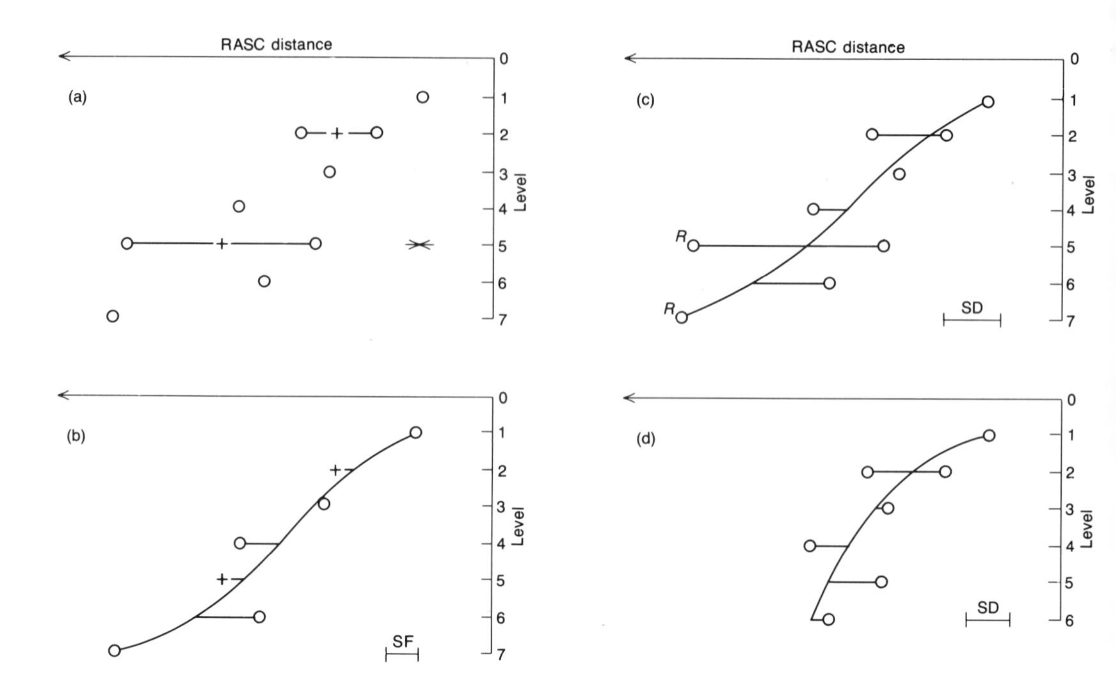

Figure 5 *Schematic illustration of preliminary computing and optional editing procedure at beginning of CASC computer program;*

(a) Events found to be anomalous with a probability of over 99 per cent (asterisk) may be omitted from spline curve fitting and later plots; RASC distances of two (or more) coeval events are averaged;

(b) Cubic spline curve is fitted using RASC distance as the dependent variable; smoothing factor (SF) representing standard deviation of differences between RASC distances assigned to levels and curve is chosen in advance;

(c) Standard deviation (SD) is computed from differences between original values and curve after curve fitting; original values (e.g., those labelled R) can be deleted; and

(d) New curve with new standard deviation (SD) is obtained without use of deleted values.

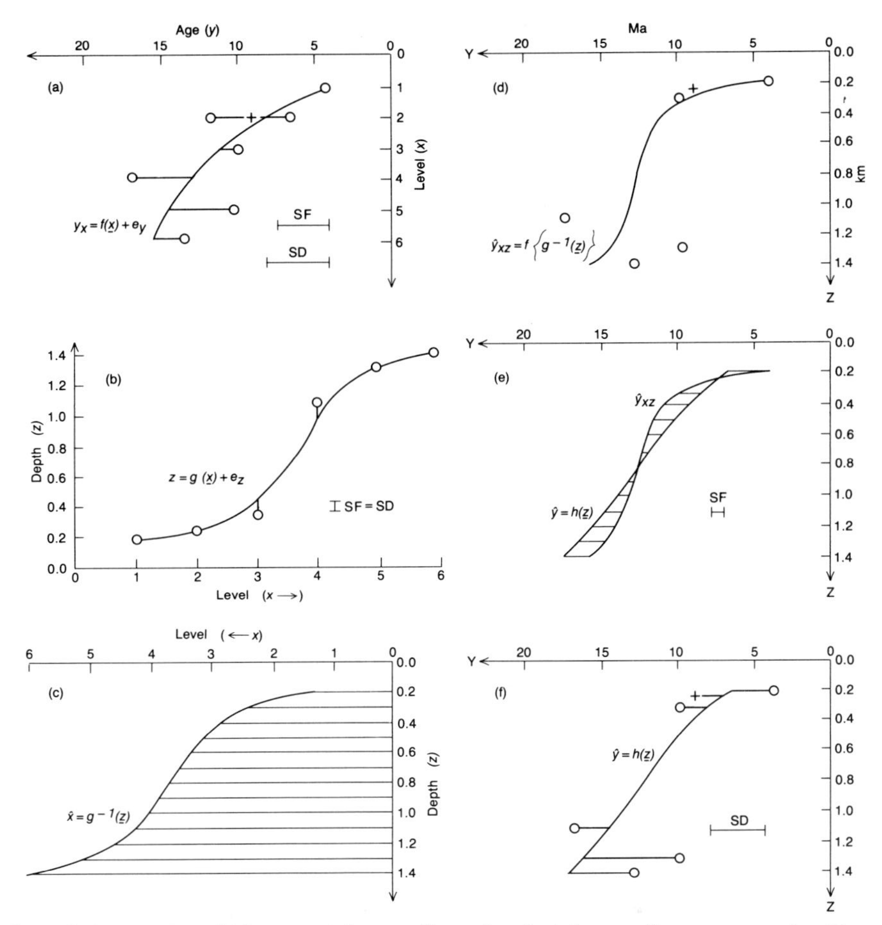

Figure 6 *Schematic illustration of calculation of an age-depth curve from RASC output for a well.*

(a) RASC distances have been replaced by ages using relation illustrated in Figure 4d; new spline curve $f(\underline{x})$ is fitted; bar in $\underline{x}$ denotes use of regular sampling interval for x; smoothing factor (SF), which was selected before curve fitting using one age per level, is smaller than standard deviation (SD) for all original values;

(b) spline curve $g(x)$ is fitted to express depth as a function of level x; bar in $\underline{x}$ denotes use of regular sampling interval for x; SM = SD is equal to some small value;

(c) $\hat{x}$ represents spline curve $g(\underline{x})$ in Figure 6b now coded as set of values for x at regular interval of z;

(d) $\hat{y}_{xz}$ denotes curve passing through set of values of y at regular interval of z obtained by combining spline curve of Figure 6a with that of Figure 6c;

(e) $\hat{y}$ is spline curve fitted to values y_{zx} of Figure 6d using new smoothing factor SF; and

(f) standard deviation SD is computed after curve fitting, using one age per level.

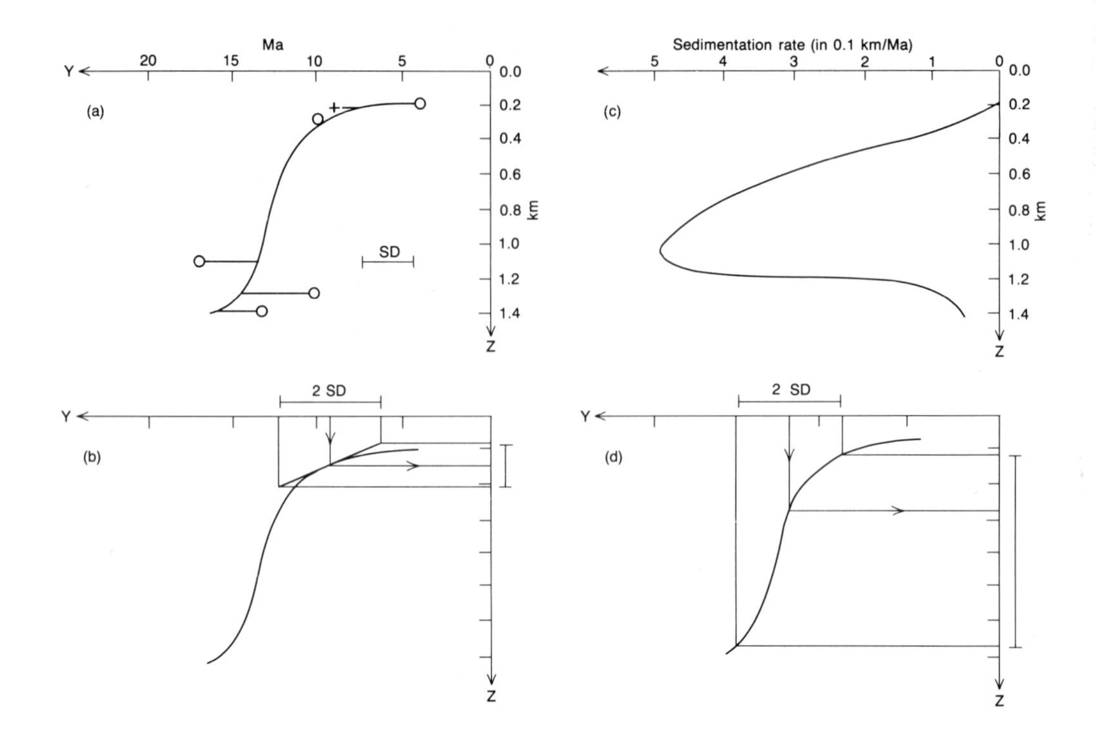

Figure 7 *Schematic illustration of estimation of local error bar and modified local error bar.*

(a) Standard deviation SD was computed after curve fitting, using one age per level;

(b) Error bar of age value plus or minus SD along Y-axis is transformed into error bar along Z-axis using first derivative (dz/dy) of age-depth curve;

(c) Rate of sedimentation (=dz/dy) can be displayed on screen during CASC interactive session; and

(d) Modified local error bar is asymmetrical with respect to depth value for a given age.

Next, the CASC user can display and edit the RASC distances for any well from among the set of wells used. Editing options are schematically shown in Figure 5, which displays preliminary data analysis. The scale in the vertical direction is relative. It shows successive levels for the stratigraphic events in the well considered. RASC distances of 2 or more coeval events are averaged (see Figure 5a) before cubic spline fitting (Figure 5b). The user has the option of omitting events for which the second-order differences were anomalously high, showing two asterisks in the RASC normality test. Such anomalous events are then displayed by use of a special symbol (asterisk in Figure 5a) and are not employed for curve fitting. The deviations are measured in the horizontal direction. SF and SD serve the same purpose as in Figure 4. The user may wish to remove other events considered to be anomalous, for example, those labelled R in Figure 5c. Then a new cubic spline-curve will be fitted for the reduced data set (Figure 5d). If extreme values are deleted, SD will probably be decreased in value. The original RASC model is based on the assumption that positions of events in a well are distributed around their expected value, according to a normal probability distribution, with standard deviation set equal to $1/\sqrt{2}$ = 0.7071. We, therefore, would expect SD to be approximately equal to 0.7 if the number of events in the stratigraphic section is sufficiently large.

For further analysis in preparation of automated correlation, RASC distance is replaced by age (see Figure 6a) using the earlier derived relationship between RASC distance and age (see Figure 4d). In the following discussion, the variables for event level, age and depth will be denoted as x, y and z, respectively. A spline curve can be fitted to express y as a function of x, as was done for distance in Figure 5d. It is also possible to replace the levels by their depths and fit a spline curve to express y as a function of z using depth as the independent variable. This would lead directly to a plot similar to Figure 6f.

However, the rate of sedimentation may have changed significantly during geologic time at a well site and this can result in irregular distribution of the points along the z-axis. This, in turn, may make it difficult to obtain a spline curve that extrapolates in a satisfactory manner across data gaps along the z-axis corresponding to short periods with high sedimentation rates. For this reason, the indirect method given in Figure 6 can be employed instead. Suppose that the spline-curve of Figure 6a is written as $y = f\ (\underline{x}) + e_y$ where e_y represents a random deviation in the y-direction. The bar under x indicates that y is regressed on x using data points which are regularly spaced along the x-axis. Depth (z) is plotted against x in Figure 6b, and a separate spline-curve with $z = g\ (\underline{x}) + e_z$ is obtained, using the same set of regularly spaced data points along the x-axis. The deviation e_z points in the z-direction. Obviously, the curve $g\ (\underline{x})$ cannot decrease in the x-direction.

The curve for z in Figure 6b is shown again in Figure 6c. It been rewritten in the form $\hat{x} = g^{-1}\ (\underline{z})$, to indicate that estimates $\hat{x}$

were obtained at points which are regularly spaced along the z-axis. Suppose that $\hat{y}$ is obtained for the irregularly spaced values of x in Figure 6c using f(x) shown, in Figure 6a. This results in a set of $\hat{y}_{xz} = f(g^{-1}(\underline{z}))$ for regularly spaced points along the z-axis (see Figure 6d). The function $f(g^{-1}(\underline{z}))$ is not a simple mathematical expression. For example, its first derivative is not readily available. A cubic spline $\hat{y} = h(\underline{z})$ can be fitted to the values $\hat{y}_{xz}$ (see Figure 6e). In Figure 6e, $\hat{y}$ is considerably smoother than $\hat{y}_{xz}$. By using a smaller smoothing factor (SF), the difference between $\hat{y}$ and $\hat{y}_{xz}$ may be kept negligibly small (see curve to be used for example in Figure 7a). The standard deviation SD for points used for fitting in Figure 6a with respect to the curve $\hat{y}$ is provided in Figure 6f. The deviations from $\hat{y}$ are measured in the y-direction. A similar age-depth diagram is shown in Figure 7a where less smoothing was applied. The spline-curve $\hat{y}$ = h(z) can be used to assign a probable age to any point along the well.

Figure 8 shows a so-called multiwell comparison for five wells. It was based on a 7-2-4 RASC run on 21 wells. Points with estimated ages of 10, 30, and 50 Ma along the five wells are connected by lines of correlation in Figure 8. The uncertainty in the position of these isochron contours is indicated by error bars, constructed according to one of three methods further explained in Figures 7 and 9. The displays of Figures 8a and 8c were redrafted from displays on the Tektronix terminal obtained during an interactive CASC session; the error bars in Figure 8b were obtained from age-depth plots according to the method explained in Figure 7d.

The local error bar in Figure 8a is obtained by multiplying s(y) (=SD) along the y-axis by rate of sedimentation to obtain a modified error s(z) along the z-axis, as shown in Figure 7b. The rate of sedimentation (Figure 7c) is the first derivative dz/dy for z in $\hat{y} = h(z)$. In general, a cubic spline curve y, fitted to n data points, consists of (n-1) successive cubic polynomials

$$\hat{y} = y_i + c_{1i}\, d + c_{2i}\, d^2 + c_{3i} d^3$$

with $d = z - z_i$, $z_i < z < z_{i+1}$, where z_i and z_{i+1} (i = 1, 2,.., n-1) represent the n depths used to convert $\hat{y}_x = f(x)$ into $\hat{y} = h(z)$. The coefficients c_{1i}, c_{2i} and c_{3i} can be used to calculate

$$dy/dz = c_{1i} + 2c_{2i}\, d + 3c_{3i}\, d^2$$

at any point. Inversion of this expression gives dz/dy. The new standard error s(z) = (dz/dy) s(y) can be displayed for any z as the local error bar z ± s(z), (see Figure 8a). This propagation of error is based on local rate of sedimentation, which is assumed to remain approximately constant over the interval y ± s(y). The latter condition is frequently not satisfied, especially when $\hat{y}$ has many inflection points (between

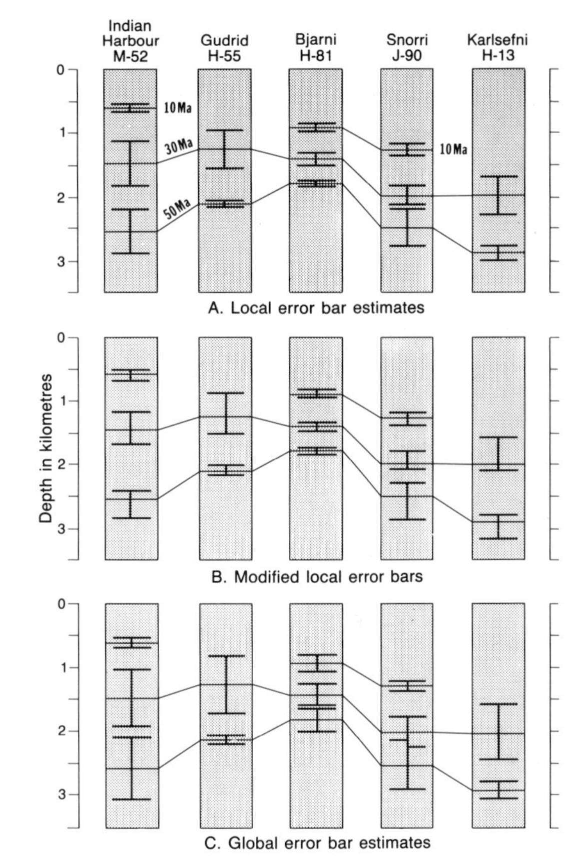

Figure 8

Example of multi-well comparison with error bar. For locations of wells, see Figure 5 in Chapter I.1. Methods used for constructing different types of error bars are illustrated in Figures 7 and 9. Local error bar (A) and global error bar (C) estimates can be displayed on screen during CASC interactive session. Modified local error bar (B) estimates were obtained from final age-depth curves for these five wells.

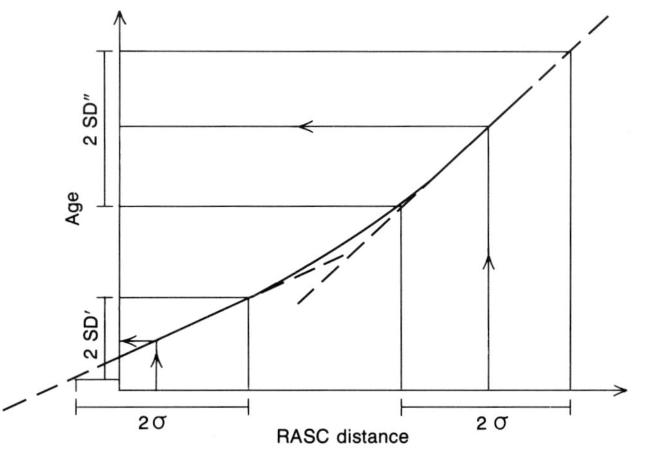

Figure 9 Schematic illustration of estimation of global error bar. Theoretical standard deviation σ(=0.7071) along RASC distance scale is assumed to remain constant. It is transformed into variable SD along age scale (e.g., SD' and SD").

local maxima and minima in sedimentation rate). Curvature of $\hat{y}$ is considered in the construction of a modified local error bar as illustrated in Figure 7d. For any point $z = h^{-1}(y)$, this bar extends from the point $h^{-1}\{y-s(y)\}$ to $h^{-1}\{y + s(y)\}$. It is asymmetrical with respect to z and is significantly shorter at places where the rate of sedimentation is high.

Finally, a global error bar (Figure 8c) can be constructed as illustrated in Figure 9. The standard deviation $\sigma = 1/\sqrt{2}$ of events along the RASC linear scale for distance is changed into a variable standard error s(y) along the age scale. This new variable standard error is changed into s(z) according to the method used for SD in Figure 7b. In global error bar estimation, it is assumed that a single RASC distance error σ can be applied to all wells. On the other hand, in local error bar estimation, use is made of a constant SD value along the age scale which was estimated from the deviations between the points used for spline fitting and the spline curve itself (cf. Figure 6e). Because of elimination of anomalous events and averaging of ages for events at the same levels, the local error bar is likely to be narrower than the global error bar.

The purpose of the error bar is to quantify the uncertainty of the observed depths of events with respect to their estimated depths in the wells. Each local or global error bar extends from the estimated depth minus one standard deviation to estimated depth plus one standard deviation. If the observed depth is normally distributed about the estimated depth, the error bar can be interpreted as a 68 percent confidence interval for single events. Then there is a 68 percent probability that the observed event falls within the range outlined by the error bar. Likewise, there is 95 per cent probability that the observed event falls within an extended error bar, which is 1.96 times as wide as the error bar shown.

The actual precision of the estimated depth of an event in a well is probably much higher than that indicated by the error bar for single events. If a value on the spline-curve would be interpreted as the arithmetic average of all (n) values used for its estimation in a well, the standard deviation of this mean would be equal to the standard deviation of the single events, divided by $\sqrt{n}$. For example, in a well with 16 event levels, the standard deviation of the difference between estimated depth and 'true' depth would be one-fourth the standard deviation of the 16 single events used to construct the error bar. The degree of validity of the assumption that a value on the spline-curve can be interpreted as an arithmetic mean of all values used for estimation is not precisely known, except when the smoothing factor is large. In the limit, which is reached when SF exceeds an upper threshold value, the spline-curve reduces to the best fitting straight line of least squares. Then the preceding assumption is known to hold approximately true. Further research would be needed to estimate the standard deviation of the difference between an estimated value on the spline-curve and its corresponding 'true' depth when the smoothing factor is smaller.

Table 1 *RASC output for 7-2-4 run on 21 wells (Grand Banks - Labrador Shelf) used as CASC input. Event levels (sequence position numbers 1-40) (A), optimum sequence of events identified by their dictionary numbers (B), modified optimum sequence after final reordering (C), and cumulative RASC distances for events in column D.*

A	B	C	RASC Distance	A	B	C	RASC Distance
1	10	10	0.000	21	261	261	4.364
2	17	17	0.391	22	263	263	4.518
3	16	16	0.912	23	40	40	4.645
4	67	67	1.204	24	29	32	4.802
5	21	18	1.647	25	32	29	4.809
6	18	71	1.799	26	41	264	5.175
7	71	21	1.919	27	264	42	5.189
8	20	20	2.108	28	42	41	5.237
9	26	15	2.442	29	30	30	5.251
10	15	26	2.486	30	86	90	5.531
11	70	70	2.513	31	36	36	5.620
12	27	29	3.198	32	90	86	5.667
13	69	25	3.358	33	45	45	5.786
14	24	27	3.418	34	57	57	5.799
15	25	69	3.499	35	50	50	6.302
16	81	81	3.722	36	46	46	6.429
17	259	259	3.738	37	54	54	6.738
18	33	33	4.295	38	56	56	7.178
19	34	34	4.311	39	55	55	7.689
20	260	260	4.342	40	59	59	8.033

Table 2 *Estimated ages for 22 events and calculation of average RASC distances for two or three events with same estimated age.*

Event No.	Age (Ma)	RASC Distance	Average Distance	Event No.	Age (Ma)	RASC Distance	Average Distance
4	3.5	-0.585	-0.385	85	38	4.456	4.456
5	3.5	-0.476		29	40	4.809	4.809
269	3.5	-0.096		90	49	5.531	5.531
17	11	0.391	0.391	57	52	5.799	5.247
179	15	1.984	1.984	93	52	5.895	
15	17	2.442	2.442	50	55	6.302	6.302
26	20	2.486	2.427	194	57	7.073	7.073
137	20	2.368		55	58	7.689	7.434
24	28	3.198	3.198	56	58	7.178	
33	37	4.295	4.017	61	63	8.228	8.039
259	37	3.738		253	63	7.849	

Table 3 *CASC input for Indian Harbour well; definition of 18 event levels; and transformation of RASC distances into ages using spline curve in Figure 10a.*

Event No.	Event Level	RASC Distance	Age (Ma)	Event No.	Event Level	RASC Distance	Age (Ma)
10	1	0.000	5.6	34	12	4.311	37.0
18	2	1.647	15.2	263	13	4.518	38.9
15	3	2.442	21.0	-36	13	5.620	47.9
-20	3	2.108	18.6	29	14	4.809	41.4
-16	3	0.912	10.9	-40	14	4.645	39.9
17	4	0.391	7.9	-41	14	5.237	44.2
24	5	3.198	27.3	-42	14	5.189	44.4
-25	5	3.358	28.7	86	15	5.667	48.2
26	6	2.486	21.4	45	16	5.786	49.1
-27	6	3.418	29.2	-46	16	6.429	53.7
259	7	3.738	32.0	57	17	5.779	49.2
261	8	4.364	37.5	-54	17	6.738	55.6
30	9	5.251	44.9	-50	17	6.302	52.9
260	10	4.342	37.3	55	18	7.689	61.5
-32	10	4.302	41.2	-56	18	7.178	58.4
33	11	4.295	36.9	(59)			

Table 4 *Data used for fitting spline-curves in Indian Harbour well example shown in Figures 10 to 12.*

Event Level	Depth (m)	Average Distance	Average Age	Event Level	Depth (m)	Average Distance	Average Age
1	546	0.000	5.6	10	1912	4.572	39.3
2	619	1.647	15.2	11	2045	4.295	36.9
3	720	1.821	16.8	12	2305	4.311	37.0
4	747	0.391	7.9	13	2335	5.069	43.4
5	1067	3.278	28.0	14	2366	4.970	42.6
6	1232	2.952	25.3	15	2396	5.667	48.2
7	1616	3.738	32.0	16	2671	6.107	51.4
8	1674	4.364	37.5	17	2884	6.280	52.6
9	1732	5.251	44.9	18	3000	7.434	59.9

Examples of CASC Runs

Output from 7-2-4 RASC run on 21 wells and a 5-2-3 run on 24 wells were used as input for examples of actual CASC runs.

Table 1 shows the optimum sequence, modified optimum sequence (after final reordering) and RASC distances for the 7-2-4 run on 21 wells. Several events, occurring in fewer than seven wells, were later inserted as unique events. Table 2 shows estimated ages of 22 events, including these unique events. Average RASC distances for events with the same age are shown in the last column of Table 2. Figure 10a shows the ages plotted against the RASC distances. The displays in Figure 10 (and Figures 11-13) were redrafted from hard copy of displays on a Tektronix terminal. A cubic spline function with smoothing factor SF = 2.0 was fitted to the 15 ages, using the average distances shown in the last column of Table 2. The smoothing factor SF is the standard deviation of differences between the 15 ages and corresponding estimated ages on the spline-curve for the same RASC distance values. The standard deviation of the 22 original ages before averaging of some RASC distances is also shown in Figure 10a.

The original sequence of events occurring in 7 or more wells is shown in Table 3 for Indian Harbour M-52 (Well No. 5), which will be used for further analysis.

As mentioned earlier, one of two different routes can be selected at the beginning of CASC. These consist of using either optimum sequence data or RASC distances for the events. In both subprograms, event levels for successive, non-coeval events are defined, as illustrated for Indian Harbour M-52, in the second column of Table 3. In the second subprogram, the RASC distances in a well are transformed into ages using the spline curve fitted in Figure 10a (see last column in Table 3). The methods used in the two subprograms are identical, except that sequence position numbers instead of ages are used in the first subprogram. As in the previous section, only the option that uses the ages (in Ma) will be illustrated in detail here.

CASC produces a number of successive plots. For each of these plots the user is asked to answer one or more questions. The plot that comes after Figure 10a during a CASC session is shown in Figure 10b. It shows the RASC distances of Table 3 plotted against their event levels. Before this plot is actually shown on the Tektronix screen, the user is asked if he wishes to exercise the option of deleting anomalous events which are out of place with a probability of 99 per cent according to the RASC normality test. Moreover, points can be deleted from Figure 11 itself by positioning the Tektronix cursor on top of them. No points were omitted in this example. Next a cubic spline curve is fitted to the average RASC distance values in the third column of Table 4. First the user is shown the default smoothing factor (SF = 0.5146 for Figure 10b) and asked if this value should be used. This default was obtained automatically, by fitting spline curves with SF increasing from 0.0

until the first curve is found for which the distance does not anywhere decrease with increasing depth (cf. Figure 5 in Chapter I.1). The default solution is shown in Figure 10b. The smoothing factor is the standard deviation of the differences between the 18 average RASC distances and the fitted spline curve. The standard deviation of residuals (=0.5664) representing differences between original RASC distances and fitted spline curve is also given in Figure 10b. It is noted that this value is only slightly less than σ = 0.7 representing the theoretical standard deviation along the RASC scale (see previous section).

Figure 10c shows a new default result, obtained after replacing RASC distance by age. It is possible to inspect the first derivative dx/dy of this graph (Figure 10d). If the slope in the direction of increasing age for the curve in Figure 10c exceeds 10, its values are not displayed in figure 10d. Because the default yields the first monotonically increasing spline curve, normally at least one interval with very high sedimentation rate is introduced with this option. By increasing SF, the user can remove artificially high sedimentation rates.

Figure 10e shows the relationship between depth and event level with fitted spline curve for SF = 0.02. It passes almost exactly through the observed values. After display of this plot, the CASC user has the option of either using this spline curve in conjunction with the age-event level plot of Fig. 10b, or to bypass the indirect procedure by directly fitting a curve to the age-depth diagram in which event levels have been replaced by their depths. The default result for the direct method is shown in Figure 10f.

The result obtained by following the indirect method is shown in Figure 11a for small SF (=0.1). The first derivative corresponding to Figure 11a is given in Figure 11b. The irregularity between 2.2 and 2.3 km in this diagram is due to lack of precision of the approximations, used in the indirect method, to obtain new values on the spline curve (see Figures 6c and 6d). The regular spacing along the depth scale, used for this purpose in CASC, is 50 m. Consequently, irregularities due to lack of precision will not extend for more than 100 m along the depth scale.

Figures 11c to 11f show new results for the indirect method, obtained after changing the value of SF from the default (SF = 3.5751 in Figure 10c) to SF = 4.0 in Figure 11c. During a CASC session, the user is also shown the unsmoothed values of $f\{g^{-1}(\underline{z})\}$, (cf. Figure 6d) connected by straight lines. An example of the latter type of display is Figure 12a which originally appeared during the CASC session just before Figure 11e where SF = 0.1. It is not possible to see differences between the curves of Figures 11e and 12a. However, when SF is enlarged to 1.0, the smoother curve in Figure 12b is generated from Figure 12a. The rate of sedimentation for Figure 12b is shown in Figure 12c.

Figure 12d represents the depth versus event level curve that replaces Figure 10e, when SF = 0.00 instead of SF = 0.02 is selected.

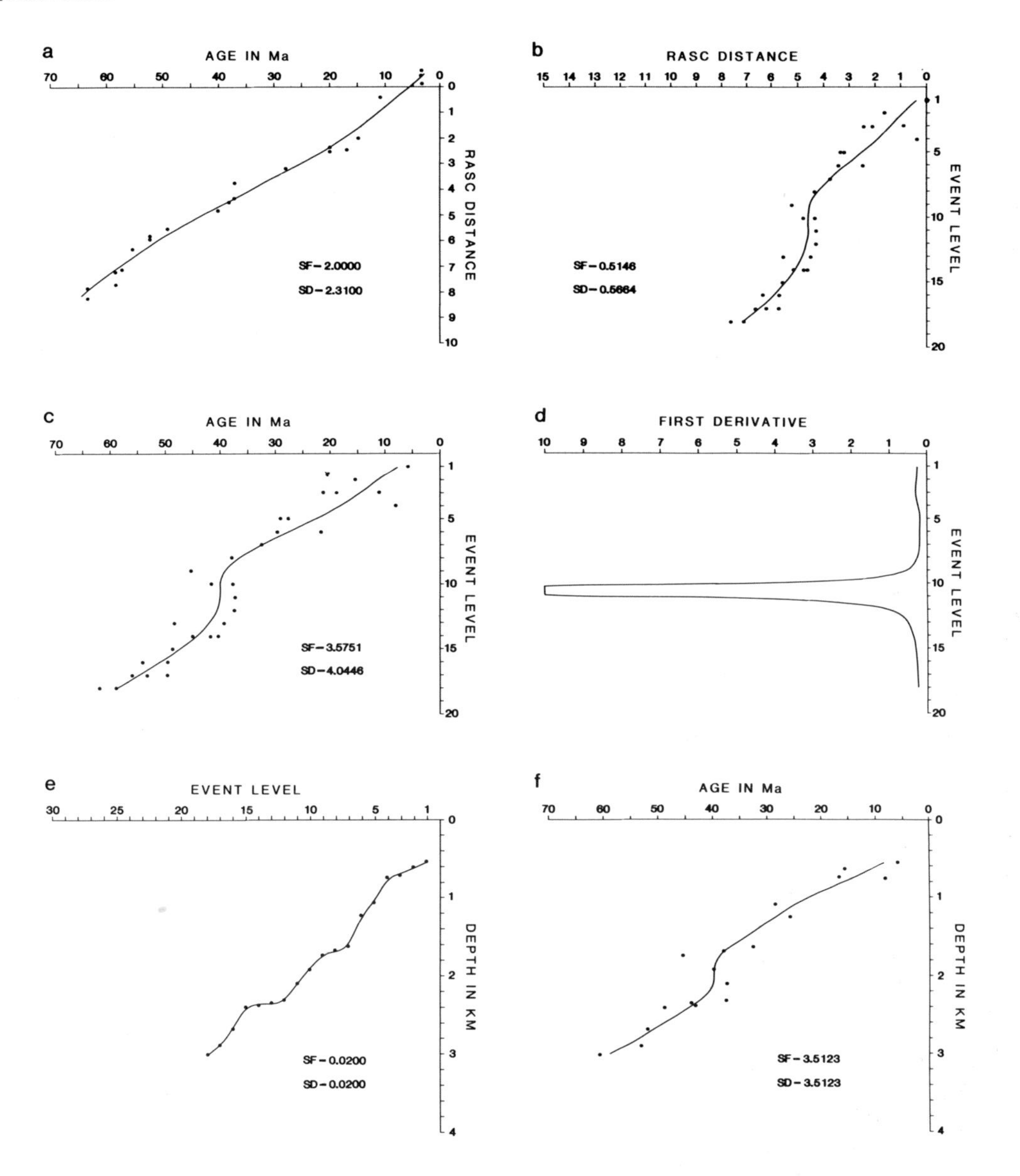

Figure 10 Example of CASC displays for Indian Harbour well based on 7-2-4 RASC results for 21 wells.
(a) Age-RASC distance relationship as derived from the 21 wells file, using method explained in Figure 4;
(b) Initial CASC plot for default smoothing factor;
(c) Age-level plot for default SF;
(d) First derivative of (c);
(e) Level-depth plot; and
(f) Age-depth plot for default SF; spline curve was fitted directly to the data, using irregularly spaced depths.

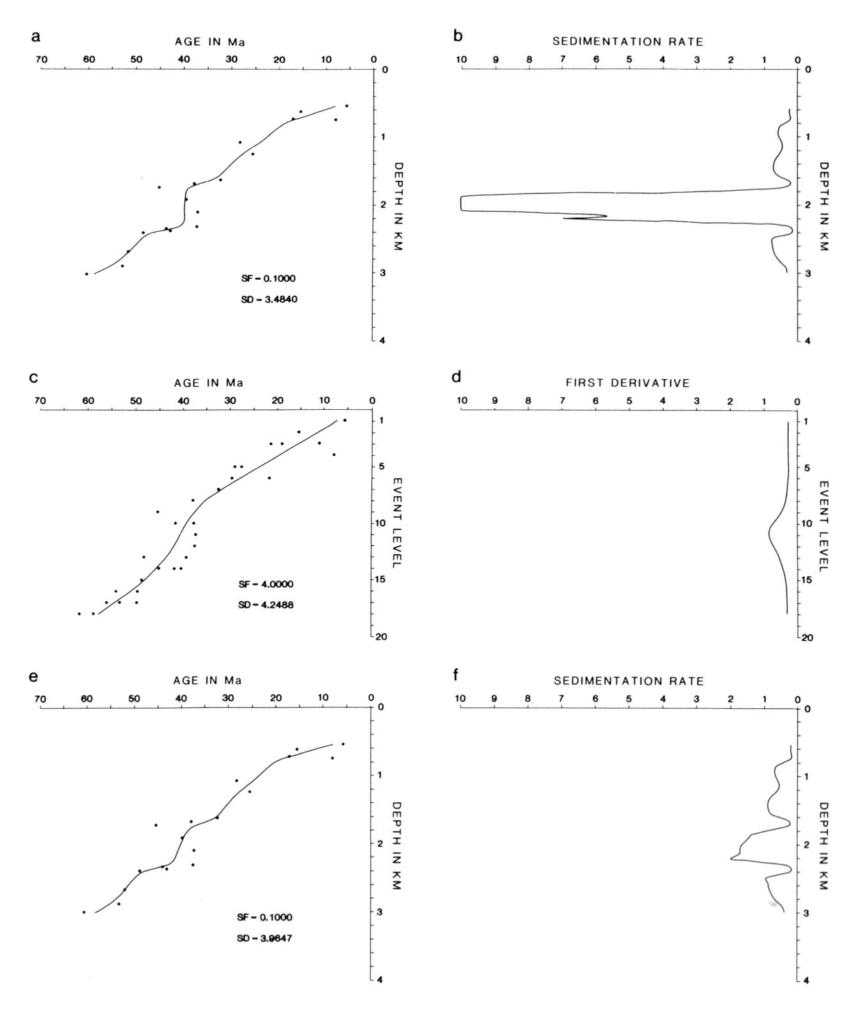

Figure 11 Example of CASC displays for Indian Harbour well (continued from Figure 10).

(a) Spline curve for small (default) SF fitted to combination of Figures 10c and 10e; indirect method explained in Figure 6 was used;

(b) Sedimentation rate in 0.1 km/my (= first derivative of spline curve in Figure 11a multiplied by 10); local maximum and minimum are due to lack of smoothness of spline curve as explained in text;

(c) Age-level plot for SF = 4.0 instead of default, used in Figure 10c;

(d) First derivative for Figure 11c; magnitude of peak in Figure 10d has been reduced;

(e) Spline curve for small (default) SF fitted to combination of Figures 10e and 11c; and

(f) Sedimentation rate in 0.1 km/my corresponding to Figure 11e.

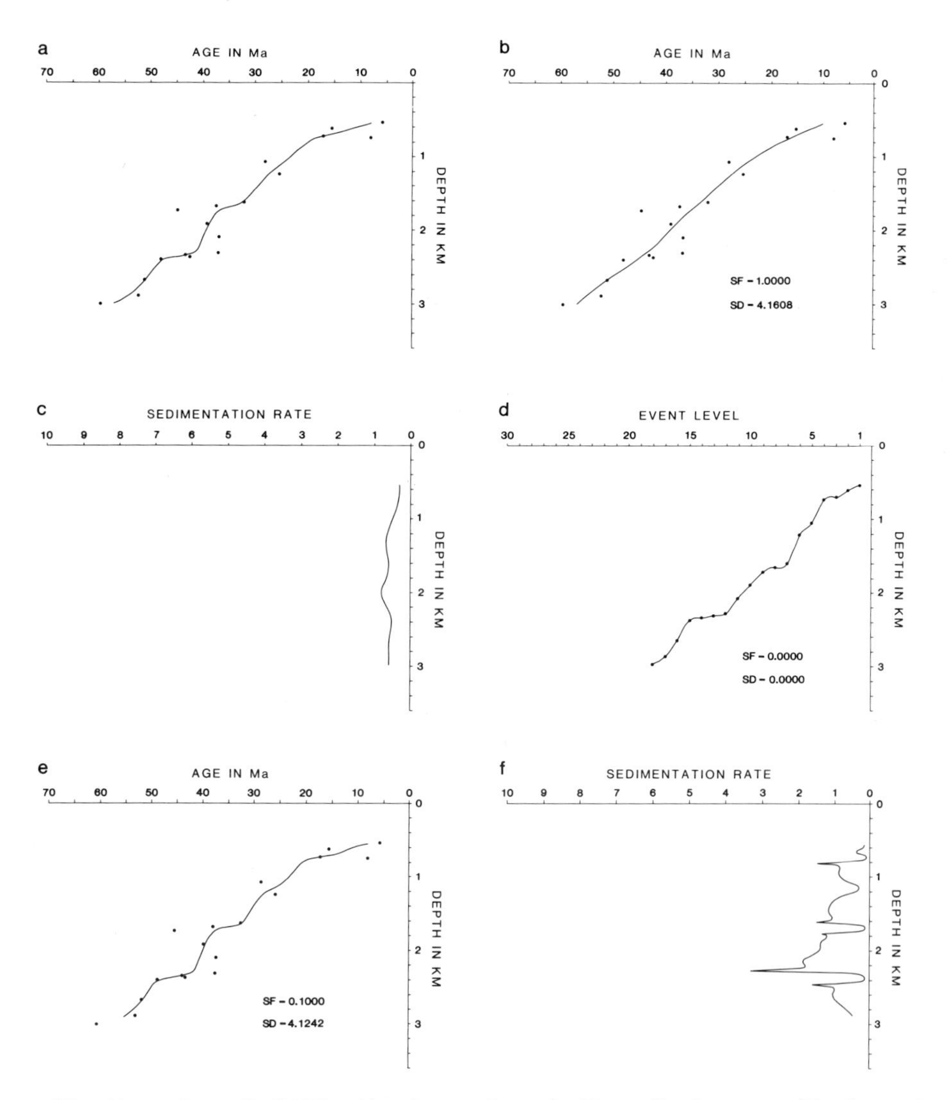

Figure 12 Example of CASC displays for Indian Harbour well (continued from Figures 10 and 11).
(a) Unsmoothed combination of Figures 10e and 11c; note similarity with spline curve in Figure 11e for SF = 0.1;
(b) Curve of Figure 12a smoothed with SF = 1.0;
(c) Sedimentation rate in 0.1 km/my corresponding to Figure 12b;
(d) Level-depth plot for SF = 0.0;
(e) Spline curve for small (default) SF fitted to combination of Figures 11c and 12d; note similarity with spline curve in Figure 11e; and
(f) Sedimentation rate in 0.1 km/my corresponding to Figure 12e; local maxima and minima are due to lack of smoothness of spline-curve as explained in text.

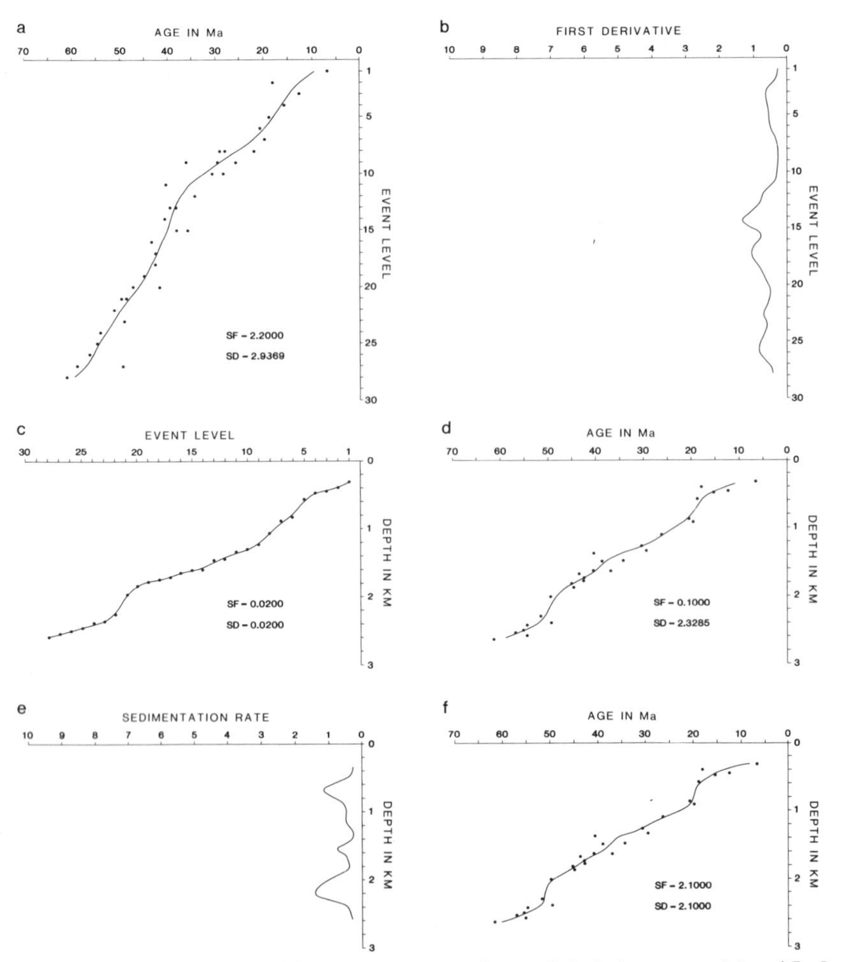

Figure 13 *Example of CASC displays for Adolphus well (5-2-3 RASC results using 24 wells).*

(a) Age-level plot for SF = 2.2;

(b) First derivative corresponding to Figure 13a; note small peak near level 14;

(c) Event-depth plot; note relatively steep slopes at depths near 0.7 km and 2.2 km, respectively;

(d) Spline curve with small (default) SF fitted to combination of Figures 13a and 13c;

(e) Sedimentation rate in 0.1 km/my corresponding to Figure 13d; two relatively high peaks correspond to steeper slopes in Figure 13c; intermediate small peak corresponds to highest first derivative in Figure 13b; and

(f) Age-depth spline curve fitted directly to the data using irregularly spaced depths; note similarity with spline curve of Figure 13d; direct method yields poorer results than indirect method when one or more intervals between successive ages are much larger than average, due to high sedimentation rate or relative lack of microfossils.

The difference between these two curves is small and when the curve of Figure 12d is combined with that of Figure 11c, the resulting plot (Figure 12e) does not differ significantly from Figure 11e. However, the first derivative of Figure 12e which is shown in Figure 12f differs significantly from Figure 11f. It shows many 100 m irregularities, which are due to lack of precision as also discussed before (see Figure 11b). In general, the final age-depth curve is less sensitive than the sedimentation rate curve to small changes in the choice of smoothing factors for successive curves during an interactive CASC session.

As a final example, Figure 13 contains various CASC displays for the Adolphus D-50 well. The input for this CASC run was output from a 5-2-3 RASC run on 24 wells. As shown in Figure 13a there are as many as 39 events on 28 levels in this well so that there is good control in the vertical direction. The first derivative of the spline-curve for SF = 2.2 (Figure 13a) remains fairly constant. It has its largest value at event level 14 (see Figure 13b). This indicates the place where the spline-curve in Figure 13a has its steepest dip. The pattern of Figure 13c suggests that rate of sedimentation was above average between events 6 and 7 and also between events 21 and 22. These two maxima also occur in Figure 13e which is the first derivative of the age-depth curve (Figure 13d), obtained by combining the spline-curves of Figures 13a and 13c with one another using the indirect method. The smaller peak in Figure 13e, which occurs at a depth of about 1600 m, represents the place (level 14) where the curve of Figure 13a has its steepest dip. The same three intervals with relatively steep slopes in the age-depth spline-curve can be observed in Figure 13f, which resulted from applying the direct method to the age-depth values. Without corroboration from other wells drilled in the immediate vicinity it may not be possible to decide with certainty whether or not small fluctuations in the rate of sedimentation, as shown in Figure 13e, are significant. Increased smoothing in the age-depth diagram (Figure 13d) would change the pattern of Figure 13e much more drastically than the pattern of Figure 13d itself.

As was illustrated in more detail for Indian Harbour M-52 in Figures 10 to 12, minor smoothing in Figure 13a for Adolphus D-50 would, in a multiwell comparison, only slightly change the position of isochrons in Adolphus D-50. However, the widths of the error bars are proportional to rate of sedimentation in both local and global error bar estimation and these widths would change drastically if smoothing is increased. This is because rate of sedimentation does depend strongly on choice of smoothing factors.

APPLICATION OF CASC

The geological use of the CASC method and its value in sedimentary basin analysis will be illustrated by means of examples drawn from exploration micropaleontology. The examples are based on the original distribution of 168 Cenozoic foraminifera in 21 Grand Banks and Labrador wells (Chapter II.4), and of 116 taxa of Mesozoic foraminifera in 16 Grand Banks wells. The latter file was largely prepared by Williamson (in press). The Cenozoic file is printed in Appendix I, with the one difference that in well Bjarni H-81, taxa 54 and 55 (**Gavelinella beccariiformis** Zone, Paleocene) have now been added. Both taxa were observed at a depth of 6660'.

The discussion will touch on the following question:

(1) What is the stratigraphical meaning of CASC-type of correlation?

(2) What is the degree of confidence, expressed in depth and in linear time units of CASC correlation, and to what extent is the error bar useful for geological interpretation?

(3) What is the difference between subsidence and sedimentation models that rely on subjective age-depth data, or more objective RASC-CASC types of age-depth interpretations?

In order to find answers to these questions, we will show conventional subjective and more objective CASC-type correlation of wells based on three related stratigraphic criteria:

(a) selected zone markers;

(b) assemblage zones; and

(c) isochrons in Ma.

In all cases, the underlying biostratigraphical zonations were defined with the RASC method, but in principle CASC can be applied to non-RASC zonations based on biostratigraphical events.

Firstly, we will trace ten selected zone markers through six wells. Starting point is the Cenozoic optimum sequence (Figure 14; 21 wells, 7-2-4 run). For the interactive spline fitting of the bivariate plots, all defaults were accepted, unless otherwise specified.

The default is the smoothing factor (SF) that defines the spline curve for which an increase in position or depth along one axis does not anywhere correspond to a decrease in position (or time) along the other axis. The default is obtained by means of an algorithm that calculates spline curve fits with SF increasing regularly, starting at SF = 0. The default satisfies the condition that the observed sedimentation rate is never negative.

In three wells the cursor option was used to delete aberrantly positioned events, including:

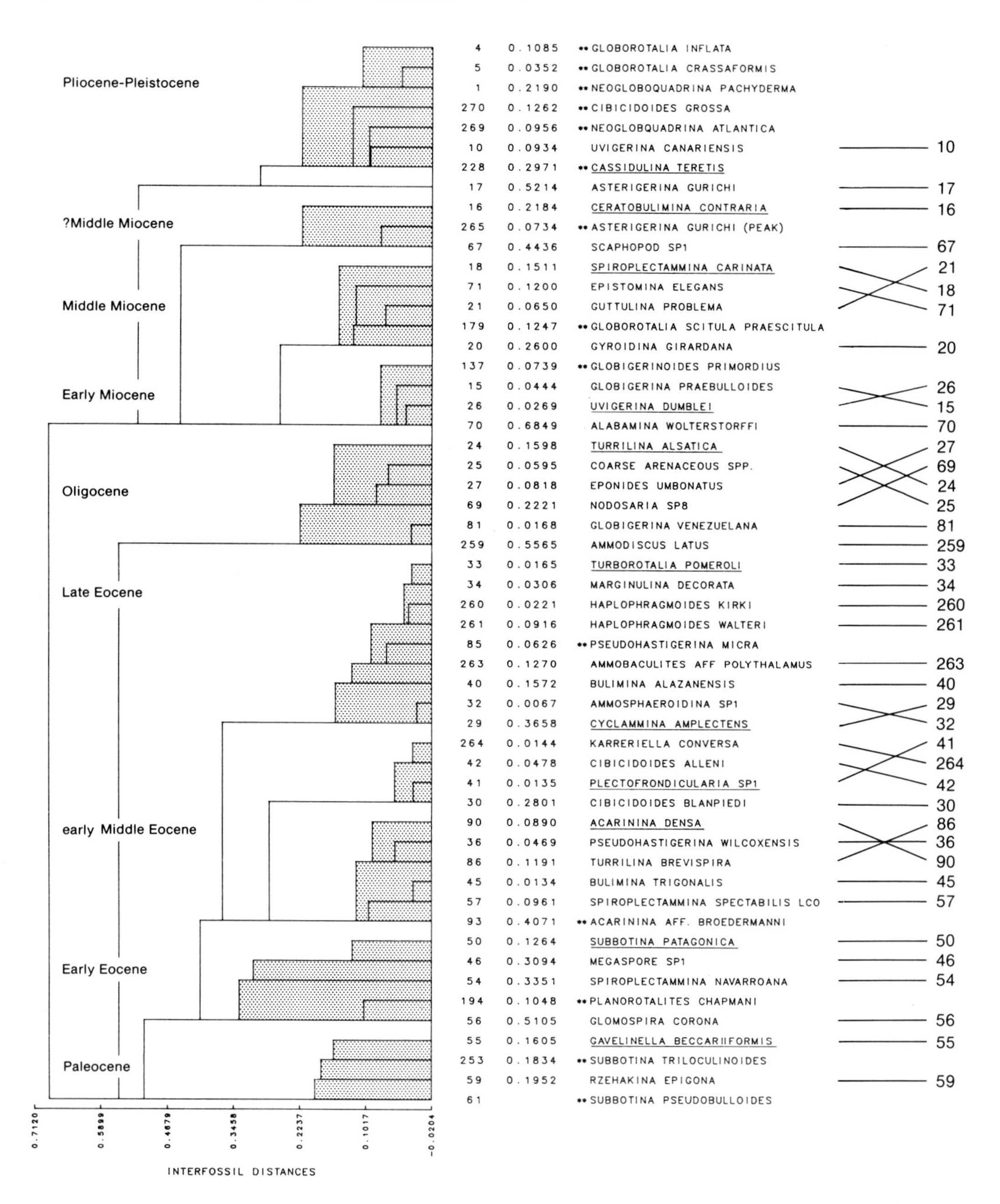

Figure 14 Cenozoic optimum sequence and scaled optimum sequence based on last occurrences of foraminifera in 21 wells, Labrador - Grand Banks (7-2-4 run). This is the same zonation as presented in Chapter II.4.

Table 5 *Observed (above line) and most likely depth (in m) of ten Eocene through Miocene zone markers in six wells. The fossil numbers are the RASC dictionary numbers. Results are based on optimum sequence CASC (21 wells; $k_c \geq 7$, $m_{c1} \geq 2$, $m_{c2} \geq 4$); * means that at that site substitute fossils (neighbours in the optimum sequence) were used.*

		Hibernia P-15	Adolphus D-50	Bonavista C-99	I. Harbour M-52	Bjarni H-81	Snorri J-90
Ceratobulimina contraria	- 16	310	485	887	750	872	1262
			350 ± 85	1130 ±381	571 ±132		
Spiroplectammina carinata	- 18	275	512	2030	649	1085	1701
		334 ± 65	481 ± 78	1681 ±363	673 ±126	1025 ± 18	1533 ±167
Uvigerina dumblei	- 26(15)	550	933	2377	1261		
		619 ±330	726 ±228	2059 ±162	786 ±155*	1094 ± 14	1756 ± 42
Turrilina alsatica	- 24	1075	1280	2377	2097	1298	
		1000 ± 75	1164 ±115	2372 ±13	1344 ±532	1307 ± 31	1910 ± 29
Ammodiscus latus	- 259	1125	1153	2316	1646		
		1083 ± 45	1291 ± 71	2571 ±135	1655 ±147	1490 ± 27	1983 ± 26
Haplophragmoides walteri	- 261(260)	1185	1509	3078	1704	1634	213
		1176 ± 40	1471 ±171	2889 ±215	1864 ±404*	1577 ± 16	2067 ± 22*
Cyclammina amplectens	- 29	1195	1890	3109	2396	1634	
			1721 ± 95	3099 ± 17	2212 ±489	1635 ± 14	
Ammosphaeroidina sp. 1	- 32	1125	1761	3386	1935	1695	2112
			1767 ± 62	3144 ± 99	2275 ±266	1653 ± 15	2155 ± 27
Acarinina densa	- 90		2062	3478			
			2293 ±249	3423 ± 96	2528 ±497	1763 ± 8	2651 ± 59
Subbotina patagonica	- 50(56)		2517		2914	2009	2932
			2501 ± 66		2861 ±327*	1812 ± 15	2798 ± 51*

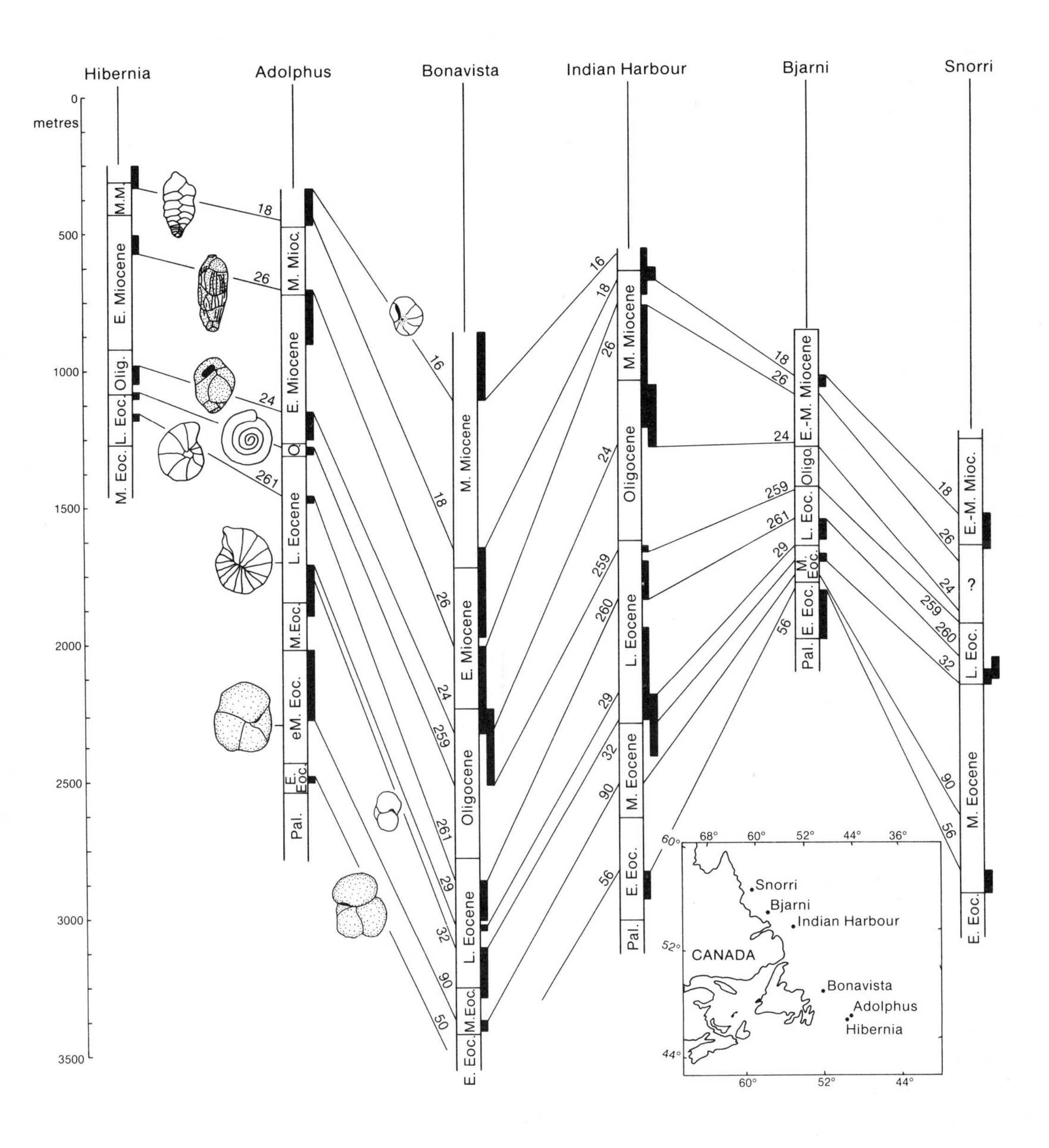

Figure 15 *Tracing of ten foraminiferal events through six wells, using the CASC (optimum sequence) method to calculate the most likely depths. Black bars show the deviation of these depths to the observed ones. The chronostratigraphic segmentation is based on observed depths only.*

Well 21 - Hibernia P-15

one point was deleted on level 12.

Well 11 - Bonavista C-99

three points were deleted at levels 4, 11 and 5 and SF for the events versus depth graph was changed from 0.02 to 0.15.

Well 9 - Snorri J-90

one point was deleted at level 6.

The results of the so-called multiwell comparison are shown in Table 5, listing both the observed and the most likely depth of the ten Paleocene through Miocene zone markers 50 (56), 90, 32, 29, 261 (260), 259, 24, 26(15), 18 and 16. In two wells substitute taxa were correlated rather than the three designated events 50, 261, and 26. The substitutes 56, 260, and 15 are neighbours of the original events in the optimum sequence.

In most instances, the observed and the most likely depth values are within half the length of the errorbar (68% probability) around the most likely value. As pointed out in the previous section, the actual precision of the estimated depth of an event in a well is probably much higher than that indicated by the local error bar for single event positions along the spline curve. Also, the local error bar at any depth is initially calculated over the time interval along the (scaled) optimum sequence scale (y), as defined by twice the standard deviation (SD) in that (y) direction. It is directly proportional to the fitted average sedimentation rate for each point (Figure 7).

Because this average sedimentation rate may frequently change through time, significant curvature may occur over the span of the original 2SD in the scaled optimum sequence (y) direction. It may therefore be better to use the so-called modified local error bar that does not, like the local error bar, calculate uncertainty by extrapolating a constant average sedimentation rate across the entire length of the error bar. Strictly speaking, the slope of the spline fit in the optimum sequence versus event depth does not represent sedimentation rate in a well, but the principle of this discussion on the modified local error bar stays valid. This new error bar estimate was not automatically obtainable at the time of writing, and only the local and global error bar options of the CASC program were used for the applications discussed below.

Figure 15 graphically correlates the ten events through the six wells. The conventional chronostratigraphic segmentation, which is shown for comparison, only uses the observed depth of events. The new, most likely, zone marker depths would lead to slight up or down adjustments of the age boundaries. It could be assumed that such a change might violate stratigraphic boundaries as adjusted for major lithology changes determined from well logs. However, using sonic and gamma logs, no evidence for this was found. In the Snorri well there is no foraminiferal evidence for the presence of events 259, 24 and 26, associated with Oligocene-Early Miocene strata, although the CASC method predicts their likely depth in this well. These depths are not unreasonable, given that Oligocene strata were thought to be present at that depth, based on spores and dinoflagellates.

The next example, shown in Table 6 and in Figure 16, correlates the RASC zones of **Spiroplectammina carinata** (events 18, 71, 21 and 20) and of **Turborotalia pomeroli - Cyclammina amplectens** (events 260, 261, 263, 32 and 29 only), Late Eocene. The six wells are as in the previous example, but now the distance CASC option is employed for 21 wells (7-2-4 run; Figure 14).

A summary of the run is as follows:

Well Hibernia P-15

deleted one point at level 12;
SF in RASC distance versus event scale plot is 0.35;
SF in RASC distance versus depth scale plot is 0.15 (default is 0.10).

Well Bonavista C-99

SF in events versus depth plot is 0.10 (default is 0.02);
SF in RASC distance versus depth plot is 0.15 (default is 0.10).

Well Indian Harbour M-52

SF in age versus depth plot is 0.15 (default 0.10).

Well Bjarni H-81

SF in age versus depth plot is 0.15 (default 0.10).

Well Snorri J-90

deleted one point at level 6.

Table 6 is a listing of the most likely depths in the six wells of taxa that characterize the two RASC zones. The error bar is the local one and with few exceptions shows that there is limited uncertainty in depth. The global error bar combines error in relative time of the events in the scaled optimum sequence (RASC distance scale) with the error in position of the events along the well (depth) scale. In few cases, shown with (*) this error approaches 20-30% or more of the actual depth value, particularly in Hibernia P-15.

Table 6 *Most likely depth (in m) of events belonging in the RASC zones of* ***Spiroplectammina carinata,*** *Miocene, and* ***Turborotalia pomeroli - Cyclammina amplectens,*** *Eocene, as calculated in six wells.*

		Hibernia P-15	Adolphus D-50	Bonavista C-99	I. Harbour M-52	Bjarni C-99	Snorri J-90
Spiroplectammina carinata Zone-Middle Miocene							
Spiroplectammmina carinata	- 18	325 ± 64*	485 ± 57*	1401 ±264	707 ± 73	1012 ± 15	1476 ±141
Epistomina elegans	- 71	347 ± 76*	512 ± 63*	1467 ±278	726 ± 76	1022 ± 14	1550 ±149
Globigerina praebulloides	- 21	371 ± 91*	535 ± 72	1522 ±296	742 ± 82	1029 ± 14	1603 ±116
Gyroidina girardana	- 20	449 ±154*	615 ±118	1696 ±475*	793 ±130	1050 ± 18	1684 ± 62
Turborotalia pomeroli - Cyclammina amplectens Zone - Late Eocene							
Ammodiscus latus	- 259	1075 ± 53	1276 ± 61	2579 ±197	1536 ±214	1476 ± 15	2000 ± 30
Haplophragmoides kirki	- 260	1170 ± 51	1409 ±111	3048 ±141	1670 ± 85	1544 ± 23	2077 ± 23
Haplophragmoides walteri	- 261	1174 ± 53	1424 ±175*	3055 ±136	1702 ± 90	1547 ± 25	2089 ± 23
Ammobaculites aff. polythalamus	- 263	1206 ± 83	1645 ±154	3106 ±144	1749 ±189	1584 ± 48	2097 ± 23
Ammosphaeroidina sp. 1	- 32		1757 ± 63	3172 ±107	2307 ± 87	1638 ± 25	2131 ± 26
Cyclammina amplectens	- 29		1761 ± 62	3176 ±107	2310 ± 83	1641 ± 25	2133 ± 26

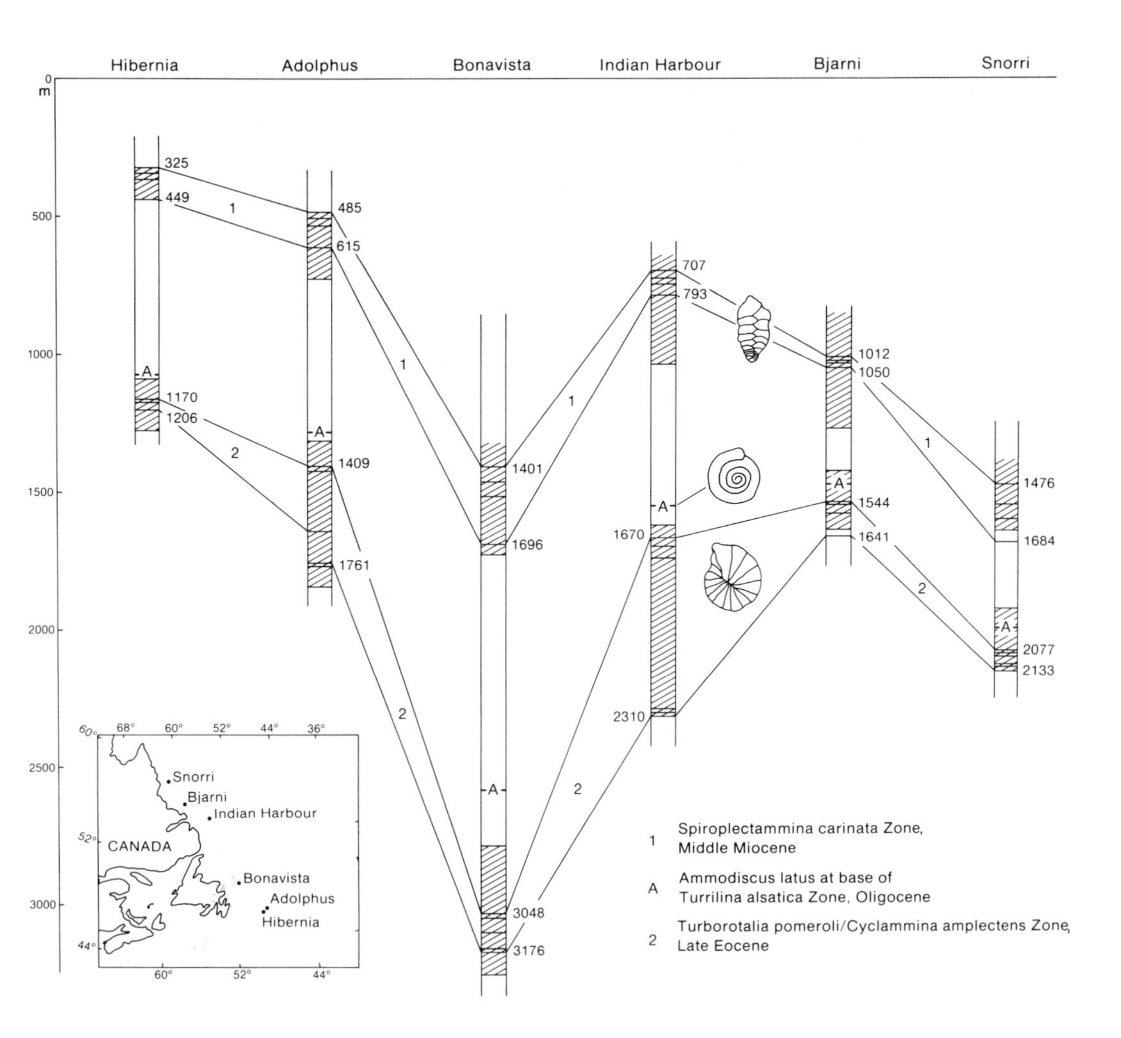

Figure 16 *Correlation of the RASC zones of* ***Spiroplectammina carinata*** *(taxa no.'s 28, 71, 21 and 20), Miocene, and of* ***Turborotalia pomeroli - Cyclammina amplectens*** *(taxa no.'s 260, 261, 263, 32 and 29), Eocene, in six wells. CASC depths for the oldest and youngest of the used zone fossils are shown in meters. Limits over which the zones would extend using the subjective, observed events depths are shown hatched. A - is the most likely depth of* ***Ammodiscus latus,*** *at the base of the RASC zone of* ***Turrilina alsatica,*** *Oligocene.*

Tracing of the two zones through the well sections (Figure 16) determines the most likely intervals of Upper Eocene and Middle Miocene strata. The same figure displays subjectively estimated thicknesses of these strata. In the Indian Harbour and Bjarni wells, the subjective analysis was not able to differentiate between the **Uvigerina dumblei** and **Spiroplectammina carinata** zones; hence, Lower and Middle Miocene were lumped. This explains why the objective thickness of Middle Miocene may seem thin. We also determined in the six wells the most likely depth of **Ammodiscus latus** (A) which occurs in the 7-2-4 run (Figure 14) at the base of the **Turrilina alsatica** zone (Oligocene). The top Eocene is below the level of A, and above the level of event 260 (**Haplophragmoides kirki**). The CASC results are entirely reasonable and define depth intervals in which the two zones are most likely to occur. The use of CASC for the tracing of zones and for stratigraphic segmentation of exploration wells, will lead to correlation of intervals in which a stratigraphic boundary is most likely to fall. The latter may then be compared to discrete wireline log parameters reflecting physical changes in the sediments.

The conversion of the scaled optimum sequence to a (local) biochronology enables the stratigrapher familiar with CASC to trace isochrons in the same way as zones were traced. The exercise starts with the designation of numerical ages in Ma to those events in the scaled optimum sequence for which literature based ages are available. In all, 23 events were dated this way, as explained below. The time scale is that of Berggren et al. (in press). The regional use of the standard planktonic zonation follows Gradstein and Srivastava (1980), and Gradstein and Agterberg (1982):

(1) 63 Ma - events 253 and 61 - **Subbotina triloculinoides** and **S. pseudobulloides** - two rare events that mark approximately the end of Danian time.

(2) 58 Ma - event 55 - **Gavelinella beccariiformis**; occurs up to standard zone P5 (Tjalsma and Lohmann, 1983), which fits well with its disappearance in the Adolphus well together with **Aragonia velascoensis** (Paleocene) and below the appearance of **Pseudohastigerina** (post P5).

(3) 57 Ma - event 194 - **Planorotalites chapmani**; disappears in standard zone P6. Specimens are often transitional between **P. chapmani** and **Pseudohastigerina**. The latter is thought to appear at the boundary of P5 and P6, or ± 57 Ma ago.

(4) 55 Ma - event 50 - **Subbotina patagonica**; this species is frequent in the Ypresian of Belgium (Muller and Willems, 1981), in the Rosnaes Clay of Denmark and in the Lower Eocene of the North Sea and Labrador Sea. The end of the **S. patagonica** peak occurs at the boundary of NP11/NP12, which coincides with the boundary of the **Morozovella formosa formosa** Zone, at the time of Anomaly 24, just after 55 Ma.

(5) 52 Ma - event 93 - **Acarinina broedermanni**; the species has its top well below **A. densa,** probably in the **A. pentacamerata - Hantkenina aragonensis** Zone, near the Early-Middle Eocene boundary at 52 Ma. In some RASC runs, **A. broedermanni** falls between Early and Middle Eocene zones.

(6) 49 Ma - event 90 - **Acarinina densa;** this is the time of the optimum climatic warming in the Labrador Sea, in early Middle Eocene time. Less common at this time are **A. senni, A. aff. pentacamerata, A. aff. broedermanni, Morozovella caucasica, M. spinulosa,** and **M. aff. aragonensis.** The event probably falls in the **Hantkenina aragonensis - Globigerinatheka subconglobata** Zone at Anomaly 21 time or 52-46 Ma, (average 49 Ma).

(7) 40 Ma - event 29 - **Cyclammina amplectens;** in RASC runs this event falls below **Turborotalia pomeroli** and **Globigerina yeguaensis** and above **Acarinina densa.** In Poland its peak occurrence is in so-called Middle Eocene; it is less frequent in Upper Eocene strata (Gradstein and Berggren, 1981). The event is tentatively placed at 40 Ma.

(8) 38 Ma - event 85 - **Pseudohastigerina micra;** same reasoning as for **Turborotalia pomeroli** (see later), but often disappears in slightly older beds, as shown also in scaling (Figure 14).

(9) 37 Ma - event 33 - **Turborotalia pomeroli;** co-occurs in southern wells with **Subbotina linaperta, Globigerina yeguaensis** and **Pseudohastigerina micra,** of the **Turborotalia cerroazulensis** Zone, late Late Eocene. The top is placed just below the inferred Eocene/Oligocene boundary.

(10) 28 Ma - event 24 - **Turrilina alsatica;** the top of this distinctive Oligocene taxon roughly equates with the top of the Boom Clay in Belgium and the top of the **Globorotalia opima opima** Zone, at ± 28 Ma.

(11) ± 20 Ma - event 26 - **Uvigerina dumblei;** slightly older than **Globigerina praebulloides** (see later).

(12) 20 Ma - event 137 - **Globigerinoides primordius trilobus;** rare Early Miocene event.

(13) 17 Ma - event 15 - **Globigerina praebulloides;** disappears locally with **Sphaeroidinella seminula** and with or just below **Globorotalia scitula praescitula,** which may equate with the **G. fohsi peripheroronda** Zone, early Middle Miocene of Scotian Shelf wells (Gradstein and Agterberg, 1982). The RASC runs (7-2-4) indicate an average disappearance in the **Uvigerina dumblei** Zone. Its local extinction is placed between 14 and 20 Ma (average 17 Ma).

(14) 15 Ma - event 179 - **Globorotalia scitula praescitula;** probably occurs in the late Early to early Middle Miocene warming event, as observed from the northern incursion of warmer water planktonic taxa.

(15) 3.5 Ma - events 266, 4, 269 and 5 - Both **Globorotalia puncticulata, G. inflata. G. crassaformis,** and **Neogloboquadrina atlantica** are thought to disappear with the onset of major glaciation in the Labrador Sea, dated at approximately 3.5 Ma.

Four other events occur at or near significant breaks in the 5-2-3 and 7-2-4 scaling solutions for 21 and 24 wells. These breaks were equated with zonal boundaries and series breaks as follows:

58 Ma - event 56 - **Glomospira corona;** Paleocene-Eocene boundary on (upper) continental margin wells.

52 Ma - event 57 - **Spiroplectammina spectabilis** LCO; Early-Middle Eocene boundary; LCO = Last Common Occurrence.

37 Ma - event 259 - **Ammodiscus latus;** Eocene - Oligocene boundary.

11 Ma - event 17 - **Asterigerina gurichi;** Middle-Late Miocene boundary.

Figure 10a is a plot of the ages of the previously listed events in a RASC distance scale (21 wells, 7-2-4 run) versus linear time scale. Smoothing of the spline curve function diminishes some of the uncertainty in subjective assignment. The spline function can now be used to convert the RASC distance scale into an age scale. This is the CASC regional biochronology, which is shown in Figure 7 of Chapter V.1.

Next, the question can be asked what is the most likely depth in the wells of the principal boundaries between RASC zones, expressed in Ma. We have traced the boundaries between the successive Cenozoic RASC zones (Figure 14), which are close approximations to the boundaries between Paleocene and Eocene (≈56 Ma), Early Eocene and early Middle Eocene (≈52 Ma), early Middle Eocene and Middle Eocene (≈49 Ma), Middle and Late Eocene (≈44 Ma), late Eocene and Oligocene (≈36 Ma), Oligocene and Miocene (≈24 Ma), Early and Middle Miocene (≈16 Ma), and top of Middle Miocene (≈12 Ma).

Table 7 lists most likely and subjectively determined (as far as known) depths for these isochrons. The same results are plotted in Figure 17, with the wells arranged latitudinally (48° - 58° N). The CASC depths are from a batch run that accepted all SF defaults. Although it yields more crude results than interactive runs, this procedure takes less time and the actual depth estimates are not influenced much. As explained in the mathematical section of this Chapter, the choice of SF has much more influence on the average rate of sedimentation and hence on the error estimation, than on the actual depth of the isochrons. In a few instances, default smoothing yielded unacceptably steep spline fits, and the local error bar estimate was deleted. In one well, Karlsefni H-13, both foraminifers and palynomorphs agree on the absence of Oligocene beds (**Turrilina alsatica** Zone). Batch CASC calculates a thin Oligocene interval (24 - 36 Ma), where in an interactive version of CASC the recently developed unconformity option should have been used. However, above the Eocene the well has only a few data points and results are crude.

The local error estimates of the most likely depths for the isochrons are within 1 to 10% of the actual depth values, and more frequently 2 to 5%. In about ten cases the subjectively assigned depths for the zonal boundaries, as converted to isochrons, are outside the 68% confidence limits (±1 SD). For geological interpretations, it should be borne in mind that the error in most likely depth is an upper limit, and that the SD is probably smaller by a factor which is, amongst others, related to the number of observations per spline curve, as explained earlier. Palynologically determined depths for these stage boundaries often are outside the depth interval (most likely depth ±1 SD), calculated by CASC. The errors in this independent biostratigraphic exercise are unknown, but the comparison suggests that multiple biostratigraphy uncertainties exceed the CASC-type of errors using one fossil discipline only. The conclusion may be drawn that the CASC program is able to predict reliable and objective well to well isochrons. The error expression which, hitherto, in conventional, subjective, correlation schemes was mysteriously vague, is conservatively large when one fossil discipline only is used.

As a further test of the use of CASC for subsurface correlation, Williamson (in press) applied RASC and CASC to Mesozoic strata of the northern Grand Banks. This author edited existing micropaleontological data and considerably expanded the file to include close to 20 wells, centered on the Hibernia oil field. Williamson determined that eleven RASC zones of Kimmeridgian through Cenomanian age are able to account for most of the biostratigraphic signal. The results are based on last occurrences of over 50 foraminiferal taxa that occur in more than four wells (4-2-3 run). Eleven taxa were defined as unique events. The eleven RASC zones supersede conventional biostratigraphical analysis based on limited data. Figure 18 presents Early Cretaceous CASC isochrons for the wells on and around the oil field. Solid lines represent subjective age interpretation; the dashed lines are the CASC ages. Both interactive and batch runs were made to arrive at the best possible results. The closeness of the subjective and the objective tielines provides another example of model verification.

A common application of exploration micropaleontological data involves elucidation of the sedimentation and subsidence history of a geological basin. A combination of geothermal and subsidence histories yields clues to the potential for oil and gas generation and entrapment.

Table 7 *Observed (above line) and most likely depth (in m) of the 56, 52, 49, 44, 36, 24, 16 and 12 Ma isochrons in 10 wells on the Grand Banks and Labrador Shelf. Results are based on scaled optimum sequence or distance - CASC (21 wells; $k_c \geq 7$, $m_{c1} \geq 2$, $m_{c2} \geq 4$).*

		Hibernia P-15	Fl. Foam I-13	Adolphus D-50	Bonavista C-99	Dominion O-23	I. Harbour M-52	Gudrid H-55	Bjarni H-81	Snorri J-90	Karlsefni H-13
12 Ma	Middle Miocene	330	576	485	585	859	676	613	871	1261	
		262 ± 42	382 ± 92	385 ± 44	521 ± 45	986 ± 46	638 ± 82	528 ± 65	952 ± 53	1288 ± 88	
16 Ma	Early Miocene	460	713	762		1767				1618	
		344 ± 77	502 ±114	508 ± 56	584 ± 84	1636 ± 56	723 ± 68	619 ±117	1019 ± 15	1546 ±159	
24 Ma	Oligocene	960	960	1310	1335	2320	1097	1066	1298		2060
		813 ± 71	809 ±147	935 ± 94	1041 ±306	2254 ± 78	1072 ±164	1025 ±241	1041 ± 20	1826 ± 43	1977 ± 40
36 Ma	Late Eocene	1125	1289	1353	1719	2804	1645	1505	1481	1975	2060
		1128 ± 41	2398 ± 98	1364 ± 46	1732 ±223	3027 ±113	1668 ± 65	1719 ±231	1524 ± 18	2041 ± 27	2137 ±118
44 Ma	Middle Eocene	1315	1585	1890	2307	2161	2377		1695	2164	
		1216	1501 ±133	2324 ±298	3274 ±184	2359 ± 50	1961 ±138	1670 ± 32	2189 ± 68	2784 ± 49	
49 Ma	Early Middle Eocene		1615	2026	2438	3265		1862		2316	2791
			1687	2311 ±133	2643 ±173	3426 ±208	2509 ±250	2091 ± 44	1779 ±178	2636 ±100	2949 ±179
52 Ma	Early Eocene		1763	2469	2764	3478	2700	2118	1847	2926	2944
			1796	2467 ± 71	2730	3564 ±20	2726 ±225	2149 ±127	1824 ± 24	2770 ± 73	3220
56 Ma	Paleocene			2569			3030	2395	2029	2993	3828
				2561 ± 29			2919 ±118	2324 ±179	2000	2995	3663 ±251
	Cretaceous	1400	2002	2667	3118	3660	3194	2400			

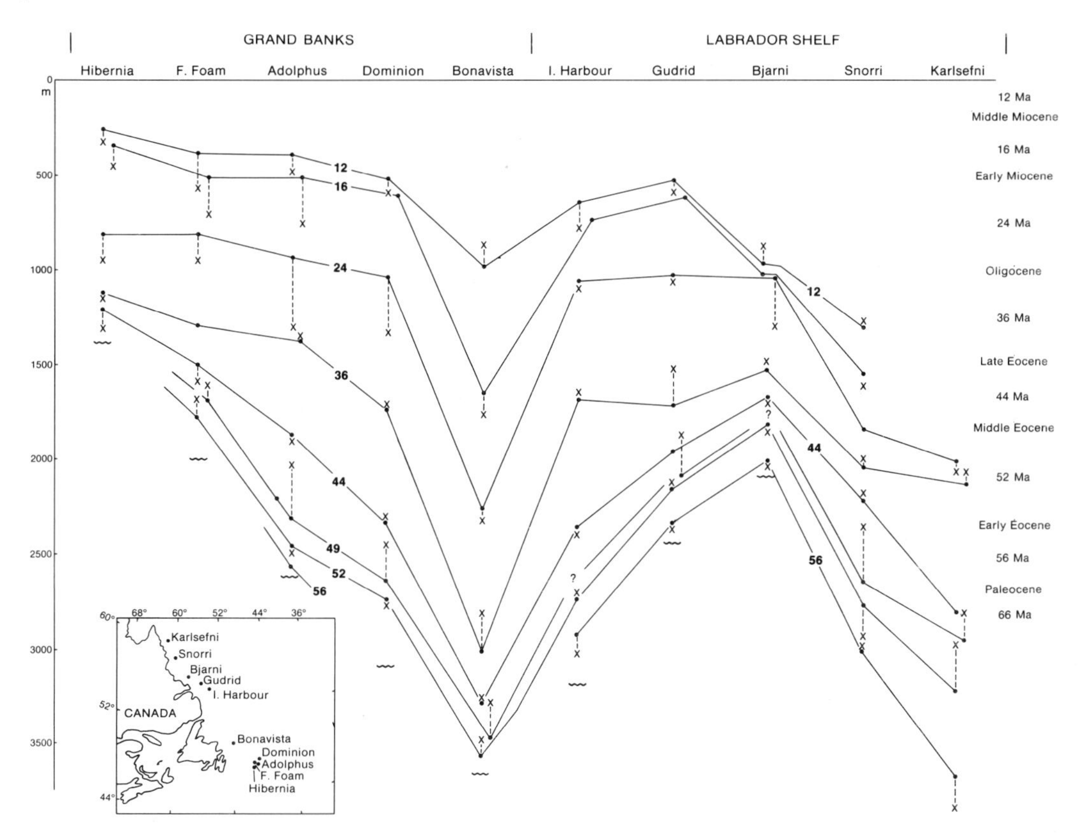

Figure 17 Correlation of 8 Cenozoic isochrons, according to their most likely depths in 10 wells on the Grand Banks and Labrador Shelf. The depths were computed by means of the RASC-CASC method explained in the text. Subjective estimates for the depths of these isochrons are shown with x.

Table 8 Age, depth (below rigfloor), paleo waterdepth and sediment data for subsidence analysis in Indian Harbour well, Labrador Shelf.

AGE (in Ma)	Depth (in m)		Waterdepth (in m)		Sediment
	Subjective	CASC	Minimum	Maximum	
0	-15		215	215	
3.5	520		100	200	sand
12	676	638 ± 82	50	100	siltstone
16		723 ± 68	50	100	siltstone
24	1097	1072 ±164	50	100	siltstone
36	1645	1668 ± 65	100	200	sand
44	2377	2359 ± 50	200	400	siltstone
49		2509 ±135	200	400	siltstone
52	2700	2726 ±113	200	400	shale
56	3030	2919 ±118	200	300	shale
66	3194		200	400	sand
70	3350		200	400	shale

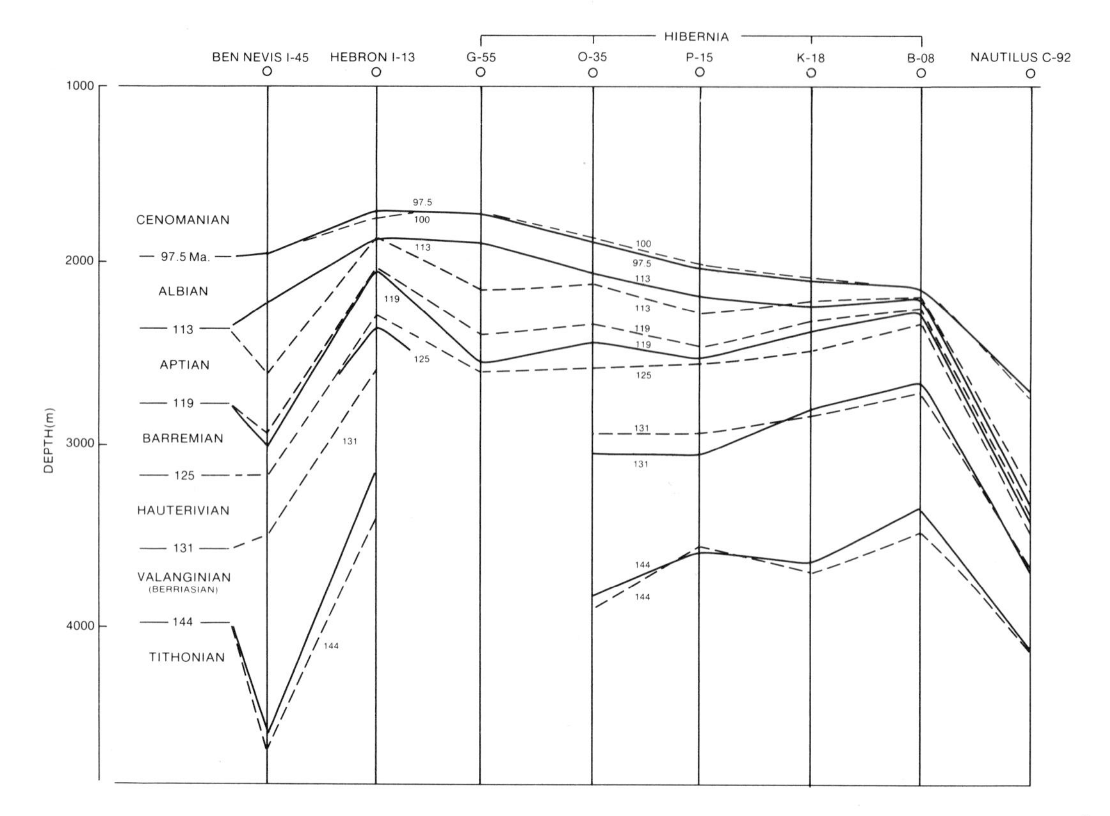

Figure 18 Comparison of subjective, and RASC-CASC derived most likely depths for Early Cretaceous isochrons (=RASC zonal boundaries) in northern Grand Banks wells, (after Williamson, in press).

The quantification of the actual basin subsidence and the rate of sedimentation through time is also of vital importance for better knowledge of tectonic history and style (Chapter V.1). In the next example, we will approach this type of analysis both from a qualitative and a quantitative stratigraphic point of view.

Briefly stated, subsidence and sediment-fill analysis runs through the following calculations, which essentially involve additions and subtractions of mass below a base line:

(1) sediment accumulation through time, starting at the oldest point in the well, using age per stratum interpretations, provided by paleontology;

(2) decompacted sediment accumulation, which estimates original sediment thickness. Observed sediment thickness is multiplied by a factor of the ratio between present and original rock densities;

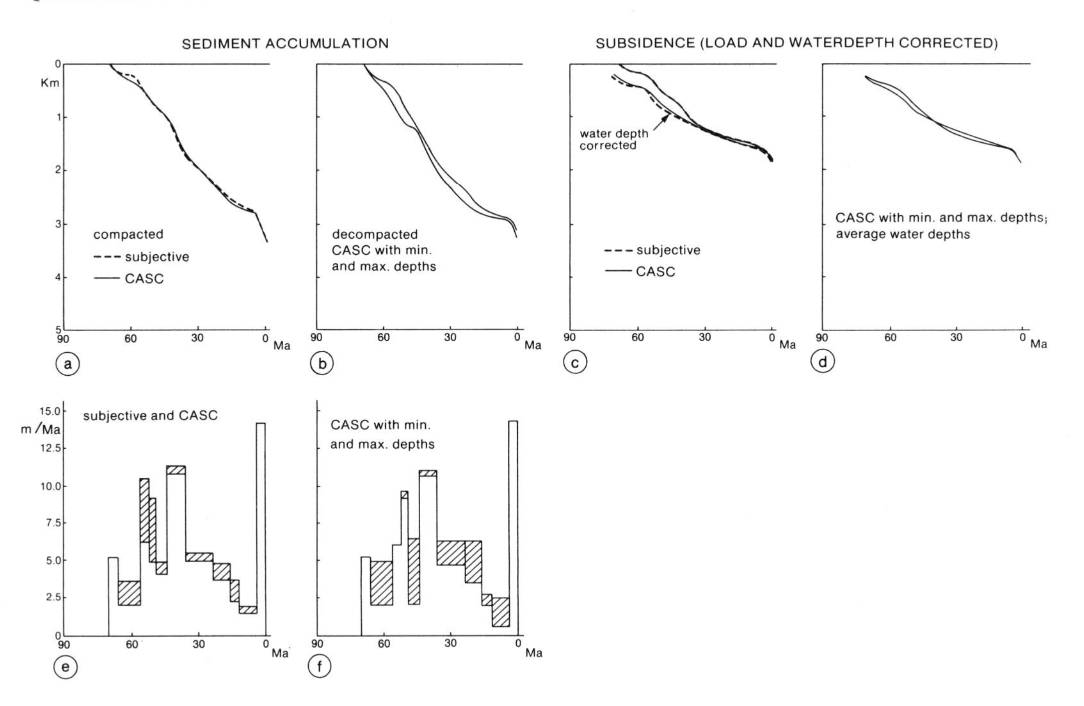

Figure 19 Subsidence and sedimentation history for the Indian Harbour well site, Labrador Shelf, based on subjective age-depth and RASC-CASC derived, most likely age-depth values.

(a) decompacted sediment accumulation;
(b) maximum and minimum estimates for CASC derived thickness under a;
(c) tectonic subsidence, load corrected;
(d) as c, using minimum and maximum - CASC derived depth values;
(e) difference between conventional and CASC calculated decompacted sedimentation rate; and
(f) difference between decompacted sedimentation rates based on minimum and maximum - CASC derived depth values.

(3) Water depth changes in addition to sediment accumulation, interpreted from the bio- and lithofacies changes in each stratum. Water depth change can be positive or negative. An increase in water depth generally means that the rate of subsidence surpassed the rate of sedimentation and vica versa. An increase in water depth is added to the sediment accumulated over the time interval considered while a decrease is subtracted. The new values yield the subsidence or so-called burial curve of points along the well;

(4) Addition of the rise and fall in sea level on a 'global' scale means that correction is in order for movement of the mentioned base line. It is widely accepted that global sea level changes of the order of 10 to 200 m do occur in time. A sea level rise is subtracted from the subsidence, a lowering added.

(5) Final calculation involves tectonic subsidence, which is the rate of subsidence of basement as if a basin had filled with (sea) water (or air), rather than with sediment. In a simple isostatic correction, the whole sediment load is taken off at once, to relieve basement. If the sediment load is taken off by stripping off one formation at a time, starting with the youngest one, tectonic (driving force) subsidence through time is found. This is the subsidence which actually took place in the upper crust.

For our example we analyze the subsidence of the Indian Harbour M-52 well site, Labrador Shelf. This well bottomed in basement on the present-day outer continental shelf, after penetrating approximately ±3200 m of Cenozoic and about ±300 m of Cretaceous (Maestrichtian) strata. Table 8 lists both the subjective and the (batch) CASC depths for successively older stratigraphic ages. The local error values at 49 and 52 Ma were arbitrarily halved. Firstly, this makes the minimum CASC depth values at 44 and 49 Ma consecutive in this order. Secondly, some smoothing of the batch-type default fits may have been preferable (Figures 10-12). Minimum and maximum paleowater depth estimates and approximate lithologies are to the right in Table 8.

Sedimentation rate (using decompacted thicknesses) in metres per my (Figure 19e) is not significantly affected by the choice of method, except that the subjective data suggest a brief high rate at 52 Ma and CASC at 49 Ma. Information (events) is sparse at this depth, but this is the time of widespread rapid sedimentation along the margin (Gradstein and Srivastava, 1980). Sedimentation rate uncertainty, calculated from all minimum and maximum CASC depths (Figure 19f), is of the order of 10 to 50%, with smaller values prevalent.

Compacted sediment accumulations through time (Figure 19a) show that CASC depths are at least as good a basis for this type of analysis as subjective data. The uncertainty in decompacted sediment accumulation is relatively minor (Figure 19b). Overall trend is for steady average sediment accumulation, which apparently kept reasonably in balance with subsidence. Both types of depth data yield an almost identical tectonic subsidence curve (Figure 19c) for basement through time, although the subjective depth curve shows a Paleocene hump. The overall curve is slightly dog-legged, with an inflection point to a lower rate around the Eocene-Oligocene boundary. This is exactly the time when seafloor spreading ceased in the adjacent Labrador Sea. The addition of the water depth estimates gives the lower pattern in Figure 19c with slightly decreased overall subsidence. It is of interest that the estimated error due to uncertainty in waterdepth is of the same order of magnitude as that predicted by the error bar for CASC depth values (compare Figures 19c and d). Detailed subsidence analysis needs to take into account both the uncertainty in waterdepth and in stratigraphic depth. Because the RASC/CASC method calculates both the biozonation, the time scale and the most likely stratigraphic segmentation of the well sites and yields conservative estimates on errors, it is a powerful tool for this type of basin analysis.

CONCLUSIONS

Observed fossil event sequences in wells or in outcrop sections, for which a RASC (scaled) optimum sequence has been calculated, can be subjected to CASC analysis. The CASC method computes the most likely depth in the wells for events or zones in this (scaled) optimum sequence. The method involves estimation of three functions per well, including:

1. Observed events sequence (x) as a function of the calculated (scaled) optimum sequence (y) in a well;

2. Observed events sequence (x) as a function of its observed depth values (z); and

3. (Scaled) optimum sequence (y) as a function of the depth of the observed events (z) in a well.

The combination of the function for x versus y(1) and x versus z(2), yields the third function of y versus z, needed to find the most likely depth in a well of the events (or zones) in the (scaled) optimum sequence.

During an interactive computer session, each function is displayed, one after another, as a bivariate scattergram on the screen of a Tectronix terminal. For each scattergram a cubic spline fit is determined, using the default or a subjectively selected smoothing factor. During a CASC batch run only defaults (=minimal smoothing factors) are used. The cubic spline fits represent the best estimate of the three functions under 1, 2 and 3. This procedure normalizes the individual well record relative to the standard. The interactive session is repeated for each well until function 3 (optimum, or scaled optimum, sequence versus normalized depth of the observed events in a well) is found for each one. Selected events or zones can now be correlated according to their most likely depth values.

If the age is known for some of the events in the scaled optimum sequence, the inter-event distance values can be used as interpolators between these tiepoints in the time scale, to make a RASC biochronology. This bivariate function of the RASC distance scale versus linear time can also be smoothed to reduce scatter in the original age data.

Error analysis in the three bivariate functions under 1, 2, and 3 makes it possible to attach confidence limits to the most likely depth values. This gives the local error bar. The standard deviation of the RASC inter-event distances in the scaled optimum sequence can be converted to a global error bar, applied in all wells. Smoothing of the three bivariate fits under 1, 2, and 3 affects the error values much more than the most likely depth values.

Examples of objective correlation involved Cenozoic foraminiferal events, zones and isochrons based on the RASC zonation in Labrador and Grand Banks wells, and Early Cretaceous isochrons based on the RASC zonation in the Hibernia oilfield, off Newfoundland. Previous analysis based on subjective age-depth data consistently confirm results obtained by the CASC model. The error for the most likely depths of the correlation lines rarely exceeds 10%; it commonly is 2 to 5%. Further research in the methods of error analysis may show the values to be relatively large. CASC-type of age/depth data contribute significantly to the analytical error analysis in tectonic subsidence and sedimentation calculations performed on a Labrador Shelf well site. RASC and CASC make subsidence analysis more objective and accurate, and easier to perform by non-paleontologists.

The combination of the RASC method for biozonations and the CASC method for the most likely segmentation and correlation of well sections is an objective and powerful stratigraphic tool in geological basin analysis.

REFERENCES

Agterberg, F.P., Oliver, J., Lew, S.N., Gradstein, F.M. and Williamson, M.A., 1985, CASC Fortran IV interactive computer program for Correlation and Scaling in Time of Biostratigraphic Events. Geol. Surv. Canada Rpt.

Berggren, W.A., Kent, D.V. and Flynn, J., in press, Paleogene Chronology and Chronostratigraphy. Geol. Soc. London, Spec. Paper

Blank, R.C., 1979, Applications of probabilistic biostratigraphy to chronostratigraphy. J. Geol., 87, 647-670.

DeBoor, C., 1978, A Practical Guide to Splines. Springer Verlag, New York, 392 p.

Drooger, C.W., 1974, The boundaries and limits of stratigraphy. Proc. Kon. Ned. Akad. Wet. Ser. 11B, 17, 159-176.

Edwards, L.E., 1984, Insights on why graphic correlation (Shaw's method) works. J. Geol. 92, 583-597.

Gradstein, F.M. and Agterberg, F.P., 1982, Models of Cenozoic foraminiferal stratigraphy - northwestern Atlantic margin. In Quantitative Stratigraphic Correlation, ed. Cubitt, J.M. and Reyment, R.A., J. Wiley and Sons Ldt., 119-170.

Gradstein, F.M. and Berggren, W.A., 1981, Flysch-type agglutinated Foraminifera and the Maestrichtian to Paleogene history of the Labrador and North Seas. Marine Micropal. 6, 211-268.

Gradstein, F.M. and Srivastava, S.P., 1980, Aspects of Cenozoic stratigraphy and paleogeography of Labrador Sea and Baffin Bay. Palaeogeogr., Palaeoclimatol., Palaeoecol., 30, 261-295.

Hay, W.W., 1972, Probabilistic Stratigraphy. Eclog. Geol. Helv., 65, 2, 255-266.

Millendorf, S.A. and Heffner, T., 1978, Fortran program for lateral tracing of time-stratigraphic units based on faunal assemblage zones. Computers and Geosciences, 4, 3, 313-318.

Miller, F.X., 1977, The graphic correlation method in biostratigraphy. In E.G. Kauffman and J.E. Hazel eds., Concepts and methods of biostratigraphy, Dowden, Hutchison and Ross, Inc., Stroudsburg, USA, 165-186.

Muller, C. and Willems, W., 1981, Nannoplankton en planktonische foraminiferen uit de Ieper-Formatie (Onder-Eoceen) in Vlaanderen (Belgie). Natuurw. Tijdschr., 62, 64-71.

Reinsch, C.H., 1967, Smoothing by spline functions. Num. Mathematik, 10, 177-183.

Shaw, A.B., 1964, Time in Stratigraphy. McGraw-Hill Book Co., New York, 365 pp.

Tjalsma, R.C. and Lohmann, G.P., 1983, Paleocene-Eocene bathyal and abyssal benthic foraminifera from the Atlantic Ocean. Micropal. Spec. Publ., no. 4, 76 pp.

Williamson, M.J., in press, Quantitative biozonation of the Late Jurassic and Early Cretaceous of the East Newfoundland Basin. Micropal.

Part IV

LITHOSTRATIGRAPHY

Contents

IV.1

PRINCIPLES OF QUANTITATIVE LITHO-STRATIGRAPHY

- The Treatment of Single Sections -

W. SCHWARZACHER

INTRODUCTION

The literal meaning of stratigraphy is the 'description of layers'. Grabau (1913) defined the subject in his fundamental work, "Principles of Stratigraphy" in the following way: "The inorganic side of historical geology or the development through the successive geologic ages of the Earth's rocky framework or lithosphere."

In spite of this definition, a great deal of the work carried out by stratigraphers is concerned with the fossils contained in sedimentary strata, so that biostratigraphy has somewhat overshadowed the lithostratigraphic approach. However, when stratigraphy is taken to be the basis of historical geology, then both bio- and lithostratigraphy are important and useful in reconstructing stratigraphic history.

Basic stratigraphic research involves three steps: one first collects the data, usually when preparing a geological map. Secondly, observations on the shape and composition of rock bodies are interpreted as lithofacies and any fossil material that has been collected is identified and summarised as biofacies. The combined facies is finally used to reconstruct depositional environments. Stratigraphic history is composed of a succession of such environments.

Although bio- and lithostratigraphy go hand in hand in this effort, the actual methods differ in a fundamental way: Whereas biostratigraphy is concerned with the order, spacing and correlation of irreversible sequences of events in time, based on the evolution of living creatures, lithostratigraphy deals with recurring events, like the presence of shales in stratigraphic sections or cyclic sedimentation.

While there is a variety of quantitative methods for (objective) lithostratigraphic analysis, the methods can be described as being of two general types. The first deals with the analysis of the data and includes such well-known statistical methods as the analysis of variance, factor analysis, trend surface analysis and so on. The second group of methods deals with the description of the sedimentary processes themselves and for this reason, these methods are often referred to as process models. The purpose of such models is to simulate or repeat the actual processes that occurred in the sedimentary environment. Naturally, this can only be achieved in the most general terms, but the use of such models as working hypotheses for geological mapping or correlation, is an essential part of the quantitative approach.

QUANTITATIVE LITHOSTRATIGRAPHY

Lithostratigraphic Data

Data must be collected before one can attempt a stratigraphic analysis and it is obviously preferable to take observations that can be expressed in numbers so that a quantitative analysis may be made. Rock properties which are derived from measurements or counts are called variables and are ideal for the purpose. However, a large proportion of geological observations consists of qualities like color, rock type or bed form, which are difficult to quantify. Such properties are known as attributes and they can be handled in two ways. Some attributes can be classified into different ranks: For example, the rock colors white, grey and black could be given the codes 1, 2, 3. Coded data of this type can be used to derive descriptive statistics. When ranking is impossible, one can consider the two alternatives, presence or absence of an attribute. By counting such incidents within defined domains, one can generate frequency data that make a very useful basis for quantitative studies.

The interpretation of a facies in terms of an environment depends largely on two sources of information; the first directly comprises sedimentary parameters, variables and attributes, while the second involves a study of the lateral and vertical variability of the sediment. The latter is particularly important because it puts the lithological observations into both a geographical and a relative time framework.

Some Statistical Properties of Sequential Observations

Most geologists are familiar with elementary statistical methods. The computation of such statistics as mean, standard deviation and correlation coefficient is frequently carried out in sedimentological and biometrical work. Judging the quality of such measurements or testing their significance (as it is more usually expressed) is just as common. It is often useful to judge the quality of measurements by establishing the significance of derived statistics. Tests of this type rely heavily on the process of random sampling which is an important technique for avoiding bias in collecting the data. Clearly unless an area is completely homogeneous, i.e. leads automatically to random samples, the outcome of any sampling experiment will depend very much on where the sample was taken.

Methods, on the other hand, that deal with measurements that were taken in an ordered sequence are applied in 'time series analysis'. This term was first used for observations that had been made in an orderly fashion and at a given time interval. Daily temperature measurements are a good example for a time series. Such a series does not represent a random sample because each successive value is somehow related to its predecessor. Furthermore, such a sequence is far from being homogeneous and one expects, for example, lower temperatures during winter in the northern hemisphere than during summer.

The dimension in which the samples in a time series analysis are taken does not necessarily have to be time; it could equally well be the vertical thickness of a sedimentary sequence or, in the horizontal direction, it could be some distance along a transect on which the measurements were taken.

In both cases, the methods of analysis are exactly the same, because it is the sequential nature of the data which is the essential feature of the analysis. The only property that is specific for time is its unidirectional sense. This means that any physical property measured in time can only be influenced by the past and not by the future. Spatial measurements along a transect can be unidirectional, but generally it is found that both neighbouring points have a physical influence upon a given sample location.

Sequential measurements of a variable x_t $(t = 0, 1, 2, ..., n)$ can be represented by their mean value:

$$\bar{x} = \{1/(n+1)\} \sum_{t=0}^{n} x_t$$

and their variance:

$$S^2 = \{1/(n+1)\} \sum_{t=0}^{n} (x_t - \bar{x})^2$$

Both of these statistics are identical with their equivalents derived from random samples. Whether they give a good representation of the data, can best be seen by making a plot. Figure 1 shows two typical cases. In Figure 1a, the sequence consists of short, possibly random fluctuations. Intuitively, one would accept the mean and also the variance as a valid description of the process. The second example in Figure 1b, illustrates a different series which shows a slower apparently predictable change in addition to the short fluctuations. The slower more predictable variation is called a trend.

In the latter case, the mean is not a very good description of the series since it would change, depending upon where it had been sampled. A series which has constant average properties is called stationary and a series which contains a trend is called non-stationary.

Two further statistics which are more specific to time series, are the 'auto-covariance' and the 'power spectrum'. The auto-covariance C_k or its standardised form, the 'auto-correlation' r_k, are functions of the lag, which is called k, and will be explained below. C_k and r_k are defined as:

$$C_k = \{1/(n-k)\} \sum_{t=1}^{n-k} (x_t x_{t+k})$$

$$r_k = C_k/\text{var } x$$

As the name implies, auto-correlation is the correlation of an ordered series with itself. One may visualise this by imagining two identical sequences that are placed side by side, and by correlating the values which are next to each other. Naturally since the sequences are identical, this would give a correlation coefficent of +1 and this is always the case when k equals zero. The coefficient r_1, $k = 1$, is calculated in the same way except that one of the sequences is shifted by one position, so that value x_0 comes opposite x_1, x_1 comes opposite x_2 and so on. A plot of r_k for $k = 0, 1, 2 \ldots$ is known as the 'correlogram' and fully describes the correlation structure of a series.

The second statistic which is frequently used in the description of a time series, is the 'power spectrum'. This function is defined mathematically as the Fourier transform of the covariance function:

$$f_\omega = \frac{1}{2\pi} \sum_{k=-\infty}^{\infty} e^{-i\omega k} C_k$$

and it analyses a time series in terms of frequencies ω. The letter i stands for the imaginary number $\sqrt{-1}$. It is convenient to think of the sequence as being made up of a great number of cosine waves with different frequencies and amplitudes. The function gives the amount of variance that each frequency contributes to the total variance. It is called a power spectrum because the power developed by an alternating electric current is proportional to the variance of the voltage.

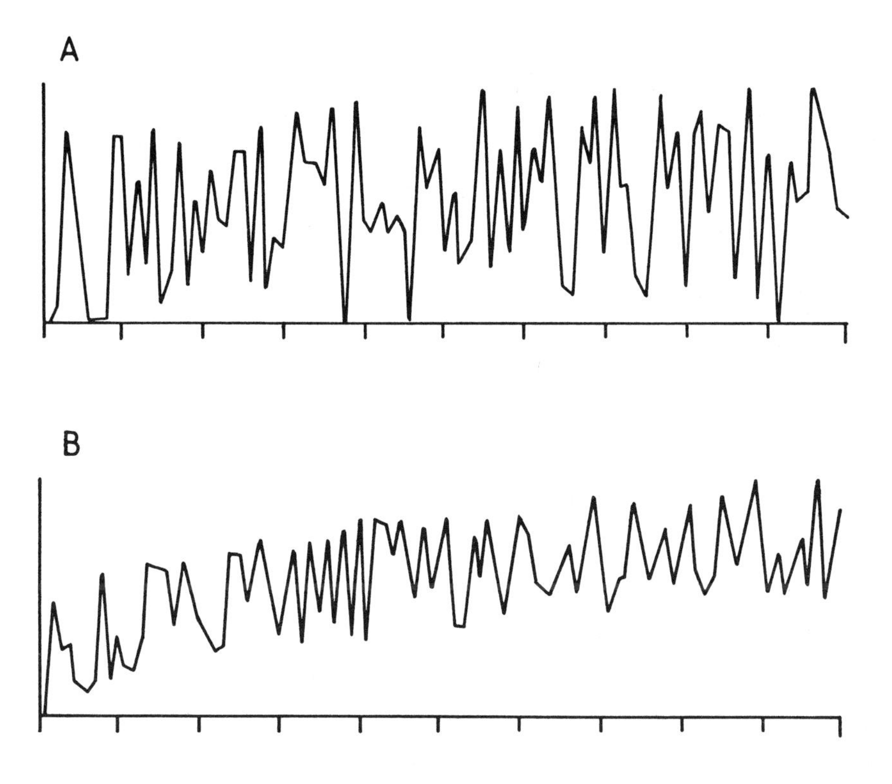

Figure 1 Illustrations to show two different types of time series.
(a) A time series consisting of random values.
(b) A series of random variations with a superimposed trend.

The decomposition of a series into its frequency elements is not necessarily restricted to sine-cosine functions; indeed any orthogonal function may serve. Walsh functions are particularily useful and consist of ordered sets of rectangular wave forms which only take the amplitudes of the values +1 and -1. An example using a Walsh power spectrum is given later and the theory of such spectra is described in Beauchamp (1975).

To a stratigrapher who is used to thinking in terms of time or stratigraphic thickness, a wavelength is usually more meaningful than a frequency. The auto-correlation function in the time domain is thus more readily interpreted than the spectrum itself. However, the spectrum has certain advantages over the correlation function, particularly if only short sequences are available. It is found that the correlogram of random processes does not damp out as it should under such circumstances and it is difficult to assess the statistical significance of a correlogram because neighbouring values are often highly correlated. Such tests are much easier to carry out on the spectrum. Furthermore, if one is concerned with cyclic sedimentation, spectral analysis is often the appropriate technique.

A sequence of completely random fluctuations is known as white noise and it consists of an infinite number of frequencies that are all contributing equally to the process. The spectrum in this case will therefore be constant for all frequencies. On the other hand, if the record is that of a regular sine wave, then there would be only a sharp spike-like maximum. Spectra of random processes usually show one or more wide maxima and if a trend is present, there will always be a strong contribution from the low frequencies.

Predictable and Unpredictable Stratigraphic Variation

The classification of sequential data into three categories:

(1) Unpredictable random values,

(2) Partly predictable fluctuations, and

(3) Predictable trend,

is of the utmost importance for both practical and theoretical considerations. Clearly, if predictability could be quantified, it would solve a great number of practical problems in geology. Before discussing such possibilities however, it is useful to examine the origin of stratigraphic variation in general terms.

It is generally assumed that sedimentation in any environment was regulated by certain conditions which somehow determined the nature of the sediment. One hopes that the examination of sediments can lead to the reconstruction of such conditions, but there are difficulties. Most sedimentary processes involve certain elements of chance, so that identical conditions do not always produce the same sediment. To take a specific example, in modern Atlantic sediments the percentage of calcium carbonate is related to water depth. This is however a statistical relationship and it would be impossible to determine an accurate water depth from the analysis of a single sediment sample. If a sequence of water depths is deduced from a sequence of carbonate percentages, then the result would contain an unpredictable variation which behaves like a random error.

The probabilistic nature of the environment-sediment relationship is not necessarily due to any breakdown of deterministic laws, but it is the result of the enormous complexity of the sedimentation processes in nature. The historical development of the environment through time will incorporate many chance fluctuations and the stratigraphic history becomes a random or stochastic process.

Any random element will lower the predictability of stratigraphic variation and it is necessary to identify the sources of such variations. The following are all potential causes of random behavior:

(a) Observational errors and misidentification by the geologist,

(b) Interference with the sedimentary record during diagenesis and possible later deformation,

(c) Ambiguities in translating environmental conditions into sediments and,

(d) The unpredictability of geological history itself.

The effect of random disturbances can vary a great deal. There are stratigraphic time intervals and areas where the sedimentation is very regular and where one can discover predictable trends clearly. In other situations, stretches of predictability seem to be very short-lived. The term trend is therefore strictly dependent upon scale and such terms as 'slow moving trend' and 'short fluctuations' are relative to the total length of the interval which is examined. Because of this relativity, it is not necessarily true that short fluctuations and trend have different explanations.

Short fluctuations are more numerous and therefore it is easier to analyse their predictability. The random nature of short fluctuations in particular, is more readily recognised.

The reliability of any prediction depends ultimately upon one being able to identify the physical processes that were responsible for such a variation. Typical geologic processes which follow a trend-like pattern through long time intervals are: The filling of sedimentary basins, the erosion of mountain areas, crustal sinking by heat loss and so on. On the other hand, typical geologic processes leading to predictable short term fluctuations are: Climatic cycles, and sedimentation controlled sequences (Bouma sequences, or sequences generated by meandering streams, for example). In order to judge the predictability of a sediment, it is necessary not only to identify the causes of the various sedimentary variations, but also to know the physical mechanisms of these causative processes. This goal can at least partly be achieved by referring to sedimentation models (see Chapter IV.2).

THE ANALYSIS OF SINGLE SECTIONS

The Models

One of the major problems in defining models for quantitative stratigraphic analysis is the separation of the predictable variation from any unpredictable random fluctuations. If the analysis is concerned with single sections, the problem is largely in determining the previously discussed components of trend, cyclic or stochastic variation, and completely unpredictable random noise. The methods of analysis that are adopted will depend very much upon the data that is available as well as on the type of model that is chosen to represent these.

Writing x for any variable at time t, the following are some possible models for single section analysis:

$$x_t = \varepsilon_t \tag{1}$$

Model 1 consists simply of a series of independent random numbers ε. In Model 2 a stochastic process results by adding a random variable to an earlier value x_{t-1}:

$$x_t = x_{t-1} + \varepsilon_t \tag{2}$$

In Model 3 the sediment record consists of a deterministic cycle with frequency ω and amplitude A, that is disturbed by random events:

$$x_t = A \sin \frac{2\pi t}{\omega} + \varepsilon_t \tag{3}$$

And finally Model 4 represents an x_t stochastic process which is added to a function m_t which is a slow-moving trend:

$$x_t = m_t + x_{t-1} + \varepsilon_t \tag{4}$$

This list of models represents only a very simple approach to the problems and it is by no means complete. It is important that the basic models should be well understood and for this reason a more detailed discussion of them follows.

A common feature of all stochastic models and indeed of all models that are relevant to the observed data, is the independent random component ε_t. Independent means that in a sequence of data, no correlation exists between successive values ε_t, ε_{t+1}, ε_{t+2} ... This can be expressed by stating that knowing the value ε_t at time t+1, does not help in any way to predict the value ε_{t+1} at time t+1. It is generally assumed that most observational errors behave in this way.

In spite of their usefulness in the analysis of stochastic processes, independent random values are rarely found in sequential data.

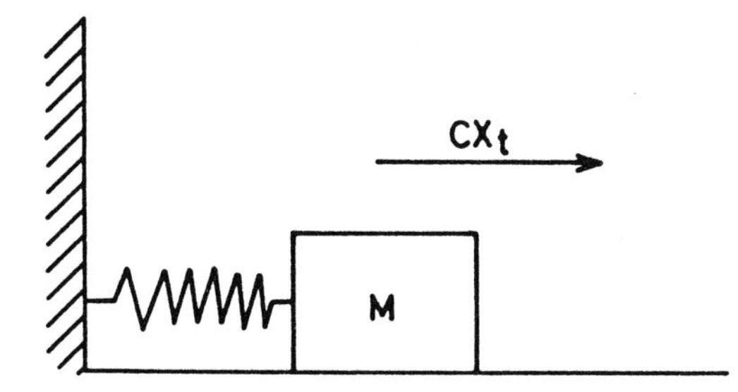

Figure 2 A mechanical analogy, used to illustrate auto-regressive processes of the first and of the second order. M is a body, Cx_t is the force required to pull it.

The reason for this is that in actual time-sequences there is always a certain amount of causal connection between successive time events. In geological models, uncorrelated random numbers are substituted only when nothing is known about the process.

A number of simple stochastic processes can be derived by introducing a certain amount of correlation between successive values. For example, the process:

$$x_t = a\, x_{t-1} + \varepsilon_t \qquad (5)$$

is called a Markov process, or written in this particular form, an auto regressive process of the first order. The value of x_t at time t is random, but it also depends upon its predecessor x_{t-1} and a constant a. For the series to be stable the coefficient a must be smaller than one.

A second order process results by extending this correlation structure:

$$x_t = a_1\, x_{t-1} - a_2\, x_{t-2} + \varepsilon_t \qquad (6)$$

In order to appreciate the two models fully, it is useful to look at the following physical analogy (Figure 2). A body M is on a horizontal surface and it is attached by a spring to a vertical wall. Assume at first, that the body has no mass. If it is pulled to the right hand side, a force C will act upon it and return it to its original position. This is counteracted by a frictional force which is proportional to the speed dx/dt and one can write for the system:

$$C\ x_t = a\ \frac{\Delta x}{\Delta t},$$ with the solution:

$$x_t = C' + e^{-(C/a)t} \qquad (7)$$

As a difference equation, this is identical with the first order autoregressive system whereby ε_t has been replaced by Cx_t. The solution shows that after M has been released, it will return to its equilibrium with an exponentially declining speed. Similarily, if there are occasional random disturbances in a time-series, they will be damped and equilibrium will be maintained until a new disturbance arrives. In the Markov process, disturbances occur at each moment in time so that new random values of ε_t are continuously incorporated into the system. In such a process, the mechanism implies that consecutive values are correlated and it is easy to show that the correlogram r_k of the process in equation (5) is given by:

$$r_k = a^k, \qquad k = 0, 1, 2, \ldots, n \qquad (8)$$

or in words an exponentially declining curve because a is smaller than one.

The analogy for the second order process is obtained by giving the sliding body some mass M. This introduces inertia which is proportional to the acceleration or second derivative of x. The difference equation can be written as:

$$C\ x_t = a_1\ \frac{\Delta x}{\Delta t} - a_2\ \frac{\Delta^2 x}{\Delta t^2}$$

with the solution:

$$x_t = p^t\ (A \cos \omega + B \sin \omega t) \qquad (9)$$

where A, B and p are constants.

The system is now capable of oscillating, but once again there is exponential damping, and if the system is excited only once it will produce regular oscillations which gradually approach zero. In the auto-regressive process given in equation (6), a new excitation is provided by each random value ε_t and the fluctuations persist but they are very irregular. However, the correlogram of the process is quite a regular damped cosine wave that is similar to the solution of equation (9).

The second order process is a typical example of a feedback mechanism which means that it is a physical process which is somehow self-regulating. This is a property of a great number of geological and environmental processes. For example the filling of a basin may lead to crustal sinking which may in its turn result in increased erosion with an eventual decreased sediment supply. An important factor in climatological arguments is the so-called albedo feedback: cold climates encourage snowfall, this increases the albedo (reflectivity) so that less energy is absorbed and therefore the amount of cooling increases, resulting in positive feedback (more snow). The so-called predator-prey relationship in the biosphere gives rise to the well-studied oscillation system where two species live in competition.

The fluctuations that are produced by systems of this type, are not very regular when taken individually but regularities are found in the wavelength and amplitude-variations when long runs are analysed.

The nature of oscillating stochastic models is probably best understood by contrasting them with the third model that was given in the previous list. This model (equation 3) consists of harmonic movement that is disturbed by superimposed random fluctuations. The mechanical equivalent of this process would be a system that is driven only by the spring force and the inertia. The equation of the system is:

$$C \frac{\Delta^2 x}{\Delta t^2} = - ax. \tag{10}$$

The process could continue indefinitely since there is no friction and so it would not need the random pulses to keep it going. The random disturbances which are present are actually due to observational errors and it is sometimes impossible to fix the precise maximum or minimum of the cycle in this system. However, one finds that the time-keeping quality is unimpaired by the errors if one regards this process as a clock. This is in distinct contrast to the auto regressive model where oscillations change in wavelength as well as in phase. The latter is the cause of the random errors being incorporated into the record and they are not merely superimposed, as in the harmonic model. A friction-free oscillating system may sound unrealistic but very large systems like the planet earth or the movement of the earth in the solar system, can be treated as if they were practically free of friction. Fluctuating environmental conditions that are based on daily, seasonal or possibly longer astronomical cycles, are examples for which the harmonic model is indicated.

The final model given by equation (4), can be applied to a series that contains a slow drift or trend and which is therefore non-stationary. It is difficult to estimate the statistical properties of such sequences without removing the trend component. The method that is normally used is to fit a polynomial (or any other mathematically defined function) to the data and to subtract it from them. If the trend consists of a gradually-changing mean, then the series will be stationary after this treatment. Unfortunately, trends that are restricted to means are rare in sedimentary successions and changes in almost any sedimentary parameter will also produce a change in the sedimentation rates. This in turn induces changes in the variance and autocorrelation that will follow the trend. For example, in a fining-upwards series, sedimentation rates usually decrease and any fluctuation will therefore not only increase in amplitude, but also apparently increase in frequency. Such trends are difficult to eliminate unless one can formulate a reasonable sedimentation model which can be justified on physical grounds.

Caution is also necessary when fitting the trend. If the usual least squares procedure is adopted, and if it is assumed that an actual physical process can be represented exactly by some polynomial, then this physical process can only be recovered if the residuals are independent random numbers. Normally data consists of a combination of trend and autocorrelated random values and in this case, parts of the stochastic component will be incorporated into the trend which is likely to be changed by this contamination. Furthermore, if such a trend is subtracted from the original data the correlation structure of the residuals will be altered unavoidably.

Fortunately the effects of trend removal can be very closely controlled if the analysis is carried out in the frequency domain and if power spectra rather than correlograms are used. If one has no prior knowledge of the trend or the random variable, then the data can be analysed by applying various filters. A filter transforms a series x_t into a series y_t by a linear operations of the type:

$$y_t = \sum_{k=0}^{m} b_k x_{t-k} - \sum_{i=1}^{m} a_i y_{t-2}.$$

in which b_k and a_i are weights. A first order filter would be given by:

$$y_t = b_0 x_t - a_1 y_{t-1}. \qquad (11)$$

The values of the weights b_0 and a_1 determine whether the operator acts as a high pass filter or a low pass filter. A high pass filter will pass high frequencies and suppress low frequencies and can be used for removing trends. A low pass filter may be used to remove random noise. Associated with each filter, is its transfer function H_f

$$H_f = \frac{y_f}{x_f}$$

where y_f and x_f are the Fourier transforms of y_t and x_t. The transfer function enables one to calculate frequency gains and phase shifts that are introduced by the filters, so that the incorrect interpretation of the treated data can be avoided. Filter methods and filter designs are used extensively in time series analysis and a number of useful textbooks exist (cf. Otnes and Enochson, 1978).

Examples of Single Section Analysis

The analysis of single sections will be illustrated by three examples that were taken from a Carboniferous section in NW Ireland (Figure 3).

Two different variables were used. In examples 1 and 3 (Figure 3a and 3c), a limestone percentage was calculated for each 20 cm interval of the section and in example 2 (Figure 3b) a bed thickness index was used. Only two lithologies were differentiated during the logging of the section, marl and limestone, which makes the two percentages complementary.

It is well known that percentages constitute a closed number system and that they must make reference, either to a unit volume or a chosen unit length of section (as in this case). The length of the sampling interval determines some of the statistics that are extracted from such data in this way. For example, if it is assumed that consecutive percentages values are not correlated, then an increased sample interval will decrease the variance. If x_{40} symbolises values based on a 40 cm interval and x_{20} denotes values that are based on a 20 cm interval, then:

$$\text{var } x_{40} = (1/2) \text{ var } x_{20}$$

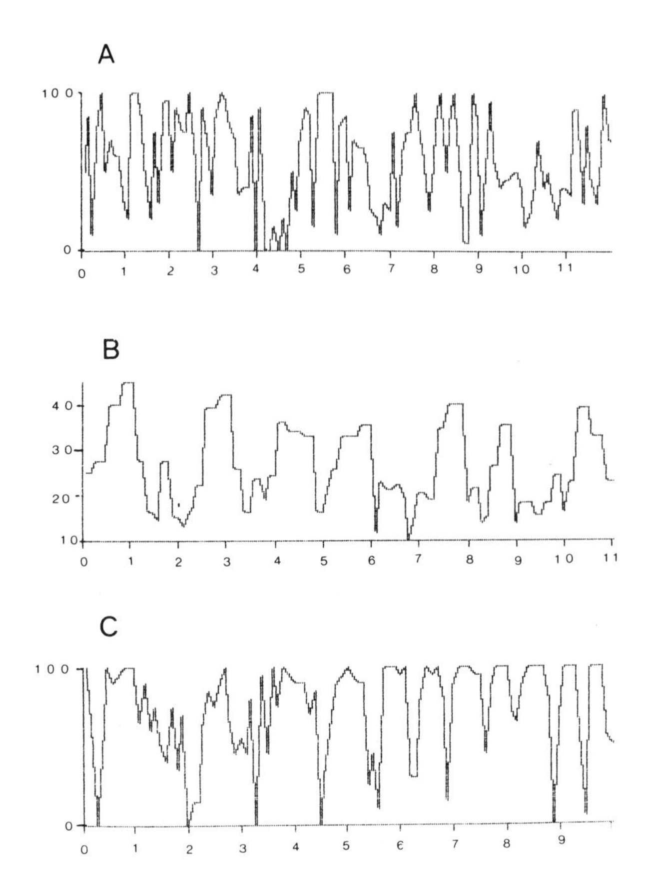

Figure 3 Three examples of stratigraphic time series;

(a) Lower Carboniferous limestone. The vertical scale shows limestone percentages, the horizontal scale is the stratigraphic position in meters.

(b) Carboniferous Dartry limestone. The vertical scale gives the thickness of limestone beds in centimetres, the horizontal scale is the stratigraphic position in metres.

(c) Middle Glencar limestone. The vertical scale shows limestone percentages, the horizontal scale is the stratigraphic position in metres.

It can be seen that doubling the interval or section length, halves the variance. However, if consecutive values are correlated, then:

$$\text{var } x_{40} = \frac{1+2r_1}{2} \text{ var } x_{20}$$

where r_1 is the autocorrelation for $k = 1$ (lag one). The relationship implies that the correlation structure is also affected by the length of the sample interval so that it would be difficult to make quantitative comparisons between sections that are based on different unit volumes or lengths.

A different variable was used in example 2 (Figure 3b). The data are from a section of well-bedded limestones with only very thin layers of marl intercalated between the beds. In this example, the limestone percentages are all very nearly one hundred percent and so they are not an instructive variable. However, the actual thicknesses of the limestone bands varied, and the thickness of any bed that coincided with a sample point was taken as a thickness index. Again, the sample points were spaced at 20 cm intervals.

It can be shown from comparative studies that shale-rich parts of a section have generally thinner beds and that with decreasing amounts of shale, the bed thickness increases. The bed thickness measurements and the limestone percentages are thus two variables which are roughly comparable.

After some experimentation with different intervals, a sample interval of 20 cm was chosen. It is probably the most efficient way of describing the section because the average thickness of the limestone bands is of this order. Since the sections were measured to the nearest centimeter, the lithologies in the 20 cm intervals can be determined with an accuracy of five percent. This degree of accuracy is quite adequate for the present study.

One can note the following from a preliminary study of the graphed data. Example 1 in Figure 3a, shows some very irregular fluctuations but the amplitudes and frequencies seem to be approximately constant and no obvious trend is present. In contrast with this, example 2 (Figure 3b) shows some fairly regular fluctutations and some cyclic sedimentation is immediately suggested by the data. The last example (Figure 3c) was chosen because it shows a fairly clear trend both in its amplitudes and in its variance and correlation structure.

The fitting of trends is always a useful starting point for an analysis. In its simplest form, this may involve fitting either a straight line or a low order polynomial to the data. A useful alternative approach is to express the trend by a trigonometric series, using

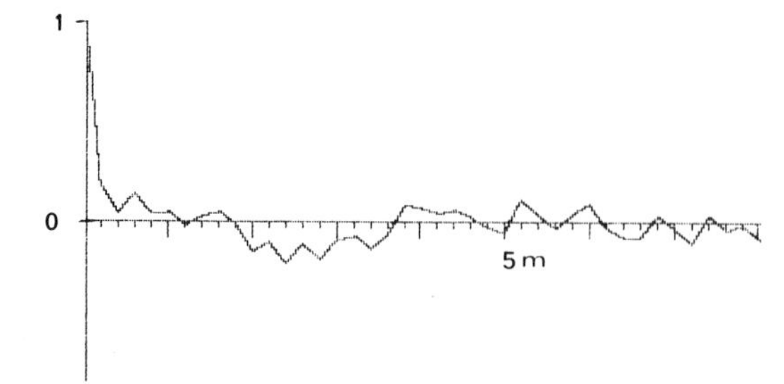

Figure 4 Correlogram of data shown in Figure 3a.

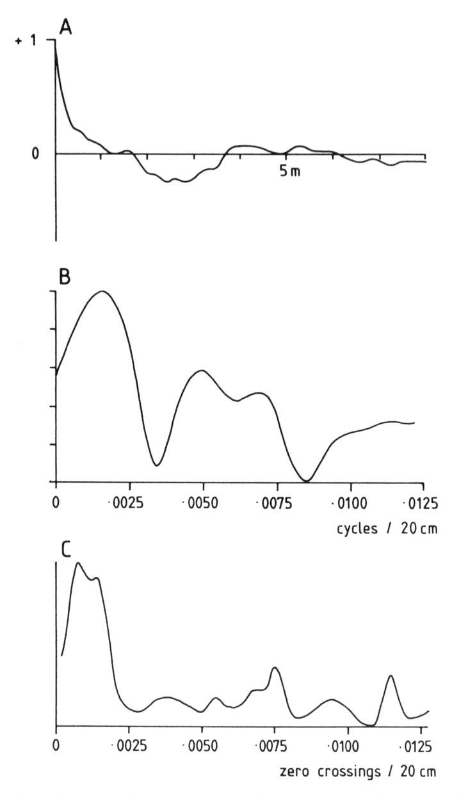

Figure 5 (a) Correlogram of data shown in Figure 3a after filtering.
(b) Powerspectrum of data in Figure 3a. The frequency units are in cycles/20 cm. The vertical scale represents variance in arbitrary units.
(c) Walsh power spectrum of the same stratigraphical interval after coding limestone = +1 and marl = -1. The frequency is given as zero crossings (of the square wave)/20 cm and the power is shown vertically in arbitrary units.

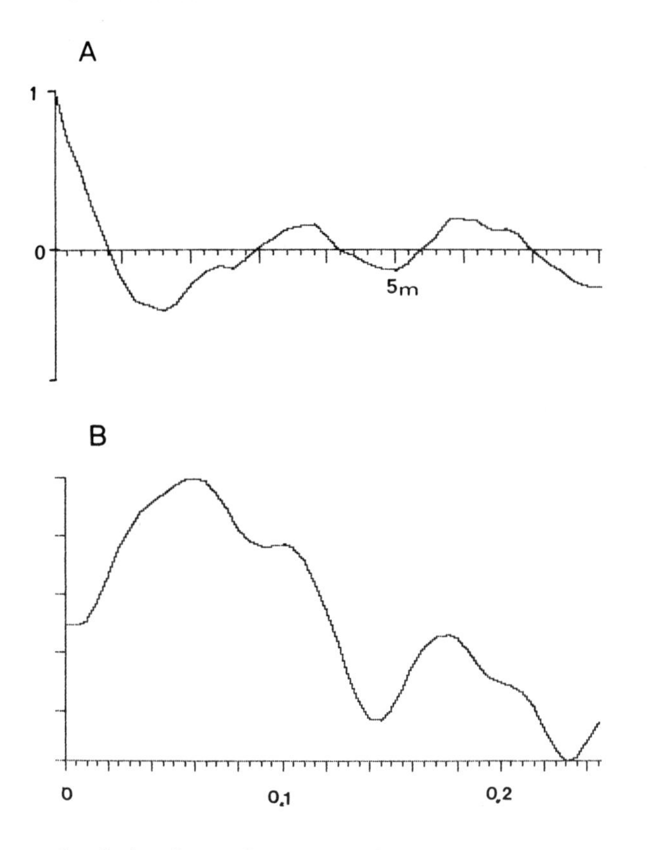

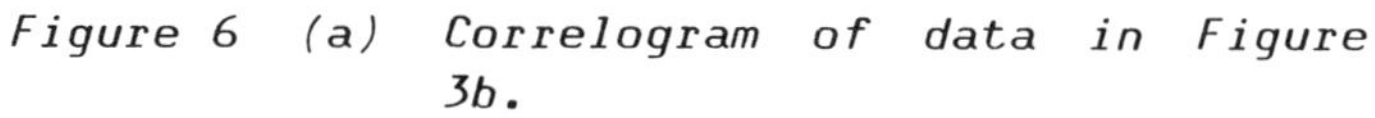

Figure 6 (a) *Correlogram of data in Figure 3b.*

(b) *Powerspectrum of data in Figure 3b. The horizontal axis shows frequencies in cycles/20 cm. The vertical axis gives the power in arbitrary units.*

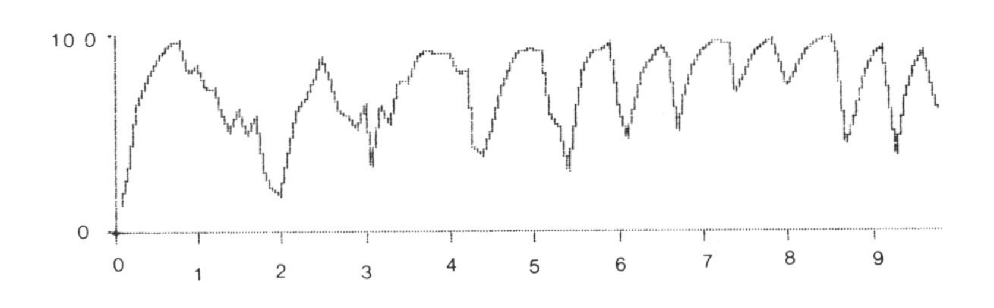

Figure 7 Filtered record of data shown in Figure 3c.

only a very restricted number of terms. This was done for the three samples and it was found that the first term of a Fourier series expresses the trend quite adequately.

Next, the correlation structure of the series was investigated. The correlation for example 1 is shown in Figure 4. The graph initially shows a very rapid damping which is followed by more persistent fluctuations. As explained earlier, theory requires such irregular records to lead to zero correlation for the higher lags. This would indeed be the case for a series of infinite length, but in samples of finite length the correlation does not disappear completely, which makes the interpretation of correlograms quite difficult.

The appearance of the correlogram for example 1 may be improved considerably by applying a simple, first order, low pass, filter:

$$y_t = (1 - \alpha)\, x_t - \alpha\, y_{t-1}$$

The resulting correlogram is given in Figure 5a. After treating the sequence with this filter and setting $\alpha = 0.3$ the correlogram becomes clearer with the presence of at least two frequencies showing.

The covariance of the filtered data were used to calculate a power spectrum. The result of this calculation is shown in Figure 5b which shows at least two well-developed frequency maxima, one at a wavelength of about 533-457 cm. and the other at 200 cm. There are further maxima in the higher frequencies but these have been altered by the low pass filter. This example is also used for demonstrating the use of Walsh power spectra. Since the data consist of two lithologies only, these can be coded as +1 and -1 and the resulting wave tract is analyzed in terms of Walsh functions. The resulting spectrum is shown in Figure 5c which shows a more or less similar frequency composition to Figure 5b.

Example 2 was much more regular and did not need any filtering, so the correlogram and power spectrum were calculated directly from the raw data (see Figures 6a and 6b). It is interesting to compare the results of these calculations with the correlogram and spectrum of example 1. One is struck immediately, by the similarities in all four graphs. The damping in the two correlograms is very similar and the spectrum for example 2 gives once again two maxima with one at about 355 cm and the other one between 118-114 cm. The thickness ratio corresponding to the high and low frequency maxima is 0.33 in example 2 and 0.36 in example 1.

All sections which are used as examples here, come from the same sedimentary series but example 2 is stratigraphically very much higher than examples 1 and 3. The similarity of the spectra suggests that the mechanism responsible for the lithological fluctuations was the same in all instances. Higher sedimentation rates in the first example could account for the greater thickness (wavelength) of the 'cycles'. It appears likely that the increased sedimentation rates go hand in hand with increased non-carbonate (marl) material. This would suggest that the cyclic fluctuations are due to variable influx of clay material, and not to varying rates of calcite dissolution. If the latter were to be the case, then one would have to assume reduced sedimentation rates in the marl-rich intervals.

The relation between cycle length and marl content becomes particularly obvious from studying example 3. The first order harmonic was fitted as an approximate trend and it already indicates an increase in limestone with stratigraphic position. At the same time, one can also see that the wavelength of the oscillations decreases systematically in the higher positions. Once more, this is more likely to be the effect of decreasing sedimentation rates. The data of example 3 have been processed by using the same low pass filter as before and the result is a much-improved record, shown in Figure 7. A new trend was fitted to this refined version and two terms of a Fourier series gave a reasonable representation of the mean trend. The correlogram of this section is extremely damped and the spectrum approaches that of white noise. This is as it should be, since the sequence is clearly non-stationary and is generated by frequencies that are continuously variable.

However, the variation of the frequencies is not random and the wave length of the oscilliations decreases exponentially with increasing stratigraphic position. If one accepts that consecutive cycles mark equal time intervals, then this type of sedimentary record would be the result of a constant and linear decrease in sedimentation rates. One may test this hypothesis by applying a uniformly increasing stretch to the data in example 3 and the result of such a transformation is shown in Figure 8. It can be seen that a record with very regular cycles is produced by correcting for uniformly decreasing sedimentation rates.

ADDITIONAL CONSIDERATIONS IN SINGLE SECTION ANALYSIS

The Sedimentation-Time Relationship

The last example in the previous section highlights one of the major problems of quantitative stratigraphic analysis. The stratigrapher is used to substituting vertical thickness for time and he uses stratigraphic measurements for comparing the duration of time-spans in relative terms; under favourable conditions, this can be done in absolute terms (see Chapter V). The transition from a depositional thickness to a time scale can only be made however, when it is safe to assume that the average rate of deposition was constant. The difficulties of actually establishing an average rate of deposition were discussed by Tipper (1983).

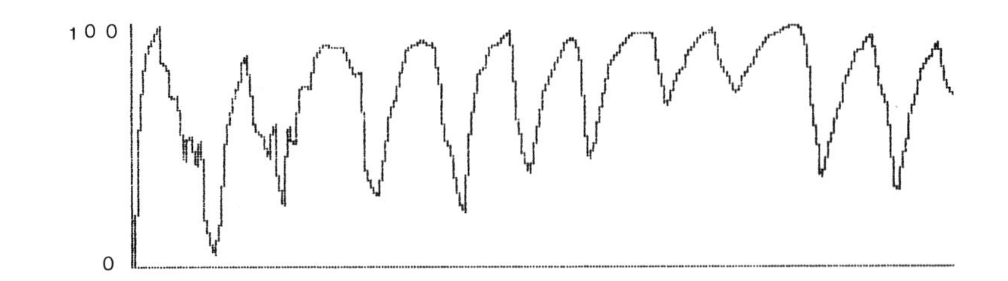

Figure 8 Exponentially stretched record of data from Figure 7.

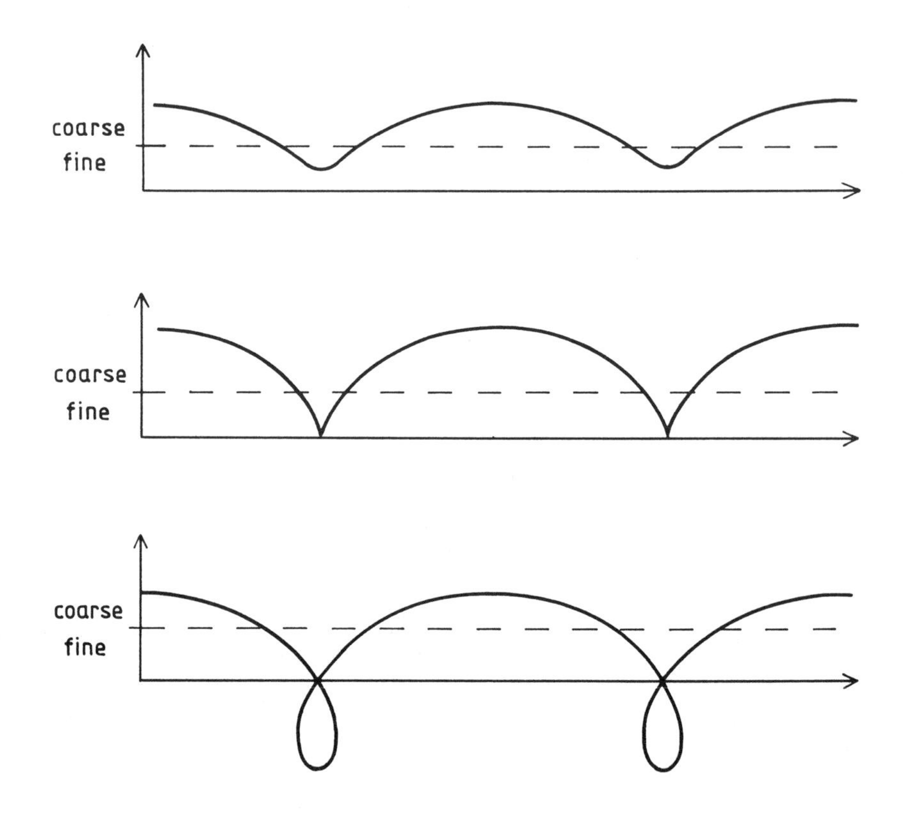

Figure 9 'Cycloids' as a model for stratigraphic position (vertical axis) in relation to a depositional history (horizontal axis) of cyclic increase and decrease of grain size;
(a) deposition slows down during production of fine material
(b) deposition comes to a standstill
(c) erosion causes the removal of some fine material

Each sediment has an absolute limit to the amount of time resolution that is possible. There is a minimum thickness below which there is no relationship between time and deposition. This minimum thickness is the smallest stratigraphic unit that is possible theoretically and it may consist of a sand grain, a lamina or a bed that is several metres thick. This unit is called the stratigraphic quantum or 'stratquant' and the stratigraphic features of this unit indicate that it was laid down extremely fast, compared with the average sedimentation rate in the area. When such a process leads to the formation of sedimentary beds, one speaks of event stratification (Seilacher 1982). Tempestites and turbidites are typical examples of event stratification. In contrast, other beds formed slowly and with reasonably steady rates of sedimentation, so that the thickness of the bed bears some relationship to the time it took to form. Such beds must consist of a sufficiently large number of stratigraphic quanta to make the concept of average sedimentation meaningful. Similarily, for larger stratigraphic intervals, it is important that the time spent in forming the bed must have been at least commensurate with the time spent in producing the bedding plane. Slowly formed beds that alternate with even considerably longer intervals of non-deposition, carry no time information and must be regarded as being at the limit of stratigraphic resolution.

If the beds which carry time information, are of approximately equal thickness then one may call them 'periodites' (Einsele, 1982).

The recognition of the various types of deposition and processes of bed formation depends to a great extent on detailed lithological and palaeontological examinations. The importance of such detailed sedimentological analysis cannot be over-stressed, and it is assumed that it always accompanies any stratigraphic analysis.

It is valuable to collect data on bed thicknesses under any circumstances, but if the beds carry time-information, then measurements of this type are even more important. Difficulties are sometimes encountered in collecting such material because of the unfortunately vague definition of sedimentary beds and bedding planes. A bedding plane is due normally to an abrupt change in sedimentation but its appearance in the outcrop can depend on the degree of weathering. For example it is well known that bedding in a certain formation may be well-developed in surface exposures and yet it may be completely absent in core material. It is therefore essential to use consistent material as far as possible in such studies.

It is unlikely that the shape of a bed-thickness distribution will give definite evidence about its origin, but one should always investigate whether an observed distribution is not contradictory to the distribution one would expect from some hypothesis. Various sedimentation models have been investigated (Schwarzacher, 1975) and the thickness-distributions that are generated by those models include normal distributions, various gamma distributions and compound Poisson distributions. The latter occur particularily frequently. The sedimentation models range from simple random walks to quite complex schemes of deposition and erosion. Because the sediment-time relationship is a statistical one, stochastic models are indicated. Therefore, if one is confronted with a sequence of beds that are all of approximately the same thickness, one will conclude correctly that they represent approximately equal time intervals.

At least a part of any observed thickness variation must be due to fluctuations in the sedimentation rates and only the remainder may be a true variation in the amount of time it records. It follows that the time fluctuation must always be less than that indicated by the thickness variation of the beds. This situation led Sander (1936) to make the following statements: "Equally spaced beds represent equal time intervals and are therefore time-periodic; beds which are not of equal thickness do not permit the exclusion of rhythmic events as their cause." This is known as the Sander Rule and it is based on sound probabilistic argument. It is recognized that sedimentation rates change at random and it seems unlikely the irregular fluctuations will combine invariably in such a way that beds of equal thickness are formed.

In the previous section, a trend was determined by plotting the stratigraphic position of sedimentary cycles against a scale with equal divisions. Here again, one may apply the argument inherent in Sander's rule and regular time intervals emerge once again as the simplest explanation for the production of such simple trends as straight lines or exponential curves. This argument is only valid if the beds carry time-information which also means that the bed thicknesses must be multiples of the minimum time resolution.

Markov Chain Analysis

The correlation structure of a stratigraphic sequence can also be examined when the data are restricted to such descriptive attributes as, for instance, rock types. Each attribute is called a state and the stratigraphic sequence can be regarded as a chain that moves from one state to the next. A transition probability matrix can be constructed by counting the transitions between the various states. The method for carrying this out has been described several times and is not difficult to apply (see Krumbein, 1967).

Structuring is the term that is applied to the process of dividing a stratigraphic section into states and two methods can be used. Lithological boundaries can be recorded as transitions. Alternatively, states can be defined by dividing the section into equal intervals and coding each interval according to the lithology it contains. Obviously, the latter method carries more information and could be used as an analytical method for very complex stratigraphic sequences which contain a great number of lithological states. However, so far no example for such an analysis is known, perhaps because it is very difficult to obtain sufficiently consistent material for such an investigation.

The method of structuring stratigraphic sections by lithologies has been both used and abused a great deal. It must be stressed that by abandoning any vertical scale, one loses almost all the information that is of stratigraphic value. The method can only be used to find some model 'sequence of lithologies' which may lead to some, necessarily primitive, sedimentation model. To assess the significance of such 'findings' is often difficult and even impossible, if no data on the stationarity and completeness of the sequences are provided.

Cyclic Sedimentation

Geologists have not always used the term cyclic sedimentation with a precise definition. However, it seems to be generally accepted that cyclic sedimentation is a pattern of sedimentation that is repeated over and over again. Considering the very nature of geological processes, one cannot expect the repetition to be invariable in the same sequence; one might well ask two questions if the process is to be quantified. How accurate must the repetition be? How many repetitions must there be for it to qualify as a sedimentary cycle?

There are three major reasons why cyclic sedimentation should arouse a stratigrapher's special interest. Any regularity in sedimentation patterns needs an explanation and this should contribute to our knowledge of the environmental history, during which the cycles formed. Secondly, any reference to real time, even it not very accurate, is important. Thus if it can be established that one is dealing with cycles carrying time information, then it is likely that similar sedimentary sequences represent similar time intervals. Thirdly there is the possibility that some sedimentary cycles really do represent accurate time cycles so that it may be possible to use these for absolute time determinations within given intervals.

The difficulty lies in deciding which of the three situations is appropriate in a specific case. Although models for the three basic types of fluctuation were given earlier in this chapter, their application is severely restricted by the inaccuracy of the stratigraphic time record. This inaccuracy is due to the statistical nature of sedimentation which makes it often a question of chance as to whether an environmental event is recorded in the sedimentary history. In addition to this, there is another factor which is important. Sedimentation is never uniform and constant in time and many environmental factors can influence sedimentation rates. Consequently, it is often found that sedimentation rates are linked to the type of sediment that is produced. A well-known example is that coarse, clastic sediments usually deposit much faster than finer ones. The significance of such a dependence can be demonstrated best by using the following imaginary example. Assume that a certain environmental history causes the deposition of a sandstone with a grain size that changes precisely as a sine wave in time. Assume further that the depositional velocity is strictly proportional to the grain size and slows down when the grain size becomes smaller. Plotting the grain size against stratigraphical thickness, would result in curves known as cycloids, which show three possible situations in the example illustrated in Figure 9. In Figure 9a, deposition during the production of fine material has slowed down. In Figure 9b, deposition came to a momentary standsill. In Figure 9c, erosion occurred and some of the finer material is missing.

The inevitable consquence of such a relationship between sedimentation rates and the variables which are being investigated is that any regularity in those variables will be incorporated into the vertical scale of the sequence. If one considers sedimentation to be based on the model given by equation 3 (see 'Analysis of Single Sections' earlier in this Chapter), then the effect of a random disturbance will be to transform the model's deterministic process into a stochastic process that follows the type of development given in equation 2. If it is taken into account that completely unrelated random numbers (as indicated by the model defined by equation 1) are very unlikely to occur, one must conclude that all stratigraphic records will follow the model given by equation 2. Therefore the three types of fluctuation which can be differentiated by their genetic models become identical or at least very similar.

Any interpretation of cycles will depend largely on determining at which scale the time information is carried by sedimentary cycles. If it can be established that one is dealing with time related stratification, the possibility exists that each cycle is triggered at quite irregular intervals, so that the time between cycles does not bear any relation to the real time that produced them. This situation is analogous to the one already discussed in relation to bedding, with the difference that in the present case each cycle may represent quite a considerable time span. However, interpretation of cycles is, stratigraphically speaking useless when the duration of a cycle bears no relationship to the total interval of time of the cyclic sedimentation itself.

A relatively homogenous time scale can only be deduced when sedimentary cycles were generated by oscillating systems of the type given in either equation 9 or equation 10. The various systems differ in the amount of damping and unfortunately, this damping is always increased by irregularities in the sedimentation rates in a way that has been discussed earlier.

Very little is known for certain about the type of oscillating systems which might cause cycles that would be of sufficient duration to be important in a geological context. Climatologists have proposed several mechanisms to explain climatic cycles, particularily during the Pleistocene. Such theories involve the interaction of such factors as

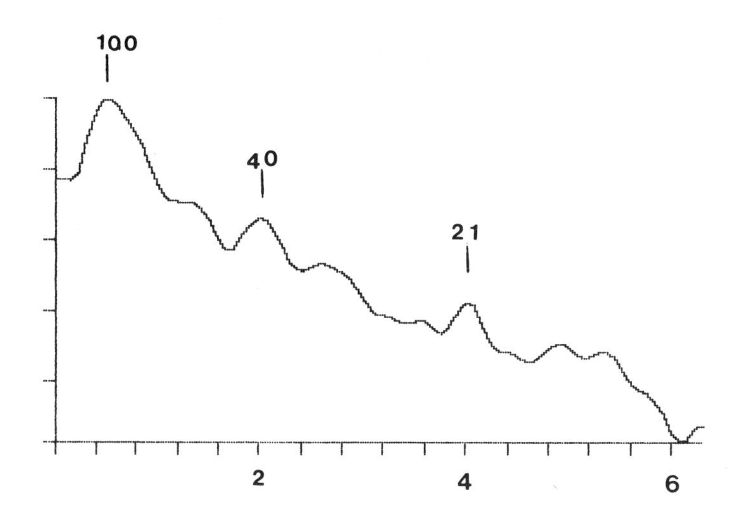

Figure 10 *Powerspectrum of ^{18}O data in core V28-239. Frequency units along horizontal axis are given in cycles/metres. The vertical axis gives the power in arbitrary units. Numbers are the duration of predominant cycles in thousands of years.*

ocean circulation, precipitation or absorbed solar radiation and so on. In explaining coal measures, some hypotheses invoke the interaction between sedimentary and ecological factors and others make reference to tectonic processes. Any such oscillating scheme is based on feed-back processes and will be damped considerably, as far as one can judge.

The only system that is capable of producing friction-free, long oscilliations that come close to undamped, harmonic movement appears to be the solar system. The shorter daily or annual cycles, are of limited importance in stratigraphy. Longer cycles are produced by the variations of the Earth's orbital elements which have durations of approximately 21, 40, 100 and 400 thousand years respectively.

Cycles of this magnitude are recognised in the oxygen isotope record of the Pleistocene deep sea sediments and power spectra show the presence of the 100, 40 and 21 thousand year periods (Figure 10). The identification of these cycles is particularly relevant because the sediment cores which provided the data can be dated with considerable accuracy (Shackleton and Opdyke, 1976).

The identification of similar cycles in older sediments is more problematic because the dating becomes less accurate and because the precise effect of the orbital variations on climate and on sedimentation in particular, is unknown. Nevertheless, some sedimentary cycles may be explained by such astronomical factors because of their length and frequency composition and it is obvious that such cycles could be of the utmost importance for stratigraphy and sedimentology.

Whether or not one accepts the astronomical hypothesis does not influence the fact that any sedimentary cycle from a series with continuous sedimentation can be regarded as an approximate time unit. As such, it can be very useful in the practical study of trends and rates of sedimentation.

REFERENCES

Einsele, G., 1982, Limestone-marl cycles (periodites): diagnosis, significance, causes - a review. In Einsele and Seilacher (Editors), Cyclic and Event Stratification, Springer, Berlin, 532 pp.

Grabau, A.W., 1913, Principles of stratigraphy. Seiler, New York, 1185 pp.

Krumbein, W.C., 1967, Fortran IV computer programs for Markov chain experiments. Kans. Geol. Surv., Comput. Contrib. 13, 1-8.

Otnes, R.K. and Enochson, L. Applied Time Series Analysis, 1. Wiley, New York, 449 pp.

Tipper, J.C. 1983, Rates of sediments and stratigraphic completeness. Nature, 302 (5910), 696-698.

Schwarzacher, W. 1975, Sedimentation Models and quantitative stratigraphy. Elsevier, Amsterdam, 382 pp.

Seilacher, A. 1982, General Remarks about Event Deposits. In Einsele and Seilacher (Editors), Cyclic and Event Stratification, Springer, Berlin, 532 pp.

Shackleton, N.J. and Opdyke, N.D., 1976, Oxygene-Isotope and Paleomagnetic Stratigraphy of Pacific Core V28-239, Late Pliocene to Latest Pleistocene. Geol. Soc. Am. Mem. 175, 449-463.

IV.2

LITHOSTRATIGRAPHIC CORRELATION AND SEDIMENTATION MODELS

W. SCHWARZACHER

INTRODUCTION

The previous Chapter restricted lithostratigraphic analysis to the study of single sections. For the analysis of entire facies, methods must be introduced to carry out stratigraphic correlation.

Lithostratigraphic correlation is best defined as the correct identification of lithological boundaries; an identification is correct when the correlated points reproduce the true shape of the rock-body (lithosome).

Two aspects of this definition are of particular importance. Firstly, it should be remembered that geological mapping is nothing else but the repeated application of lithostratigraphic correlation and this makes the importance of the procedure self-evident. Secondly, lithostratigraphic correlation is not a measureable quality in itself, since it can only be either right or wrong. Proving the correctness of a correlation can only be done in either of two ways. Attempts may be made to establish that the horizon which was observed in several localities is absolutely unique in the stratigraphic column. Alternatively, individual horizons may be traced continuously from locality to locality where there is complete exposure.

In most practical situations the similarity between two geographically separate sections can be established using so called matching procedures. In contrast to stratigraphic correlation, similarity can be measured and so a degree of matching can be established. It must be noted however that two sections can be perfectly matched without being necessarily correlated in the stratigraphical sense. For instance sections containing cyclically repeated lithologies have corresponding repeated match positions but of course only one valid lithostratigraphic correlation.

LITHOSTRATIGRAPHIC CORRELATION

Measures of Stratigraphic Similarity

Any measure of similarity depends largely on the type of data which is available. Thus a very common way of describing stratigraphic sections is the use of attribues such as fossil content or lithology, as was considered in the previous chapter. Under such circumstances, similarity can be expressed quantitatively by giving the proportion of positive matches. This is done by placing two sections side by side and counting how often two attributes which are opposite one another are identical. The so called matching coefficient S_m (Sackin et al. 1965) is defined as

$$S_m = \frac{\text{number of matches}}{\text{total number of comparisons}}$$

and can be determined for a number of different match positions that are systematically changed by keeping one section stationary and sliding the second section past the standard. The procedure is shown diagrammatically in the following example

```
    A B C E F G
          |
Z B C F E G
```

$$S_m = \frac{1}{5} = 0.2$$

```
A B C E F G
  | |     |
Z B C F E G
```

$$S_m = \frac{3}{6} = 0.5$$

```
A B C E F G
        |
    Z B C F E G
```

$$S_m = \frac{1}{5} = 0.2$$

By definition the matching coefficient varies between 1.0, for a perfect match, to zero for no match at all. The actual value of the matching coefficient depends very much on the type of attribute which is used for describing the section. Thus, if the attribute is a common property of sediments, all coefficients will be high. If relatively rare properties are used the overall matches will be low, but the maximum of the best match will be relatively sharp if some correlation exists. The problem is not defined well enough to make any statistical assessment of the significance of such matching procedures.

One of the disadvantages of this cross association technique is that sequences containing gaps, which are due either to incomplete exposures or to non-deposition, cannot be compared. Smith & Waterman (1980) proposed a method to overcome this and which therefore permits gaps to occur anywhere within the sections. The method involves calculating a distance matrix for any combination of the observed sequences and interspersed gaps. The matrix is then used to find the alignment which provides the maximum number of matches or the minimum number of miss-matches, which amounts to the same thing. In order to obtain some numerical values for the distance, different weights are given to the three events: match, miss-match and gap.

This modified cross-association method of Smith & Waterman was originally developed to match genetic sequences in molecular biology, where one is dealing with chains of well defined items. Lithological units in stratigraphy rarely are so well defined and independent of each other and for this reason none of the cross-association methods is very attractive.

Yet a further modification of the pairwise comparison was introduced by Gordon & Reyment (1979) by their slotting method. In contrast to the previous methods, this technique attempts to combine two sections into one series in such a way that the successive data points show a minimum of dissimilarity. For example two sections $A_1 \ldots A_n$ and $B_1 \ldots B_n$ are to be slotted with this principle in mind. A possible result may be the following series:

$$A_1, A_2, B_1, A_3, B_2, A_4, B_3 \ldots$$

The method has the advantage that a limited number of gaps as well as the thinning out or the change in character of some beds, would not upset the slotting process.

Possibly more satisfactory results can be obtained if the lithological variation is given by one or more numerical variables. Ideally, such variables should be given at equal intervals to provide a time series as was described in the previous chapter. The problem in this case amounts to comparing two sections x_t and y_t which refer to some lithological variable which was measured at the stratigraphic positions, $t = 0, 1, 2 \ldots$ The conventional statistic which is used for such a purpose is the cross-covariance, which is defined as:

$$\text{cov}(xy)_k = \{1/(n-k)\} \sum_{t=1}^{n-k} x_t \, y_{t+k}$$

Like the autocovariance, it can be standardised to provide a cross-correlation coefficient. The procedure for calculating the cross-correlation is exactly the same as that for calculating the autocorrelation, with the sole difference that two different sections are compared. Clearly, the lag, k, depends on the reference point from which the sections are measured, which is usually x_o and y_o and which may be the, more or less arbitrary, base of each section. By systematically changing the lag one can search for the highest correlation which will have a maximum in the position of the best overall match.

An alternative way of measuring similarity can be derived from the cross-variogram, which can be written as:

$$\gamma(xy)_k = \{1/2(n-k)\} \sum (x_t - y_{t+k})^2$$

Since the cross-variogram is based on the difference between the two values being compared rather than their product, it can be applied in situations where the mean of the lithology is not constant (non stationarity).

The cross-correlation or the cross-variogram determines the similarity of two sections in the time or space domain as the case may be. An alternative but mathematically equivalent way is correlation in the frequency domain. It was pointed out in the previous chapter that the power spectrum which analyses a time series in terms of its frequency composition, is the Fourier transform of the autocorrelation function. If one takes the Fourier transform of the cross-correlation function, one obtains the so-called cospectrum. The cospectrum is a complex function which can be represented by two real functions; the coherence and the phase-spectrum. The former is directly related to the correlation coefficient.

Like the correlation coefficient, the coherence can vary between +1.0 and -1.0 but it is frequency dependent and this also applies to the phase-spectrum. Both functions can provide some very valuable insight into the origin of stratigraphically produced time series.

The Correlation Problem in Geology

Before going any further in developing quantitative methods which can be applied to the correlation problem it is useful to consider the practical requirements of geology first.

Lithostratigraphic correlation problems can be divided into two groups:

Group 1 concerns exclusively bedded sedimentary rocks in which sedimentation rates were constant over large distances. This results in the so-called 'layer cake stratigraphy' and it should be possible to correlate extensive stratigraphic sequences based on the correlation of a single marker horizon.

Group 2 refers to any other correlation problem, in particular with sedimentary strata which have different thicknesses in different localities due to variable sedimentation rates. Included also are non stratiform deposits such as reefs, mounds or the result of erosion or non-deposition. Obviously the most difficult problem is the reconstruction of lithosomes which have undergone rapid facies changes.

With any problem concerning either of the two groups one will attempt to make the correlation as complete as possible. This means that one will try to correlate as many lithostratigraphic units as possible. To achieve a really detailed result, it is necessary to differentiate a maximum number of lithological events in a section and to correlate every single horizon. An overall best match of two sections is fairly useless to the geologist unless it leads to a detailed correlation.

Once a correlation has been achieved, it would be useful to have some measure of reliability which could be the probability that a certain correlation is correct. Unfortunately, such a statistical assessment of correlation results is extremely difficult to obtain and is only possible if some very restrictive assumptions are made, as will be shown.

SOME PRACTICAL METHODS APPLICABLE TO GROUP 1 SITUATIONS

If a situation conformed strictly to the condition of constant sedimentation rates over wide areas, then it should be possible to correlate any two sections, provided that at least one marker horizon can be identified in each locality. In this case, the problem of estimating the reliability of such correlations is reduced to the question of whether the marker horizon was identified correctly. However, absolutely constant sedimentation rates are not found in real situations and errors will also be made in measuring the section. The combined effect of this is that the reliability of the correlation is lowered.

The Graphical Correlation Method

A useful graphical representation of any correlation problem is to plot the stratigraphic positions of identified horizons along the x and y axis of a Cartesian coordinate system. Assume that the following sequence is found in locality 1: marker horizon m, rocktype A, rocktype B, rocktype C and A. The respective lithological boundaries can be plotted at the positions x_1, x_2, x_3, x_4. A similar sequence is observed in locality 2 and here the corresponding positions y_1 to y_4 are plotted. If strict group 1 conditions apply the connected points x_1 y_1; x_2 y_2;...should fall on a straight line, which is inclined at an angle of 45 degrees to the x and y axis. Such a line is called a mapping function of sequence y on sequence x. An actual example of such a correlation is shown in Figure 1. Because of the slight variations in sedimentation rates and because of possible errors in correlation, points will not fall exactly on to the straight line. It would be useful to have some confidence limits which would permit one to decide which correlations are acceptable and which fall outside the expected range. The problem was investigated by Odell (1975), who considered a situation where two correlation points were fixed by two correctly identified marker horizons in both localities. The two points determine the mapping function which represents the mean of the expected correlation points. The model used by Odell assumes that the variance of the deviations increases with increasing distance from the marker points. Therefore the maximum deviations are half way between the two. If one standardizes sections x and y to run from 0 to 1 and if one writes d for the deviation of a point from the correlation line (see Figure 2), then one can write for a fixed x:

mean of y = x, variance of y = kx (1-x), in which k is a constant.

The deviations d are approximately normally distributed and

$$\text{Prob}\ \{|d(x)|\} < \sqrt{kx(1-x)} = 0.95$$

therefore represents a 95% confidence line. An estimate of the constant k can be obtained by estimating the variance of d, $\hat{\sigma}^2$. For the 95% confidence limit, $\hat{k} = (1.96)^2\ \hat{\sigma}^2$. The variance itself is estimated from the actual data points on the plot: For n points the variance is:

$$\sigma^2 = (6/n) \sum_{i=1}^{n} d_i^2$$

The method developed by Odell is of course only applicable in an area in which there is an already well established stratigraphy and where there are a number of reliable correlations available. Odell himself applied the method to the East Midland coalfields of England using rare marine bands as marker horizons. He was able to show that some coal seams which hitherto had been thought to be well-correlated, fell outside the confidence limits. These coal seams are now recognised as being wrongly correlated.

If Odells model is applied to two localities which have only one marker horizon in common, one would have to assume that the variance of d increases linearly above and below the marker horizon and it is difficult to find a theoretical justification for this. In this situation one might make the alternative assumption that the deviations from the mapping function are constant. This would lead to confidence limits which are symmetric and parallel to the correlation line. At the end points which are possibly marker points one can postulate either a finite or a zero variance. A finite variance could represent observational errors or errors of measurements which of course also applies to the positions of the marker horizons. Like Odell's model this is probably a vast over-simplification and it is necessary to collect a great deal more evidence before the method can be elaborated.

The Reliability of Marker Horizons

The correlations which have been discussed so far, rely on the identification of marker horizons and it is therefore necessary to question the reliability of such identifications. Clearly, if any stratigraphic horizon has unique properties, then it can be positively identified. If one is dealing with sedimentary rocks however, one finds the same lithologies again and again and even if there is such a thing as a unique rock type, it would be impossible to prove. It should be noted that this is in pronounced contrast to the situation in biostratigraphy where the first and last appearance of a species is by definition a unique event. It is true that it may be difficult to pinpoint such biological events but it is relatively easy to estimate the probability that such points have been found. In the absence of this principle of unique lithostratigraphic units one tends to use relatively rare lithologies as marker horizons. Typical examples are boulder beds, bentonites, or exceptionally thick turbidites, or in fact anything which is unusual in a given sedimentary series. Of course this method works well in most cases but it is impossible to assess such methods in quantitative terms, without knowing a great deal about the local stratigraphy already. For example, if use is made of an ash layer to connect two sections then one has to know (a) that the two localities cover the same stratigraphic interval or at least that some stratigraphic overlap exists, and (b) that this particular interval contains only one ash bed in both localities. This is the type of information for which the experienced geologist may develop a certain feel, but it is not something which can easily be quantified.

The Correlation of Sections Without Marker Horizons

If it is fairly certain that one is dealing with an area in which sedimentation rates are constant and in which facies changes and

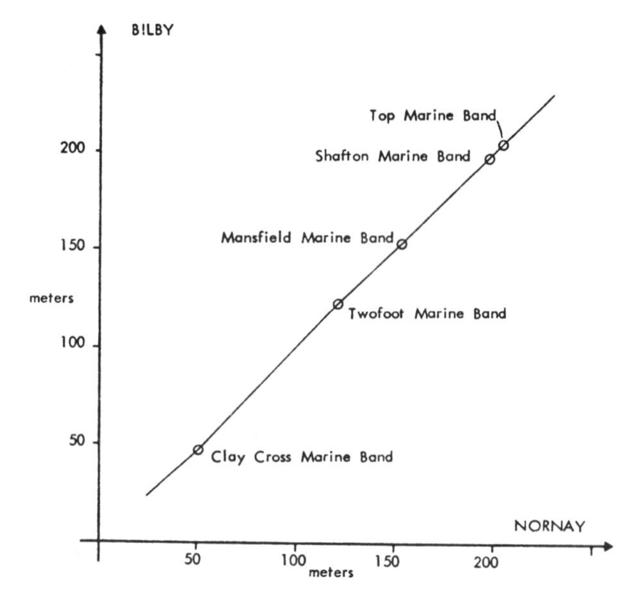

Figure 1 *Example of a graphical correlation. Horizontal and vertical axes represent the stratigraphic sequence in two boreholes (Nornay and Bilby, East Midlands, England). Correlated rock types are indicated with circles and connected (after Odell 1975).*

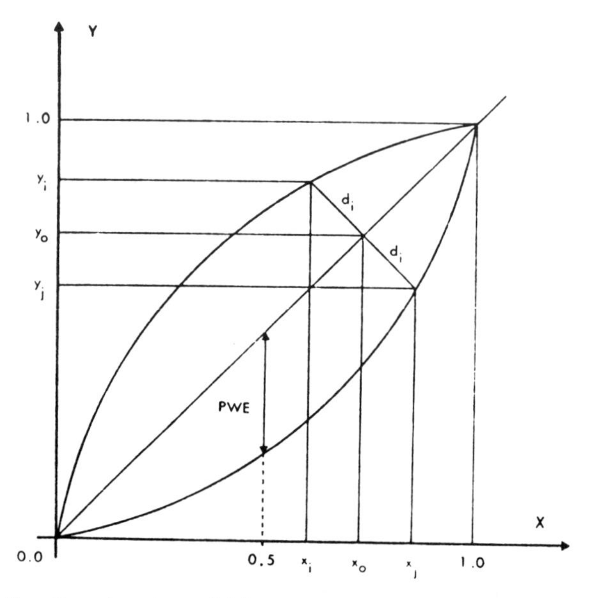

Figure 2 *Graph to illustrate the derivation of functions which describe the confidence limits of correlation (after Odell 1975).*

partial erosion are unlikely, then it should be possible to correlate two sections by finding the position of greatest similarity between the two. Genuine stratigraphic correlation should be possible, unless the sequences have a tendency towards cyclic sedimentation. For such matching, one may either use the previously discussed matching coefficient or the cross-correlation coefficient if numerical data are available.

The statistical assessment of such an analysis is again very difficult. In the case of the matching coefficient, it has been proposed that a recorded section could be compared with a random pattern, but a Markov chain analysis will probably show that this is not an acceptable procedure. Similar problems arise when cross-correlation coefficients are used. As was pointed out before, most conventional statistical tests are based on random sampling and systematic sampling along a section does hardly ever lead to a random pattern. This non randomness is best indicated by the presence of autocorrelation. Clearly, if two series are cross correlated and one of them shows auto-correlation, then the second sequence too must have auto-correlation. Similarly, if two sequences have auto-correlation then they have something in common and must show cross-correlation. In order to test the significance of any correlation between two sections one must know the number of independent observations from which the correlation coefficient was derived. This can be done approximately as follows. Assume two sections are compared by N couples which were observed, let the first k ($k < N$) positive auto-correlation coefficients of section 1 be $r_1, r_2, \ldots, r_k$ and the coefficients of section 2 $r'_1, r'_2, \ldots, r'_k$, then the number of independent observations N_e is given approximately by:

$$N_e = N/(1 + 2\, r_1 r_2' + 2\, r_2 r_2' + \ldots + 2\, r_k r_k')$$

An important condition for this test being applicable is that the auto-correlation coefficients should converge towards zero. Such a convergence will not occur if the series is too short, or if it shows a tendency towards cyclicity or if there is any trend present. The latter, in particular, can lead to misleading results. For the convergence to be recognized, N clearly must be larger than k.

Despite these difficulties, cross-correlation coefficients and similarily cross-covariograms are good qualitative guides for the similarity between sections and can therefore be used for finding good matching positions. A qualitative check on the reliability of a correlation that is based on such matches, can again be obtained by a correlation plot. If the ratio of sedimentation rates from two localities remains reasonably constant, then the correlated points should fall near a straight line. One can of course calculate the correlation coefficient for such data, but the meaning of this quantity is extremely difficult to interpret since the deviations from the straight line are due to such a variety of different factors and errors.

CORRELATION IN GROUP 2 SITUATIONS

Problems in Areas with Variable Sedimentation Rates

Stratigraphic correlation has to deal with geological situations which range from the simple 'layer cake' stratigraphy to areas which originate in the most complex environments and which contain rapid facies changes and often quite incomplete sedimentation. In the following paragraphs such more complex situations will be considered. However, it will be assumed that the sedimentary sequences are still approximately stratiform. Lithostratigraphic problems involving structures like reefs, igneous intrusions and strongly deformed rocks are outside the scope of the present discussion.

To obtain correlations in geological settings which are not strict layer cake stratigraphy one has to deal with three frequently occurring complications.

(1) Unequal sedimentation rates will lead to sections of different overall length in different localities.

(2) Some horizons are not persistent and will not be represented in all the sections.

(3) Facies changes will substitute different lithologies for the same horizon in different localities.

The last two points imply of course that the lithological boundaries which one attempts to find are, strictly speaking, no longer stratiform. In this connection, one should remember that practically every stratigraphic survey provides very detailed information about the vertical variation but comparatively little data about the horizontal variation of sedimentary rocks. Profiles or well logs are often measured in centimeters but the spacing between bore holes or measured sections is often measured in kilometers. This practice reflects the inherent variability of most sediments to a large extent and it is caused by the enormous difference between the vertical sedimentation rates and the rates of lateral spreading. As long as this pattern of variability is present one can use correlation methods which treat sediments as stratiform and which compare essentially two profiles horizon by horizon, as in the previous section. Differing sedimentation rates and indeed gaps in the sedimentary record can be neutralized by adjusting the vertical scale of the two profiles to be compared. One of the first techniques therefore to be investigated is the shrinking or stretching of stratigraphic sections.

Uniform Stretch of Stratigraphic Sections

The artifice of giving a uniform stretch to sections can be achieved graphically and has always been used by stratigraphers. Haites

(1963) discussed the geometry of such a procedure. The comparison of two sections of unequal length involves the projection of the shorter section on to the longer using straight lines which intersect in a single point or vertex. For example, Figure 3 gives three sections from a Carboniferous limestone in which the horizons A to M have been identified and correlated. The condition that the rays meet in a single point is that the ratios of the thickness intervals are identical i.e.:

$$\overline{A_1B_1} / \overline{A_2B_2} = \overline{B_1C_1} / \overline{B_2C_2} = \overline{C_2D_2} \dots$$

This condition is clearly given for horizons A to D in sections one and two and also for horizons B to G in sections three and two. Haites calls such a system of connected stratigraphic levels a perspectivity and Figure 3 shows five perspectivities altogether.

If sections are represented in digitized form and a stratigraphic variable is given at equal intervals, stretching can cause problems which were first discussed by Neidell (1969). Changing the vertical scale (usually by stretching) the shorter section will in most instances cause the data points of the two sections to be no longer at identical intervals. To restore this situation, equally spaced points have to be found by interpolation. Furthermore, stretching a sequence has an effect which is best understood if the series is considered in the frequency domain. It has been shown in the previous chapter that a time series can be considered as the sum of harmonic functions with different frequencies. Because data are collected at finite sampling intervals, any frequency which is higher than one divided by twice the sampling interval will be missed. The highest frequency which can be represented is called the Nyquist frequency and frequencies beyond it are ignored and can be regarded as noise. Stretching a series, however, lowers the Nyquist frequency and in this way part of the noise, previously cut off, enters the data. In some situations this may cause difficulties and Neidell recommends that the stretched series be filtered; his computer program (Neidell 1969) provides for this.

Neidell gives particular attention to interpolation and uses the so-called Whittaker interpolation which is given by:

$$F_x = \sum_{i=1}^{n} F(i\Delta x) \frac{\sin \pi \, (x/\Delta x - i)}{\pi \, (x/\Delta x - i)}$$

Where x stands for the value at which the function F is to be interpolated, i indicates data points, Δx is the sampling interval and n the length of the series. It is possible to truncate the sum but it should at least be extended over 10% of the length n. The trigonometric

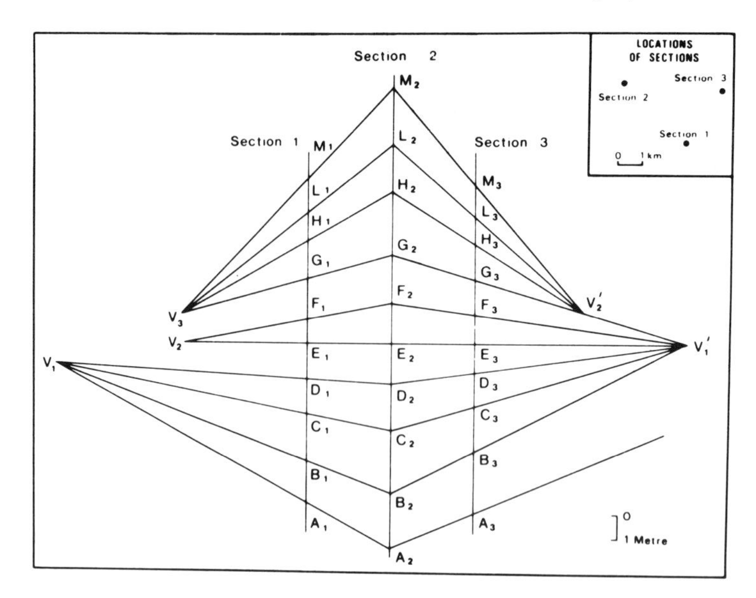

Figure 3 Perspective correlation of three sections. Horizons A to M in each section have been identified and correlated. Lines originating from points V connect horizons which can be correlated by uniform stretch (after Schwarzacher 1975).

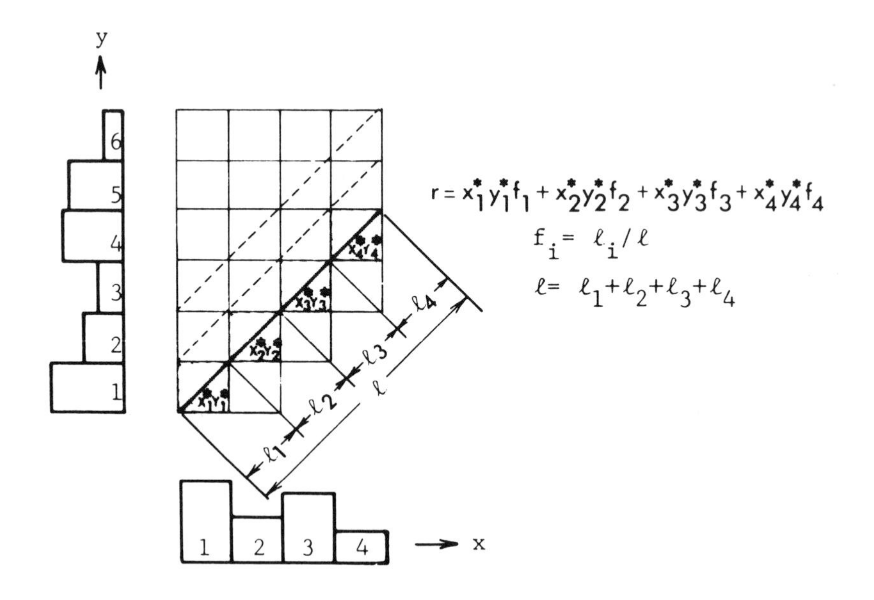

Figure 4 Graph illustrating the derivation of Kemp's correlation coefficient without stretch. For details see text (after Kemp 1982).

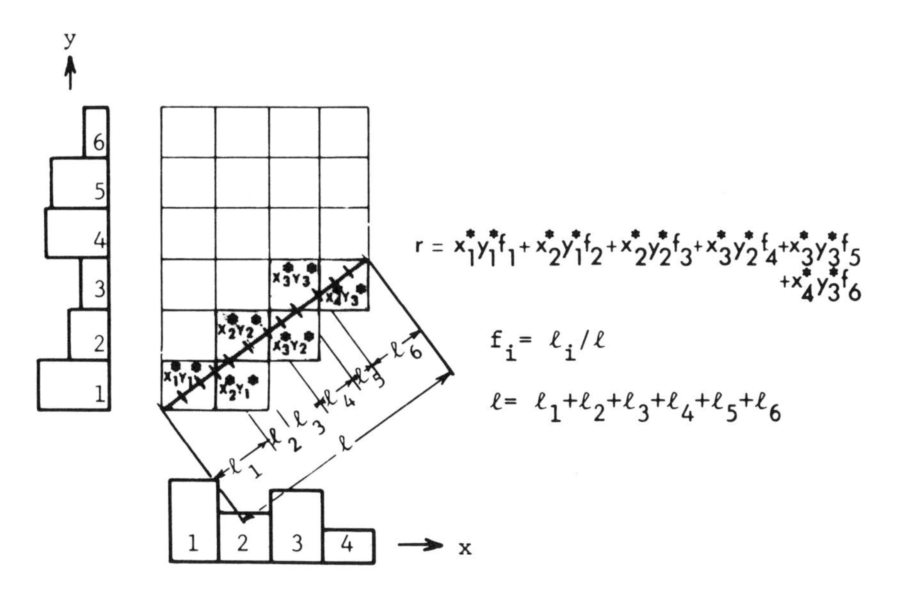

Figure 5 *Graph to illustrate the derivation of Kemp's correlation when uniform stretch is applied. For details see text (after Kemp 1982).*

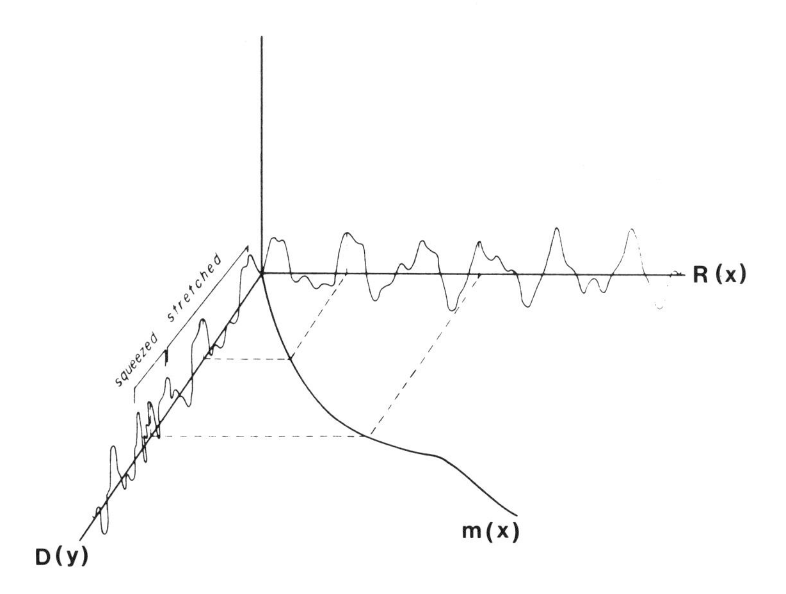

Figure 6 *The correlation of two time series using a variable stretch. The function R(x) is taken as standard and the function D(y) is squeezed and stretched to make it similar to the standard R(x). The curve m(x) records the amount of stretching, (after Martinson et al. 1982).*

interpolation is only suitable for data changed by uniform stretch. If data points are sampled at unequal intervals, as for example happens with variable stretch (see later), other methods must be used. Experiments in interpolation have shown that good results can be obtained with spline functions (see previous Chapters), and if a great number of interpolations has to be performed one may use even simpler techniques such as linear interpolation.

A computational method which avoids the necessity of direct interpolation has been given by Kemp (1982). The method is best understood by making reference to the correlation diagram of two sections x and y. It is noted that the correlation coefficient r can be defined as:

$$r = \sum_{i=1}^{n} x_i^* y_i^* f_i \quad ; \quad f_i = i/n$$

where n is the length of the section, x_i^* y_i^* are the normalized data pairs which have been correlated, and f_i is weight (see below). The products x_i y_i are geometrically represented as heights above the shaded squares in Figure 4. If the sedimentation rates are the same in both localities the points fall on to a straight line which is inclined 45 degrees to the x axis and passes through the origin. This line is called the mapping function between the sections x and y. Shift correlation, for example r_{01}, r_{12}...or r_{02}, r_{13}..., is represented by the dashed straight lines in Figure 4.

These lines are parallel to the original mapping function but intercept the y axis at 1 and 2 when x equals zero. The weights f_i are geometrically represented by the intercepts l_1, l_2, ... in Figure 4. If the mapping function is less than 45 degrees, section x is relatively stretched compared with section y. If the mapping function forms an angle larger than 45 degrees, then section x has to be shrunk in relation to section y. The correlation coefficient for stretched (or shrunk) sections can directly be calculated by adding all products through which the mapping function passes and multiplying them by the weights which are proportional to the lengths of intercepts, ℓ, shown in Figure 5.

Kemp's method requires approximately the same computational effort as the previously discussed procedures with the advantage that no special interpolation is needed. Other advantages will be pointed out later.

In order to make the stretching procedure useful for the purpose of stratigraphic correlation, two operations have to be performed. The best stretch has to be found by an iterative process which keeps one section constant and gradually increases the stretch of the second sequence. Secondly, to find the greatest similarity between the two sections it is necessary to shift the two sections at each stretch increment against each other. The two procedures will be referred to as stretch and shift correlation.

In order to identify the best value of stretch correlation, Rudman & Lankston (1973) compare the cross-correlogram between two sections with the auto-correlogram of the standard, meaning the non-stretched section. The comparison can be made by averaging the squares of differences between the two functions. Any of these methods requires a great deal of computer time which can be considerably reduced if either the required shift or stretch is at least approximately known.

Correlations Involving Variable Stretch of Sections

In many situations one finds that sedimentation rates change either continuously or abruptly during the deposition of sequences which have to be correlated and such problems require variable stretch. Using the methods of cross-correlation, Rudman and Lankston (1972) compared well-log data which were subdivided into different subsections, whereby each of these 'windows' was given a different stretch or shrink factor until a good fit between two sections was obtained. Kemp (1982) showed that if one calculates a correlation surface rather than a single correlogram one can identify areas of particularly high correlation in it. The connection of such areas of high correlation by a series of straight lines yields a mapping function which reflects the variable sedimentation rates.

An elegant method which can handle much more complex problems was introduced by Martinson et al. (1982). Only the basic principles of this method have been published and a complete computer program which will probably be quite extensive is not yet available. The Martinson method involves finding the mapping function m(x) that relates a stretch sequence D(y) to a standard series R(x), which has a length of $n\Delta x$. The sections again are plotted on a Cartesian system and the mapping function connects the correctly correlated points x and y. In contrast to the previous examples the mapping function is no longer restricted to straight lines and can take on any shape according to whether section y has to be shrunk or expanded (see Figure 6). The mapping function can be approximated by a truncated Fourier series, containing r coefficients:

$$m(x) = a_0 x + \sum_{i=1}^{r-1} a_i \sin\left(\frac{i\pi x}{n\Delta x}\right)$$

in which the coefficient a_0 determines the initial inclination of m(x), and represents the best uniform stretch which is necessary between two

sections. Any added coefficient will modify the mapping function. The successive Fourier coefficients are determined by finding the maximum correlation between the two series after applying the mapping function at each stage. The number of coefficients to be used in determining the mapping function depends very much on the resolution which is required and of course also on the type of record which is used. For example if some cyclical record is to be correlated and if each peak in this sequence is to be tuned, then m(x) must be represented by the number of harmonics corresponding to the record. Thus, if $n\Delta x$ is the length of the record and Xmin the shortest significant period the number of required harmonics r is given by:

$$r = 2n\Delta x/X_{min}.$$

The maximum number of coefficients which Martinson et al. (op. cit.) used was around 30. The iterative process of determining the coefficients is begun by making an initial guess at the mapping function. The success of the method often depends on the quality of this initial estimate and this is particularly true for fairly noisy data. Martinson et al. recommend that a first estimate should be obtained after treating the data with a low pass filter and the analysis of the unfiltered series can then be based on the initial mapping function. The method is capable of dealing with quite abrupt changes in the mapping function, as would occur if there were gaps or hiatuses in one of the sections. Furthermore, the method also deals reasonably well with records which are cyclic in nature, but here of course multiple solutions are again possible.

Martinson et al. applied the method to correlate ^{18}O data from deep sea cores and to the correlation of marine magnetic profiles. In both of the examples the starting point of the sections was known and it was not necessary to have a search procedure for the best shift correlation. It would be possible, of course, to incorporate such a search procedure.

The Reliability of Correlations Involving Stretch

In assessing the reliability of correlations based on stretching procedures, it is particularly important to differentiate between the degree of matching and stratigraphic correlation. The degree of matching is based on the similarity between two sections and it must be noted that the technique, discussed previously, makes it possible to change any sedimentary record in such a way that it becomes similar to a possibly arbitrary standard section. Therefore, given similar parameters, there is no reason why one should not alter the mapping function to such a degree that all sections become identical. A very similar problem arises when one is concerned with establishing the regularity of cyclic sedimentation. Once it is admitted that sedimentation rates vary and that it is therefore legitimate to adjust the vertical scale of sections, it is possible to make cycles as regular as one desires. In this and similarly in the correlation problem, one has to somehow decide that the assumed sedimentation rates are reasonable. This type of assessment is almost entirely based on geological arguments and is not easily quantified.

One might suggest two general rules. The first is that sedimentation rates are more likely to be simple and steady functions. The second rule is that changes in sedimentation rates are almost invariably somehow connected to changes in lithology.

The most likely situations are either constant or constantly declining sedimentation rates. The condition of constantly increasing sedimentation seems to be much rarer. Constant sedimentation leads to a linearly increasing cumulative sedimentation curve and any constantly decreasing or increasing sedimentation leads to exponential accumulation curves. In most sedimentary basins which permit lithostratigraphic correlation one would expect the relative conditions to be the same in different localities and this (apart from minor fluctuations) should lead to straight line mapping functions. An exponential mapping function for example could only occur if sedimentation is constant in one locality and constantly decreasing in another. The latter is of course possible but would certainly require some geological explanation. When variations in relative sedimentation rates occur, the mapping function is most likely to consist of a number of near straight line segments with different inclinations. Following the second proposed rule, one would expect any change of slope to coincide with changes in lithology in one or both of the sections to be compared. Indeed, if two sections can only be made similar by frequently changing the relative sedimentation rates without any sign of associated lithological changes, one may not accept such matches as stratigraphic correlations. On the other hand, if one finds that good matches can be obtained by approximately straight line mapping functions which change their slope at lithological boundaries one may argue that this is unlikely to be an accidential result. The same argument was used earlier in judging straight line relative sedimentation curves. As before, the value of this argument cannot be quantified.

SEDIMENTATION MODELS AND QUANTITATIVE CORRELATION

Sedimentation models can be defined as the quantitative equivalent of the working hypothesis, which is an essential part of any geological activity. Any hypothesis which relates to the distribution and extent of lithological units is of potential use in lithostratigraphic correlation problems. Ideally, the sedimentation model should be capable of predicting the three dimensional shape of lithostratigraphic units. In practice, one is far removed from this ideal, and only simplified systems can be treated as yet.

Because lithostratigraphy is primarily concerned with the geometry of rock bodies, various trend-surface techniques are useful in describing the shape of lithological units. Any such geometrical model, however, has to have predictive properties and for this reason must be somehow linked to the mechanism of sediment formation. It has already been stressed that sedimentation processes involve stochastic elements and the geometry of the rock units formed in this way will reflect this. It is this stochastic component in particular which makes the correlation by matching difficult and uncertain.

The difficulties of analysing the stochastic element of shape variation are very similar to the problems of time series analysis and in particular it is very difficult to separate the stochastic variation from an assumed well-defined trend. However, as in time series analysis, it is possible to formulate stochastic processes which operate in space and time and which generate structures which are at least similar to actually observed configurations. The theory of such temporal spatial processes is still fairly incomplete but more complex situations have been solved by Monte Carlo methods.

The link of models with reality can be achieved by comparing the correlation structure of the observed and theoretical processes. The description of the correlation structure of stratigraphic variation can be derived from a series of profiles and this reduces the problem to two dimensions. It will be assumed that in any profile a stratigraphic variable $x_{s,t}$ is given at points s, t where s is measured along the horizontal axis and t along the vertical or time axis. In practice a profile of this kind will be made up of a number of measured vertical sections which are spaced at intervals s. The covariance (C)-matrix of such data can be calculated by using the formula:

$$C_{s,t} = \{1/(m-s)(n-t)\} \sum_{i=0}^{m-s-1} \sum_{j=0}^{n-t-1} x_{i,j}\, x_{i-s,j-t}$$

In this m is the number of vertical sections and n is the number of observations in each section giving a total of m times n observations per profile.

A particularly simple process, which will be taken as an example for a sedimentation model, is the unilateral first order Markov process which operates in space and time. The model can be written as:

$$x_{s,t} = \alpha x_{s-1,t} + \beta x_{s,t-1} + \varepsilon_{s,t}$$

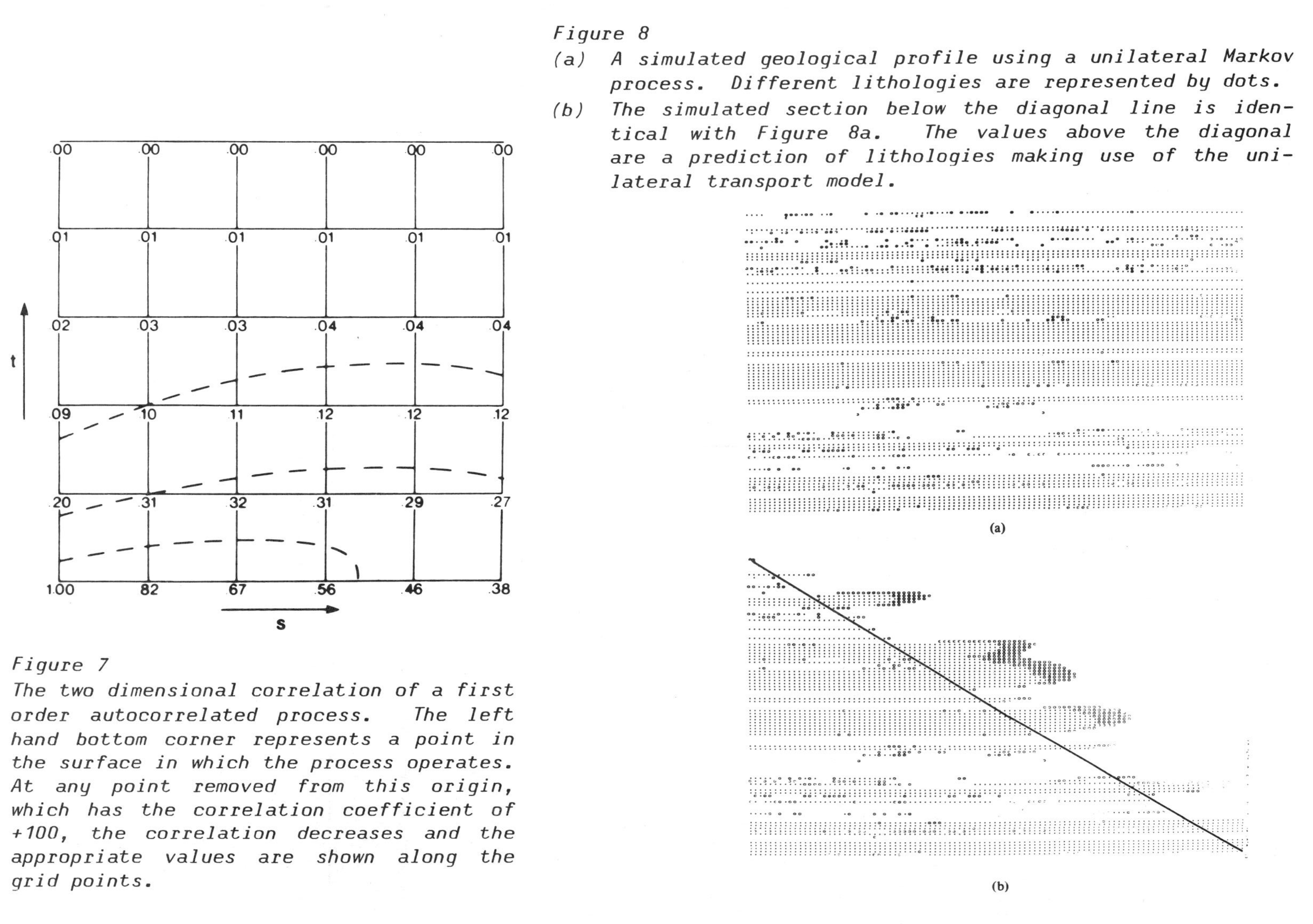

Figure 7
The two dimensional correlation of a first order autocorrelated process. The left hand bottom corner represents a point in the surface in which the process operates. At any point removed from this origin, which has the correlation coefficient of +100, the correlation decreases and the appropriate values are shown along the grid points.

Figure 8
(a) A simulated geological profile using a unilateral Markov process. Different lithologies are represented by dots.
(b) The simulated section below the diagonal line is identical with Figure 8a. The values above the diagonal are a prediction of lithologies making use of the unilateral transport model.

In this α and β are constants and ε is a stochastic element. Expressed in words this model assumes that the type of sedimentation at site s depends on what happens at the neighbouring site s-1 and also what happened at site s at time t-1. The model therefore is a somewhat primitive representation of undirectional transport, the deposition of sediment depending to a limited extent on the past history of each site.

The correlation structure of this process can be calculated (Besag 1972) and an example of such a two dimensional correlogram is shown in Figure 7.

To illustrate the procedure, a simulated geological profile was generated using the above equation; the value of x was taken to vary from -1 to +1 and in Figure 8 all positive values are shown. Dots represent x values greater than 0.5 and zeros represent values greater than 0 and smaller than 0.5

The lower half of Figure 8 shows the identical profile, but in this case the simulation was stopped along the indicated diagonal line, and the values found along this line were used to predict the profile to the right of the line. It can be seen that after a short distance the predictions become zero, which is the mean value of x and which is shown blank. One can use the variogram for measuring the similarity of sections which are h units apart:

$$\gamma_n = (m/2) \sum_t (x_{s,t} - x_{s+h,t})^2$$

The theoretical variogram for this particular process is:

$$\gamma_n = \text{var } x \; (1-\lambda^{|h|})$$

where λ is a constant which can be calculated from the covariance structure (see Schwarzacher 1980). The variogram shows that the similarity between sections decreases exponentially with distance and this relationship may be used to measure the realiability of a proposed correlation.

The above example demonstrates two important points. Firstly to assess the reliability of a stratigraphic correlation in quantitative terms one has to know the correlation structure of the field data at least in parts of the area to be investigated. Secondly, one has to formulate a sedimentation model which is valid in the sense that it can generate a correlation structure which conforms to the observed one. Furthermore the model must also be representative of an unknown area so that predictions can be made.

One of the major difficulties in applying the method in practice is in finding the spatial correlation structure of an area. For instance in trying to calculate a correlogram for some Carboniferous limestone (NW Ireland), a 10 km long profile was available and any amount of vertical thickness data. Analysis of single sections showed that the vertical variation could be reasonably portrayed by data points at 2 meter intervals. To obtain a comparable variation in the horizontal sites, sections had to be taken at least 1 km apart. The reason for this disparity is the already-mentioned difference between the horizontal and vertical variability of most sediments. Naturally, with an available exposure of 10 km, only very few lags can be calculated (see Figure 9). Clearly, what is needed for such studies is a large area in which the stratigraphy and correlation is well established, otherwise the correlation structure cannot be obtained. Of course, it is precisely in such areas that testing the significance of correlations becomes unnecessary.

The choice of the sedimentation model which is the second requirement for the method, does not appear to be too critical. Experiments have shown that relatively simple two dimensional stochastic models can produce correlation structures which come close to actually observed situations. This again is largely limited by the accuracy of the observations. The attempt to construct sedimentation models largely on theoretical grounds (see for example Jacod & Joathon 1971) may be an interesting academic exercise, but unless their correlation structure is tested against reality they are of little use in practical correlation. On the other hand, using the ability of a model to provide predictive correlations which can be checked by other methods, is an excellent way of testing sedimentation models.

SUMMARY OF CORRELATION AND MATCHING METHODS

It has been claimed for each of the reviewed methods that they can be applied to solve stratigraphic problems; it is however difficult to compare and judge the various methods.

The most difficult correlation problems are either concerned with complex facies changes or sequences which contain sedimentary cycles. None of the methods which have been proposed can deal with such situations properly. The earlier methods of matching that are based on comparing lithological states, are rather restricted in their applications and they certainly have not been taken up enthusiastically by stratigraphers. Gross correlation methods which permit the use of variable sedimentation rates, can be useful in correlating well log data. A potentially important use for such methods is the correlation of magnetic profiles and the correlation of magnetic profiles with stratigraphic profiles (Napoleone et al., 1983). The analytical methods of Martinson et al. (1982) should provide particularly valuable results in these fields.

However, all the methods which have been discussed here should be regarded as auxiliary aids to stratigraphic correlation and they cannot be used automatically without additional geological evidence. The ideal way of using the methods is in an environment of interactive computing. Stretching, shrinking and filtering are all procedures which can be displayed graphically and are very useful, providing that they are easily available. Since most of the procedures can be handled by microcomputers, there is no reason why geologists should not have access to such methods.

LITHOSTRATIGRAPHY AND TIME CORRELATION

Chapter IV.1 discussed some of the difficulties in relating a sedimentary record to absolute time. It was found that the discrete nature of the sedimentation process invariably determines a limit to the amount of time resolution that is possible, and stratigraphic intervals below this limit can no longer be used for time correlations. However, the conditions for time correlation do not only depend upon the accuracy of the time scale but largely on the ability to recognize synchronous surfaces in the sediment. As with most geological problems, one cannot expect complete accuracy and therefore a synchronous surface will be only an approximately accurate time plane.

The accuracy with which the position of a synchronous surface can be established depends upon two factors: The rate of sediment accumulation and the rate with which a marker horizon expands in a horizontal direction. In an ideal situation this lateral rate may be zero, which in practice is the case when layers are produced by ash falls or when deep sea sediments show stratification that has been induced by rapid climatic changes. However, the more common sedimentary processes involve lateral transport and any marker horizon will therefore take a certain time to be completed. It has been estimated that the lateral spreading of sediments is between 10^6 to 10^7 times faster than the vertical accumulation rate (Schwarzacher, 1975) and so the diachronism of surfaces which have developed by lateral spreading can often be ignored. Furthermore, many lithological boundaries are either erosional or are followed by periods of nondeposition. Since this reduces the vertical accumulation rate, such boundaries often contain time planes which unfortunately cannot be dated accurately. Also, it must be recognized that lithological markers can spread with different speeds. For example, it is not unusual to see bedding planes that cut obliquely into larger lithological units. One might generalize and say that large scale lithological variation is caused by relatively slow moving facies belts and that phenomena like cyclic or event bedding are caused by much faster environmental changes. Of course there are exceptions.

The actual position of a time transgressive lithological boundary can be a complicated function of the basin geometry, movement of facies belts and sedimentation rates. A simple example is shown in

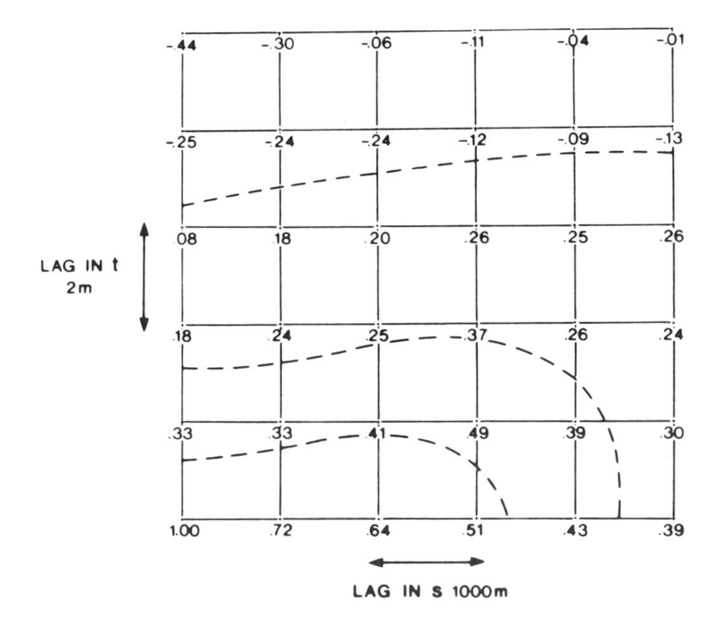

Figure 9 *Two dimensional correlation plot of a 10 km long Carboniferous limestone section in NW Ireland. Note the large difference in the vertical and horizontal lags.*

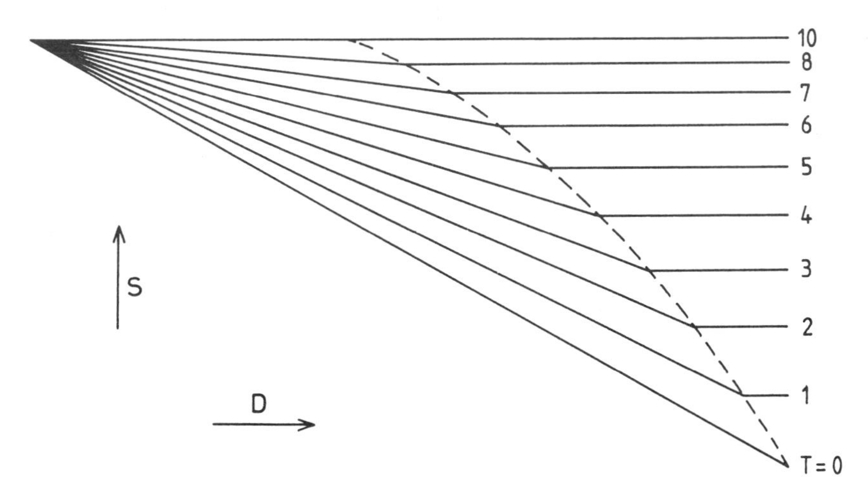

Figure 10 *Position of a lithological boundary (dashed line) which moved at a constant rate from right to left. It is assumed that sedimentation rates to the right of this boundary were constant but increased from left to right on the other side. The numbered lines represent time planes.*

Figure 10. It is assumed that the sedimentation rates dS/dt = a increase linearly in a direction from left to right. A lithological boundary moves with constant rate dD/dt = λ in a direction from right to left. The position of the lithological boundary is then given in terms of the distance D and the stratigraphic position S as $D = \sqrt{s\lambda/a}$. The diagram assumes that the sedimentation rates remain constant after they have crossed the lithological boundary. The successive stages in the diagram are time planes and these could well be realized in the field as bedding planes. The figure also illustrates that the frequently quoted 'Walther's Law' makes little sense when the vertical variation is caused by phenomena which have different lateral spreading rates.

Real situations are considerably more complicated. The migration of a lithological boundary is often determined by facies developments and these in turn have an effect on sedimentation rates. Discontinuities along shelf margins for example, and large areas of nondeposition can confuse the picture. The time interpretation of lithostratigraphic data can therefore only be attempted after a careful analysis of the reconstructed sedimentation conditions.

A Practical Example

To illustrate some of the previously discussed methods, a quantitative comparison has been made of three cores. Two cores (V28 - 283 and V28 - 239) are from the Pacific and oxygen analyses are available (Shackleton & Opdyke, 1976). The third core is from the Atlantic and its stratigraphy is described in a paper by Herterich and Sarnthein (1984). The ^{18}O values found in the cores are plotted at 5 cm intervals using an arbitrary vertical scale (Figure 11).

In general, oxygen isotope records of deep-sea sediments do not pose difficult problems of correlation. The tops of such cores generally represent the present time and this level therefore, can be correlated fairly accurately, while the maxima and minima representing the glacial and interglacial periods can be matched without difficulty. In these paticular cores, the position of the Bruhnes-Matuyama magnetic reversal is also known and this provides another time level which can be correlated.

Taking the date of the magnetic reversal as 730 ka bp (= 730,000 year before present), one can calculate the average sedimentation rates in the cores as:

0.995 cm/ka in V28-239
1.644 cm/ka in V28-238
1.349 cm/ka in Meteor 13519

Since core V28-238 has the highest sedimentation rates, core V28-239 would have to be expanded by a factor of 1.652 and core 13519 by a factor of 1.218 to make them all of equal length. A test was made to see whether such relative sedimentation rates could be recovered without prior knowledge of the magnetic reversal datum. Core 13519 and core V28-239 were expanded by increasing amounts and after each expansion, the mean squared differences were calculated. The results of this experiment are shown in Table 1. As may be seen, minima in the differences occur at expansions 1.20 and 1.62 respectively and this is in good agreement with data derived from the magnetic stratigraphy.

Variable expansion rates can be applied to the problem by using the approach of Martinson et al. (1982), discussed earlier. The programme used by these authors has not yet been published. In this study, the Fourier coefficients which determine the mapping function were computed by an iterative process which uses the cross variogram as a criterion for the best fit at each stage. In order to obtain equally spaced data, Newton's interpolation formula was used after each expansion. Since no particular criteria for convergence were used, the method relied heavily on a reasonable initial guess being made. This was provided by fixing the mapping function in such a way that the sharp minima after each interglacial coincided.

Figure 12 shows core 13591 and core V28-239 expanded using core V28-238 as a template. From the figure it can be seen that a remarkable similarity has been achieved but a study of the mapping function between the cores is more valuable still. The most informative way of interpreting such curves is to show the first derivative of the functions, since this represents relative sedimentation rates. A plot of such rate curves is shown in Figures 13a and 13b, where the expansion (the reciprocal of the sedimentation rates) is plotted against depth in the unexpanded core.

The mean rate is shown as a reference line and from this one can calculate the average expansion of core 28-239 as 1.80 and the expansion of core 13519 as 1.22. This should be compared with the earlier derived values of 1.62 and 1.20.

A useful application of this approach would be in comparing a large number of cores, since this should bring out irregularities in the sedimentation rates of the individual cores.

If the records are compared with a Milankovitch solar insolation curve (for example July 65°N), it is found that there is a high coherence in the frequencies corresponding to the 41 and 21 ka cycles (Pisias & Moore, 1981; Herterich & Sarnthein, 1984). On the other hand, it should be obvious even without analysis that the highest power of the ^{18}O records corresponds to frequencies of the 100 ka cycle. The reason why the coherence between the Milankovitch insolation curve and the 100 ka cycle is insignificant, is because the former function has very little power in this band. If the ^{18}O record is maximised against the 100 ka eccentricity cycle, then naturally, a high coherence is obtained.

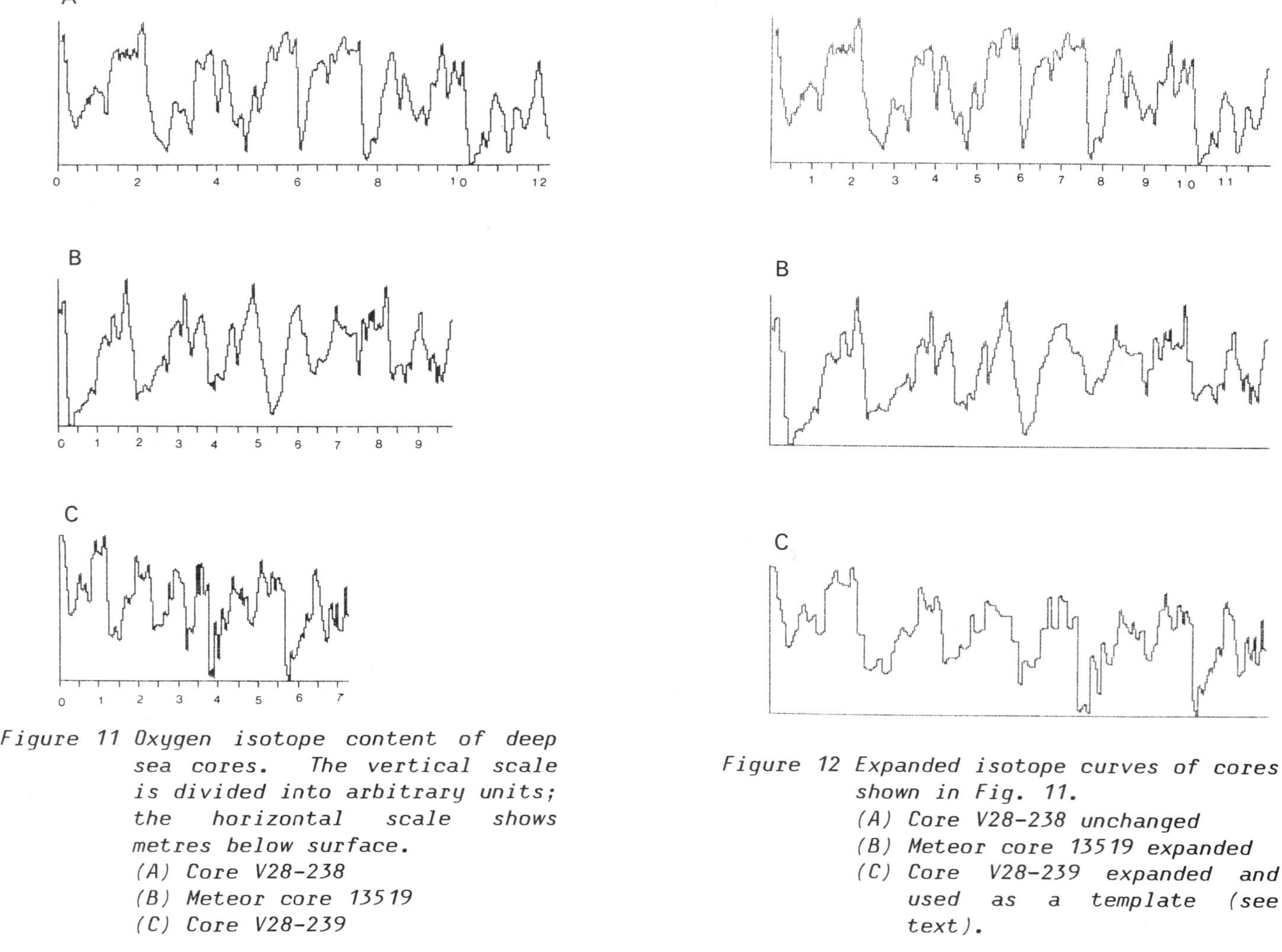

Figure 11 Oxygen isotope content of deep sea cores. The vertical scale is divided into arbitrary units; the horizontal scale shows metres below surface.
(A) Core V28-238
(B) Meteor core 13519
(C) Core V28-239

Figure 12 Expanded isotope curves of cores shown in Fig. 11.
(A) Core V28-238 unchanged
(B) Meteor core 13519 expanded
(C) Core V28-239 expanded and used as a template (see text).

Table 1 Comparison of expanded cores with core V28-238.

Core 13519 against V28-238		V28-239 against V28-238	
Expansion	Mean squared differences	Expansion	Mean squared differences
		1.16	51.78
1.58	28.10		
1.17	48.52	1.59	27.66
1.18	46.00	1.60	27.20
1.19	45.12	1.61	26.46
1.20	44.06	1.62	26.22
1.21	45.14	1.63	26.64
1.22	44.28	1.64	27.66
1.23	51.28	1.65	28.36
1.24	55.22	1.66	29.26
1.25	58.74	1.67	29.28

Table 2 Standardized depths of stage boundaries.

Boundary	depth cm Core 13519	Standardized V28-238	Standardized V28-239
	35	37.5	43.5
3 - 4	95	100.5	108.5
5 - 6	200	230	195.5
7 - 8	395	397	347.5
9 - 10	540	510	485
11 - 12	645	648	550.5
13 - 14	755	748	702.5
15 - 16	840	840	840
17 - 18	930	940.5	970.5

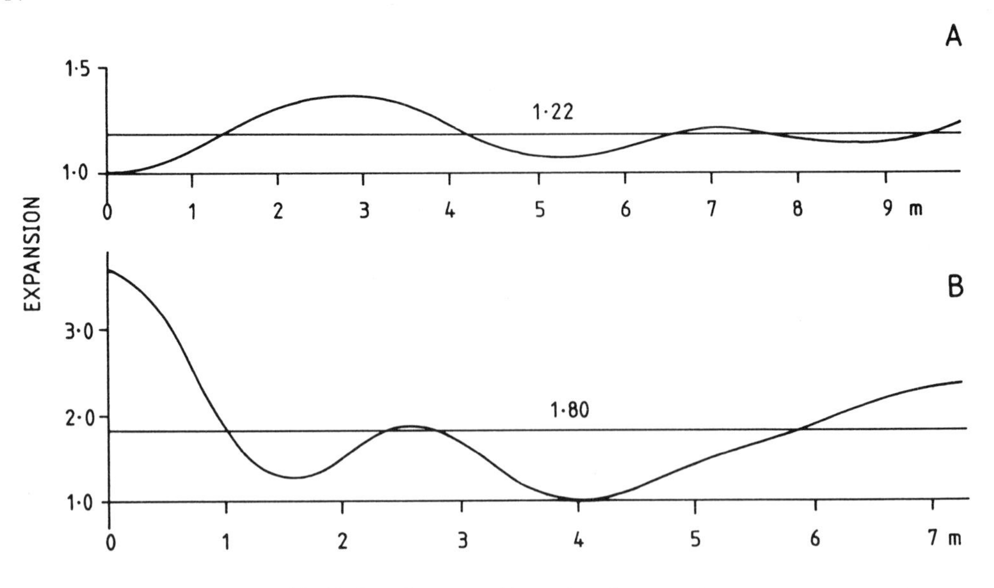

Figure 13 *Relative sedimentation rates comparing:*
(A) Meteor core 13519 against core V28-238
(B) Core V28-239 against core V28-238
The vertical scale gives the stretch factor (expansion), the horizontal scale represents depth below surface before stretching.

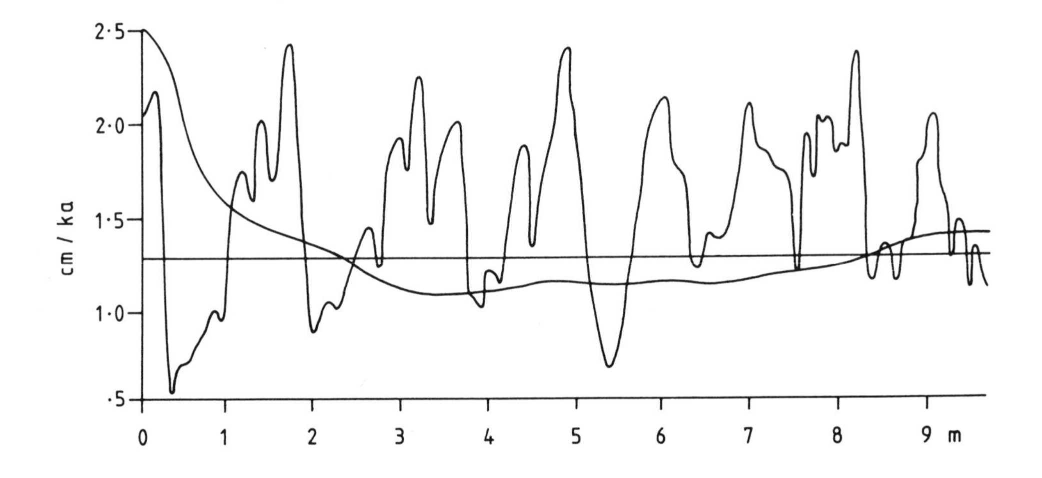

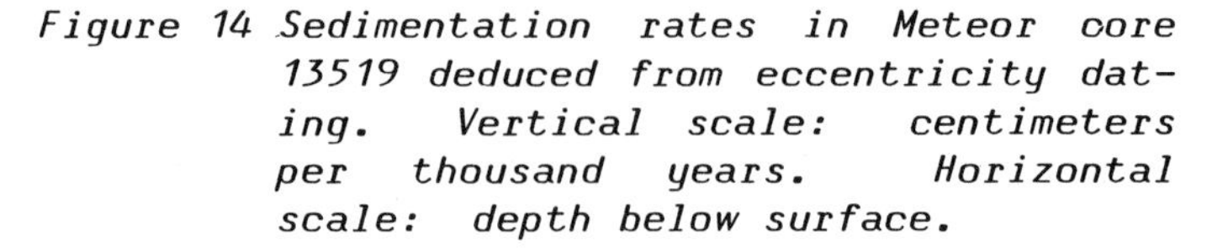

Figure 14 *Sedimentation rates in Meteor core 13519 deduced from eccentricity dating. Vertical scale: centimeters per thousand years. Horizontal scale: depth below surface.*

In order to test the capability of the method, the three cores were fitted to the eccentricity data as obtained from Berger's programme (Berger, 1978). In each case, the initial mapping function was fixed by choosing eight points where the minima in the eccentricity curve were made to coincide with the minima in the ^{18}O record. No use was made in this attempt, of the known datum of the B-M reversal.

A sedimentation rate diagram was calculated for core 13518, (Figure 14) which shows that the deviations occur at the top of the core and this may be for two possible reasons. Any error which is made by neglecting a phase shift between the eccentricity function and the ^{18}O record is particularly noticeable at the beginning. It is also possible that the top has been disturbed or that an effect of compaction is responsible. For simplicity, the top of the core was equated with the eccentricity which occurred in the year 1950. This is undoubtedly wrong and the initial discrepancy could be reduced. The average sedimentation rates which were found by comparing the three cores with the eccentricity curve gave the following values:

0.738 in V28-239
1.695 in V29-238
1.375 in Meteor 13519

These rates are in good agreement with the rates deduced from the B-M reversal and by assuming constant sedimentation rates.

It is possible that Figure 14 indicates slightly higher sedimentation rates during stage 6; also, a slight tendency towards a 100 ka cyclicity was noted in all the cores. However, considerably more material will be needed for studying such variations.

It is of some interest to compare the eccentricity time mapping function with the time depth relationship found by Herterich & Sarnthein. The authors published two curves. Scale 'STUNE' is based on fitting the records to the Milankovitch insolation curve and scale 'CARPOR' was calculated from the carbonate production and from porosity. Figure 15 shows that the eccentricity scale compares well with scale STUNE. Sarnthein's calculated curve CARPOR seems to indicate slightly lower sedimentation rates. It should be noted that in matching the eccentricity curve, no use was made of the palaeomagnetic datum. This probably explains the deviation of the eccentricity curve from the Herterich-Sarnthein data in the lowest part of the core. The exercise shows that tuning the isotope data to the relatively simple and straight-forward function of the eccentricity, can provide a very reasonable time scale. It also has the advantage that one does not have to get involved with hypothetical arguments as to which insolation curve is best.

The correlation of astronomical data with isotope records cannot prove that there is a causal connection between the two time series. However, it is possible to estimate the errors which are generated by

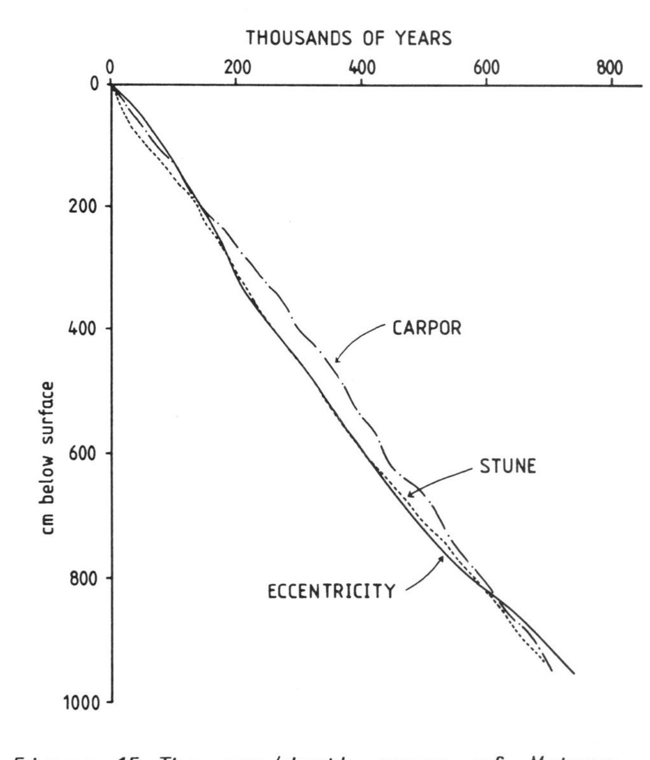

Figure 15 *The age/depth curve of Meteor core 13519. The eccentricity data compared with the scales STUNE and CARPOR. Vertical scale is depth below surface. Horizontal scale is age in thousands of years.*

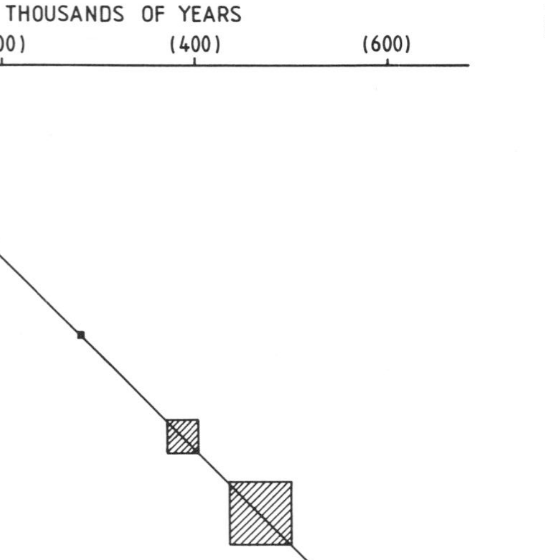

Figure 16 *The correlation plot of the three cores shown in Figure 11. The vertical scale is depth below surface after standardizing (see Table 2, in text). The horizontal scale in brackets is a time scale which strictly is only applicable to Meteor core 13519. The correlated stage boundaries fall within the shaded squares.*

variable sedimentation rates and which must enter diagrams like Figure 15. If one has several profiles in which a number of levels can be correlated with certainty, then the points on the correlation plot must fall within a square whose size is given by the largest difference in levels.

Comparing the three cores, core 13519 was chosen as a standard and the remaining two cores were standardised in such a way that the boundary of stages 15-16 (see Table 2) was at a depth of 840 cm in each case. Assuming that all the correlations were correct, then the differences in depth of the various points, given in Table 2, are due to variations in sedimentation rates.

The variability of the three cores is shown in Figure 16, which has been drawn to the same scale as Figure 15, so that the two are directly comparable. One might have reservations in substituting lateral variation for time errors which are caused by variable sedimentation rates within one and the same core. One might also question whether it is legitimate to compare two cores from the Pacific Ocean with a core from the Atlantic Ocean. However, if the isotope record is determined by a periodic signal, then differences in a single core can only represent variations in sedimentation rates. A study of Table 2 shows that the distance between the cores does not seem to determine the amount of discrepancy. Thus it seems reasonable to use plots of the type given in Figure 16, as an aid in judging the accuracy of the derived time scales. Similarily, one can thus assess the accuracy of the stratigraphic correlation.

REFERENCES

Berger, A.L., 1978, A simple algorithm to compute long term variations of daily or monthly Isolation. Inst. D'Astronomie et de Geophysique, Université Catholique de Louvain, Contribution 18.

Besag, J.E., 1972, On the correlation structure of some two-dimensional processes. Biometrika, 59, 43-48.

Gordon, A.D., and Reyment, R.A., 1979, Slotting of borehole sequences. J. Math. Geology, 11, 309-327.

Haites, T.B., 1963, Perspective Correlation, Bull. Am. Assoc. Pet. Geol., 47, 553-574.

Herterich, K., and Sarnthein, M., 1984, Bruhnes Time Scale: Tuning by rates of Calcium-Carbonate Dissolution and Cross Spectral Analysis with Solar Radiation. A.L. Berger et al. (eds.) Milankovitch and Climate, Part 1, 447-460, D. Reidel Pub. Co.

Jacod, J., and Joathon, P., 1971, Use of random genetic models in the study of sedimentary processes. J. Math. Geol., 3, 265-279.

Kemp, F., 1982, An Algorithm for the Stratigraphic Correlation of Well Logs. J. Math. Geol. 14, 271-284.

Martinson, D.G., and Menke, W., and Stoffa, P., 1982, An inverse Approach to Signal Correlation. J. Geophys. Res. 87, 4807-4818.

Napoleone, G., Corezzi, S., Premoli-Silva, I, Heller, F, Chili, P. and Fischer, A.G., 1983, Eocene magnetic stratigraphy at Gubbio, Italy, and its implication for palaeogene geochronology. Geol. Soc. Am. Bull., 94, 181-191.

Neidell, N.S., 1969, Ambiguity functions and the concept of geological correlation. Kansas Geol. Surv., Comput. Contrib., 40, 19-29.

Odell, J., 1975, Error estimation in stratigraphic correlation. J. Math. Geol., 7, 167-182.

Pisias, N.G., and Moore, T.C., 1981, The evolution of Pleistocene climate: a time series approach. Earth and Planet. Sc. Let., 52, 450-458.

Rudman, A.J., and Lankston, R.W., 1973, Stratigraphic correlation of well logs by computer techniques. Amer. Assoc. Petrol. Geol., 57, 577-588.

Sackin, M.J., and Sneath, P.H.A., and Merriam, D.F., 1965, Algol program for cross-association of non-numeric sequences using a medium sized computer. Kansas Geol. Surv., Spec. Publ., 23, 33 pp.

Schwarzacher, W., 1980, Models for the study of stratigraphic correlation. J. Math. Geol., 12, 213-234.

Shackleton, N.J., and Opdyke, N.D., 1976, Oxygene-Isotope and Paleomagnetic Stratigraphy of Pacific Core V28-239 Late Pliocene to Latest Pleistocene. Geol. Soc. Amer., Memoirs 145, 449-464.

Smith, T.F., and Waterman, M.S., 1980, New Stratigraphic Correlation Techniques. J. Geol., 88, 451-457.

Part V

TIMESCALES, SUBSIDENCE AND SEDIMENTATION

Contents

V.1

TIMESCALES AND BURIAL HISTORY

F.M. GRADSTEIN

'The historical succession of phenomena, their correlation and meaning form the most attractive and interesting feature of geology.' <u>In</u>, *"The Pulse of the Earth", J.H.F. Umbgrove, 1942*

INTRODUCTION

The accurate portrayal of geological history demands that relative and subjective judgement be modified to include models in which observations are expressed numerically, preferably with an estimate of uncertainty. When events or processes are organized along a linear timescale, expressed in millions of years (Ma), rates of subsidence and sedimentation, or the duration of a hiatus in a geological basin may be calculated. Such numerical geological hypotheses, particularly when also expressed graphically, can elegantly summarize much information.

The nineteenth century philosopher and poet Christian Morgenstern once wrote a poem in which his tragicomic hero Korf expresses his uneasiness with lengthy, descriptive stories, where all could have been said in a few words. Korf, therefore, invented eyeglasses with the power to condense such stories to the point that the particular poem itself is only a question mark:

Die Brille

Korf liest gerne schnell und viel;
darum widert ihn das Spiel
all des zwölfmal unertbetnen
Ausgewaltzten, Breitgetretnen

Meistes ist in sechs bis acht
Wörtern völlig abgemacht
und in ebensoviel Sätzen
lässt sich Bandwurmweisheit schwätzen

Er erfindet drum sein Geist
etwas, was in dem entreisst:

Brillen, deren Energieen
ihm den Text - zusammenziehen

Beispielweise dies Gedicht
läse, so bebrillt, man - nicht!
Dreiunddreissig seines gleichen
gäben erst - Ein - Fragezeichen!!

Burial curves, i.e., time-depth diagrams that combine estimates of numerical age and paleo waterdepth for successive strata in oceanic and continental boreholes, are a geological equivalent of Korf's concentrating and revealing eyeglasses. No single quantitative stratigraphic technique is more powerful in unravelling and portraying geological history at a glance. Burial curves provide a direct means of reading the timing and magnitude of geologic events and calculating subsidence and sedimentation in basins. When combined with information on the temperature gradient through time, the burial curve reveals how long deposits resided in a potential oil or gas window. The method also helps to decide if data make geological sense, and this is a key to understanding basin evolution.

This Chapter introduces the construction and use of burial curves and the components involved, including multiple biozonations, timescales, paleobathymetric estimates (including backtracking) and corrections for sediment decompaction, eustatic sealevel changes and sediment loading of basement. Special attention is given to methods for constructing numerical timescales and the DNAG (Decade of North American Geology) scale for the last 200 Ma is reproduced. Examples of subsidence and sedimentation are largely drawn from boreholes in the North Atlantic Ocean and margins.

STRATIGRAPHY AND TIMESCALES

The construction of modern geological timescales is an intricate process that involves interrelation of biostratigraphy and radiometric dates with magnetostratigraphy and the sea-floor spreading record of reversals of geomagnetic polarity.

The derivation of a numerical geological timescale ultimately depends on the availability of radiometric ages. The use of radiometric dates to determine the age of stratigraphic boundaries is emphasized in the approaches of Odin et al. (1982) and Armstrong (1978). Radiometrics involves measuring the ratio of the original element and its isotopic daughter products. Depending on the half life of the element, several radiometric clocks are available; $^{87}Rb \rightarrow {}^{87}Sr$ and $^{40}K \rightarrow {}^{40}Ar$ are suitable in the Phanerozoic. Laboratories work with agreed-upon decay constants (Steiger and Jager, 1978) and isotopic abundances.

Dating of sedimentary rocks follows several geological strategies:

1. Dating of igneous intrusions within sediments records the time of primary cooling, when the igneous rocks were emplaced and had cooled sufficiently (to a few hundreds of degrees centigrade) to set the radiometric clock. This critical temperature is often referred to as the Curie point. Because of uncertainty in the relation of the intrusion to the sediments, such dates may be of limited stratigraphic use.

2. Dating of volcanic flows and pyroclastics (tuffs, bentonites, etc.) as part of stratified sedimentary successions. Dates under 1 and 2 are often referred to as being of a high-temperature nature.

3. Dating of authigenic sedimentary minerals. This mainly involves glauconite, although in principle any Rubidium or Potassium bearing mineral may be used. Mild heating or overburden pressure after burial may lead to loss of Argon, the daughter product in the $^{40}K \rightarrow ^{40}Ar$ decay series. As a result, the radiometric clock was seemingly set more recently and calculated ages are too young. Glauconite dates are sometimes referred to as being of a low-temperature nature.

Inter-laboratory and analytical differences (errors) of up to 3% are commonly encountered in radiometrics. This means that the radiometric error (but not accuracy) exceeds the biostratigraphic resolution for all but the youngest Phanerozoic time.

A special problem is the estimation of the age of a stratigraphic boundary based on (few or many) radiometric ages in the vicinity of this boundary and the calculation of the accompanying error. This statistical appraisal of timescales is introduced in the section on 'Stratigraphic Resolution'.

Unfortunately, radiometric age data, particularly those derived from high-temperature minerals as opposed to authigenic glauconites, are not sufficiently numerous in time to adequately and directly define the ages of most stage boundaries, and more indirect techniques must be utilized.

Since there are many more combined fossil events and magnetic reversals than datable horizons in the Phanerozoic sediments, the ordinal framework of bio- and magnetochronology provides the principal fabric for geochronology. Until recently, it was difficult to relate this relative framework to a continuous linear scale, because many of the datums can not be closely related to a radiometrically dated level.

Van Hinte (1976b) has eloquently expressed what happened next; 'The breakthrough came with recognition of a continuous record of geomagnetic reversals recorded in a linear magnetic-anomaly pattern in oceanic sea floors and subsequent paleontologic dating by deep-sea drilling of oceanic basement at key anomalies. The assumption of constancy of sea-floor-spreading rate between the dated anomalies then allowed for a linear age interpretation of the pattern between these points and, with additional extrapolation beyond the calibration points, yielded a linear geomagnetic-reversal timescale for the entire Tertiary. Because geomagnetic reversals also can be recognized in sediments, the reversal timescale now makes possible the assignment of numeric ages to biohorizons which are not radiometrically dated, by interrelating the magnetostratigraphy and biostratigraphy of sedimentary sections. As a result, it has become possible to place a linear timescale beside the biostratigraphic zonal scheme. The addition of new and better data to this feedback system continues to improve it'.

The introduction of many new and better data has spurred presentation of a new Mesozoic and Cenozoic timescale by Berggren et al. (in press) and Kent and Gradstein (in press), and the following discussion is taken from their work. The new scale has been accepted as the standard timescale for the Decade of North American Geology (DNAG) (Palmer, 1983). The well-known scale by Harland et al. (1982) incorporates an older version of the new Cenozoic scale by Berggren et al. (op. cit.), but is conceptually different from the Mesozoic scale by Kent and Gradstein (op. cit.), (see Table 1).

Cenozoic

The Cenozoic geomagnetic reversal timescale is based on radiometrically dated magnetic polarity data on lavas for 0 to 4 Ma (Mankinen and Dalrymple, 1979), and extended in time by age calibration of the polarity sequence inferred from marine magnetic anomalies. The polarity sequence compiled by LaBrecque et al. (1977) is taken as representative of the sea-floor-spreading record for the late Cretaceous and Cenozoic. In order to satisfy six selected high-temperature age calibration tie-points (Table 2) and to minimize apparent acceleration in sea-floor-spreading history, Berggren et al. (in press) consider this polarity sequence as three linear segments on an - age versus distance calibration plot of the marine magnetic anomalies on the seafloor (Figure 1). An initial segment is defined by the origin, a 3.40 Ma age for the Gauss/Gilbert boundary and a 8.87 Ma age for the younger boundary of Anomaly 5, yielding an extrapolated age of 10.42 Ma for the older boundary of Anomaly 5. Available radiometric age estimates for magnetozones in land sections correlated to the younger boundaries of Anomalies 12, 13 and 21 (32.4, 34.6 and 49.5 Ma, respectively) are used to extend the chronology by a linear best fit anchored to the estimated age for the older boundary of Anomaly 5; an estimated age of 56.14 Ma for the older boundary of Anomaly 24 is extrapolated for this segment. Interpolation between this estimated age for Anomaly 24 and a radiometric age estimate of 84 Ma for Anomaly 34 (which is based on correlation of the magnetozone equivalent with the Campanian/Santonian boundary), completes the reversal chronology to the younger end of the Cretaceous Long Normal Polarity Interval, corresponding to the oceanic Cretaceous Quiet Zone.

Table 1 Different methods used to construct modern versions of the Mesozoic and Cenozoic geological time scale.

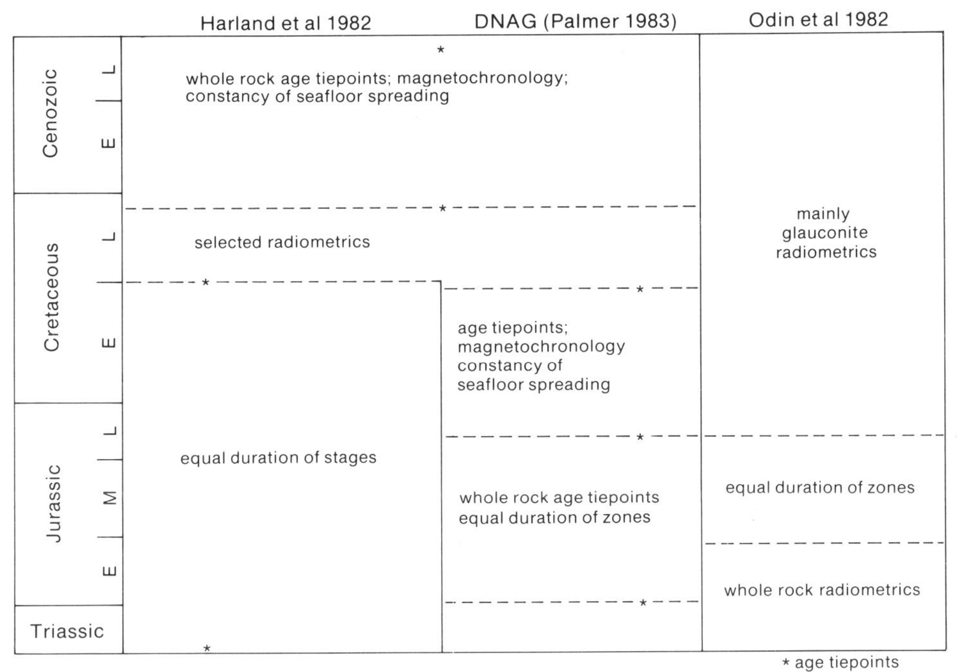

Table 2 Mesozoic and Cenozoic calibration ages to construct the time scale reviewed in this chapter; (1) Berggren et al. (in press), (2) Kent and Gradstein (in press), (3) Harland et al. (1982).

Calibration Level	Age (Ma)	
Gauss/Gilbert, Anomaly 2A	3.40 Ma	(1)
Younger end of Anomaly 5	8.87 Ma	(1)
Younger end of Anomaly 12	32.4 Ma	(1)
Younger end of Anomaly 13	34.6 Ma	(1)
Younger end of Anomaly 21	49.5 Ma	(1)
Santonian/Campanian boundary, younger end of Anomaly 34	84.0 Ma	(1)
Santonian/Coniacian boundary	87.5 Ma	(2)
Coniacian/Turonian boundary	88.5 Ma	(2)
Turonian/Cenomanian boundary	91.0 Ma	(2)
Cenomanian/Albian boundary	97.5 Ma	(2)
Albian/Aptian boundary	113 Ma	(2)
Aptian/Barremian boundary, Anomaly M0	119 Ma	(3)
Oxfordian/Kimmeridgian boundary, Anomaly M25	156 Ma	(3)
Norian/Sinemurian boundary	208 Ma	(3)

Ages for magnetic polarity intervals are calculated according to the linear regression equations for these three segments. Relative precision of the reversal sequence depends on the spatial resolution of the magnetic anomaly data and the assumption that sea-floor spreading occurred somewhere at a constant rate over tens of million years. The accuracy of the reversal chronology ultimately depends on the radiometric age data set used for calibration and on the method of interpolation.

The first extended geomagnetic reversal timescale was presented by Heirtzler et al. (1968). A magnetic profile from the South Atlantic was chosen as representative for the geomagnetic reversals record in the last 80 Ma. The chronology was derived by a correlation of the mid-ocean ridge-axis magnetic anomalies to the 0 to 4 Ma, radiometrically-dated magnetic reversal timescale of Cox et al. (1963) and by extrapolation to the oldest then recognized polarity interval (anomaly 32). This twenty-plus fold extrapolation assumed that the rate of sea-floor spreading in this area of the South Atlantic was constant over about 1400 km or 80 Ma, at the rate calculated over the time interval from 0 to 3.35 Ma (anomaly 2A). Remarkably, the absolute age estimates for the magnetochrons are within 10 per cent of the numerical age estimates in the much more detailed scale of Berggren et al. (in press), (Figure 1). This agreement indicates that the constant spreading rate assumption applied to selected areas of the world ocean provides a very good first-order approximation in the derivation of a geomagnetic reversal chronology.

For biochronologic calibration of the new timescale, interrelation of bio- and magnetostratigraphy was carried out in European Paleogene and Neogene stratotype sections. An assessment was made of some 200 Cenozoic and late Cretaceous calcareous plankton datum events that area directly tied to magnetic polarity stratigraphy in deep-sea sediment cores. Both procedures improved identification of the stage boundaries and their duration in terms of bio- and magnetochronology.

An assessment by Berggren et al. (in press) of published radiometric dates and their relation to the above bio- magnetochronology suggests biochronologic age estimates virtually identical to the ones based on the marine magnetic reversal chronology in Figure 1 (see Table 3). The radiometric dates favoured are based on whole rock, high temperature minerals, rather than glauconites. The latter tend to give ages for Paleogene stage boundaries up to 7 my younger (e.g. Odin et al., 1982). The high internal consistency of this integrated magneto-biochronology supports the use of the age-calibrated magnetic reversal sequence as a vernier (analogous to use of age-calibrated stratigraphic thickness) to obtain precise age estimation for various boundaries, in accordance with magneto-biostratigraphic correlations. In effect, Berggren et al. (in press) have extended the philosophy and approach to the Neogene timescale by Ryan et al. (1974) to the Paleogene. Numerical ages in the Cenozoic timescale therefore are based on the magnetochronology of Figure 1. The actual timescale is shown in Figure 4.

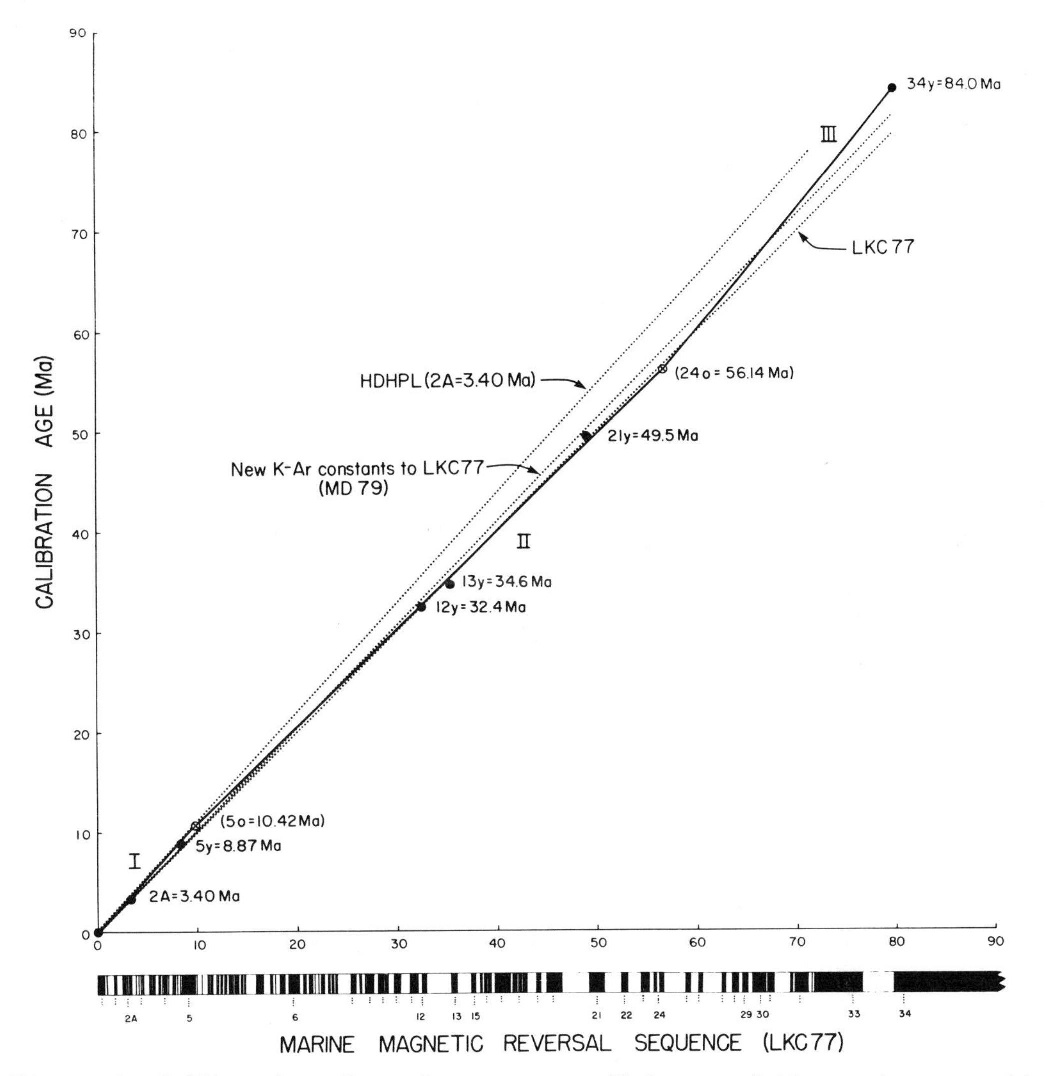

Figure 1 *Calibration plot of age versus distance of the marine magnetic anomalies on the seafloor for the magnetic reversals over the last 84 Ma (after Berggren et al., in press). Solid lines are three linear apparent age-calibration segments (I, II and III) which satisfy calibration tie-points indicated by solid circles (Table 2). The two open circles with X's at anomalies 5 and 24 are the inferred inflection points whose ages are derived by extrapolation from linear segments I and II, respectively. Shown for comparison by dotted lines are the geomagnetic reversal time scales of Heirtzler et al. (1968) (HDHPL68 with anomaly 2A set for 3.40 Ma to conform with current estimate) and LaBreque et al. (1977) (LKC77 in original form and modified (MD79) to account for new K-Ar constants. Anomaly numbers are indicated below bar graph of geomagnetic reversal sequence (filled for normal, open for reversed polarity).*

Table 3 Comparison of ages in Ma for the Cenozoic epochs using (a) the interrelation of biochronology and (mainly) high-temperature radiometrics and (b) magnetochronology and selected high-temperature radiometric tiepoints (after Berggren et al., in press).

Biochronology/assorted high temperature radiometrics		Magnetochronology/selected radiometric tiepoints
<2 Ma	Pleistocene	1.6 Ma
ca 5	Pliocene	5.3
ca 23.5	Miocene	23.7
ca 37	Oligocene	36.6
ca 56.5	Eocene	57.8
ca 66	Paleocene	66.4

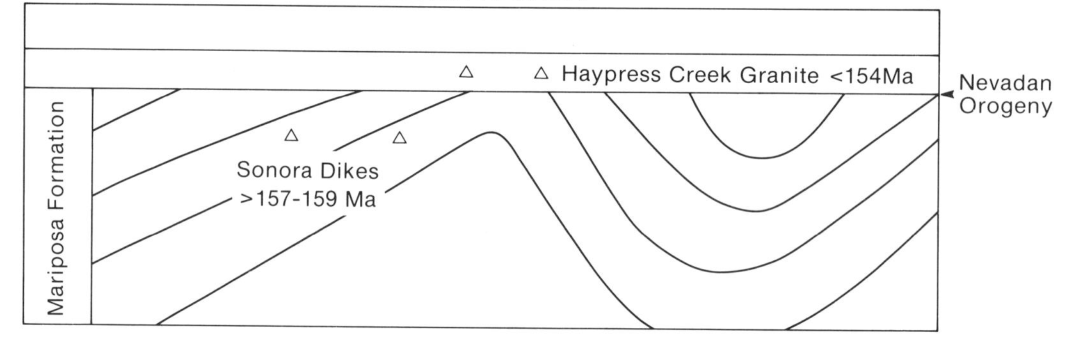

Figure 2 High-temperature, radiometric ages of 157-159 Ma on the folded Sonora Dikes in the folded Mariposa Formation of late Oxfordian-Early Kimmeridgian age, contrast with ages younger than 154 Ma for the non-folded Haypress granite on top (after Schweikert et al., 1984). The Oxfordian-Kimmeridgian boundary age can thus be approximated at 156 Ma.

Mid-Cretaceous (Santonian to Aptian)

The stratigraphic interval from the lower Aptian to the top of the Santonian generally records predominantly normal geomagnetic polarity (Lowrie et al., 1980), which accounts appropriately for the Cretaceous Quiet Zone in the oceans. Consequently, there are no well-documented magnetozones or anomalies that can be correlated and used for interpolation. Sufficient radiometric ages are, however, available to allow direct chronometric age estimates for stage boundaries in this interval.Numerical ages (Table 2) for Santonian to Aptian stage boundaries are therefore taken directly from the chronometric estimates of Harland et al. (1982) which are largely based on assessment of radiometric ages from the list of Armstrong (1978).

Early Cretaceous and Late Jurassic (Barremian to Kimmeridgian)

The older end of the Cretaceous Quiet Zones in the oceans typically is bounded by lineated magnetic anomalies referred to as the M-sequence. These anomalies are best defined over the higher spreading rate systems in the Pacific, but they are correlatable to the Keathley sequence of the North Atlantic (Larson and Chase, 1972; Larson and Pitman, 1972). The standard magnetic reversal model for the M-sequence (M0 to M25 from youngest to oldest, designating key anomalies which are interpreted to correspond usually to reversed polarity) was derived from the Hawaiian lineations (Larson and Hilde, 1975).

Recent magneto-biostratigraphic correlations, primarily from studies of land-sections and supported by additional basal ages and magneto-biostratigraphy in DSDP holes provide a basis for a refined chronology as discussed below.

Marine magnetic anomaly M0 is early Aptian in age; M3 essentially straddles the Hauterivian/Barremian boundary according to magneto-biostratigraphic studies in land sections from the Mediterranian region (Lowrie et al., 1980). A detailed study of pelagic limestones in DSDP Sites 534 and 603 indicates that the Hauterivian-Barremian boundary is at M7 (top of dinoflagellate **Druggidium apicopaucicum**) or M5 (top of nannofossil **Speetonia colligata**), (J. Ogg, pers. comm. 1985). In the same study M10N is assigned to the Valanginian-Hauterivian boundary.

From the magnetostratigraphic studies summarized by Channell et al. (1982), M10N also correlates with the Valanginian/Hauterivian boundary and M14 to the Berriasian/Valanginian boundary. M16, based on the age of basal sediments at DSDP Site 387, occurs at or just below this latter boundary; the correlation is supported by magnetostratigraphic correlations at DSDP Sites 534 and 603 which suggests that M16 is about middle Berriasian (Ogg, 1983 and pers. comm. 1985). Channell et al. (1982) correlate M17 to earlier Berriasian. The Tithonian/Berriasin boundary is poorly defined and as a result there is no consensus on criteria for definition of the Jurassic-Cretaceous boundary by means of ammonites, calpionellids, nannofossils or magnetic reversals. Ogg et al. (1984) find that in southern Spain the boundary (defined at the base of the **Grandis-Jacobi** ammonites Zone) falls between M18 and M19. Using nannofossils in the Maiolica limestones in Italy, the boundary correlates close to the older part of M17 (Lowrie and Channell, 1984). The same study places M18 and M19 in the late Tithonian.

Correlation using the co-occurrence of the benthic foraminifers **Epistomina** aff. **uhligi** and **Lenticulina quenstedti** in sediments immediately above basement of approximately M25 age in DSDP Sites 105 and 367, and correlations of this event to DSDP Site 534 suggests that M24 and M25 are older than late Kimmeridgian (Sheridan, Gradstein et al., 1983). Early Kimmeridgian is a minimum age, based on the palynological age of basal sediments in DSDP Site 105 and its correlation to DSDP Site 534 (Habib and Drugg, 1983). Ogg et al. (1984), estimate the Kimmeridgian/ Tithonian boundary in deep water limestones in southern Spain, to be at the base of M22; the correlation to M25, the oldest globally recognized magnetic anomaly is just below, or at the Oxfordian/Kimmeridgian boundary.

Despite the improved magneto-biostratigraphic correlations, numerical age estimates for late Jurassic and early Cretaceous stages are still poor due to a lack of reliable radiometric data. To be consistent with our use of the Harland et al. (1982) chronology for the Santonian to Albian and to avoid an artificial discontinuity, we use their age estimate of 119 Ma for the Barremian/Aptian boundary to calibrate the younger end of the M-sequence. The Barremian/Aptian boundary falls within the M-sequence and is the next (older) boundary from the Aptian/Albian, which is considered by Harland et al. (1982) to be the only chronometrically well-constrained tie-point (113 Ma) in the Early Cretaceous and the Jurassic. An isochron age of 120 Ma was determined for basalt overlain by lower Aptian sediment at DSDP Hole 417D, which was drilled on anomaly M0 in the western North Atlantic (Ozima et al., 1979). This age is admittedly poor but it is nevertheless consistent with the 119 Ma age estimate of Harland et al. (1982) for the base of the Aptian.

Armstrong (1978) interpolates whole-rock K/Ar ages to arrive at approximately 156 Ma for the Oxfordian/Kimmeridgian boundary. This is identical to the broadly interpolated age derived by Harland et al. (1982) who used the equal duration of stages between Anisian/Ladinian at 238 Ma and the Aptian/Albian at 113 Ma. Additional evidence for a 154-158 Ma age range for the Oxfordian/Kimmeridgian boundary comes from the Sierra Nevada (California), where numerous igneous intrusions that have been radiometrically dated, are found (Schweikert et al., 1984). Of particular stratigraphic significance is the Mariposa Formation, dated by **Buchia concentrica, Dichotomosphinctes, Discosphinctes** and **Amoebites (Amoeboceras),** to be late Oxfordian-early Kimmeridgian in age (R.W. Imlay in Clark, 1964, 1976). The Mariposa Formation shows a so-called Nevadan-type of regional deformation and is the youngest unit affected by the Nevadan orogeny. Nevadan cleavage and folds affect dikes and plutons as young as 157-159 Ma (Sonora dikes) but do not affect plutons as old as 154 Ma (Haypress Creek granite) (Figure 2). The Nevadan orogeny is therefore constrained to the interval 154 and 157-169 Ma, which can be interpreted as an indirect minimum age range for the Mariposa Formation and the Oxfordian-Kimmeridgian boundary. The latter in turn equates to M25.

For these reasons, we have accepted the ages of 119 Ma (Barremian/Aptian) and 156 Ma (Oxfordian/Kimmeridgian) as reasonable tie-points for calibration of the M-sequence. We note, however, that these age estimates are older, by as much as 14-16 my in the case of the Oxfordian/Kimmeridgian boundary, than age estimates of Van Hinte (1976b) and of Odin et al. (1982), (Figure 3) which authors rely on ages determined by glauconites.

We used the magnetostratigraphic correlations, the assumption of a constant spreading rate on the Hawaiian lineations (Larson and Hilde, 1975) and the above calibration tie-points to derive age estimates for Kimmeridgian to Barremian stage boundaries (Figure 3) and for the M-sequence geomagnetic reversals. This is simply another case of using the magnetic reversal sequence as a vernier to estimate ages of correlated boundaries between points of known or assumed age, and provides an independent means to assess the relative duration of stages.

In comparison to the equal stage duration model of Harland et al. (1982), (Table 1) the magnetochronological method also gives approximately equal (ca. 6 my) durations for the oldest four stages of the Early Cretaceous; however, it results in an apparently longer Tithonian (8 my) and a shorter Kimmeridgian (4 my). This factor of two ratio in relative duration of the Tithonian and Kimmeridgian stages is in good agreement with the relative duration inferred from assuming equal duration of ammonite zones (see below). Differences with previously published magnetochronologies for the M-sequence vary according to the degree of magneto-biostratigraphic data incorporated (e.g., Cox, 1982) and the geologic timescale used in its calibration (Figure 3).

Jurassic (pre-Kimmeridgian)

The older end of the M-sequence of anomalies is bounded by the Jurassic Quiet Zone which is represented by sea-floor areas of subdued magnetic signature. Unlike the generally well-defined boundary with the Cretaceous Quiet Zone, the boundary of the M-sequence with the Jurassic Quiet Zone is typically gradational and indistinct, characterized by a decreasing anomaly amplitude envelope from at least M22 to M25 and complicated by small-scale lineated anomalies extending the sequence.

Magnetostratigraphic studies on land-sections in Spain and Poland, summarized by Ogg and Steiner (1985), indicate frequent reversals in the Callovian-Oxfordian. There are no reliable correlations to the marine magnetic anomaly record, which is either quiet ('Jurassic

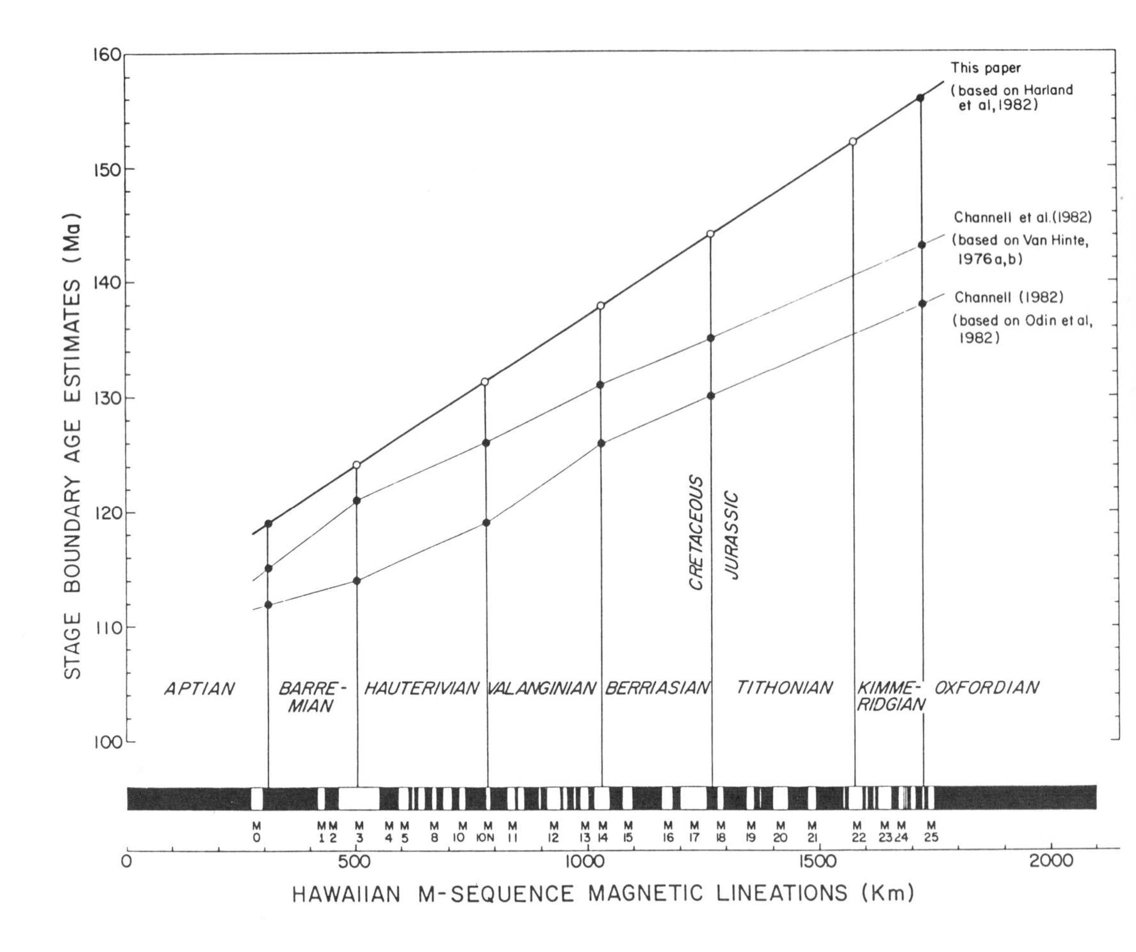

Figure 3 *Calibration plot of age versus distance of the marine magnetic anomalies on the seafloor for the standard magnetic M- reversal sequence as derived from Hawaiian lineations (Larson and Hilde, 1975). A constant spreading rate of 3.836 cm/yr is assumed here for the spacing and age of the Hawaiian lineations based on age estimates of 119 Ma and 156 Ma for the Barremian/Aptian and Oxfordian/Kimmeridgian boundaries, respectively (solid circles); age estimates of intervening stage boundaries (open circles) are based on interpolation. For comparison are shown the spreading rates using the magnetochronology of Channell et al., 1982 and the timescales by Van Hinte (1976a, b) and Odin et al. (1982), (after Kent and Gradstein, in press).*

Quiet Zone'), or displays weak amplitude lineations on (Pacific) plate sequences with uncertain paleolatitudes.

Preliminary magnetostratigraphic work on Middle and Lower Jurassic land-sections primarily from the Mediterranean area suggests that the Sinemurian to Bathonian stages are characterized by frequent geomagnetic reversals even though correlation of magnetozones between sections is difficult (Channell et al., 1982; Steiner and Ogg, 1983). No marine magnetic anomalies record this interval of frequent reversals at the older end of the Jurassic Quiet Zone, suggesting that the present Atlantic oceanic crust is of post-Bathonian age (ca. 169 Ma).

For pre-Kimmeridgian Jurassic geochronology, the general lack of correlatable lineated magnetic anomalies means that it is not possible to utilize magnetochronology to calibrate the timescale as we did for the Kimmeridgian-Aptian interval. We therefore apply bio- and radiochronologic methods for Early and Middle Jurassic stage age estimation, according to the following arguments.

We proceed from the age estimate of 156 Ma for the Oxfordian/ Kimmeridgian boundary. Harland et al. (1982) show that few radiometric dates exist to allow radiochronologic estimates of stage boundaries between the base of the Kimmeridgian and the base of the Jurassic. However, according to Armstrong (1978), at least part of the Sinemurian should be older than 203 Ma and some part of the Toarcian should fall in the age bracket of 185-189 Ma. The Triassic/Jurassic boundary age is inferred from another series of whole rock cooling ages from Triassic and Jurassic beds in volcanogenic and sedimentary complexes in British Columbia, Canada (Armstrong, 1978). Armstrong, taking into account all world data, suggests that the best available evidence places the base of the Jurassic at 208 Ma. This figure is just within the 200-208 Ma range of Odin and Letolle (1982) that is also based on high-temperature mineral radiometric dates. Harland et al. (1982) estimate an age of 213 Ma for the Triassic/Jurassic boundary based on interpolation between middle Triassic and middle Cretaceous using equal duration of stages. Agterberg (in press) applied spline curve fitting on Harland et al.'s (1982) original data for this same interpolation and arrived at an age of 208 Ma for the Triassic/Jurassic boundary. Clearly, the 208 Ma estimate preferred by us is within the chronometric uncertainty.

The Jurassic chronology is then built on the radiometric age constraints of 208 Ma for the Jurassic/Triassic boundary and an Oxfordian/Kimmeridgian boundary age of 156 Ma. We use an interpolation mechanism that also was adopted by Van Hinte (1976b), the equal duration of zones, but we use the updated ammonite zonation advocated by Hallam (1975). There are 50 zones between the base of the Hettangian (208 Ma) and the top of the Oxfordian (156 Ma), which is 1.04 Ma per zone. The equal zone duration method is in our opinion less crude a vernier than the equal duration of stages assumed by Harland et al. (1982). In fact, Harland et al. (1982) justify their equal-duration criterion by evolutionary turnover; thus it might be better to assume equal duration of zones which are the shortest (bio)stratigraphic building blocks.

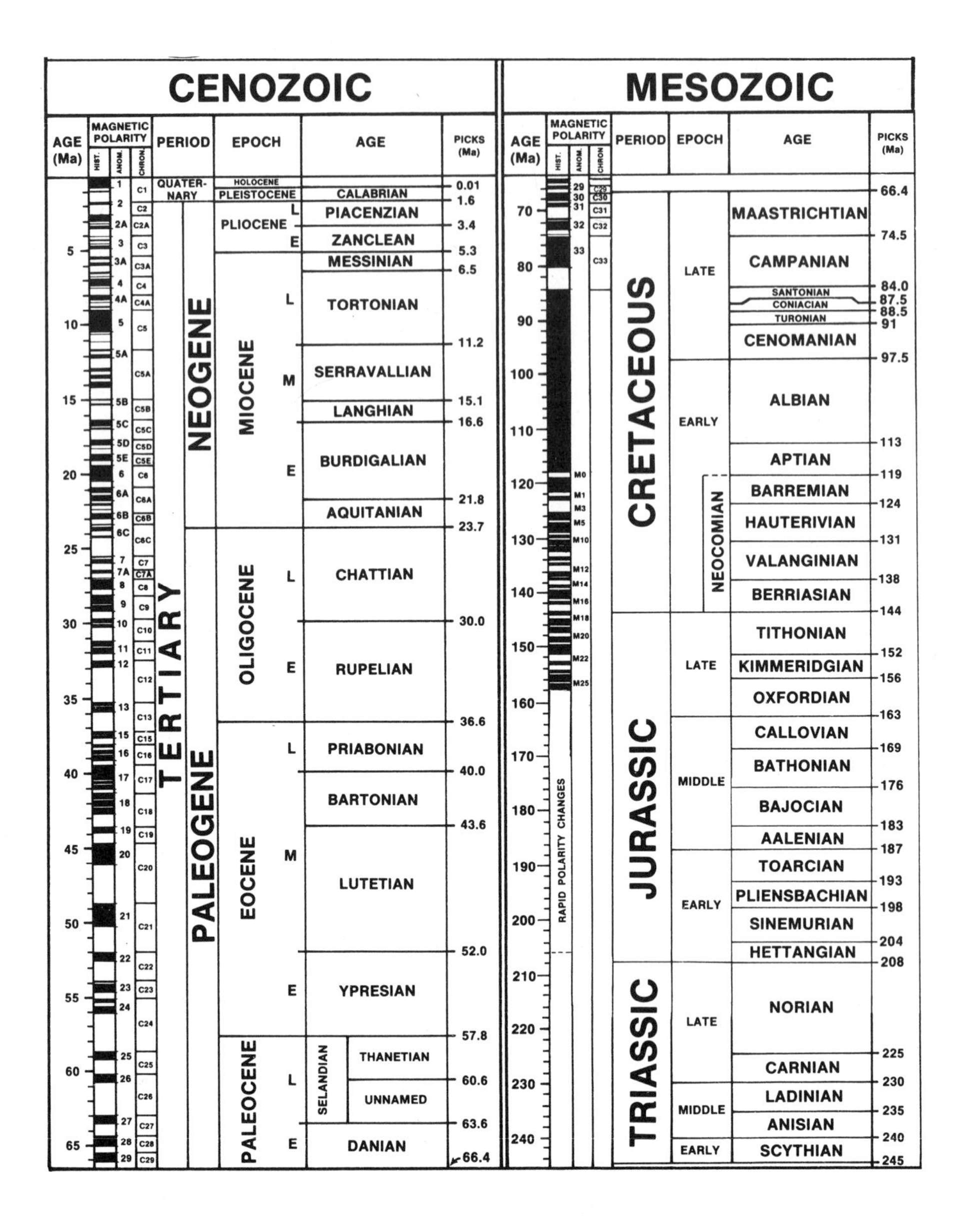

Figure 4 The geological timescale for the Mesozoic and Cenozoic discussed in the text. This is the time scale adopted by the Decade of North American Geology (DNAG), (modified after Palmer, 1983).

From this information, we derive boundary age estimates for Jurassic stages as shown in Figure 4. The older age limit of the Sinemurian (204 Ma) falls just within the age constraint of Armstrong (1978), noted earlier, and at least part of the Toarcian (193-187 Ma) is in Armstrong's 185-189 Ma range. According to our equal zone duration method of interpolation, Jurassic stages vary in duration by a factor of 2, but of course the average duration of the 11 Jurassic stages (5.8 my) is still very near to the average (ca. 6 my) assumed by Harland et al. (1982) in their interpolation. Our Jurassic stage boundary age estimates therefore tend to differ by at most 5 my (at the Triassic/Jurassic boundary), and usually not more than 2 my, from Harland et al. (1982). Recently, Westermann (in press) has scaled the Jurassic stages based on equal duration of subzones, also using the radiometric tie-points of Kent and Gradstein (in press). The main difference is that the Middle Jurassic (Aalenian) starts about 4 my younger.

Above the Oxfordian the ammonite zonation is less well established, particularly because latitudinal provincialism creates more of a problem for correlation. However, the Kimmeridgian may have 4-6 zones and the Tithonian 7-9 zones (Hallam, 1975); this ostensibly requires a Tithonian stage twice as long as the Kimmeridgian and is consistent with the magnetochronological estimate given above.

The question of the accuracy in time that this new scale achieves will be answered only when many more well spaced and stratigraphically meaningful radiometric ages become available.

STRATIGRAPHIC RESOLUTION

The typical micropaleontological biochronologic unit in the Cenozoic timescale is of about 1 my duration. For the Mesozoic this is on the order of 2-5 my, where the larger values pertain to the Jurassic. This resolution is generally adequate for the purpose of geological basin analysis, although it says little of the uncertainties in the data. Errors that pertain to stratigraphic resolution include:

a. uncertainty in estimates of the age of stratigraphic boundaries, and

b. uncertainty in correlation of the zonal units to the timescale

The errors under (b) are difficult to assess, but clearly decrease with increasing number of zones or events along the scale. If for example such errors are half the duration of each zone, this gives uncertainties of 0.5 my for the Cenozoic and 1-2.5 my for the Cretaceous-Jurassic. Both Harland et al. (1982) and Agterberg (in press) have given thought to the error in estimating the age of stratigraphic boundaries from radiometric dates, spaced along the linear scale.

Table 4 Ages and estimated standard deviation (S.D.) used for fitting of the spline curve in Figure 5.

Lower Boundary of Stage		Age	S.D.
1	Maastrichtian (Maa)	72	1.41
2	Campanian (Cmp)	84	1.59
3	Santonian (San)	87.5	1.59
4	Coniacian (Con)	88.5	0.88
5	Turonian (Tur)	91	0.88
6	Cenomanian (Cen)	97.5	0.70
7	Albian (Alb)	113	1.41
8	Aptian (Apt)	} 122	3.18
9	Barremian (Brm)		
10	Hauterivian (Hau)	124	2.83
11	Valanginian (Vlg)	} 135	1.77
12	Berriasian (Ber)		
13	Tithonian (Tth)	145	4.24
14	Kimmeridgian (Kim)	151	2.12
15	Oxfordian (Oxf)	} 158	5.30
16	Callovian (Clv)		
17	Bathonian (Bth)		
18	Bajocian (Baj)		
19	Aalenian (Aal)		
20	Toarcian (Toa)		
21	Pliensbachian (plb)		
22	Sinemurian (Sin)		
23	Hettangian (Het)	212	4.95
24	Rhaetian (Rht)	213	6.36
25	Norian (Nor)	218	2.83
26	Carnian (Crn)	228	7.78
27	Ladinian (Lad)	238	3.54
28	Anisian (Ans)	} 242	7.43
29	Scythian (Scy)		
30	Tatarian (Tat)	246	7.07
31	Kazanian/Ufimian (Kaz-Ufi)	} 253	8.13
32	Kungurian (Kun)		
33	Artinskian (Art)	268	4.24

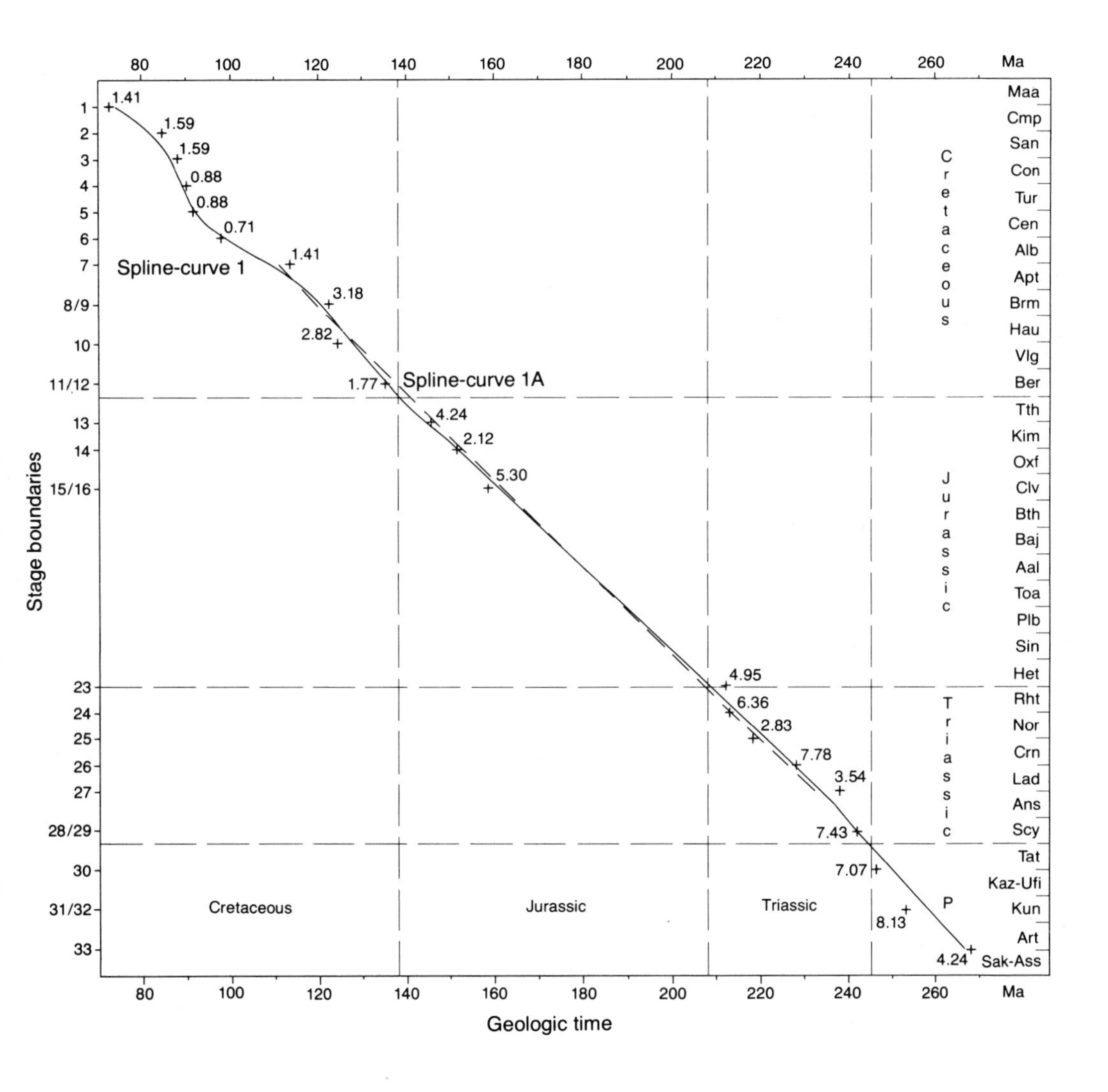

Figure 5 Spline curve fitted to ages of stage boundaries listed in Table 4. Spline curve 1A was fitted to data for stage boundaries numbered 7 to 27 only (after Agterberg, in press).

Maximum likelihood estimates and error range in time (standard deviation) were employed by Agterberg (in press) to calculate the age of a stratigraphic boundary using radiometric ages assigned to or close to this boundary. This method works well both with large and small samples, i.e. numbers of dates per boundary. Table 4 shows age estimates with calculated uncertainty in time. The original data were derived from Harland et al. (1982) with minor modification:

a. if there are not sufficient dates for the two boundaries of a stage, the estimate was assigned to a single point mid-way between the stage boundaries

b. imprecise estimates for six successive Jurassic stage boundaries were not used

As can be seen, the standard deviation increases irregularly from around 1.5 my for the late Cretaceous stage boundaries to 7 or 8 my in the Triassic.

When the ages of a number of successive chronostratigraphic boundaries have been estimated, they can be further improved by smoothing with the aid of cubic spline functions. A spline curve is a very smooth interpolation, based on the principle of least squares. Deviation between observed values (crosses in Figure 5) and spline curve are permitted to exist, but the sum of squares of these deviations can be regulated. The weight assigned to each observed value is inversely proportional to its variance.

Figure 5 shows that the fitted spline curve no. 1 follows the stage boundaries in the Late Cretaceous more closely because they are relatively precise. In places where the uncertainty is great, the spline curve tends to become a straight line. Spline curve no. 1A shown in Figure 5 was fitted to the available estimated ages between the Anisian (Triassic) and Cenomanian (Late Cretaceous).

Because the intervals between stage boundaries in the vertical direction of Figure 5 are equally spaced, a straight line in this type of plot would agree with the hypothesis of equal duration of stages. Harland et al. (1982) applied this linear interpolation between relatively precise stage boundaries defined as tie-points, i.e. no. 47 (Aptian-Albian boundary, 113 Ma) and no. 27 (Ladinian-Anisian boundary, 238 Ma). Because the crosses for boundaries nos. 7 and 27 fall slightly to the right of the fitted spline curve, the two estimates obtained by spline interpolation are younger than those by Harland et al. (1982) (see Figure 6). For example, the Triassic-Jurassic boundary is 208 Ma, rather than 212 Ma and the Jurassic-Cretaceous boundary is close to 140 Ma instead of 144 Ma. The interpolation for the Triassic-Jurassic boundary agrees with the DNAG Mesozoic scale (Figure 4) and the Jurassic-Cretaceous estimate is only slightly younger. The latter is mainly due to the effect of a relatively young Oxfordian glauconite age of 148 Ma and four other relatively young glauconite ages for the

Tithonian, as listed in Harland et al. (1982). If these 5 glauconite dates were not used, the spline curve would also give an age of approximately 144 Ma for the top Jurassic. Odin et al. (1982) use more glauconite dates and arrive at an age of 130 Ma for this boundary.

The same estimates of 208 Ma for the Triassic-Jurassic and of ±140 Ma for the Jurassic-Cretaceous boundaries are found (Figure 6) when the concept of the equal duration of Jurassic ammonite zones is used as a meter to guide the spline curve. This concept underlies the DNAG scale, discussed earlier. Also, when tie-points of 208 Ma (Triassic-Jurassic boundary) and of 156 Ma for the Oxfordian-Kimmeridgian boundary are used (spline curve no. 2 with tie-points, Figure 6), the DNAG and the spline curve timescale are virtually identical, except for the younger (glauconite) top Jurassic age, as discussed earlier. No error analysis along these lines of reasoning is available for the Cenozoic timescale.

The conclusion from the previous discussion is that stratigraphic resolution (but not necessary accuracy) of the biochronologic and magnetochronologic scale exceeds the resolution using radiometric interpolation. The most pressing need for the Jurassic and Early Cretaceous timescale is reliable radiometric dates.

LOCAL TIMESCALES

In many sedimentary basins, particularly in higher latitudes or with restricted environments, standard biozonations may be difficult to apply. Limited correlation to a standard timescale may be achieved for selected stratigraphic horizons, but in general local biostratigraphic correlations are employed. Sometimes it is possible to use extrapolation of sedimentation rates, or tectono-eustatic or radiometric ties to calibrate the local scale.

The consistent use of such a local timescale in a sedimentary basin, even if the timescale calibration is weak, is valuable for quantitative interpretation. After all, when rates of subsidence and sedimentation are calculated using the same numeric standard, the numbers will show the trends, if not the exact magnitude.

In Chapter III.1, on the CASC method of quantitative correlation, mention was made of the so-called RASC timescale. Suppose that a detailed scaled optimum sequence (Chapter II.4) of fossil events has been obtained for a basin. Each event in this sequence has a known interfossil distance to its neighbors. In the CASC chapter it was shown that, if for some of these events an age in Ma is known, the interfossil distance can be used to stretch the whole scaled sequence in linear time. If there are sufficient calibration points on the scale, the new timescale is reliable for isochron correlation and subsidence and sedimentation rates (Figures 10a, 18 and 19 in Chapter III.1). Smoothing by means of a spline curve of the plot of scaled optimum sequence (distance) against linear time reduces the effect of error in individual key ages.

Figure 10a in Chapter III.1 is the age-distance plot of the scaled optimum sequence for Cenozoic foraminifers in 21 Labrador-Grand Banks wells (7-2-4 thresholds). The smoothing factor is 2.0. Figure 7 shows the new RASC timescale for this region, which has a resolution of 1-3 Ma. Accuracy is considerably better than crude timescales that relate broad regional interpretation of Paleocene or Early Oligocene etc. age to a numerical geological scale. The same interpolation approach as for RASC timescales can be applied with the graphic correlation method, using the composite standard.

PALEOBATHYMETRY

The interpretation of the original depth below sealevel at which sediments were laid down is a complex art. The position of a paleo-seafloor relative to sealevel can be estimated from sedimentary or seismic facies and most commonly is attempted from the fossil record. In deltas or large fan deposits the foreset height of prograding clastic fans may be a useful indicator of the water-depth in the basin, which depth may be compared to estimates based on microfossils. In Mesozoic and Cenozoic basins ratios of planktonic versus benthic foraminifers and assemblage composition of benthic foraminifers are often excellent indicators of depth of deposition. This is particularly true when trends can be analyzed in (a) planktonic/benthic ratios, (b) assemblage composition or (c) test morphology, as a function of shallowing or deepening basins. Most studies are actualistic in the sense that the modern depth distribution of taxa is taken as representative also for the past. The Recent as a Key to the Past has been shown to be particularly reliable for photic zone type of organisms in water depth of 200 m or so and less. Certainly, the error in paleo/water-depth interpretations may be large when deeper bathyal and the abyssal realms are involved. Also, the uncertainty in deep water estimates increases with the age of the deposits, and often only minimum water depths are given. Fortunately, in areas of detailed exploration, local geological factors like paleogeography, paleoslopes from seismic sections and paleontological trends may considerably enhance the accuracy of observations. Table 5 lists selected studies of use in paleobathymetric analysis, arranged by geological period.

If paleobathymetric reconstructions are sufficiently detailed, short term eustatic sealevel changes may be deduced. Since the magnitude of such sealevel changes probably did not exceed 100-200 m and frequently was much less (e.g. Hardenbol et al., 1981) a well or a section has to be available in a portion of a sedimentary basin that was filled within 100-200 m of sealevel during eustatic highstand.

Table 5 Selected references of use with paleobathymetric analysis of benthic foraminiferal assemblages in Mesozoic-Cenozoic strata.

Cenozoic	Cretaceous	Jurassic
Poag 1981	Butt 1982	Hallam 1975
Murray 1973	Nyong & Olsson 1984	(paleoenvironment in
Boltovskoy & Wright 1976	Haig 1979	general
Berggren & Aubert 1975	Sliter 1972	Gradstein 1983
Lipps et al. 1979		Munk 1980
Tjalsma & Lohmann 1983		Wall 1960
Gradstein & Berggren 1981		

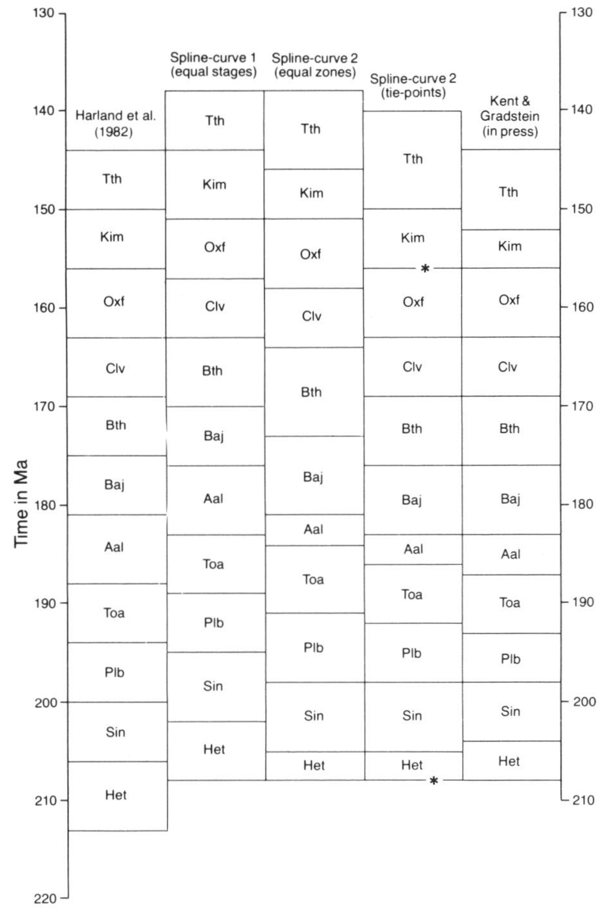

Figure 6 Comparison of spline curve ages (rounded off to nearest interger values) to ages estimated by Harland et al. (1982) and Kent and Gradstein (in press), (after Agterberg, in press).

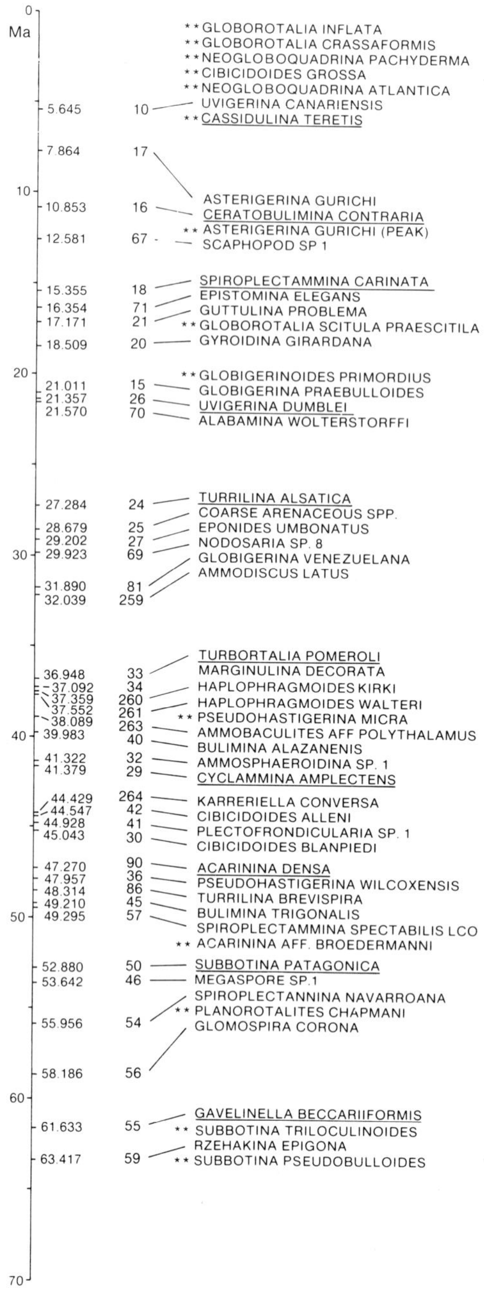

0 Ma

**GLOBOROTALIA INFLATA
**GLOBOROTALIA CRASSAFORMIS
**NEOGLOBOQUADRINA PACHYDERMA
**CIBICIDOIDES GROSSA
**NEOGLOBOQUADRINA ATLANTICA
5.645 10 UVIGERINA CANARIENSIS
**CASSIDULINA TERETIS
7.864 17

10
ASTERIGERINA GURICHI
10.853 16 CERATOBULIMINA CONTRARIA
**ASTERIGERINA GURICHI (PEAK)
12.581 67 SCAPHOPOD SP 1

15.355 18 SPIROPLECTAMMINA CARINATA
16.354 71 EPISTOMINA ELEGANS
17.171 21 GUTTULINA PROBLEMA
**GLOBOROTALIA SCITULA PRAESCITILA
18.509 20 GYROIDINA GIRARDANA

20
**GLOBIGERINOIDES PRIMORDIUS
21.011 15 GLOBIGERINA PRAEBULLOIDES
21.357 26 UVIGERINA DUMBLEI
21.570 70 ALABAMINA WOLTERSTORFFI

27.284 24 TURRILINA ALSATICA
COARSE ARENACEOUS SPP.
28.679 25 EPONIDES UMBONATUS
29.202 27 NODOSARIA SP. 8
30 29.923 69 GLOBIGERINA VENEZUELANA
AMMODISCUS LATUS
31.890 81
32.039 259

TURBORTALIA POMEROLI
36.948 33 MARGINULINA DECORATA
37.092 34 HAPLOPHRAGMOIDES KIRKI
37.359 260 HAPLOPHRAGMOIDES WALTERI
37.552 261 **PSEUDOHASTIGERINA MICRA
38.089 263 AMMOBACULITES AFF POLYTHALAMUS
40 39.983 40 BULIMINA ALAZANENIS
41.322 32 AMMOSPHAEROIDINA SP. 1
41.379 29 CYCLAMMINA AMPLECTENS
44.429 264 KARRERIELLA CONVERSA
44.547 42 CIBICIDOIDES ALLENI
44.928 41 PLECTOFRONDICULARIA SP. 1
45.043 30 CIBICIDOIDES BLANPIEDI
47.270 90 ACARININA DENSA
47.957 36 PSEUDOHASTIGERINA WILCOXENSIS
48.314 86 TURRILINA BREVISPIRA
49.210 45 BULIMINA TRIGONALIS
50 49.295 57 SPIROPLECTAMMINA SPECTABILIS LCO
**ACARININA AFF. BROEDERMANNI
52.880 50 SUBBOTINA PATAGONICA
53.642 46 MEGASPORE SP.1
SPIROPLECTANNINA NAVARROANA
**PLANOROTALITES CHAPMANI
55.956 54 GLOMOSPIRA CORONA
58.186 56

60
61.633 55 GAVELINELLA BECCARIIFORMIS
**SUBBOTINA TRILOCULINOIDES
63.417 59 RZEHAKINA EPIGONA
**SUBBOTINA PSEUDOBULLOIDES

70

Figure 7 Local timescale for Grand Banks-Labrador biostratigraphy. This scaled optimum sequence for 21 Labrador-Grand Banks wells was calibrated in linear time by means of the CASC (7-2-4 RASC run) age-depth plot.

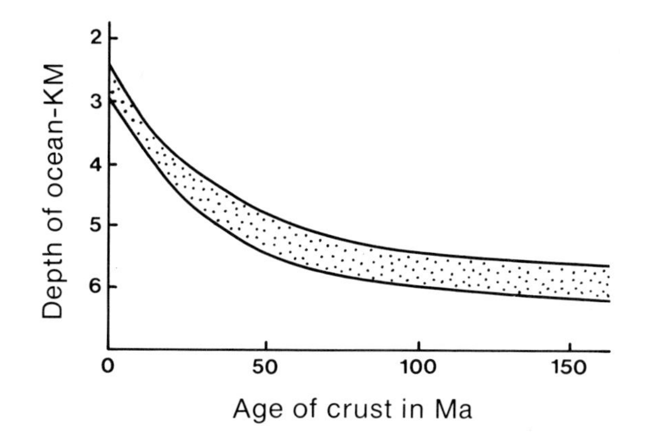

Figure 8 The relationship between depth and age for normal oceanic crust (after Parsons and Sclater, 1977).

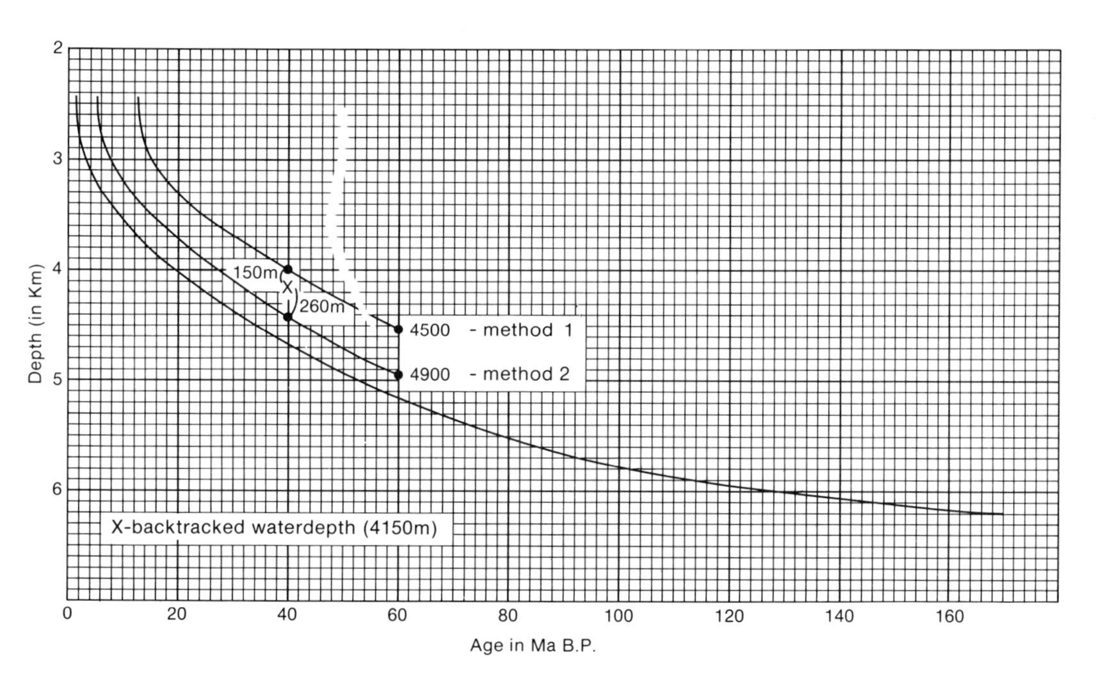

Figure 9 Graphical methods to determine the paleo waterdepth of sediments deposited on oceanic crust, following the empirical relationship between depth and age in Figure 8. For explanation see text.

In recent years, the method of backtracking oceanic sites to the original depth at which the site formed on the spreading ridge, has received considerable attention by marine geologists and paleontologists, interested in accurate abyssal water depth estimates.

BACKTRACKING

It is now widely recognized that there is a stable age-depth relationship of the seafloor on spreading ridge flanks through time (e.g. Parsons and Sclater, 1977). The empirical relationship goes back to three significant observations:

1. rigid, basaltic plates, move at constant velocity away from a hot boundary at an oceanic ridge crest

2. ocean depth increases away from these ridges

3. ocean ridge crests show a constant elevation of approximately 2.7 km, over almost all ocean basins, independent of slow or fast spreading.

On this basis, Sclater et al. (1971) showed that there are empirical relationships between heat flow and age, and depth and age that are similar in all oceans. The relationship represents a simple cooling curve for a basaltic slab of crust that is relatively buoyant and high in elevation when hot, but cools, contracts and sinks in isostatic equilibrium when it moves away from the ridge. The sinking curve is found as follows:

1. Basement depths at oceanic sites with a known age (preferably from marine magnetics) are corrected for loading due to sediment cover. The amount of unloading is 30-50% of the thickness of the cover (for each 10 m of sediment cover, the site is depressed by ± 5 m).

2. The corrected basement depths under (1) are plotted in an age-depth diagram and fitted by means of an empirical curve (Figure 8). The curve is dependent on reliable bio- and magnetostratigraphy for the ocean sites and reliable numerical estimates of basement ages. The curve would be slightly different and more drawn-out, when the new DNAG timescale is used. This change is probably within the error in water depth, which has been estimated for this method at ± 300 m.

Young crust sinks according to the square root of time ($\sqrt{t}$), and older crust according to a negative exponential function. For 0-70 Ma old crust

$$d(t) = 2500 + 350\ (t)^{1/2}\ \text{m} \qquad (1)$$

and for 70-160 Ma old crust

$$d(t) = 6400 - 3200\ e^{(-t/62.8)}\ \text{m} \qquad (2)$$

where d(t) is the depth increase per time increment in m per Ma.

The backtracking curve gives an excellent idea of the paleobathymetic history through time for oceanic sites. If one knows the age of basement, the stratigraphy of the overlying sediment and the present water depth, the paleodepth for each interval may be obtained. A simple example illustrates the procedure as described by Berger and Von Rad (1972), and by Van Andel et al. (1977), (Figure 9). Assume a site with

age of basement - 60 Ma
water depth - 4500 m
age sample in site - 40 Ma
depth of sample below seafloor - 300 m
total sediment thickness - 600 m

Berger and Von Rad (1972) proceed as follows:

1. In the (empirical) subsidence curve diagram find the intersection of 60 Ma and 4500 m.

2. Draw a line parallel to the best fit curve, passing through this point.

3. Find the intersection of this line with the 40 Ma age point when the site had subsided for 20 Ma.

4. The new depth should have been deeper due to loading and shallower due to sediment accumulation. From the intersection under (3), move downward by one-half of the distance between present water depth at the site and the depth of the sample in the hole.

The paleo waterdepth at 40 Ma age was 4150 m.

Van Andel et al. (1977) obtain the paleodepth in an ocean site by a method which reduces the depth of the standard curve by the amount of sediment accumulated up to that time. Only where sediment thickness exceeded 100 m is isostatic compensation for loading necessary.

1. Reduce the present basement depth by 1/3 of the sediment (5100 - 200 = 4900 m).

2. Fit the standard curve through the new basement depth.

3. Reduce paleodepth at each time by 2/3 of the accumulated sediment, rather than the entire value (2/3 x 400 = 260 m).

The paleo waterdepth at 40 Ma age was 4140 m, or almost exactly equal to that found by the previous method.

These two graphical methods for backtracking are shortcuts to obtain oceanic paleo waterdepth. The full calculation involves applying corrections for thermal subsidence, sediment loading, and decompaction.

The seafloor depth at time t_1 is the present depth of basement minus

1. the increment in metres due to thermal cooling for the $t_1 - t_0$ (=Recent) time interval

2. the increment in metres due to sedimentary loading by the sediments deposited in the $t_1 - t_0$ interval, and

3. the decompacted sediment thickness of the older than time t_1 non-backstripped column.

The thermal subsidence increment can be found with equations (1) and (2), cited previously. The loading correction is

$$\ell = S_1^* \frac{\rho_{s'} - \rho_w}{\rho_m - \rho_w} \qquad (3)$$

where ℓ is the loading correction in m, S_1* is the stripped off and decompacted sediment thickness for the time interval $t_1 - t_0$ and ρ_w, ρ_s' and ρ_m are the densities in g/cm^3 of seawater (1.028), the decompacted sediment and the mantle (3.23) respectively. The decompaction correction is given by

$$d = S_0 \frac{\rho_s - \rho_{s'}}{\rho_1 - \rho_w} \qquad (4)$$

where d is the decompaction correction in m, S_0 is the measured thickness in m, ρ_s is the measured density of the sediment in g/cm^3, ρ_w is the density of seawater and ρ_1 is the estimated original density of the sediment (≈1.8). Residual errors may be of the order of 10%.

After the corrected seafloor depth is found with the youngest unit stripped off, the calculations are repeated with successively older units removed. The restored seafloor depths through time are connected to form the subsidence pathway.

Practical Application of Backtracking

The backtracking relationship has seen a pleothora of applications, including reconstruction of ocean paleobathymetry in time and space (Sclater et al., 1977), changes in depth of the deep sea carbonate ooze/clay dissolution interval through time (Ramsay, 1977), and changes through time in bathyal and abyssal benthic communities as a function of increasing water depth (Tjalsma and Lohmann, 1983). By means of the empirical curve it can be established if a newly drilled ocean site conforms to the expected subsidence pattern or is on too hot (higher) or too cold (lower) ocean crust. Application of the model may help to deduce the approximate paleobathymetry of such sites through time.

In 1981 Deep Sea Drilling Project Leg 76 completed Site 534 in the Blake Bahama Basin (Sheridan, Gradstein et al., 1983). This Site, in 4990 m of water, bottomed at 1635 m in oceanic crust in the Jurassic marine magnetic quiet zone. The immediately overlying sediments are Callovian in age or about 165 Ma. Figure 10 shows the stratigraphy at the Site as determined from multiple biostratigraphy, lithostratigraphy and the intersection of the borehole with the regional seismic markers. Site 534 was the first to reach oceanic crust of that age and determined that Atlantic opening was 2-3 stages or almost 20 Ma younger than postulated. Important questions for paleoceanography are whether the basement followed the average back-track subsidence pattern and what the depth transect of the seafloor was through time.

Both backtrack method no. 2 and the backstripping procedure were applied at Site 534. Basement is at 1625 m below seafloor. Unloading of basement by an amount in m of 1/3 of the sedimentary cover gives a depth of 6080 m for 165 Ma old crust. This is almost exactly on the average backtrack curve of Figure 8, which establishes that the Site in Callovian time was on the ridge crest at 2.7 km below sealevel. In order to find the depth of the seafloor back through time, the youngest formation (Blake Ridge + Abaco Member of Neogene age) is first stripped off (696 m) and the backtrack procedure repeated for the 23.5 Ma point and 1635 - 696 = 939 m of sedimentary section. This procedure is repeated for successively older units and the new depth points connected to show the position of the paleo seafloor through time. Figure 11 illustrates the transect of subsidence for each of the 7 lithological units. The seafloor subsided at an almost linear rate of 1 cm in 690 y. Superimposed on the plotted paleobathymetry is the fluctuating Carbonate Compensation Depth curve for the western North Atlantic. The CCD controls the formation of shale versus limestone lithofacies in the basin, and these boundaries form the key seismic reflectors β, C and D.

BURIAL HISTORY

In contrast to age-depth predictions for oceanic sites, there is no simple empirical relationship between subsidence and paleo waterdepth on continents and margins. Mechanisms for subsidence include cooling of (deep) crust following rifting, extension due to stretching, and loading due to sediment and water fill. As a result of a better understanding of the fundamental processes that affect vertical motion, drilling of many deep (exploration) wells and improved reliability of biostratigraphy and timescales, a great deal of attention has recently been focussed on continental subsidence. An excellent introduction to the intricacies of theoretical modelling and geological analysis is found in the studies presented in Blanchet and Montadert (editors, 1981).

Subsidence and burial curves are an analytical rather than predictive tool to study the tectonic and depositional history of continental margins or other sedimentary basins. Van Hinte (1978) coined the term **"geohistory analysis"** for this method, but in this Chapter I prefer the more direct and older term **"burial history".** The technique shows the rate of burial of a geological horizon as a function of basement subsidence, sediment fill and sediment water loading. Excellent introductions to the subject and its geological applications are in Van Hinte (1978), Hardenbol et al. (1981), Wood (1981) and Brunet and Le Pichon (1980).

As already briefly stated in Chapter III.1 on the CASC program, burial analysis runs through a series of stepwise calculations, which I review below.

1. Sediment accumulation through time, starting at the oldest point in a well, using age per stratum interpretations provided by paleontology, is defined as

$$R = \frac{S}{10t} \text{ cm}/10^3\text{y} \qquad (5)$$

where R is the sedimentation rate. S is the sediment thickness and t is the time interval in my.

2. Rates of deposition for two stratigraphic units of equal duration and thickness, but buried at different depths are not the same, because the more deeply buried one generally is more consolidated. Decompacted sediment accumulation is based on the thickness restored to the time of deposition. Thus, the equation for R is modified to express the restored or decompacted sedimentation rate per unit time

$$R_s = \frac{S^*}{10t} \text{ cm}/10^3\text{y} \qquad (6)$$

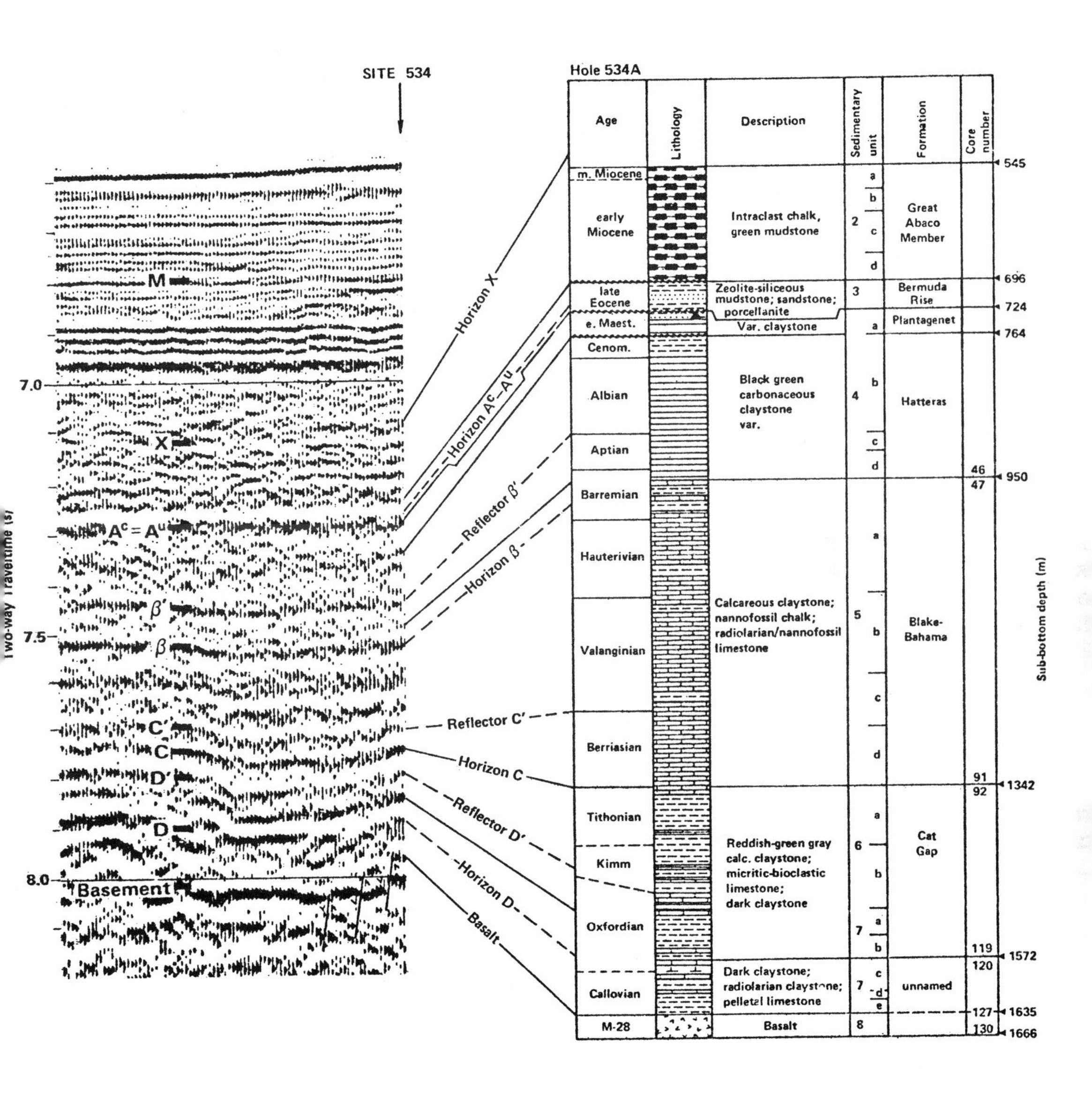

Figure 10 Stratigraphic summary of Site 534, DSDP Leg 76, Blake Bahama Basin, western North Atlantic Ocean.

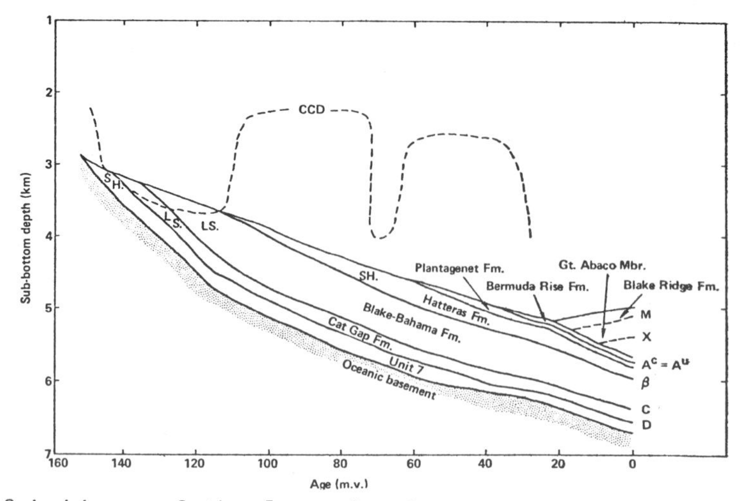

Figure 11 Subsidence of the Jurassic, Cretaceous and Cenozoic formations on oceanic crust in Site 534, DSDP Leg 76 (after Sheridan, Gradstein, et al., 1983).

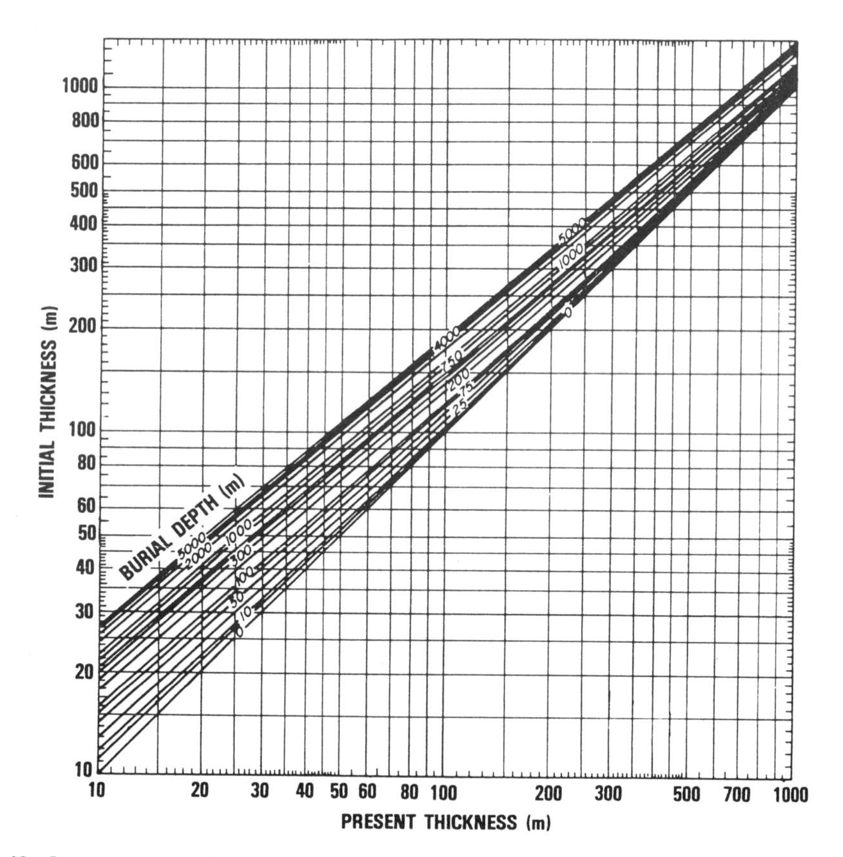

Figure 12 Relation between present thickness, burial depth and initial thickness of shales (after Perrier and Quiblier, 1974).

where R_S is the decompacted (original) sedimentation rate and S^* is the restored thickness.

In order to perform decompaction, each borehole is partitioned according to its main lithologies, with emphasis on shale units that undergo most consolidation.

Perrier and Quiblier (1974) and Van Hinte (1978) give functions that (a) estimate shale porosity at increasing burial depth and relate present thickness, original porosity and initial thickness, and (b) show the relation between present thickness, burial depth and initial thickness of shales, here reproduced as Figure 12. For example, a 100 m thick shale now buried at 1 km, initially was more than 150 m thick and at 4 km almost 200 m.

Reef carbonates may have undergone little compaction and the same may be true for sands. Grain carbonates are assumed to compact like sand containing 30% shale, and micrites like shale containing 35% sand (Hardenbol et al., 1981).

When sonic or density logs are available for a well, the original decompacted thickness may be calculated from

$$S^* = \frac{1 - \phi_1}{1 - \phi_1'} S \qquad (7)$$

where ϕ_1 is the measured porosity, ϕ_1' is the original porosity and S is the interval thickness as measured. The original porosity is found by sliding each package of lithologies up its own porosity versus depth curve (Sclater and Christie, 1980; program DEPOR, p. 567).

3. In order to construct a burial curve, a time/depth plot is made of the observed and restored thicknesses versus linear time. However, first the paleobathymetric trends through time in a borehcle are plotted against linear time in Ma along the horizontal axis and water depth in feet or metres along the vertical axis. Estimates of paleobathymetric error can be added to this graph using error bars. The ancient seafloor is now used as a baseline to plot cumulatively the thickness of sediment determined for each stratigraphic increment recognized. The resulting curve is a first approximation of relative movements at well sites and represents a burial curve relative to present day sealevel.

In an early application of this principle Bandy (1953) used micro-paleontological age and paleodepth data to demonstrate aspects of the tectonic history of a site in the Ventura Basin, California (Figure 13). Initially, subsidence surpassed sedimentation and water depth increased, but in early or middle Pliocene time subsidence either slowed consider-ably or some uplift occurred because sediment filling led to rapidly

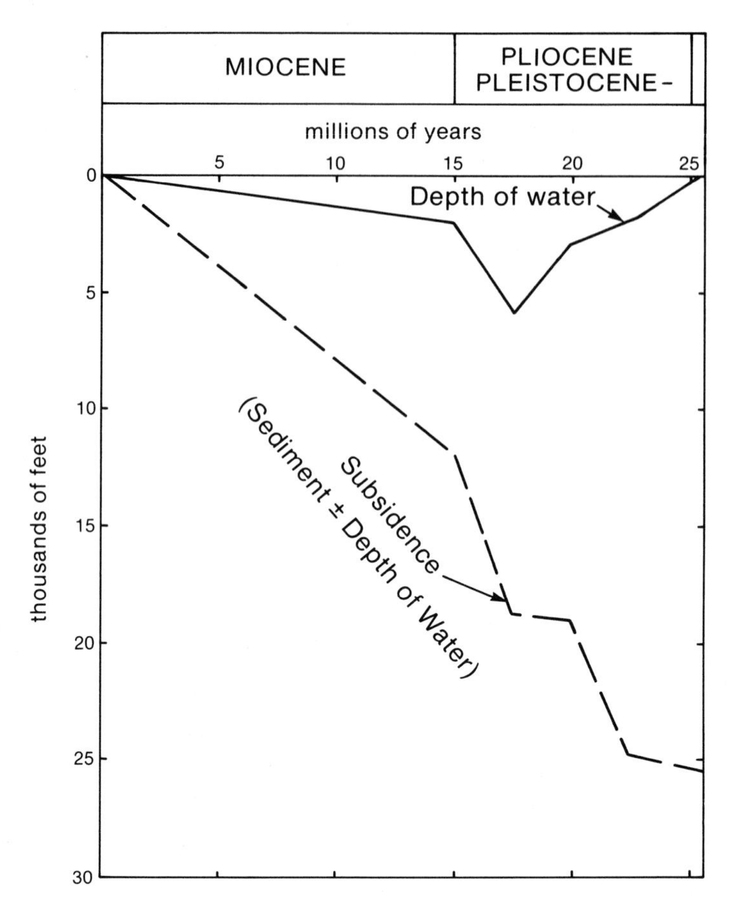

Figure 13 Burial history of the Wheeler Ridge area in the Ventura Basin (after Bandy, 1953).

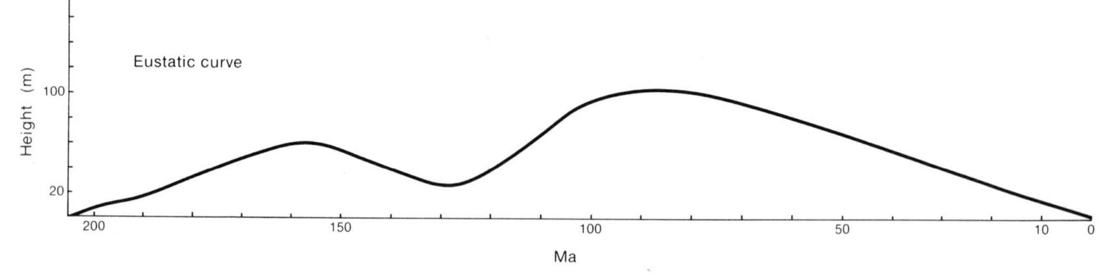

Figure 14 Schematic curve of eustatic changes in sea level through time. This eustatic correction, which is a conservative estimate, is incorporated in the input data (Table 6) for tectonic subsidence calculations of Figures 15, 16 and 17.

decreasing water depth. Renewed subsidence again was followed by a slowdown up to Recent time.

The waterdepth changes through time added to the (compacted or decompacted) sedimentation rate give the burial rate R_S per unit time

$$R_s = \frac{s + d_w}{10t} \text{ cm}/10^3\text{y} \qquad (8)$$

where d_W is the change in water depth per time interval. An increase in water depth (deepening of the sea) is added to the sediment accumulation, and a decrease in water depth (shallowing of the sea) is subtracted (d_W assumes a negative value).

4. It is widely accepted in the geological literature that sealevel has not been stable through time. As to the causes for such long-term changes in elevation and the magnitude of the phenomenon, opinions diverge. Relative movement of sealevel on the order of 100 or more metres over 50 my or so, is a conservative estimate. Figure 14 shows such a schematic and conservative sealevel curve (compiled) from various sources (Hallam, 1975; Hardenbol et al., 1981; Watts and Steckler, 1979). This curve will be applied as a correction in a practical example later in the Chapter. Only the long-term, gradual changes have been entered, and not the higher frequency changes.

The equation for the rate of burial R_S per unit time now is

$$R_s = \frac{S + d_w - d_E}{10t} \text{ cm}/10^3\text{y} \qquad (9)$$

where d_E is the change in sealevel elevation through time. A rise in sealevel is subtracted from the subsidence or burial rate (d_E assumes a negative value) and a drop in sealevel is added.

An interesting question is whether long-term eustatic sealevel changes are not accounted for in the observed paleobathymetric changes in a well section indicated by (bio)facies. This would be true if a basin filled only with water, and if the paleobathymetric estimates through time were sufficiently accurate to record these slow changes. In reality a basin fills both with water and with sediment and a sealevel change also influences sedimentation rate. Assume a situation where sedimentation surpasses subsidence while sealevel rises. Depending on the relative rates, paleo waterdepth may either

(a) increase (sealevel rise surpasses sediment fill rate minus subsidence rate),

(b) remain constant (sealevel rise equals sediment fill rate minus subsidence rate),

(c) decrease (sealevel rise is less than sediment fill rate minus subsidence rate).

Clearly, paleobathymetric interpretations using biofacies changes through time do not necessarily reflect long term changes in eustatic sealevel.

5. The final calculation in the reconstruction of burial history involves tectonic or driving force subsidence, which is the rate of basement subsidence if the basin had only filled with water or air. For this calculation it is necessary to think in terms of weight balancing. With a sediment density of 2.7 g/cm^3 (at 0% porosity) the mantle displacement as a result of sediment loading is often approximately 0.7 x the thickness of the solid sediment column, which means that thermotectonic subsidence is about 30% of the burial depth. The same rule of thumb number was earlier used for the loading correction in backtracking.

Following Watts and Steckler (1981) the unloaded basement depth y is given by

$$y = S^* \frac{\rho_m - \rho_s}{\rho_m - \rho_w} - \Delta S\ell \frac{\rho_w}{\rho_m - \rho_w} + w_d + \Delta S\ell \qquad (10)$$

where ρ_m, ρ_s and ρ_w are the mean density of the mantle, sediment and (sea)water respectively, $\Delta S\ell$ is the eustatic height of sealevel above present day level, w_d is the paleo waterdepth and S^* is decompacted sediment thickness.

The equation is applied using the backstripping method, where for each package of sediment y is calculated, starting with the youngest sediment. First, the youngest sediment package is stripped off and its weight (load) subtracted from basement. The relaxation results in a shallower depth point for basement, as if the last sediment package had not depressed it. This procedure is repeated stepwise back through time for each discrete lithological unit and the new depth points thus obtained show the subsidence of the basement if no sediments had been deposited, and only water had filled the basin.

Practical Application of Burial History

Burial curves provide a direct means of reading the timing and magnitude of geologic events and calculating subsidence and sedimentation in basins. The inclusion of such diagrams in a paleontological report on a borehole or in a regional stratigraphic publication has the following obvious advantages (Van Hinte, 1978).

1. The construction itself forces the biostratigrapher to be aware of the geologic implications of his analytical conclusions. He sees, for example, where more data are needed, what the consequences are when an unconformity is reported, and which of his interpretations do and which do not make geological sense. For example, when a displaced shallow-water fossil assemblage is not recognized as being displaced. it will plot too high on the diagram, and consequently, the subsidence curve will show an uplift "kick". Uplift would also be seen when, for example, a Miocene fossil assemblage in a well site is reported as representing a water depth of 2 km, while there is only 1 km of sediment fill since that time to arrive at the present water depth at the site of 100 m. If 900 m uplift is needed and there is no tectonic evidence for it, chances are the paleo waterdepth estimate is excessive.

2. The diagram forms a convenient linear time-depth frame for the geologist to plot other parameters, such as heat-flow gradients, porosity changes and mineralization.

3. The diagrams are excellent first-order models of basin evolution and tectonic style, continental margin history, etc.

4. Extrapolation of the curves fills gaps in cases of fragmentary stratigraphic information.

Many of the curves, except for the ones depicting tectonic subsidence, can be easily produced graphically on a pre-formatted time-depth diagram. A time scale is plotted along one axis of the graph and the borehole depth along the other.

A computer program can calculate all results and the following example, using three Grand Banks wells, was generated on a microcomputer with graphics output (Programs Bursub and Depor, p. 567). The input is shown in Table 6. The Grand Banks of Newfoundland is a passive margin region underlain by several sedimentary subbasins. Each of three wells selected for burial- and subsidence-history analysis reflects a different tectonic setting (Figures 15, 16) in each of these basins. The Puffin well is near the transform margin of the Grand Banks, where large-scale lateral movement since Jurassic time may have stretched the rifted crust. The Hibernia well is on the edge of a continental graben basin, which experienced steady downward movement since early Jurassic time. The Gabriel site, in the 'deep water' passage between Grand Banks and the continental extension Flemish Cap, is closest to the oceanic regime between Grand Banks and Iberia. Here rifting started in late

Table 6 Original data input for the burial analysis of three Grand Banks wells, Hibernia O-35 (top), Puffin B-90 (middle) and Gabriel C-90 (bottom).

Age (Ma)	Depth below rigfloor (m)	Water depth min	Water depth max	Sea level min	Sea level max	
0.0	107.0	80.0	80.0	0.0	0.0	
17.5	555.0	50.0	100.0	0.0	20.0	Shale
23.5	640.0	100.0	400.0	0.0	25.0	Shale
37.0	1000.0	300.0	500.0	0.0	45.0	Shale
40.0	1300.0	300.0	500.0	0.0	50.0	Shale
76.0	1400.0	0.0	30.0	0.0	95.0	Siltstone
88.0	1720.0	200.0	300.0	0.0	100.0	Sand
91.0	1790.0	100.0	200.0	0.0	100.0	Limestone
97.0	1900.0	100.0	200.0	0.0	95.0	Shale
105.0	2050.0	50.0	100.0	0.0	90.0	Shale
119.0	2300.0	10.0	50.0	0.0	40.0	Siltstone
124.0	2700.0	10.0	50.0	0.0	30.0	Siltstone
131.0	2870.0	5.0	30.0	0.0	30.0	Siltstone
135.0	3200.0	5.0	30.0	0.0	30.0	Siltstone
144.0	3730.0	5.0	30.0	0.0	50.0	Siltstone
156.0	4375.0	50.0	100.0	0.0	60.0	Sand
160.0	4788.0	50.0	100.0	0.0	60.0	Mudstone
0.0	0.0	106.0	106.0	0.0	0.0	
1.6	358.0	10.0	100.0	0.0	5.0	Sand
5.3	579.0	50.0	100.0	0.0	10.0	Siltstone
11.2	1116.0	50.0	100.0	0.0	10.0	Siltstone
16.2	1585.0	200.0	500.0	0.0	20.0	Siltstone
23.7	1699.0	200.0	500.0	0.0	25.0	Shale
36.6	2023.0	200.0	500.0	0.0	50.0	Shale
40.0	2170.0	200.0	500.0	0.0	50.0	Shale
52.0	2206.0	200.0	500.0	0.0	65.0	Shale
57.8	2249.0	200.0	500.0	0.0	70.0	Shale
59.0	2258.0	200.0	300.0	0.0	80.0	Shale
71.0	2259.0	100.0	200.0	0.0	90.0	Shale
74.5	2301.0	100.0	200.0	0.0	95.0	Sandy marl
84.0	2423.0	100.0	200.0	0.0	100.0	Sandy marl
87.5	2633.0	100.0	200.0	0.0	100.0	Sandy marl
88.5	2737.0	200.0	300.0	0.0	100.0	Limestone
97.5	2738.0	0.0	0.0	0.0	95.0	Sand
113.0	2880.0	5.0	50.0	0.0	50.0	Sand
119.0	3000.0	0.0	10.0	0.0	40.0	Siltstone
124.0	3276.0	0.0	10.0	0.0	30.0	Sand
131.0	3703.0	10.0	50.0	0.0	30.0	Sand
138.0	4500.0	50.0	150.0	0.0	38.0	Mudstone
141.0	4701.0	50.0	100.0	0.0	40.0	Mudstone
0.0	0.0	1134.0	1134.0	0.0	0.0	
3.0	612.0	1000.0	1500.0	0.0	8.0	Shale
15.0	613.0	1000.0	1500.0	0.0	15.0	Sand
23.0	842.0	1000.0	1500.0	0.0	25.0	Sand
30.0	843.0	1000.0	1500.0	0.0	40.0	Sand
36.0	892.0	1000.0	1500.0	0.0	45.0	Shale
40.0	1282.0	1000.0	1500.0	0.0	50.0	Shale
52.0	1283.0	1000.0	1500.0	0.0	65.0	Shale
55.0	1358.0	800.0	1000.0	0.0	70.0	Sandy marl
72.0	1359.0	500.0	500.0	0.0	90.0	Sandy marl
78.0	1372.0	200.0	500.0	0.0	95.0	Sandy marl
113.0	1373.0	200.0	500.0	0.0	55.0	Sandy marl
116.0	1652.0	100.0	300.0	0.0	50.0	Sandy marl
121.0	2642.0	50.0	100.0	0.0	40.0	Sand
124.0	3402.0	10.0	50.0	0.0	30.0	Sand
131.0	3822.0	100.0	200.0	0.0	30.0	Mudstone
133.0	4013.0	100.0	200.0	0.0	30.0	Mudstone

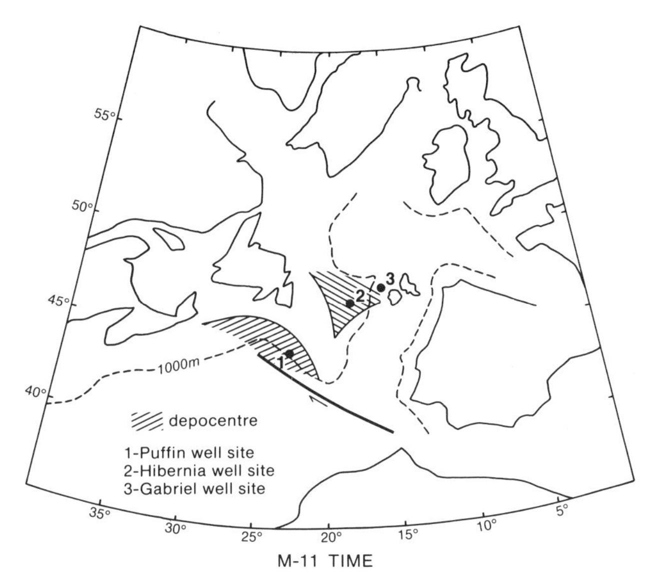

Figure 15 Location of the three wells used for burial history analysis; the base map shows the approximate paleogeography at M-11 (Early Cretaceous) time.

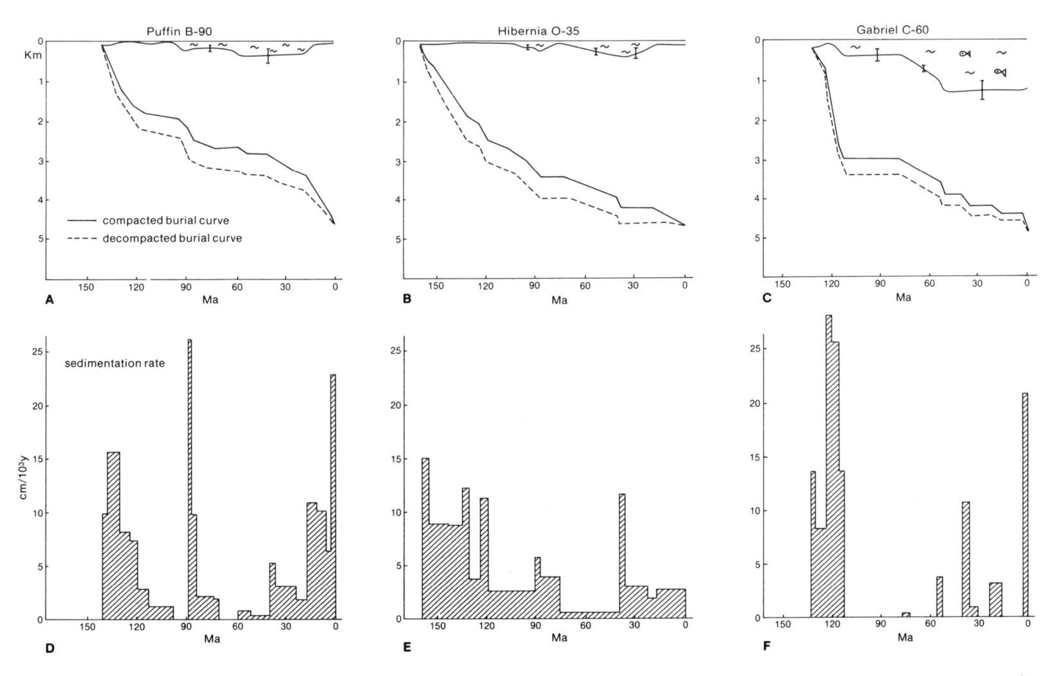

Figure 16 a,b,c Water depth trends and burial history for three passive continental margin wells. Location in Figure 15.

d,e,f Histograms of the (decompacted) sediment accumulation in the three wells in cm/10^3 y.

Jurassic- early Cretaceous time. None of the wells went to (continental) basement, which is probably of Paleozoic age.

At all three sites, a major Early Cretaceous clastic wedge formed (Figure 16). at a rate of 15-25 cm/10^3y, when shallow water sedimentation more or less kept up with subsidence. This rate uses restored thicknesses. A major sand influx (Hibernia sands - G. Williamson, pers. comm., 1984) occurred around 120 Ma ago. During late Early Cretaceous, Late Cretaceous and Tertiary subsidence slowed, but nevertheless the sea deepened because sedimentation slowed even more. Gabriel, the most distal site, experienced more sediment starvation and hence experienced the deepest water conditions, with only intermittent deposition during the major sedimentary cycles in late Eocene (≈40 Ma) and Neogene (<20 Ma). The Quaternary peak sedimentation is due to glacially induced debris fill.

At all three sites, late Miocene, Oligocene, and Senonian-middle Eocene were periods of low sedimentation rates or even may have witnessed erosion as a result of the rapid Messinian, mid Oligocene and latest Cretaceous eustatic sealevel drops respectively. Erosion may also explain why subsidence (Figure 17) in Gabriel has an upward kick as if late Cretaceous rebound occurred. It is worth investigating if there is evidence for abnormally high Lower Cretaceous compaction, which would indicate that mid-Cretaceous section was stripped. The upward bump in the burial curves and the subsidence curve for the Puffin site around 90 Ma (Turonian-Coniacian), may reflect the relative buoyancy of the salt structure on which the well is drilled.

All three subsidence curves are compared to thermal contraction models in Figure 17, using either oceanic crust, or a thinned continental crust with 60% lithospheric injection following rifting (see Royden et al., 1980, and Hardenbol et al., 1981).

Gabriel, closest to the late Jurassic-early Cretaceous oceanic rift between Grand Banks and Iberia, shows a strong, presumably fault controlled, early subsidence, which later levelled off and may then be largely thermal. Mantle and crustal cooling can explain most of the almost perfect thermal curve for Hibernia's subsidence, whereas the continued downward incline of the subsidence curve in Puffin may be the result of stretching. A systematic investigation of adjacent well sites, using equally detailed paleobathymetry and stratigraphy, in combination with seismically deduced thicknesses for the sedimentary sections down to basement, will shed further light on the trends outlined.

The general trend in burial history of these three continental edge wells is rapid early subsidence of a tectonic-thermal nature, that decreases following the thermal contraction pattern of the mantle and crust. A second downward cycle, evidenced in increased water depth and renewed high sedimentation is largely the result of sedimentary loading.

The burial and subsidence history is entirely different for sites drilled on an active margin, as illustrated by an example (Figure 18) shown in Van Hinte (1978). Here deep water sediments accumulated on a normally subsiding oceanic seafloor until about 58 Ma ago, when they were uplifted abruptly as a result of underthrusting that accompanied subduction. The rate of uplift cannot be interpretated from the well data because of a hiatus between 58 and 43 Ma. Three possibilities exist (Figure 18):

(1) Uplift occurred almost instantaneously 58 Ma ago and was followed by a period of non-deposition because of increased current velocities over the new seafloor high.

(2) Uplift was spasmodic or gradual during the entire period of the hiatus (with a minimum average rate of about 10 $cm/10^3y$, and again non-deposition (due to erosion?) was experienced.

(3) Uplift occurred rapidly but later than 58 Ma ago, and pre-event sediments have been eroded.

Whatever actually occurred, the paleontologic conclusions point out that uplift (due to underthrusting) apparently took place within a specific time span, during which the location had become part of the continent, and thereafter subsided slowly under an increasing sediment load. The two younger hiatuses in this well site (stippled parts in Figure 18) may be the result of erosion caused by eustatic sealevel drops or by changes in current regimes.

A final example of the use of burial history for geological basin analysis is drawn from a study by Gradstein and Srivastava (1980) on the tectonic-stratigraphic trends in the Labrador Sea. If margin subsidence can be (partly) interpreted in terms of a simple thermal cooling model, as empirically established for basins on oceanic crust, then regional (passive) continental margin transgression and regression and seafloor rifting and incipient spreading in the adjacent oceanic basin can be expected to be correlative. Although this simplistic notion should be applied with caution and can be easily overprinted by eustatic changes and local tectonics, it holds approximately true for the Labrador Sea. Figure 19a shows the Late Mesozoic and Cenozoic depositional history, (well locations are in Figure 10, Chapter II.4). The marine magnetic anomalies 32, 28, 24 and 13 are as recognized over the ocean floor. The main transgressions of the Labrador Shelf coincide with the stratigraphic interval of anomalies 32-13 (i.e. the period of actual seafloor spreading). Figures 19a and b show that following generally insignificant, slow sedimentation (0-3 $cm/10^3$ years) in late Cretaceous time, rapid subsidence and rapid sedimentation (often exceeding 10 $cm/10^3$ years) was initiated at the close of Cretaceous time, shortly after active seafloor spreading started in the southern Labrador Sea. Eocene sedimentation rates did not keep up with the subsidence, which allowed the ocean margin area to deepen. On the northern Labrador

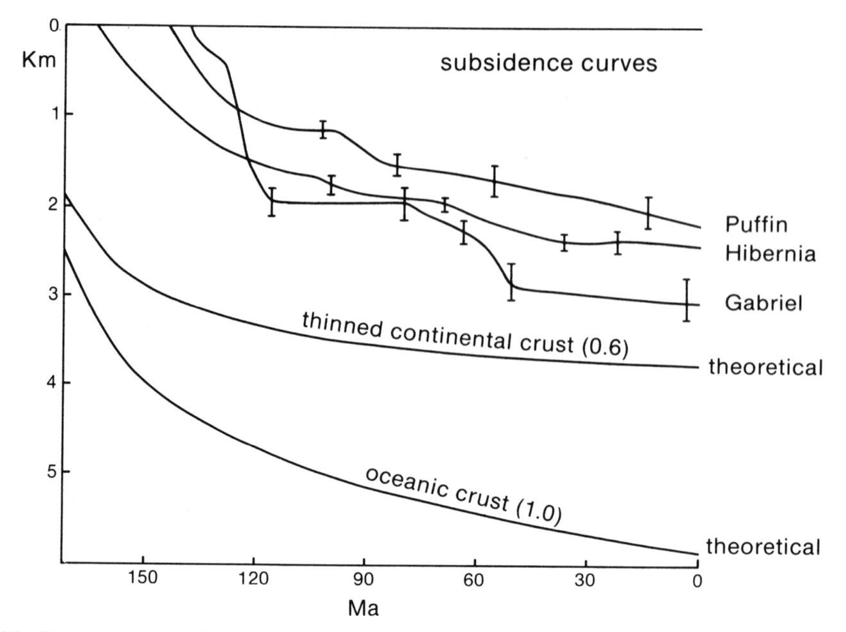

Figure 17 Tectonic subsidence component of the burial history in the three wells of Figure 16. For comparison are shown subsidences curves of oceanic crust and of thinned continental crust with 60% dike injection, using data in Royden et al. (1980), and Hardenbol et al. (1981).

Shelf, an abrupt decline in sedimentation coincides in time with termination of opening of the Labrador Sea as if subsidence is of the (exponential) crustal cooling type (see Keen, 1979 and Figure 19 in Chapter III.1). In several wells the Oligocene section is either very thin, or absent and I postulate that the major mid-Tertiary eustatic sea level lowering event (Vail et al., 1977) may have accentuated the often abrupt shallowing and/or non-deposition, or possibly even erosional trend in the region.

Subsequent young tectonic movements, probably in combination with the sediment loading, may have caused the renewed late Cenozoic high sedimentation rate in the northern wells, a trend that from seismostratigraphic studies (Grant, 1980) appears to be enhanced northward along the Labrador Shelf.

CONCLUSIONS

The interrelation of whole rock, high-temperature radiometric dates and magneto-biochronology provide the detailed timescale for the last 208 Ma, Jurassic-Cenozoic. Stratigraphic resolution, but not necessarily accuracy, of the magneto-biochronologic scale exceeds the

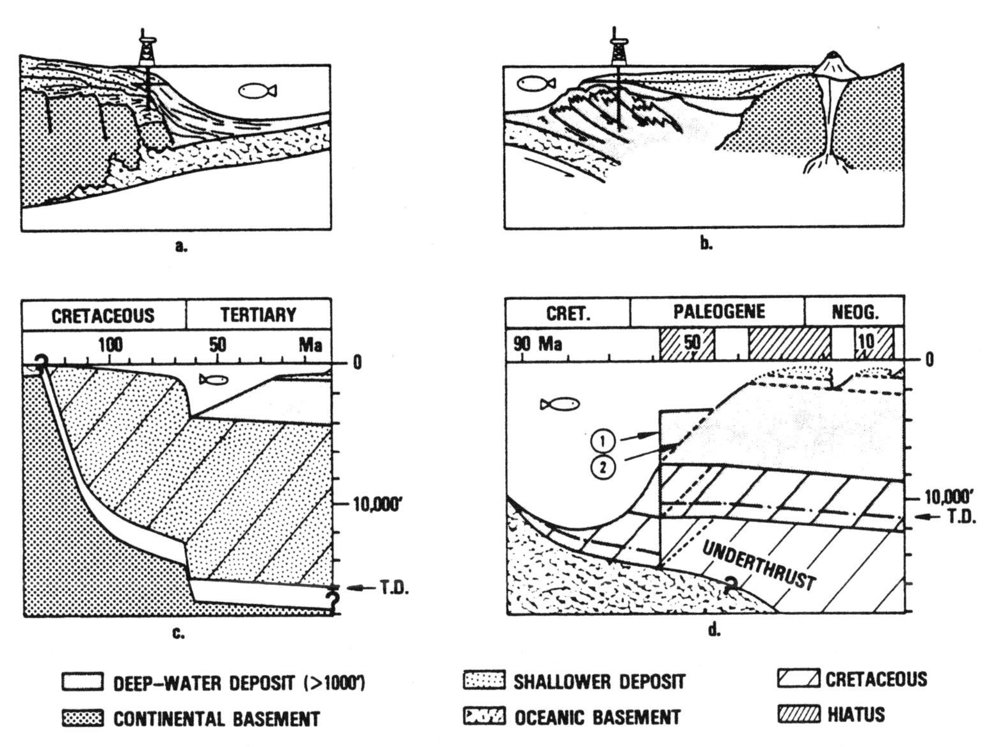

Figure 18 Tectonostratigraphic models: (a) for passive continental margin; (b) for active continental margin. Schematic burial history (after Van Hinte, 1978) for typical well sections: (c) on passive continental margin (since well at a did not reach basement, vertical and horizontal scales are interrupted to show its supposed subsidence path); (d) on active continental margin illustrates two of three alternative interpretations for uplift and shoaling path during first major hiatus: (1) "instantaneous" uplift; (2) gradual uplift. Stippled bathymetry lines at right suggest one possibility of what may have happened during times not represented because of later hiatuses.

resolution using stratigraphically meaningfull radiometric dates. Until recently, it was difficult to relate this relative magnetic and biochronologic framework to a continuous linear scale, because many of the relative datums cannot be closely tied to a radiometrically dated level. Constancy of seafloor spreading interpolation assists in making the Cenozoic (magnetic anomalies 34-3A) and late Jurassic - Early Cretaceous (magnetic anomalies M_{25}-M_0) timescales. The time-successive magnetic polarity changes are considered to be virtually linear segments on an - age versus distance calibration plot of the marine magnetic anomalies on the seafloor. Knowing the age of a few key anomalies allows interpolation of the age in Ma of the other ones. The new magnetochronology is

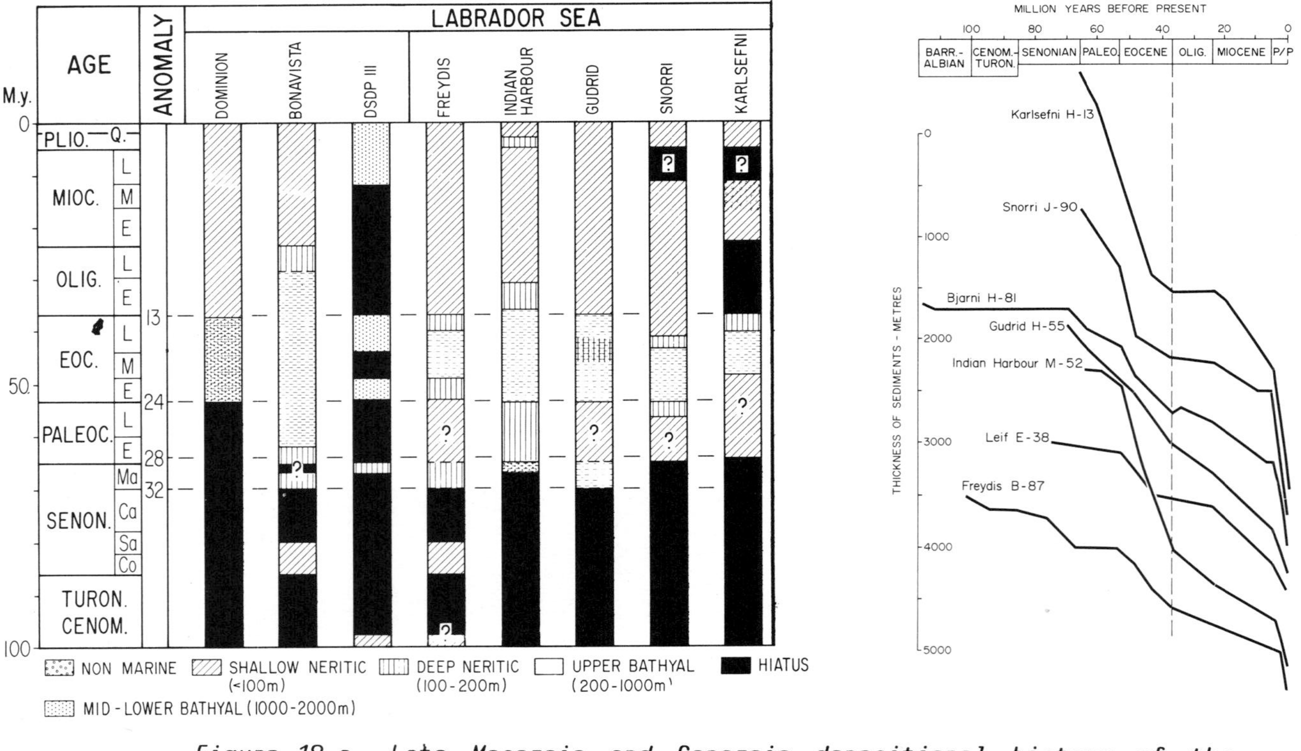

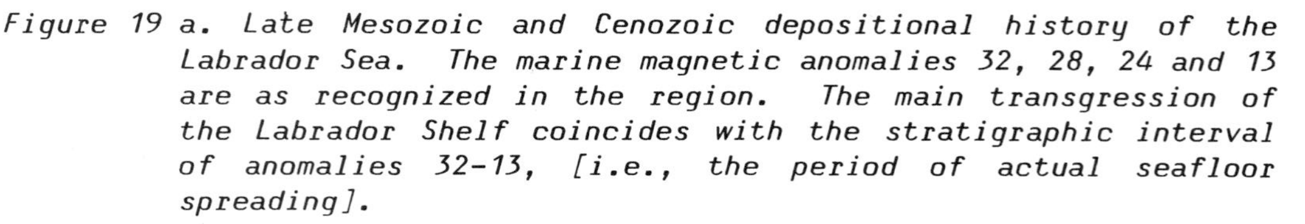

Figure 19 a. *Late Mesozoic and Cenozoic depositional history of the Labrador Sea. The marine magnetic anomalies 32, 28, 24 and 13 are as recognized in the region. The main transgression of the Labrador Shelf coincides with the stratigraphic interval of anomalies 32-13, [i.e., the period of actual seafloor spreading].*

b. *First-order approximation of relative movement since mid-Cretaceous time on the Labrador Shelf. The main phase of subsidence occurs in Maastrichtian - Eocene time and coincides with the Labrador Sea opening.*

used to scale the observed bio-magneto-chronostratigraphy in linear time. Equal duration of the 50 standard ammonite zones in the Hettangian-Oxfordian (Jurassic) presents another vernier to reconstruct geochronology. The new Mesozoic-Cenozoic scale is used as a standard for the Decade of North American Geology (DNAG).

Maximum likelihood interpolation of (stratigraphic) clusters of radiometric dates and spline fits to model the verniers show that geochronological uncertainty in the Jurassic-Cretaceous time scale varies from 2-6 my. No such uncertainty estimate is available for the Cenozoic, but comparison of dates using both the seafloor spreading vernier and the (in)direct ties between magneto-biochronologic datums and radiometric dates shows accuracies between 0.3 and 1.5 my. The larger uncertainty values pertain to the older stage boundaries.

The use of a few key index fossil ages in the RASC scaled optimum sequence makes it possible to construct a local (basin) timescale, of use for regional isochron correlations and burial history.

As a result of (a) better understanding of the fundamental processes that affect vertical motion, (b) drilling of many deep ocean and continental wells and, (c) improved reliability of biostratigraphy, paleobathymetry and timescales, the backtrack curves of oceanic sites and the burial curves for continental (margin) sites are now easily accessible and powerful analytical tools. Corrections for decompaction and eustatic sealevel changes improve the use of age-depth curves in well sites. Backstripping of successively older stratigraphic units helps to calculate and graphically display tectonic - or driving force subsidence.

Advantages to paleontologists and geologists of age-depth curves are:

(1) The construction itself forces the biostratigrapher to be aware of the geological implications of his analytical conclusions.

(2) The diagrams form convenient linear time-depth frames to plot other parameters, such as heat-flow gradients, porosity changes and mineralizations.

(3) The diagrams are excellent first-order models of basin evolution and tectonic style.

(4) Extrapolation of the age-depth curves fills gaps in cases of fragmentary stratigraphic information.

Error analysis on paleobathymetry and timescales enhances the reliability of age-depth trends.

REFERENCES

Agterberg, F.P., in press. Quality of time scales - a statistical appraisal. In "Computers and Geology" 5, D.F. Merriam, ed., Pergamon Press, Oxford.

Armstrong, R.L., 1978. Pre-Cenozoic Phanerozoic Time Scale - computer file of critical dates and consequences of new and in-progress decay - constant revisions. In The Geologic-Time Scale, eds. G.V. Cohee, M.F. Glaessner, and H.D. Hedberg, Studies in Geol. 6, Am. Assoc. Petrol. Geol., 73-93.

Bandy, O.L., 1953. Ecology and paleoecology of some California Foraminifera. Pt. II. Foraminiferal evidence of subsidence rates in the Ventura Basin. J. Paleont. 27, 2, 200-203.

Berger, W.H. and von Rad, U., 1972. Cretaceous and Cenozoic sediments from the Atlantic Ocean. In, Hayes, D.E. et al., Init. Rept. Deep Sea Drilling Project (U.S. Govern. Printing Office, Washington, D.C.), 14, 787-984.

Berggren, W.A. and Aubert, J., 1975. Paleocene benthonic foraminiferal biostratigraphy, paleobiogeography and paleoecology of Atlantic-Tethyan regions: Midway-type faunas. Palaeogeogr. Palaeoclimat., Palaeoecol. 18, 73-192.

Berggren, W.A., Kent, D.V., Flynn, J.J. and Van Couvering, J.A., in press. Cenozoic Geochronology. Bull. Geol. Soc. Am.

Blanchet, R. and Montadert, L. eds., 1981. Geology of Continental Margins. Coll. C3 26th Intern. Geol. Congress, Paris 1980, Oceanol. Acta 4, 294 pp.

Boltovskoy, E. and Wright, R., 1976. Recent Foraminifera. W. Junk, The Hague, 515 pp.

Brunet, M.F. and LePichon, X., 1980. Effet des variations eustatiques sur la subsidence dans le Bassin de Paris. Bull. Soc. geol. France 7, XXII, 4, 631-637.

Butt, A., 1982. Micropaleontological bathymetry of the Cretaceous of western Morocco. Palaeogeogr., Palaeoclimat., Palaeoecol., 37, 235-275.

Channell, J.E.T., Ogg, J.G. and Lowrie, W., 1982. Geomagnetic polarity in the early Cretaceous and Jurassic. Phil. Trans. R. Soc. Lond., A306, 137-146.

Clark, L.D., 1964. Stratigraphy and structure of part of the Western Sierra Nevada Metamorphic Belt, California. U.S. Geological Survey Prof. Paper 410, 70 pp.

Clark, L.D., 1976. Stratigraphy of the northern half of the western Sierra Nevada, Metamorphic belt, California. U.S. Geological Survey Prof. Paper 923, 26 pp.

Cox, A., Doell, R.R. and Dalrymple, G.B., 1963. Geomagnetic polarity epochs and Pleistocene geochronometry. Nature, 198, 1049-1051.

Cox, A.V., 1982. Magnetic reversal time-scale. In: A Geologic Time Scale, Harland et al., Cambridge Univ. Press, Cambridge, 128 pp.

Gradstein, F.M. 1983. Paleoecology and stratigraphy of Jurassic abyssal foraminifera in the Blake-Bahama Basin, Deep Sea Drilling Project Site 534. In Sheridan, R.E., F.M. Gradstein et al., Initial Reports of the Deep Sea Drilling Project, 76. (U.S. Government Printing Office), Washington, D.C., 537-559.

Gradstein, F.M. and Srivastava, S.P., 1980. Aspects of Cenozoic stratigraphy and paleoceanography of the Labrador Sea and Baffin Bay. Palaeogeogr. Palaeoclim., Palaeoecol. 30, 261-295.

Gradstein, F.M. and Berggren, W.A., 1981. Flysch-type agglutinated Foraminifera and the Maestrichtian to Paleogene History of the Labrador and North Seas. Mar. Micropal. 6, 211-268.

Grant, A.C., 1980. Problems with plate tectonics: The Labrador Sea. Bull. Can. Petrol. Geol. (2), 252-273.

Habib, D. and Drugg, W.S., 1983. Dinoflagellate age of Middle Jurassic-Early Cretaceous sediments in the Blake Bahama Basin. In Sheridan, R.E., and others, Initial Reports of the Deep Sea Drilling Project, 76, (U.S. Government Printing Office), Washington, D.C., 623-639.

Haig, D.W., 1979. Global distribution patterns for mid-Cretaceous foraminiferids. J. Foram. Res. 9, 1, 29-40.

Hallam, A., 1975. Jurassic Environments. Cambridge University Press, Cambridge, ix + 269 p.

Hardenbol, J., Vail, P.R. and Ferrer, J., 1981. Interpreting paleoenvironments; subsidence history and sea-level changes of passive margins from seismics and biostratigraphy. Oceanologica Acta 1981, no SP, 33-44.

Harland, W.B., Cox, A.V., Llewellyn, P.G., Pickton, C.A.G., Smith, A.G. and Walters, R., 1982. A Geologic Time Scale. Cambridge University Press, Cambridge, 128 pp.

Heirtzler, J.R., Dickson, G.O., Herron, E.M., Pitmman III, W.C. and X. Le Pichon, 1968. Marine magnetic anomalies, geomagnetic field reversals, and motions of the ocean floor and continents. J. Geophys. Res., 73, 2119-2136.

Heller, F., 1977. Palaeomagnetism of Upper Jurassic Limestones from Southern Germany. J. Geophys., 42, 475-488.

Keen, C.E., 1979. Thermal history and subsidence of rifted continental margins - evidence from wells on the Nova Scotian and Labrador Shelves. Can. J. Earth Sciences 16, 505-522.

Kent, D.V. and Gradstein, F.M., in press. A Cretaceous and Jurassic Geochronology. Bull. Geol. Soc. Am.

LaBrecque, J.L., Kent, D.V. and Cande, S.C., 1977. Revised Magnetic Polarity Time Scale for the Late Cretaceous and Cenozoic Time. Geology, 5, 330-335.

Larson, R.L. and Pitmann, W.C. III, 1972. Worldwide correlation of Mesozoic magnetic anomalies, and its implications. Bull. Geol. Soc. America, 83, 3645-3662.

Larson, R.L. and Hilde, T.W.C., 1975. A revised time scale of magnetic anomalies for the Early Cretaceous and Late Jurassic. J. Geophys. Res., 80, 2586-2594.

Larson, R.L. and Chase, C.G., 1972. Late Mesozoic evolution of the western Pacific. Geol. Soc. Amer. Bull., 83, 3627-3644.

Lipps, J.H., Berger, W.H., Buzas, M.A., Douglas, R.G. and Ross, C.A., 1979. Foraminiferal Ecology and Paleoecology. SEPM Short Course No. 6, Houston 1979, 198 pp.

Lowrie, W. and Channell, J.E.T., 1984. Magnetostratigraphy of the Jurassic-Cretaceous boundary in the Maiolica Limestone (Umbria, Italy). Geology, 12, 44-47.

Lowrie, W., Channell, J.E.T., and Alvarez, W., 1980. A review of magnetic stratigraphic investigations in Cretaceous pelagic carbonate rocks. J. Geophys. Res., 89, 3597-3605.

Mankinen, E.A. and Dalrymple, G.B., 1979. Revised geomagnetic polarity time scale for the interval 0-5 m.y. B.P. J. Geophys. Res., 84, 615-626.

Munk, C., 1980. Foraminiferen aus dem unteren Kimmeridge ***(Platynota-*** *Schichten) der Nordlichen und Mittleren Frankenalb-Faunen bestand und Palökologie. Facies 2, 149-218.*

Murray, J.W., 1973. Distribution and Ecology of living Benthic Foraminiferids. Heinemann Books Ltd., pp. 273.

Nyong, E.E. and Olsson, R.K., 1984. A paleoslope model of Campanian to Lower Maestrichtian Foraminifera in the North American Basin and adjacent continental margin. Mar. Micropal. 8, 437-477.

Odin, G.S. and Letolle, R., 1982. The Triassic time scale in 1982. In Numerical Dating in Stratigraphy, G.S. Odin, ed., John Wiley and Sons, Ltd., 523-533.

Odin, G.S., Curry, D., Gale, N.H. and Kennedy, W.J., 1982. The Phanerozoic time scale in 1981. In Numerical Dating and Stratigraphy, G.S. Odin, ed., John Wiley and Sons, Ltd., New York, 957-960.

Ogg, J.G., 1983. Magnetostratigraphy of Upper Jurassic and Lower Cretaceous sediments, DSDP Site 534, Western North Atlantic. In Sheridan, R.E. and others, Initial Reports of the Deep Sea Drilling Project, 76 (U.S. Government Printing Office, Washington D.C.), 685-699.

Ogg, J.G. and Steiner, M.B., 1985. Jurassic magnetic polarity timescale-current status and compilation. In Michelson, O. and Zeiss, A. (eds.). Jurassic Stratigraphy. Spec. Publ. Danish Geol. Survey, p. 1-16.

Ogg, J.C., Steiner, M.B., Oloriz, F. and Tavera, J.M., 1984. Jurassic magnetostratigraphy, 1. Kimmeridgian-Tithonian of Sierra Gorda and Carcabuey, Southern Spain. Earth and Planetary Science Letters 71, 147-162.

Ozima, M., Kaneoka, I. and Yanagisawa, M., 1979. ^{40}Ar - ^{39}Ar geochronological studies of drilled basalts from Legs 51 and 52. In Init. Rept. Deep Sea Drilling Prog. 51, 52, 53, (U.S. Govern. Printing Office, Washington, D.C.), 1127-1128.

Palmer, A.R., 1983. The Decade of North American Geology 1983. Geologic Time Scale, Geology, 11, 503-504.

Parsons, B. and Sclater, J.G., 1977. An analysis of the variation of ocean floor bathymetry and heat flow with age. J. Geophys. Res. 82, 5, 803-827.

Perrier, R. and Quiblier, J., 1974. Thickness changes in sedimentary layers during compaction history; methods for quantitative evaluation. Bull. Am. Ass. Petr. Geol. 58, 507-520.

Poag, C.W., 1981. Ecologic atlas of benthic Foraminifera of the Gulf of Mexico. Stroudsburg et al., 125 pp.

Ramsay, A.T.S., 1977. Sedimentological clues to Palaeo-oceanography. In Oceanic Micropalaeontology, 2, Ramsay, A.T.S. ed., Academic Press, 1371-1453.

Royden, L., Sclater, J.G. and Von Herzen, R.P., 1980. Continental margin subsidence and heat flow: Important parameters in formation of petroleum hydrocarbons. Bull. Am. Assoc. Petr. Geol., 64, 2, 173-187.

Ryan, W.B.F., Cita, M.B., Rawson, M.D., Burckle, L.H. and Saito, T., 1974. A paleomagnetic assignment of Neogene stage boundaries and the development of isochronous datum planes between the Mediterranean, the Pacific and Indian Oceans in order to investigate the response of the world ocean to the Mediterranean "Salinity Crisis". Riv. Ital. Paleont., 80, 631-688.

Schweickert, R.A., Bogen, N.L., Girty, G.H., Hanson, R.E. and Merguerian, C., 1984. Timing and structural expression of the Nevadan orogeny, Sierra Nevada, California. Geol. Soc. Am. Bull., 95, 967-979.

Sclater, J.G., Anderson, R.N. and Bell, M.L., 1971. The elevation of ridges and evolution of the central eastern Pacific. J. Geophys Res. 76, 7888-7915.

Sclater, J.G., Hellinger, S. and Tapscott, C., 1977. The paleobathymetry of the Atlantic Ocean from the Jurassic to the present. J. Geol. 85, 5, 509-552.

Sclater, J.G. and Cristie, P.A.F., 1980. Continental stretching: an explanation of the post mid-Cretaceous subsidence of the central North Sea basin. J. Geophys. Res. 85, 371-379.

Sheridan, R.E., Gradstein, F.M. et al., 1983. Initial Reports of the Deep Sea Drilling Project, 76 (U.S. Government Printing Office, Washington, D.C.), 947 pp.

Sliter, W.V., 1972. Cretaceous Foraminifers, depth habitats and their origin. Nature 239, 514-515.

Steckler, M.S. and Watts, A.B., 1978. Subsidence of the Atlantic-type continental margin off New York. Earth Planet. Sci. Lett. 41, 1-13.

Steiger, R.H. and Jager, E., 1978. Submission on Geochronology: Convention on the use of decay constants in geochronology and cosmochronology. In The Geologic Time Scale, eds. G. Cohee, M. Glaessner and H. Hedberg, Studies in Geol. 6, Am. Assoc. Petrol. Geol., 67-73.

Steiner, M.B. and Ogg, J.G., 1983. Jurassic Magnetic Polarity Time Scale. EOS 45, 689.

Tjalsma, R.C. and Lohmann, G.P., 1983. Paleocene-Eocene bathyal and abyssal benthic Foraminifera from the Atlantic Ocean. Micropal. Spec. Publ. 4, 90 pp.

Umbgrove, J.H.F., 1942. The Pulse of the Earth. Nijhoff Publ. Co., The Hague, 179 pp.

Vail, P.R., Mitchum, R.M., and Thompson, S., 1977. Seismic stratigraphy and global changes of sea level, Part 4: Global cycler of relative change of sea level. In Seismic stratigraphy: Application to Hydrocarbon exploration, Ed. C.E. Payton, American Association of Petroleum Geologists, Memoir 26, 83-97.

Van Andel, T.H., Thiede, J., Sclater, J.G., and Hay, W.W., 1977. Depositional history of the South Atlantic Ocean during the last 125 million years. J. Geol. 85, 651-698.

Van Hinte, J.E., 1976a. A Cretaceous Time scale, Am. Assoc. Petrol. Geol. Bull. 60, 498-516.

Van Hinte, J.E., 1976b. A Jurassic Time Scale, Am. Assoc. Petrol. Geol. Bull. 60, 489-497.

Van Hinte, J.F., 1978. Geohistory analysis, application of micropaleontology in exploration geology. Am. Assoc. Petrol. Geol. Bull. 62, 201-227.

Wall, J.H., 1960. Jurassic microfaunas from Saskatchewan. Dept. of Mineral. Res. Saskatchewan, Rept. 53, 1-229.

Watts, A.B. and Steckler, M.S., 1981. Subsidence and tectonics of Atlantic-type continental margins. In Geology of Continental Margins, eds. R. Blanchet and L. Montadert. Oceanol. Acta 1981, 143-155.

Westermann, G.E.G., in press. Duration of Jurassic Ammonite zones and stages based on scaled-equal subzone method. Proc. Int. Symp. Quant. Strat. Kharagpur, India 1983; Assoc. Expl. Geoph., Hyderabad.

Wood, R.I., 1981. The subsidence history of the Conoco well 15/30-1, central North Sea. Earth and Planet. Sci. Lett. 54, 306-312.

APPENDIX

Contents

NINE DATA BASES WITH APPLICATIONS OF RANKING AND SCALING OF STRATIGRAPHIC EVENTS

F.P. AGTERBERG, F.M. GRADSTEIN,
S.N. LEW AND F.C. THOMAS

INTRODUCTION

In the course of IGCP Project 148, a number of statistical techniques for stratigraphic correlation were developed and applied to different data bases. (A data base is a collection of data on a computer, used by several people, Rumble, 1984, p. 4.) It would not be possible to gather in a systematic way all these data bases and the codes of the computer programs applied to them. The nine data bases listed in this Appendix were originally developed by participants in IGCP Project 148 and treated by them using different techniques, most of which are described in the March 1984 Special Issue of Computers and Geosciences, vol. 10, no. 1, entitled "Theory, Application and Comparison of Stratigraphic Correlation Techniques" (Proceedings, IGPP-148 Symposium held in Geneva, Switzerland, November 1982; F.P. Agterberg, Editor). On the following pages, each data base was ranked and scaled using the RASC computer program for ranking and scaling of stratigraphic events. In all runs, the presorting technique was used for ranking, in most instances followed by the modified Hay method. RASC output was redrawn automatically on the CALCOMP plotter using DISSPLA (Jackson et al., 1984).

The formats of the nine data bases are basically of two types:

1. RASC input format;

2. Unitary Association (UA) input format.

Each listing of sequence data for stratigraphic events in wells or outcrop sections is preceded by a dictionary. In the RASC input format (also see discussion of computer programs in Appendix II), events are listed in order of occurrence in the stratigraphically downward direction (=direction of drilling in wells). Events preceded by hyphens were observed to be coeval with events preceding them in the sequence.

In the UA input format the code number of each fossil is followed by two codes: respectively for its lowest and highest occurrences in a section. Computer programs for converting UA input into RASC input and vice versa have been developed (see Appendix II). In two situations, (Gradstein-Thomas data base, and Rubel's Silurian Baltic data bases) the depths of the samples (in metres or feet) are also listed. For RASC/CASC input (Gradstein-Thomas data base) this is performed by addition of a separate depth file with values that correspond to the codes of events in the listing of the sequence data. For Rubel's data bases the UA format was modified by providing distances in metres for the lowest and highest occurrences of the fossils in each section.

The nine data bases are identified as follows:

(1) Gradstein-Thomas data base - Labrador Shelf and Grand Banks Cenozoic Foraminifers;

(2) Gradstein data base - Grand Banks Late Cretaceous Foraminifers;

(3) Doeven data base - Scotian Shelf Cretaceous Nannofossils;

(4) Baumgartner data base - Tethyan Jurassic Radiolarians;

(5) Blank data base - Mesozoic/Cenozoic Nannofossils in Atlantic Deep Sea Drilling Project (DSDP) sites;

(6) Rubel's data bases - Silurian Fossils of the Baltic;

(7) Sullivan-Bramlette data base - Californian Paleogene Nannofossils;

(8) Corliss data base - Paleogene DSDP Benthic Foraminifers;

(9) Agterberg-Lew data base - Computer Simulation Experiments.

On the next pages, a brief description of each data base is followed by the optimum sequence (ranking output) and dendrogram (scaling output) obtained for it using the RASC program (Figures 1-27).

These descriptions are followed by computer print-outs of the original information in the nine data bases which was also used as input for the optimum sequences and dendrograms shown in Figures 1-27.

REFERENCES

Agterberg, F.P., and Rubel, M., in press, Quantitative stratigraphic correlation in the Baltic Silurian-comparison of results obtained by probabilistic and deterministic methods. In Oleynikov, N.A., and Rubel, M., eds., Quantitative Stratigraphy - Retrospective Evaluation and Future Development; Institute of Geology, Acad. Sc. Estonian SSR, Tallinn, U.S.S.R.

Baumgartner, P.D., 1984, Comparison of Unitary Associations and probabilistic ranking and scaling as applied to Mesozoic Radiolaria. Computers and Geosciences 10, 167-183.

Blank, R.G., 1984, Comparison of two binomial models in probabilistic biostratigraphy. Computers and Geosciences 10, 59-67.

Blank, R.G., and Ellis, C.H., 1982, The probable range concept applied to the biostratigraphy of marine microfossils. J. Geol. 90, (4), 415-433.

Bramlette, M.N., and Sullivan, F.R., 1961, Cocolithophorids and related nannoplankton of the Early Tertiary in California. Micropal. 7, 129-188.

Davaud, E., and Guex, J., 1978, Traitement analytique "manuel" et algorithmique de problèmes complexes de correlations biochronologiques. Eclogae Geol. Helv. 71, (3), 581-610.

Doeven, P.H., 1983, Cretaceous nannofossil stratigraphy and paleoecology of the Canadian Atlantic Margin. Geol. Survey Canada Bull. 356, 70 pp.

Doeven, P.H., Gradstein, F.M., Jackson, A., Agterberg, F.P., and Nel, D., 1982, A quantitative nannofossil range chart. Micropal. 28, (1), 85-92.

Gradstein, F.M., 1984, On stratigraphic normality. Computers and Geosciences 10, (1), 43-57.

Gradstein, F.M., and Agterberg, F.P., 1982, Models of Cenozoic foraminiferal stratigraphy - Northwestern Atlantic Margin. In Cubitt, J.M., and Reyment, R.A., eds., Quantitative Stratigraphic Correlation, John Wiley, Chichester, 119-173.

Heller, M., Gradstein, W.S., Gradstein, F.M., and Agterberg, F.P., 1983, RASC-Fortran IV Computer Program for ranking and scaling of biostratigraphic events. Geol. Survey of Canada, Rept. 922, 54 pp.

Jackson, A., Lew, S.N., and Agterberg, F.P., 1984, DISSPLA program for display of dendrograms from RASC output. Computers and Geosciences 10, (1), 59-165.

Rubel, M., 1982, The succession of brachiopods as a tool for correlation in the Silurian of the East Baltic. In Ecostratigraphy of the East Baltic Silurian, Vagus, Tallinn, 51-61.

Rubel, M., and Pak, D.N., 1984, Theory of stratigraphic correlation by means of ordinal scales. Computers and Geosciences, 10, 97-105.

Rumble, J.R., Jr., 1984, Data, computers and data management systems, I. In J.R. Rumble, Jr. and V.E. Hample (eds.), 'Database Management in Science and Technology', Elsevier (North Holland), 1-13.

Sullivan, F.R., 1965, Lower Tertiary nannoplankton from the California Coast Ranges, II, Eocene. Univ. Calif. Publ. Geol. Sc., 53, 1-52.

(1) Gradstein-Thomas data base

Labrador Shelf Cenozoic Foraminifers

RASC input format for sequence data (with depths).

Total number of exploration wells = 24.

Total number of events in dictionary = 275.

Type of event = last occurrences.

Largest number of events in single section = 77.

Only events that occur in k_c = 7 or more sections were used in the RASC run. For ranking (presorting followed by modified Hay method), only pairs of events that occur in m_{c1} = 2 or more sections were used. For scaling (dendrogram after final reordering), only pairs of events that occur in m_{c2} = 4 or more sections were used. Optimum sequence (ranking output) and dendrogram (scaling output) are shown in Figures 1 and 2.

In this data base, as well as in others, the numbering of the events in the dictionary is not consecutive. Dummy lines for missing numbers should be inserted when the RASC program is used for processing. There should be a one-to-one correspondence between the sequence data and the dictionary.

This is a revised and enlarged version of previous Labrador Shelf and Grand Banks data bases. Two of these earlier versions were previously published in Gradstein and Agterberg (1982) and Gradstein (1984), respectively. A detailed discussion of zonations using this data file is in Chapter II.4.

OPTIMUM FOSSIL SEQUENCE

SEQUENCE POSITION	FOSSIL NUMBER	RANGE	FOSSIL NAME
1	10	0 — 2	UVIGERINA CANARIENSIS
2	17	1 — 3	ASTERIGERINA GURICHI
3	16	2 — 4	CERATOBULIMINA CONTRARIA
4	67	3 — 5	SCAPHOPOD SP1
5	18	4 — 6	SPIROPLECTAMMINA CARINATA
6	71	5 — 7	EPISTOMINA ELEGANS
7	26	6 — 8	UVIGERINA DUMBLEI
8	21	7 — 9	GUTTULINA PROBLEMA
9	20	8 — 10	GYROIDINA GIRARDANA
10	15	9 — 11	GLOBIGERINA PRAEBULLOIDES
11	70	10 — 13	ALABAMINA WOLTERSTORFFI
12	27	10 — 13	EPONIDES UMBONATUS
13	69	12 — 14	NODOSARIA SP8
14	24	13 — 15	TURRILINA ALSATICA
15	81	14 — 16	GLOBIGERINA VENEZUELANA
16	25	15 — 17	COARSE ARENACEOUS SPP.
17	259	16 — 18	AMMODISCUS LATUS
18	147	17 — 19	CATAPSYDRAX AFF. DISSIMILIS
19	33	18 — 20	TURBOROTALIA POMEROLI
20	34	19 — 21	MARGINULINA DECORATA
21	260	20 — 23	HAPLOPHRAGMOIDES KIRKI
22	261	20 — 23	HAPLOPHRAGMOIDES WALTERI
23	263	22 — 24	AMMOBACULITES AFF POLYTHALAMUS LOEBLICH
24	29	23 — 25	CYCLAMMINA AMPLECTENS
25	32	24 — 26	AMMOSPHAEROIDINA SP1
26	40	25 — 27	BULIMINA ALAZANENSIS
27	41	26 — 28	PLECTOFRONDICULARIA SP1
28	264	27 — 29	KARRERIELLA CONVERSA
29	42	28 — 30	CIBICIDOIDES ALLENI
30	30	29 — 31	CIBICIDOIDES BLANPIEDI
31	86	30 — 32	TURRILINA BREVISPIRA
32	57	31 — 33	SPIROPLECTAMMINA SPECTABILIS LCO
33	90	32 — 35	ACARININA DENSA
34	36	32 — 35	PSEUDOHASTIGERINA WILCOXENSIS
35	50	34 — 36	SUBBOTINA PATAGONICA
36	46	35 — 37	MEGASPORE SP1
37	230	36 — 38	BULIMINA OVATA
38	45	37 — 40	BULIMINA TRIGONALIS
39	52	37 — 41	ACARININA SOLDADOENSIS
40	54	38 — 41	SPIROPLECTAMMINA NAVARROANA
41	56	40 — 42	GLOMOSPIRA CORONA
42	55	41 — 43	GAVELINELLA BECCARIIFORMIS
43	59	42 — 44	RZEHAKINA EPIGONA

Figure 1 Optimum sequence (ranking output) for Gradstein-Thomas data base.

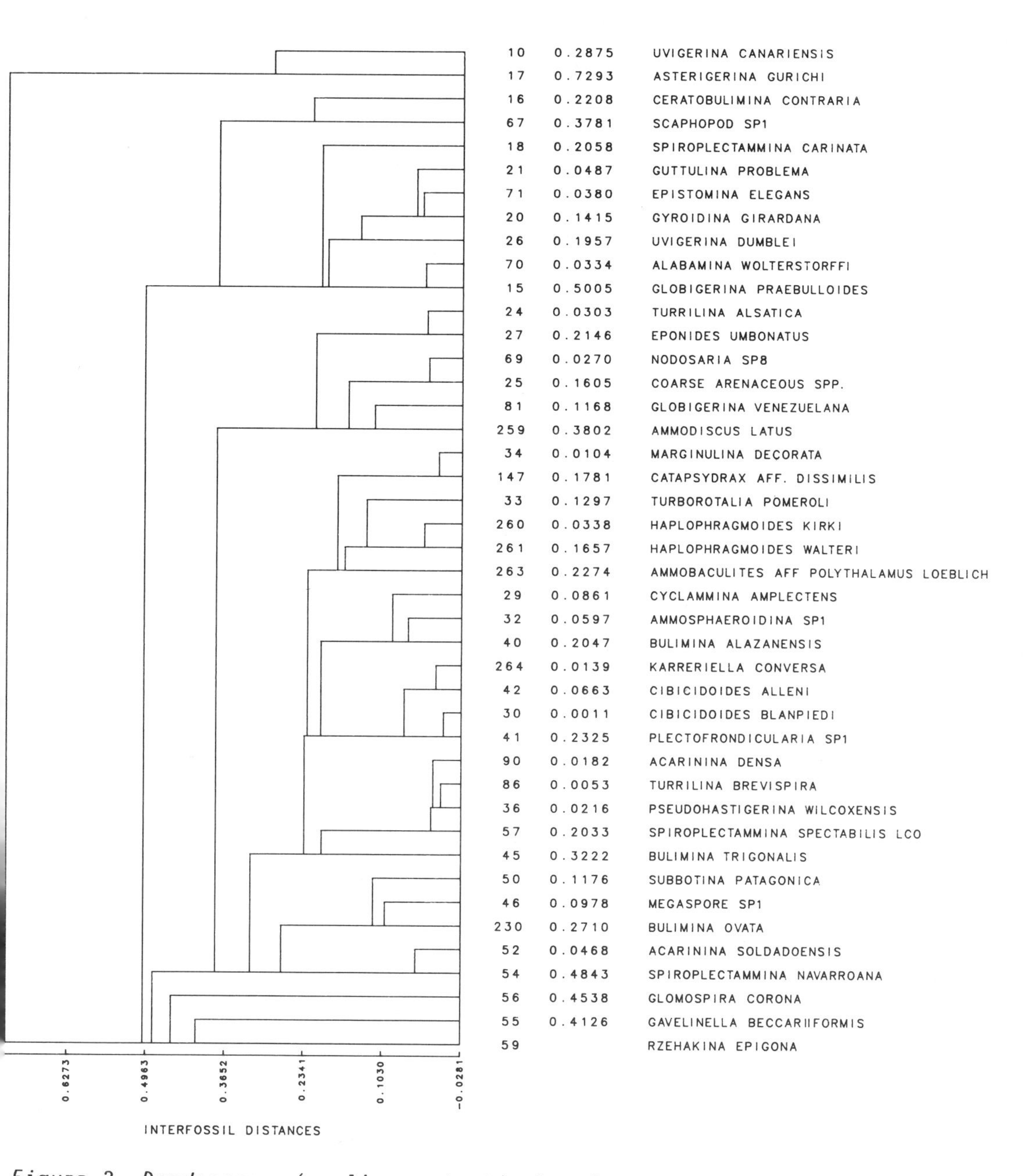

Figure 2 Dendrogram (scaling output) for Gradstein-Thomas data base. This output can be compared with scaled optimum sequences for 16 wells and 24 wells (see Figure 17 in Chapter II.4). Note that seismic event no. 2 occurs near the top of the Oligocene.

(2) Gradstein data base

Late Cretaceous Grand Banks Foraminifers

RASC input format for sequence data.

Total number of sections = 20

Total number of events in dictionary = 204

Type of event = last occurrences

Largest number of events in single section = 34

Only events that occur in k_c = 5 or more sections were used in RASC run. For ranking (presorting followed by modified Hay method) only pairs of events that occur in m_{c1} = 2 or more sections were used. For scaling (dendrogram after final reordering), only pairs of events that occur in m_{c2} = 3 or more sections were used. Optimum sequence and dendrogram are shown in Figures 3 and 4. An earlier version of this data base was published in Heller et al. (1983).

OPTIMUM FOSSIL SEQUENCE

SEQUENCE POSITION	FOSSIL NUMBER	RANGE	FOSSIL NAME
1	5	0 – 5	STENSIOINA POMMERANA
2	4	0 – 6	GLOBOTRUNCANA ARCA
3	3	0 – 6	GLOBOTRUNCANA STUARTI
4	105	0 – 5	LOXOSTOMA GEMMUM
5	6	4 – 6	GLOBOTRUNCANELLA HAVANENSIS
6	8	5 – 8	GLOBIGERINELLOIDES MESSINAE
7	16	5 – 8	RUGOGLOBIGERINA RUGOSA
8	13	7 – 9	GLOBOTRUNCANA LINNEIANA
9	12	8 – 10	GLOBOTRUNCANA STUARTIFORMIS
10	11	9 – 11	GLOBOTRUNCANA FORNICATA
11	14	10 – 12	GLOBOTRUNCANA CRETACEA
12	15	11 – 13	GLOBOTRUNCANA MARGINATA
13	26	12 – 16	GLOBOTRUNCANA ANGUSTICARINATA
14	108	12 – 19	GAVELINELLA MINIMA
15	56	12 – 16	GAUDRYINA AUSTINANA
16	20	15 – 17	GLOBOTRUNCANA CORONATA
17	23	16 – 18	STENSIOINA EXCULPTA
18	55	17 – 19	GLOBOTRUNCANA CARINATA
19	109	18 – 20	COARSE AGGLUTINATED SPP
20	24	19 – 21	HEDBERGELLA AMABILIS
21	22	20 – 22	SIGALIA DEFLAENSIS
22	54	21 – 23	GLOBOTRUNCANA CONCAVATA
23	72	22 – 24	GLOBOTRUNCANA RENZI
24	110	23 – 25	GLOBOTRUNCANA IMBRICATA
25	70	24 – 26	HEDBERGELLA BOSQUENSIS
26	112	25 – 28	GLOBOTRUNCANA HELVETICA
27	30	25 – 29	PRAEGLOBOTRUNCANA STEPHANI
28	71	26 – 29	GLOBOTRUNCANA SCHNEEGANSI
29	31	28 – 30	PRAEGLOBOTRUNCANA TURBINATA
30	27	29 – 32	ROTALIPORA CUSHMANI
31	131	29 – 32	GAVELINOPSIS CENOMANICA

Figure 3 Optimum sequence for Gradstein data base.

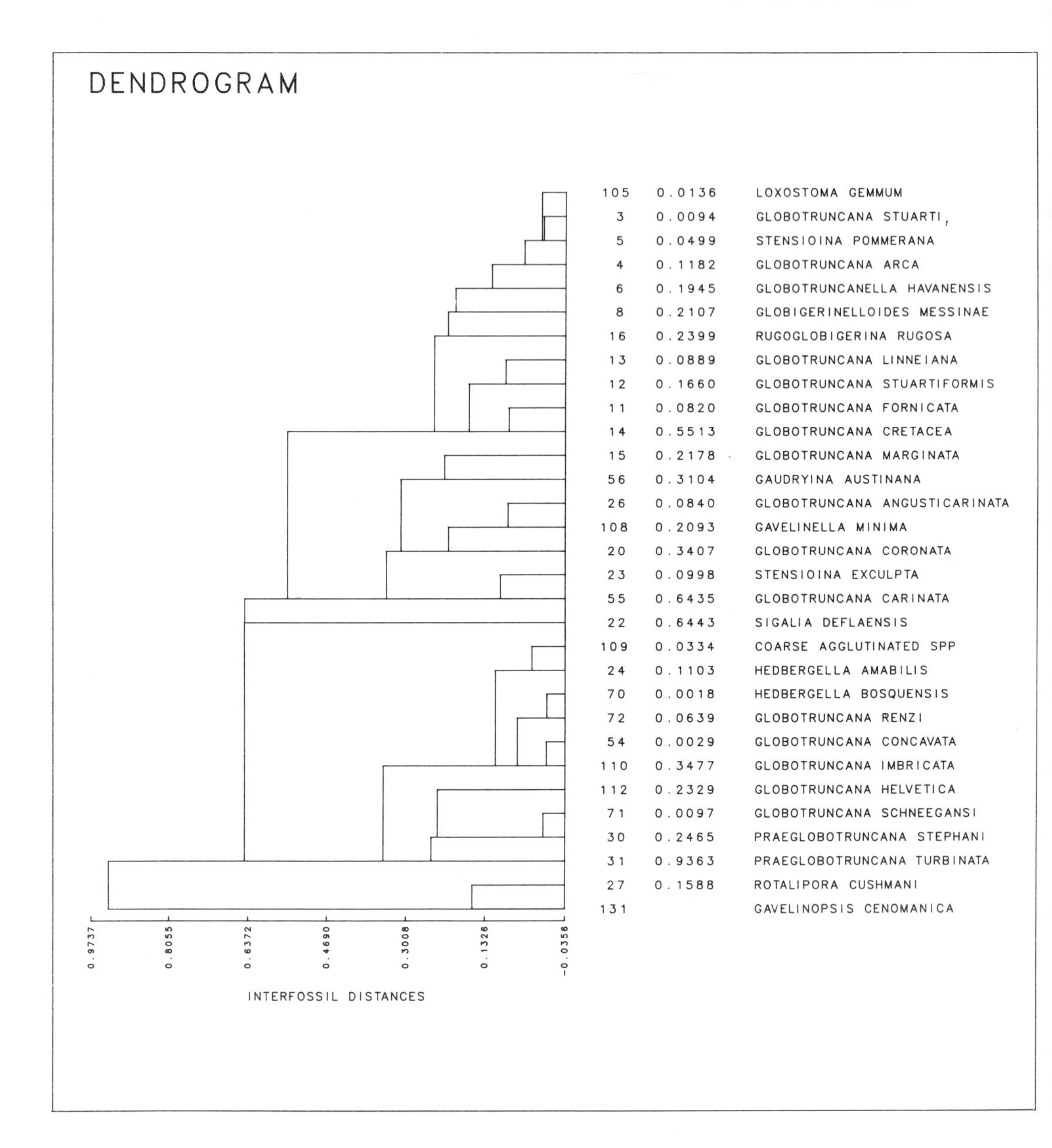

Figure 4 Scaled optimum sequence for Gradstein data base.

(3) Doeven data base

Scotian Shelf Cretaceous Nannofossils

RASC input format for sequence data.

Total number of sections = 10.

Total number of events in dictionary = 223.

Types of events = Highest occurrence (HI)
Highest consistent occurrence (subtop, ST)
Lowest consistent occurrence (subbottom, SB)
Lowest occurrence (LO), (see Chapter II.4.).

Largest number of events in single section = 109

Only events that occur in k_c = 7 or more sections were used in RASC run. For ranking (presorting followed by modified Hay method), only pairs of events that occur in m_{c1} = 3 or more sections were used. For scaling (dendrogram after final reordering), only pairs of events that occur in m_{c2} = 5 or more sections were used. Optimum sequence and dendrogram are shown in Figures 5 and 6. Background material on this data base is provided in Doeven (1983), Doeven et al. (1982) and in Chapter II.4.

OPTIMUM FOSSIL SEQUENCE

SEQUENCE POSITION	FOSSIL NUMBER	RANGE	FOSSIL NAME
1	3	0 – 2	HI ARKHANGELSKIELLA CYMBIFORMIS
2	43	1 – 3	HI LUCIANORHABDUS CAYEUXI
3	28	2 – 4	HI EIFFELLITHUS EXIMIUS
4	10	3 – 8	HI BROINSONIA PARCA
5	62	3 – 6	HI PHANULITES SPP
6	26	5 – 7	HI CYLINDRALITHUS SP
7	73	6 – 8	HI REINHARDTITES LEVIS
8	9	7 – 9	HI BROINSONIA MATALOSA
9	44	8 – 11	HI LUCIANORHABDUS MALEFORMIS
10	45	8 – 11	HI LUCIANORHABDUS SCOTUS
11	71	10 – 13	HI REINHARDTITES ANTHOPHORUS
12	80	10 – 13	HI TRANOLITHUS PHACELOSUS
13	173	12 – 14	LO REINHARDTITES LEVIS
14	49	13 – 15	HI MICRORHABDULUS BELGICUS
15	203	14 – 16	SB ARKHANGELSKIELLA CYMBIFORMIS
16	111	15 – 17	LO CERATOLITHOIDES ACULEUS
17	205	16 – 18	ST EIFFELLITHUS EXIMIUS
18	90	17 – 19	HI BROINSONIA SP
19	75	18 – 20	HI RUCINOLITHUS HAYI
20	64	19 – 21	HI PODORHABDUS CORONADVENTIS
21	16	20 – 22	HI COROLLITHON SIGNUM
22	202	21 – 23	ST RUCINOLITHUS HAYI
23	48	22 – 25	HI MARTHASTERITES FURCATUS
24	86	22 – 25	HI ZYGODISCUS BIPERFORATUS
25	210	24 – 28	SB BROINSONIA PARCA
26	42	23 – 27	HI LUCIANORHABDUS ARCUATUS
27	37	26 – 28	HI LITHASTRINUS GRILLI
28	36	27 – 29	HI LITHASTRINUS FLORALIS
29	216	28 – 31	SB PHANULITES OBSCURUS
30	201	27 – 31	ST LITHASTRINUS GRILLI
31	145	30 – 32	LO LUCIANORHABDUS SCOTUS
32	103	31 – 33	LO ARKHANGELSKIELLA CYMBIFORMIS
33	110	32 – 35	LO BROINSONIA PARCA
34	209	32 – 35	SB LUCIANORHABDUS CAYEUXI
35	142	34 – 36	LO LUCIANORHABDUS ARCUATUS
36	212	35 – 37	SB MICRORHABDULUS STRADNERI
37	221	36 – 38	SB RUCINOLITHUS HAYI
38	175	37 – 40	LO RUCINOLITHUS HAYI
39	204	37 – 41	SB ARKHANGELSKIELLA SPECILLATA
40	213	38 – 41	SB MICULA CONCAVA
41	160	40 – 42	LO PHANULITES OBSCURUS
42	220	41 – 43	SB REINHARDTITES ANTHOPHORUS
43	182	42 – 45	LO VEKSHINELLA BOCHOTNICAE
44	104	42 – 46	LO ARKHANGELSKIELLA SPECILLATA
45	152	43 – 48	LO MICRORHABDULUS STRADNERI
46	57	44 – 47	HI NANNOCONUS SP
47	153	46 – 48	LO MICULA CONCAVA
48	143	47 – 49	LO LUCIANORHABDUS CAYEUXI
49	222	48 – 51	SB BROINSONIA SPP
50	33	46 – 51	HI GARTNERAGO STRIATUM
51	190	50 – 52	LO BROINSONIA SP
52	171	51 – 53	LO REINHARDTITES ANTHOPHORUS
53	218	52 – 55	SB PHANULITES SPP
54	217	52 – 55	SB PHANULITES OVALIS
55	186	54 – 58	LO ZYGODISCUS BIPERFORATUS
56	211	54 – 57	SB MARTHASTERITES FURCATUS
57	148	56 – 58	LO MARTHASTERITES FURCATUS
58	13	57 – 59	HI COROLLITHION ACHYLOSUM
59	101	58 – 60	LO AHMUELLERELLA OCTORADIATA
60	135	59 – 61	LO KAMPTNERIUS SP
61	74	60 – 62	HI RHAGODISCUS ASPER
62	126	61 – 63	LO CYLINDRALITHUS SP
63	215	62 – 65	SB MICULA DECUSSATA
64	102	62 – 65	LO AHMUELLERELLA REGULARIS
65	154	64 – 66	LO MICULA DECUSSATA
66	206	65 – 67	SB EIFFELLITHUS EXIMIUS
67	161	66 – 70	LO PHANULITES OVALIS
68	162	66 – 70	LO PHANULITES SSP
69	137	66 – 71	LO LITHASTRINUS GRILLI
70	144	68 – 72	LO LUCIANORHABDUS MALEFORMIS
71	219	69 – 72	SB QUADRUM GARTNERI
72	168	71 – 74	LO QUADRUM GARTNERI
73	149	71 – 75	LO MICRORHABDULUS BELGICUS
74	132	72 – 75	LO GARTNERAGO OBLIQUUM
75	150	74 – 76	LO MICRORHABDULUS DECORATUS
76	128	75 – 77	LO EIFFELLITHUS EXIMIUS
77	130	76 – 78	LO EIFFELLITHUS TURRISEIFFELI

Figure 5 Optimum sequence for Doeven data base. Other optimum sequences were previously used to construct a range chart, (Figure 19 in Chapter II.4 and Doeven, 1983).

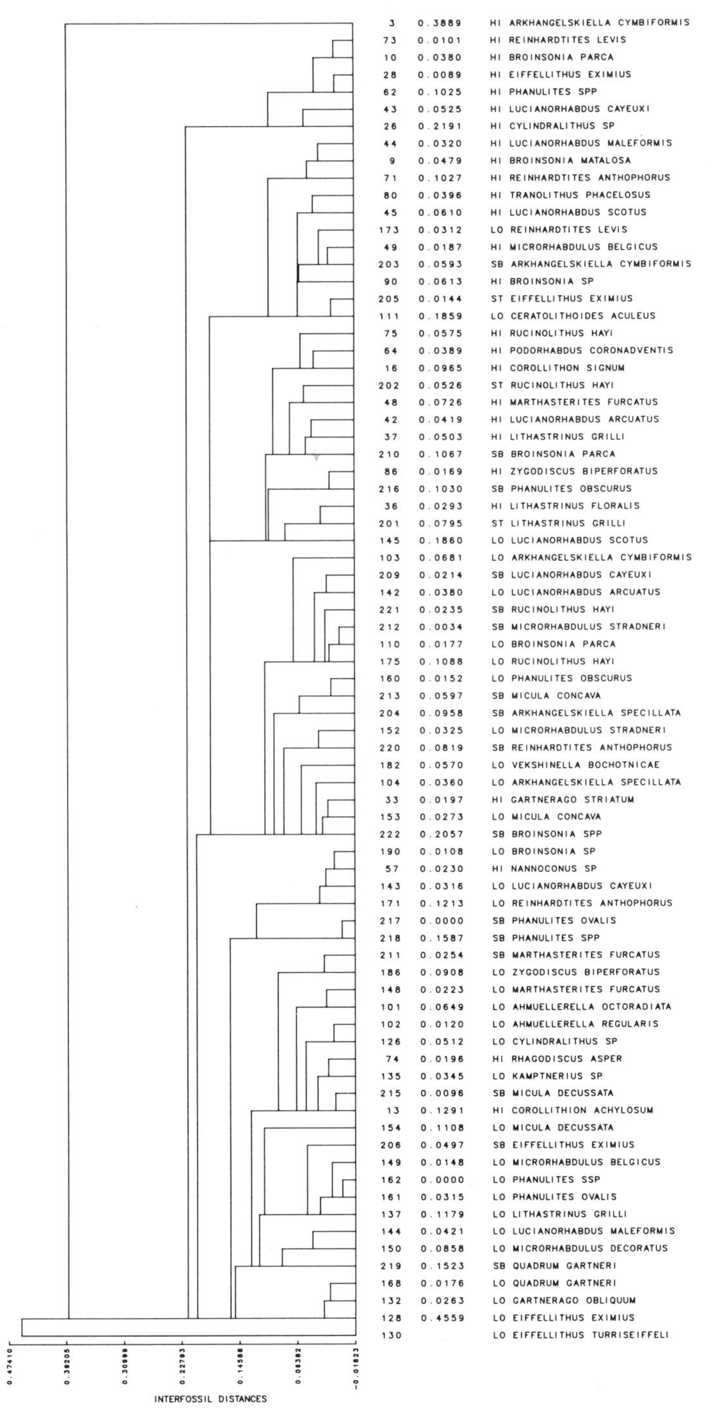

Figure 6 Scaled optimum sequence for Doeven data base.

(4) Baumgartner data base

Tethyan Jurassic Radiolarians

UA input format for sequence data.

Total number of sections = 43.

Total number of events in dictionary = 220.

Type of events = LO (lowest occurrence), HI (highest occurrence).

Largest number of events in single section = 160.

Only events that occur in k_c = 13 or more sections were used in the RASC run. For ranking (presorting only), all pairs of events were used. Scaling (dendrogram after final reordering), is based on pairs of events that occur in m_{c2} = 8 or more sections. Optimum sequence and dendrogram are shown in Figures 7 and 8.

The data base listed in the computer print-out part of this appendix was used as input for the RASC computer program. The largest number of events in a single section should not exceed 160 in the version of RASC used. Two sections in Baumgartner's data base (24 and 30) have over 160 events. In order to meet the input requirement, a number of events occurring in fewer than k_c = 13 sections were omitted, so that the optimum sequence (Figure 7) and dendrogram (Figure 8) were not affected. Only pairs of events were omitted. Using U.A. format these were, for Section 24: 2.2-2, 14.3-7, 34.2-2, 35.9-21, 63.2-5, 66.8-8, 87.3-10, 92.4-4, 93.24-24, 94.24-24, 95.24-24, and 101.24-24. The missing events for Section 30 are: 5.1-2, 9.1-23, 87.1-23, 92.2-12, 93.25-25, 94.25-28, 95.25-25, and 96.25-25.

This data base which was previously published in Baumgartner (1984), is analyzed in some detail in Chapter II.8.

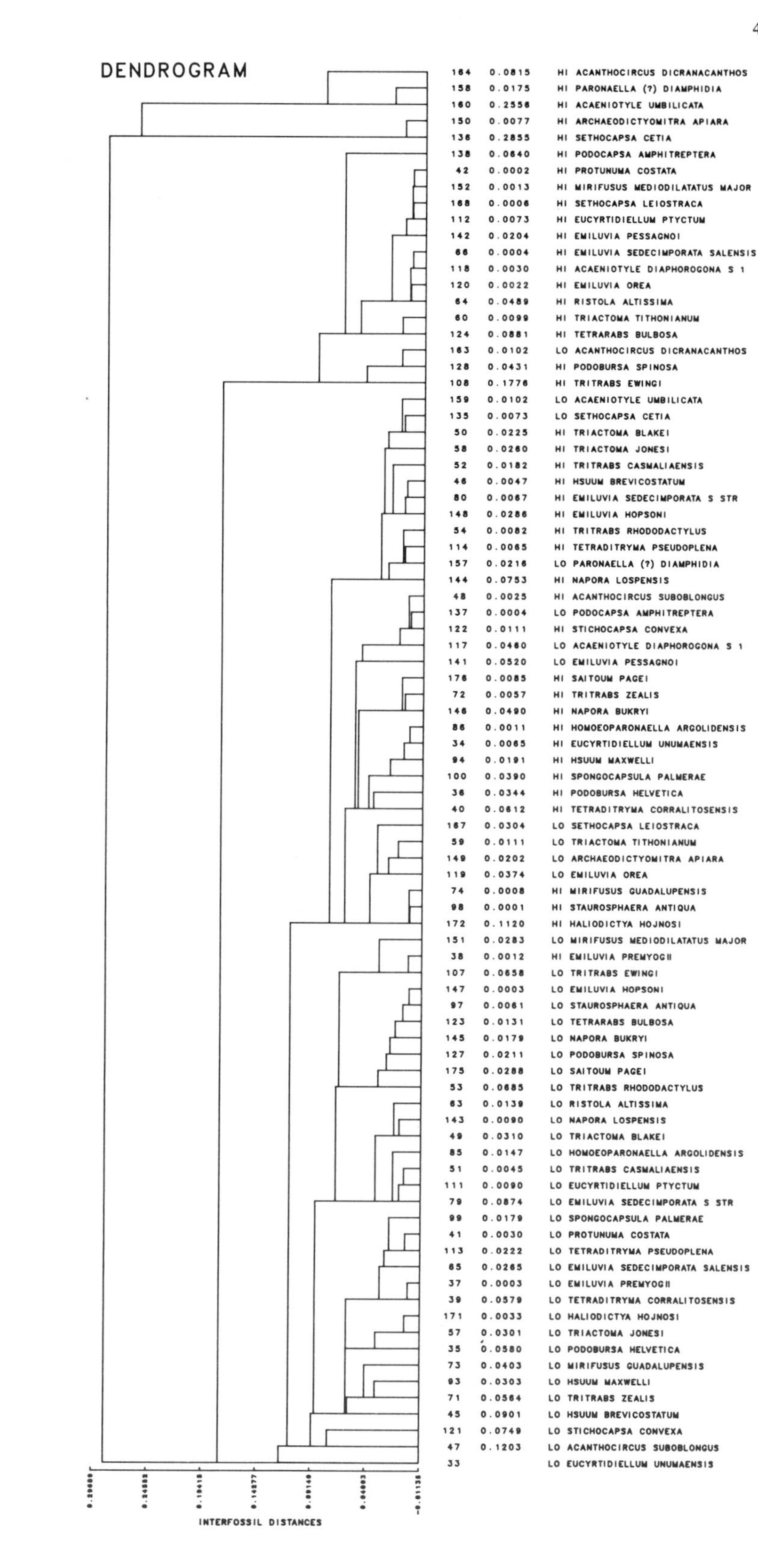

Figure 7 Optimum sequence for Baumgartner data base. For processing results with $k_c = 8$ and comparison to Unitary Association output, see Chapter II.8 and Baumgartner (1984, Figure 4, p. 173).

OPTIMUM FOSSIL SEQUENCE

SEQUENCE POSITION	FOSSIL NUMBER	RANGE	FOSSIL NAME
1	164	0 – 2	HI ACANTHOCIRCUS DICRANACANTHOS
2	160	1 – 3	HI ACAENIOTYLE UMBILICATA
3	158	2 – 4	HI PARONAELLA (?) DIAMPHIDIA
4	150	3 – 5	HI ARCHAEODICTYOMITRA APIARA
5	136	4 – 6	HI SETHOCAPSA CETIA
6	138	5 – 7	HI PODOCAPSA AMPHITREPTERA
7	152	6 – 10	HI MIRIFUSUS MEDIODILATATUS MAJOR
8	159	5 – 9	LO ACAENIOTYLE UMBILICATA
9	135	8 – 15	LO SETHOCAPSA CETIA
10	112	7 – 12	HI EUCYRTIDIELLUM PTYCTUM
11	142	8 – 13	HI EMILUVIA PESSAGNOI
12	168	10 – 15	HI SETHOCAPSA LEIOSTRACA
13	64	11 – 16	HI RISTOLA ALTISSIMA
14	42	7 – 16	HI PROTUNUMA COSTATA
15	163	12 – 17	LO ACANTHOCIRCUS DICRANACANTHOS
16	124	14 – 17	HI TETRARABS BULBOSA
17	120	16 – 18	HI EMILUVIA OREA
18	66	17 – 19	HI EMILUVIA SEDECIMPORATA SALENSIS
19	108	18 – 23	HI TRITRABS EWINGI
20	148	17 – 23	HI EMILUVIA HOPSONI
21	128	18 – 23	HI PODOBURSA SPINOSA
22	118	18 – 23	HI ACAENIOTYLE DIAPHOROGONA S 1
23	60	22 – 26	HI TRIACTOMA TITHONIANUM
24	50	21 – 26	HI TRIACTOMA BLAKEI
25	58	22 – 26	HI TRIACTOMA JONESI
26	54	25 – 31	HI TRITRABS RHODODACTYLUS
27	48	24 – 28	HI ACANTHOCIRCUS SUBOBLONGUS
28	52	27 – 31	HI TRITRABS CASMALIAENSIS
29	80	27 – 33	HI EMILUVIA SEDECIMPORATA S STR
30	157	23 – 32	LO PARONAELLA (?) DIAMPHIDIA
31	46	28 – 35	HI HSUUM BREVICOSTATUM
32	137	30 – 36	LO PODOCAPSA AMPHITREPTERA
33	141	29 – 36	LO EMILUVIA PESSAGNOI
34	122	27 – 35	HI STICHOCAPSA CONVEXA
35	114	34 – 40	HI TETRADITRYMA PSEUDOPLENA
36	72	34 – 37	HI TRITRABS ZEALIS
37	86	36 – 39	HI HOMOEOPARONAELLA ARGOLIDENSIS
38	144	36 – 40	HI NAPORA LOSPENSIS
39	146	37 – 45	HI NAPORA BUKRYI
40	176	38 – 42	HI SAITOUM PAGEI
41	117	35 – 47	LO ACAENIOTYLE DIAPHOROGONA S 1
42	100	40 – 43	HI SPONGOCAPSULA PALMERAE
43	36	42 – 45	HI PODOBURSA HELVETICA
44	94	40 – 45	HI HSUUM MAXWELLI
45	40	44 – 51	HI TETRADITRYMA CORRALITOSENSIS
46	34	40 – 50	HI EUCYRTIDIELLUM UNUMAENSIS
47	119	42 – 50	LO EMILUVIA OREA
48	59	43 – 49	LO TRIACTOMA TITHONIANUM
49	149	48 – 50	LO ARCHAEODICTYOMITRA APIARA
50	74	49 – 52	HI MIRIFUSUS GUADALUPENSIS
51	98	47 – 53	HI STAUROSPHAERA ANTIQUA
52	172	50 – 53	HI HALIODICTYA HOJNOSI
53	167	52 – 54	LO SETHOCAPSA LEIOSTRACA
54	123	53 – 55	LO TETRARABS BULBOSA
55	107	54 – 56	LO TRITRABS EWINGI
56	151	55 – 58	LO MIRIFUSUS MEDIODILATATUS MAJOR
57	145	55 – 60	LO NAPORA BUKRYI
58	175	56 – 59	LO SAITOUM PAGEI
59	38	58 – 62	HI EMILUVIA PREMYOGII
60	127	58 – 62	LO PODOBURSA SPINOSA
61	97	58 – 66	LO STAUROSPHAERA ANTIQUA
62	147	60 – 64	LO EMILUVIA HOPSONI
63	143	60 – 64	LO NAPORA LOSPENSIS
64	85	63 – 65	LO HOMOEOPARONAELLA ARGOLIDENSIS
65	51	64 – 66	LO TRITRABS CASMALIAENSIS
66	63	65 – 70	LO RISTOLA ALTISSIMA
67	53	65 – 68	LO TRITRABS RHODODACTYLUS
68	111	67 – 71	LO EUCYRTIDIELLUM PTYCTUM
69	99	64 – 73	LO SPONGOCAPSULA PALMERAE
70	49	67 – 71	LO TRIACTOMA BLAKEI
71	39	70 – 76	LO TETRADITRYMA CORRALITOSENSIS
72	79	68 – 73	LO EMILUVIA SEDECIMPORATA S STR
73	113	72 – 75	LO TETRADITRYMA PSEUDOPLENA
74	37	70 – 78	LO EMILUVIA PREMYOGII
75	171	73 – 78	LO HALIODICTYA HOJNOSI
76	65	73 – 77	LO EMILUVIA SEDECIMPORATA SALENSIS
77	93	76 – 78	LO HSUUM MAXWELLI
78	41	77 – 79	LO PROTUNUMA COSTATA
79	73	78 – 81	LO MIRIFUSUS GUADALUPENSIS
80	35	78 – 82	LO PODOBURSA HELVETICA
81	121	79 – 84	LO STICHOCAPSA CONVEXA
82	71	80 – 83	LO TRITRABS ZEALIS
83	57	82 – 85	LO TRIACTOMA JONESI
84	45	82 – 85	LO HSUUM BREVICOSTATUM
85	33	84 – 86	LO EUCYRTIDIELLUM UNUMAENSIS
86	47	85 – 87	LO ACANTHOCIRCUS SUBOBLONGUS

Figure 8 Scaled optimum sequence for Baumgartner data base. Total RASC distance from first to last event amounts to 2.9611 only, indicating very strong clustering of the 86 events retained. Baumgartner (1984) has discussed limitations of scaling output for 156 events with $k_c = 8$ and $m_{c2} = 6$. Then the total RASC distance amounted to 2.3395 only.

(5) Blank data base

Atlantic Mesozoic/Cenozoic Nannofossils

RASC input format for sequence data.

Total number of deep sea drilling sites = 81.

Total number of events in dictionary = 590.

Type of event = LO (lowest occurrence), HI (highest occurrence).

Largest number of events in single section = 156.

Only events that occur in k_c = 15 or more sections were used in RASC run. For ranking (presorting followed by modified Hay method), only pairs of events that occur in m_{c1} = 4 or more sections were used. For scaling (dendrogram after final reordering), only pairs of events that occur in m_{c2} = 8 or more sections were used. Optimum sequence and dendrogram are shown in Figures 9 and 10.

This data base was originally compiled and used by R.G. Blank for testing and comparing binomial models in probabilistic biostratigraphy (see Blank and Ellis, 1982; Blank, 1984).

OPTIMUM FOSSIL SEQUENCE

SEQUENCE POSITION	FOSSIL NUMBER	RANGE	FOSSIL NAME
1	151	0 — 3	HI CYCLOCOCCOLITHUS LEPTOPORUS
2	301	0 — 3	HI GEPHYROCAPSA OCEANICA
3	319	2 — 4	HI HELICOSPHAERA KAMPTNERI
4	59	3 — 6	HI CERATOLITHUS CRISTATUS
5	555	2 — 6	HI UMBILICOSPHAERA SIBOGAE
6	103	5 — 8	HI COCCOLITHUS PELAGICUS
7	447	5 — 8	HI RHABDOSPHAERA CLAVIGERA
8	429	7 — 9	HI PSEUDOEMILIANIA LACUNOSA
9	271	8 — 10	HI DISCOLITHINA JAPONICA
10	302	9 — 11	LO GEPHYROCAPSA OCEANICA
11	155	10 — 12	HI CYCLOCOCCOLITHUS MACINTYREI
12	329	11 — 13	HI HELICOSPHAERA SELLII
13	117	12 — 14	HI CRENALITHUS DORONICOIDES
14	448	13 — 15	LO RHABDOSPHAERA CLAVIGERA
15	187	14 — 16	HI DISCOASTER BROUWERI
16	60	15 — 17	LO CERATOLITHUS CRISTATUS
17	243	16 — 18	HI DISCOASTER PENTARADIATUS
18	173	17 — 20	HI DISCOASTER ASYMMETRICUS
19	255	17 — 21	HI DISCOASTER SURCULUS
20	61	18 — 21	HI CERATOLITHUS RUGOSUS
21	118	20 — 23	LO CRENALITHUS DORONICOIDES
22	556	19 — 23	LO UMBILICOSPHAERA SIBOGAE
23	439	22 — 24	HI RETICULOFENESTRA PSEUDOUMBILIC
24	477	23 — 25	HI SPHENOLITHUS ABIES
25	272	24 — 27	LO DISCOLITHINA JAPONICA
26	263	24 — 27	HI DISCOASTER VARIABILIS
27	11	26 — 28	HI AMAUROLITHUS TRICORNICULATUS
28	191	27 — 29	HI DISCOASTER CHALLENGERI
29	430	28 — 30	LO PSEUDOEMILIANIA LACUNOSA
30	330	29 — 31	LO HELICOSPHAERA SELLII
31	174	30 — 32	LO DISCOASTER ASYMMETRICUS
32	62	31 — 33	LO CERATOLITHUS RUGOSUS
33	247	32 — 34	HI DISCOASTER QUINQUERAMUS
34	12	33 — 35	LO AMAUROLITHUS TRICORNICULATUS
35	256	34 — 36	LO DISCOASTER SURCULUS
36	244	35 — 37	LO DISCOASTER PENTARADIATUS
37	188	36 — 38	LO DISCOASTER BROUWERI
38	207	37 — 39	HI DISCOASTER EXILIS
39	248	38 — 40	LO DISCOASTER QUINQUERAMUS
40	192	39 — 41	LO DISCOASTER CHALLENGERI
41	156	40 — 43	LO CYCLOCOCCOLITHUS MACINTYREI
42	478	40 — 43	LO SPHENOLITHUS ABIES
43	97	42 — 44	HI COCCOLITHUS EOPELAGICUS
44	195	43 — 45	HI DISCOASTER DEFLANDREI
45	264	44 — 46	LO DISCOASTER VARIABILIS
46	491	45 — 47	HI SPHENOLITHUS MORIFORMIS
47	320	46 — 48	LO HELICOSPHAERA KAMPTNERI
48	208	47 — 50	LO DISCOASTER EXILIS
49	152	47 — 50	LO CYCLOCOCCOLITHUS LEPTOPORUS
50	440	49 — 51	LO RETICULOFENESTRA PSEUDOUMBILIC
51	492	50 — 53	LO SPHENOLITHUS MORIFORMIS
52	104	50 — 53	LO COCCOLITHUS PELAGICUS
53	443	52 — 54	HI RETICULOFENESTRA UMBILICA
54	196	53 — 55	LO DISCOASTER DEFLANDREI
55	175	54 — 56	HI DISCOASTER BARBADIENSIS
56	249	55 — 57	HI DISCOASTER SAIPANENSIS
57	77	56 — 58	HI CHIASMOLITHUS GRANDIS
58	98	57 — 59	LO COCCOLITHUS EOPELAGICUS
59	444	58 — 60	LO RETICULOFENESTRA UMBILICA
60	250	59 — 61	LO DISCOASTER SAIPANENSIS
61	78	60 — 62	LO CHIASMOLITHUS GRANDIS
62	176	61 — 63	LO DISCOASTER BARBADIENSIS
63	425	62 — 64	HI PREDISCOSPHAERA CRETACEA
64	131	63 — 65	HI CRIBROSPHAERELLA EHRENBERGI
65	281	64 — 66	HI EIFFELLITHUS TURRISEIFFELI
66	563	65 — 67	HI WATZNAUERIA BARNESAE
67	573	66 — 68	HI ZYGODISCUS DIPLOGRAMMUS
68	407	67 — 69	HI PARHABDOLITHUS EMBERGERI
69	403	68 — 71	HI PARHABDOLITHUS ANGUSTUS
70	347	68 — 71	HI LITHRAPHIDITES CARNIOLENSIS
71	507	70 — 72	HI STEPHANOLITHION LAFFITTEI
72	132	71 — 73	LO CRIBROSPHAERELLA EHRENBERGI
73	426	72 — 74	LO PREDISCOSPHAERA CRETACEA
74	282	73 — 75	LO EIFFELLITHUS TURRISEIFFELI
75	404	74 — 76	LO PARHABDOLITHUS ANGUSTUS
76	508	75 — 77	LO STEPHANOLITHION LAFFITTEI
77	574	76 — 78	LO ZYGODISCUS DIPLOGRAMMUS
78	348	77 — 79	LO LITHRAPHIDITES CARNIOLENSIS
79	408	78 — 80	LO PARHABDOLITHUS EMBERGERI
80	564	79 — 81	LO WATZNAUERIA BARNESAE

Figure 9 Optimum sequence for Blank data base.

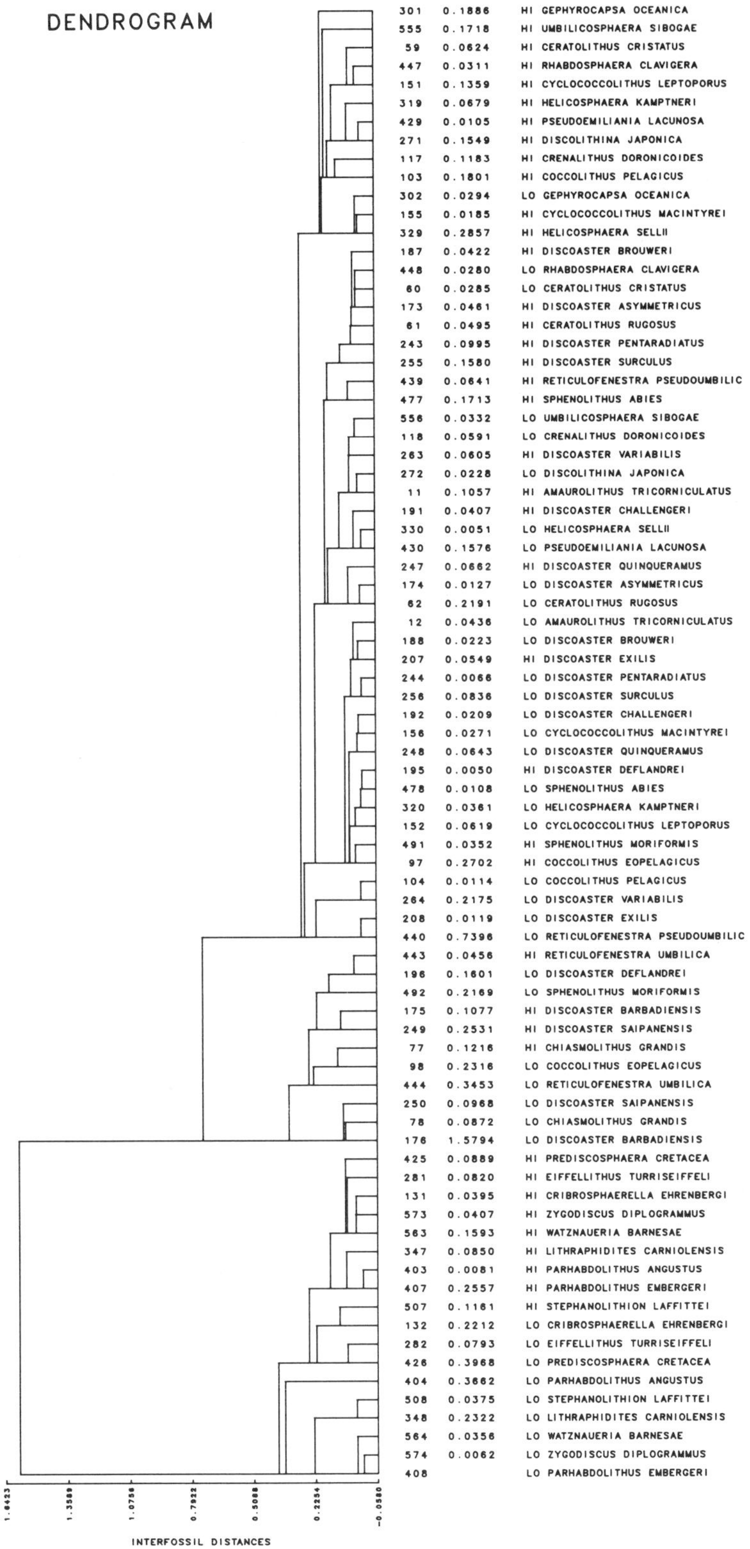

DENDROGRAM

301	0.1886	HI GEPHYROCAPSA OCEANICA
555	0.1718	HI UMBILICOSPHAERA SIBOGAE
59	0.0624	HI CERATOLITHUS CRISTATUS
447	0.0311	HI RHABDOSPHAERA CLAVIGERA
151	0.1359	HI CYCLOCOCCOLITHUS LEPTOPORUS
319	0.0679	HI HELICOSPHAERA KAMPTNERI
429	0.0105	HI PSEUDOEMILIANIA LACUNOSA
271	0.1549	HI DISCOLITHINA JAPONICA
117	0.1183	HI CRENALITHUS DORONICOIDES
103	0.1801	HI COCCOLITHUS PELAGICUS
302	0.0294	LO GEPHYROCAPSA OCEANICA
155	0.0185	HI CYCLOCOCCOLITHUS MACINTYREI
329	0.2857	HI HELICOSPHAERA SELLII
187	0.0422	HI DISCOASTER BROUWERI
448	0.0280	LO RHABDOSPHAERA CLAVIGERA
60	0.0285	LO CERATOLITHUS CRISTATUS
173	0.0461	HI DISCOASTER ASYMMETRICUS
61	0.0495	HI CERATOLITHUS RUGOSUS
243	0.0995	HI DISCOASTER PENTARADIATUS
255	0.1580	HI DISCOASTER SURCULUS
439	0.0641	HI RETICULOFENESTRA PSEUDOUMBILIC
477	0.1713	HI SPHENOLITHUS ABIES
556	0.0332	LO UMBILICOSPHAERA SIBOGAE
118	0.0591	LO CRENALITHUS DORONICOIDES
263	0.0605	HI DISCOASTER VARIABILIS
272	0.0228	LO DISCOLITHINA JAPONICA
11	0.1057	HI AMAUROLITHUS TRICORNICULATUS
191	0.0407	HI DISCOASTER CHALLENGERI
330	0.0051	LO HELICOSPHAERA SELLII
430	0.1576	LO PSEUDOEMILIANIA LACUNOSA
247	0.0662	HI DISCOASTER QUINQUERAMUS
174	0.0127	LO DISCOASTER ASYMMETRICUS
62	0.2191	LO CERATOLITHUS RUGOSUS
12	0.0436	LO AMAUROLITHUS TRICORNICULATUS
188	0.0223	LO DISCOASTER BROUWERI
207	0.0549	HI DISCOASTER EXILIS
244	0.0066	LO DISCOASTER PENTARADIATUS
256	0.0836	LO DISCOASTER SURCULUS
192	0.0209	LO DISCOASTER CHALLENGERI
156	0.0271	LO CYCLOCOCCOLITHUS MACINTYREI
248	0.0643	LO DISCOASTER QUINQUERAMUS
195	0.0050	HI DISCOASTER DEFLANDREI
478	0.0108	LO SPHENOLITHUS ABIES
320	0.0361	LO HELICOSPHAERA KAMPTNERI
152	0.0619	LO CYCLOCOCCOLITHUS LEPTOPORUS
491	0.0352	HI SPHENOLITHUS MORIFORMIS
97	0.2702	HI COCCOLITHUS EOPELAGICUS
104	0.0114	LO COCCOLITHUS PELAGICUS
264	0.2175	LO DISCOASTER VARIABILIS
208	0.0119	LO DISCOASTER EXILIS
440	0.7396	LO RETICULOFENESTRA PSEUDOUMBILIC
443	0.0456	HI RETICULOFENESTRA UMBILICA
196	0.1601	LO DISCOASTER DEFLANDREI
492	0.2169	LO SPHENOLITHUS MORIFORMIS
175	0.1077	HI DISCOASTER BARBADIENSIS
249	0.2531	HI DISCOASTER SAIPANENSIS
77	0.1216	HI CHIASMOLITHUS GRANDIS
98	0.2316	LO COCCOLITHUS EOPELAGICUS
444	0.3453	LO RETICULOFENESTRA UMBILICA
250	0.0968	LO DISCOASTER SAIPANENSIS
78	0.0872	LO CHIASMOLITHUS GRANDIS
176	1.5794	LO DISCOASTER BARBADIENSIS
425	0.0889	HI PREDISCOSPHAERA CRETACEA
281	0.0820	HI EIFFELLITHUS TURRISEIFFELI
131	0.0395	HI CRIBROSPHAERELLA EHRENBERGI
573	0.0407	HI ZYGODISCUS DIPLOGRAMMUS
563	0.1593	HI WATZNAUERIA BARNESAE
347	0.0850	HI LITHRAPHIDITES CARNIOLENSIS
403	0.0081	HI PARHABDOLITHUS ANGUSTUS
407	0.2557	HI PARHABDOLITHUS EMBERGERI
507	0.1161	HI STEPHANOLITHION LAFFITTEI
132	0.2212	LO CRIBROSPHAERELLA EHRENBERGI
282	0.0793	LO EIFFELLITHUS TURRISEIFFELI
426	0.3968	LO PREDISCOSPHAERA CRETACEA
404	0.3662	LO PARHABDOLITHUS ANGUSTUS
508	0.0375	LO STEPHANOLITHION LAFFITTEI
348	0.2322	LO LITHRAPHIDITES CARNIOLENSIS
564	0.0356	LO WATZNAUERIA BARNESAE
574	0.0062	LO ZYGODISCUS DIPLOGRAMMUS
408		LO PARHABDOLITHUS EMBERGERI

1.6423 1.3589 1.0756 0.7922 0.5088 0.2254 -0.0580

INTERFOSSIL DISTANCES

Figure 10 Scaled optimum sequence for Blank data base. RASC dendrogram for 170 events ($k_c = 9$, $m_{c1} = m_{c1} = 5$) was published by Blank (1984, Figure 2, p. 62-63) and compared to several other optimum sequences.

(6) Rubel's Silurian data bases

A. Baltic Brachiopods

UA input format for sequence data (with depths).

Total number of sections = 20.

Total number of events in dictionary = 398 (112 Brachiopod events).

Largest number of events in single section = 80.

Only events that occur in k_c = 8 or more sections were used in RASC run. For ranking (presorting followed by modified Hay method) only pairs of events that occur in m_{c1} = 3 or more sections were used. For scaling (dendrogram after final reordering), only pairs of events that occur in m_{c2} = 5 or more sections were used. Optimum sequence and dendrogram, based on lowest (LO) and highest (HI) occurrences, are shown in Figures 11 and 12.

Rubel's Silurian Baltic data bases are part of material collected by the USSR Working Group of IGCP Project 148 (Rubel, 1982) and processed by the mathematical method of constructing 'datum planes' bases on assemblages of fossils for biostratigraphic correlation (Rubel and Pak, 1984). For a comparison of results obtained by the latter method and RASC results for these data bases, see Agterberg and Rubel (in press).

OPTIMUM FOSSIL SEQUENCE

SEQUENCE POSITION	FOSSIL NUMBER	RANGE	FOSSIL NAME
1	50	0 — 2	HI MICROSPHAERIDIORHYNCUSNUCULA
2	52	1 — 3	HI PROTOCHONETES PILTENENSIS
3	54	2 — 4	HI HOMOEOSPIRA BAYILEI
4	38	3 — 5	HI ISORTHIS CANALICULATA
5	58	4 — 6	HI DELTHYRIS MAGNA
6	30	5 — 7	HI DELTHYRIS ELEVATA
7	60	6 — 8	HI SHALERIA DZWINOGRODENSIS
8	10	7 — 9	HI ATRYPA RETICULARIS
9	56	8 — 10	HI ATRYPOIDEA PRUNUM
10	18	9 — 11	HI DALEJINA HYBRIDA
11	62	10 — 12	HI MORINORHYNCHUS ORBIGNYI
12	40	11 — 13	HI DAYIA NAVICULA
13	57	12 — 14	LO DELTHYRIS MAGNA
14	68	13 — 15	HI HOWELLELLA SP. SP.
15	59	14 — 16	LO SHALERIA DZWINOGRODENSIS
16	2	15 — 17	HI SHALERIA SP. SP.
17	53	16 — 18	LO HOMOEOSPIRA BAYILEI
18	37	17 — 20	LO ISORTHIS CANALICULATA
19	51	17 — 20	LO PROTOCHONETES PILTENENSIS
20	55	19 — 22	LO ATRYPOIDEA PRUNUM
21	29	19 — 22	LO DELTHYRIS ELEVATA
22	61	21 — 23	LO MORINORHYNCHUS ORBIGNYI
23	49	22 — 24	LO MICROSPHAERIDIORHYNCUSNUCULA
24	16	23 — 25	HI ISORTHIS CRASSA
25	28	24 — 26	HI CYPIDULA GALEATA
26	36	25 — 27	HI STROPHONELLA EUGLYPHA
27	1	26 — 28	LO SHALERIA SP. SP.
28	27	27 — 29	LO CYPIDULA GALEATA
29	12	28 — 30	HI RESSERELLA SP. SP.
30	66	29 — 31	HI EOSPIRIFER RADIATUS
31	42	30 — 32	HI PROTOCHONETES MINIMUS
32	20	31 — 33	HI CYRTIA EXPORRECTA
33	39	32 — 34	LO DAYIA NAVICULA
34	67	33 — 35	LO HOWELLELLA SP. SP.
35	6	34 — 36	HI GLASSIA OBOVATA
36	14	35 — 37	HI DICOELOSIA BILOBA
37	22	36 — 38	HI SKENIDIOIDES LEWISI
38	96	37 — 39	HI EOPLECTODONTA SP. SP.
39	65	38 — 40	LO EOSPIRIFER RADIATUS
40	35	39 — 41	LO STROPHONELLA EUGLYPHA
41	4	40 — 42	HI AEGIRIA GRAYI
42	41	41 — 43	LO PROTOCHONETES MINIMUS
43	26	42 — 44	HI CLORINDA SP. SP.
44	11	43 — 45	LO RESSERELLA SP. SP.
45	13	44 — 46	LO DICOELOSIA BILOBA
46	17	45 — 47	LO DALEJINA HYBRIDA
47	19	46 — 48	LO CYRTIA EXPORRECTA
48	9	47 — 49	LO ATRYPA RETICULARIS
49	25	48 — 50	LO CLORINDA SP. SP.
50	15	49 — 52	LO ISORTHIS CRASSA
51	21	49 — 52	LO SKENIDIOIDES LEWISI
52	3	51 — 53	LO AEGIRIA GRAYI
53	5	52 — 54	LO GLASSIA OBOVATA
54	95	53 — 55	LO EOPLECTODONTA SP. SP.

Figure 11 Optimum sequence for Rubel's Silurian Brachiopods, Baltic region.

50	0.2122	HI MICROSPHAERIDIORHYNCUSNUCULA
52	0.0718	HI PROTOCHONETES PILTENENSIS
54	0.0212	HI HOMOEOSPIRA BAYILEI
38	0.2458	HI ISORTHIS CANALICULATA
60	0.1152	HI SHALERIA DZWINOGRODENSIS
58	0.1799	HI DELTHYRIS MAGNA
30	0.0191	HI DELTHYRIS ELEVATA
10	0.1360	HI ATRYPA RETICULARIS
62	0.0188	HI MORINORHYNCHUS ORBIGNYI
56	0.1169	HI ATRYPOIDEA PRUNUM
18	0.0267	HI DALEJINA HYBRIDA
40	0.1392	HI DAYIA NAVICULA
68	0.1547	HI HOWELLELLA SP. SP.
2	0.0781	HI SHALERIA SP. SP.
57	0.1237	LO DELTHYRIS MAGNA
59	0.1781	LO SHALERIA DZWINOGRODENSIS
51	0.0115	LO PROTOCHONETES PILTENENSIS
53	0.1529	LO HOMOEOSPIRA BAYILEI
37	0.1540	LO ISORTHIS CANALICULATA
55	0.0168	LO ATRYPOIDEA PRUNUM
61	0.1883	LO MORINORHYNCHUS ORBIGNYI
16	0.0141	HI ISORTHIS CRASSA
29	0.0199	LO DELTHYRIS ELEVATA
28	0.0736	HI CYPIDULA GALEATA
49	0.0116	LO MICROSPHAERIDIORHYNCUSNUCULA
12	0.0029	HI RESSERELLA SP. SP.
66	0.0897	HI EOSPIRIFER RADIATUS
42	0.1121	HI PROTOCHONETES MINIMUS
27	0.0050	LO CYPIDULA GALEATA
1	0.0179	LO SHALERIA SP. SP.
36	0.0572	HI STROPHONELLA EUGLYPHA
39	0.0066	LO DAYIA NAVICULA
6	0.0121	HI GLASSIA OBOVATA
67	0.0981	LO HOWELLELLA SP. SP.
14	0.0756	HI DICOELOSIA BILOBA
65	0.0183	LO EOSPIRIFER RADIATUS
22	0.0084	HI SKENIDIOIDES LEWISI
96	0.0158	HI EOPLECTODONTA SP. SP.
20	0.0877	HI CYRTIA EXPORRECTA
35	0.1177	LO STROPHONELLA EUGLYPHA
41	0.0125	LO PROTOCHONETES MINIMUS
4	0.1534	HI AEGIRIA GRAYI
26	0.1886	HI CLORINDA SP. SP.
17	0.0031	LO DALEJINA HYBRIDA
11	0.0555	LO RESSERELLA SP. SP.
19	0.0332	LO CYRTIA EXPORRECTA
9	0.0580	LO ATRYPA RETICULARIS
13	0.0980	LO DICOELOSIA BILOBA
25	0.1369	LO CLORINDA SP. SP.
15	0.1629	LO ISORTHIS CRASSA
21	0.1408	LO SKENIDIOIDES LEWISI
5	0.1059	LO GLASSIA OBOVATA
3	0.2586	LO AEGIRIA GRAYI
95		LO EOPLECTODONTA SP. SP.

0.28880 0.22278 0.17676 0.13073 0.08471 0.03869 -0.00733

INTERFOSSIL DISTANCES

Figure 12 Dendrogram for Rubel's Silurian Brachiopods, Baltic region.

(6) Rubel's Silurian data bases

B. Baltic Ostracods

UA input format for sequence data (with depths).

Total number of sections = 12.

Total number of events in dictionary = 398 (204 ostracod events).

Largest number of events in single section = 110.

Only events that occur in k_c = 8 or more sections were used in RASC run. For ranking (presorting followed by modified Hay method), only pairs of events that occur in m_{c1} = 3 or more sections were used. For scaling (dendrogram after final reordering), only pairs of events that occur in m_{c2} = 5 or more sections were used. Optimum sequence and dendrogram are shown in Figures 13 and 14.

OPTIMUM FOSSIL SEQUENCE

SEQUENCE POSITION	FOSSIL NUMBER	RANGE	FOSSIL NAME
1	168	0 — 2	HI CUTHERELLINA MAGNA
2	120	1 — 3	HI A. SUBCLUSA
3	148	2 — 4	HI C. CIRCULATA
4	196	3 — 5	HI H. MACCOYIANA
5	144	4 — 6	HI CAVELLINA ANGULATA
6	226	5 — 7	HI N. BUCHIANA
7	304	6 — 8	HI SLEIA INERMIS
8	218	7 — 9	HI MACRYPSILON SALTERIANA
9	122	8 — 10	HI A. SOLIDA
10	228	9 — 11	HI NODIBEYRICHIA TUBERCULATA
11	118	10 — 12	HI AMYGDALELLA NASUTA
12	298	11 — 13	HI S. PROFUNDIGENUS
13	114	12 — 14	HI AECHMINA MOLENGRAAFFI
14	227	13 — 15	LO NODIBEYRICHIA TUBERCULATA
15	194	14 — 16	HI H. MARGARITAE
16	306	15 — 17	HI S. EQUESTRIS
17	217	16 — 19	LO MACRYPSILON SALTERIANA
18	184	15 — 19	HI H. TRIVIALE
19	121	18 — 20	LO A. SOLIDA
20	182	19 — 21	HI HEBELLUM TETRAGONA
21	290	20 — 24	HI S. SIMPLEX
22	113	20 — 23	LO AECHMINA MOLENGRAAFFI
23	208	22 — 24	HI LEIOCYAMUS LIMPIDUS
24	195	23 — 25	LO H. MACCOYIANA
25	254	24 — 26	HI "OCTONARIA" PERPLEXA
26	117	25 — 27	LO AMYGDALELLA NASUTA
27	225	26 — 28	LO N. BUCHIANA
28	297	27 — 31	LO S. PROFUNDIGENUS
29	183	27 — 30	LO H. TRIVIALE
30	305	29 — 31	LO S. EQUESTRIS
31	303	30 — 32	LO SLEIA INERMIS
32	181	31 — 33	LO HEBELLUM TETRAGONA
33	143	32 — 34	LO CAVELLINA ANGULATA
34	193	33 — 35	LO H. MARGARITAE
35	289	34 — 36	LO S. SIMPLEX
36	207	35 — 37	LO LEIOCYAMUS LIMPIDUS
37	253	36 — 38	LO "OCTONARIA" PERPLEXA
38	119	37 — 39	LO A. SUBCLUSA
39	147	38 — 40	LO C. CIRCULATA
40	167	39 — 41	LO CUTHERELLINA MAGNA

Figure 13 Optimum sequence for Rubel's Silurian Ostracods, Baltic region.

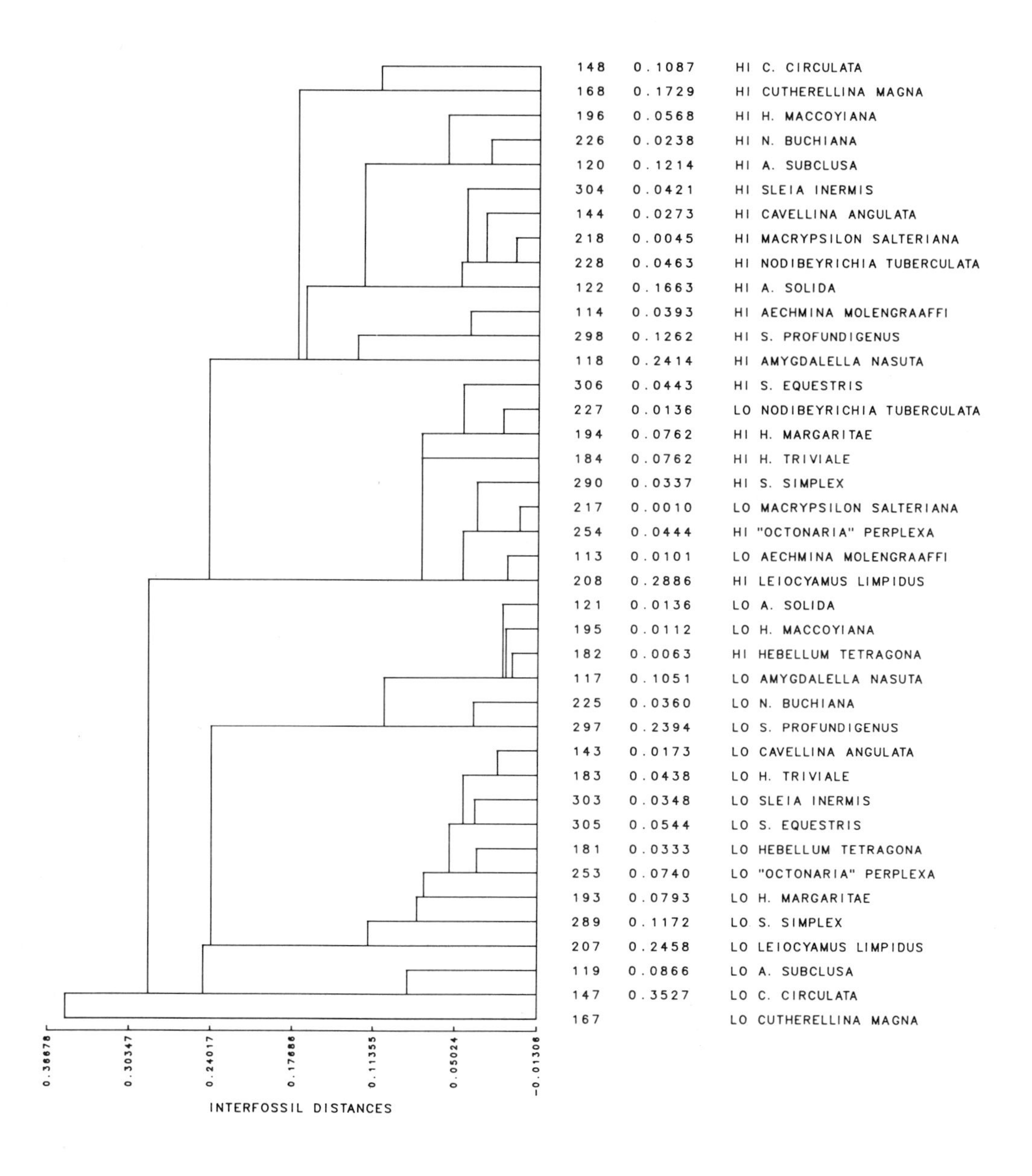

Figure 14 Dendrogram for Rubel's Silurian Ostracods, Baltic region.

(6) Rubel's Silurian data bases

C. Baltic Thelodonts

UA input format for sequence data (with depths).

Total number of sections = 20.

Total number of events in dictionary = 398 (82 Thelodont events).

Largest number of events in single section = 54.

Only events that occur in k_c = 8 or more sections were used in RASC run. For ranking (presorting followed by modified Hay method), only pairs of events that occur in m_{c1} = 3 or more sections were used. For scaling (dendrogram after final reordering), only pairs of events that occur in m_{c2} = 5 or more sections were used. Optimum sequence and dendrogram are shown in Figures 15 and 16.

OPTIMUM FOSSIL SEQUENCE

SEQUENCE POSITION	FOSSIL NUMBER	RANGE	FOSSIL NAME
1	328	0 — 3	HI GOMPHONCHUS SANDELENSIS
2	348	0 — 3	HI N. STRIATA
3	344	2 — 4	HI NOSTOLEPIS GRACILIS
4	336	3 — 6	HI LOGANIA CUNEATA
5	366	3 — 6	HI T. PARVIDENS
6	334	5 — 7	HI K. TRICAVUS
7	370	6 — 8	HI T. TRAQUAIRI
8	343	7 — 9	LO NOSTOLEPIS GRACILIS
9	326	8 — 10	HI G. HOPPEI OR PARACANTHODES POROSUS
10	360	9 — 11	HI THELODUS ADMIRABILIS
11	368	10 — 12	HI T. SCULPTILIS
12	359	11 — 13	LO THELODUS ADMIRABILIS
13	322	12 — 14	HI CYATHASPIDINAE (ARCHEOGONASPIS?)
14	398	13 — 15	HI OSTEOOSTRACI (GEN. INDET.)
15	318	14 — 16	HI ANASPIDA GEN. N. ET SP.A
16	369	15 — 17	LO T. TRAQUAIRI
17	333	16 — 18	LO K. TRICAVUS
18	317	17 — 19	LO ANASPIDA GEN. N. ET SP.A
19	335	18 — 20	LO LOGANIA CUNEATA
20	325	19 — 21	LO G. HOPPEI OR PARACANTHODES POROSUS
21	367	20 — 22	LO T. SCULPTILIS
22	321	21 — 23	LO CYATHASPIDINAE (ARCHEOGONASPIS?)
23	365	22 — 24	LO T. PARVIDENS
24	347	23 — 25	LO N. STRIATA
25	327	24 — 26	LO GOMPHONCHUS SANDELENSIS
26	350	25 — 27	HI PHLEBOLEPIS ELEGANS
27	356	26 — 28	HI SAAROLEPIS OESELENSIS
28	338	27 — 29	HI L. MARTINSSONI
29	362	28 — 30	HI T. LAEVIS
30	397	29 — 31	LO OSTEOOSTRACI (GEN. INDET.)
31	349	30 — 32	LO PHLEBOLEPIS ELEGANS
32	355	31 — 33	LO SAAROLEPIS OESELENSIS
33	337	32 — 34	LO L. MARTINSSONI
34	361	33 — 35	LO T. LAEVIS

Figure 15 Optimum sequence for Rubel's Silurian Thelodonts, Baltic region.

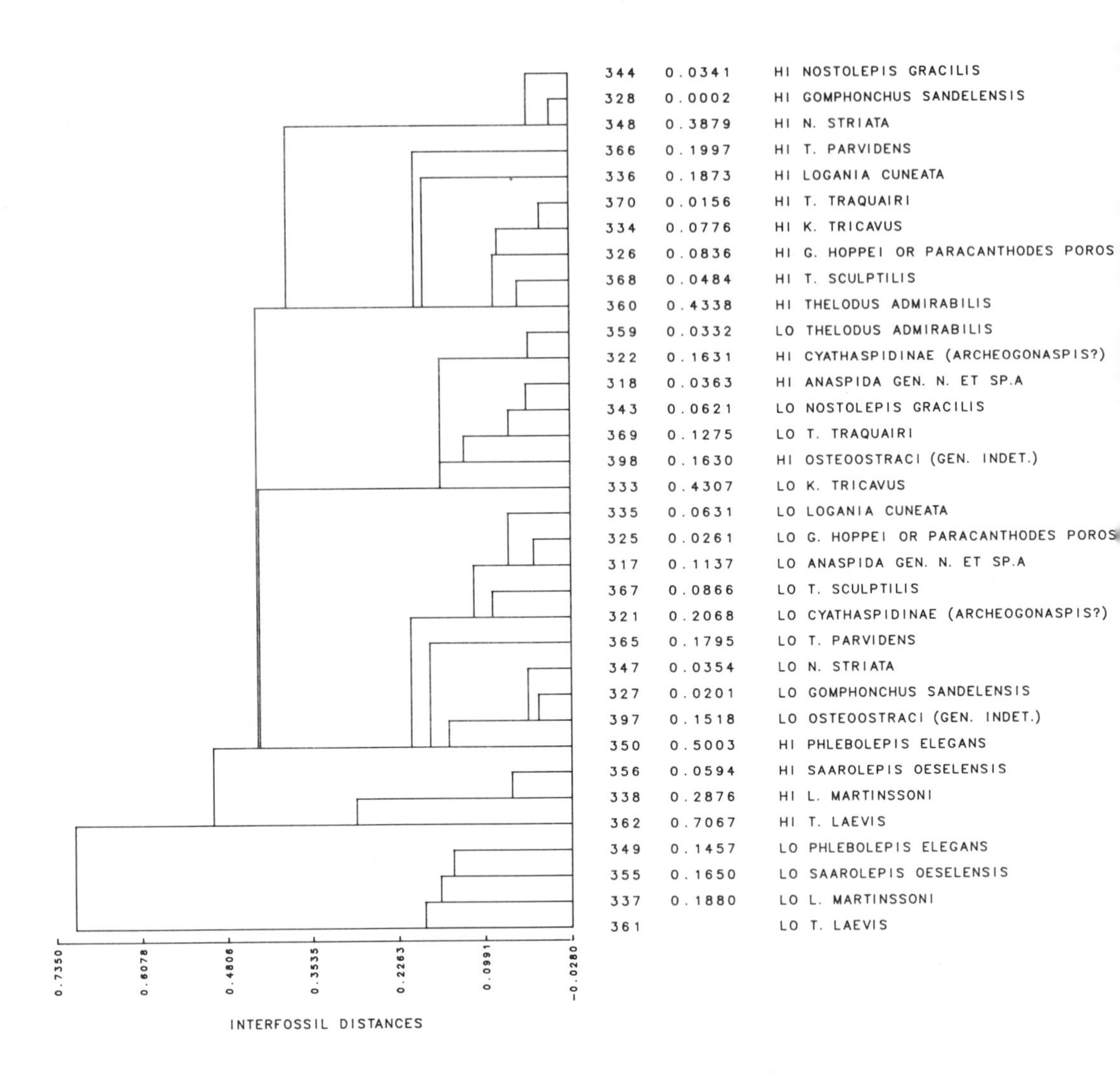

Figure 16 Dendrogram for Rubel's Silurian Thelodonts, Baltic region.

(6) Rubel's Silurian data bases

D. Three data bases (A, B and C) combined

UA input format for sequence data (with depths).

Total number of sections = 35.

Total number of events in dictionary = 398.

Largest number of events in single section = 160.

Only events that occur in k_c = 13 or more sections were used in RASC run. For ranking (presorting only), all pairs of events were used. For scaling (dendrogram after final reordering), only pairs of events that occur in m_{c2} = 3 or more sections were used. Optimum sequence and dendrogram are shown in Figures 17 and 18.

OPTIMUM FOSSIL SEQUENCE

SEQUENCE POSITION	FOSSIL NUMBER	RANGE	FOSSIL NAME
1	328	0 — 3	HI GOMPHONCHUS SANDELENSIS
2	348	0 — 3	HI N. STRIATA
3	50	2 — 4	HI MICROSPHAERIDIORHYNCUSNUCULA
4	52	3 — 5	HI PROTOCHONETES PILTENENSIS
5	54	4 — 6	HI HOMOEOSPIRA BAYILEI
6	38	5 — 7	HI ISORTHIS CANALICULATA
7	58	6 — 8	HI DELTHYRIS MAGNA
8	366	7 — 9	HI T. PARVIDENS
9	30	8 — 11	HI DELTHYRIS ELEVATA
10	336	7 — 11	HI LOGANIA CUNEATA
11	334	10 — 12	HI K. TRICAVUS
12	10	11 — 13	HI ATRYPA RETICULARIS
13	370	12 — 15	HI T. TRAQUAIRI
14	18	12 — 19	HI DALEJINA HYBRIDA
15	326	13 — 16	HI G. HOPPEI OR PARACANTHODES POROSUS
16	56	15 — 17	HI ATRYPOIDEA PRUNUM
17	368	16 — 19	HI T. SCULPTILIS
18	40	12 — 20	HI DAYIA NAVICULA
19	333	17 — 20	LO K. TRICAVUS
20	57	19 — 21	LO DELTHYRIS MAGNA
21	369	20 — 27	LO T. TRAQUAIRI
22	51	20 — 26	LO PROTOCHONETES PILTENENSIS
23	53	20 — 24	LO HOMOEOSPIRA BAYILEI
24	37	23 — 26	LO ISORTHIS CANALICULATA
25	68	20 — 26	HI HOWELLELLA SP. SP.
26	55	25 — 27	LO ATRYPOIDEA PRUNUM
27	335	26 — 28	LO LOGANIA CUNEATA
28	325	27 — 29	LO G. HOPPEI OR PARACANTHODES POROSUS
29	367	28 — 30	LO T. SCULPTILIS
30	29	29 — 31	LO DELTHYRIS ELEVATA
31	365	30 — 33	LO T. PARVIDENS
32	39	25 — 38	LO DAYIA NAVICULA
33	347	31 — 34	LO N. STRIATA
34	327	33 — 36	LO GOMPHONCHUS SANDELENSIS
35	16	33 — 37	HI ISORTHIS CRASSA
36	49	34 — 37	LO MICROSPHAERIDIORHYNCUSNUCULA
37	20	36 — 38	HI CYRTIA EXPORRECTA
38	67	37 — 39	LO HOWELLELLA SP. SP.
39	6	38 — 40	HI GLASSIA OBOVATA
40	17	39 — 41	LO DALEJINA HYBRIDA
41	19	40 — 42	LO CYRTIA EXPORRECTA
42	9	41 — 43	LO ATRYPA RETICULARIS
43	15	42 — 44	LO ISORTHIS CRASSA

Figure 17 Optimum sequence for combined Silurian data bases.

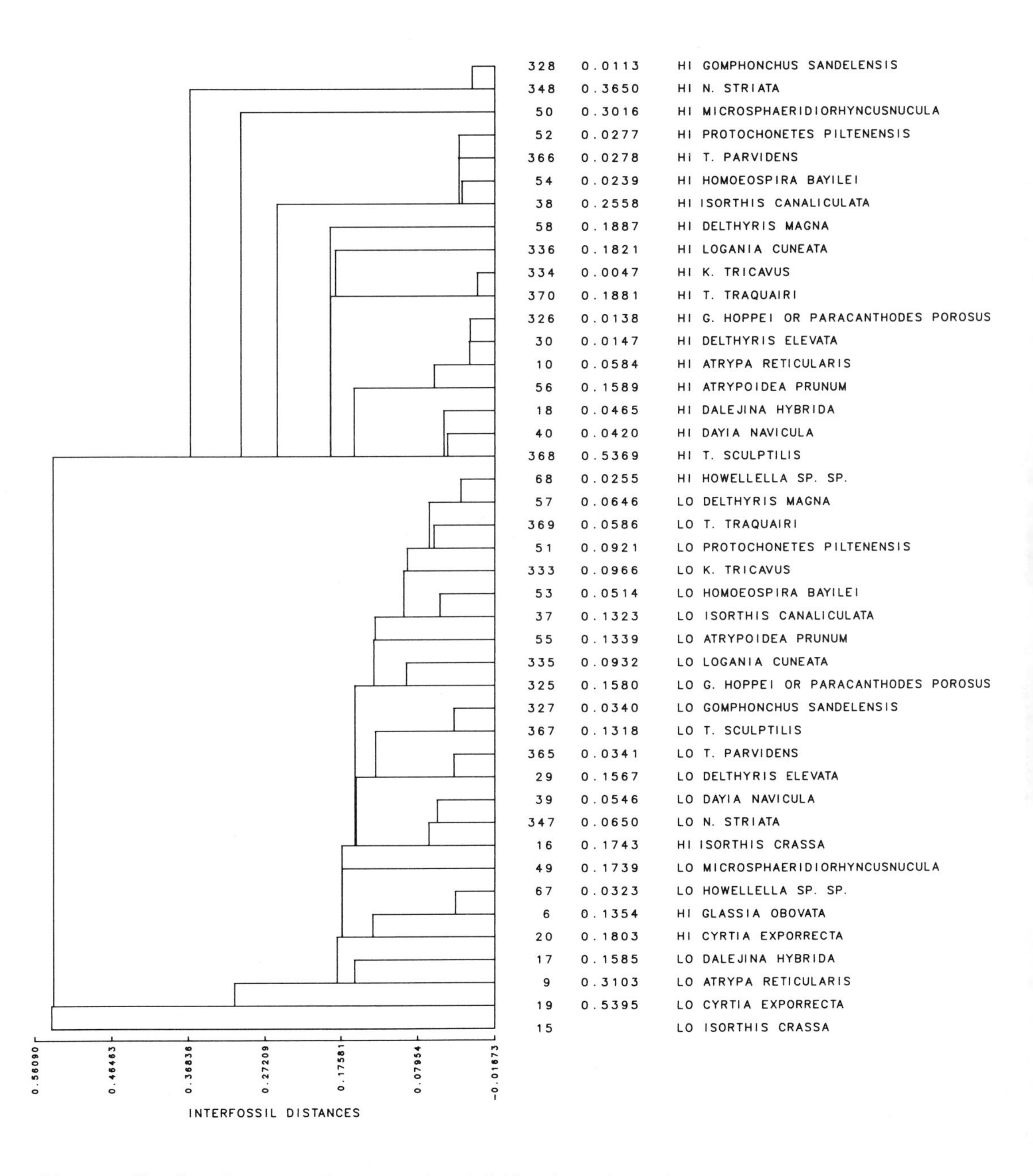

Figure 18 Dendrogram for combined Silurian data bases.

(7) Sullivan - Bramlette data base

Californian Paleogene Nannofossils

UA input format for sequence data.

Total number of sections = 10.

Total number of events in dictionary = 208.

Largest number of events in single section = 176.

Only events that occur in k_c = 9 or more sections were used in RASC run. For ranking (presorting followed by modified Hay method), only pairs of events that occur in m_{c1} = 3 or more sections were used. For scaling (dendrogram after final reordering), only pairs of events that occur in m_{c2} = 5 or more sections were used. Optimum sequence and dendrogram are shown in Figures 19 and 20.

This data base was originally coded and analyzed using the Unitary Association method by Davaud and Guex (1978) from information published in Bramlette and Sullivan (1961) and Sullivan (1965).

OPTIMUM FOSSIL SEQUENCE

SEQUENCE POSITION	FOSSIL NUMBER	RANGE	FOSSIL NAME
1	20	0 — 3	HI COCCOLITHUS GRANDIS
2	84	0 — 3	HI HELICOSPHAERA LOPHOTA
3	152	2 — 4	HI SPHENOLITHUS RADIANS
4	32	3 — 5	HI COCCOLITHUS CRASSUS
5	156	4 — 6	HI DISCOASTER BARBADIENSIS
6	8	5 — 7	HI CHIPHRAGMALITHUS DUBIUS
7	170	6 — 8	HI DISCOASTER LODOENSIS
8	28	7 — 9	HI COCCOLITHUS DELUS
9	38	8 — 10	HI CYCLOCOCCOLITHUS GAMMATION
10	166	9 — 11	HI DISCOASTER DISTINCTUS
11	202	10 — 12	HI DISCOASTEROIDES KUEPPERI
12	44	11 — 13	HI DISCOLITHUS PLANUS
13	160	12 — 14	HI DISCOASTER DEFLANDREI
14	48	13 — 15	HI DISCOLITHUS PULCHEROIDES
15	128	14 — 16	HI ZYGRHABLITHUS BIJUGATUS
16	46	15 — 17	HI DISCOLITHUS PULCHER
17	82	16 — 18	HI HELICOSPHAERA SEMILUNUM
18	30	17 — 19	HI COCCOLITHUS CONSUETUS
19	50	18 — 20	HI DISCOLITHUS RIMOSUS
20	130	19 — 21	HI BAARUDOSPHAERA BIGELOWI
21	86	20 — 22	HI LOPHODOLITHUS NASCENS
22	158	21 — 23	HI DISCOASTER BINODOSUS
23	178	22 — 24	HI DISCOASTER TRIBRACHIATUS
24	6	23 — 26	HI CHIPHRAGMALITHUS CALATUS
25	88	23 — 26	HI LOPHODOLITHUS RENIFORMIS
26	78	25 — 28	HI ELLIPSOLITHUS MACELLUS
27	83	25 — 28	LO HELICOSPHAERA LOPHOTA
28	37	27 — 29	LO CYCLOCOCCOLITHUS GAMMATION
29	165	28 — 31	LO DISCOASTER DISTINCTUS
30	159	28 — 32	LO DISCOASTER DEFLANDREI
31	43	29 — 32	LO DISCOLITHUS PLANUS
32	129	31 — 33	LO BAARUDOSPHAERA BIGELOWI
33	81	32 — 34	LO HELICOSPHAERA SEMILUNUM
34	7	33 — 37	LO CHIPHRAGMALITHUS DUBIUS
35	87	33 — 36	LO LOPHODOLITHUS RENIFORMIS
36	31	35 — 41	LO COCCOLITHUS CRASSUS
37	157	35 — 39	LO DISCOASTER BINODOSUS
38	49	34 — 41	LO DISCOLITHUS RIMOSUS
39	201	37 — 40	LO DISCOASTEROIDES KUEPPERI
40	19	39 — 44	LO COCCOLITHUS GRANDIS
41	169	38 — 46	LO DISCOASTER LODOENSIS
42	5	38 — 46	LO CHIPHRAGMALITHUS CALATUS
43	177	38 — 46	LO DISCOASTER TRIBRACHIATUS
44	47	40 — 45	LO DISCOLITHUS PULCHEROIDES
45	127	44 — 47	LO ZYGRHABLITHUS BIJUGATUS
46	155	44 — 48	LO DISCOASTER BARBADIENSIS
47	45	45 — 51	LO DISCOLITHUS PULCHER
48	85	46 — 51	LO LOPHODOLITHUS NASCENS
49	27	46 — 51	LO COCCOLITHUS DELUS
50	151	45 — 51	LO SPHENOLITHUS RADIANS
51	29	50 — 52	LO COCCOLITHUS CONSUETUS
52	77	51 — 53	LO ELLIPSOLITHUS MACELLUS

Figure 19 Optimum sequence for Sullivan-Bramlette data base.

DENDROGRAM

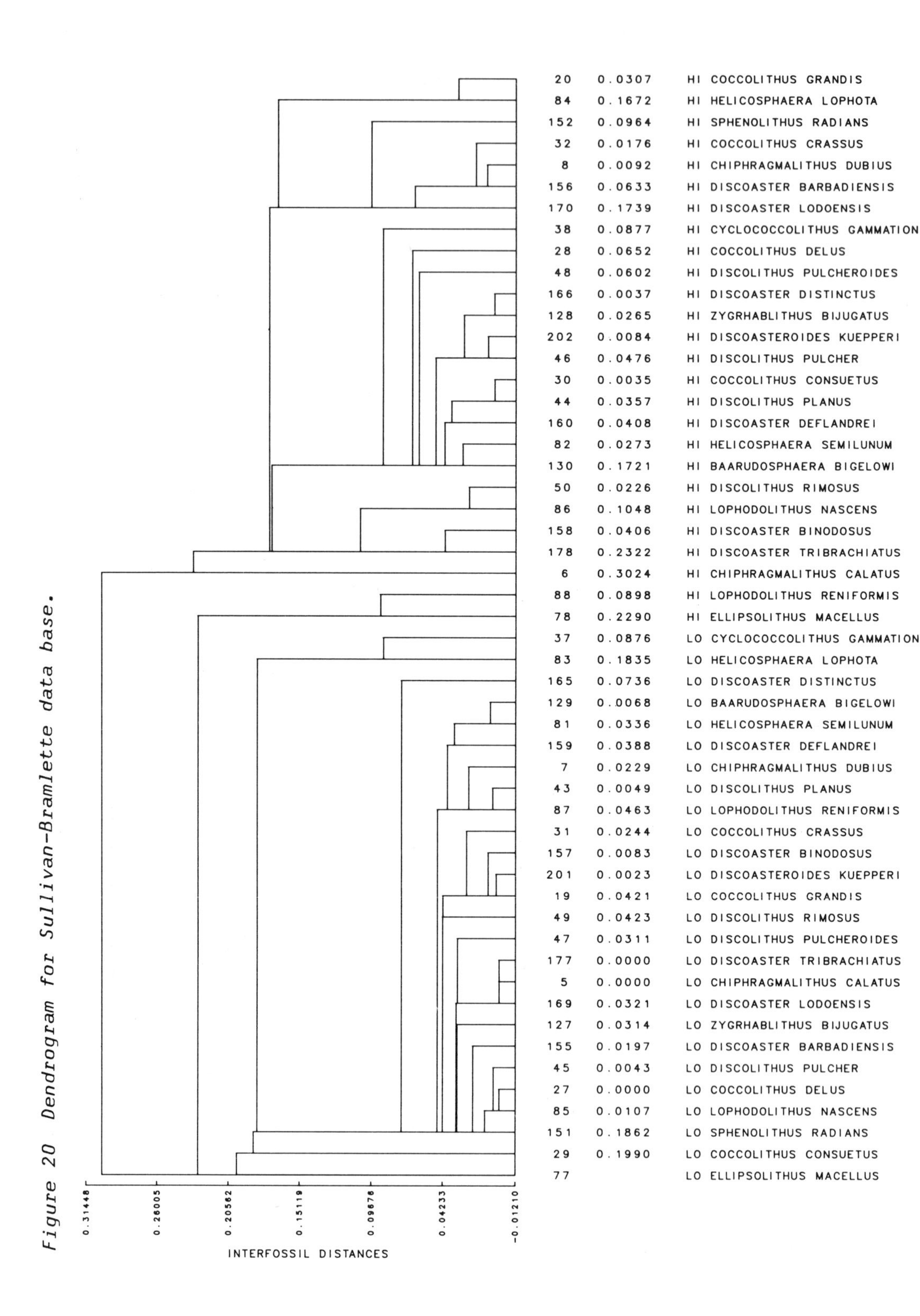

Figure 20 Dendrogram for Sullivan-Bramlette data base.

(8) Corliss data base

Paleogene DSDP Benthic Foraminifers

A. Highest occurrences only

RASC input format for sequence data.

Total number of sections = 6.

Total number of events in dictionary = 68.

Largest number of events in single section = 23.

Only events that occur in k_c = 3 or more sections were used in RASC run. For ranking (presorting followed by modified Hay method), all pairs of events were used. For scaling (dendrogram after final reordering), only pairs of events that occur in m_{c2} = 2 or more sections were used. Optimum sequence and dendrogram are shown in Figures 21 and 22.

Tops (exits) and bottoms (entries) were analyzed separately for this data set because the combined data set produced biased results. (Several exits were located too low with respect to the entries in their immediate vicinity; cf. discussion of bias in mixed data sets in 'Presorting and Ranking by Harper', Chapter II.5.)

OPTIMUM FOSSIL SEQUENCE

SEQUENCE POSITION	FOSSIL NUMBER	RANGE	FOSSIL NAME
1	25	0 — 2	CIBICIDOIDES LAURISAE
2	61	1 — 3	TURRILINA BREVISPIRA
3	55	2 — 4	OSANGULARIA SP 1
4	50	3 — 7	OOLINA
5	12	3 — 6	BULIMINA IMPENDENS
6	35	5 — 8	GAVELINELLA MICRA
7	6	5 — 8	ASTRONONION PUSILLUM
8	17	7 — 9	BULIMINA TUXPAMENSIS
9	49	8 — 10	NUTTALLIDES TRUEMPYI

Figure 21 Optimum sequence for tops (exits) in Corliss data base.

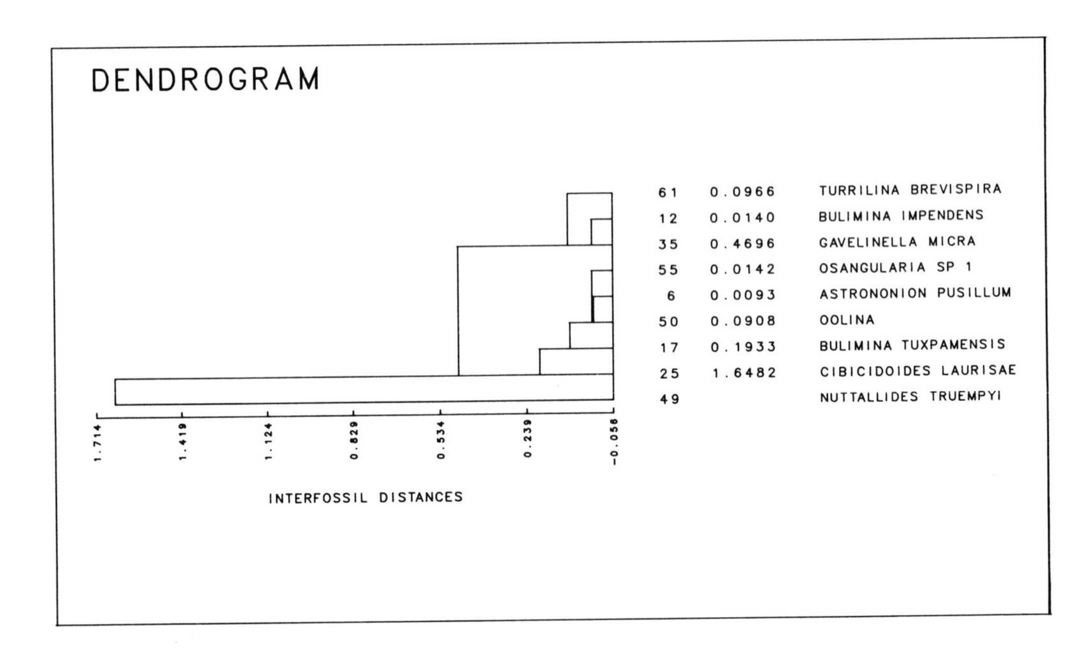

Figure 22 Dendrogram for tops (exits) in Corliss data base.

(8) Corliss data base

Paleogene DSDP Benthic Foraminifers

B. Lowest occurrences only

RASC input format for sequence data.

Total number of sections = 6.

Total number of events in dictionary = 68.

Largest number of events in single section = 32.

Only events that occur in $k_c = 4$ or more sections were used in RASC run. For ranking (presorting followed by modified Hay method), only pairs of events that occur in $m_{c1} = 2$ or more sections were used. For scaling (dendrogram after final reordering), only pairs of events that occur in $m_{c2} = 3$ or more sections were used. Optimum sequence and dendrogram are shown in Figures 23 and 24.

OPTIMUM FOSSIL SEQUENCE

SEQUENCE POSITION	FOSSIL NUMBER	RANGE	FOSSIL NAME
1	6	0 — 2	ASTRONONION PUSILLUM
2	62	1 — 3	UVIGERINA ELONGATA
3	8	2 — 4	BOLIVINA SP 1
4	9	3 — 5	BULIMINA ALAZANENSIS
5	41	4 — 8	GYROIDINOIDES SP 2
6	12	4 — 8	BULIMINA IMPENDENS
7	22	3 — 9	CASSIDULINA HAVANENSE
8	35	6 — 9	GAVELINELLA MICRA
9	44	8 — 10	LAGENA
10	39	9 — 11	GYROIDINOIDES PERAMPLA
11	54	10 — 12	OSANGULARIA MEXICANA
12	16	11 — 14	BULIMINA TRINITATENSIS
13	25	11 — 14	CIBICIDOIDES LAURISAE
14	17	13 — 15	BULIMINA TUXPAMENSIS
15	34	14 — 16	FISSURINA

Figure 23 Optimum sequence for bottoms (entries) in Corliss data base.

DENDROGRAM

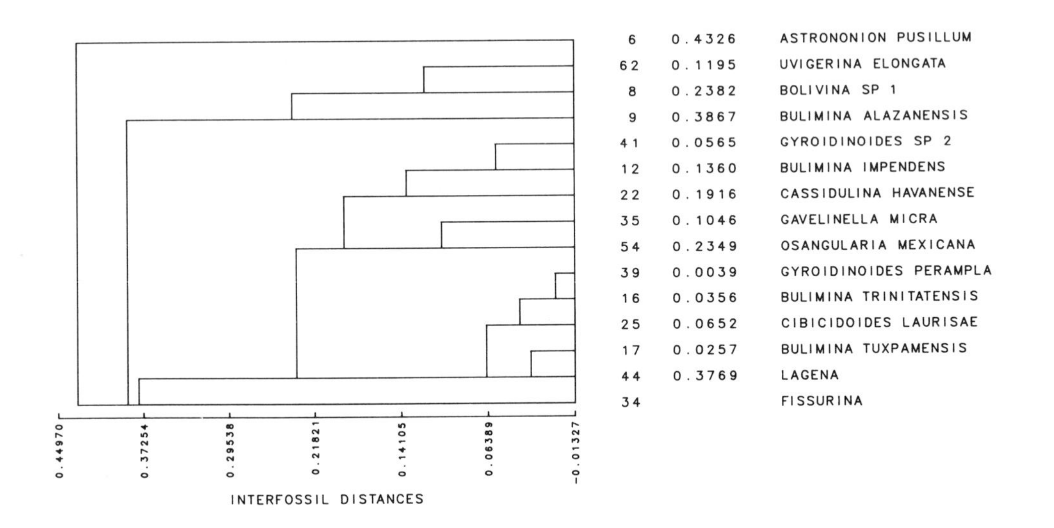

Figure 24 Dendrogram for bottoms (entries) in Corliss data base.

(9) Agterberg - Lew data base

Input for computer simulation experiments

RASC input format for sequence data.

Total number of sections = 25.

Total number of events in dictionary = 20.

Largest number of events in single section = 20.

All events in all sections were used in RASC run for ranking (presorting followed by modified Hay method), and scaling (dendrogram after final reordering). Events were not named in dictionary. Three experiments were performed with interval along RASC scale set equal to (A) 0.5, (B) 0.3, and (C) 0.1, respectively. Optimum sequence and dendrogram for each of these series are shown in Figures 25, 26 and 27.

This data base is part of a larger data base generated on the computer using a pseudo-random normal number generator (cf. 'Computer Simulation Experiments' in Chapter II.6, and 'Revised Normality Test' in Chapter II.7). An identical sequence of random normal numbers was used in all three experiments. Consequently, there is a general resemblance of input and output for the three series. However, the deviation from the theoretical model, consisting of equally spaced events ordered 1 to 20, increases when E(D), representing spacing between successive events in the theoretical model, decreases.

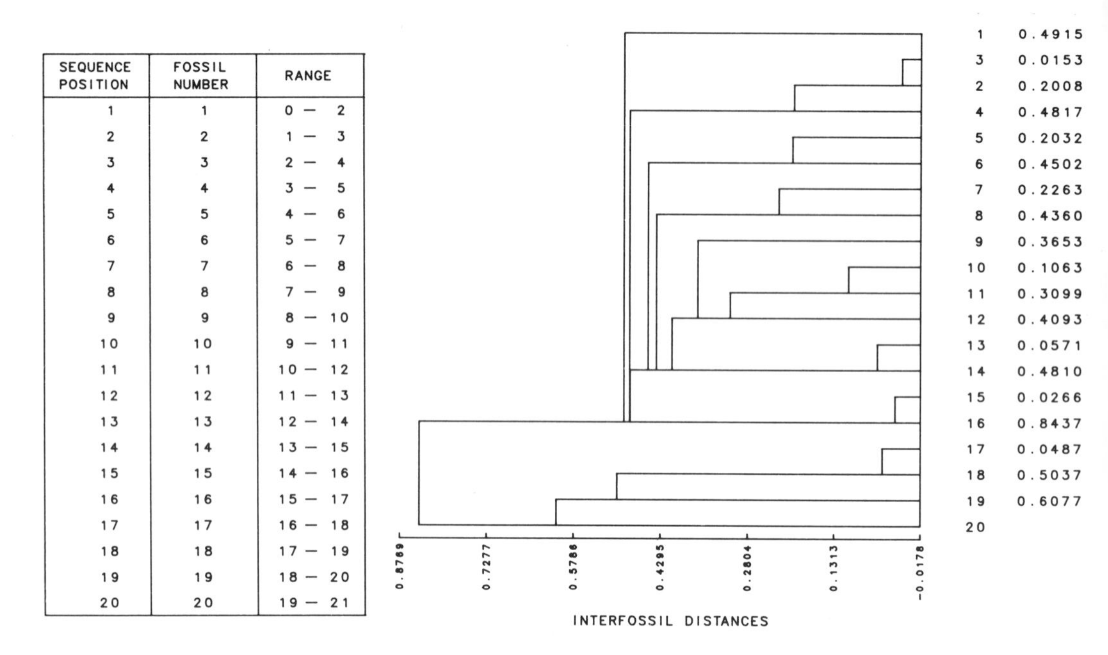

SEQUENCE POSITION	FOSSIL NUMBER	RANGE
1	1	0 — 2
2	2	1 — 3
3	3	2 — 4
4	4	3 — 5
5	5	4 — 6
6	6	5 — 7
7	7	6 — 8
8	8	7 — 9
9	9	8 — 10
10	10	9 — 11
11	11	10 — 12
12	12	11 — 13
13	13	12 — 14
14	14	13 — 15
15	15	14 — 16
16	16	15 — 17
17	17	16 — 18
18	18	17 — 19
19	19	18 — 20
20	20	19 — 21

Figure 25 Optimum sequence and dendrogram for sample drawn at random from population (theoretical model of equally spaced events labelled 1 to 20 along RASC scale with E(D) = 0.5).

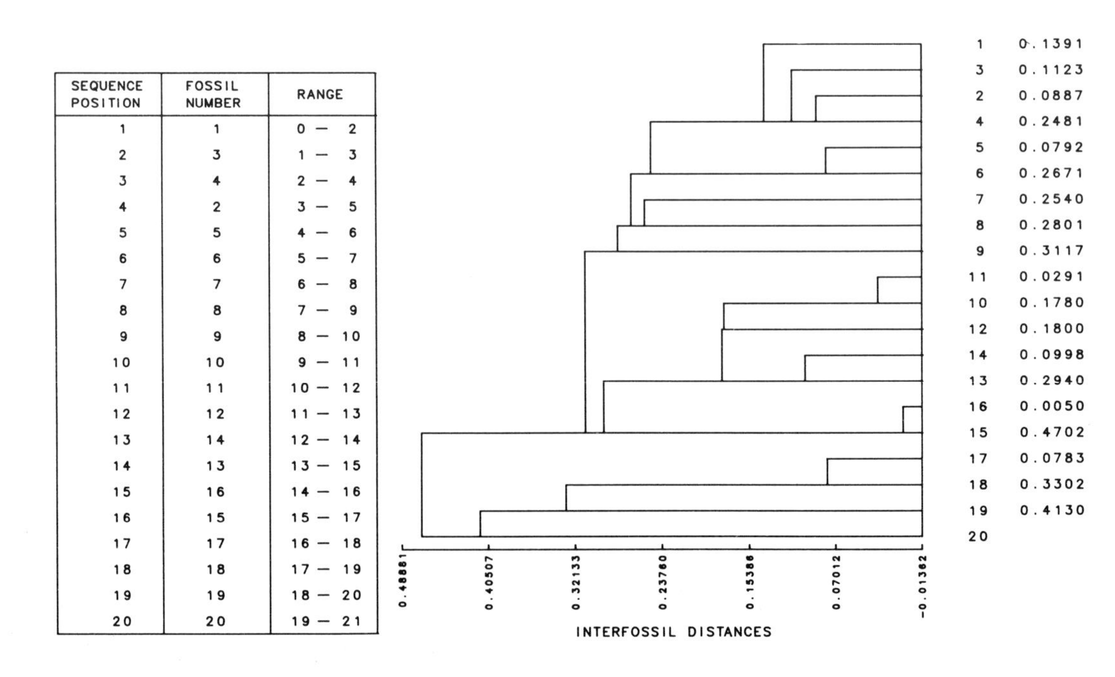

SEQUENCE POSITION	FOSSIL NUMBER	RANGE
1	1	0 — 2
2	3	1 — 3
3	4	2 — 4
4	2	3 — 5
5	5	4 — 6
6	6	5 — 7
7	7	6 — 8
8	8	7 — 9
9	9	8 — 10
10	10	9 — 11
11	11	10 — 12
12	12	11 — 13
13	14	12 — 14
14	13	13 — 15
15	16	14 — 16
16	15	15 — 17
17	17	16 — 18
18	18	17 — 19
19	19	18 — 20
20	20	19 — 21

Figure 26 Optimum sequence and dendrogram for computer simulation experiments of Figure 25 repeated with E(D) = 0.3.

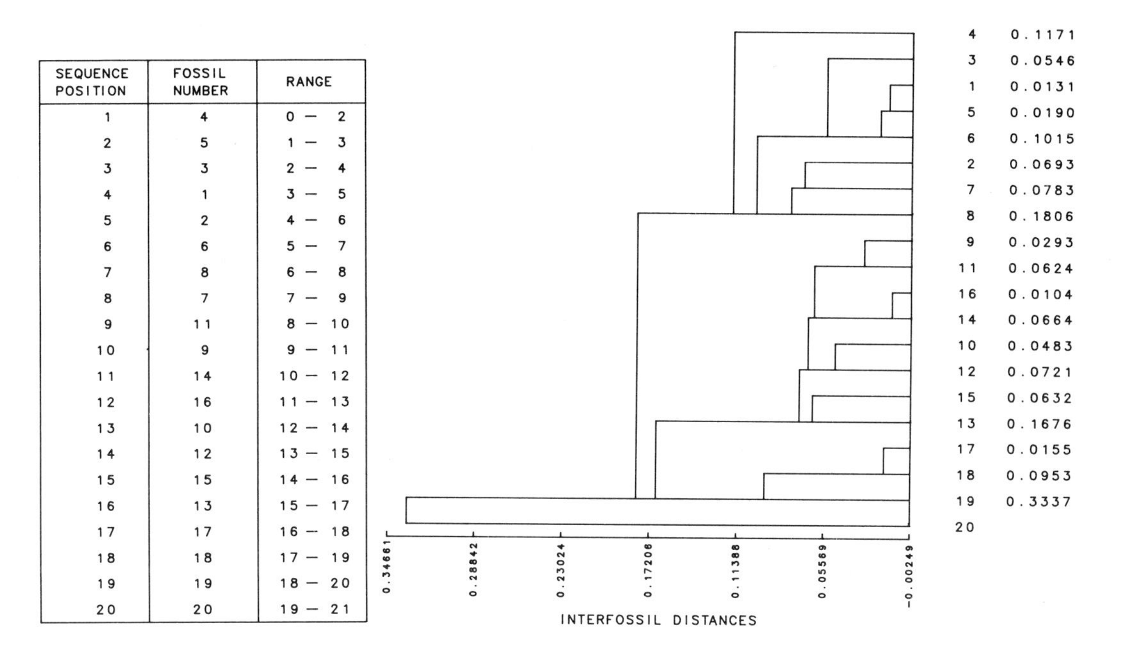

SEQUENCE POSITION	FOSSIL NUMBER	RANGE
1	4	0 — 2
2	5	1 — 3
3	3	2 — 4
4	1	3 — 5
5	2	4 — 6
6	6	5 — 7
7	8	6 — 8
8	7	7 — 9
9	11	8 — 10
10	9	9 — 11
11	14	10 — 12
12	16	11 — 13
13	10	12 — 14
14	12	13 — 15
15	15	14 — 16
16	13	15 — 17
17	17	16 — 18
18	18	17 — 19
19	19	18 — 20
20	20	19 — 21

Figure 27 *Optimum sequence and dendrogram for computer simulation experiment of Figure 25 repeated with E(D) = 0.1.*

```
1************************
**
**
** (1) GRADSTEIN - THOMAS DATA BASE
**     CENOZOIC LABRADOR SHELF FORAMINIFERS
**
**
************************
```

DICTIONARY

1 NEOGLOBOQUADRINA PACHYDERMA
2 GLOBIGERINA APERTURA
3 GLOBIGERINA PSEUDOBESA
4 GLOBOROTALIA INFLATA
5 GLOBOROTALIA CRASSAFORMIS
6 NEOGLOBOQUADRINA ACOSTAENSIS
7 GLOBIGERINOIDES RUBER
8 ORBULINA UNIVERSA
9 FURSENKOINA GRACILIS
10 UVIGERINA CANARIENSIS
11 NONIONELLA PIZARRENSE
12 EHRENBERGINA SERRATA
13 HANZAWAIA CONCENTRICA
14 TEXTULARIA AGGLUTINANS
15 GLOBIGERINA PRAEBULLOIDES
16 CERATOBULIMINA CONTRARIA
17 ASTERIGERINA GURICHI
18 SPIROPLECTAMMINA CARINATA
19 GLOBIGERINOIDES SP
20 GYROIDINA GIRARDANA
21 GUTTULINA PROBLEMA
22 COSCINODISCUS SP3
23 COSCINODISCUS SP4
24 TURRILINA ALSATICA
25 COARSE ARENACEOUS SPP.
26 UVIGERINA DUMBLEI
27 EPONIDES UMBONATUS
28 CIBICIDOIDES SP5
29 CYCLAMMINA AMPLECTENS
30 CIBICIDOIDES BLANPIEDI
31 PTEROPOD SP1
32 AMMOSPHAEROIDINA SP1
33 TURBOROTALIA POMEROLI
34 MARGINULINA DECORATA
35 SPIROPLECTAMMINA DENTATA
36 PSEUDOHASTIGERINA WILCOXENSIS
37 ACARININA AFF PENTACAMERATA
38 LENTICULINA SUBPAPILLOSA
39 ALABAMINA WILCOXENSIS
40 BULIMINA ALAZANENSIS
41 PLECTOFRONDICULARIA SP1
42 CIBICIDOIDES ALLENI
43 BULIMINA MIDWAYENSIS
44 CIBICIDOIDES AFF WESTI
45 BULIMINA TRIGONALIS
46 MEGASPORE SP1
47 PLANOROTALITES PLANOCONICUS
48 ANOMALINA SP5
49 OSANGULARIA EXPANSA
50 SUBBOTINA PATAGONICA
51 ACARININA PRIMITIVA
52 ACARININA SOLDADOENSIS
53 UVIGERINA BATJESI
54 SPIROPLECTAMMINA NAVARROANA
55 GAVELINELLA BECCARIIFORMIS
56 GLOMOSPIRA CORONA
57 SPIROPLECTAMMINA SPECTABILIS LCO
58 EPONIDES SP8
59 RZEHAKINA EPIGONA
60 PLANOROTALITES COMPRESSUS
61 SUBBOTINA PSEODOBULLOIDES
62 GAVELINELLA DANICA
63 NODOSARIA SP11
64 CASSIDULINA ISLANDICA
65 COSCINODISCUS SP1
66 COLEITES RETICULOSUS
67 SCAPHOPOD SP1
68 SPIROPLECTAMMINA SPECTABLIS LO
69 NODOSARIA SP8
70 ALABAMINA WOLTERSTORFFI
71 EPISTOMINA ELEGANS
72 CYCLOGYRA SP3
73 EPONIDES SP3
74 EPONIDES SP5
75 LENTICULINA ULATISENSIS
76 CASSIDULINA SP
77 ELPHIDIUM SP
78 UVIGERINA PEREGRINA
79 GLOBIGERINA TRIPARTITA
80 CYCLAMMINA CANCELLATA
81 GLOBIGERINA VENEZUELANA
82 GLOBIGERINA LINAPERTA
83 PLANOROTALITES PSEUDOSCITULUS
84 GLOBIGERINA YEGUAENSIS
85 PSEUDOHASTIGERINA MICRA
86 TURRILINA BREVISPIRA
87 BULIMINA AFF. JACKSONENSIS
88 SIPHOGENEROIDES ELEGANTA
89 MOROZOVELLA SPINULOSA
90 ACARININA DENSA
91 RADIOLARIANS
92 MOROZOVELLA CAUCASICA
93 ACARININA AFF. BROEDERMANNI
94 GLOBIGERINATHEKA KUGLERI
95 ARAGONIA VELASCOENSIS
96 ACARININA INTERMEDIA WILCOXENSIS
100 GLOBIGERINA RIVEROA
109 CASSIDULINA CURVATA
110 GLOBIGERINA BULLOIDES
111 PARAROTALIA SP1
112 MARGINULINA BACHEI
113 GLOBOROTALIA MENARDII GROUP
114 GLOBIGERINOIDES SACCULIFER
115 GLOBOROTALIA OBESA

116 ORBULINA SUTURALIS
117 SPHAEROIDINA BULLOIDES
118 EPISTOMINA SP5
119 SPHAEROIDINELLA SUBDEHISCENS
120 GLOBOROTALIA SIAKENSIS
121 GLOBIGERINA NEPENTHES
122 SPHAEROIDINELLOPSIS SEMINULINA
123 GLOBIGERINOIDES TRILOBUS
124 GLOBOQUADRINA DEHISCENS
125 "GLOBOROTALIA" CONTINUOSA
126 GLOBIGERINOIDES OBLIQUUS
127 GLOBIGERINITA NAPARIMAENSIS
128 GLOBOROTALIA PRAEMENARDII
130 SIPHONINA ADVENA
131 CIBICIDOIDES TENELLUS
132 "GLOBOROTALIA" OPIMA NANA
133 LENTICULINA SP3
134 LENTICULINA SP4
135 GLOBIGERINA SP40
136 MELONIS BARLEANUM
137 GLOBIGERINOIDES PRIMORDIUS
138 GLOBIGERINA ANGUSTIUMBILICATA
139 "GLOBOROTALIA" OPIMA OPIMA
140 ROTALIATINA BULIMINOIDES
141 PLANULINA RENZI
142 GYROIDINA SOLDANII MAMILLIGERA
143 UVIGERINA GALLOWAY
144 GLOBOROTALIA CERROAZULENSIS
145 ANOMALINOIDES ALLENI
146 SUBBOTINA EOCAENA
147 CATAPSYDRAX AFF. DISSIMILIS
148 GLOBIGERINATHEKA INDEX
149 GLOBIGERINATHEKA TROPICALIS
150 GLOBIGERINA GORTANII
151 BULIMINA BRADBURYI
153 BULIMINA COOPERENSIS
154 ANOMALINOIDES MIDWAYENSIS
155 ANOMALINOIDES GROSSERUGOSA
156 SUBBOTINA FRONTOSA
157 TRITAXIA SP3
158 SUBBOTINA INAEQUISPIRA
159 MOROZOVELLA ARAGONENSIS
160 ACARININA PSEUDOTOPILENSIS
161 PLANOROTALITES AUSTRALIFORMIS
162 MOROZOVELLA AEQUA
164 NUTTALIDES TRUMPYI
166 MOROZOVELLA SUBBOTINAE
167 MOROZOVELLA FORMOSA GRACILIS
169 EPISTOMINELLA TAKAYANAGII
172 PSEUDOHASTIGERINA SP
173 ANOMALINA SP1
175 ALLOGROMIA SP
176 ALLOMORPHINA SP1
177 BOLIVINA DILATATA
179 GLOBOROTALIA SCITULA PRAESCITULA
180 GYROIDINA SP4
181 CYCLOGYRA INVOLVENS
182 PLECTOFRONDICULARIA SP3
184 GYROIDINA OCTOCAMERATA
187 CIBICIDOIDES GRANULOSA
188 PLEUROSTOMELLA SP1
190 ANOMALINOIDES ACUTA
191 "GLOBIGERINA" AFF. HIGGINSI
194 PLANOROTALITES CHAPMANI
196 OSANGULARIA SP4
201 SEISMIC EVENT #1
202 SEISMIC EVENT #2
203 SEISMIC EVENT #3
204 SEISMIC EVENT #4
206 EPONIDES POLYGONUS
210 LOXOSTOMOIDES APPLINAE
211 HANTKENINA SP
213 ARENOBULIMINA SP2
216 GLOBIGERINOIDES SICANUS
217 GLOBOROTALIA SCITULA
218 MARGINULINA AMERICANA
219 MARTINOTIELLA COMMUNIS
220 CIBICIDOIDES WUELLERSTORFFI
221 GLOBIGERINOIDES SUBQUADRATUS
222 GLOBOQUADRINA ALTISPIRA
223 GLOBIGERINA CIPEROENSIS
224 UVIGERINA MEXICANA
225 GLOBIGERINA AFF. AMPLIAPERTURA
226 GLOBIGERINA SENNI
227 CIBICIDOIDES AFF. TUXPAMENSIS
228 CASSIDULINA TERETIS
230 BULIMINA OVATA
231 UVIGERINA RUSTICA
232 GLOBIGERINOIDES IMMATURUS
233 CATAPSYDRAX UNICAVUS
234 TRUNCAROTALOIDES AFF. ROHRI
235 SUBBOTINA BOLIVARIANA
236 EPONIDES SP4
237 LENTICULINA SP8
238 CIBICIDOIDES SP7
239 NONIONELLA LABRADORICA
240 ELPHIDIUM CLAVATUM
241 GLOBOROTALIA TRUNCATULINOIDES
242 GLOBOROTALIA FOHSI GROUP
243 GLOBIGERINA DECAPERTA
244 GAUDRYINA SP10
245 PRAEORBULINA GLOMEROSA
246 GLOBIGERINATELLA INSUETA
247 GLOBIGERINOIDES ALTIAPERTURA
248 "GLOBOROTALIA" AFF. INCREBESCENS
249 GLOBIGERINATHEKA SEMIINVOLUTA
250 VULVULINA JARVISI
251 ANOMALINA SP4
252 MOROZOVELLA AFF. QUETRA
253 SUBBOTINA TRILOCULINOIDES
254 PLANOROTALITES PSEUDOMENARDII
255 MOROZOVELLA CONICOTRUNCATA
256 "MOROZOVELLA" AFF. PUSILLA
257 CHILOGUEMBELINA SP
258 TAPPANINA SELMENSIS
259 AMMODISCUS LATUS
260 HAPLOPHRAGMOIDES KIRKI
261 HAPLOPHRAGMOIDES WALTERI
262 KARRERIELLA APICULARIS
263 AMMOBACULITES AFF POLYTHALAMUS
264 KARRERIELLA CONVERSA
265 ASTERIGERINA GURICHI (PEAK)
266 GLOBOROTALIA PUNCTICULATA
267 GLOBOROTALIA HIRSUTA
268 GLOBOROTALIA AFF KUGLERI
269 NEOGLOBQUADRINA ATLANTICA
270 CIBICIDOIDES GROSSA
271 GLOBOROTALIA INCREBESCENS
272 GLOBOQUADRINA BAROEMOENSIS
273 BULIMINA GRATA
274 GAUDRYINA AFF HILTERMANNI
275 PARAROTALIA SP2

```
1************************
**
**
**  (1)  GRADSTEIN - THOMAS DATA BASE
**       CENOZOIC LABRADOR SHELF FORAMINIFERS
**
**
************************

SEQUENCE DATA

BJARNI H-81
  16   67   20  -21   18  -69  -70  -71   15   24   25   34   29 -261   42  -74  -41  -32   30 -264
 -75   57   46   56

CARTIER D-70
  16   18   15   21  -70   67   69   24 -172   25  259   34  260 -261  118 -173  -85  -29 -263   46
 -42  -32   35   41  -57   54   56  175  -59

FREYDIS B-87
  16  181  -67  -21  -18   20   69  -27   15  -70   25  190  -34 -206  -42  -74 -173  260   29 -261
 -45   33  -81  -41  -75 -210  -32  211  -85  -94   57  -88  -86  -30  -46  -35   56   54  213  -55
  59

GUDRID H-55
  10  -17  265   20  -21  -18  -16   24   15  -25   33  259   40  -34   84  -90  -36   37 -260 -261
  29   35   45  -74   42   57  -88  -30   32   46  -50   56  -59  -54   55

INDIAN HARBOUR M-52
   1   -3   -4   -5   -8    9  -10  269    2   -7    6  -18   15  -20  -16   17   24  -25   26  -27
 -28  259  261   30  260  -32   33   34  -35  263  -36  -39   29  -40  -41  -42   86   37  -38   44
  45  -46  -47   49   57  -54  -50  -52   55  -56   59   60  -61  -62

KARLSEFNI H-13
 228   67   25   41 -118   69  260 -261   68  -39   53 -206 -173   29   86  -30  -63  -34   46 -264
 230  -44  -42   96  -36  164  -50   52   45  -54   56   55  -62   61 -253  258

LEIF M-48
 228  -77  -10  181   16  -67   15   20  -21  -18   70   69   85  -24   25 -238   42   29  260  -34
  57  -74 -118 -263   30  -41   46  -56  -54

LEIF E-38
 228  -77 -270   17   67  -16   18  -21   20

SNORRI J-90
  77  228   16   67   15  -21   18   25   57 -263  -32  -34   29 -260  -53  -41  -30  -36 -173   27
 -46  118  264  230   86  -63   42   45   56   59  -54

HERJOLF M-92
  67   18  -15  -20  -16   78   70   25 -259   85 -145  -71  -40   45  -35 -263 -261  -34   29   41
 -53  -30  -32 -264   86   57   54   46  190   47 -154  -56   55   60   59

BONAVISTA C-99
  76  -77   10   17  -16   21   25  -20   18   79  -15  259   24  -26   81  -33   82   83   40   84
 -27   29 -261   32 -263   85  -86  -87 -264   41  -34   57   88  -42  -90   89  159  -92  -93  -94
  56  -50  -30   47  -96  -36   46

DOMINION O-23
 177 -109 -169   11   -9   17   10 -117  -78  112   18  179  -16  -15  -71  122  180   26 -123 -137
  14 -136   27   20   21 -181  201   24   25   34  264 -260  -38  259  142  -81  184  -82  -30 -146
  69 -263  202   32   68  187   49 -188 -147 -190 -140   29  -40  191 -156  151  250 -226   36  -44
 194  -90  -57  203   50  -47 -158  161  -52  -46   37 -159 -162  196   45 -230  164
```

EGRET K-36
17 26 16 20 -21 -18 -71 -15 24 27 -42 202 69 82

OSPREY H-84
17 18 -20 15 -16 26 -181 81 82 84 -147 -69 -148 90 -89 -33 -187 -234 -34 -244
52 -51 -162 -159 -166 -50 -93

CUMBERLAND B-55
76 228 -1 17 10 -11 -9 -109 -71 265 -16 -20 18 15 -119 117 219 26 24 25
-259 132 42 261 41 84 29 32 226 144 49 57 -36 90 52 -54 161 -93 -96 -151
-164 -157 46 -50 -159 55 -56 -254 -194

EGRET N-46
11 -16 -18 14 -27 -71 26 -20 202 15 -24 172

ADOLPHUS D-50
10 71 218 16 -136 18 20 179 201 26 15 -81 -69 24 -33 -202 259 -25 263 82
85 -261 203 147 -260 68 32 40 30 49 -29 144 -90 -156 -37 -89 234 160 -93 36
161 -164 50 -230 54 57 -56 55 194 -95

HIBERNIA O-35
17 201 26 18 -20 16 275 24 -71 72 27 140 202 34 -81 203 259 -29 -25 15
-28 57 -260 -261 204 40 -32 91

FLYING FOAM I-13
9 -10 16 71 17 275 -265 18 -110 70 26 -15 -81 201 24 -20 -27 25 259 202
263 -32 -34 260 -261 264 29 -57 -203 54 46 36 41 230

BLUE H-28
77 1 4 267 269 110 -10 -64 266 124 -125 -6 -113 122 26 -71 268 -2 147 -27
29 -261 -81 -150 82 -15 -118 -138 146 -84 32 -79 -172 -53 -68 164 -190 42 86 -151
33 -94 -57 37 90 -52

HARE BAY H-31
228 -270 77 1 10 136 16 70 -15 24 18 -20 -25 260 -263 259 29 -233 -69 -118
-32 -81 68 49 41 227 93 -42 -96 50 57 66 -54 55 -161 -56 59 253 -255 -46

HIBERNIA K-18
201 16 -18 -20 -71 -72 24 -27 15 -34 81 202 259 147 25 -29 -260 30 -57 -203
32 263 36 -40 -63 45 -91 -155 -230 204

HIBERNIA B-08
17 26 18 -20 16 15 -27 -71 72 81 -25 24 146 -259 32 -57 -147 -260 -261 -263
36 -40 45 63 47 -144 -194 -54 -91 -230 56 55 -61 52 -59 -61 -96 -253

HIBERNIA P-15
17 18 -265 16 20 -100 26 201 15 71 72 69 202 81 27 147 24 25 -32 -57
-259 -260 261 29 203 53 -263 40 45 204

```
1************************
 **
 **
 **  (1)  GRADSTEIN - THOMAS DATA BASE
 **       CENOZOIC LABRADOR SHELF FORAMINIFERS
 **
 **
 ************************

 DEPTH FILE     M = METRES,   F = FEET
                FIRST NUMBER AFTER 'M' OR 'F' IS ROTARY TABLE HEIGHT
                SECOND NUMBER IS WATER DEPTH

 BJARNI H-81
 F 40.0  456.0
   2860.  3360.  3460.  3560.  4060.  4260.  4860.  5060.  5360.  5560.  5660.  5960.  6060.  6590.

 CARTIER D-70
 F 41.0 1017.0
   2270.  2520.  2960.  3360.  3460.  3660.  3870.  4060.  4170.  4760.  4860.  4960.  5160.  5260.
   5470.  5670.  5870.  6070.

 FREYDIS B-87
 F 41.0  586.0
   1000.  1840.  2470.  2560.  2770.  3600.  3700.  3810.  3910.  4180.  4270.  4360.  4960.  5050.
   5170.  5230.

 GUDRID H-55
 F 40.0  982.0
   1660.  1960.  2010.  3500.  4460.  4940.  5810.  6020.  6110.  6200.  6410.  6530.  6590.  6680.
   6800.  6860.  6950.  7860.  8550.

 INDIAN HARBOUR M-52
 F 98.0  649.0
   1740.  1890.  1950.  2040.  2130.  2460.  2550.  3600.  4140.  5400.  5590.  5780.  6370.  6970.
   7660.  7760.  7860.  7960.  8140.  8230.  8860.  9130.  9560.  9940. 10090. 10230.

 KARLSEFNI H-13
 F 40.0  573.0
   1760.  6260.  6760.  7060.  7380.  8660.  8760.  8960.  9060.  9160.  9260.  9460.  9660. 10260.
  10952. 11663. 12460. 12560. 12760. 12960.

 LEIF M-48
 F 40.0  542.0
   1300.  2380.  2470.  3290.  3380.  3480.  3790.  3880.  4630.  4720.  4850.  5150.  5250.  5340.
   5620.

 LEIF E-38
 F 40.0  550.0
   1210.  1310.  2710.  3230.  3520.

 SNORRI J-90
 F 37.0  462.0
   1260.  2940.  4140.  4410.  5310.  5580.  6480.  6930.  7020.  7110.  7290.  7310.  7650.  7740.
   8100.  9020.  9620.  9820.

 HERJOLF M-92
 F 88.0  456.0
   3750.  3840.  4020.  4110.  4300.  4650.  5640.  5740.  6000.  6180.  6450.  6540.  6630.  6720.
   7060.  7250.  7610.  7790.

 BONAVISTA C-99
 F 42.6 1080.0
   1860.  2820.  2910.  5820.  5827.  6660.  7380.  7600.  7800.  8400.  8730.  9400.  9800. 10000.
  10100. 10200. 10686. 10910. 11010. 11110. 11210. 11310. 11410. 11510. 11910.
```

```
DOMINION O-23
F 98.0  530.0
  1380.  1560.  1650.  1740.  1830.  1920.  2010.  2100.  2670.  2940.  3030.  3120.  3210.  3330.
  4250.  4380.  4560.  5280.  5640.  5820.  5910.  6090.  6270.  6800.  6900.  7080.  7270.  7320.
  7570.  7670.  7820.  7920.  8320.  8770.  8800.  9070.  9170.  9320.  9570.  9770. 10230.

EGRET K-36
F 98.0  223.0
   860.  1040.  1340.  1520.  1580.  1610.  1640.  1940.  2240.

OSPREY H-84
F 85.0  201.0
  1190.  1370.  1640.  1820.  1910.  2180.  2360.  2450.  2540.

CUMBERLAND B-55
F 98.0  639.0
   920.  2190.  2280.  2370.  2550.  2910.  3450.  3540.  3810.  4890.  6330.  6600.  7700.  7800.
  7900.  8710.  9100.  9700.  9800.  9900. 10140. 10440. 10560. 10680. 10830. 10930. 11030. 11830.

EGRET N-46
F 98.0  222.0
  1080.  1320.  1410.  1555.  1590.  1950.

ADOLPHUS D-50
F 98.0  377.0
  1140.  1410.  1500.  1590.  1680.  1980.  2700.  2900.  3060.  3660.  4200.  4440.  4562.  4920.
  4950.  5400.  5420.  5550.  5778.  5896.  6018.  6200.  6646.  6975.  7596.  7917.  8020.  8258.
  8384.  8520.  8700.  8726.

HIBERNIA P-15
M 11.3   80.2
   255.   275.   310.   410.   550.   620.   695.   720.   915.   945.   960.   975.  1005.  1035.
  1075.  1125.  1185.  1195.  1200.  1315.  1345.  1375.  1400.

FLYING FOAM I-13
F 98.0  300.0
   990.  1170.  1260.  1530.  1620.  1890.  2340.  2520.  2700.  3150.  3700.  4230.  4400.  4500.
  4770.  4950.  5300.  5490.  5650.  6200.  6390.  6570.

BLUE H-28
M 15.   1486.0
  2090.  2150.  2240.  2540.  2600.  2720.  2870.  3170.  3220.  3310.  3550.  3640.  3830.  3920.
  4070.  4340.  4490.  4635.  4670.  4700.  4730.  4760.

HARE BAY H-13
M 24.2  239.1
   393.   420.   450.   870.   990.  1110.  1140.  1170.  1430.  1550.  1910.  2060.  2210.  2400.
  2630.  2660.  2840.  2870.  2960.  3020.  3080.  3140.  3200.

HIBERNIA K-18
M 28.5   78.3
   700.   920.   950.   980.  1035.  1040.  1065.  1185.  1245.  1275.  1305.  1395.  1485.  1515.
  1580.

HIBERNIA B-08
M 27.4   83.0
   540.   600.   690.   720.   810.  1020.  1105.  1165.  1255.  1345.  1525.  1565.  1595.  1655.
  1685.  1710.  1745.

HIBERNIA O-35
M 24.0   80.0
   520.   540.   550.   580.   610.   640.   670.   700.   730.   790.   840.   910.  1025.  1090.
  1120.  1150.  1240.  1265.  1295.
```

```
1*************************
 **
 **
 **  (2)  GRADSTEIN DATA BASE
 **       LATE CRETACEOUS GRAND BANKS FORAMINIFERS
 **
 **
 *************************
```

DICTIONARY

1 GLOBOTRUNCANA CONTUSA
2 FRUCTICOSE PSEUDOTEXTULARIA
3 GLOBOTRUNCANA STUARTI
4 GLOBOTRUNCANA ARCA
5 STENSIOINA POMMERANA
6 GLOBOTRUNCANELLA HAVANENSIS
7 PSEUDOTEXTULARIA ELEGANS
8 GLOBIGERINELLOIDES MESSINAE
9 GLOBOROTALITES MICHELINIANUS
10 ARENOBULIMINA AMERICANA
11 GLOBOTRUNCANA FORNICATA
12 GLOBOTRUNCANA STUARTIFORMIS
13 GLOBOTRUNCANA LINNEIANA
14 GLOBOTRUNCANA CRETACEA
15 GLOBOTRUNCANA MARGINATA
16 RUGOGLOBIGERINA RUGOSA
17 GLOBOROTALITES AFF. MULTISEPTUS
18 PRAEBULIMINA SP
19 ARAGONIA MATERNA KUGLERI
20 GLOBOTRUNCANA CORONATA
21 REUSSELLA SZAJNOCHAE
22 SIGALIA DEFLAENSIS
23 STENSIOINA EXCULPTA
24 HEDBERGELLA AMABILIS
25 HEDBERGELLA SPP
26 GLOBOTRUNCANA ANGUSTICARINATA
27 ROTALIPORA CUSHMANI
28 ROTALIPORA GREENHORNENSIS
29 ROTALIPORA DECKERI
30 PRAEGLOBOTRUNCANA STEPHANI
31 PRAEGLOBOTRUNCANA TURBINATA
32 GRANDES HEDBERGELLES
33 ROTALIPORA APPENNINICA
34 PRAEGLOBOTRUNCANA DELRIOENSIS
35 DOROTHIA AFF. FILIFORMIS
48 HEDBERGELLA PLANISPIRA
50 AGGLUTINATED SPP ABUNDANT
53 GLOBOTRUNCANA CONICA
54 GLOBOTRUNCANA CONCAVATA
55 GLOBOTRUNCANA CARINATA
56 GAUDRYINA AUSTINANA
57 GLOBOTRUNCANA VENTRICOSA
58 GLOBIGERINELLOIDES ASPERA
59 RUGOGLOBIGERINA ROTUNDIDORSATA
70 HEDBERGELLA BOSQUENSIS
71 GLOBOTRUNCANA SCHNEEGANSI
72 GLOBOTRUNCANA RENZI
95 GLOBOTRUNCANA FALSO STUARTI
104 GLOBOTRUNCANA ROSETTA
105 LOXOSTOMA GEMMUM
106 ARENOBULIMINA DORBIGNYI
107 KYPHOPYXA CHRISTNERI
108 GAVELINELLA MINIMA
109 COARSE AGGLUTINATED SPP
110 GLOBOTRUNCANA IMBRICATA
111 GLOBOTRUNCANA PRIMITIVA
112 GLOBOTRUNCANA HELVETICA
113 GLOBOTRUNCANA SIGALI
114 GLOBOTRUNCANA MARIANOSI
115 LINGULOGAVELINELLA TURONICA
116 HEDBERGELLA PARADUBIA
128 GAVELINELLA TOURAINENSIS
129 PRAEGLOBOTRUNCANA HAGNI
130 PRAEGLOBOTRUNCANA DIFFORMIS
131 GAVELINOPSIS CENOMANICA
139 GAUDRYINA PYRAMIDATA
140 MARSONELLA OXYCONA
141 HEDBERGELLA SIMPLEX
168 GLOBOTRUNCANA GANSSERI
169 FAVUSELLA WASHITENSIS
170 PRAEGLOBOTRUNCANA SP
173 PLANULINA TAYLORENSIS
174 EPISTOMINA STELLIGIRA ALVEOLATA
175 VAGINULINA TEXANA
178 NEOFLABELLINA RUGOSA
184 PETREL LIMESTONE
204 SEISMIC EVENT #4

```
1************************
 **
 **
 **  (2)  GRADSTEIN DATA BASE
 **       LATE CRETACEOUS GRAND BANKS FORAMINIFERS
 **
 **
*************************

SEQUENCE DATA

BONAVISTA-C66
   4  -12  -13  -53  -11   -6  -16   54  -55  -17  -56  -23

CUMBERLAND B-55
  21  -16   -2  -13  -11   -6   -5   70  -24  -71  -20  -54  -72

ADOLPHUS D-50
   1   -2   -3   -4   -5   -6   -7   -8   -9  -10   11  -12  -13  -14  -15  -16  -17   18   19  -20
 -21   22  -23  -24  -25  -26  184   27  -28  -29  -30  -31  -32   33

BLUE H-28
   4  -13  -12  -57   -3  -21   -7  -58   -6  -59   33  -30  -48  -50  -35

A GABRIEL C-60
   4  -11  -12  -13  -57  -95   -3   16

HARE BAY E-21
   6  -16  -53   -4   -3  -12  -13 -104   -7   -5 -105   -8  106  107

EGRET K-36
 204    4   55  108   72  -30  112 -111   71  -15 -109 -128 -110 -113 -114   31 -129 -130 -115  -54

HIBERNIA O-35
  93   13  -14  -20  -15  -72  -56 -108  113 -111  112 -109   30  -71 -141  131  -31

EGRET N-46
   9   -8  -13   -4   -6  204   11   14   22  -15  108  -54  109 -110  -72 -111  112  -71 -113  -25
-114  -70  -31 -115 -116   30

PUFFIN B-90
   3   -4 -105   -5   12   57  -11   13   14   20  -15  -26   23   54  -55   72 -113  109

KITTIWAKE P-11
   4   -6  -13   -5   -1  -16    8  168   12  -14   11  -15   23  -24   54  -22   72  -30  131  -34
  27  -28  169

HERMINE E-94
   4   -6   -3 -168 -105   -5   -7   16   -8   13  -14   26  -15   12   20   70   54   71  170

PETREL A-62
   4   14   15   13   72   27   33 -170  131  169
```

HERON H-73
8 14 -11 -107 17 -26 23 -20 24 70 55 72 184 27 -31 131 169

BITTERN M-62
16 105 -5 6 14 22 54 72

EIDER M-75
8 14 -107 -173 174 15 -175 56 -13 9 20 -26 -11 -70 54 112 -110 -30 -31

JAEGER A-49
4 -105 -16 -5 107 -173 -15 -11 -13 -178 -174 175 56 54

CORMORANT K-83
14 -13 -11 112 -30 -31 -110 -115 -24 -108 27 -28 -131

MURRE G-67
204 11 -13 -15 -14 108 -109 -110 115 -70 30 -31 112 -71 -24 131 27 -28

OSPREY H-84
173 -10 -9 -17 -11 -13 -4 -12 -5 26 -57 -53 -14 55 -56 -109 20 -15 72 -71
-54 -22 -70 -24 111 -30 131

```
1************************
**
**
**  (3)  DOEVEN DATA BASE
**       CRETACEOUS SCOTIAN SHELF NANNOFOSSILS
**
**          HI = HIGHEST OCCURRENCE
**          ST = SUBTOP (LAST CONSISTENT OCCURRENCE)
**          SB = SUBBOTTOM (FIRST CONSISTENT OCCURRENCE)
**          LO = LOWEST OCCURRENCE
**
************************
```

DICTIONARY

```
  1  HI AHMUELLERELLA OCTORADIATA
  2  HI AHMUELLERELLA REGULARIS
  3  HI ARKHANGELSKIELLA CYMBIFORMIS
  4  HI ARKHANGELSKIELLA SPECILLATA
  5  HI BRAARUDOSPHAERA AFRICANA
  6  HI BRAARUDOSPHAERA BIGELOWI
  7  HI BROINSONIA FURTIVA
  8  HI BROINSONIA LACUNOSA
  9  HI BROINSONIA MATALOSA
 10  HI BROINSONIA PARCA
 11  HI CERATOLITHOIDES ACULEUS
 12  HI CERATOLITHOIDES KAMPTNERI
 13  HI COROLLITHION ACHYLOSUM
 14  HI COROLLITHION COMPLETUM
 15  HI COROLLITHION EXIGUUM
 16  HI COROLLITHON SIGNUM
 17  HI CRETARHABDUS LORIEI
 18  HI CRIBROCORONA GALLICA
 19  HI CRIBROSPHAERELLA CIRCULA
 20  HI CRUCIELLIPSIS CHIASTA
 21  HI CYCLAGELOSPHAERA MARGERELI
 22  HI CYCLAGELOSPHAERA SP
 23  HI CYLINDRALITHUS ASYMMETRICUS
 24  HI CYLINDRALITHUS BIARCUS
 25  HI CYLINDRALITHUS CORONATUS
 26  HI CYLINDRALITHUS SP
 27  HI CYLINDRALITHUS SERRATUS
 28  HI EIFFELLITHUS EXIMIUS
 29  HI EIFFELLITHUS PARALLELUS
 30  HI EIFFELLITHUS TURRISEIFFELI
 31  HI FLABELLITES BIFORAMINIS
 32  HI GARTNERAGO OBLIQUUM
 33  HI GARTNERAGO STRIATUM
 34  HI HELICOLITHUS TRABECULATUS
 35  HI KAMPTNERIUS SP
 36  HI LITHASTRINUS FLORALIS
 37  HI LITHASTRINUS GRILLI
 38  HI LITHRAPHIDITES ACUTUM
 39  HI LITHRAPHIDITES ALATUS
 40  HI LITHRAPHIDITES PRAEQUADRATUS
 41  HI LITHRAPHIDITES QUADRATUS
 42  HI LUCIANORHABDUS ARCUATUS
 43  HI LUCIANORHABDUS CAYEUXI
 44  HI LUCIANORHABDUS MALEFORMIS
 45  HI LUCIANORHABDUS SCOTUS
 46  HI LUCIANORHABDUS SP
 47  HI MARKALIUS CIRCUMRADIATUS
 48  HI MARTHASTERITES FURCATUS
 49  HI MICRORHABDULUS BELGICUS
 55  HI NANNOCONUS AFF. MULTICADUS
 56  HI NANNOCONUS TRUITTI
 57  HI NANNOCONUS SP
 58  HI NEPHROLITHUS FREQUENS
 59  HI PARHABDOLITHUS ACHYLOSTAURION
 60  HI PHANULITES OBSCURUS
 61  HI PHANULITES OVALIS
 62  HI PHANULITES SPP
 63  HI PODORHABDUS ALBIANUS
 64  HI PODORHABDUS CORONADVENTIS
 65  HI PODORHABDUS DECORUS
 66  HI PREDISCOSPHAERA COLUMNATA
 67  HI PREDISCOSPHAERA CRETACEA
 69  HI QUADRUM GOTHICUM
 70  HI QUADRUM TRIFIDUM
 71  HI REINHARDTITES ANTHOPHORUS
 72  HI REINHARDTITES BROOKSII
 73  HI REINHARDTITES LEVIS
 74  HI RHAGODISCUS ASPER
 75  HI RUCINOLITHUS HAYI
 76  HI SCAPHOLITHUS FOSSILIS
 77  HI SOLLASITES HORTICUS
 78  HI STEPHANOLITHION LAFFITTEI
 79  HI TETRALITHUS COPULATUS
 80  HI TRANOLITHUS PHACELOSUS
 81  HI VEKSHINELLA GAUSORHETHIUM
 83  HI WATZNAUERIA BIPORTA
 84  HI WATZNAUERIA BRITANNICA
 85  HI WATZNAUERIA OVATA
 86  HI ZYGODISCUS BIPERFORATUS
 87  HI ZYGODISCUS PSEUDANTHOPHORUS
 89  HI ZYGODISCUS XENOTUS
 90  HI BROINSONIA SP
101  LO AHMUELLERELLA OCTORADIATA
102  LO AHMUELLERELLA REGULARIS
103  LO ARKHANGELSKIELLA CYMBIFORMIS
104  LO ARKHANGELSKIELLA SPECILLATA
105  LO BRAARUDOSPHAERA AFRICANA
106  LO BRAARUDOSPHAERA BIGELOWI
107  LO BROINSONIA FURTIVA
108  LO BROINSONIA LACUNOSA
109  LO BROINSONIA MATALOSA
110  LO BROINSONIA PARCA
111  LO CERATOLITHOIDES ACULEUS
112  LO CERATOLITHOIDES KAMPTNERI
113  LO CORLLITHIOPN ACHYLOSUM
114  LO COROLLITHION COMPLETUM
115  LO COROLLITHION EXIGUUM
```

116 LO COROLLITHION SIGNUM
117 LO CRETARHABDUS LORIEI
118 LO CRIBROCORONA GALLICA
119 LO CRIBROSPHAERELLA CIRCULA
120 LO CRUCIELLIPSIS CHIASTA
121 LO CYCLAGELOSPHAERA MARGERELI
122 LO CYCLAGELOSPHAERA SP
123 LO CYLINDRALITHUS ASYMMETRICUS
124 LO CYLINDRALITHUS BIARCUS
125 LO CYLINDRALITHUS CORONATUS
126 LO CYLINDRALITHUS SP
127 LO CYLINDRALITHUS SERRATUS
128 LO EIFFELLITHUS EXIMIUS
129 LO EIFFELLITHUS PARALLELUS
130 LO EIFFELLITHUS TURRISEIFFELI
131 LO FLABELLITES BIFORAMINIS
132 LO GARTNERAGO OBLIQUUM
133 LO GARTNERAGO STRIATUM
134 LO HELICOLITHUS TRABECULATUS
135 LO KAMPTNERIUS SP
136 LO LITHASTRINUS FLORALIS
137 LO LITHASTRINUS GRILLI
138 LO LITHRAPHIDITES ACUTUM
139 LO LITHRAPHIDITES ALATUS
140 LO LITHRAPHIDITES PRAEQUADRATUS
141 LO LITHRAPHIDITES QUADRATUS
142 LO LUCIANORHABDUS ARCUATUS
143 LO LUCIANORHABDUS CAYEUXI
144 LO LUCIANORHABDUS MALEFORMIS
145 LO LUCIANORHABDUS SCOTUS
146 LO LUCIANORHABDUS SP
147 LO MARKALIUS CIRCUMRADIATUS
148 LO MARTHASTERITES FURCATUS
149 LO MICRORHABDULUS BELGICUS
150 LO MICRORHABDULUS DECORATUS
151 LO MICRORHABDULUS HELICOIDEUS
152 LO MICRORHABDULUS STRADNERI
153 LO MICULA CONCAVA
154 LO MICULA DECUSSATA
155 LO NANNOCONUS AFF. MULTICADUS
156 LO NANNOCONUS TRUITTI
157 LO NANNOCONUS SP
158 LO NEPHROLITHUS FREQUENS
159 LO PARHABDOLITHUS ACHYLOSTAURION
160 LO PHANULITES OBSCURUS
161 LO PHANULITES OVALIS
162 LO PHANULITES SSP
163 LO PODORHABDUS ALBIANUS
164 LO PODORHABDUS CORONADVENTIS
165 LO PODORHABDUS DECORUS
166 LO PREDISCOSPHAERA COLUMNATA
167 LO PREDISCOSPHAERA CRETACEA
168 LO QUADRUM GARTNERI
169 LO QUADRUM GOTHICUM
170 LO QUADRUM TRIFIDUM
171 LO REINHARDTITES ANTHOPHORUS
172 LO REINHARDTITES BROOKSII
173 LO REINHARDTITES LEVIS
174 LO RHAGODISCUS ASPER
175 LO RUCINOLITHUS HAYI
176 LO SCAPHOLITHUS FOSSILIS
177 LO SOLLASITES HORTICUS
178 LO STEPHANOLITHION LAFFITTEI
179 LO TETRALITHUS COPULATUS
180 LO TRANOLITHUS PHACELOSUS
181 LO VEKSHINELLA GAUSORHETHIUM
182 LO VEKSHINELLA BOCHOTNICAE
183 LO WATZNAUERIA BIPORTA
184 LO WATZNAUERIA BRITANNICA
185 LO WATZNAUERIA OVATA
186 LO ZYGODISCUS BIPERFORATUS
187 LO ZYGODISCUS PSEUDANTHOPHORUS
188 LO ZYGODISCUS SPIRALIS
189 LO ZYGODISCUS XENOTUS
190 LO BROINSONIA SP
191 LO MICULA MURA
201 ST LITHASTRINUS GRILLI
202 ST RUCINOLITHUS HAYI
203 SB ARKHANGELSKIELLA CYMBIFORMIS
204 SB ARKHANGELSKIELLA SPECILLATA
205 ST EIFFELLITHUS EXIMIUS
206 SB EIFFELLITHUS EXIMIUS
207 SB BROINSONIA FURTIVA
208 SB BROINSONIA LACUNOSA
209 SB LUCIANORHABDUS CAYEUXI
210 SB BROINSONIA PARCA
211 SB MARTHASTERITES FURCATUS
212 SB MICRORHABDULUS STRADNERI
213 SB MICULA CONCAVA
214 ST COROLLITHION COMPLETUM
215 SB MICULA DECUSSATA
216 SB PHANULITES OBSCURUS
217 SB PHANULITES OVALIS
218 SB PHANULITES SPP
219 SB QUADRUM GARTNERI
220 SB REINHARDTITES ANTHOPHORUS
221 SB RUCINOLITHUS HAYI
222 SB BROINSONIA SPP
223 ST BROINSONIA LACUNOSA

```
1*************************
 **
 **
 **  (3)  DOEVEN DATA BASE
 **       CRETACEOUS SCOTIAN SHELF NANNOFOSSILS
 **
 **
 *************************

 SEQUENCE DATA

 PRIMROSE A41
    3  -40  -10  -28  -73  -43  -45  140  -44  -64  -80 -122   62  -86   42   70   -9   69  -71   26
  169  -37  -79 -179  -15 -115  170  173 -145 -205  103 -203    8 -223  -90   75 -202  111  110 -210
  -48   -7  201   16  -36  -49  182  104 -204 -208 -152 -212  -57 -142 -221   33 -153 -213  175 -220
 -160  143 -209  108 -107 -207 -190 -222  -13 -154 -215  171 -217 -218 -102 -186   46   56  137  161
 -162 -150   -5  -74  128 -206 -148 -211 -168 -219 -132 -101 -135 -144  -59 -126   63  -38  -14 -214
  -20   39  138 -149 -114 -146

 ADOLPHUS D50
   58 -158   -3  -43   -9  141  -40  -44  -12 -112   28  -26 -122  -62  140   10  -73  -80   70  -69
  170 -173  -71  -75  169  -49  205 -111 -103 -203   16   -8 -223  -64  -86  -90    7 -216  209  202
  -48  110 -210  -37 -201  -36   57  -42  142  207 -204 -182 -213 -212  175 -221  102  -56  107 -160
 -153 -149  171 -220 -108 -208 -104 -143 -152 -190 -222  148 -211  -33   13 -154 -215 -144 -161 -162
 -217 -218  206 -137 -101 -135 -186   63 -168 -219  -74  -14 -214   38  132   20  -39   66  138  150
    5 -126  128 -114  -84  130

 TRIUMPH P50
   10   -3  -28  -70  -69  -73  -62  -43  -49  103 -203 -170  -75  -71   -7   -8 -223  -90  -44  -14
  -26  173  -80  169 -111 -205 -152 -212  202   -9  110 -210  -37 -201  -36  -86   48 -101  -64   16
 -160 -216 -153 -213 -102  148 -211  -42 -142  -45 -145  175 -221 -171 -220 -190 -222 -108 -208 -107
 -207 -161 -217 -162 -218 -154 -215 -135 -104 -204 -143 -209 -182 -151  186  206  -57   74   63 -168
 -219 -132 -144   -5   13  -38 -138 -149  214 -114  -59   20  -66  126  137  150  128   56  -84  130

 ONONDAGA E84
   58 -158   -3  -12   -9  -26   28  141  -40 -112 -129  -15   45  -44   10  -73  -62  -80  145  140
 -111 -173 -122  205  -16  -75 -202  -71  -90  -43  -64  -42  -49  110 -210  -48  -37 -201  -36 -216
 -203  212  182  175 -221 -143 -209 -213  220 -160  -57  190 -222  -33  -56  148 -211  -86 -186 -126
 -217 -218   13  102 -152 -149 -153  206 -101 -135  161 -162  -74  -14 -214 -114  171  -63 -168 -219
  144  132  -39  103  150  130 -128 -154 -215   -5 -137

 WYANDOT E53
   10   -3 -103 -203  -70 -170 -111  -73 -173  -71  -62  -43  -45  -26  -49  110 -210  -28 -205  -75
 -202   -8 -223  -90 -160 -216  -44  -80   -9  175 -221  -48  -36 -171 -220   -7 -104 -204 -143 -209
  -64  -42 -142 -152 -212  -86   16  -37 -201   57  -15 -145   56  107 -207 -217 -218 -182 -151  148
 -211 -108 -208 -190 -222  -33 -153 -213  135 -186  154 -215 -128 -206  -74   13 -161 -162 -126  137
 -168 -219 -144  -46 -115   38  -59  -39  102 -101 -150  -66 -130 -146 -132 -149 -138  -20

 MICMAC H86
   58 -158  -91 -191 -141  -40 -140   -3  -12 -112 -122 -129   10  -28  -70  -69  -16  -48  -37  -73
   -8   -7  -90  -62  -43  -64  -45  -15  -80  -49  -26  170 -111  -71 -203  -44  -42  -79 -179  173
   86  205 -169   -9   75 -202 -223  210  -36  201  160 -216 -145  175 -221 -209 -142   33  153 -213
 -103  -74  110 -171 -220 -108 -208 -107 -207 -190 -222 -217 -150 -143 -104 -204 -152 -212 -151 -149
 -218  148 -211 -102 -101 -115 -186 -126  128 -206 -137  -13 -154 -215 -135 -144 -182  -59  168 -219
 -132 -161 -162  -57  130
```

MISSISSAUGA H54

```
  58 -158  -40   -3   -9  -43  -15  -12 -112  -14  -26  141 -140  -10 -203  -28  -70  -69  -73  -62
 -44 -122  -79 -179   16  -71  -42  -45  170 -173  -80  205  169   49    8 -223  -90  111   75 -202
 -48   -7  -86  210  -36 -216   37 -201  -64  175 -221 -220 -204 -209  208 -182 -142 -151 -126 -213
 207 -222  217 -218   33  -46  171 -160 -145 -186  211  108  -13 -104  206 -148 -107 -190  -57  -59
 115  137 -215 -135  -56  -74  219 -146  132 -150 -101 -144   63   20  -38 -138  110 -103 -161 -162
-153 -102 -143  128 -168 -154 -149  130  -39
```

INDIAN HARBOUR M52

```
   3  -58  -86  158
```

HERON H73

```
  10  -28 -205  -16  -62   -3  -71  -26  -15   75 -202  -90   -9  210  -48  -37 -201   44   36  -43
 -64   49  -80  216 -209  175 -221  -45 -145  190 -222 -110   42 -142  -86  171 -220 -153 -213  104
-204 -151  160 -101 -182 -186  137   57  -74  135 -115 -152 -212 -126  128 -206 -148 -211 -168 -219
-132 -161 -217 -162 -218 -154 -215 -150 -102 -144 -143 -130 -149  -39
```

OSPREY H84

```
  10  -28 -205  -16  -48  -90  -62   -3  -44  -43  -64  -45 -145  -71   -9  -42  -49  210  -37 -201
  75 -202 -175 -221  -36 -153 -213 -101 -142  -26 -126   33 -160 -154 -215 -135 -103 -104 -143 -182
 110 -128 -206 -148 -137 -171 -190  -13 -168 -219 -132 -161 -162 -150 -102 -144  -74 -214 -114  -14
-130 -149  -59
```

```
1************************
 **
 **
 **  (4)  BAUMGARTNER DATA BASE
 **       JURASSIC TETHYAN RADIOLARIANS
 **
 **
 ************************
```

DICTIONARY

1 TRILLUS SP CF T SEIDERSI
2 UNUMA ECHINATUS
3 ANDROMEDA PRAEPODBIELENSIS
4 STICHOCAPSA SP AFF S JAPONICA
5 TETRADITRYMA PRAEPLENA
6 STYLOCAPSA OBLONGULA
7 LITHOCAMPE NUDATA
8 HAGIASTRID SP A
9 TRICOLOCAPSA PLICARUM
10 ANDROMEDA PRAECRASSA
11 GORGANSIUM PULCHRUM
12 NAPORA PYRAMIDALIS
13 HIGUMASTRA IMBRICATA
14 MIRIFUSUS FRAGILIS
15 THEOCAPSOMMA CORDIS
16 ANDROMEDA PODBIELENSIS
17 EUCYRTIDIELLUM UNUMAENSIS
18 PODOBURSA HELVETICA
19 EMILUVIA PREMYOGII
20 TETRADITRYMA CORRALITOSENSIS
21 PROTUNUMA COSTATA
22 STICHOCAPSA JAPONICA
23 HSUUM BREVICOSTATUM
24 ACANTHOCIRCUS SUBOBLONGUS
25 TRIACTOMA BLAKEI
26 TRITRABS CASMALIAENSIS
27 TRITRABS RHODODACTYLUS
28 TRITRABS HAYI
29 TRIACTOMA JONESI
30 TRIACTOMA TITHONIANUM
31 PERISPYRIDIUM ORDINARIUM
32 RISTOLA ALTISSIMA
33 EMILUVIA SEDECIMPORATA SALENSIS
34 DIACANTHOCAPSA NORMALIS
35 BERNOULLIUS DICERA
36 TRITRABS ZEALIS
37 MIRIFUSUS GUADALUPENSIS
38 PARONAELLA MULLERI
39 BERNOULLIUS CRISTATUS
40 EMILUVIA SEDECIMPORATA S STR
41 EMILUVIA(?) SP P
42 MONOTRABS PLENOIDES
43 HOMOEOPARONAELLA ARGOLIDENSIS
44 TRITRABS EXOTICA
45 RISTOLA PROCERA
46 NAPORA DEWEVERI
47 HSUUM MAXWELLI
48 PARONAELLA KOTURA
49 STAUROSPHAERA ANTIQUA
50 SPONGOCAPSULA PALMERAE
51 PSEUDOCRUCELLA SANFILIPPOAE
52 PSEUDOCRUCELLA ADRIANI
53 PARONAELLA BROENNIMANNI
54 TRITRABS EWINGI
55 DIBOLOACHRAS CHANDRIKA
56 EUCYRTIDIELLUM PTYCTUM
57 TETRADITRYMA PSEUDOPLENA
58 PARONAELLA BANDYI
59 ACAENIOTYLE DIAPHOROGONA S 1
60 EMILUVIA OREA
61 STICHOCAPSA CONVEXA
62 TETRARABS BULBOSA
63 EUCYRTID GEN ET SP INDET
64 PODOBURSA SPINOSA
65 HOMOEOPARONAELLA ELEGANS
66 HIGUMASTRA SP AFF H INFLATA
67 ANGULOBRACCHIA PURISIMAENSIS
68 SETHOCAPSA CETIA
69 PODOCAPSA AMPHITREPTERA
70 HOMOEOPARONAELLA GIGANTHEA
71 EMILUVIA PESSAGNOI
72 NAPORA LOSPENSIS
73 NAPORA BUKRYI
74 EMILUVIA HOPSONI
75 ARCHAEODICTYOMITRA APIARA
76 MIRIFUSUS MEDIODILATATUS MAJOR
77 MIRIFUSUS CHENODES
78 PARONAELLA (?) HIPPOSIDERICUS
79 PARONAELLA (?) DIAMPHIDIA
80 ACAENIOTYLE UMBILICATA
81 TRIACTOMA ECHIODES
82 ACANTHOCIRCUS DICRANACANTHOS
83 OBESACAPSULA ROTUNDA
84 SETHOCAPSA LEIOSTRACA
85 SPONGOCAPSULA PERAMPLA
86 HALIODICTYA HOJNOSI
87 PRAECONOCARYOMMA (?) HEXACUBICA
88 SAITOUM PAGEI
89 TRIACTOMA CORNUTA
90 PARVICINGULA DHIMENAENSIS
91 EUCYRTIDELLUM PUSTULATUM
92 ARCHAEOHAGIASTRUM MUNITUM
93 PANTANELLIUM (?) BERRIASIANUM
94 ACAENIOTYLE DIAPHOROGONA S STR
95 OBESACAPSULA RUSCONENSIS
96 SYRINGOCAPSA LUCIFER
97 PSEUDODICTYOMITRA DEPRESSA
98 ANGULOBRACCHIA (?) PORTMANNI
99 MIRIFUSUS MEDIODILATATUS MINOR
100 ACHAEODICTYOMITRA EXECELLENS
101 RISTOLA CRETACEA
102 PARVICINGULA COSMOCONICA
103 DITRABS SANSALVATORENSIS
104 ALIEVUM HELENAE
105 SYRINGOCAPSA AGOLARIUM
106 XITUS SP CF X SPICULARIUS
107 PSEUDODICTYOMITRA CARPATICA
108 HOLOCRYPTOCANIUM BARBUI
109 THANARLA ELEGANTISSIMA
110 CECROPS SEPTEMPORATUS

```
**************************
**
**
**  (4)  BAUMGARTNER DATA BASE
**       JURASSIC TETHYAN RADIOLARIANS
**
**
**************************

SEQUENCE DATA

1. DHIMAINA, ARGOLIS PENINSULA, PELOPONNESUS, GREECE: 9
 17 7 9  18 3 3  20 3 3  21 1 8  25 1 1  27 1 3  28 3 3  29 3 6  32 3 5  40 1 3  46 2 2  49 1 1
 56 2 9  57 1 1  60 3 3  62 3 3  64 1 4  72 3 3  73 3 3  75 1 8  76 2 9  77 3 3  78 2 9  79 2 3
 84 1 5  88 2 3

2. ANGELOKASTRON, ARGOLIS PENINSULA, PELOPONNESUS, GREECE: 8
 16 2 2  18 2 7  19 2 4  20 4 6  23 1 8  24 2 8  25 1 6  26 2 8  27 1 6  28 2 4  29 1 8  30 4 4
 31 2 4  32 1 8  33 2 8  35 2 4  36 1 8  37 1 6  38 2 6  40 2 8  43 2 5  44 2 4  45 1 4  46 2 6
 47 1 8  48 1 5  49 2 4  50 1 4  51 2 5  52 2 4  53 2 6  54 2 4  55 2 4  56 2 8  57 1 8  58 2 4
 59 2 4  60 1 8  62 2 8  64 2 8  65 1 4  66 4 4  67 2 4  70 2 4  72 2 4  73 2 5  74 2 4  75 1 8
 76 1 6  77 4 8  78 5 5  79 2 5  84 2 4  86 2 4  88 6 6  89 1 5  90 2 8  91 1 1

3. PROSIMNI, ARGOLIS PENINSULA, PELOPONNESUS, GREECE: 3
 21 1 3  25 1 1  29 1 1  32 3 3  33 1 1  37 1 1  40 1 3  49 3 3  56 1 3  60 1 3  62 1 1  64 1 3
 71 1 3  73 3 3  75 1 3  76 1 3  77 1 1  84 1 1  86 1 1

4. TAXIARCHIS, ARGOLIS PENINSULA, PELOPONNESUS, GREECE: 3
 18 2 3  20 2 2  21 3 3  23 1 3  24 1 2  25 3 3  27 1 3  29 1 3  30 1 3  32 1 3  33 1 3  36 2 2
 37 1 3  40 1 1  43 1 1  44 1 1  47 1 3  49 1 1  50 1 3  53 1 1  54 3 3  56 1 3  60 1 3  62 1 3
 64 1 1  71 3 3  72 1 1  73 1 2  75 1 3  76 1 3  77 1 2  85 3 3  88 1 1  89 1 1  90 1 2

5. KANDHIA, ARGOLIS PENINSULA, PELOPONNESUS, GREECE: 2
 18 1 2  21 1 2  23 1 2  24 2 2  25 1 2  26 2 2  27 2 2  29 1 2  30 1 2  32 1 2  33 1 2  40 1 2
 44 1 1  46 1 1  54 2 2  56 1 2  57 2 2  59 1 1  60 1 2  62 1 2  64 1 2  69 1 2  71 1 1  72 2 2
 73 1 2  75 1 2  76 1 2  86 1 1  90 1 2

6. SERRADA, TRENTO PLATEAU, NORTHERN ITALY: 1
 24 1 1  25 1 1  30 1 1  33 1 1  55 1 1  60 1 1  64 1 1  69 1 1  75 1 1  76 1 1  84 1 1

7. KOLIAKI, ARGOLIS PENINSULA, PELOPONNESUS, GREECE: 7
  1 1 2   2 1 2   3 1 1   4 1 2   6 3 3   9 2 3  11 3 3  12 3 3  14 3 3  15 3 3  17 1 4  18 3 3
 20 3 3  21 3 4  23 2 4  24 1 3  27 4 4  28 4 4  29 4 4  32 3 3  34 2 3  35 3 3  36 2 4  37 3 3
 39 3 3  40 4 4  43 3 3  47 3 3  49 3 3  56 3 4  57 3 4  60 4 4  61 3 3  63 1 1  71 4 4  75 4 4
 76 4 4  88 3 3  90 3 3

8. THEOKAFTA, ARGOLIS PENINSULA, PELOPONNESUS, GREECE: 1
 20 1 1  23 1 1  24 1 1  25 1 1  27 1 1  28 1 1  29 1 1  30 1 1  31 1 1  33 1 1  36 1 1  38 1 1
 40 1 1  43 1 1  54 1 1  56 1 1  57 1 1  58 1 1  59 1 1  60 1 1  62 1 1  65 1 1  67 1 1  68 1 1
 69 1 1  71 1 1  73 1 1  75 1 1  76 1 1  78 1 1  79 1 1  80 1 1  81 1 1  86 1 1  88 1 1  90 1 1

9. RHADON, ARGOLIS PENINSULA, PELOPONNESUS, GREECE:
 21 1 1  27 1 1  29 1 1  37 1 1  47 1 1  88 1 1  91 1 1

10. PINDOS, CENTRAL GREECE: 3
 25 2 2  27 2 2  32 2 2  36 2 2  37 2 2  40 2 2  44 2 2  45 2 2  48 1 1  54 1 1  56 1 1  60 2 2
 62 2 2  64 2 2  76 2 2  86 2 2  97 3 3 106 3 3 109 3 3

11. MARATHOS, CENTRAL GREECE: 6
 23 2 2  32 3 4  33 2 3  47 1 1  53 1 1  62 3 4  68 4 4  69 2 3  76 2 4  99 5 5 106 5 6 109 5 6
```

12. POJORITA, RARAU MOUNTAINS, ROUMANIA: 2
18 2 2 19 2 2 21 1 2 23 1 2 24 2 2 25 2 2 26 2 2 27 2 2 28 2 2 29 2 2 32 2 2 33 2 2
35 2 2 36 1 2 37 2 2 43 1 2 45 2 2 47 2 2 54 1 1 56 2 2 57 1 2 60 2 2 62 2 2 64 2 2
67 1 1 75 2 2 76 2 2 91 2 2

13. LAEN ROSU, HAGHIMAS MOUNTAINS, ROUMANIA: 1
16 1 1 23 1 1 29 1 1 36 1 1 37 1 1 46 1 1 47 1 1 88 1 1 91 1 1

14. PIATRA SOIMULUI, RARAU MOUNTAINS, ROUMANIA: 1
21 1 1 23 1 1 27 1 1 29 1 1 32 1 1 35 1 1 36 1 1 43 1 1 44 1 1 56 1 1 60 1 1 62 1 1
75 1 1 76 1 1 85 1 1 87 1 1 91 1 1

15. GOMIELOR VALLEY, DROCEA MOUNTAINS, ROUMANIA: 1
8 1 1 35 1 1 36 1 1 37 1 1 57 1 1

16. SVINITA, BANAT, DANUBE SECTION, S-CARPATHIANS, ROUMANIA: 9
29 1 1 54 1 9 59 5 5 68 1 2 69 1 1 71 1 1 75 1 7 79 1 8 80 1 8 81 1 8 82 1 8 83 1 8
84 3 8 94 1 8 95 1 8 97 2 7 98 2 8 99 1 9 100 2 8 101 1 3 102 1 1 103 1 8 104 2 8 105 1 8
106 2 7 108 2 8 109 2 8 110 7 9 111 6 8

17. BESOZZO II, PROV. VARESE, LOMBARDY, NORTHERN ITALY: 3
6 2 2 7 2 2 11 2 2 12 2 2 13 1 2 17 2 2 18 2 2 19 2 2 21 3 3 24 1 3 26 3 3 27 2 3
29 3 3 31 2 2 32 3 3 33 2 3 36 2 3 37 2 2 41 2 2 42 2 2 44 3 3 47 3 3 50 3 3 55 2 2
56 3 3 57 3 3 60 3 3 61 2 2 64 3 3 65 2 2 67 3 3 69 3 3 71 3 3 73 2 2 74 2 2 76 3 3
84 3 3 85 2 2 86 1 1

18. MONTE GENEROSO, TICINO, SOUTHERN SWITZERLAND: 3
6 1 1 15 1 1 17 2 2 18 2 2 23 2 2 24 2 2 25 2 2 26 2 2 27 2 2 31 2 2 36 2 3 37 1 2
38 2 2 40 2 2 43 2 2 48 2 2 49 2 2 50 2 2 51 2 2 53 2 2 55 2 2 56 2 3 57 2 2 60 2 2
64 2 2 65 2 2 69 3 3 71 2 3 72 2 2 73 2 2 74 2 2 75 2 2 76 2 2 85 2 2 88 2 2

19. TORRE DE BUSI, PROV. COMO, LOMBARDY, NORTHERN ITALY: 4
1 1 1 2 1 1 3 1 1 4 1 1 5 1 1 6 4 5 7 3 6 10 1 1 11 2 2 12 1 6 15 3 5 16 6 6
17 1 6 21 7 7 23 1 6 24 1 6 29 1 1 32 6 8 33 3 4 35 4 6 36 1 1 37 6 6 38 4 4 40 1 1
43 2 6 46 4 6 47 2 2 50 6 6 51 7 7 56 7 7 60 7 7 61 2 9 63 1 1 66 1 2 69 7 9 70 7 7
71 7 7 72 7 9 75 9 9 76 7 7 77 7 7 82 9 9 86 2 6 88 4 9

20. VALMAGGIORE, BRENTA, PROV. VARESE, NORTHERN ITALY: 4
6 1 1 7 1 1 11 1 1 15 1 1 17 1 1 19 1 1 22 1 1 23 1 1 24 1 3 26 3 3 27 2 2 29 2 2
33 3 3 37 1 1 40 2 2 41 3 3 47 1 1 49 2 2 55 2 2 60 3 3 61 1 1 62 3 4 64 2 3 67 2 2
69 3 4 74 1 1 75 4 4 82 4 4 84 3 4 85 3 3 88 2 2

21. BESOZZO I, BESOZZO SUP., PROV. VARESE, NORTHERN ITALY: 5
17 1 1 19 1 1 21 1 5 22 5 5 25 2 2 27 2 5 30 5 5 32 4 5 33 1 5 41 1 2 44 2 2 48 4 5
49 1 1 50 1 1 51 5 5 52 5 5 53 2 5 54 2 5 55 1 5 56 1 5 60 1 5 61 4 4 64 4 5 69 1 5
71 1 5 72 1 1 73 1 1 74 1 1 75 1 5 76 2 5 77 5 5 82 5 5 84 1 1

22. SANGIANO, PROV. VARESE, NORTHERN ITALY: 7
7 1 3 8 1 3 11 1 3 12 1 3 13 2 3 15 3 4 16 1 1 17 1 4 18 1 4 19 1 3 20 1 3 21 1 7
23 1 4 24 1 4 25 4 4 26 1 5 27 2 4 29 2 4 31 4 4 32 4 7 33 2 7 35 4 4 36 1 4 37 1 4
41 6 6 42 2 3 43 2 2 44 4 4 46 4 4 47 1 4 48 4 4 50 1 2 54 2 4 55 4 7 56 1 6 57 3 4
58 3 6 60 4 6 61 1 7 62 4 7 64 3 3 68 7 7 69 5 7 71 5 7 72 3 7 74 2 7 75 4 7 76 6 6
77 2 7 80 7 7 82 6 7 84 6 6 85 1 4 86 1 3 88 4 4 89 4 4

23. CAVA RUSCONI, CITTIGLIO, PROV. VARESE, NORTHERN ITALY: 1
27 1 1 29 1 1 30 1 1 33 1 1 50 1 1 54 1 1 68 1 1 69 1 1 71 1 1 72 1 1 73 1 1 74 1 1
75 1 1 78 1 1 79 1 1 80 1 1 81 1 1 82 1 1 83 1 1 84 1 1 93 1 1 94 1 1 95 1 1 96 1 1
97 1 1 98 1 1 99 1 1 101 1 1 102 1 1 103 1 1 106 1 1

24. BREGGIA GORGE, TICINO, SOUTHERN SWITZERLAND: 24
1042424 1002424 6 3 9 7 113 8 310 9 7 7 11 314 12 513 13 512 15 3 9 16 314 17 115
18 313 19 515 20 315 21 520 221212 23 314 24 121 25 521 962424 26 320 271218 28 3 5
29 219 301821 31 518 32 523 33 323 35 921 972424 36 312 37 314 38 216 40 519 41 419
42 515 43 120 441014 461212 47 717 481518 491919 501419 51 419 531216 54 724 561023
57 218 58 410 982424 992424 592424 601518 61 420 62 819 641420 651417 682424 691924
711024 72 812 731818 74 724 751524 76 723 771221 802424 812424 822224 1022424 1032424
832424 841224 85 119 86 120 881717 891418 90 2 3 91 4 4

25. SALTRIO, PROV. VARESE, NORTHERN ITALY: 12
2 1 1 5 3 5 6 612 7 112 8 212 9 311 10 1 2 11 112 12 212 13 111 14 111 15 112
16 211 17 112 18 112 19 211 20 211 21 111 221010 23 111 24 112 25 311 26 612 271212
28 3 4 29 111 31 5 5 32 212 33 212 35 712 36 212 371111 38 711 39 212 40 210 42 711
43 412 46 611 47 511 49 311 50 1 1 52 111 57 2 8 58 111 61 112 63 1 2 66 3 9 67 4 4
701111 73 511 74 612 84 311 86 411 87.111 88 511 89 6 7 91 5 9 92 511

26. FIUME BOSSO, NEAR PIANELLO, UMBRIA, CENTRAL ITALY: 16
2 1 1 6 2 2 8 4 4 9 1 2 11 6 6 12 5 5 13 3 3 15 2 2 16 2 5 17 1 4 18 1 5 19 2 6
20 2 2 21 113 22 1 2 23 1 7 24 1 9 25 214 26 6 6 27 1 8 29 115 30 7 8 31 6 7 32 2 7
33 310 35 5 5 36 2 8 37 210 38 6 6 39 2 2 41 6 6 431414 47 2 4 49 5 5 50 6 8 54 715
55 7 8 56 414 57 213 59 7 7 60 714 61 1 4 62 714 64 710 65 6 6 67 2 2 681515 69 715
701515 711415 72 5 7 73 5 5 741314 75 714 76 614 791515 801515 811515 821416 831515
84 9 9 85 311 86 3 3 87 2 5 88 6 6 89 5 5 99 7 7 1021516 1031515

27. MONTE CETONA, TUSCANY, CENTRAL ITALY: 9
6 1 8 7 8 8 8 2 2 11 2 8 12 4 4 13 8 8 15 3 8 16 8 8 17 2 9 18 8 8 19 6 6 20 3 9
22 6 6 23 4 4 24 1 9 26 1 7 29 3 4 32 8 8 33 1 8 36 8 8 38 2 3 39 8 8 42 5 8 43 3 8
47 4 8 50 3 8 56 9 9 57 8 8 61 2 9 65 2 2 66 3 3 86 1 6

28. SANTA ANNA, NEAR CALTABELLOTTA, SICILY, ITALY: 4
16 2 3 18 3 3 19 4 4 21 4 4 23 4 4 24 1 1 25 4 4 27 4 4 29 1 1 30 3 3 32 1 4 33 4 4
36 2 3 37 1 3 38 4 4 40 4 4 50 3 4 54 3 3 56 1 4 57 1 3 59 4 4 60 3 4 62 3 4 64 1 4
65 4 4 69 1 3 71 2 3 72 1 3 73 4 4 74 3 3 75 4 4 76 1 4 77 4 4 78 3 3 84 4 4 85 4 4
86 4 4 90 3 3 91 4 4

29. DSDP LEG 41, SITE 367, CAPE VERDE BASIN, E-ATLANTIC: 7
16 3 3 18 1 3 21 1 6 23 1 6 24 1 1 25 1 5 30 2 6 31 1 3 32 4 4 33 1 3 37 1 1 40 1 3
43 4 4 47 1 1 52 1 1 54 6 6 55 1 6 56 1 5 57 3 3 59 3 3 60 1 4 62 1 1 64 1 7 68 4 7
69 2 6 70 4 4 71 3 5 73 1 3 74 2 4 75 1 5 76 1 6 79 1 7 82 6 6 86 1 6 89 4 4 90 6 6
99 7 7

30. DSDP LEG 76, SITE 534, BLAKE BAHAMA BASIN, W-ATLANTIC: 28
1022526 1002628 6 523 7 123 8 123 1022526 11 223 12 219 13 222 14 218 15 223 16 220
17 223 18 223 19 221 20 220 21 223 22 220 23 223 24 123 25 124 26 219 27 222 281718
29 224 30 224 31 224 32 224 33 224 34 323 35 323 36 323 37 520 38 720 39 312 40 324
411111 421220 431420 441919 451620 47 123 48 420 492424 501920 531618 541624 551624
561721 571818 581819 591826 602424 61 120 63 3 3 642424 651212 692425 712424 742525
752428 762424 792525 802428 812425 822528 84 228 85 820 861219 1032525 88 224 902424
91 223 1052528 1062628 1072628 1082628 1092626 972528 982525

31. DSDP LEG 17, SITE 167, MAGELLAN RISE, CENTRAL PACIFIC: 6
27 2 2 30 3 3 64 1 1 68 6 6 69 1 2 78 2 2 79 2 3 80 3 3 82 2 5 85 1 4 86 1 5 88 1 2
99 1 6 100 2 6 102 2 3 103 3 3 107 2 6 109 3 6 110 3 6 111 5 5

32. DSDP LEG 32, SITE 306, SHATSKY RISE, E-CENTRAL PACIFIC: 7
30 1 4 54 4 4 59 1 7 68 1 5 69 1 4 75 1 7 79 1 7 80 1 7 81 3 7 82 1 4 83 1 4 84 1 4
93 2 2 97 2 3 98 2 3 99 1 6 100 3 3 101 2 3 102 1 4 103 2 2 104 6 7 106 2 3 107 2 7

33. DSDP LEG 17, SITE 307, SHATSKY RISE, E-CENTRAL PACIFIC: 6
30 1 2 59 2 5 68 1 4 75 1 5 78 5 5 79 5 5 80 2 6 81 3 6 82 1 6 83 2 4 84 2 4 94 4 4
99 1 4 100 4 5 102 1 4 104 3 6 106 4 5 107 1 6 109 5 5 110 3 6 111 4 5

34. DSDP LEG 20, SITE 195, SE-JAPAN ABYSSAL PLAIN, W-PACIFIC: 4
59 1 3 79 4 4 80 1 3 83 1 2 84 1 3 102 1 4 104 3 4 105 3 3 107 3 4 110 1 3

35. DSDP LEG 20, SITE 196, SE-JAPAN ABYSSAL PLAIN, W-PACIFIC: 3
30 1 1 54 1 1 59 1 3 68 1 1 69 1 1 71 1 1 75 1 1 78 2 2 79 1 3 80 1 3 81 2 3 82 1 3
83 1 3 84 1 3 93 1 1 98 1 2 99 1 1 100 1 1 102 1 1 104 2 3 105 2 3 106 2 2 107 1 3 110 2 3

36. GLASENBACH GORGE, NEAR SALZBURG, AUSTRIA: 2
6 1 2 8 2 2 11 2 2 12 2 2 13 1 2 15 2 2 17 1 2 19 2 2 20 2 2 23 2 2 24 1 2 26 1 1
27 1 2 28 1 1 29 1 1 31 1 2 33 1 2 35 1 2 36 1 1 37 2 2 38 1 2 43 2 2 46 1 2 50 1 1
53 1 2 54 2 2 61 1 2 65 1 1 66 1 1 67 2 2 74 2 2 86 1 2

37. POINT SAL, SANTA BARBARA COUNTY, CALIFORNIA, USA: 3
16 1 1 18 1 1 19 1 1 20 1 2 24 1 2 25 1 2 26 1 3 27 1 1 28 1 2 29 1 3 30 2 2 31 1 2
32 1 2 33 1 2 36 1 1 37 1 1 38 2 2 42 1 1 43 1 1 44 1 2 45 1 2 47 1 3 48 1 1 49 1 1
50 1 3 51 1 2 53 1 3 54 1 2 56 1 2 57 1 2 58 2 2 64 1 2 65 1 2 67 1 1 70 1 1 71 2 2
72 1 3 73 1 2 74 1 3 75 1 2 76 1 3 77 1 1 78 1 1 85 1 1 86 1 1 88 1 2 89 1 1

38. VEVEYSE DE CHATEL S. DENIS, CANT. VAUD, SWITZERLAND: 1
59 1 1 77 1 1 80 1 1 83 1 1 84 1 1 97 1 1 98 1 1 99 1 1 104 1 1 106 1 1 109 1 1 110 1 1
111 1 1

39. DSDP LEG1, SITE 5, BLAKE BAHAMA BASIN, W-ATLANTIC: 1
54 1 1 69 1 1 71 1 1 '80 1 1 83 1 1 98 1 1 101 1 1 103 1 1

40. IN 7, NEAR UNUMA, INUYAMA AREA, CENTRAL JAPAN: 1
2 1 1 3 1 1 5 1 1 9 1 1 10 1 1 14 1 1 17 1 1 18 1 1 19 1 1 20 1 1 22 1 1 24 1 1
27 1 1 28 1 1 29 1 1 34 1 1 58 1 1 61 1 1 86 1 1

41. GUATEMALA, NEAR SANTA ROSA, NICOYA PENINSULA, COSTA RICA: 1
7 1 1 9 1 1 18 1 1 23 1 1 24 1 1 37 1 1 61 1 1 87 1 1

42. OM 191, OM200, NEAR SUR, HAWASINA COMPLEX, SE-OMAN: 2
29 2 2 68 1 1 69 1 1 71 1 1 80 2 2 81 2 2 82 1 2 99 1 1 100 1 2 104 2 2 105 1 2 106 2 2
109 2 2 110 1 2 111 2 2

43. TRATTBERG, SALZBURG, AUSTRIA: 2
30 1 1 50 1 1 54 1 1 59 2 2 68 1 2 69 1 1 71 1 1 75 2 2 79 1 2 80 1 2 81 2 2 82 1 2
84 2 2 93 1 2 94 1 2 95 1 1 96 1 1 97 1 2 98 1 1 99 1 2 100 2 2 101 1 1 102 1 2 104 1 1
106 2 2 107 1 2 108 2 2

```
1************************
 **
 **
 **  (5)  BLANK DATA BASE
 **       MESOZOIC / CENOZOIC ATLANTIC DSDP NANNOFOSSILS
 **
 **             IN THE DICTIONARY USED FOR THE RASC SEQUENCE
 **          DATA, THE HIGHEST OCCURRENCES (ODD NUMBERS) ARE
 **          FOLLOWED BY THE LOWEST OCCURRENCES (EVEN NUMBERS)
 **          NOT PRINTED HERE.
 **
 ************************
```

DICTIONARY

```
  1 HI AHMUELLERELLA ASPER
  3 HI AHMUELLERELLA OCTORADIATA
  5 HI AMAUROLITHUS AMPLIFICUS
  7 HI AMAUROLITHUS DELICATUS
  9 HI AMAUROLITHUS PRIMUS
 11 HI AMAUROLITHUS TRICORNICULATUS
 13 HI APERTAPETRA GRONOSA
 15 HI ARKHANGELSKIELLA CYMBIFORMIS
 17 HI ARKHANGELSKIELLA ERRATICA
 19 HI ARKHANGELSKIELLA STRIATA
 21 HI BIDISCUS IGNOTUS
 23 HI BIDISCUS ROTATORIUS
 25 HI BISCUTUM CONSTANS
 27 HI BISCUTUM TESTUDINARIUM
 29 HI BRAARUDOSPHAERA AFRICANA
 31 HI BRAARUDOSPHAERA BIGELOWI
 33 HI BRAARUDOSPHAERA DISCULA
 35 HI BRAMLETTEIUS SERRACULOIDES
 37 HI BROINSONIA BEVIERI
 39 HI BROINSONIA LATA
 41 HI BROINSONIA ORTHOCANCELLATA
 43 HI BROINSONIA PARCA
 45 HI CALCIALATHINA OBLONGATA
 47 HI CAMPYLOSPHAERA DELA
 49 HI CAMPYLOSPHAERA EODELA
 51 HI CATINASTER CALYCULUS
 53 HI CATINASTER COALITUS
 55 HI CERATOLITHOIDES KAMPTNERI
 57 HI CERATOLITHUS ARMATUS
 59 HI CERATOLITHUS CRISTATUS
 61 HI CERATOLITHUS RUGOSUS
 63 HI CHIASMOLITHUS ALTUS
 65 HI CHIASMOLITHUS BIDENS
 67 HI CHIASMOLITHUS CALIFORNICUS
 69 HI CHIASMOLITHUS CONSUETUS
 71 HI CHIASMOLITHUS EXPANSUS
 73 HI CHIASMOLITHUS EXPANSUS SP. CF
 75 HI CHIASMOLITHUS GIGAS
 77 HI CHIASMOLITHUS GRANDIS
 79 HI CHIASMOLITHUS OAMARUENSIS
 81 HI CHIASMOLITHUS SOLITUS
 83 HI CHIASTOZYGUS AMPHIPONS
 85 HI CHIASTOZYGUS CUNEATUS
 87 HI CHIASTOZYGUS LITTERARIUS
 89 HI CHIPHRAGMALITHUS CRISTATUS
 91 HI CHIPHRAGMALITHUS QUADRATUS
 93 HI COCCOLITHUS CRASSUS
 95 HI COCCOLITHUS DEFLANDREI
 97 HI COCCOLITHUS EOPELAGICUS
 99 HI COCCOLITHUS FENESTRATUS
101 HI COCCOLITHUS MIOPELAGICUS
103 HI COCCOLITHUS PELAGICUS
105 HI COCCOLITHUS STAURION
107 HI COROLLITHION ACHYLOSUM
109 HI COROLLITHION SIGNUM
111 HI CORONOCYCLUS NITESCENS
113 HI CORONOCYCLUS SERRATUS
115 HI CRASSAPONTOSPHAERA JONESII
117 HI CRENALITHUS DORONICOIDES
119 HI CRETARHABDUS CONICUS
121 HI CRETARHABDUS CORONADVENTIS
123 HI CRETARHABDUS CRENULATUS
125 HI CRETARHABDUS LORIEI
127 HI CRETARHABDUS SURIRELLUS
129 HI CRETATURBELLA ROTHII
131 HI CRIBROSPHAERELLA EHRENBERGI
133 HI CRUCIELLIPSIS CHIASTA
135 HI CRUCIELLIPSIS CUVILLIERI
137 HI CRUCIPLACOLITHUS TENUIS
139 HI CYCLAGELOSPHAERA MARGERELII
141 HI CYCLICARGOLITHUS ABISECTUS
143 HI CYCLICARGOLITHUS FLORIDANUS
145 HI CYCLICARGOLITHUS PSEUDOGAMMAT
147 HI CYCLOCOCCOLITHUS FORMOSUS
149 HI CYCLOCOCCOLITHUS GAMMATION
151 HI CYCLOCOCCOLITHUS LEPTOPORUS
153 HI CYCLOCOCCOLITHUS LEPTOPORUS VA
155 HI CYCLOCOCCOLITHUS MACINTYREI
157 HI CYCLOCOCCOLITHUS NEOGAMMATION
159 HI CYCLOLITHELLA ROTUNDA
161 HI CYLINDRALITHUS ASYMMETRICUS
163 HI CYLINDRALITHUS GALLICUS
165 HI DIADORHOMBUS RECTUS
167 HI DIAZOMATOLITHUS LEHMANII
169 HI DICTYOCOCCITES BISECTUS
171 HI DICTYOCOCCITES SCRIPPSAE
173 HI DISCOASTER ASYMMETRICUS
175 HI DISCOASTER BARBADIENSIS
177 HI DISCOASTER BELLUS
179 HI DISCOASTER BERGGRENII
181 HI DISCOASTER BINODOSUS
183 HI DISCOASTER BOLLII
185 HI DISCOASTER BRAARUDII
187 HI DISCOASTER BROUWERI
189 HI DISCOASTER CALCARIS
191 HI DISCOASTER CHALLENGERI
193 HI DISCOASTER DECORUS
195 HI DISCOASTER DEFLANDREI
```

197 HI DISCOASTER DELICATUS
199 HI DISCOASTER DIASTYPUS
201 HI DISCOASTER DISTINCTUS
203 HI DISCOASTER DIVARICATUS
205 HI DISCOASTER DRUGGII
207 HI DISCOASTER EXILIS
209 HI DISCOASTER HAMATUS
211 HI DISCOASTER ICARUS
213 HI DISCOASTER INTERCALARIS
215 HI DISCOASTER KUGLERI
217 HI DISCOASTER LODOENSIS
219 HI DISCOASTER LOEBLICHII
221 HI DISCOASTER MEDIOSUS
223 HI DISCOASTER MOHLERI
225 HI DISCOASTER MULTIRADIATUS
227 HI DISCOASTER NEOHAMATUS
229 HI DISCOASTER NEORECTUS
231 HI DISCOASTER NOBILIS
233 HI DISCOASTER NODIFER
235 HI DISCOASTER NONARADIATUS
237 HI DISCOASTER OBTUSUS
239 HI DISCOASTER ORNATUS
241 HI DISCOASTER PANSUS
243 HI DISCOASTER PENTARADIATUS
245 HI DISCOASTER PREPENTARADIATUS
247 HI DISCOASTER QUINQUERAMUS
249 HI DISCOASTER SAIPANENSIS
251 HI DISCOASTER STELLULUS
253 HI DISCOASTER SUBLODOENSIS
255 HI DISCOASTER SURCULUS
257 HI DISCOASTER TAMALIS
259 HI DISCOASTER TANII
261 HI DISCOASTER TRIRADIATUS
263 HI DISCOASTER VARIABILIS
265 HI DISCOASTER WEMMELENSIS
267 HI DISCOASTER WOODRINGII
269 HI DISCOASTEROIDES KUEPPERI
271 HI DISCOLITHINA JAPONICA
273 HI DISCOLITHINA MULTIPORA
275 HI DISCOLITHUS PHASEOLUS
277 HI EIFFELLITHUS EXIMIUS
279 HI EIFFELLITHUS TRABECULATUS
281 HI EIFFELLITHUS TURRISEIFFELI
283 HI ELLIPSOLITHUS DISTICHUS
285 HI ELLIPSOLITHUS MACELLUS
287 HI EMILIANIA ANNULA
289 HI EMILIANIA HUXLEYI
291 HI ERICSONIA SUBPERTUSA
293 HI FASCICULITHUS TYMPANIFORMIS
295 HI GARTNERAGO OBLIGUUM
297 HI GEPHYROCAPSA APERTA
299 HI GEPHYROCAPSA CARIBBEANICA
301 HI GEPHYROCAPSA OCEANICA
303 HI GEPHYROCAPSA SP.
305 HI HAYASTER PERPLEXUS
307 HI HAYESITES ALBIENSIS
309 HI HELICOSPHAERA AMPLIAPERTA
311 HI HELICOSPHAERA COMPACTA
313 HI HELICOSPHAERA EUPHRATIS
315 HI HELICOSPHAERA GRANULATA
317 HI HELICOSPHAERA INTERMEDIA
319 HI HELICOSPHAERA KAMPTNERI
321 HI HELICOSPHAERA LOPHOTA
323 HI HELICOSPHAERA PARALLELA
325 HI HELICOSPHAERA RECTA
327 HI HELICOSPHAERA RETICULATA
329 HI HELICOSPHAERA SELLII
331 HI HELICOSPHAERA SEMINULUM
333 HI HELICOSPHAERA TRUNCATA
335 HI HELIOLITHUS KLEINPELLI
337 HI ISTHMOLITHUS RECURVUS
339 HI KAMPTNERIUS MAGNIFICUS
341 HI LANTERNITHUS MINUTUS
343 HI LITHASTRINUS FLORALIS
345 HI LITHASTRINUS GRILLII
347 HI LITHRAPHIDITES CARNIOLENSIS
349 HI LITHRAPHIDITES QUADRATUS
351 HI LOPHODOLITHUS NASCENS
353 HI LUCIANORHABDUS CAYEUXII
355 HI MANIVITELLA PEMMATOIDEA
357 HI MARKALIUS CIRCUMRADIATUS
359 HI MARKALIUS INVERSUS
361 HI MARTHASTERITES FURCATUS
363 HI MARTHASTERITES TRIBRACHIATUS
365 HI MICRANTHOLITHUS HOCHSCHULZII
367 HI MICRANTHOLITHUS OBTUSUS
369 HI MICRORHABDULUS BELGICUS
371 HI MICRORHABDULUS DECORATUS
373 HI MICULA MURA
375 HI MICULA STAUROPHORA
377 HI MINYLITHA CONVALLIS
379 HI NANNOCONUS BUCHERI
381 HI NANNOCONUS COLOMII
383 HI NANNOCONUS DOLOMITICUS
385 HI NANNOCONUS GLOBULUS
387 HI NANNOCONUS KAMPTNERI
389 HI NANNOCONUS STEINMANNII
391 HI NANNOCONUS TRUITTII
393 HI NEOCHIASTOZYGUS CHIASTUS
395 HI NEOCHIASTOZYGUS CONCINNUS
397 HI NEOCHIASTOZYGUS JUNCTUS
399 HI NEPHROLITHUS FREQUENS
401 HI OOLITHOTUS FRAGILIS
403 HI PARHABDOLITHUS ANGUSTUS
405 HI PARHABDOLITHUS ASPER
407 HI PARHABDOLITHUS EMBERGERI
409 HI PARHABDOLITHUS INFINITUS
411 HI PARHABDOLITHUS SPLENDENS
413 HI PEMMA STRADNERII
415 HI PODORHABDUS DECORUS
417 HI PODORHABDUS DIETZMANNII
419 HI PODORHABDUS ORBICULOFENESTRUS
421 HI PONTOSPHAERA DISCOPORA
423 HI PONTOSPHAERA SCUTELLUM
425 HI PREDISCOSPHAERA CRETACEA
427 HI PREDISCOSPHAERA SPINOSA
429 HI PSEUDOEMILIANIA LACUNOSA
431 HI REINHARDTITES FENESTRATUS
433 HI RETICULOFENESTRA DICTYODA
435 HI RETICULOFENESTRA GARTNERI
437 HI RETICULOFENESTRA HILLAE
439 HI RETICULOFENESTRA PSEUDOUMBILIC
441 HI RETICULOFENESTRA SAMODUROVI I
443 HI RETICULOFENESTRA UMBILICA

445 HI RHABDOLITHINA SPLENDENS
447 HI RHABDOSPHAERA CLAVIGERA
449 HI RHABDOSPHAERA INFLATA
451 HI RHABDOSPHAERA PROCERA
453 HI RHABDOSPHAERA SPINULA
455 HI RHABDOSPHAERA STYLIFER
457 HI RHOMBOASTER CUSPIS
459 HI RUCINOLITHUS IRREGULARIS
461 HI RUCINOLITHUS WISEI
463 HI SCAPHOLITHUS FOSSILIS
465 HI SCYPHOSPHAERA AMPHORA
467 HI SCYPHOSPHAERA APSTEINII
469 HI SCYPHOSPHAERA GLOBULATA
471 HI SCYPHOSPHAERA INTERMEDIA
473 HI SCYPHOSPHAERA PULCHERRIMA
475 HI SCYPHOSPHAERA RECURVATA
477 HI SPHENOLITHUS ABIES
479 HI SPHENOLITHUS BELEMNOS
481 HI SPHENOLITHUS CIPEROENSIS
483 HI SPHENOLITHUS DISSIMILIS
485 HI SPHENOLITHUS DISTENTUS
487 HI SPHENOLITHUS FURCATOLITHOIDES
489 HI SPHENOLITHUS HETEROMORPHUS
491 HI SPHENOLITHUS MORIFORMIS
493 HI SPHENOLITHUS NEOABIES
495 HI SPHENOLITHUS PREDISTENTUS
497 HI SPHENOLITHUS PSEUDORADIANS
499 HI SPHENOLITHUS RADIANS
501 HI STAUROLITHITES COMPACTUS
503 HI STAUROLITHITES CRUX
505 HI STEPHANOLITHION BIGOTI
507 HI STEPHANOLITHION LAFFITTEI
509 HI SYRACOSPHAERA HISTRICA
511 HI SYRACOSPHAERA PULCHRA
513 HI SYRACOSPHAERA SP.
515 HI TEGUMENTUM STRADNERI
517 HI TETRALITHUS ACULEUS
519 HI TETRALITHUS NITIDUS
521 HI TETRALITHUS OBSCURUS
523 HI TETRALITHUS PYRAMIDUS
525 HI TETRALITHUS TRIFIDUS
527 HI THORACOSPHAERA IMPERFORATA
529 HI TOWEIUS CRATICULUS
531 HI TOWEIUS EMINENS
533 HI TRANOLITHUS EXIGUUS
535 HI TRANOLITHUS GABALUS
537 HI TRANOLITHUS ORIONATUS
539 HI TRANSVERSOPONTIS PULCHER
541 HI TRANSVERSOPONTIS PULCHEROIDES
543 HI TRIBRACHIATUS CONTORTUS
545 HI TRIBRACHIATUS ORTHOSTYLUS
547 HI TRIQUETRORHABDULUS CARINATUS
549 HI TRIQUETRORHABDULUS INVERSUS
551 HI TRIQUETRORHABDULUS MILOWII
553 HI TRIQUETRORHABDULUS RUGOSUS
555 HI UMBILICOSPHAERA SIBOGAE
557 HI VAGALAPILLA MATALOSA
559 HI VAGALAPILLA STRADNERI
561 HI WATZNAUERIA ACTINOSA
563 HI WATZNAUERIA BARNESAE
565 HI WATZNAUERIA BIPORTA
567 HI WATZNAUERIA BRITANNICA
569 HI WATZNAUERIA COMMUNIS
571 HI WATZNAUERIA MARTELAE
573 HI ZYGODISCUS DIPLOGRAMMUS
575 HI ZYGODISCUS ELEGANS
577 HI ZYGODISCUS ERECTUS
579 HI ZYGODISCUS PHACELOSUS
581 HI ZYGODISCUS PONTICULUS
583 HI ZYGODISCUS SALILLUM
585 HI ZYGODISCUS SIGMOIDES
587 HI ZYGOLITHUS DUBIUS
589 HI ZYGRHABLITHUS BIJUGATUS

```
1************************
 **
 **
 **  (5)  BLANK DATA BASE
 **       MESOZOIC / CENOZOIC ATLANTIC DSDP NANNOFOSSILS
 **
 **
 ************************

 SEQUENCE DATA

 SEQUENCE NUMBER  1:  DSDP SITE 4;  HOLES 4, 4A
    61  -62 -151 -152 -187 -188 -243 -244 -255 -256 -439 -440   15  -43 -427 -428 -517 -523 -525   16
   -44 -518 -524 -526  131 -132 -277 -278 -361 -362 -371 -372 -533 -534 -561   13 -281 -343 -344 -403
  -425 -426 -581   14  -17  -18  -19  -20 -109 -110 -282 -404 -407 -415 -416 -562 -577 -578 -582   33
   -34 -347 -367 -368 -389 -390  348 -408 -563 -564

 SEQUENCE NUMBER  2:  DSDP SITE 4;  HOLES 4, 4A
   151 -152 -187 -188   43 -131 -281 -347 -371 -375 -425 -517   44 -132 -348 -426 -518  361 -362 -372
  -376 -523 -524  282  139 -140 -167 -168 -407 -408 -563 -564

 SEQUENCE NUMBER  3:  DSDP SITE 5;  HOLES 5, 5A
    59  -60 -151 -152 -301 -302 -319 -447 -448  157 -195 -196 -317 -318 -320 -333 -334   61  -62 -243
  -244 -255 -256  158 -169 -170 -259 -260 -311 -312 -495 -496  131 -132 -281 -282 -343 -375 -425 -561
  -563  361 -362 -376 -562  344 -426 -581   13  -14  -19  -20 -109 -110 -403 -404 -407 -415 -416 -577
  -578 -582   33 -347 -367 -368 -385 -507  139 -348 -386 -389   34 -140 -167 -168 -383 -384 -390 -408
  -505 -506 -508 -564 -567 -568

 SEQUENCE NUMBER  4:  DSDP SITE 5
   301 -302   97  -98 -305 -306  281 -282 -375 -376  139 -140 -167 -168 -389 -390 -407 -408 -563 -564

 SEQUENCE NUMBER  5:  DSDP SITE 6
    71  -81 -175 -249 -259 -587   72  -82 -176 -250 -260 -588

 SEQUENCE NUMBER  6:  DSDP SITE 6;  HOLES 6, 6A
   117 -118 -153 -154 -187 -188 -255 -256 -319 -320   71  -77  -91  -92 -147 -175 -249 -443  149 -487
   148 -150 -331 -332 -341 -342 -444   72  -75  -76  -78  -81  -82 -176 -250 -488

 SEQUENCE NUMBER  7:  DSDP SITE 7;  HOLES 7, 7A
    59 -151 -299 -300  187 -188 -243   65  -66 -175   60 -152 -244 -261 -262 -401 -402 -555 -556   77
   -78  -97  -98 -176

 SEQUENCE NUMBER  8:  DSDP SITE 7;  HOLES 7, 7A
    61 -117 -151   62 -118 -152 -187 -188   71  -72  -75  -76  -77  -78  -89  -90 -321 -322 -443 -444
    81  -82 -147 -148 -149 -150 -175 -176 -453 -454

 SEQUENCE NUMBER  9:  DSDP SITE 8;  HOLE 8A
   103 -104 -151 -152 -301 -302 -319 -320

 SEQUENCE NUMBER 10:  DSDP SITE 9;  HOLES 9, 9A
    59 -103 -289 -303 -555  151 -271 -290 -447  104  272 -275 -276 -301 -302 -319 -320 -448   60 -187
  -304 -556  243 -255 -263 -439 -477  152  264   11  -61   12  -62 -188 -244 -440 -478  256   55 -163
  -373 -425 -517 -563   56 -164 -374 -426 -518 -564

 SEQUENCE NUMBER 11:  DSDP SITE 9;  HOLES 9, 9A
   103 -104 -153 -154 -301 -302 -447 -448   59  -60 -117 -118 -455 -456   61 -191 -192 -243 -255 -439
    11  -12  -62 -244 -256 -263 -264 -440  131 -132 -375 -425 -426 -517 -523 -524  376 -518 -563 -564
```

SEQUENCE NUMBER 12: DSDP SITE 10
61 -117 -118 -187 -188 -243 -255 11 -191 -439 -440 12 -97 -192 62 -98 -157 -195 -203 -204
-205 -206 -207 -208 -244 -256 -263 -264 481 -482 -491 -492 158 -169 -196 -485 -486 -495 -496 35
-147 -259 -443 -497 -498 175 -249 -337 -338 47 -77 -148 -260 36 -48 -78 -170 -250 -444 -499
-500 93 -94 -176 -217 -218 -269 -270 -351 -352 15 -83 -84 -131 -349 -350 -425 -426 -563 -564
43 -371 -517 -525 281 -372 -519 163 -164 -353 -354 526 16 -523 277 -520 44 -132 -278 -282
-518 -524

SEQUENCE NUMBER 13: DSDP SITE 10
187 -243 -255 61 -151 -275 -439 -465 276 -466 263 11 -477 62 191 -192 207 -215 12 -203
-209 -210 -256 -264 -440 152 -188 -189 -190 -244 -478 195 -204 -216 -491 159 -196 -208 -489 -490
205 -206 -237 -238 -547 -548 481 103 -104 97 -160 -492 98 -443 482 -495 -496 -497 35 -259
-337 249 175 36 -250 -260 -338 -444 -498 77 -149 -217 -269 150 -176 78 -218 -270 -499 -500
15 -119 -123 -131 -163 -281 -347 -349 -371 -373 -375 -413 -425 -427 -445 -517 -563 164 -350 -374
-428 43 16 -44 -120 -124 -132 -282 -348 -372 -376 -414 -426 -446 -518 -564

SEQUENCE NUMBER 14: DSDP SITE 11; HOLES 11, 11A
61 -117 -118 -151 -319 62 -153 -154 -320 -447 -448 97 -187 -188 -191 -192 -243 -244 -439 98
-152 -157 -158 -440

SEQUENCE NUMBER 15: DSDP SITE 11; HOLES 11, 11A
303 -555 556 61 -473 -474 187 -188 -304 -439 -477 11 -62 -151 12 203 -478 152 -204 -440
489 -490

SEQUENCE NUMBER 16: DSDP SITE 12; HOLES 12C, 12D
151 -303 -319 -447 -555 320 59 -61 -187 -275 60 -243 103 -473 276 -448 -474 304 255 439
104 -152 191 556 62 -192 -244 -256 -440 188 -263 -264

SEQUENCE NUMBER 17: DSDP SITE 12; HOLES 12B, 12C
61 -117 -151 -301 -455 -456 187 243 -255 319 302 -320 101 -118 102 -439 11 -12 -62 -152
-188 -191 -192 -244 -256 -440

SEQUENCE NUMBER 18: DSDP SITE 13; HOLES 13, 13A
151 -187 -243 -319 -429 303 304 61 320 -430 62 -152 -188 -191 -192 -244 175 -176 -443 -444
15 -119 -131 -281 -347 -371 -425 -563 -575 16 -120 -132 -282 -348 -372 -426 -564 -576

SEQUENCE NUMBER 19: DSDP SITE 13; HOLES 13, 13A
61 -117 -118 -153 -154 -187 -243 -244 62 -188 -191 -192 -439 -440 81 -82 -89 -90 -433 -434
375 -376 -519 -520 -525 -526

SEQUENCE NUMBER 20: DSDP SITE 26
301 -319 302 -320

SEQUENCE NUMBER 21: DSDP SITE 26
59 -151 -243 -301 -319 60 -244 -299 297 555 429 477 -478 300 298 -430 -556 152 -320 302

SEQUENCE NUMBER 22: DSDP SITE 27
157 -158 -169 -170 -323 -324 -333 -334 -481 -482 -491 -492 -547 -548 77 -78 -147 -148 -175 -176
-233 -234 -249 -250 -443 -444 -487 -488

SEQUENCE NUMBER 23: DSDP SITE 27
547 143 103 -111 -112 -169 -195 -323 -485 -486 -491 311 -481 97 -482 175 -176 -196 -312 -492
98 -104 -144 -170 -324 -548

SEQUENCE NUMBER 24: DSDP SITE 27
97 -103 -169 -175 -249 -443 170 195 -196 77 33 -34 -78 -98 -104 -176 -233 -234 -250 -444

SEQUENCE NUMBER 25: DSDP SITE 28
35 -36 -47 -77 -97 -98 -175 -201 -202 -249 -443 -444 -487 -488 -587 -588 33 -34 -48 -75
-76 -78 -176 -217 -218 -250 -253 -254 -433 -434 -499 -500

```
SEQUENCE NUMBER 26:  DSDP SITE 28
  33  -77  -78  -97 -103 -175 -239   34 -499   47  -48  -98 -104 -176 -240 -500  281 -282 -347 -348
-371 -372 -375 -376 -425 -426 -563 -564

SEQUENCE NUMBER 27:  DSDP SITE 28
 103 -143 -443   75  175 -195   77  -97 -249 -589  233  323 -324   91   92 -144  234 -250   76  -78
 -98 -104 -176 -196 -444 -539 -540 -590

SEQUENCE NUMBER 28:  DSDP SITE 98
  59  -60 -103 -151 -271 -301 -302 -439 -477 -511 -555   11 -155 -173 -174 -243 -255 -263 -305 -319
  61  -62 -187 -191 -512 -556  115 -247 -465 -467  152 -244 -256 -272 -473   12  116 -156  188  192
-248 -264 -306 -440 -466 -474 -478  111 -157 -317 -323 -333 -481 -491  104 -318 -320 -324 -492  112
-158 -169 -334 -482   31  -33  -97 -147 -175 -249 -443 -499  170   75 -468   77  -81  -82 -321 -322
-549   76 -217 -444 -550   34  -69  -78  -98 -176 -218 -250   47  -65 -148 -199 -285 -293 -351 -363
-364   67 -395  181 -182 -197 -500  200 -225 -231 -286  137 -198 -352 -529  138 -396   66 -530 -531
-532   32  -48  -68  -70 -226 -232 -294   15  -43 -119 -131 -281 -371 -375 -425 -519 -563  523  120
-520 -561   44 -339 -340   16 -132 -282 -372 -376 -426 -524 -562 -564 -573 -574

SEQUENCE NUMBER 29:  DSDP SITE 99;  HOLE 99A
 103 -151 -287 -289 -290 -301 -319 -455 -463 -511 -555   59 -187 -429 -556   60 -152 -302 -512   61
-155 -173 -243 -255 -430 -439   62 -104 -156 -174 -188 -244 -256 -288 -320 -440 -456 -464 -477 -478
  13  -33 -277 -278 -367 -371 -372 -379 -381 -382 -389 -407 -563   14 -380    1  -95 -135 -385 -386
-507  383 -508  347    2  -34 -139 -167 -368 -567  136 -348  384 -390 -408  577   96 -140 -168 -505
-506 -564 -568 -578 -583 -584

SEQUENCE NUMBER 30:  DSDP SITE 100
  13  -19 -347 -383 -384 -385 -387 -389 -407 -507 -561 -563  348 -388    1   -2  -14  -20  -33  -34
-381 -382 -386 -390 -508 -562 -573   95 -139 -167 -567 -577 -583  408  135 -505 -574  136  578  506
 140   96 -168 -564 -568 -584

SEQUENCE NUMBER 31:  DSDP SITE 101;  HOLES 101, 101A
  61 -103 -155 -173 -187 -243 -255 -263 -305 -319 -429 -439   59  -60 -156 -174 -329 -430 -455 -456
-473 -477   11 -207 -247 -474   62 -189 -191 -215  330   12 -256  190 -216 -306 -320  104 -188 -192
-208 -244 -248 -264 -440 -478   13  -19 -281 -407 -425 -561 -563 -573    1  -17  -18  -20 -282 -426
-577 -581  403   14 -123 -124 -404 -562 -578 -582   95 -347 -381 -389    2  -33 -139 -367 -382 -383
-384 -385 -386   34  -96 -135 -167 -368  136 -140 -168 -348 -390 -408 -564 -574

SEQUENCE NUMBER 32:  DSDP SITE 102
  59 -103 -151 -301 -319 -455 -527 -555  511  299 -463 -556  429  271  300  302 -465 -467  187  243
-512   61 -173 -329   60 -263 -439 -456  191 -430 -477  255 -272   11  528   62  330  174 -207 -256
 215 -244 -247  188 -464  466   12 -104 -152 -192 -208 -216 -248 -264 -320 -440 -468 -478

SEQUENCE NUMBER 33:  DSDP SITE 103
  11  -61 -103 -155 -173 -187 -243 -255 -263 -319 -439 -477   62 -174  247  207 -256   12 -156 -191
 215 -317  183 -188 -209 -216 -244 -248 -318 -478  184 -210 -465  104 -192 -208 -264 -320 -440 -466

SEQUENCE NUMBER 34:  DSDP SITE 104
 247 -477  207 -263  209  103 -187 -188 -210 -243 -248 -319 -439  151 -191 -244   97 -183 -192 -215
 473  216  264 -478  152   98 -104 -157 -158 -184 -208 -320 -440 -474 -489 -490

SEQUENCE NUMBER 35:  DSDP SITE 105
  59  -60 -151 -152 -301 -319 -455 -456 -463 -464 -511 -512 -527 -555  103 -187 -439 -556  155 -302
-467 -477   11  -12  -61 -173 -191 -243 -255 -263   62 -104 -156 -174 -188 -192 -244 -256 -264 -320
-440 -468 -478 -528    1  -13  -19 -123 -281 -347 -403 -407 -425 -561 -563 -573 -577  426  282  404
 387 -507  388   33  124 -574   20   14  139    2 -508  135 -167 -383 -567   34 -348  389  384  390
-408  583  505  562  578  136 -140 -168 -506 -564 -568 -584

SEQUENCE NUMBER 37:  DSDP SITE 106;  HOLES 106, 106B
  59 -103 -151 -271 -287 -288 -289 -290 -301 -319 -455 -463 -511 -527 -555  464  429  528  272 -302
-512   60  -61 -155 -187 -243 -329 -467  430 -456  330   62 -156 -173 -174 -188 -244 -255 -263 -439
-477  468   11  -12 -207 -247 -248 -256 -556  157 -183 -489  490   53  -54  -97  -98 -152 -215 -216
-440 -478  104 -208  158 -184 -264 -320
```

SEQUENCE NUMBER 39: DSDP SITE 108
31 -47 -71 -75 -77 -97 -147 -175 -331 -453 -477 -589 32 -321 -332 -443 -499 76 -199 48
-72 -78 -98 -148 -176 -195 -196 -200 -201 -202 -322 -444 -454 -478 -500 -590

SEQUENCE NUMBER 40: DSDP SITE 108
81 -82 -299 -300 -319 -320 -321 -322 -587 -588

SEQUENCE NUMBER 41: DSDP SITE 111; HOLES 111, 111A
103 151 -287 117 -155 -271 -319 -329 -513 288 447 -448 118 -207 -208 61 -243 -255 11 -62
-187 -188 -263 -272 -273 -274 -439 -451 -452 -471 -472 -493 -494 12 -104 -152 -156 -179 -180 -185
-186 -244 -247 -248 -256 -264 -315 -316 -320 -440 47 -77 -147 -148 -175 -176 -201 -202 -217 -253
-254 -265 -266 -441 -442 81 -82 -93 -94 -235 -236 -269 -499 -500 -514 -541 -542 -587 48 -78
-218 -270 -330 -351 -352 -539 -540 -545 -546 -588 13 -15 -27 -31 -32 -119 -120 -123 -131 -163
-164 -281 -339 -349 -350 -353 -354 -371 -375 -413 -414 -425 -563 16 -28 -43 -44 -124 -132 -340
-372 -376 -426 -517 -518 -525 -526 14 -282 -343 -344 -407 -408 -564

SEQUENCE NUMBER 42: DSDP SITE 112; HOLES 112, 112A
103 -151 -155 -299 -300 -319 -329 -447 117 -271 -287 118 -187 272 -288 -330 -448 -513 -514 61
-62 -104 -188 -191 -192 -241 -242 -243 -244 -255 -256 -320 -439 -451 -452 -477 -478 156 -177 -183
-184 -185 -189 -190 -227 -228 -237 -238 -245 -246 -263 -493 -494 -553 178 -186 97 -152 -195 -207
-208 -264 -440 -554 63 -141 -142 -157 -169 -171 -196 -435 -491 436 64 -79 -233 -234 -327 -328
-337 -437 -443 -453 -454 -492 80 -175 -249 -338 -438 -541 47 -48 -73 -74 -98 -99 -100 -147
-158 -170 -172 -201 -202 -311 -312 -359 -360 -413 -414 -441 -442 -444 -495 -496 -539 -542 -587 71
-72 -81 -82 -105 -106 -145 -146 -148 -176 -235 -236 -250 -253 -254 -449 -450 -540 -588 -589 -590

SEQUENCE NUMBER 43: DSDP SITE 113
103 -301 -302 -513 -514 155 187 -188 151 -243 -255 -439 104 -152 -156 -244 -256 -440

SEQUENCE NUMBER 44: DSDP SITE 114
151 -287 -288 -299 -300 -319 103 -155 -329 104 -117 -118 -152 -156 -273 -274 -320 -330

SEQUENCE NUMBER 45: DSDP SITE 115
155 -156 -271 -272 -319 -320

SEQUENCE NUMBER 46: DSDP SITE 116; HOLES 116, 116A
103 -151 -319 117 -155 -271 -273 -274 -287 -299 -300 -509 -510 329 -447 -448 -513 -514 272 -288
-320 118 -152 -156 -187 -188 -263 -264 -330 -467 -468 -511 -512 97 -157 -315 -316 -489 -490 141
-171 -195 -323 -324 -491 -589 63 -64 -99 -142 -169 -196 -435 -436 -492 79 -100 -147 -233 -234
-337 -437 -443 98 -158 -249 -338 -341 -342 -438 73 -74 -80 -104 -148 -170 -172 -250 -444 -590

SEQUENCE NUMBER 47: DSDP SITE 117; HOLES 117, 117A
33 -34 -63 -97 -141 -157 -169 -171 -195 -491 -589 142 -196 -435 -481 -482 -492 98 -170 -172
-333 -334 64 -158 -436 77 -78 -93 -94 -181 -182 -359 -541 -545 -587 269 -270 -351 -352 539
-540 31 -65 -225 -226 -239 -240 -283 -284 -393 -394 -397 -398 -588 32 -66 -199 -200 -285 -286
-360 -542 -543 -544 -546 -590

SEQUENCE NUMBER 48: DSDP SITE 118
103 -117 -151 -155 -271 -272 -273 -299 -300 -319 -329 -330 -447 -448 104 -118 11 -12 -243 -244
-439 97 -185 -191 -195 -207 -263 -320 -553 156 -192 -264 -274 98 -152 -186 -196 -208 -440 -493
-494 -554 47 -48 -71 -72 -75 -76 -77 -81 -89 -91 -92 -105 -145 -147 -175 -249 -250 -265
-266 73 -74 -82 -90 -106 -146 -549 -550 93 -148 -217 -218 -269 -270 -545 49 -69 -78 -94
-176 -199 -200 -225 -239 -285 -286 -546 50 -65 -66 -70 -226 -240 -457 -458 -531 -532 -543 -544

SEQUENCE NUMBER 49: DSDP SITE 119
103 -104 -151 -319 97 -152 -207 -208 -315 -316 -320 -439 -440 -493 -494 -553 -554 63 -99 -141
-142 -157 -169 -171 -195 -435 -436 -485 -486 -491 -492 -495 -496 64 -79 -80 -98 -100 -158 -170
-172 -196 -233 -259 -260 -311 -312 -337 -338 -437 -438 -443 -444 47 -48 -75 -76 -77 -147 -148
-175 -234 -249 -250 -265 -441 -442 78 -89 -90 -105 -106 -145 -146 -253 -254 -266 -449 -450 -499
-500 -549 -550 217 -235 -269 176 -218 -236 -270 -545 -546 31 -49 -50 -65 -67 -69 -197 -198
-225 -226 -283 -393 -397 -398 -531 -585 -589 -590 66 -68 -223 -224 -284 -285 -293 -335 70 -137
-286 -294 -336 -394 -532 32 -138 -291 -292 -586

SEQUENCE NUMBER 50: DSDP SITE 120
45 -46 -129 -130 -139 -140 -347 -348 -357 -358 -459 -460 -563 -564 119 -120 -135 -136 -167 -168
-507 -508 -567 -568

SEQUENCE NUMBER 51: DSDP SITE 120
103 -104 -159 -160 -191 -192 -195 -196 -203 -204 -237 -238 19 -27 -109 -110 -123 -347 -355 -357
-403 -407 -445 -507 -563 -573 -577 356 -404 -446 -508 29 -135 -358 -367 -368 -381 -382 -387 -388
-389 -390 -391 -392 -574 30 -578 139 -567 20 -28 -124 -136 -140 -345 -346 -348 -408 -564 -568

SEQUENCE NUMBER 52: DSDP SITE 120
103 -104 -159 -160 -191 -192 -195 -196 -203 -204 -237 -238 -477 -478 27 -347 -355 -357 -403 -407
-445 -563 -573 19 -404 -577 109 -110 -123 -507 356 -446 -508 343 -344 135 -381 -387 -388 29
-358 -367 -368 -382 -389 -390 -391 -392 -574 -579 -580 578 30 139 -567 20 -28 -345 -346 -348
124 -136 -140 -408 -564 -568

SEQUENCE NUMBER 53: DSDP SITE 135
3 -37 -38 -39 -40 -41 -42 -87 4 -15 -16 -281 -282 -425 -426 23 -25 -121 -123 -127
-133 -139 -167 -307 -347 -355 -357 -365 -381 -405 -407 -409 -411 -412 -415 -417 -431 -501 -503 -507
-515 -565 -567 -569 -573 122 -129 -366 -379 -380 -535 107 -119 -165 -166 24 -120 -128 -130 -134
-136 -140 -348 -356 -358 -368 -382 -406 -408 -410 -418 -432 -502 -504 -508 -516 -566 -570 -574 135
-367 -536 108 -279 -280 88 26 -124 -168 -308 -416 -568

SEQUENCE NUMBER 54: DSDP SITE 135
103 -151 -301 -319 -421 -423 -447 -509 -555 302 -448 -510 -556 11 -61 -117 -155 -187 -243 -255
-329 -424 -429 -439 -477 191 -263 430 422 62 12 -256 179 -180 -189 -227 -553 118 -156 -188
-190 -192 -228 -244 -330 -478 -554 97 -143 -195 -203 -207 -489 -491 113 -152 -208 -320 204 -264
-440 -490 141 -492 -547 98 -104 -114 -142 -144 -169 -170 -196 -479 -480 -548 181 -182 -217 -218
-221 -222 -363 -364 -457 -458

SEQUENCE NUMBER 55: DSDP SITE 136
9 -11 -103 -151 -155 -187 -191 -243 -263 -439 -553 173 -174 247 -248 10 -12 -188 -192 -244
97 -143 -195 -203 -207 -489 -491 113 -156 -554 264 114 152 -208 267 -309 -310 -479 -490 -547
480 141 -319 -320 -440 144 -204 -205 -268 -492 206 98 -104 -142 -196 -548

SEQUENCE NUMBER 56: DSDP SITE 136
23 -25 -87 -109 -110 -121 -123 -127 -131 -132 -277 -278 -279 -280 -281 -282 -347 -355 -371 -372
-425 -427 -507 -515 -533 -534 -573 107 -119 -133 -139 -165 -166 -167 -307 -357 -403 -405 -407 -409
-415 -417 -431 -459 -501 -503 -535 -565 -567 -569 125 -129 -135 -136 -411 -412 308 26 -108 -122
-124 -126 -140 -345 -346 -348 -356 -358 -408 -418 -460 -502 -504 -508 -516 -568 24 -29 -30 -88
-120 -128 -130 -134 -168 -367 -368 -379 -380 -381 -382 -391 -392 -404 -406 -410 -416 -419 -420 -426
-428 -432 -536 -566 -570 -574

SEQUENCE NUMBER 57: DSDP SITE 137
85 -87 -107 -119 -121 -123 -125 -127 -133 -281 -343 -347 -355 -357 -403 -405 -407 -411 -415 -419
-425 -501 -515 -533 -535 -557 -565 -569 -573 23 -25 -131 -427 -431 -503 109 507 537 -567 279
37 -39 -41 17 -409 18 110 -410 86 132 307 -459 -460 108 428 38 -40 -42 -280 -420
-432 24 -26 -88 -120 -122 -124 -126 -128 -134 -282 -308 -344 -348 -356 -358 -404 -406 -408 -412
-416 -426 -502 -504 -508 -516 -534 -536 -538 -558 -566 -568 -570 -574

SEQUENCE NUMBER 58: DSDP SITE 139
59 -61 -103 -117 -151 -155 -161 -187 -243 -255 -319 -421 -423 -429 -439 -451 60 -62 -118 -162
-452 11 -191 -247 -263 -477 -553 273 -422 12 -554 188 -192 -244 -248 -256 -424 -430 -478 97
-143 -195 -207 -215 -264 -329 -330 152 -156 -208 -216 -320 -491 267 203 -268 -274 -547 440 141
-313 -314 98 -104 -142 -144 -196 -492 -548 204

SEQUENCE NUMBER 59: DSDP SITE 140
59 -61 -103 -117 -151 -155 -187 -243 -255 -257 -319 -329 -401 -423 -429 -447 448 60 -173 402
62 -118 -152 -156 -174 -188 -244 -256 -258 -330 -421 -422 -424 -430 97 -113 -143 -195 -203 -263
-309 -313 -317 -439 -489 -491 264 -314 -479 547 318 98 -104 -114 -144 -196 -204 -267 -268 -310
-320 -440 -480 -490 -492 -548

SEQUENCE NUMBER 60: DSDP SITE 141
59 -103 -117 -151 -155 -319 -329 -423 -429 -447 -509 -555 421 -513 -514 401 187 61 60 422
-448 -510 243 -255 -271 -556 257 -424 272 161 118 -263 -477 439 162 -258 -430 11 -402 62
-191 -330 5 -247 -553 6 554 12 -192 -256 -320 -478 152 -156 -183 -189 -209 -227 440 190
-248 104 -184 -188 -210 -228 -244 -264

SEQUENCE NUMBER 61: DSDP SITE 142
151 -301 -319 -429 -509 -555 117 510 302 59 423 60 -155 -421 -447 -448 329 61 -173 -187
-191 -243 -255 -257 -263 -439 -477 330 273 258 -430 424 118 -422 11 -103 179 -247 -553 189
-227 9 305 10 -12 -180 -192 -244 -248 -256 156 -203 -306 -478 -554 62 -174 -188 -190 -228
-264 -491 97 -195 -207 143 -489 490 -492 98 -104 -144 -152 -196 -204 -208 -274 -320 -440 -556

SEQUENCE NUMBER 62: DSDP SITE 143
25 -26 -31 -32 -87 -88 -107 -108 -109 -110 -121 -122 -127 -128 -355 -356 -403 -404 -411 -412
-417 -418 -419 -420 -501 -502 -503 -504 -507 -508 -573 -574

SEQUENCE NUMBER 63: DSDP SITE 144; HOLES 144, 144A
15 -25 -37 -39 -41 -85 -119 -127 -131 -161 -371 -375 -407 -425 -515 -573 87 -427 -565 3
-281 -339 -355 -411 347 503 -507 -517 -523 -533 345 -403 -405 -415 -525 43 -44 -521 -522 -537
277 -340 569 16 -518 -526 23 -109 -343 -369 -557 107 359 -360 278 -370 4 -121 -346 -376
-524 -566 123 -132 -162 -279 -538 419 -501 428 38 -40 -42 -86 -110 -280 -282 -420 -426 29
-88 -122 -133 -134 -372 -416 -535 31 -348 -412 -534 357 -358 -502 26 -128 -408 -508 -536 570
30 -356 -406 32 -108 -124 -404 -516 24 -120 -344 -504 -558 -574

SEQUENCE NUMBER 64: DSDP SITE 144; HOLES 144, 144A
97 -143 -169 -195 -233 -311 -313 -317 -437 -485 -491 -495 497 333 11 -12 -103 -104 -151 -152
-155 -156 -187 -188 -191 -192 -243 -244 -247 -248 -255 -256 -263 -264 -319 -320 -423 -424 -477 -478
334 113 327 -486 -589 341 -443 293 147 337 -338 175 35 273 -314 -438 -498 342 196 -318
47 -77 -249 36 -114 -274 499 -549 -590 331 81 -550 75 71 -76 -234 48 -72 -78 -82
-98 -144 -148 -170 -176 -250 -312 -328 -444 -492 -496 49 -65 -67 -69 -137 -221 -225 -283 -285
-332 -500 -529 -585 -587 -588 335 222 -395 50 -66 -226 -393 -394 138 -223 224 -284 336 -586
286 68 -70 -294 -396 -530

SEQUENCE NUMBER 65: DSDP SITE 332; HOLE 332A
59 -60 -103 -151 -299 -301 -319 -329 -447 -475 271 -300 -302 -429 -448 -467 117 -173 -185 -187
-243 -251 -255 -451 -465 -473 61 115 -257 -471 439 330 263 474 430 244 104 -320 62 -116
-118 -152 -174 -186 -188 -252 -256 -258 -264 -272 -440 -452 -466 -468 -472 -476

SEQUENCE NUMBER 66: DSDP SITE 332; HOLE 332B
103 -117 -151 -173 -185 -187 -251 -255 -271 -319 -329 -429 -451 -465 -467 -471 115 -193 -194 -243
-257 -439 430 475 61 -330 -452 -472 62 -104 -116 -118 -152 -174 -186 -188 -244 -252 -256 -258
-263 -264 -272 -320 -440 -466 -468 -476 -477 -478

SEQUENCE NUMBER 67: DSDP SITE 333; HOLES 333, 333A
103 -151 -289 -299 -301 -319 -329 -475 59 -60 -447 -448 290 -300 -302 115 -117 -173 -185 -187
-243 -251 -255 -257 -271 -429 -465 -467 473 451 263 61 -439 471 193 194 452 477 478 62
-474 -476 104 -116 -118 -152 -174 -186 -188 -244 -252 -256 -258 -264 -272 -320 -330 -430 -440 -466
-468 -472

SEQUENCE NUMBER 68: DSDP SITE 333
173 -187 -243 -255 -257 174 -258 -261 188 -244 -256 -262

SEQUENCE NUMBER 69: DSDP SITE 334
9 -103 -151 -179 -185 -187 -241 -243 -251 -263 -319 -439 -467 -473 -475 -477 5 -191 -213 -255
-271 -451 -471 -553 247 6 -10 256 211 219 244 -248 -452 180 242 476 177 245 212 -220
-468 472 246 183 -214 51 -272 52 -474 -554 104 -152 -178 -184 -186 -188 -192 -252 -264 -320
-440 -478

SEQUENCE NUMBER 70: DSDP SITE 335
59 -103 -151 -173 -271 -319 -329 -429 -467 -469 -473 -475 -477 185 -187 -243 -251 -255 60 257
465 -553 213 61 -191 -471 193 194 -330 174 -263 62 -256 -264 -320 -466 -470 -472 -476 179
-211 -241 -247 252 104 -180 -186 -212 -242 -244 -468 -474 101 -188 -214 102 -152 -192 -207 -208
-248 -258 -272 -430 -478 -554

SEQUENCE NUMBER 71: DSDP SITE 353
59 -151 -289 -290 -299 -301 -319 61 -187 -263 -325 -326 -329 -330 -447 -463 -509 -555 62 -101
-102 -103 -111 -112 -143 -188 -191 -192 -195 -205 -206 -207 -208 -243 -255 -257 -258 -264 -429 -479
-480 -489 -490 -493 -494 -495 -496 -547 -548 -589 -590 511 196 104 -510 60 -144 -152 -244 -256
-300 -302 -320 -430 -448 -464 -491 -492 -512 -556

SEQUENCE NUMBER 72: DSDP SITE 354
59 -151 -289 -299 -301 -319 -447 -463 -509 -555 511 290 -455 -456 61 -287 -297 -298 -300 -302
-429 103 -155 -185 -187 -213 -243 -255 -257 -451 9 173 11 -263 -288 -439 -553 191 -258 -430
-469 7 -57 -62 -174 -477 -493 60 -512 247 -510 229 219 309 58 310 177 -178 -214 491
5 12 -179 -189 -452 6 -230 8 -111 -183 -256 10 -180 -227 -248 -377 305 464 -556 209
51 -53 -186 378 184 470 52 -54 -190 -210 -220 -228 -244 -478 207 -211 -215 -315 101 -195
-212 -264 448 188 -192 -216 -320 -551 313 -316 -489 -547 -554 143 208 -490 156 -440 -479 -480
205 -206 306 102 -104 -112 -144 -152 -196 -314 -492 -494 -548 -552

SEQUENCE NUMBER 73: DSDP SITE 354
97 -101 -111 -141 -143 -195 -313 -331 -483 -491 -547 -589 321 484 481 169 102 -325 -548 551
485 -495 -552 259 -497 311 -482 327 486 112 -142 171 -443 437 147 -337 326 144 175 -181
-196 -249 -312 -314 -433 -438 499 35 -170 338 -487 -496 172 -434 -444 -498 47 -75 -77 -81
-265 -587 -588 328 36 250 260 -488 -492 48 -76 -78 -82 -98 -148 -176 -182 -266 -322 -332
-500 -590

SEQUENCE NUMBER 74: DSDP SITE 354
49 -65 -69 -97 -223 -225 -231 -283 -397 -529 -531 -585 359 293 181 182 285 137 226 -291
398 221 -222 -232 50 -66 224 -335 395 360 -396 286 -336 138 98 -284 70 -292 -294 -530
-532 -586

SEQUENCE NUMBER 75: DSDP SITE 354
3 -4 -15 -87 -119 -131 -281 -295 -347 -349 -353 -375 -399 -400 -403 -411 -415 -425 -521 -563
339 350 -585 427 -522 120 -348 428 404 -586 88 -412 282 -416 132 -296 -340 -354 -376 -426
16 -564

SEQUENCE NUMBER 76: DSDP SITE 366; HOLE 366A
173 -187 -193 -194 -243 -255 -257 -305 174 -188 -244 -256 -258 -261 -262 -306

SEQUENCE NUMBER 77: DSDP SITE 366
83 -84 -131 -132 -281 -282 -291 -292 -349 -350 -371 -372 -375 -376 -407 -408 -563 -564

SEQUENCE NUMBER 78: DSDP SITE 367
173 -187 -243 -255 -257 -261 174 -188 -193 -194 -244 -256 -258 -262

SEQUENCE NUMBER 79: DSDP SITE 367
521 -522 25 -87 -109 -123 -281 -343 -419 -425 -463 -464 -507 -557 -559 -563 -573 -577 295 -420
353 -354 296 -405 -407 -565 83 -127 -575 558 21 -84 -107 -347 -355 -403 22 -110 -119 -279
-280 -417 -418 -576 108 -426 26 282 -533 -534 404 -578 508 88 -167 -344 -571 409 367 139
-365 -381 128 -366 120 -368 -574 135 124 -410 560 356 165 -166 -406 -461 168 462 136 -348
-382 408 140 -564 -566 -572

SEQUENCE NUMBER 80: DSDP SITE 368
339 -340 563 39 -343 -355 -375 -376 -573 281 -565 21 -22 -25 -419 -420 -425 295 -407 -408
-557 -558 26 -296 -403 -404 -566 40 -87 -88 -131 -132 -282 -344 -356 -426 -564 -574 -575 -576

SEQUENCE NUMBER 81: DSDP SITE 369; HOLE 369A
3 -15 -83 -127 -131 -281 -339 -347 -349 -371 -373 -375 -413 -425 -523 -563 -571 -573 355 -374
350 -517 -565 119 -295 -353 -407 -537 -559 4 -414 525 16 -354 -518 -526 43 -44 -277 84
-85 -107 -109 -121 -123 -343 -361 -411 87 -372 25 -86 -279 -280 -296 -340 -362 -369 -370 -376
-403 -507 -524 -557 21 -132 -405 278 -282 -535 -575 29 307 -417 39 -463 -538 419 -577 40
409 -410 -418 22 -110 -120 -124 -308 -412 -420 -464 -508 -572 26 -30 -88 -108 -122 -128 -344
-348 -356 -404 -406 -408 -426 -536 -558 -560 -564 -566 -574 -576 -578

SEQUENCE NUMBER 82: DSDP SITE 370
173 -187 -213 -214 -243 -255 -257 174 -188 -244 -256 -258 -261 -262

SEQUENCE NUMBER 83: DSDP SITE 370
21 -25 -87 -107 -109 -119 -121 -123 -127 -131 -281 -343 -347 -355 -403 -405 -407 -411 -419 -425
-507 -559 -563 -565 -573 -577 417 535 -557 -571 -575 459 -460 279 108 -110 -122 -132 -412 -463
-464 280 -282 -420 -426 -536 -558 88 344 -404 381 367 135 23 -24 129 -365 409 -410 167
45 139 576 26 22 46 -366 -368 -382 120 -140 -418 -508 -574 -578 124 -128 -130 -136 -168
-348 -356 -406 -408 -560 -564 -566 -572

```
1************************
 **
 **
 **  (6)  RUBEL'S BALTIC SILURIAN DATA BASES
 **
 **
 ************************
```

DICTIONARY

1 SHALERIA SP. SP.
2 AEGIRIA GRAYI
3 GLASSIA OBOVATA
4 ANASTROPHIA DEFLEXA
5 ATRYPA RETICULARIS
6 RESSERELLA SP. SP.
7 DICOELOSIA BILOBA
8 ISORTHIS CRASSA
9 DALEJINA HYBRIDA
10 CYRTIA EXPORRECTA
11 SKENIDIOIDES LEWISI
12 ATRYPINA BARRANDI
13 CLORINDA SP. SP.
14 CYPIDULA GALEATA
15 DELTHYRIS ELEVATA
16 STRIISPIRIFER PLICATELLUS
17 VISBYELLA VISBYENSIS
18 STROPHONELLA EUGLYPHA
19 ISORTHIS CANALICULATA
20 DAYIA NAVICULA
21 PROTOCHONETES MINIMUS
22 STROPHOCHONETES CINGULATUS
23 PROTOCHONETES STONISHKENSIS
24 BRACHYPRION KURZEMENSIS
25 MICROSPHAERIDIORHYNCUSNUCULA
26 PROTOCHONETES PILTENENSIS
27 HOMOEOSPIRA BAYILEI
28 ATRYPOIDEA PRUNUM
29 DELTHYRIS MAGNA
30 SHALERIA DZWINOGRODENSIS
31 MORINORHYNCHUS ORBIGNYI
32 EOMARTINIOPSIS LUDLOVENSIS
33 EOSPIRIFER RADIATUS
34 HOWELLELLA SP. SP.
35 SPHAERIRHYNCHIA WILSONI
36 STEGERHYNCHUS DIODONTUS
37 PENTAMERIID
38 COELOSPIRA PUSILLA
39 DIDYMOTHYRIS DIDYMA
40 STEGERHYNCHUS PSEUDOBIDENTATUS
41 PROTOCHONETES STRIATELLUS
42 SALOPINA CONSERVATRIX
43 SEPTATRYPA SUBAEQUALIS
44 RESSERELLA SAWDDENSIS
45 PROTOZEUGA BICARINATA
46 PHOLIDOSTROPHIA LAEVIGATA
47 NUCLEOSPIRA PISUM
48 EOPLECTODONTA SP. SP.
49 DICOELOSIA OSLOENSIS
50 ATRYPA HEDEI
51 LEPTOSTROPHIA FILOSA
52 PLAGIORHYNCHA DEPRESSA
53 LEANGELLA SP. SP.
54 PTYCHOPLEURELLA SP. SP.
55 ESTONIRHYNCHIA SP. SP.
56 MERISTINA OBTUSA
57 AECHMINA MOLENGRAAFFI
58 ALANELLA TECTUMIFORMIS
59 AMYGDALELLA NASUTA
60 A. SUBCLUSA
61 A. SOLIDA
62 BEYRICHIA GLOBIFERA
63 B. CF. GLOBIFERA
64 B. SIMPLICIB
65 B. SNODERNIANA
66 B. VENUSTA
67 BOLLIA AMABILIS
68 B. PARVA
69 BORUSSULUS RETICULIFER
70 BEROLINELLA STEUSLOFFI
71 CALCARIBEYRICHIA ALTONODOSA
72 CAVELLINA ANGULATA
73 C. BALTICA
74 C. CIRCULATA
75 CLAVOFABELLA ATTRITA
76 C. HETEROSA
77 C.? LATIVELATA
78 C. MAXIMA
79 C. NODOSA
80 C. RELIQUA
81 CRASPEDOBOLBINA EZERENSIS
82 C. LIETUVENSIS
83 C. PERCURRENS
84 CUTHERELLINA MAGNA
85 DELOSIA CUNEATA
86 DIZYGOPLEURA OPPORTUNA
87 FROSTIELLA GROENVALLIANA
88 F. LEBIENSIS
89 F. PLICULATA
90 HAMMARIELLA PULCHRIVELATA
91 HEBELLUM TETRAGONA
92 H. TRIVIALE
93 HEMSIELLA ANTEROVELATA
94 H. HEMSIENSIS
95 H. LATVIENSIS
96 HEMSIELLA LOENSIS
97 H. MARGARITAE
98 H. MACCOYIANA
99 HOBURGIELLA ANTEROVELATA
100 JUVIELLA JUVENSIS
101 J. PILTENENSIS
102 KURESAARIA SP.
103 KLOEDENIA LEPTOSOMA
104 LEIOCYAMUS LIMPIDUS

105 LIMBINARIELLA MACRORETICULATA
106 L. MALORNATA
107 LOPHOCTENELLA ANGUSTILAQUEATA
108 MICROCHEILINELLA LACRIMA
109 MACRYPSILON SALTERIANA
110 M. PARVISULCATUM
111 NOVIBEYRICHIA BALTICIVAGA
112 NEOBEYRICHIA ALIA
113 N. BUCHIANA
114 NODIBEYRICHIA TUBERCULATA
115 N. SCISSA
116 N. BIFIDA
117 N. GEDANENSIS
118 N. JURASSICA
119 N. SALDUSENSIS
120 NEOBEYRICHIA BULBATA
121 N. CTEROPHORA
122 N. INCERTA
123 N. LAUENSIS
124 N. NUTANS
125 OCHESAARINA SP.
126 O. VARIOLARIS
127 "OCTONARIA" PERPLEXA
128 ORCOFABELLA ARANEOSA
129 O. ARGUTA
130 O. OBSCURA
131 O. TESTATA
132 PARABOLBINA BALTICA
133 PLICIBEYRICHIA CALCARISPINOSA
134 PLICIBEYRICHIA NUMEROSA
135 POLENOVULA REDA
136 PRIMITIOPSELLA RECTELLAFORMIS
137 PRIMITIOPSIS EZERENSIS
138 PSEUDORAYELLA ACUTA
139 P. SCALA
140 RETISACCULUS SEMICOLONATUS
141 R. SULCATUS
142 SACCELATIA BIMARGINATA
143 S. OLESKOENSIS
144 S. PERSONATA
145 S. SIMPLEX
146 SCIPIONIS AMPLUS
147 S.? ASSUCTUS
148 S. PRAECEPS
149 S. PROFUNDIGENUS
150 S. VAGUS
151 SIGNETOPSIS DECORATA
152 SLEIA INERMIS
153 S. EQUESTRIS
154 UNDULIRETE BALTICUM
155 VENZAVELLA COSTATA
156 V. DICOSTATA
157 V. MULTICOSTATA
158 V. SUBCOSTATA
159 ANASPIDA GEN. N. ET SP.A
160 ANDREOLEPIS HEDEI
161 CYATHASPIDINAE (ARCHEOGONASPIS?)
162 GOMPHONCHUS HOPPEI
163 G. HOPPEI OR PARACANTHODES POROSUS
164 GOMPHONCHUS SANDELENSIS
165 GONIPORUS ALATUS
166 KATOPORUS TIMANICUS
167 K. TRICAVUS
168 LOGANIA CUNEATA
169 L. MARTINSSONI
170 L. TAITI
171 LOPHOSTEUS SUPERBUS
172 NOSTOLEPIS GRACILIS
173 N. N. SP.
174 N. STRIATA
175 PHLEBOLEPIS ELEGANS
176 P. ORNATA
177 PORACANTHODES PUNCTATUS
178 SAAROLEPIS OESELENSIS
179 STROSIPHERUS INDETATUS
180 THELODUS ADMIRABILIS
181 T. LAEVIS
182 T. MARGINATUS
183 T. PARVIDENS
184 T. SCULPTILIS
185 T. TRAQUAIRI
186 TOLYPELEPIS UNDULATA
187 TRAQUAIRASPIS SP.
188 TREMATASPIS SP. SP.
189 TYLODUS DELTOIDES
191 ANASPIDA GEN ET SP. N. B
192 LOGANIA KUMMEROWI
193 L. LUDLOWIENSIS
194 PORACANTHODES CF. PUNCTATUS
195 TESSERASPI SP.
196 TURINIA PAGEI
197 THYESTES SP.
198 OESELASPIS SP. ?
199 OSTEOOSTRACI (GEN. INDET.)

```
1*************************
 **
 **
 **  (6A)  RUBEL'S BALTIC SILURIAN DATA BASES
 **
 **
 *************************

 SEQUENCE DATA   (BRACHIOPODS)
```

```
01
052 0327.10 0278.38
044 0319.05 0295.70
006 0314.90 0160.58
022 0307.36 0160.22
003 0306.44 0229.23
008 0299.20 0160.42
048 0297.85 0187.14
013 0296.85 0281.75
002 0292.55 0287.70
011 0292.55 0273.70
005 0285.04 0058.00
053 0284.38 0273.70
033 0272.83 0236.90
055 0272.83 0160.00
056 0252.85 0206.80
018 0237.00 0206.48
009 0236.90 0005.10
016 0210.65 0198.25
010 0183.10 0182.40
037 0182.45 0182.40
034 0164.31 0095.38
025 0160.00 0001.75
031 0159.26 0059.10
035 0150.27 0116.93
041 0117.70 0095.40
039 0116.43 0095.17
042 0116.20 0001.75
015 0098.08 0004.30
027 0091.95 0001.75
026 0089.73 0001.75
029 0085.35 0001.75
028 0079.25 0047.67
019 0074.10 0001.75
030 0062.20 0004.20
040 0015.90 0015.90
```

```
02
039 0075.60 0057.20
041 0057.20 0057.20
034 0057.20 0057.20
042 0057.20 0012.70
031 0057.20 0009.30
025 0057.20 0004.35
027 0057.20 0004.00
029 0049.95 0004.35
028 0049.50 0013.90
036 0048.00 0029.60
015 0045.55 0006.00
030 0044.20 0009.80
009 0039.30 0004.35
019 0038.10 0009.30
005 0033.00 0015.10
026 0028.00 0013.00
020 0019.00 0016.25
040 0013.00 0012.85
```

```
03
022 0462.50 0418.00
010 0460.00 0429.60
008 0460.00 0379.00
048 0460.00 0417.50
003 0460.00 0417.00
011 0457.10 0377.50
045 0447.30 0385.00
043 0444.50 0444.50
025 0433.90 0173.90
005 0432.30 0198.60
009 0429.60 0252.00
021 0417.80 0377.30
013 0416.80 0415.20
006 0394.00 0382.00
034 0391.30 0237.00
035 0386.80 0386.80
033 0376.80 0376.80
028 0369.45 0245.90
026 0282.90 0175.00
030 0281.00 0199.40
019 0280.70 0173.90
036 0279.90 0279.90
027 0268.90 0173.90
015 0268.90 0192.70
029 0257.10 0191.00
037 0239.70 0239.60
031 0266.00 0210.00
047 0404.20 0393.20
020 0261.30 0253.60
```

```
04
002 0638.00 0577.20
028 0638.00 0275.10
011 0630.60 0555.40
008 0630.60 0513.40
020 0630.60 0390.00
031 0616.00 0324.20
034 0616.00 0275.10
021 0597.10 0586.90
010 0594.80 0594.80
013 0594.80 0577.20
018 0588.10 0555.40
033 0577.20 0577.20
005 0561.50 0396.00
014 0531.20 0520.40
001 0513.40 0473.20
035 0513.00 0488.80
026 0488.80 0275.10
025 0488.80 0270.20
019 0473.70 0317.80
015 0470.00 0291.50
036 0450.20 0449.60
030 0450.20 0320.10
027 0450.20 0275.50
029 0450.20 0275.50
037 0418.10 0418.10
042 0306.80 0275.10
```

```
05
003 0826.60 0671.00
002 0795.30 0644.00
008 0795.30 0634.93
038 0757.00 0606.80
043 0748.50 0740.00
034 0723.50 0540.70
020 0698.50 0519.70
010 0697.80 0697.80
005 0684.50 0469.60
032 0672.50 0644.00
026 0634.93 0434.50
028 0644.00 0523.00
001 0634.93 0538.70
015 0634.90 0443.50
019 0630.00 0473.50
004 0630.00 0624.00
018 0630.00 0610.50
009 0630.00 0594.00
014 0624.00 0609.20
025 0621.50 0418.50
031 0609.20 0456.50
033 0608.40 0608.10
023 0601.00 0598.20
024 0594.00 0473.50
030 0551.80 0443.50
027 0545.00 0443.10
047 0473.50 0473.50
040 0472.90 0472.90
029 0516.90 0448.00
```

```
06
003 1310.50 1247.80
020 1310.50 0979.30
023 1269.70 1073.50
019 1266.80 1053.00
005 1187.20 1103.00
034 1166.00 1092.00
025 1187.20 0984.00
001 1164.00 1164.00
030 1103.00 0979.30
026 1069.50 1001.50
027 1069.50 0984.00
029 1001.50 0984.00
```

```
07
003 1124.40 1042.40
032 1110.80 1035.90
020 1097.00 0927.90
008 1087.00 0978.20
018 1082.50 1077.60
009 1082.50 0962.75
038 1082.00 1056.00
001 1081.70 0975.80
015 1080.80 1059.00
005 1080.00 0927.90
010 1079.60 1042.00
007 1078.95 1052.00
033 1078.70 1077.30
014 1078.70 1078.50
025 1030.40 0927.90
023 1030.00 0971.85
030 0998.50 0927.90
026 0991.60 0927.90
029 0984.00 0980.30
024 0998.50 0927.90
031 0980.40 0979.00
019 0976.30 0927.50
034 0976.00 0930.80
028 0963.20 0963.20
027 0960.00 0960.00
```

```
08
017 1150.42 1125.30
013 1150.42 1005.50
048 1150.42 1009.45
002 1150.42 0882.10
049 1149.85 1122.65
003 1140.75 0986.40
011 1140.75 0965.90
044 1120.80 1018.10
022 1115.85 1009.35
052 1113.80 1087.90
045 1035.40 1035.40
008 1012.05 0919.45
043 1001.50 0972.70
005 1001.10 0753.00
034 0986.40 0754.90
028 0984.00 0743.80
018 0983.00 0969.80
010 0978.55 0978.55
021 0978.55 0978.35
020 0978.35 0742.90
009 0971.20 0747.75
014 0944.25 0944.25
031 0887.45 0757.20
019 0885.20 0713.30
015 0881.00 0732.20
001 0881.00 0802.15
025 0862.10 0730.80
026 0862.10 0705.50
029 0851.40 0802.15
030 0847.00 0738.40
023 0811.30 0762.30
027 0807.00 0729.70
037 0765.10 0765.10
```

```
09
011 0515.50 0511.30
007 0515.50 0507.90
002 0515.50 0499.60
003 0515.50 0499.60
008 0515.50 0499.60
005 0515.50 0251.60
009 0515.50 0213.40
006 0515.50 0499.60
021 0499.60 0499.60
046 0445.00 0445.00
019 0445.00 0221.50
051 0430.00 0415.20
055 0430.00 0414.80
018 0429.00 0395.60
033 0422.50 0422.50
001 0419.60 0317.60
026 0401.50 0183.00
025 0401.50 0181.00
014 0396.40 0396.40
027 0337.10 0188.00
029 0331.90 0188.00
031 0328.60 0181.00
015 0328.40 0181.80
028 0324.60 0279.30
034 0318.20 0190.50
030 0305.60 0210.80
020 0297.80 0251.60
024 0242.80 0242.80
042 0208.80 0184.10
```

```
10
039 0009.90 0000.12
042 0009.40 0000.20
025 0007.05 0007.05
041 0004.70 0000.60
```

```
11
041 0025.00 0021.70
042 0025.00 0021.55
039 0024.70 0021.40
025 0024.70 0003.40
031 0021.55 0012.00
034 0021.55 0021.40
026 0015.50 0004.60
029 0013.50 0003.40
028 0007.80 0004.00
019 0006.85 0003.08
```

```
12
002 0799.80 0799.80
003 0799.80 0669.50
004 0739.00 0739.00
005 0738.00 0537.10
006 0737.50 0731.50
007 0737.50 0731.50
008 0737.50 0695.50
009 0737.50 0663.00
010 0736.50 0736.50
011 0736.50 0695.50
012 0732.20 0732.20
013 0731.50 0717.50
014 0730.60 0695.50
015 0728.00 0516.30
016 0717.50 0717.50
001 0703.10 0585.40
018 0696.70 0695.90
019 0688.00 0516.30
020 0687.00 0546.80
021 0686.90 0646.00
023 0638.70 0595.60
024 0638.00 0594.00
025 0631.00 0505.80
026 0594.40 0489.85
027 0594.00 0503.90
028 0551.70 0551.70
029 0551.70 0489.85
030 0549.00 0489.85
```

```
13
002 1007.70 0909.80
048 1007.70 0870.70
003 1007.70 0847.80
053 1007.40 0994.50
017 1007.40 0984.90
013 1005.80 0858.60
049 1005.20 0993.70
020 1005.20 0826.00
052 0990.70 0856.80
022 0984.40 0857.80
044 0976.00 0875.00
010 0974.80 0901.50
008 0944.00 0826.70
021 0936.80 0847.80
005 0934.10 0647.10
043 0931.00 0903.70
033 0923.80 0835.70
032 0919.00 0916.80
037 0910.70 0910.70
018 0907.80 0907.80
025 0888.60 0644.20
007 0884.50 0854.50
045 0877.20 0874.40
015 0877.20 0692.40
004 0876.50 0876.50
009 0873.70 0826.00
027 0873.70 0644.20
056 0870.30 0816.50
034 0855.80 0692.40
020 0836.50 0836.50
```

```
14
014 1102.90 1102.90
038 1102.90 1098.60
034 1102.90 0959.00
008 1102.90 1079.70
016 1102.90 1079.70
023 1102.90 1078.40
001 1102.90 0959.00
005 1102.90 0959.00
009 1102.90 0959.00
025 1102.90 0950.10
011 1098.60 1098.60
027 1098.60 0974.00
015 1098.60 1078.40
028 1033.60 1033.60
024 1030.00 1030.00
020 1024.20 0959.00
029 1004.20 0974.00
019 1004.20 0950.10
026 0959.00 0959.00
```

```
15
017 0847.30 0841.60
049 0847.30 0837.50
048 0847.30 0720.00
011 0847.30 0710.60
050 0846.30 0837.50
013 0844.20 0835.20
009 0844.20 0696.00
003 0844.00 0729.60
002 0835.20 0788.40
054 0835.20 0738.80
006 0835.20 0710.60
053 0835.20 0740.90
005 0835.20 0702.50
044 0835.20 0800.80
007 0835.20 0699.00
010 0834.20 0757.10
008 0825.00 0699.00
047 0818.00 0729.80
052 0816.80 0780.30
021 0805.10 0735.50
045 0801.30 0710.70
043 0794.30 0756.50
033 0770.40 0729.80
034 0759.70 0703.70
055 0732.50 0732.50
```

```
014 0835.30 0835.30
006 0830.40 0829.60
036 0816.50 0816.50
035 0808.10 0806.90
019 0801.10 0650.00
030 0799.30 0690.80
028 0746.10 0725.50
029 0682.70 0664.00
026 0680.40 0650.00
031 0650.00 0650.00
```

```
16
012 1268.00 1268.00
008 1224.00 1200.40
038 1222.00 1219.90
005 1222.00 1117.00
043 1219.20 1219.20
014 1218.80 1198.10
007 1218.00 1217.00
009 1218.00 1115.90
013 1217.00 1217.00
018 1216.00 1209.00
015 1216.00 1113.00
003 1204.90 1192.50
001 1204.90 1119.10
032 1196.50 1196.50
019 1196.50 1105.30
051 1193.50 1193.50
021 1192.50 1142.00
025 1189.60 1105.30
020 1183.50 1100.50
024 1147.60 1147.60
023 1117.50 1100.50
027 1115.90 1115.90
```

```
17
055 0088.38 0087.50
025 0093.21 0077.32
004 0092.83 0078.38
013 0111.60 0092.50
008 0077.60 0073.35
054 0114.60 0085.17
011 0122.22 0098.49
022 0121.18 0115.13
046 0107.77 0087.50
051 0091.83 0091.83
031 0087.50 0087.50
052 0121.18 0073.35
034 0094.00 0076.50
033 0112.28 0082.45
016 0104.25 0085.41
010 0121.49 0075.40
012 0120.34 0085.36
050 0123.20 0120.38
005 0118.51 0042.55
047 0094.15 0074.65
003 0121.49 0074.50
056 0095.93 0082.10
007 0116.00 0082.10
049 0124.00 0119.35
006 0115.20 0072.20
017 0120.52 0118.41
009 0119.95 0074.78
053 0121.49 0083.71
048 0120.92 0072.60
043 0112.53 0112.53
```

```
18
002 0603.00 0525.20
011 0600.00 0528.70
008 0575.80 0482.20
013 0575.80 0575.80
028 0432.30 0306.40
021 0564.50 0483.80
010 0560.20 0525.00
009 0539.00 0409.00
034 0528.00 0480.30
020 0527.20 0480.60
018 0525.00 0525.00
005 0522.00 0311.70
055 0521.50 0521.50
031 0509.00 0413.30
035 0480.60 0480.60
041 0452.00 0451.50
025 0441.00 0312.20
027 0439.60 0399.00
029 0432.30 0412.70
015 0432.20 0399.00
042 0432.20 0432.20
026 0431.00 0395.30
019 0427.60 0308.80
024 0412.60 0412.60
037 0405.80 0405.80
```

```
19
050 0566.00 0566.00
048 0552.00 0508.80
005 0537.00 0504.50
007 0537.00 0537.00
047 0537.00 0537.00
010 0531.40 0498.40
006 0517.20 0508.80
052 0517.20 0516.00
008 0517.20 0516.00
003 0516.50 0516.50
046 0516.00 0516.00
018 0510.80 0510.80
056 0510.80 0503.80
034 0503.80 0503.80
034 0503.80 0503.80
```

```
20
011 0120.10 0120.10
048 0120.10 0120.10
050 0120.10 0120.10
052 0120.10 0120.10
006 0094.50 0083.60
007 0094.50 0083.60
009 0094.50 0083.60
053 0094.50 0094.50
005 0094.50 0083.60
012 0094.50 0083.60
010 0092.40 0092.40
047 0092.40 0083.60
054 0092.40 0083.60
004 0092.40 0092.40
056 0092.40 0083.60
```

```
1*************************
 **
 **
 ** (6B) RUBEL'S BALTIC SILURIAN DATA BASES
 **
 **
 *************************


 SEQUENCE DATA   (OSTRACODES)

 01
 124 0115.00 0081.50
 060 0115.00 0003.00
 084 0115.00 0005.00
 076 0110.80 0110.80
 130 0110.80 0048.40
 094 0106.60 0085.50
 090 0102.80 0102.80
 127 0102.80 0067.70
 107 0098.00 0098.00
 074 0098.00 0004.50
 111 0094.00 0092.12
 104 0094.00 0005.00
 071 0092.10 0067.70
 072 0089.20 0006.10
 152 0085.50 0004.50
 097 0082.00 0044.60
 140 0080.00 0073.90
 066 0080.00 0067.70
 134 0080.00 0067.70
 133 0080.00 0067.70
 126 0080.00 0065.40
 145 0080.00 0023.90
 141 0076.20 0067.70
 110 0076.20 0067.70
 062 0076.20 0067.70
 120 0076.20 0058.50
 079 0076.20 0067.70
 078 0076.20 0076.20
 105 0076.20 0076.20
 106 0076.20 0073.90
 091 0076.20 0058.50
 096 0073.90 0067.70
 149 0073.90 0050.70
 077 0072.30 0067.70
 142 0067.70 0065.40
 085 0067.70 0067.70
 059 0065.40 0005.00
 153 0064.10 0048.40
 113 0058.50 0003.00
 087 0058.50 0034.40
 061 0058.50 0012.70
 109 0056.20 0006.10
 092 0056.20 0043.20
 098 0048.40 0004.50
 100 0048.40 0017.30
 116 0044.60 0034.40
 114 0029.80 0003.00
 057 0017.30 0012.70
 089 0014.30 0011.40
 129 0005.00 0005.00
 119 0004.50 0004.50

 02
 105 0053.48 0039.93
 130 0053.00 0003.70
 134 0053.00 0034.20
 127 0053.00 0003.70
 071 0052.37 0032.60
 072 0052.37 0006.60
 074 0052.37 0006.60
 104 0052.37 0005.00
 084 0051.70 0006.60
 060 0051.70 0011.65
 091 0051.70 0046.25
 151 0051.70 0005.00
 096 0051.70 0044.40
 079 0051.70 0051.70
 140 0051.70 0006.60
 078 0051.70 0039.93
 126 0046.25 0003.70
 139 0045.58 0003.70
 097 0044.40 0003.70
 145 0044.40 0005.00
 142 0039.93 0003.70
 092 0039.93 0014.40
 066 0038.30 0028.60
 120 0035.80 0027.95
 128 0034.20 0005.00
 100 0034.80 0014.40
 129 0034.20 0003.70
 149 0034.20 0005.00
 113 0034.20 0012.35
 152 0032.60 0012.35
 073 0030.62 0003.70
 141 0030.62 0009.40
 061 0028.60 0003.70
 157 0028.60 0028.60
 077 0027.95 0006.60
 059 0027.95 0018.90
 109 0023.25 0023.25
 153 0018.90 0014.40
 068 0014.40 0003.70
 137 0014.40 0010.30
 098 0011.65 0010.30
 116 0011.65 0010.30
 155 0011.65 0003.70
 132 0010.30 0003.70
 058 0009.40 0003.70
 158 0006.60 0005.00

 03
 084 0283.20 0197.70
 074 0283.20 0174.60
 060 0283.20 0174.60
 130 0274.50 0197.70
 096 0270.00 0260.50
 071 0270.00 0256.60
 104 0270.00 0197.70
 076 0270.00 0270.00
 128 0270.00 0197.70
 145 0270.00 0256.60
 127 0270.00 0259.00
 097 0270.00 0250.00
 066 0264.00 0260.50
 152 0260.50 0200.90
 100 0260.50 0256.60
 061 0260.50 0237.00
 105 0260.50 0260.50
 149 0260.50 0232.80
 079 0260.50 0260.50
 144 0260.50 0260.50
 153 0259.00 0259.00
 113 0237.00 0211.00
 098 0237.00 0174.60
 129 0237.00 0192.50
 114 0236.70 0211.00
 109 0236.70 0230.00
 057 0234.50 0234.50
 089 0221.00 0190.00
 072 0209.00 0209.00
 141 0209.00 0197.70
 118 0207.00 0177.00
 086 0207.00 0207.00
 119 0205.30 0205.30
 157 0200.90 0177.00
```

04
082 0645.00 0645.00
143 0645.00 0589.00
094 0617.50 0476.00
138 0605.00 0510.00
108 0605.00 0525.00
081 0567.50 0515.00
115 0487.50 0476.00
123 0476.00 0476.00
075 0462.00 0433.00
074 0452.50 0295.00
060 0452.50 0271.00
154 0447.50 0447.50
071 0445.50 0417.50
062 0440.00 0420.00
084 0428.00 0319.50
097 0422.50 0384.00
112 0422.50 0384.00
122 0422.50 0384.00
145 0417.50 0417.50
072 0416.50 0387.00
153 0416.50 0303.50
105 0413.00 0413.00
059 0413.00 0412.00
091 0407.50 0407.50
109 0402.00 0295.00
089 0398.50 0305.00
113 0386.50 0315.00
087 0374.00 0374.00
095 0367.00 0304.00
114 0366.00 0325.50
098 0352.00 0304.00
149 0340.00 0340.00
128 0332.50 0332.50
061 0319.50 0319.50
155 0319.50 0319.50
135 0312.50 0312.50
118 0302.50 0271.00
086 0302.50 0302.50
117 0302.50 0302.50
158 0278.00 0278.00

05
143 0733.00 0687.00
138 0725.00 0664.60
090 0722.00 0694.00
074 0714.00 0425.00
104 0687.00 0639.40
108 0681.00 0590.50
094 0663.70 0651.50
123 0660.00 0639.40
121 0660.00 0639.40
084 0647.50 0462.30
147 0639.40 0639.40
099 0630.00 0589.00
144 0628.00 0568.00
069 0628.00 0472.50
075 0624.00 0624.00
153 0623.00 0623.00
060 0619.50 0418.50
107 0618.50 0589.00
150 0615.50 0484.00
091 0615.50 0545.00
072 0615.50 0433.00
154 0614.80 0590.50
068 0594.30 0455.00
098 0589.00 0455.00
145 0568.00 0555.40
152 0568.00 0419.00
080 0568.00 0568.00
119 0567.20 0455.20
057 0555.40 0510.60
092 0553.00 0550.00
097 0550.00 0524.50
073 0545.00 0441.00
059 0540.00 0460.50
113 0538.20 0455.00
061 0537.00 0455.00
088 0534.70 0484.00
122 0524.50 0524.50
155 0519.70 0519.70
114 0495.20 0457.60
146 0484.00 0455.00
089 0484.00 0484.00
135 0469.50 0469.50
095 0466.00 0455.00
109 0466.00 0433.00
157 0465.00 0455.00
103 0455.00 0432.50
101 0455.00 0433.00
128 0455.00 0455.00
129 9455.00 0430.00
131 0455.00 0419.00
158 0455.00 0441.00
127 0455.00 0455.00
086 0441.00 0441.00
118 0441.00 0419.00
117 0439.00 0439.00

06
093 1310.40 1291.40
125 1310.40 1310.40
084 1267.00 0979.00
097 1267.00 1187.00
071 1267.00 1199.00
064 1267.00 1248.00
144 1267.00 1248.00
102 1267.00 1248.00
149 1260.00 1053.00
074 1254.00 1002.00
068 1254.00 1055.00
069 1248.00 1055.00
145 1248.00 1187.00
098 1226.00 0984.00
112 1226.00 1155.00
152 1218.00 1055.00
092 1218.00 1218.00
110 1204.50 1187.00
072 1199.00 1009.00
060 1187.00 1002.00
153 1187.00 1055.00
091 1187.00 1103.00
095 1166.00 1092.00
088 1166.00 1155.00
057 1160.00 1055.00
073 1155.00 1055.00
113 1147.50 1055.00
061 1103.00 1055.00
104 1103.00 1086.00
109 1069.50 1009.00

07

090 1170.00 1138.00
143 1170.00 1138.00
065 1120.00 1120.00
081 1120.00 1120.00
137 1112.00 0994.00
069 1087.00 0970.00
104 1087.00 1056.00
144 1076.00 0994.00
092 1064.00 0971.50
084 1064.00 0926.85
108 1064.00 1064.00
154 1064.00 1032.50
075 1056.00 1042.00
099 1038.20 1032.50
098 1035.00 0973.80
112 1035.00 0950.00
085 1032.50 1032.50
074 1032.50 0926.85
072 1032.50 0950.00
097 1032.50 0970.00
152 1032.50 0950.00
091 1032.50 0973.80
145 1032.50 0981.00
080 1032.00 0970.00
060 1030.30 0926.85
113 1030.30 0926.85
132 1003.30 0984.00
059 1001.50 0981.00
057 0995.50 0926.85
153 0995.00 0980.50
135 0995.00 0995.00
150 0995.00 0995.00
068 0981.00 0981.00
061 0981.00 0981.00
087 0973.80 0973.80
114 0935.50 0935.50

08

067 1048.00 1048.00
083 1048.00 0887.00
136 1046.50 1032.50
082 1032.50 0964.00
108 0971.50 0871.00
137 0964.00 0871.00
081 0932.00 0892.00
084 0927.00 0742.00
138 0905.00 0888.50
090 0897.00 0871.00
093 0896.00 0892.00
148 0892.00 0892.00
121 0887.00 0871.10
123 0876.00 0876.00
099 0871.00 0871.00
145 0871.00 0792.00
122 0867.00 0800.00
154 0862.00 0862.00
060 0862.00 0862.00
153 0862.00 0742.00
142 0827.20 0807.50
097 0827.20 0814.00
134 0827.20 0827.20
133 0827.20 0827.20
066 0827.20 0827.20
063 0827.20 0827.20
130 0827.20 0827.20
104 0827.20 0827.20
105 0827.20 0827.20
126 0827.20 0827.20
139 0827.20 0792.00
127 0827.20 0827.20
141 0816.50 0816.50
155 0814.00 0814.00
069 0814.00 0814.00
098 0807.50 0807.50
112 0807.50 0807.50
061 0807.50 0742.00
068 0807.50 0807.50
072 0807.50 0807.50
074 0807.50 0792.00
092 0807.50 0792.00
057 0806.00 0735.00
147 0784.00 0774.50
149 0742.00 0742.00

09

082 0506.00 0459.00
065 0506.00 0475.00
081 0460.00 0403.00
060 0460.00 0403.00
084 0336.00 0188.00
154 0336.00 0188.00
127 0334.00 0314.00
129 0334.00 0334.00
106 0334.00 0334.00
112 0332.00 0292.00
122 0332.00 0292.00
153 0315.00 0240.00
145 0314.00 0314.00
091 0310.00 0310.00
097 0305.00 0294.00
113 0305.00 0222.00
109 0298.00 0188.00
087 0297.00 0293.00
089 0295.00 0212.00
061 0294.00 0212.00
098 0292.00 0258.00
114 0274.00 0234.00
152 0270.00 0270.00
095 0270.00 0222.00
071 0242.00 0242.00
155 0240.00 0240.00
135 0214.00 0212.00
141 0212.00 0212.00
075 0212.00 0212.00
103 0188.00 0188.00
118 0187.00 0185.00

21

114 0021.90 0006.30
113 0021.90 0001.20
060 0021.90 0000.50
098 0021.90 0000.50
084 0021.90 0000.50
102 0021.90 0001.20
100 0021.00 0020.40
095 0021.00 0021.00
089 0021.00 0002.00
130 0021.00 0000.80
129 0021.00 0001.20
061 0020.40 0001.60
152 0020.40 0003.00
059 0019.40 0007.20
149 0015.90 0013.20
070 0012.30 0001.60
119 0009.90 0001.60
077 0006.60 0006.60
127 0006.60 0000.80
104 0003.30 0001.60
157 0003.30 0000.80
101 0003.30 0000.80
118 0003.30 0001.60
064 0003.30 0001.60
086 0003.00 0003.00
128 0003.00 0000.80
155 0003.00 0001.20
141 0003.00 0001.60
109 0003.00 0001.20
103 0003.00 0001.60
126 0003.00 0001.60
068 0002.00 0001.60
151 0002.00 0002.00
156 0001.60 0001.60
158 0001.60 0001.20
145 0001.60 0001.60
131 0001.20 0001.20

22

079 0528.60 0528.00
153 0528.60 0453.00
154 0528.60 0528.60
069 0528.60 0383.50
145 0528.60 0386.50
092 0528.60 0449.00
060 0528.60 0386.50
084 0528.60 0383.50
074 0528.60 0383.50
130 0492.80 0386.50
127 0492.80 0459.50
063 0492.80 0480.20
125 0484.00 0480.20
097 0466.70 0449.00
152 0466.70 0386.50
059 0466.70 0383.50
113 0466.70 0383.50
122 0463.40 0414.50
061 0463.40 0383.50
075 0463.40 0453.00
109 0459.50 0389.50
072 0459.50 0383.50
120 0453.00 0453.00
057 0453.00 0414.50
149 0453.00 0386.50
089 0451.00 0392.50
098 0449.00 0383.50
114 0428.50 0386.50
095 0414.50 0383.50
070 0389.00 0389.00
119 0386.50 0386.50
141 0386.50 0386.50
129 0386.50 0383.50
155 0386.50 0386.50
067 1204.00 1204.00

23

136 1204.00 1204.00
083 1204.00 1088.00
065 1195.50 1167.70
082 1195.50 1158.60
108 1133.20 1068.90
137 1117.10 1094.50
090 1117.10 1074.00
124 1103.20 1094.50
138 1100.00 1013.60
085 1100.00 1100.00
148 1094.50 1088.00
084 1094.50 0902.50
060 1088.00 0902.50
062 1088.00 1020.20
074 1073.90 0907.50
105 1073.90 1053.80
099 1073.90 1073.90
121 1073.90 1068.90
144 1073.90 1023.50
130 1073.90 1053.80
126 1053.80 1013.60
104 1053.80 1013.60
096 1038.00 1038.00
142 1023.50 1013.40
111 1020.20 1020.20
061 1009.30 0970.80
128 1006.00 0926.80
091 1006.00 0997.60
068 1006.00 0989.30
069 1006.00 0972.70
153 1006.00 0948.20
152 1006.00 0919.90
149 1003.40 0929.90
113 1001.40 0929.90
057 0999.20 0968.50
092 0999.20 0957.60
098 0989.30 0919.90
109 0964.60 0961.90
146 0948.20 0948.20
151 0926.80 0926.80
058 0919.90 0919.90
059 0907.50 0902.50

```
1************************
 **
 **
 **  (6C)  RUBEL'S BALTIC SILURIAN DATA BASES
 **
 **
 ************************
```

SEQUENCE DATA (THELODONTS)

01	02	03
159 0095.17 0092.27	159 0052.05 0052.05	159 0284.50 0284.20
160 0098.95 0098.90	161 0052.05 0044.90	165 0205.90 0158.30
161 0098.80 0093.15	164 0056.55 0002.10	164 0158.30 0040.31
164 0098.95 0002.00	163 0053.40 0002.10	162 0202.30 0158.30
163 0095.17 0041.46	167 0019.21 0013.40	163 0284.50 0205.60
167 0092.31 0006.95	168 0053.40 0013.40	167 0245.50 0164.00
168 0095.17 0090.73	169 0099.95 0063.45	168 0284.50 0158.30
169 0153.20 0105.03	174 0056.55 0002.10	169 0397.00 0342.00
170 0174.10 0163.30	175 0068.22 0063.45	171 0205.90 0158.30
172 0038.15 0002.00	180 0042.25 0013.40	172 0236.80 0160.20
174 0098.95 0002.00	183 0053.40 0007.65	174 0324.40 0158.30
176 0112.25 0111.80	184 0056.55 0036.00	173 0173.00 0160.20
175 0108.30 0098.90	194 0013.40 0013.40	177 0198.00 0158.30
178 0151.70 0150.40		179 0205.90 0158.30
180 0083.10 0073.60		183 0284.50 0166.80
182 0105.03 0105.03		184 0284.50 0242.00
183 0098.95 0014.45		185 0276.40 0205.60
184 0095.17 0064.65		186 0217.80 0161.30
185 0094.48 0073.60		189 0181.00 0161.30
186 0006.95 0004.30		181 0412.40 0402.80
181 0153.20 0105.03		180 0279.90 0242.00
193 0094.48 0094.45		199 0245.50 0160.20
188 0150.53 0106.25		166 0161.60 0161.30
		194 0245.50 0242.00

```
04
160 0484.50 0472.40
161 0481.40 0472.40
165 0321.80 0266.80
164 0604.80 0266.80
162 0304.20 0266.80
163 0456.00 0456.00
167 0343.70 0298.30
168 0343.70 0266.80
192 0268.70 0266.80
171 0325.20 0271.30
172 0379.40 0266.80
174 0604.80 0266.80
173 0289.90 0266.80
199 0474.80 0474.80
175 0552.50 0474.80
177 0325.20 0268.40
179 0336.10 0266.80
180 0418.50 0390.80
183 0456.00 0280.00
184 0456.00 0421.60
185 0336.10 0283.40
186 0376.30 0268.40
195 0268.40 0268.40
187 0269.70 0268.40
196 0268.90 0260.00
189 0289.90 0271.30
166 0273.30 0271.30
```

```
10
169 0011.75 0001.30
188 0011.75 0011.75
181 0011.75 0001.30
175 0003.90 0001.30
178 0011.75 0001.30
170 0049.48 0045.00
198 0011.75 0011.75
```

```
11
159 0018.38 0014.50
164 0022.52 0001.50
163 0018.38 0003.66
167 0018.25 0014.50
168 0018.38 0009.23
174 0021.30 0001.50
199 0018.38 0010.40
175 0035.05 0029.47
178 0034.35 0030.82
183 0022.52 0001.50
185 0018.10 0001.50
181 0035.05 0030.82
182 0035.05 0030.82
184 0018.38 0003.66
170 0092.99 0091.35
188 0031.86 0030.82
198 0031.86 0030.82
197 0031.41 0030.82
161 0018.38 0003.66
180 0012.45 0006.82
```

```
12
164 0716.20 0492.50
163 0604.40 0570.40
172 0540.40 0492.50
174 0626.40 0492.50
180 0648.60 0604.00
183 0645.00 0492.50
184 0705.50 0625.00
```

```
24
159 0011.60 0007.60
160 0011.60 0011.60
161 0010.00 0007.60
163 0010.00 0004.70
164 0011.60 0004.70
168 0010.00 0004.70
167 0010.00 0004.70
174 0010.00 0004.70
199 0011.60 0007.60
185 0007.80 0004.70
180 0004.70 0004.70
183 0011.60 0004.70
184 0010.00 0004.70
```

```
25
161 0090.10 0088.40
163 0090.10 0086.80
164 0090.10 0086.80
168 0090.10 0086.80
167 0088.90 0088.90
174 0090.10 0086.80
185 0088.90 0088.90
180 0086.80 0086.80
183 0090.10 0086.80
182 0093.20 0093.20
169 0093.20 0093.20
188 0093.20 0093.20
181 0093.20 0093.20
184 0090.10 0086.80
```

```
26
159 0022.00 0019.10
161 0022.00 0019.10
163 0022.00 0019.10
164 0022.00 0019.10
168 0022.00 0019.10
167 0022.00 0019.10
174 0022.00 0019.10
183 0022.00 0019.10
182 0032.30 0023.30
169 0032.30 0023.30
188 0032.30 0031.80
184 0022.00 0019.10
181 0032.30 0031.80
175 0032.30 0023.30
178 0032.30 0023.30
```

```
27
159 0014.50 0014.00
164 0018.20 0010.65
163 0014.50 0014.00
167 0014.50 0014.00
168 0014.50 0010.65
174 0014.50 0010.65
183 0014.50 0010.65
185 0014.50 0010.65
184 0014.50 0010.65
```

```
28
164 0039.15 0038.95
169 0008.80 0008.65
174 0039.15 0038.95
178 0008.80 0003.20
181 0008.80 0008.65
170 0046.15 0023.70
188 0008.80 0008.65
198 0008.80 0008.65
197 0008.80 0008.65
```

29
159 0144.40 0031.10
164 0144.40 0030.60
162 0034.60 0030.60
163 0144.40 0134.55
167 0144.40 0030.60
168 0144.40 0033.50
171 0041.20 0031.10
174 0144.40 0030.60
172 0094.80 0031.10
177 0055.70 0034.60
179 0050.20 0030.60
183 0144.40 0030.60
185 0143.50 0031.10
186 0094.80 0030.60
184 0144.40 0130.80
189 0033.70 0033.50
180 0130.90 0130.80

30
164 0035.50 0008.60
162 0023.10 0012.00
167 0032.30 0020.60
168 0032.30 0013.30
171 0032.30 0017.30
174 0035.50 0008.60
177 0023.10 0023.10
179 0032.30 0012.00
172 0035.50 0012.00
183 0035.50 0017.30
185 0029.30 0017.30
186 0029.30 0013.30

31
191 0172.60 0163.20
164 0190.80 0148.80
162 0174.50 0157.15
165 0163.20 0149.10
167 0174.50 0149.10
168 0177.40 0149.10
171 0174.50 0148.00
174 0190.80 0148.00
199 0165.20 0163.20
177 0174.50 0148.00
179 0177.40 0148.00
172 0189.40 0148.00
183 0174.50 0149.10
185 0163.20 0162.40
189 0157.15 0150.20
186 0181.90 0149.10

32
164 0012.13 0005.20
174 0012.13 0005.20
168 0005.30 0005.20
199 0021.95 0021.82
183 0012.13 0005.20
185 0005.30 0005.20
181 0021.95 0021.85
182 0021.95 0017.82
169 0021.95 0021.82
175 0021.95 0017.82
178 0021.95 0021.82
184 0005.30 0005.20
163 0005.30 0005.20

33
164 0060.20 0007.00
162 0050.30 0039.10
165 0025.50 0007.00
167 0046.50 0007.00
168 0057.50 0007.00
171 0042.40 0009.20
174 0060.20 0007.00
177 0055.10 0025.50
179 0048.30 0024.80
172 0060.20 0007.00
183 0046.50 0014.00
185 0042.40 0029.80
189 0039.10 0009.20
186 0060.20 0007.00

34
160 0014.65 0014.47
161 0014.65 0006.20
164 0014.65 0006.20
163 0013.40 0006.20
159 0013.20 0012.80
169 0025.70 0023.92
168 0013.40 0006.20
167 0013.40 0006.20
174 0013.40 0006.20
199 0025.70 0012.80
175 0024.04 0020.45
178 0025.70 0020.45
185 0008.07 0006.20
180 0008.07 0006.20
183 0019.90 0006.20
184 0013.40 0006.20
182 0025.70 0025.60
181 0025.70 0023.92
188 0037.17 0037.05
170 0076.98 0058.20

35
170 0072.03 0057.06

36
169 0004.60 0002.70
199 0004.60 0002.70
175 0009.75 0002.70
178 0003.60 0002.70
181 0016.04 0002.70
182 0003.10 0002.70

```
1*************************
 **
 **
 **  (7)  SULLIVAN - BRAMLETTE DATA BASE
 **       PALEOGENE CALIFORNIAN NANNOFOSSILS
 **
 **
 *************************
```

DICTIONARY

1 CHIPHRAGMALITHUS CRISTATUS
2 CHIPHRAGMALITHUS ACANTHODES
3 CHIPHRAGMALITHUS CALATUS
4 CHIPHRAGMALITHUS DUBIUS
5 CHIPHRAGMALITHUS PROTENUS
6 CHIPHRAGMALITHUS QUADRATUS
7 COCCOLITHUS BIDENS
8 COCCOLITHUS CALIFORNICUS
9 COCCOLITHUS EXPANSUS
10 COCCOLITHUS GRANDIS
11 COCCOLITHUS SOLITUS
12 COCCOLITHUS STAURION
13 COCCOLITHUS GIGAS
14 COCCOLITHUS DELUS
15 COCCOLITHUS CONSUETUS
16 COCCOLITHUS CRASSUS
17 COCCOLITHUS CRIBELLUM
18 COCCOLITHUS EMINENS
19 CYCLOCOCCOLITHUS GAMMATION
20 CYCLOCOCCOLITHUS LUMINIS
21 DISCOLITHUS PECTINATUS
22 DISCOLITHUS PLANUS
23 DISCOLITHUS PULCHER
24 DISCOLITHUS PULCHEROIDES
25 DISCOLITHUS RIMOSUS
26 DISCOLITHUS DISTINCTUS
27 DISCOLITHUS FIMBRIATUS
28 DISCOLITHUS OCELLATUS
29 DISCOLITHUS PANARIUM
30 DISCOLITHUS PUNCTOSUS
31 DISCOLITHUS SOLIDUS
32 DISCOLITHUS VESCUS
33 DISCOLITHUS VERSUS
34 DISCOLITHUS PERTUSUS
35 DISCOLITHUS EXILIS
36 DISCOLITHUS DUOCAVUS
37 DISCOLITHUS INCONSPICUUS
38 CYCLOLITHUS ROBUSTUS
39 ELLIPSOLITHUS MACELLUS
40 ELLIPSOLITHUS DISTICHUS
41 HELICOSPHAERA SEMILUNUM
42 HELICOSPHAERA LOPHOTA
43 LOPHODOLITHUS NASCENS
44 LOPHODOLITHUS RENIFORMIS
45 LOPHODOLITHUS MOCHOLOPHORUS
46 RHABDOSPHAERA CREBRA
47 RHABDOSPHAERA MORIONUM
48 RHABDOSPHAERA PERLONGA
49 RHABDOSPHAERA RUDIS
50 RHABDOSPHAERA SCABROSA
51 RHABDOSPHAERA SEMIFORMIS
52 RHABDOSPHAERA TENUIS
53 RHABDOSPHAERA TRUNCATA
54 RHABDOSPHAERA INFLATA
55 ZYGODISCUS SIGMOIDES
56 ZYGODISCUS ADAMAS
57 ZYGODISCUS HERLYNI
58 ZYGODISCUS PLECTOPONS
59 ZYGOLITHUS CONCINNUS
60 ZYGOLITHUS CRUX
61 ZYGOLITHUS DISTENTUS
62 ZYGOLITHUS JUNCTUS
63 ZYGRHABLITHUS SIMPLEX
64 ZYGRHABLITHUS BIJUGATUS
65 BAARUDOSPHAERA BIGELOWI
66 BAARUDOSPHAERA DISCULA
67 MICRANTHOLITHUS FLOS
68 MICRANTHOLITHUS INAEQUALIS
69 MICRANTHOLITHUS VESPER
70 MICRANTHOLITHUS BASQUENSIS
71 MICRANTHOLITHUS CRENULATUS
72 MICRANTHOLITHUS AEQUALIS
73 CLATHROLITHUS ELLIPTICUS
74 RHOMBOASTER CUSPIS
75 POLYCLADOLITHUS OPEROSUS
76 SPHENOLITHUS RADIANS
77 FASCICULOLITHUS INVOLUTUS
78 DISCOASTER BARBADIENSIS
79 DISCOASTER BINODOSUS
80 DISCOASTER DEFLANDREI
81 DISCOASTER DELICATUS
82 DISCOASTER DIASTYPUS
83 DISCOASTER DISTINCTUS
84 DISCOASTER FALCATUS
85 DISCOASTER LODOENSIS
86 DISCOASTER MULTIRADIATUS
87 DISCOASTER NONARADIATUS
88 DISCOASTER STRADNERI
89 DISCOASTER TRIBRACHIATUS
90 DISCOASTER CRUCIFORMIS
91 DISCOASTER GERMANICUS
92 DISCOASTER LENTICULARIS
93 DISCOASTER MARTINII
94 DISCOASTER MINIMUS
95 DISCOASTER SEPTEMRADIATUS
96 DISCOASTER SUBLODOENSIS
97 DISCOASTER HELIANTHUS
98 DISCOASTER LIMBATUS
99 DISCOASTER MEDIOSUS
100 DISCOASTER PERPOLITUS
101 DISCOASTEROIDES KUEPPERI
102 DISCOASTEROIDES MEGASTYPUS
103 HELIOLITHUS KLEINPELLI
104 HELIOLITHUS RIEDELI

```
1************************

 **
 **
 **  (7)  SULLIVAN - BRAMLETTE DATA BASE
 **       PALEOGENE CALIFORNIAN NANNOFOSSILS
 **
 **              CF., APPENDIX 3 IN:  DAVAUD, E., ET GUEX, J., 1978, TRAITEMENT ANALYTIQUE
 **                      <<MANUEL>> ET ALGORITHMIQUE DE PROBLEMES COMPLEXES DE CORRELATIONS
 **                      BIOCHRONOLOGIQUES:  ECLOGAE GEOL. HELV., VOL.71, PP. 581-610
 **
 ************************


 SEQUENCE DATA

 LOCALITY A:  PACHECO
   1 6 8   2 3 6   3 1 4   4 1 8   6 7 8   9 4 8  10 1 8  11 1 8  12 1 3  13 8 8  14 1 8  15 2 8
  16 1 8  17 1 1  19 1 7  20 1 1  21 1 8  22 1 7  23 1 6  24 1 6  25 1 6  26 1 8  27 1 7  28 1 5
  29 1 8  33 1 6  35 5 5  39 1 1  41 1 6  42 1 8  43 1 5  44 1 2  45 1 6  46 1 6  47 1 6  48 1 5
  50 1 3  51 1 6  52 1 8  59 1 1  64 1 6  65 2 6  67 1 4  68 3 3  69 1 8  71 1 5  73 1 3  76 1 6
  78 1 8  79 1 4  80 1 8  83 1 8  85 1 8  87 1 8  88 1 8  89 1 2  91 1 6  94 1 4  95 1 8 101 1 7

 LOCALITY B:  VACA VALLEY
   3 1 2   1 8 9   4 1 9   6 8 9  14 1 8  15 1 6  16 1 9  17 1 7   9 6 6  10 1 9  11 1 9  12 9 9
  19 1 7  20 1 8  26 4 4  27 3 9  28 2 6  29 1 8  21 1 9  22 1 7  23 1 9  24 1 9  25 1 6  31 3 7
  33 1 7  32 3 3  39 1 1  41 1 7  42 1 9  45 1 8  43 1 7  44 1 7  46 1 7  47 3 6  48 1 7  50 1 7
  51 8 9  52 1 9  53 3 7  60 1 5  64 1 9  65 2 9  66 5 6  70 3 6  71 3 7  67 2 7  68 3 7  69 1 9
  73 1 6  78 1 8  79 3 6  90 3 7  80 4 6  83 1 9  91 1 7  92 1 1  85 1 9  93 7 7  94 3 4  87 8 8
  95 1 9  88 1 9  96 3 7  89 1 7 101 1 8  75 3 6  76 1 9

 LOCALITY C:  NEW IDRIA
   3 2 9   4 2 9   5 2 9  14 2 9   7 1 1  15 1 9  16 2 9  17 2 8  18 1 1  10 2 9  11 7 9  19 2 9
  36 2 7  37 2 5  28 4 9  21 2 9  22 2 9  23 2 9  24 2 9  30 2 9  25 2 9  33 2 9  40 1 1  39 1 9
  41 2 9  42 2 9  43 2 9  44 2 9  46 2 9  47 2 9  48 2 9  49 2 9  50 8 9  53 2 9  55 1 1  59 1 9
  64 2 9  65 1 9  66 2 3  71 9 9  67 3 9  68 2 9  69 2 9  73 2 9  78 2 9  79 2 9  90 6 9  80 2 7
  82 2 6  83 3 9  91 5 6  92 2 3  85 2 9  94 5 5  86 2 2  95 4 4  88 2 9  89 2 9 101 2 9  77 1 1
 104 1 1  76 2 9

 LOCALITY D:  UPPER CANADA DE SANTA ANITA
   2 4 8   3 1 3   1 4 8   4 1 8   5 1 1  14 1 8  15 1 8  16 1 8  17 2 3   9 8 8  10 2 8  11 1 8
  12 4 8  19 2 8  26 3 8  27 5 6  37 1 5  28 6 7  29 3 7  21 1 8  22 1 8  23 1 8  24 1 8  25 1 7
  33 4 5  39 1 8  41 2 5  42 2 8  45 4 4  43 1 8  44 2 4  46 6 6  54 4 6  47 5 5  48 1 5  49 2 2
  51 4 7  52 5 6  53 4 5  59 1 5  64 1 8  65 1 6  66 6 6  70 6 6  71 5 6  69 3 6  73 3 3  78 1 8
  79 1 7  90 3 3  80 1 8  82 1 2  83 3 7  91 1 6  85 1 8  94 1 5  86 2 2  87 7 7  88 1 7  96 5 8
  89 1 8 101 1 6  76 1 8
```

```
LOCALITY E:  LAS CRUCES
  1 4 4   2 2 3   3 1 3   4 2 4   6 4 4  10 1 4  11 3 3  14 1 3  15 1 2  16 1 3  19 2 4  22 1 3
 23 1 3  24 1 3  25 1 3  26 2 3  41 1 3  42 1 4  43 1 1  44 1 1  45 2 3  47 3 3  51 3 3  52 2 4
 59 2 3  64 1 3  65 2 3  76 1 4  78 1 3  79 1 2  80 1 4  82 1 1  83 2 3  85 1 3  87 3 3  88 2 3
 89 1 2  94 1 1  96 2 3 101 1 3

LOCALITY F:  SIMI VALLEY
  4 6 6   5 4 7   7 1 6   8 1 6  101111  14 410  15 112  16 811  18 2 7  19 8 8  22 4 7  24 712
 25 1 5  28 9 9  31 9 9  34 211  37 210  38 2 4  39 1 5  40 1 7  43 411  55 1 7  56 2 7  58 2 7
 59 112  61 1 5  62 1 5  63 1 7  64 2 7  57 2 7  65 212  66 2 3  67 3 7  721010  73 2 9  77 1 7
 78 610  81 2 7  821011  84 2 7  92 4 6  98 2 7  99 5 6 100 4 7 1011212 102 2 2 103 1 1 104 2 2

LOCALITY G:  TRES PINOS
  11010   3 4 8   4 4 8   5 4 7   6 910  10 310  14 4 7  15 1 8  16 410  17 4 4  18 2 2  19 710
 21 5 7  22 6 6  23 4 5  24 4 8  25 4 6  39 1 7  41 4 8  42 810  43 4 8  44 4 5  55 1 1  59 4 7
 64 4 6  65 5 8  67 5 5  76 310  77 1 2  78 3 9  79 4 8  80 4 9  81 2 3  82 3 7  83 3 8  84 2 3
 85 4 8  86 2 3  88 4 8  89 4 8 101 4 8 103 1 1 104 1 2

LOCALITY H:  MEDIA AQUA CREEK
  11012   2 4 9   3 1 1   4 110   61012   9 8 9  10 111  11 111  131112  14 1 9  15 1 9  16 112
 17 1 8  19 1 9  21 110  22 1 9  23 1 9  24 1 9  25 3 9  26 4 9  29 3 8  31 9 9  33 2 9  37 1 3
 39 1 1  41 1 9  42 112  43 1 6  44 1 1  45 4 9  46 6 6  48 5 5  51 4 9  52 5 9  59 1 4  64 1 9
 65 1 8  66 1 2  67 1 7  68 1 7  69 7 9  70 6 6  71 6 9  76 112  78 110  79 1 8  80 1 9  83 1 9
 85 1 9  87 8 8  88 3 9  89 1 5  90 3 9  91 1 3  931212  94 3 3  96 4 9 101 1 9

LOCALITY I:  UPPER RELIZ CREEK
  3 1 7   4 211   5 2 4  10 111  121111  14 111  15 111  16 111  17 211  19 511  21 211  22 411
 23 111  24 1 9  25 211  26 911  28 1 1  291011  30 1 2  33 6 6  39 1 5  41 211  42 811  43 110
 44 1 8  46 6 6  48 2 8  49 6 8  53 1 3  64 111  65 2 9  66 7 7  691111  76 111  78 111  79 111
 80 4 9  82 1 6  83 2 2  85 111  88 311  89 1 8  90 811  94 3 3 101 111

LOCALITY J:  LODO
  1 611   2 6 8   3 4 5   4 410   5 3 3   6 911   7 1 2   9 610  10 411  11 511  121010  131111
 14 3 9  15 1 5  16 5 6  17 4 5  18 1 1  19 5 6  21 410  22 4 9  23 3 8  26 6 8  27 7 8  28 5 8
 29 6 9  30 4 5  31 3 6  33 4 6  35 7 8  36 4 5  38 2 2  39 1 5  40 1 5  41 5 8  42 5 9  43 3 6
 44 5 5  45 6 8  46 5 6  47 5 6  48 4 6  49 5 5  50 5 6  51 6 9  52 610  53 4 5  54 7 9  55 1 5
 56 3 5  58 3 5  60 5 7  61 3 5  62 1 1  63 1 5  64 4 9  65 3 6  66 3 6  67 5 6  69 5 6  71 3 6
 73 3 6  74 3 5  75 5 6  76 4 9  77 1 5  78 411  79 4 5  80 4 5  81 2 5  82 4 5  83 410  84 2 5
 85 411  86 3 5  87 610  89 4 5  90 5 6  91 4 6  92 3 5  931111  95 510  97 1 5  98 3 3  99 3 4
101 4 6 102 2 5 104 1 5  25 2 8
```

```
1************************
 **
 **
 **  (8)  CORLISS DATA BASE
 **        PALEOGENE DSDP BENTHIC FORAMINIFERS
 **
 **
 ************************
```

DICTIONARY

```
  1  ALABAMINA DISSONATA
  2  ANOMALINOIDES ALAZANENSIS
  3  ANOMALINOIDES INTERMEDIA
  4  ARAGONIA ARAGONENSIS
  5  ARAGONIA SEMIRETICULATA
  6  ASTRONONION PUSILLUM
  7  BOLIVINA HUNERI
  8  BOLIVINA SP 1
  9  BULIMINA ALAZANENSIS
 10  BULIMINA BRADBURYI
 11  BULIMINA GLOMARCHALLENGERI
 12  BULIMINA IMPENDENS
 13  BULIMINA JARVISI/SEMICOSTATA
 14  BULIMINA MACILENTA
 15  BULIMINA PUPOIDES
 16  BULIMINA TRINITATENSIS
 17  BULIMINA TUXPAMENSIS
 18  BULIMINA SP 2
 19  BULIMINA SP 3
 20  BULIMINA SP 4
 21  BULIMINELLA GRATA
 22  CASSIDULINA HAVANENSE
 23  CIBICIDOIDES GRIMSDALEI
 24  CIBICIDOIDES HAITIENSIS
 25  CIBICIDOIDES LAURISAE
 26  CIBICIDOIDES SUBSPIRATUS
 27  CIBICIDOIDES TUXPAMENSIS
 28  CIBICIDOIDES UNGERIANUS
 29  CIBICIDOIDES SP 6
 30  CIBICIDOIDES SP 9
 31  CIBICIDOIDES SP 12
 32  CIBICIDOIDES SP 14
 33  EPISTOMINELLA UMBONIFERA
 34  FISSURINA
 35  GAVELINELLA MICRA
 36  GLOBOCASSIDULINA SUBGLOBOSA
 37  GLOBOCASSIDULINA SP 1
 38  GYROIDINOIDES COMPLANATA
 39  GYROIDINOIDES PERAMPLA
 40  GYROIDINOIDES ZEALANDICA
 41  GYROIDINOIDES SP 2
 42  HANZAWAI CUSHMANI
 43  KARRERIELLA SUBGLABRA
 44  LAGENA
 45  LENTICULINA
 46  NONION HAVANENSE
 47  NONION SP 1
 48  NONIONELLA SP 1
 49  NUTTALLIDES TRUEMPYI
 50  OOLINA
 51  ORIDORSALIS TENER
 52  ORIDORSALIS SP 1
 53  ORIDORSALIS SPIRAL
 54  OSANGULARIA MEXICANA
 55  OSANGULARIA SP 1
 56  PLANULINA AMMOPHILA
 57  PLEUROSTOMELLIDAE
 58  PULLENIA EOCENICA
 59  PULLENIA QUINQUELOBA
 60  STILOSTOMELLA
 61  TURRILINA BREVISPIRA
 62  UVIGERINA ELONGATA
 63  UVIGERINA RIPPENSIS/SPINICOSTATA
 64  UVIGERINA SP 1
 65  UVIGERINA SP 1
 66  UVIGERINA SP 3
 67  UVIGERINA-BULIMINA
 68  VULVULINA SPINOSA
```

```
1************************
 **
 **
 **  (8A)  CORLISS DATA BASE
 **        PALEOGENE DSDP BENTHIC FORAMINIFERS
 **
 **
 ************************


 SEQUENCE DATA   (TOPS)

 CORE 77B
   27  -35  -52   16  -25   17  -22    9

 CORE 219
   15  -19  -25  -39  -42  -57  -58  -59  -66    2  -12  -14  -38    6  -33  -52   35  -61   53   26
   11   64    7

 CORE 253
    2  -17  -21  -24  -44  -65   11  -12  -50   19  -35   34   18   63   49   23   10    3   53

 CORE 292
   35  -54  -55   50    6  -27  -43   46   17  -25  -41   49

 CORE 363
   23  -61   22  -55   50    1    6  -16    8   49   10    5   -7  -26

 CORE E128
   20  -63   35  -61   13  -49  -55  -56  -66    4   12   40    5   15


1************************
 **
 **
 **  (8B)  CORLISS DATA BASE
 **        PALEOGENE DSDP BENTHIC FORAMINIFERS
 **
 **
 ************************


 SEQUENCE DATA   (BOTTOMS)

 CORE 77B
   12   35  -54   16  -17  -25  -44

 CORE 219
   66   19  -37    6  -65   56   30   53   11  -14    9  -62   26   23  -25  -29  -64    8  -39  -63
   12  -13  -17  -33  -34  -44  -50  -52  -55  -67

 CORE 253
   62   11   66    6   18   23    8   13   16    9   17  -22  -33  -41  -44  -55  -65   34  -38  -39
  -49

 CORE 292
   35   22  -67    6  -27  -44   59   39  -47    9   34  -41   11  -13  -25  -40   43  -52  -54

 CORE 363
    8    1    6  -33   55    9  -22  -41  -54  -63  -67    7  -10  -12  -16  -21  -25  -26  -35  -42
  -44  -50  -58  -62

 CORE E128
   62   20  -48    1   32   31   12   41    5   61    8  -27  -42   16  -17  -22  -24  -25  -29  -34
  -35  -38  -39  -40  -43  -54  -56  -58  -59  -63  -65  -66
```

```
1************************
 **
 **
 **  (9A)  AGTERBERG - LEW DATA BASE
 **        COMPUTER SIMULATION EXPERIMENTS
 **
 **
 ************************

 SEQUENCE DATA   (STEP SIZE = 0.5)

   1   2   5   4   3   6  10   8   9  11  13  14  12  15   7  17  16  18  19  20
   1   4   3   2   7   8   9   6  11   5  12  13  10  15  18  19  16  14  17  20
   3   1   2   4   5   6  10   8   7   9  12  11  13  15  16  14  17  18  19  20
   5   3   1   2   4   7   6   8   9  10  12  11  13  14  18  19  16  15  17  20
   2   1   3   5   6   4   7   8   9  12  10  13  11  14  15  16  19  17  20  18
   3   4   5   2   1   6  11   9   7  10  12   8  16  15  14  13  17  18  20  19
   2   3   4   1   7   6   9  10   5  12   8  13  14  15  11  16  18  17  19  20
   1   3   5   4   9   6   2   7  11  12   8  10  13  16  15  14  17  19  18  20
   1   8   3   2   4   6   9   5  12   7  10  11  14  13  15  16  18  17  20  19
   2   3   4   1   8   7   6   5  10  12  14  16  11  13   9  15  17  18  19  20
   1   5   6   2   3   4   8   7   9  13  10  14  16  11  12  15  17  18  19  20
   1   4   6   2   3   5   8   7   9  13  11  14  10  12  15  17  18  16  19  20
   2   4   1   5   3  11   6   7   9   8  10  13  14  12  16  15  17  18  19  20
   6   3   1   4   2   5   7   8  14   9  11  12  15  16  10  13  17  18  19  20
   3   4   2   1   5   7   6   8   9  12  10  11  14  13  16  17  15  19  18  20
   3   1   7   6   2   5   4   8  10  15  12   9  13  14  11  17  16  20  19  18
   1   2   4   5   7   3   8   6  14  10   9  11  16  12  13  19  18  17  15  20
   2   1   4   3   8   6   5   7   9  11  15  14  12  13  10  16  17  20  18  19
   1   2   4   7   3   5   6   9  10  11   8  18  13  12  14  15  16  17  19  20
   2   1   4   3   6   5   7  11  10   9   8  14  15  16  12  13  18  17  19  20
   3   1   5   4  10   6   2   7   8  11   9  12  14  16  13  17  15  18  19  20
   1   2   5   3   4   6   8   7   9  11  10  15  14  13  12  16  19  17  18  20
   1   5   4   3   6   2   8   7  11   9  12  10  16  14  17  15  18  13  19  20
   2   1   7   3   6   5   4   8  13  12   9  10  11  16  18  20  14  15  19  17
   4   1   3   2   8   6   5   7  11   9  13  10  12  16  14  15  17  18  20  19
```

```
1************************
 **
 **
 **  (9B)  AGTERBERG - LEW DATA BASE
 **        COMPUTER SIMULATION EXPERIMENTS
 **
 **
 ************************
```

SEQUENCE DATA (STEP SIZE = 0.3)

5	1	4	2	10	3	6	8	11	9	15	13	14	17	12	16	7	19	18	20
1	4	3	7	2	8	9	11	6	12	13	18	15	5	10	19	16	20	17	14
3	1	2	4	5	6	10	12	8	9	11	7	16	15	13	17	14	18	19	20
5	3	1	7	6	4	2	9	8	10	12	13	11	14	18	19	17	20	16	15
2	1	3	5	6	8	7	12	9	4	10	14	13	19	15	11	16	17	20	18
3	4	5	11	9	2	6	1	7	10	12	16	15	14	8	17	13	18	20	19
2	3	4	7	1	10	9	6	12	13	15	14	5	8	16	18	11	17	19	20
1	9	3	5	4	6	2	11	7	12	10	16	8	13	15	14	19	17	18	20
8	3	1	2	4	6	9	12	5	10	7	14	11	15	13	16	18	17	20	19
2	3	4	8	7	1	6	10	5	14	12	16	15	13	11	17	9	18	19	20
1	5	6	2	3	8	7	13	4	9	16	14	10	11	12	17	15	18	19	20
1	4	6	5	3	2	8	7	14	13	9	17	11	15	10	12	18	20	19	16
2	4	5	11	3	1	9	6	7	8	13	10	14	16	12	15	17	18	19	20
6	3	4	1	2	5	14	7	8	11	9	16	12	15	17	13	10	18	19	20
3	4	2	1	5	7	12	9	8	6	11	10	14	16	13	17	19	15	18	20
3	1	7	6	5	15	2	10	8	4	14	12	13	9	11	17	16	20	19	18
1	4	7	2	5	14	8	6	3	10	16	11	9	19	12	18	13	17	15	20
2	8	4	1	3	6	7	5	9	11	15	14	12	13	20	18	16	17	19	10
7	1	4	2	5	3	6	9	18	10	11	13	8	12	14	15	16	17	19	20
2	4	1	6	7	3	5	11	14	10	9	8	16	15	18	17	12	13	19	20
3	10	1	5	6	4	7	2	8	11	9	14	12	16	17	13	15	18	19	20
1	2	5	3	4	8	6	7	9	11	15	10	14	13	19	12	16	17	18	20
5	1	4	6	3	11	2	8	7	9	12	16	17	18	14	10	15	13	19	20
2	7	6	1	3	5	13	8	12	4	16	9	10	20	18	11	14	19	15	17
4	3	8	1	2	6	5	11	7	9	13	12	10	16	14	17	15	18	20	19

```
1************************
 **
 **
 **  (9C)  AGTERBERG - LEW DATA BASE
 **        COMPUTER SIMULATION EXPERIMENTS
 **
 **
 ************************


 SEQUENCE DATA    (STEP SIZE = 0.1)

     5  10   4   2   1  11  17  15  14   8   9  13   6   3  16  12  19  20  18   7

     1   4   7  18  11  19   9  13   8  12   3  15  20   2   6  16  17  10  14   5

     3   4   1   2   6  10  12   5  16  11  15   8   9  13   7  17  18  19  20  14

     5   7   3   6   9   1   4   8  18  10  19   2  12  13  14  20  11  17  16  15

     2   5  12   1   3  19   8   6   7   9  10  15  14  20  16  13  17   4  11  18

    11  16   9   5   4   3  10  12   6  15   7  17   2  14  18   1  20  13  19   8

    10  15   9   3   7  12   4   2  13  14   6  16  18   1  17   8   5  11  19  20

     9   1   5   6   3   4  11  16  12  19   7  15   2  10  13  17  14  18   8  20

     8  12   3   9   6   1  15  14   4   2  16  10  13  18  11   7  17   5  20  19

     2   8   3   4   7  16  14  10   6  12  15   1  17  13   5  19  18  20  11   9

     5   6   1  13  16   8   7  14   9   2   3   4  10  17  12  18  19  11  15  20

     4   1   6   5   3   8   2  17  14  13  20  15  19  18  11   7   9  16  12  10

    11   4   5   2   9  13   8   7   3   6  10   1  14  16  17  18  19  15  12  20

     6   3  14   4  16  11  17   5  15   2   8   1   7  12   9  19  18  20  13  10

     3   4  12   5   7   2   9  14   8   1  16  19  17  11   6  10  13  15  18  20

     3   1  15   7   6  10  14   8  13   5  12  20  17   2   4  16  11  19   9  18

    14   7  16   4   1  19   8   5   2  10   6  18  11  12  17   3   9  13  20  15

     2   8  15  20   7   4   6  11  14   9  19   5  18   3  17   1  13  16  12  10

    18   7   4   5   1   9  10  11   2   6  13   3  12  14  16  17  15   8  20  19

     2   4  14   1  11   6   7  16  10  15   9   5   3   8  18  17  19  20  13  12

    10   3   5   6   7   1   4   8  11  16  14  17  12   9   2  19  18  15  13  20

     5   1   2   3   4   8  15  14  11   6   7  19  13   9  10  16  18  17  12  20

     5  11   6   4   1   3   8  18  16  17   9   7  12   2  14  15  19  20  10  13

     7  13   2  20   6  16  12  18   5   8   3   1  19  10   9   4  11  14  15  17

     8   4  11  13   3   6  16   5  17   9   1   2  18  12   7  15  14  10  20  19
```

COMPUTER PROGRAMS

SERIATION

Program to seriate a matrix of biostratigraphic events (see Chapter II.3). Method by J. Brower and W. Burrough, Syracuse University, 1979. The program to produce biostratigraphic seriation is composed of 6 FORTRAN IV routines (main program SER and 5 subroutines) and implemented on IBM 370, DEC system-10 and CDC CYBER computers. Implementation on other computers should be straightforward in that no half-word variables are used in the coding and alphanumeric data is transmitted in A1 Format or as Hollerith strings. For Source Listing see Computers and Geosciences, vol. 8, no. 2, pp. 137-148, 1982.

BIOSTRATSIM

The program Biostratsim by S. Millendorf, G. Srivastava, T. Dyman and J.C. Brower in Computers and Geosciences, vol. 4, no. 3, pp. 307-311 calculates a matrix of similarity coefficients (Chapters I.2 and II.2). The data represent taxa in a series of samples. Similarities are calculated between the columns in the data matrix, and an option permits the choice of five binary similarity coefficients, namely the Dice, Jaccard, Otsuka, Simpson and Simple Match coefficients.

UNITARY ASSOCIATION

The program to calculate unitary associations (Chapters II.7 and 8) from the overlap in range of many stratigraphically successive taxa has been documented by J. Guex and E. Davaud in Bull. Soc. Vaud. Sc. Nat. no. 361, vol. 76, 1982 and with graph theory in Computers and Geosciences, vol. 10, no. 1, pp. 69-96, 1984. It is written in Fortran ASCII and developed on a UNIVAC 1100/60 computer in 65k words.

RASC

The method and program RASC for Ranking and Scaling of biostratigraphic events (Chapters II.4, 5, 6, 7) was developed by F.P. Agterberg, F.M. Gradstein, L. Nel and S.N. Lew. It is coded in FORTRAN

IV and listed by F.P. Agterberg and L. Nel in Computers and Geosciences, vol. 8, no. 1, pp. 69-90 and vol. 8, no. 1, pp. 163-189. The program consists of about 3000 statements and has been implemented on several different main frames: for original development on CDC CYBER 74 and 730; on different types of IBM and CDC computers; UNIVAC 1108 and DEC System 10. An expanded version in FORTRAN - extended version IV is available on 5" floppy diskette with the RASC syllabus in the Geological Survey of Canada Open File No. 922, 1985 (2nd revised edition) by M. Heller, W.S. Gradstein, F.M. Gradstein, F.P. Agterberg and S.N. Lew. The diskette also contains a sample run (Late Cretaceous foraminiferal exits in 20 Grand Bank wells). An interactive microcomputer version of RASC was written by S.N. Lew. RASC-PC in PC BASIC uses 64k.

A versatile microcomputer version in language C for IBM XT (with co-processor) or other machines is marketed by Alethic Software Inc., 25 Parkhill Road, Halifax, Nova Scotia, B3P 1R5, telephone 902-423-9860.

CONVERSION PROGRAMS

Program RASCSET by R. Yang in FORTRAN IV, 458 lines.

- to convert U.A. formatted data into RASC input; levels may be reported in metres; it is possible to reverse the sequences.

Program CONVRT by S.N. Lew in FORTRAN IV, 192 lines.

- to convert Doeven-type RASC formatted data into U.A. input (for example, highest and lowest occurrences of fossil 32 are 32 and 132 in RASC input format).

ENTERDAT

Data management program of approximately 350 statements in FORTRAN V and C languages for microcomputers (IBM PC, etc.). Enterdat can be used in conjunction with routine micropaleontological sample analysis of wells and long outcrop sections to create paleontological data files and (sorted) taxonomic dictionaries. The new files and dictionaries are in format suitable for RASC and CASC runs. The program is marketed by Alethic Software Ind., 25 Parkhill Road, Halifax, Nova Scotia, B3P 1R5, telephone 902-423-9860.

DISSPLA

The DENO program listed by A.E. Jackson, S.N. Lew and F.P. Agterberg in Computers and Geosciences, vol. 10, no. 1, pp. 159-165, 1984 is written in the plotting language DISSPLA. It serves to display dendrograms of scaled optimum sequences and the optimum sequences of stratigraphic events from RASC output. For examples see Chapter II.4 and the Appendix.

CASC

The CASC method and computer program for automated correlation and scaling of events in time (Chapter III.1) was developed by F.P. Agterberg, F.M. Gradstein, J. Oliver and S.N. Lew and implemented on a CDC CYBER 730 with a TECTRONIX 4014 terminal. It is coded in FORTRAN Extended Version IV with about 6000 statements. Two libraries are required to use CASC, IMSL Library (Subroutine ICSSCU) and TECTRONIX Advanced Graphics Library (Advanced Graphing-II). Both an interactive and batch run version have been developed. At present best geological results are obtained by using the interactive version.

For details and source code of the interactive version on 5" floppy diskettes see: Agterberg, F.P., Oliver, J., Lew, S.N., Gradstein, F.M. and Williamson, W.A., 1985. Geological Survey of Canada, Open File Report 1179. Alethic Software Inc., Halifax (see under RASC) plans to market a microcomputer version.

BURIAL HISTORY

A FORTRAN V program for basement subsidence and mass rate of sedimentation was rewritten by D. Issler (Dalhousie University, Halifax) from a source code listing by R. Wood (Subsidence in the North Sea; Ph.D. thesis, 1983, Cambridge University, U.K.). Burial history, depth - porosity curves fitting, depth-porosity curves fitting, extended graphics and a more versatile input/output system for basin modelling (Chapters III.1 and V.1) were later added by B. Stam, F. Gradstein, P. Lloyd and A. Jackson, (Computers and Geosciences, in press).

The actual programs (BURSUB and DEPOR) contain about 1300 statements and were implemented on Victor 9000 and IBM-XT, both with co-processors.

MULTIVARIATE ANALYSIS

A number of multivariate analysis computer packages are available in standard statistical libraries, including:

BMDP (statistical software) (Department of Biomathematics) University of California Press 22223 Fulton Street BERKELY, California 94720 U.S.A.	and	SAS (Statistical Analysis System) SAS Institute Inc. SAS Circle Box 8000 CARY, N.C. 27511 U.S.A.

GLOSSARY

Acme Zone - a body of strata characterized by the flourishing of one or several fossil taxa.

Adjacency Matrix - two-dimensional array of ones and zeros, expressing stratigraphic relationships of co-occurrence (2 ones) or superposition (one and zero) for a group of fossil species.

Affinis - taxonomic term to indicate a degree of morphological similarity of observed specimens to a paleontological species (see confer).

Age - principal geochronological unit, e.g. Maestrichtian.

Age-Depth Diagram - displays the subsidence and sedimentation history of a well site (see burial history) in a x(age)-y(depth) graph.

Agglutinated - foreign particles bound together by cement (e.g. in agglutinated benthic Foraminifera).

Agglomerative Cluster Analysis - clustering which proceeds by successive fusions of the taxa or samples into groups.

Albedo - reflectance of sunlight on the earth surface.

Algorithm - method or device to solve specific mathematical problem usually by means of digitized computer.

Alveolinids - cigar shaped larger foraminifera (up to several mm in length) with spirally coiled, multi-chambered whorls.

Analog Image - optical image stored by physical means (e.g. photography).

Arc (in graph theory) - edge with an arrow indicating direction for an ordered pair of vertices.

Arenaceous - tests (e.g. of foraminifera) composed of sand or other foreign particles.

Assemblage Zone - a group of strata characterized by a distinctive assemblage of fossil taxa.

Authigenic - (minerals) generated on the spot by chemical or biochemical action prior to burial and consolidation of the sediment.

Auto Correlation - correlation of variable forming series with respect to itself by comparing values that are a fixed number of spaces apart in the series.

Autocorrelation Coefficient - estimate of autocovariance divided by variance.

Auto Covariance - covariance of variable with respect to itself comparing values that are the same number of spaces apart in a series.

Average Range Chart - graphic display of the average stratigraphic ranges of selected fossil taxa in geological time.

Average Zonation - produced by sequencing methods designed to estimate the average highest (or average lowest) occurrence of a fossil event.

Backstripping - stepwise removal of successively older and more deeply buried lithological units for the purpose of unloading each older unit and finding its subsidence.

Basin History - tectonic and sedimentary history of a geologically coherent sedimentary basin.

Batch Computer Program - job submitted to run without interference until output is completed.

Bentonite - a clayey rock formed by alteration of glassey igneous material like tuff or ash.

Benthic - bottom dwelling.

Biochronology - the succession of fossil taxa in their geological time content.

Biofacies - the paleontological aspects of a particular sedimentary deposit (e.g. reef facies).

Biometry - the study of the measurements of living or fossil organisms for diagnostic purposes.

Binomial Theory - the theory of the binomial distribution expressing the probability that an event occurs a specific number of times during a number of trials if the probability that the event occurs remains constant.

Biostratigraphic Attributes - consist of vertical range, facies independence and geographic persistence.

Biostratigraphic Constancy - percentage of samples within a zone containing a particular species.

Biostratigraphic Event - the presence of a taxon in its time context, derived from its position in a rock sequence.

Biostratigraphic Value - relative usefulness of fossils for correlation, incl. relative age.

Biostratigraphy - the global or regional record in rock sections of paleontological events or zones and their lower and upper limits in the succession of rocks.

Biozonation - empirical scheme that emphasizes the restriction of morphologically distinct fossil taxa in time (and frequently also in a geographical sense).

Biozone - a body of rock characterized by a particular fossil content, used to distinguish it from older and younger geological strata.

Bivariate Scatter Plot - graphical representation of the corresponding values of two variables (generally an independent and a dependent one) in two dimensional space (x-y axes).

Burial History - subsidence and sedimentation history of a well site in a geological basin, generally graphically displayed in age/depth diagrams.

Calpionellids - basket shaped calcareous skeletons of uncertain origin, up to about 100 microns in size, typically found in fine-grained limestones of Jurassic-Cretaceous boundary age.

CASC - abreviation for Correlation And SCaling in time method and computer program, that performs automated geological correlation (with error analysis) of events, zones and isochrons.

Centroid - mean, median or some other measure of the "center of gravity" of a group of samples or taxa.

Chi Square - sum of number of squared z-values, arises in chi square test for goodness of fit of assumed distribution compared with set of observed frequencies.

Chronostratigraphy - geological science that deals with the relative succession of stages (see stage).

Clique - complete subgraph of an undirected graph that contains all possible edges.

Cluster Analysis - encompasses many diverse multivariate techniques for discovering relationships between taxa and/or samples by successive partitioning or grouping.

Coded Sequence - stratigraphic sequence of fossil events coded according to the dictionary of event numbers.

Coeval - of same age.

Column - the vertical row(s) of numbers in a matrix (see also row).

Composite Section - stratigraphic section assembled from several (shorter) sections.

Composite Standard - a composite of several stratigraphic sections that combines the locally observed stratigraphic ranges of the taxa into a total stratigraphic range.

Computer - an analog or digital device capable of solving problems or manipulating data by accepting prescribed operations on the data and supplying the result of these operations.

Computer Language - the grammar, reserved words and symbols and techniques for providing instructions to a computer system.

Concurrent Range Zone - the overlapping part of the range zones of two or more selected taxa.

Confidence Belt - interval associated with an estimated value to express uncertainty, e.g. a 95% confidence belt contains the true value that is being estimated with a probability of 95%.

Confer - taxonomic term to indicate that specimens have morphological features of a paleontological species, but are probably not related to it.

Conservative Zonation - produced by sequencing methods designed to give the stratigraphically highest possible estimate of the top of a range zone and the stratigraphically lowest possible estimate of the base of the range zone.

Constrained Solution - type of seriation which employs the known stratigraphic information about the distribution of species and samples within the individual stratigraphic sections.

Core (rock) - mechanically obtained vertical section of strata.

Correlation - causal linkage of present or past processes and events; correlation of geological attributes generally expresses the hypothesis that a mutual relation exists between stratigraphic units.

Correlation Coefficient - a measure of the strength of the linear relationship between two variables.

Correspondence Analysis - an ordination technique (also termed Reciprocal Averaging) that is a variety of principal components where the data are treated as a contingency table.

Cospectrum - the power spectrum of two or more time-series.

Cross-Association - a measure of similarity between two sequences based on descriptive data (attributes).

Cross Correlation Coefficient - the correlation between two sequentially ordered sets of measurements. It is the cross-covariance divided by the variance.

Cross-Covariance - the covariance of two sequentially ordered sets of measurements.

Cross-over Frequency - the frequency of reversal in stratigraphic position of a pair of paleontological events as observed in a given number of stratigraphic sections.

Cross Variogram - the similarity between variables in two series which are shifted against each other by a given amount and measured by the mean square of their difference.

Cubic Polynomial - polynomial third degree.

Cubic Spline Function - Succession of cubic polynomials fitted to a sequence of values with continuous first and second derivatives.

Curvilinear Data - fall along a curved line or surface.

Cycling - irreconcilable differences in the order of three or more stratigraphic events as observed in a given number of stratigraphic sections.

Cycle Destruction - the process of eliminating stratigraphic inconsistencies, corresponding to 3,4,5, or 6 event cycles; the actual destruction is based on suppression of the weakest link in each cycle, where such a link is defined as the pair of events with the least number of observations.

Datum Plane - geological time plane based on dated events.

Dendrogram - tree like diagram for displaying the results of hierarchical cluster analysis.

Deterministic Approach - attempt to explain observed data by causal processes or mathematical equations allowing for little or no uncertainty in the data.

Diachronous - correlation across time lines.

Diagenesis - process involving physical and chemical changes in sediment after deposition.

Dice Coefficient - a similarity coefficient which compares two samples with respect to the taxa present or the distributions of two taxa in terms of samples. This coefficient usually generates intermediate similarities.

Dictionary(RASC) - alphabetical and numerical listings of all taxa used in a quantitative zonation (see also index); the unique taxon number replaces the original name.

Difference Matrix - a square matrix containing coefficients which illustrate the lack of similarity between all possible pairs of taxa or samples.

Digital Image - optical image stored by means of binary codes on (magnetic) tape or disc.

Dinoflagellate - one celled organic-walled organism in possession of a flagellum for free movement in the water; uses photosynthesis for energy.

Disconformity - an unconformity in which the beds on opposite side of the unconformable contact are parallel (see unconformity).

Discriminant Analysis - multivariate techniques which measure the statistical separation between two or more a priori groups. The probability that a particular individual belongs to a certain group can also be calculated. The technique is often used to identify unknown samples.

Dissimilarity - the lack of resemblance between taxa or samples, i.e. the reverse of similarity.

Distance Coefficient - type of dissimilarity coefficient which measures the distance between two samples or taxa and satisfies the properties of a metric such as the triangular inequality.

Divisive Clustering - clustering which separates the samples or taxa into progressively smaller groups.

Directed Graph - a graph consisting of arcs only.

Dual Space Technique - group of multivariate methods based on eigenvalues and eigenvectors, such as principal components and correspondence analysis, which allows the plotting of taxa and samples on the same graph.

Ecozone - body of strata characterized by a fossil assemblage characteristic for a particular geological enviroment, (also named facies-zone).

Edge (in graph theory) - undirected line segment connecting two vertices of a graph.

Eigenvalue - one of the latent or characteristic roots of a square matrix.

Eigenvector - one of the latent or characteristic vectors of a square matrix.

Embedded Absence - is located between the highest and lowest presences of a seriated data matrix.

Entry - first occurrence of a fossil in a body of strata.

Entropy Function - measures the uncertainty in predicting the identity of a randomly selected item in a population.

Error Bar - graphical representation of uncertainty of an estimated value usually ranging from this value plus or minus its standard deviation.

Eurytopic - a species which can tolerate a wide range of enviromental conditions.

Eustacy - world wide simultaneous change in sea level.

Evaporite - sediments deposited from evaporation of an aqueous solution.

Exit - last occurrence of a fossil in a body of strata.

F-Matrix - two-dimensional array of cross-over frequencies f_{ij} of paleontological events; consists of N rows labeled i and N columns labeled j.

Facies - part of a rock body, as contrasted with other parts by appearance, composition, sedimentological or lateral stratigraphic position.

Factor Analysis - attempts to estimate the structure of the covariance or correlation matrix of the population from which the sample was drawn. Often incorrectly confused with principal components.

FAD - first appearance datum (evolutionary).

Fence Diagram - three-dimensional correlation diagram of outcrops or well sections, generally arranged in stratigraphic order.

F_i - cumulative number of events occurring in i or more sections.

f_{ij} - cross-over frequency of paleontological events labelled i and j; represents number of times i is observed to occur above j or vice versa.

Final Reordering - procedure in RASC method of replacing optimum sequence by succession of estimated scaled optimum sequences before new scaled optimum sequence is estimated.

First Order Difference - difference between two RASC-distance values of successive events in a section.

First Appearance (Datum) - first appearance of a fossil taxon in geological time.

Foraminifera - single-celled marine animals, consisting of a protoplasma body that secretes an often elaborate calcareous test or collects an arenaceous one; both floating and bottom dwelling families occur, each with an extensive fossil record.

Forbidden Structure (in graph theory) - arrangement of arcs and edges in a biostratigraphic graph which presents its representation by means of an interval graph.

Formation - rock unit used to map strata of a distinct lithology.

Fossil - skeletal remains of organism buried and preserved in geological strata.

Fourier Series - Series of sine and cosine terms used for approximating a mathematical function or series of discrete observations.

Fractile - value of cumulative frequency distribution corresponding to a specific percentage.

Gamma Distribution - A type of frequency distribution of which the mathematical expression is an incomplete gamma-function.

Gaussian Distribution - see Normal Distribution.

Generated Subgraph - subgraph of a graph containing all edges in the original graph for the vertices in the subgraph.

Geochronology - classification system of geological time units.

Geographical Range - the area inhabited by a particular taxon during its geological life span.

Geographical Persistence - the proportion of stratigraphic sections within an area in which a selected taxon is known to occur.

Geohistory Analysis - see Burial History.

Global Error Bar - estimate of uncertainty of points on fitted age-depth curve in CASC as based on standard deviation of biostratigraphic events along RASC distance scale.

Graph Theory - the theory of graphs which are geometrical representations of ordered or non-ordered relationships between objects or events.

Highest Occurrence - last occurrence of a fossil in a body of strata.

Homeomorph - of same shape or building-plan, but not necessarily (evolutionary) related.

Independent Random Component - a random variable which is independent of either preceeding or subsequent values in a series of measurements.

Index - alphabetical and numerical listings of all taxa used in a quantative zonation (see also Dictionary); the unique taxon number replaces the original name.

Index Fossil - fossil deemed of particular use for (chrono)stratigraphic correlation.

Information Function - see Entropy Function.

Initial Unitary Association (IUA) - clique of interval graph obtained after eliminating forbidden structures from biostratigraphic graph.

Interactive Computer Program - menu driven program in which the user "steers" the execution.

Interval Graph - undirected graph representing overlapping or non-overlapping line segments on the real line.

Inter-event Distance - ('interfossil' distance) interval between two successive events along RASC distance scale.

Interval Zone - zone in which one of the two boundaries is defined by the immediately under- or overlying zone; zones based on tops using well cuttings are interval zones.

Isochronous - (strata of) equal age.

Isochron - a line connecting points of similar age in different (well) sections.

Iteration - step in a repetitive process leading to a final solution.

IUA - (see Initial Unitary Association).

Jaccard Coefficient - a similarity coefficient which compares two samples with respect to the taxa present or the distribution patterns of two taxa in series of samples. The Jaccard Coefficient stresses the differences between the items involved.

Kendall Rank Correlation Coefficient - measure of correspondence of orders of elements in two series.

LAD - last appearance datum (evolutionary).

Lag - fixed number of spaces between values forming a sequence for which an auto correlation coefficient is estimated.

Last Appearance (datum) - last appearance of a taxon in geological time.

Lateral Tracing - a multivariate technique which links or joins the most similar samples from adjacent stratigraphic sections in order to form a line of sections or a fence diagram.

Lateral Linkage - a category of multivariate techniques which forces samples to pair across rather than within adjacent stratigraphic sections.

Line of Correlation - best fit of scatterpoints representing lowest and/or highest occurrences of fossils in two stratigraphic sections, each section is plotted along one axis of a bivariate graph.

Line of Observation - bivariate fit of events in a section plotted according to order(x-axis) and depth(y-axis).

Lineage - morphological series of taxa organised according to geological time.

Lithofacies - part of a rock body with distinct composition or appearance.

Lithosome - the geometrical shape of a defined rock body.

Lithostratigraphy - the framework of successive and lateral rock units (formations) in a geological basin, recognized mainly by their physical character.

Loading - compaction and subsidence at depth as a result of the weight of overlying strata.

Local Error Bar - estimate of uncertainty based using CASC method on deviations from fitted spline curve as observed in a section using CASC method.

Lower Triangle (of a matrix) - elements located below the principal diagonal of a matrix which connects the first element in the top row with the last element in the bottom row.

Lowest Occurrence - first occurrence of a fossil in a body of strata.

MA - age in millions of years before present.

My - duration in millions of years.

Magnetostratigraphy - the global or regional record of magnetic reversals in rock sections used for correlation.

Marker Event - a geological event of stratigraphic (correlation) value.

Marker Horizon - stratigraphic event such as volcanic ash layer or seismic marker which can be located with certainty within a section and scaled between the biostratigraphic events by means of the RASC method.

Markov Process - a time series in which a random variable is partly independent and partly dependent on its predecessors. First order Markov processes are sequences in which the random variable depends only on its immediate predecessor.

Matching Coefficient - measure of similarity based on counting identical items in two comparable sequences.

Matrix (pl. matrices) - two-dimensional array of numbers with rows of equal length and columns of equal length.

Matrix Permutation - Changing a matrix by one or more interchanges of two of its columns and the rows corresponding to these columns.

Maximal Clique - clique which is not contained in a larger clique.

Maximum Likelihood Interpolation - interpolation by using the maximum likelihood method which consists of estimating one or more parameters of a population by maximizing the probability of appearance of a sample which is assumed to be at random from this population.

Modified Hay Method - technique of ranking biostratigraphic events proposed by W.W. Hay in 1972 and modified to allow ranking solutions when cycles occur.

Monothetic Association Analysis - a form of divisive clustering which divides samples into successive groups based on the presence or absence of one taxon at each step.

Morphometry - study of morphology (of taxa) using physical measurements of the body.

Morphotype - (species) taxonomic type based on morphological features of the body, (all fossils are morphotypes).

Morphology - the study of the form and structure of organisms or physical units.

Multivariate Analysis - simultaneous statistical analysis with respect to more than two variables.

(N x N) Matrix - square matrix with N rows and N columns.

Nannoplankton - unicellular, autotrophic, marine algae, 1-15 microns in size with a skeleton of systematically arranged calcite crystals; the group includes coccolithophores, discoasters, and several other families; Jurassic-Recent.

Nannoconid - cone-shaped nannofossil with internal canal; Mesozoic only.

Noisy Data - data of relatively low quality due to contamination.

Nonmetric Multidimensional Scaling - a multivariate ordination technique which can handle nonlinear data because it only attempts to preserve the rank order of samples or taxa.

Normality Test - method in RASC computer program that compares the position of an observed event in a given section with those of its neighbours in the same section, using second order differences.

Normal Distribution - frequencies satisfying the normal or Gaussian curve.

Numerical Time Scale - geological time scale expressed in numerical time units, like years or millions of years.

Numerical Taxonomy - system using a combination of code numbers to differentiate the taxa in the plant and animal kingdoms.

Oppel Zone - a body of strata characterized by the co-ocurrence (co-existence) of a group of fossil taxa.

Optimum Clustering - see Scaled Optimum Sequence.

Optimum Sequence - most likely ('average') stratigraphic sequence of fossil events based on ranking as used in RASC computer program.

Ordination - the ordering of taxa and/or samples with respect to one or more axes, for example those calculated by principal components.

Ostracods - subdivision of crustaceans that grow a test consisting of two interlocking calcareous shells inhabiting both salt and fresh water. Shells were molted during growth of animal and range from 0.15 - 2.0 mm in size; known since Paleozoic time.

Otsuka Coefficient - the binary equivalent of the cosine theta coefficient for comparing taxa or samples.

Paleoecology - science that reconstructs the original living enviroment of fossil assemblages containing plants and animals.

Paleontological Events - the presence of a taxon in its time context, derived from its position in a rock sequence.

Palynomorph - organic walled microfossils like acritarchs,pollen, spores and dinoflagellates.

Paratype - formally designated specimens, other than the holotype, on which an original description of a species is based.

Peak Zone - a body of strata characterized by the maximum development or greatest abundance of some fossils.

Pelagic - floating.

Periodites - bedded sediments in which each layer represents an approximately equal time interval.

Permutation - possible arrangement of objects or events forming a sequence.

Phase Spectrum - part of cospectrum which gives the phase difference between fluctuations of a given frequency in two time-series.

Phi Coefficient - the binary equivalent of the Pearson Product Moment Correlation Coefficient for comparing taxa or samples.

Phylo Zone - a body of strata containing a taxon representing a segment of morphological-evolutionary lineage defined between the predecessor and the successor.

Planktonic - floating (animal or plant).

Poisson Probability - Probability satisfying the Poisson distribution for occurrence of a discrete event. For example, if points are randomly distributed in space, then the number of ponts falling within a volume of specific size and shape is Poisson distributed.

Polar Ordination - an ordination scheme which uses samples to define the ordination axes.

Polynomial - linear combination of powers of a variable.

Power Spectrum - a function which analyses a time series by the frequency content of its fluctuations. Alternatively the frequency may be expressed in terms of wave length.

Presorting - initial ordering of stratigraphic events from different sections using cross-over frequencies; usually followed by modified Hay method.

Principal Components - a multivariate statistical technique which employs the eigenvectors of a correlation or covariance matrix as axes for ordination.

Probabilistic Approach - attempt to explain observed data by means of relatively simple models allowing for considerable uncertainty in the data.

Probit - z-values plus 5 to avoid appearance of negative values.

Process Model - a hypothesis which attempts to explain a given geological situation through the action of physical processes.

Q-Mode - analysis of relationships between samples.

Quantitative Correlation - geological correlation with confidence limits, using mathematical and statistical methods.

Quantitative Stratigraphy - use of relative simple or complex mathematical/statistical methods to calculate stratigraphic models that with a minimum of data provide a maximum of predictive potency and include formulation of confidence limits.

R-matrix - (N x N) matrix consisting of r_{ij} - values.

R-Mode - analysis of relations between taxa or attributes.

Radiolarian - unicellular, planktonic, marine animals, 60 micron - 2.0 mm in size, with amorphous silicious skeleton often displaying a radial symmetry; Paleozoic to Recent.

Random Number Generator - computer algorithm to generate pseudo-random numbers from one another beginning with a seed or initial number. For example, in the multiplicantive - congruence method, each new number is the remainder of the previous number multiplied by one constant and divided by another constant.

Random Process - a process which involves the action of random variables.

Range Chart - graphical arrangement of the distribution of fossil taxa in geological time as expressed in depth-, zone-, stage- or age units.

Range, Stratigraphical - geological interval between the lowest and highest occurrences of a fossil taxon.

Range, Statistical - interval for stratigraphic event in optimum sequence expressing uncertainty in its position due to missing data and other reasons.

Range-Through Method - records a taxon as present in all samples within its local range zone.

Range Zone - a body of strata corresponding to the stratigraphic range of a fossil taxon.

Rank - relative position in a sequence.

Ranking - organization of items in a sequence, using one or more attributes.

Ranking Model - see Optimum Sequence.

RASC Distance - see Inter-Event Distance.

RASC (program) - abbreviation for RAnking and SCaling computer program of biostratigraphic events.

Reciprocal Averaging - see Correspondence Analysis.

Regional Marker - geological feature, like the presence of a fossil or sediment type, useful in regional correlation.

Regression (geological) - relative retreat of the shoreline, resulting in emergence of land above sealevel.

Regression Analysis (statistical) - method of estimating value of one variable from that of another variable by using the principle of least squares.

Relative Biostratigraphic Value - an index which measures the amount of biostratighraphic information derived from observing the presence of a particular taxon.

Relative Order Frequency - the number of times that a certain event occurs above (or below) another one.

Reworking - movement of sediment (incl. fossils) after preliminary deposition.

r_{ij} (RASC) - sample size or number of times two stratigraphic events labelled i and j occur together in a section.

Row - the horizontal row(s) of numbers in a matrix (see also column).

S-Matrix - (N x N) matrix of scores (s_{ij}).

s-Ratio - number of times arc occurs in a cycle within the strong component of biostratigraphical graph divided by number of times this arc occurs in the same strong component.

Scaled Most Likely Sequence - see Scaled Optimum Sequence.

Scaling - method of estimating distances (intervals) between successive events in an optimum sequence.

Scaled Optimum Sequence - optimum sequence based on RASC computer program, which has been scaled such that the stratigraphically successive groups of fossil events in the dendogram reflect natural clusters in relative (or linear) geological time.

Scatter Diagram - graphical representation of the corresponding (numerical) values of two or more related parameters.

Scatter Ellipse - bivariate cluster of points (in a scatter diagram) that fit an ellipse.

Score (RASC) - see s_{ij}.

Second Order Difference - difference between two successive first order differences for RASC distances of events in a section; used in normality test.

Section, Stratigraphic - an exposure of successive geological strata.

Sediment - solid material, both mineral and organic, that has been move by air, water or ice and has come to rest on the earth's surface, either above or below sealevel.

Sedimentary Basin - geologically depressed area with thicker sediments in the interior and thinner sediment at the edges.

Sedimentation - process of separation, movement and (re)deposition of rock particles and organisms and precipitation of suspended minerals.

Sedimentation Rate - quantity of sediments deposited in time.

Seismic Record (Seismogram) - the record made by a seismograph.

Seismograph - instrument which records earth vibrations due to elastic waves.

Seismostratigraphy - the study of the vertical and lateral record of subsurface strata obtained from shock wave experiments.

Sequence, Composite - see Composite Standard.

Sequence, Optimum - see Scaled Optimum Sequence.

Sequence, Scaled Optimum - see Optimum Clustering.

Seriation - to arrange taxa and/or samples in a one-dimensional sequence which hopefully represents "time or evolution".

SF - see Smoothing Factor.

Shape Analysis - Digital (computer-based) method to discriminate between different shapes (e.g. sand grains, rock pores or microfossil tests).

s_{ij} (RASC) - score for two events labelled i and j in S-matrix ($S_{ij} = f_{ij} + 0.5\ r_{ij}$).

Similarity Coefficient - measures the amount of resemblance between two samples or taxa.

Similarity Matrix - a square matrix containing similarity coefficients which compare all possible pairs of taxa or samples.

Simpson Coefficient - a similarity coefficient which compares two samples with respect to the taxa present or the distribution patterns of two taxa in terms of samples. The Simpson Coefficient stresses the similarities between the items being compared.

Slotting - method for pairwise comparison of successions of data points by forming single series with minimum of dissimilarity between successive data points.

Smoothing Factor (SF) - standard deviation for deviations from spline-curve (see below) entered before fitting in CASC.

Spearman Coefficient - measure of correspondence of orders of elements in two series.

Spectral Analysis - the computation of the power spectrum of a time series which is particularly useful in the analysis of cyclic sediments.

Spline Curve - very smooth interpolation of a scatter diagram based on the principle of least squares.

Squared Multiple Correlation - ratio of the sum of squares explained by regression divided by the total sum of squares of the dependent variable.

Stage - well-delimited body of rocks, of an assigned and agreed upon age.

Standard Deviation - square root of variance.

Stenotopic - limited to a narrow range of environmental conditions.

Step-Model - comparison of the events in individual section with those in the optimum sequence in RASC method by scoring penality points for events that are out of place in the section.

Stochastical - involving element of random variation.

Strata (plural of stratum) - sedimentary beds or layers, independent of thickness.

Stratotype - the designated and agreed upon type-section of the rock body representing a stage or containing a formal stratigraphic boundary.

Stratquant - see Stratigraphic Quantum.

Stratigraphic Correlation - causal linkage of strata or events in time as found in different rock sections.

Stratigraphic Event - the presence of a fossil taxon or physical item in its (relative) time context as derived from its position in a rock sequence.

Stratigraphic Horizon - any definite horizon in a sedimentary sequence.

Stratigraphic Normality - the degree of correspondence between the individual stratigraphic record and the designated standard which summarizes the consistency in the order, the duration and the spacing of events in time.

Stratigraphic Quantum - the minimum thickness of sediment containing time information.

Stratigraphic Record - the geological record on which stratigraphy (see below) is based.

Stratigraphy - the science dealing with the succession and chronology of stratified rocks, with the life of past ages and the evolutionary changes from age to age as recorded by fossils, and with successive changes in the distribution of land and sea as interpreted by the character and fossil content of the sedimentary rocks.

Strong Component - generated subgraph of a biostratigraphical graph which is strongly connected and has the maximum number of vertices.

Student's t-Test - method of testing the hypothesis that two simple means of normally distributed data are equal to one another.

Sub Bottom - a designated, recognizable stratigraphic occurrence above the first occurrence.

Sub Top - a designated, recognizable stratigraphic occurrence below the last occurrence.

Subsidence - (as opposite to Uplift), downward vertical movement of portions of the earth crust.

Superposition - the order in which rocks are placed above one another.

Superpositional Relationship - relative stratigraphic order.

T-Matrix - (N X N) matrix of ties (t_{ij}).

T-Zonation (RASC) - Tertiary Zonation.

tau-value - see Kendall Rank Correlation Coefficient.

Taxa (taxon) - a stable unit consisting of all individuals (for example fossils) considered to be (morphologically) sufficient alike to be given the same (Linnean) name.

Taxonomy - the systematic (natural) classification of plants and animals.

Tectonic Subsidence - downward vertical movement of strata due to crustal cooling or deformation and dislocation in the earth crust.

Tempestites - sedimentary layers which have been deposited practically instantaneously.

Test Criterion (Index) - used in seriation of an original data matrix which ignores the known information about the stratigraphic distribution of taxa and samples within the individual stratigraphic sections.

Tethys - ancient (Mesozoic) east-west seaways that separated Europe and Africa and extended across Southern Asia and Central America.

Threshold Parameter - in trimonial model of Glenn and David, a tie of two events is assumed to occur if the distance between them is less than a given threshold parameter.

t_{ij} - number of ties where two stratigraphic events labelled i and j are observed to be coeval in a section.

Time Series Analysis - the analysis of ordered sequences. A time series may either consist of measurements taken at progressive time intervals or along a geological profile.

Top Occurrence - highest occurrence.

Transgression (geological) - relative advance of the shore line, resulting in the immersion of land below sealevel.

Transitively Oriented Graph - directed graph with the property that a vertex w can be reached from u if w can be reached from v and v from u.

Trend-Fitting Technique - representation of the trend component of a time series by a particular curve-form or by using polynomial series.

Trinomial Model - extension of binomial distribution model to possible occurrrence of three instead of two events.

Truncated Normal Distribution - normal distribution of which the values above or below a given fractile are not know.

Turbidites - deposit laid down by a density current that moves sediment in suspension down slope in standing water or air.

UA - see Unitary Association.

Unconstrained Solution - type of seriation of an original data matrix which ignores the known information about the stratigraphic distribution of taxa and samples within the individual stratigraphic sections.

Unconformity - lack of continuity in deposition between strata in contact.

Undirected Graph - graph consisting of vertices and edges only.

Unique Event - important stratigraphic (paleontologic) event that has been observed in one or few sections.

Unitary Association - The largest group of compatible species, where two species are classified as compatible if they have lived together at least once in the same stratigraphic horizon, or have never been simultaneously observed in the same horizon but stratigraphic order is reversed from one section to another, indicating chronological coexistence (see Oppel Zone).

UPGM - unweighted-pair-group-method of cluster analysis.

Upper Triangle (of a matrix) - elements located above the principal diagonal of a matrix which connects the first element in the top row with the last element in the bottom row.

Variance - square of standard deviation representing measure of the dispersion of values around their mean.

Variogram - the similarity between variables which are a given distance apart, measured by the mean square of their differences.

Vertex (plural vertices) - point of a graph usually connected to other vertices by edges or arcs.

Vertical Range - one of the biostratigraphic attributes of fossils, calculated by maximum value of (thickness of sediment in section m occupied by Species i)/(total thickness of section m).

Vertical Weighting Factor - the weight required to eliminate crossing linkages in lateral tracing.

Virtual Co-Occurrence - when two taxa exchange order in lateral sections, without actual co-occurring, or when both taxa co- exist with a third taxon, without actually co-occurring. Used in Unitary Association method.

Walther's Law - states that the vertical sequence of sediments reflects the lateral facies changes in time.

Weight - measure of relative importance of individual observation which is inversely proportional to its variance.

Weighted Distance Analysis - method of estimating distances between successive events forming and optimum sequence in which cross over frequencies are weighted according to the total numbers of pairs of events on which they are based.

Z-Matrix - (N x N) matrix consisting of z-values.

z-Value - fractile of the normal distribution corresponding to a given percentage value.

Z4-Structure - graph consisting of 4 vertices connected by 4 edges only, (with 2 edges per vertex)

Zonal Marker - fossil typical for a zone.

Zonation - empiral scheme, which emphasises the temporal and spatial restriction of morphologically distinct fossil taxa.

INDEX